Mechatronics

Dynamical systems approach and theory of holors

Mechatronics

Dynamical systems approach and theory of holors

B T Fijalkowski
Cracow University of Technology and
State Higher Vocational School in Nova Sandec, Poland

IOP Publishing, Bristol, UK

ISBN 978-0-7503-1350-6 (ebook)
ISBN 978-0-7503-1351-3 (print)
ISBN 978-0-7503-1352-0 (mobi)

DOI 10.1088/978-0-7503-1350-6

Version: 20161201

IOP Expanding Physics
ISSN 2053-2563 (online)
ISSN 2054-7315 (print)

British Library Cataloguing-in-Publication Data: A catalogue record for this book is available from the British Library.

Published by IOP Publishing, wholly owned by The Institute of Physics, London

IOP Publishing, Temple Circus, Temple Way, Bristol, BS1 6HG, UK

US Office: IOP Publishing, Inc., 190 North Independence Mall West, Suite 601, Philadelphia, PA 19106, USA

Once again, to Madeleine with love

Contents

Preface xvii

Acknowledgements xxvi

Author biography xxvii

Acronyms xxviii

Part I Towards a unified theory exploiting a dynamical systems approach for physical heterogeneous continuous dynamical hypersystems

1 General considerations 1-1

1.1 Dynamical hypersystem definition 1-1
1.2 Conclusions 1-4
1.3 Summary 1-6
Reference 1-7

2 Model definition 2-1

2.1 Introduction 2-1
2.2 Physical model 2-4
2.3 Deterministic mathematical model 2-5
2.3.1 Synthetic deterministic mathematical model 2-5
2.3.2 Chaotic deterministic mathematical model 2-6
2.4 Statistical mathematical model 2-6
2.5 Stochastic mathematical model 2-6
References 2-6

3 Dynamical systems approach 3-1

3.1 Introduction 3-1
3.2 The dynamical systems approach to physical heterogeneous continuous dynamical hypersystems 3-8
3.3 Summary 3-10
References 3-10

4 Algorithm for a formulation of the mathematical model 4-1

5 Developmental systems approach 5-1

Reference 5-5

6 Synthetic mathematical model of the abstract functional heterogeneous continuous dynamical hypersystem **6-1**

6.1 Abstract functional heterogeneous continuous dynamical hypersystem 6-2
6.2 Independent generalised coordinates 6-3
6.3 Unconstrained and constrained nonlinear functional heterogeneous continuous dynamical hypersystems—classification of the constraints 6-4
6.4 Principle of stationary action 6-6
References 6-16

Part II Towards a unified theory exploiting the theory of holors for physical heterogeneous continuous dynamical hypersystems

7 Physical heterogeneous continuous dynamical hypersystems as generalised physical commutation matrixers **7-1**

7.1 Introduction 7-1
7.2 The generalised physical commutation matrixer's input- and output-signal holor functions 7-6
7.3 Multi-port, multi-input/multi-output, generalised physical commutation matrixers 7-13
7.4 The principal physical commutation matrixer dynamical components 7-26
7.5 Input- and output-signal holors of physical commutation matrixers 7-27
7.6 Physical and mathematical modelling methodologies 7-31
7.7 Signal holors of physical commutation matrixers 7-32
References 7-33

8 Physical commutation matrixer dynamical components **8-1**

8.1 Introduction 8-1
8.2 Electrical commutation matrixer dynamical components 8-3
8.2.1 Electrical-signal holors 8-4
8.2.2 Electrical-resistance holors 8-6
8.2.3 Electrical-inductance holor 8-8
8.2.4 Electrical-capacitance holor 8-11
8.3 Magnetic commutation matrixer dynamical components 8-14
8.3.1 Magnetomotive-force and magnetic-flux holors 8-16
8.3.2 Magnetic-reluctance holor 8-16
8.3.3 Magnetic-energy-source holors 8-17
8.4 Mechanical commutation matrixer dynamical components 8-19

	8.4.1 Mechanical-signal holors	8-19
	8.4.2 Mechanical-resistance or damping holor	8-23
	8.4.3 Mechanical inductance holor of the spring	8-26
	8.4.4 Mechanical-capacitance holor of the mass	8-34
8.5	Fluidic commutation matrixer components	8-37
	8.5.1 Fluidic-signal holors	8-38
	8.5.2 Fluidic-resistance holor	8-41
	8.5.3 Fluidic-inertance holor	8-45
	8.5.4 Fluidic-capacitance holor	8-48
	8.5.5 Fluidic energy sources	8-57
8.6	Thermal commutation matrixer dynamical components	8-57
	8.6.1 Thermal-signal holors	8-58
	8.6.2 Thermal-resistance holor	8-60
	8.6.3 Thermal-inertance holor	8-62
	8.6.4 Thermal-capacitance holor	8-63
	8.6.5 General thermal dynamical components	8-65
8.7	Energy sources and loads	8-66
	8.7.1 Energy-potential-difference sources	8-67
	8.7.2 Energy-transfer sources	8-69
	8.7.3 Loads	8-71
	8.7.4 Coulomb friction	8-72
	8.7.5 Weight	8-74
8.8	Physical dynamical components as mathematical operators	8-75
8.9	Summary	8-75
	Reference	8-78
9	**Principal physical commutation matrixers as physical models**	**9-1**
9.1	Introduction	9-1
9.2	Creation of physical commutation matrixers	9-2
9.3	Principal physical commutation matrixer laws	9-2
10	**Generalised impedance and admittance holors of physical commutation matrixers**	**10-1**
10.1	Sinusoidal responses to energy-potential-difference holor and energy-transfer holor functions	10-1
10.2	Impedance and admittance of principal physical commutation matrixer dynamical components	10-1
10.3	Generalised impedance and admittance holors in combination	10-4
	10.3.1 Driving-point impedance and/or admittance holors	10-7

10.4 Physical heterogeneous continuous dynamical hypersystems' transmittance or transfer-function holors 10-9
10.5 General formulation of physical-heterogeneous-continuous-dynamical-hypersystem holor equations 10-11
10.6 Some properties of linear physical heterogeneous continuous dynamical hypersystems 10-13
10.6.1 Two-port single-input/single output physical commutation matrixers 10-13
Summary 10-24
10.6.2 Physical commutation matrixer having three independent nodes 10-25
10.6.3 Superposition theorem 10-27
10.6.4 Reciprocity theorem 10-29
10.6.5 Thévenin's theorem 10-31
10.6.6 Norton's theorem 10-37
10.6.7 Millman's theorem 10-40
Reference 10-42

11 Electrical homogeneous continuous dynamical systems as DC and AC electrical commutation matrixers 11-1

11.1 Introduction 11-1
11.2 Electrical homogeneous continuous dynamical systems' holor algebra, steady-state DC analysis 11-6
11.2.1 DC holor algebra, steady-state analysis 11-6
11.2.2 Node-voltage holor method 11-9
11.2.3 Loop-current holor method 11-10
11.2.4 Superposition method 11-14
11.2.5 The DC energy source 11-15
11.2.6 The generation of DC energy 11-16
11.3 Electrical homogeneous continuous dynamical systems holor algebra, steady-state AC analysis 11-23
11.3.1 Sinusoidal alternating-current holor algebra, steady-state analysis 11-23
11.3.2 The sinusoidal AC energy source 11-23
11.3.3 The generation of AC 11-23
11.3.4 Sinusoidal alternating current 11-28
11.3.5 Sinusoidal voltage and current waveforms—holor representation 11-29
11.3.6 The passive AC electrical-homogenous-continuous-dynamical-system components in the holor domain 11-31

11.3.7	An impedance-holor representation	11-36
11.3.8	An admittance-holor representation	11-38
11.3.9	AC power holor representation	11-39
11.3.10	A geometrical interpretation of the voltage, current and power holors	11-42
11.3.11	Conclusion	11-44
11.4	The single-phase AC electrical homogeneous continuous dynamical system as the single-phase AC electrical commutation matrixer	11-45
11.4.1	Introduction	11-45
11.4.2	Series and/or parallel transformations	11-47
11.4.3	Delta-to-wye or pi-to-tee equivalent	11-50
11.5	The polyphase AC electrical homogeneous continuous dynamical system as the polyphase AC electrical commutation matrixer	11-54
11.5.1	Introduction	11-54
11.6	The two-phase AC electrical homogeneous continuous dynamical system as the two-phase AC electrical commutation matrixer	11-56
11.6.1	Introduction	11-56
11.7	The three-phase AC electrical homogeneous continuous dynamical system as the three-phase AC electrical commutation matrixer	11-61
11.7.1	Introduction	11-61
11.7.2	Three-phase energy sources	11-63
11.7.3	Balanced three-phase AC electrical homogeneous continuous dynamical systems	11-65
11.7.4	Wye connection of the AC electrical homogenous continuous dynamical system	11-70
11.7.5	Delta connection of the AC electrical homogeneous continuous dynamical system	11-73
11.7.6	Unbalanced wye-connected load of the AC electrical homogeneous continuous dynamical system	11-76
11.7.7	Unbalanced delta-connected load of the AC electrical homogeneous continuous dynamical system	11-78
11.7.8	Power holor computations in balanced three-phase AC electrical homogeneous continuous dynamical systems	11-79
11.7.9	Delta three-phase AC electrical homogeneous continuous dynamical systems	11-84
11.7.10	Wye three-phase AC electrical homogeneous continuous dynamical systems	11-87
11.7.11	Holor algebra analysis of the wye–wye three-phase AC electrical homogenous continuous dynamical system	11-88

11.7.12 Holor algebra analysis of the wye-delta three-phase AC electrical homogeneous continuous dynamical system 11-94
11.7.13 Holor algebra analysis of the delta–wye three-phase AC electrical homogenerous continuous dynamical system 11-96
11.7.14 Holor algebra analysis of the delta–delta three-phase AC electrical homogeneous continuous dynamical system 11-97
11.7.15 Delta-to-wye transformations 11-98
11.7.16 Repetitive example I 11-99
11.7.17 Repetitive example II 11-106
11.8 Parallel-series resistance-inductance-capacitance AC electrical commutation matrixer 11-109
11.9 Application of Thévenin's theorem 11-111
11.10 Application of superposition theorem 11-115
References 11-115

12 Mechanical homogenous continuous dynamical systems as mechanical commutation matrixers 12-1

12.1 Simple plane-motion mechanical homogenous continuous dynamical system 12-1
12.2 Simple pendulum mechanical homogeneous continuous dynamical system (approximate solution) 12-2
12.3 Bicycle mechanical homogeneous continuous dynamical system 12-4
12.4 Damper–spring–mass mechanical homogeneous continuous dynamical system (I) 12-4
12.5 Damper–spring–mass mechanical homogeneous continuous dynamical system (II) 12-6
12.6 Automotive vehicle's suspension mechanical commutation matrixer 12-9
12.7 Determination of the analogous impedance holor of mechanical commutation matrixers 12-10
12.8 Damper–spring–mass mechanical homogeneous continuous dynamical system (III) 12-11
12.9 Damper–mass mechanical commutation matrixer 12-14
References 12-18

13 Fluidic homogeneous continuous dynamical systems as fluidic commutation matrixers 13-1

13.1 Fluidic-transmission line 13-1
13.2 Node-to-datum holor equations for the fluidic commutation matrixer 13-2
13.3 Loop and mesh holor equations for the fluidic commutation matrixer 13-4

Part III Physical Matrixers as Physical Commutators

14 Mechanical commutation matrixer commutators for conventional DC and AC magneto-mechano-dynamical electrical machines **14-1**

14.1 Introduction 14-1
14.2 MCM AC–DC/DC–AC commutator dynamotors 14-7
14.3 Schrage MCM AC–AC commutator motor 14-9
14.4 Exemplary applications of mechanical commutation matrixer (MCM) commutators 14-11
14.4.1 AC commutatorless motor—an MCM AC–DC commutator generator for converting AC–DC 14-11
14.4.2 MCM AC–DC commutator amplidyne 14-12
14.4.3 MCM AC–DC/DC–AC ring-commutator single-armature frequency converter 14-13
14.4.4 MCM DC–AC/AC–DC ring-commutator single-armature frequency converter and adjustable-ratio EE transformer connected to an AC commutatorless wound-rotor motor for rotational-speed-control purposes 14-16
14.4.5 MCM DC–AC/AC–DC ring-commutator single-armature frequency converter and MCM DC–AC commutator motor coupled to the shaft of the AC commutatorless wound-rotor motor for speed-control purposes 14-18
References 14-21

15 Electrical commutation matrixer commutators for modern DC and AC magneto-mechano-dynamical electrical machines **15-1**

15.1 Introduction 15-1
15.2 Status and trends 15-5
15.3 Physical and mathematical models of a generalised ECM AC–AC and/or AC–DC/DC–AC commutator 15-19
15.3.1 Introduction 15-19
15.3.2 Hybrid electrical commutation matrixer commutators 15-35
15.3.3 Monolithic electrical commutation matrixer commutators 15-36
Summary 15-49
15.4 MCM and ECM AC–AC and AC–DC/DC–AC commutator dynamotors—a basic application 15-51

15.5 New-concept ECM AC–AC and AC–DC/DC–AC commutator dynamotors 15-53
15.6 Exemplary applications of electrical commutation matrixer commutators 15-61
15.6.1 A 2 × 2 ECM AC–DC/DC–AC commutator 15-61
15.6.2 A 3 × 3, 3 × 5 or 5 × 5 ECM AC–AC commutator 15-61
15.6.3 A 3 × 2 ECM AC–DC commutator 15-64
15.6.4 A 3 × 3 ECM AC–AC commutator 15-65
15.6.5 A single-phase ECM AC–AC and/or AC–DC/DC–AC commutator 15-66
15.6.6 A 2 × 3/3 × 2 ECM DC–AC/AC–DC commutator and 2 × 2 ECM DC–DC commutator 15-67
15.6.7 A 2 × 5/5 × 2 ECM DC–AC/AC–DC commutator 15-70
15.6.8 A 2 × 3 ECM DC–AC/AC–DC commutator and 2 × 2 DC–DC commutator 15-70
15.6.9 A 2 s(2 × 5)/2 s(5 × 2) ECM DC–AC/AC–DC commutator 15-71
15.7 Electrical commutation matrixer commutators—a look into the future 15-72
15.8 Conclusion 15-74
References 15-78

16 Electrical commutation matrixer keyboards for computers 16-1
16.1 Introduction 16-1
16.2 How electrical commutation matrixer keyboards for computers work 16-4
16.3 Electrical commutation matrixer keyboard fundamentals 16-4
16.4 Electrical commutation matrixer keyboard's electrical valves 16-6
16.5 Unconventional electrical commutation matrixer keyboards 16-15
16.6 Virtual electrical commutation matrixer keyboards 16-16
16.7 Canesta keyboard 16-16
16.8 Samsung's Scurry keyboard 16-17
16.9 Exemplary electrical commutation matrixer keyboard 16-17
16.10 Conclusions 16-21
References 16-22

17 Programmable-logic and/or generic-logic electrical commutation matrixers for digital devices **17-1**
17.1 Introduction 17-1

17.2 Programmable-logic electrical-commutation-matrixers and classifications 17-2
17.2.1 The OR electrical commutation matrixer 17-2
17.2.2 The AND electrical commutation matrixer 17-2
17.2.3 Classifications of programmable-logic electrical commutation matrixers 17-3
17.3 Programmable ROMs (PROMs and EPROMs) 17-3
17.4 Programmable matrixer logic (PML) 17-5
17.5 Generic matrixer logic (GML) 17-6
17.6 Electrical commutation matrixer crossbar 17-8
17.7 Electrical commutation matrixer tactile sensor 17-10
17.8 Electrical commutation matrixer seven-segment display 17-11
17.9 Fluidic commutation matrixer combinational-chemistry microreaction 17-13
17.10 Exemplary applications of programmable-logic electrical commutation matrixers for digital devices 17-14
17.10.1 PCM implementing a 2048 × 8 EPROM 17-14
17.10.2 PCM implementing a sum-of-products (SOP) expression 17-14
17.10.3 GML implementing a sum-of-products (SOP) expression 17-17
17.10.4 Electrical commutation matrixer seven-segment display to put on show ‘5’ 17-18
17.11 Conclusion 17-18
References 17-20

18 Nano-magneto-rheological fluido-mechanical commutation matrixer for internal combustion engines 18-1

18.1 Introduction 18-1
18.2 The Fijalkowski engine concept 18-5
18.2.1 Magnetic-field-exciter’s electric-current sequencing for rotary motion 18-12
18.2.2 Engine output-shaft angular velocity control 18-14
18.2.3 Engine output-shaft deceleration and reverse 18-15
18.3 Fijalkowski engine cooling 18-15
18.4 The Fijalkowski engine advantages versus conventional internal combustion engines 18-16
18.5 Conclusions 18-18
References 18-19

19 Conclusion and future trends **19-1**

19.1 Concluding remarks 19-1

19.2 Future work 19-4

References 19-7

Glossary 20-1

Preface

To make progress, science should not accept the limitations placed on discovery by traditional methods, conventional approaches, or existing infrastructure. Moselio Schaechter, Roberto Kolter and Merry Buckley 2004 Microbiology in the 21st Century: Where Are We and Where Are We Going? (Washington DC: AMS)

The real voyage of discovery consists not in seeking new landscapes but in having new eyes.

Marcel Proust

A particular new scientific result does not usually gain a victory in a way that the opponents suffer a defeat and declare that they are converted but much rather the opponents gradually die out and the new generations grow ab ovo familiar with the truth.

Max Planck

The resistance to a new idea increases as the square of its importance.

Bertrand Russell

The title of this textbook requires some explanation. 'Mechatronics' is the integration of physics—mechanics, fluidics, electrics and electronics, etc—and computer technologies into the research and development of physical heterogeneous continuous complex and/or simple dynamical hypersystems (Fijalkowski 2010, 2011). Thus, mechatronics consists of the synergistic combination of different physical disciplines like mechanics, fluidics, electrics and electronics, as well as informatics, etc. Synergy is a very creative and therefore dynamic process, far more so than the usual co-operation of the different physical disciplines, and even more so than a close integration of machine hardware and software.

Mechatronics offers new solutions and unprecedented flexibility in transportation systems, industrial production processes, and aerospace, automotive and traction components, etc. Its economic success is based on functional integration, i.e. the multiple use of mechano-mechanical (MM), fluido-mechanical (FM) and/or electro-mechanical (EM) actuators (machines); the decentralisation of intelligence into physical heterogeneous continuous dynamical hypersystems, e.g. aerospace, automotive and/or traction single-shaft propulsion mechatronical control systems; and the inherent options for sensorless self-monitoring and system protection.

The 'dynamical systems approach' is a holistic description of observable physical reality. *The whole of the dynamical systems approach is not composed of the sum of its elementary units—the multiplicity of elementary units is a result of the diversification of the whole.*

The idea of a *holor* (from the Greek $\acute{o}\lambda o\sigma$, as in the English *holistic*, pertaining to an entity or a whole) seems to have been originated by Wallis (1673) to describe a mathematical entity that is made up of one or more *merates* (from the Greek $\mu\acute{\epsilon}\rho o\sigma$, part of an entity) or *coordinates* (independent quantities) of the holor. It includes complex numbers, scalars, vectors, matrices, tensors, quaternions and other hypernumbers. The works of Grassmann (1862, 1844, 1894) strongly influenced the subsequent development of holors, including the works of Gibbs (1881, 1884) and Heaviside (1893, 1925, 1950).

Holors, thus defined, have been known for centuries, but each type has been developed more or less independently, accompanied by a separate nomenclature and theory (Moon and Spencer 1986).

The 'theory of holors' demonstrates how these complicated subjects can be made simple by using a single notation that applies to all holors, both tensors and non-tensors. Moon and Spencer (1965, 1986) consider all possible types of holors and develop holor algebra and holor calculus in the most general sense.

Thus, the reader will learn to develop a new holor that fits the application, rather than forcing an application onto a holor representation that is known but does not perfectly describe it.

The discussion includes non-tensors that have no transformation and holors that transform in more complicated ways than are allowed with ordinary tensors. This opens up the possibility of devising a holor for a new physical application without being limited to a few conventional types of holor. The theory of holors should establish a method through which students and teachers can learn vector and tensor analysis via a uniform treatment (Moon and Spencer 1963, 1965, 1986; Fijalkowski (1967)).

This textbook deals with the identification of natural Lagrangian dynamical systems (Arnold, 1978), that is, the formulation of the mathematical description, i.e. the mathematical model of natural Lagrangian dynamical hypersystems. They have inputs and outputs that are related by means of dynamical relationships in a form of the Euler–Lagrange second-order differential equation of dynamics in the original dynamical systems approach and holor notation. Physical heterogeneous continuous complex and/or simple dynamical hypersystems with holonomic and unholonomic constraints are generalised physical heterogeneous continuous dynamical objects and devices. This book takes into consideration a single 'identification method', which uses the dynamical systems approach and the matrix notation of the Euler–Lagrange second-order differential equations of dynamics. Identification is defined as the *formulation of a mathematical description*, i.e. it is the *mathematical model of the physical heterogeneous continuous dynamical hypersystem.* As the mathematical model is a generalised presentation of a significant feature of real or designed physical heterogeneous continuous dynamical hypersystems, it is in a convenient form for the researcher. A mathematical model is as a rule a simplification of reality.

In physics and engineering, the author discriminates three fundamental aims for a formulation of mathematical models of physical heterogeneous continuous dynamical hypersystems: analytical study, design and control.

In an identification of some real (existing) physical heterogeneous continuous complex and/or simple dynamical hypersystems there is as a rule a great degree of indeterminacy, connected with the existence of stochastic disturbances and often an inaccurate knowledge of the physical heterogeneous continuous dynamical hypersystem's structure. This indeterminacy often limits statistical identification methods (Manczak and Nahorski 1983).

In this book, the author has confined himself exclusively to 'deterministic identification methods', because in the considered mathematical models of physical heterogeneous continuous dynamical hypersystems, the stochastic disturbances may be neglected.

Deterministic identification methods for physical heterogeneous continuous dynamical hypersystems are interesting for practitioners and designers.

Dynamical process identification is a complex work that comprises (Wegrzyn 1974): the formulation of mathematical models, experimental studies (collection of measurement data), analytical studies (physical-parameter computer-simulation of mathematical models) and the verification (and inspection) of mathematical models. The most important and most difficult aspect in physically continuous dynamical hypersystems is the formulation of mathematical models, which is why this is a principal aim of this textbook.

In principle, the problems that exist upon a formulation of a simple mathematical model of the dynamical system may be considered to have been sufficiently studied. In the case of a formulation of a synthetic mathematical model or a functional mathematical model of the physical heterogeneous continuous dynamical hypersystem, a suitable methodology of its formulation should be applied.

In the holor theory of abstract functional heterogeneous continuous dynamical hypersystems (including physically continuous dynamical hypersystems, among others) two modes of mathematical modelling methodology may be selected: an analytical mode with decomposition methods and a synthetic mode comprising aggregation methods.

The full advantages of both modes are taken into consideration in this book. That is why when considering physical heterogeneous continuous dynamical hypersystems two kinds of mathematical models may be created. Namely, synthetic mathematical models for a structure of the physical heterogeneous continuous dynamical hypersystem, as well as simple mathematical models for individual, functional and structural dynamical components of physical heterogeneous continuous dynamical hypersystems, that is, elemental homogeneous continuous dynamical systems, or dynamical hyposystems.

The formalisation of physical heterogeneous continuous dynamical hypersystems provides everything about the mathematical models' methodology, comprising different operations, phases and connections in a synthetic dynamical process. However, this is not the case for identification, which most often resolves itself in the identification of the homological, functional and structural dynamical components of physically continuous dynamical hypersystems, that is, physically homogenous continuous dynamical systems, or dynamical hyposystems, by means of well-known methods.

Stating a generality of knowledge on a considered problem precisely in the domain of mathematical formalism is termed 'mathematical model formulation', and concrete mathematical relationships are termed 'mathematical models'. Because the mathematical model only simulates these features, which specify a purpose for which it has been created, a given physical heterogeneous continuous dynamical hypersystem may have no single mathematical model, but instead several, simulating heterogeneous viewpoints. It may be affirmed, going out from most generalised assumptions, that each of the simple 'component' mathematical models may be led out from a certain synthetic mathematical model of the physical heterogeneous continuous dynamical hypersystem, in which they are connected to each other.

A synthetic mathematical model that is solely in the domain of abstract mathematical models and uses the Euler–Lagrange second-order differential equations of dynamics for its formation may be formatted in the dynamical systems approach and holor matrix notation. In compliance with this, it is good to determine the physical heterogeneous continuous dynamical hypersystem by means of a structure, representing the so-called 'synthetic mathematical model'.

The physical heterogeneous continuous dynamical hypersystem as a whole may be concerned or it may be divided into physical homogeneous continuous dynamical systems on the grounds of which form of energy is concerned (e.g. kinetic energy: radiant, thermal, motion, sound and electrical homogeneous continuous dynamical systems; potential energy: chemical, nuclear, stored mechanical, fluidic, and electrical and gravitational homogeneous continuous systems). Next, they may be divided as physical homogeneous continuous dynamical hyposystems and, as follows, they may be divisible as functional and structural physical homogeneous continuous dynamical components.

The author's intention in writing this textbook is to submit 'generalised physical commutation matrixer holor analyses' in the easiest practicable dynamical systems approach (systems thinking). The author supposes that this can be best realised through the application of physical commutation matrixers and the functional and structural physical homogeneous continuous dynamical components.

Nearly all mechanical and electrical engineering students have been educated concisely regarding physical commutation matrixers during their university physics courses. In view of this, the physical model represents an exceptional structure with which to form a junction between static and dynamical analysis.

It is probably particularly important that the generalised physical commutation matrixer concept lends itself systematically to analogy. Owing to the application of the physical commutation matrixer method, one may confirm that it is practicable to consider the area under discussion in mechatronics (kinetic energy—radiant, thermal, motion, sound and electrical—as well as potential energy—chemical, nuclear, stored mechanical, fluidic, electrical and gravitational). For instance, the analogy between charging an electrical capacitor and saturating a fluidic container of water is made evident by equating the physical and mathematical models of these physical processes. Naturally, one might have identified this analogy instinctively. In spite of this, one becomes aware that these physical processes are also comparable to the variation in velocity of a mass when a force is relevant. This is not practically so evident. As a result, the physical commutation matrixer concept confirmation is exceptionally convenient in the absence of physical situations in various physical heterogeneous continuous dynamical hyposystems for a generalised physical heterogeneous continuous dynamical hyposystem from which a universal type of solution may be completed.

A new definition of the 'physical commutation matrixer', based on matrix-interconnection physical heterogeneous continuous dynamical hypersystem theory concepts, is proposed. The definition may be used to evaluate physical commutation matrixers, which perform multivalent logical functions, and continuously operating physical commutation matrixers. The serial and parallel connections of two physical commutation matrixers and simplified formulae are defined, which are valid for high component physical commutation matrixers.

A comparison is made between results arrived at using the new and conventional definitions. It is found that the results differ little in the case of equal probability output letters. For instance, a physical commutation matrixer is a general term referring to a physical heterogeneous continuous dynamical hypersystem or part of a physical heterogeneous continuous dynamical hypersystem of matrixery conductive parts and their matrixery interconnections through which an energy-transfer holor is intended to flow.

A physical commutation matrixer, i.e. a configuration of physically (electrically, magnetically, optically, or radiationally) connected dynamical components or analogue and/or digital devices, is made up of active and passive physical, functional and structural dynamical components or an assemblage of physical, functional and structural dynamical components, and their matrix-interconnected row and column conductive collectors. Thus, a physical commutation matrixer is a physical device powered by physical energy.

The active physical, functional and structural dynamical components are the sources of physical energy for the physical commutation matrixer; for instance, they may be chemo-electrical/electro-chemical (ChE/ECh) storage batteries, direct current (DC) or alternating current (AC) mechano-electrical/electro-mechanical (ME/EM) dynamotors (generators/motors), photovoltaic cells, or fuel cells.

The passive physical, functional and structural dynamical components are impeders or admitters, i.e. resistors, inductors and capacitors.

The physical commutation matrixer described by a matrixer physical model (matrixery signal-flow diagram or map) shows the active and passive physical, functional and structural dynamical components and their matrixery interconnected row and column conductive collectors (conductors).

For the purposes of analysis, apparatus (equipment) and devices with an individual physical identity are often represented by equivalent physical commutation matrixers. These equivalent physical commutation matrixers are made up of the basic passive and active physical, functional and structural dynamical components listed above. For instance, electrical commutation matrixers are used to transmit electrical energy in high-voltage power lines and electro-electrical (EE) transformers or in low-voltage distribution in factories and homes. They are used to convert energy from or to its electrical form, for example, as in EM motors, ME generators, microphones, loudspeakers and lamps; to communicate information, as in telephones, radios, televisions and internet systems; to process and store data and make logical decisions, as in computers; and to form systems for the automatic control of equipment.

Physical commutation matrixer theory includes the study of all aspects of matrixers, including analysis, design and application. In it, the fundamental quantities are the 'energy-potential-difference holors' (i.e. generalised force holors) between various points, the 'energy-transfer holors' (i.e. generalised velocity holors) flowing in a number of row and column conductive collectors, and the parameters, which describe the passive physical, functional and structural dynamical components. Other important physical commutation matrixer quantities, such as power, energy and time constants, may be computed from the fundamental physical variables. For a discussion of these parameters, physical commutation matrixer theory is often divided into special topics. This can be based on how the energy-potential-difference holors (i.e. voltage holors and current holors) in the physical commutation matrixer vary with time (i.e. DC or AC, sinusoidal, non-sinusoidal, digital and transient physical commutation matrixer theory). Moreover, they can be divided based on the arrangement or configuration of the energy-transfer holors' (i.e. electrical current holors) row and column conductive collectors (series, parallel, series parallel, parallel series, coupled, open circuited and short-circuited physical commutation matrixers).

The physical commutation matrixer can be divided into special topics according to which physical devices form the matrixer, or the application and use of physical commutation matrix theory can also be divided (power, communication, macro- and microelectronic, solid state, integrated computers and programmable commutation matrixers).

This textbook provides a comprehensive and careful development of the physical commutation matrixer base and the use of holor theory formulations of analytical methods.

No attempt is made to provide the required mathematical tools: matrixer theory, the functions of a complex variable, Laplace transforms, numerical methods and programming. However, some references in these areas have been included in the bibliography.

Holor analysis has been shown beyond doubt to be of prime importance in physics and engineering. Every current scientist and engineer must be systematically well acquainted with the symbolism and methods of manipulation.

Because both the nomenclature and routine manipulation are quite uncomplicated and may be learned in a short time, the question emerges as to whether an independent one-term subject on holors is adequate for a programme of scientific study.

If the subject is on the experiential level, it may be integrated into one of the necessary physics subjects, such as electronics and magnetics, mechanics, fluidics, photonics, electronics, etc.

However, if one expects to use matrices as a systematically dependable tool, one must really understand them. This not only necessitates that an independent subject is taught in the programme of study, but also that there is a textbook that is dissimilar from the old school one.

An examination of the handbooks on matrix analysis since 2000 confirms that virtually all of them are founded on the same physical models. One considers these physical models to be inadequate on two counts (Moon and Spencer 1965):

- they do not refer to 'invariance', which is in reality the imperative factor that makes a matrix a *matrix*, a vector a *vector* and a scalar a *scalar*;
- they give too much weight to rectangular coordinates.

Matrix analysis as column matrix or vector analysis is the infant of *quaternions* and *Ausdehnungslehre*. The former are principally algebraic, and in them questions of invariance do not arise. The latter deals with geometrical figures that are tacitly assumed to be unconcerned by coordinate transformations. However, precise consideration of invariance under coordinate transformation (which has assumed such importance in the 20th century, especially after the advent of relativity) is not present in either parents or offspring. This is why some mathematicians have considered column matrix analysis or vector analysis to be a marginal subject: an ad hoc combinatory logic foundation. The inexperienced confidence that a column matrix or a vector maintains its form and size is transformed when the coordinates are not adequate. One must also inquire, 'under what group of transformations?'. Because matrix analysis has disregarded this question, it is concerned by imprecision. The unacceptable dissimilarity among free, bound and sliding column matrices or vectors is a paradigm. In addition, the old school treatment of matrices gives too much weight to orthogonal Cartesian coordinates. Even the definitions of 'gradient', 'divergence' and 'curl' are normally only provided for particular cases of rectangular coordinates, as if no other coordinates ever emerge in mathematics or physics.

The treatment presented in this interdisciplinary textbook differs from the old school one by introducing *invariance* into the theory and providing *general definitions* that are sensible in all coordinate systems. In this approach, matrix analysis is endowed with a concrete logical foundation. Irrespective of the forward-thinking aspects of the book, we do not sense that the treatment is too difficult for ordinary readers.

This interdisciplinary subject has been taught to undergraduates and postgraduates at the Cracow University of Technology in Krakow and the State Higher Vocational School in Nowy Sacz, Poland. This textbook has been written primarily for undergraduate students of physics and electrical, mechanical, fluidic and thermal engineering. Its logical structure, however, should also make it valuable for mathematicians, and it may serve as a helpful review source for graduate students.

The first few chapters of the book introduce a variety of concepts that may be unknown to some readers, but mastery of these concepts may provide a much deeper knowledge of matrix analysis than could otherwise be acquired. Index notation is employed where necessary, but applications to electronics and mechanics, etc, are performed in the well-known physical commutation matrixer nomenclature. Various problems are known to provide readers with satisfactory preparation in using holors.

I am the sole author of this textbook and all the text contained herein is of my own conception unless otherwise indicated. Any text, figures, theories, results, or designs that are not of my own devising are appropriately referenced in order to give acknowledgment to the original authors. All sources of assistance have been assigned due acknowledgement. In this book, all the information has been obtained and presented in accordance with academic rules and conduct. I have fully cited and referenced all the material and results that are not original to this book.

I am also indebted to my international and national colleagues who contributed indirectly to this book. In addition, I am grateful to all of you who have adopted this text for your interdisciplinary classes or for your own use. Without you I should not be in business.

I hope that you find this textbook to be a valuable learning tool and reference for students. The author is also grateful to his friends for helpful comments. The above declaration should not be interpreted as a suggestion that these benevolent friends concur with all the iconoclastic propositions presented in my text; quite the contrary!

For all the radicalisms and imperfections, I assume total responsibility. I would like to conclude this preface by sharing my life motto—*with focus, motivation and concerted efforts, one can make the seemingly impossible, possible.*

Bogdan Thaddeus Fijalkowski
Krakow, 21 August 2015

References

Arnold W I 1978 Mathematical methods of classical mechanics (Graduate Texts in Mathematics vol 60) (Berlin: Springer)

Cajori F 1912 Historical notes on the graphic representation of imaginaries before the time of Wessel *Am. Math. Mon.* **19** 167

Fijalkowski B 1967 Zastosowanie rachunku holorowego w teorii pradow przemiennych (Application of a holor calculus in the theory of alternating currents) *Przeglad elektrotechniczny (Electrotechnics Review)* Rok XLIII, Zeszyt 2, Warszawa, 1967 (In Polish)

Fijalkowski B T 2010 *Automotive Mechatronics: Operational and Practical Issues.*— vol I. International Series on Intelligent Control, and Automation Science and Engineering—Vol 47 (Berlin: Springer)

Fijalkowski B T 2011 *Automotive Mechatronics: Operational and Practical Issues.*— vol II. International Series on Intelligent Control, and Automotive Science and Engineering—Vol 52 (Berlin: Springer)

Gibbs J W 1881 *Elements of Vector Analysis* privately printed, New Haven pp 17–50; 1881 pp 50–90 1884; 1906 The scientific papers of J Willard Gibbs Vol. II (London: Longmans Green and Co.) 17

Gibbs J W 1884 *Elements of Vector Analysis* privately printed, New Haven, 1881 pp 17–50; 1884 pp 50–90; 1906 The scientific papers of J Willard Gibbs Vol. II (London: Longmans Green and Co.)

Grassmann H 1862 *Die Ausdehnungslehre* Enslin, Berlin, 1862 (In German)

Grassmann H 1844 *Die lineale Ausdehnungslehre, ein neuer Zweig der Mathematik* Ottp Wigand, Leipzig

Grassmann H 1894 *Grassmann's gesammelte math. u. phys. Werke* 3 vols (Leipzig: B G Teubner)

Heaviside O 1893 *Electromagnetic Theory* (London: Electrician Printing and Pub. Co.), 3 vols. 1950 (New York: Dover) Heaviside's treatment is given in vol 1, 3 132–305: The elements of vectorial algebra and analysis

Heaviside O 1925 *Electromagnetic theory* 3 vols (London: The Electrician Printing and Publishing Company Ltd.)

Heaviside O 1950 *Electromagnetic theory* 3 vols contains a critical and historical introby ed Ernst Weber (New York: Dover), see vol 11 pp 192–305 on vector analysis

Manczak K and Nahorski Z 1983 *Komputerowa identyfikacja obiektow dynamicznych (Computerised Identification of Dynamical Objects)* Panstwowe Wydawnictwo Naukowe, 1983 (In Polish)

Moon P H and Spencer D E 1963 A new mathematical representation of alternating currents *Tensor* 110

Moon P H and Spencer D E 1965 *Vectors* (Princeton, NJ: Van Nostrand)

Moon P H and Spencer D E 1986 Theory of holors *A generalization of Tensors* (Cambridge: Cambridge University Press)

Wallis J 1673 Letter to Collins, 6 May 1673 published in *Treatise of algebra* (London 1685) p 264

Wallis J 1685 Letter to Collins, 6 May 1673 Treatise of algebra (London, 1685) p 264

Wegrzyn S 1974 Systemy automatyki kompleksowej. Teoria i doswiadczenie *Prace VI Krajowej Konferencji Automatyki* 1974 (In Polish)

Acknowledgements

The author gives his sincere thanks to the many friends and students who not only inspired his initial interest in the subject, but also gave many helpful suggestions during the writing of the book. Particularly, the author wishes to express his gratitude to Dr Jozef Tutaj, who has been indispensible in the writing of this book, and whose enthusiasm, advice, and encouragement helped make all of it possible. Special thanks go to my grandson Marcel for his very helpful discussions and for the preparation of many of the figures in this book. I am grateful to the many authors referenced in this book from whom, during the course of writing, I learned so much regarding the subjects covered. I am also indebted to my international and national colleagues who contributed indirectly to this book. I would like to take this opportunity to also thank the IOP Publishing crew for their persistence in making this book a reality. In addition, I am grateful to all of you who have adopted this text for your interdisciplinary classes or for your own use. Without you I would not be in business.

Author biography

B T Fijalkowski

Professor Dr Bogdan T Fijalkowski was born in 1932 in Poland. He obtained an MSc in Electrical Engineering, PhD in Power Electronics and DSc in Physical Heterogeneous Continuous and Discrete Dynamical Systems from Szczecin University of Technology in 1959, Academy of Mining and Metallurgy in 1965 and Poznan University of Technology in 1988, respectively. He began his professional and electrical engineering carrier in 1955. He has been working 5 years in industry and 50 years in academia; he was the Director of the Electrotechnics & Industrial Electronics Institute, Faculty of Electrical and Computer Engineering and he was the Head of Automotive Mechatronics Institution, Faculty of Mechanical Engineering at the Cracow University of Technology, Poland. He was a visiting professor at several well-known universities. He serves as Consultant to several organizations in Poland and the United States. He is the signatory of the MoU for the establishment of World Electric Vehicles Association (WEVA). He was Guest Editor of *Journal of Circuits, Systems and Computers*—Special Issue on Automotive Electronics, reviewer and referee, *IEEE Transactions on Circuits and Systems*—Part I, *IEEE Transactions on Fuzzy Systems*, *International Journal of Vehicle Design*, USA, *International Journal on Advanced Robotic Systems* and *International Journal of Technology Management (IJTM)*, USA. He is a reviewer of books and book chapters on automotive electrics and electronics, electric drives systems dynamics and wireless information transmission, as well as information security, and is listed in *Who's Who in the World*, *Who's Who in Science and Engineering* and so on. He has published 29 books and book chapters as well as over 200 technical papers and 25 patents on mining and automotive electrics and electronics as well as mechatronics. Recent publications by him include books (monographs), book chapters, journal articles, and conference proceedings on topics such as mathematical models of selected aerospace and automotive discrete dynamical hypersystems as well as the civil and military, wheeled and tracked all-electric and hybrid-electric vehicles and also nanomagneto-rheological fluid (NMRF) mechatronic commutator, 'crankless' internal combustion engines with energy storage, termed the Fijalkowski engines and automotive gas turbines that are based on the Fijalkowski turbine boosting (FTB) system. Recently, Springer published two volumes of his book entitled *Automotive Mechatronics: Operational and Practical Issues* and IOP Publishing published his interdisciplinary book entitled *Mechatronics: Dynamical Systems Approach and Theory of Holors*. He made valuable contributions to several of ASME, ISATA, ISTVS, SAE, IEEE SMC, as well as EVS and WEVA, conferences as a track and/or session chair and speaker. His hobby is yachting. He has the rank of Ocean-Going Yacht Master.

Acronyms

ABEL	Advanced Boolean expression language
ABW	Absorb-by-wire
AC	Alternating current
AEV	All-electrical vehicle
AI	Artificial intelligence
AWA	All-wheel-absorbed
AWB	All-wheel-braked
AWD	All-wheel-driven
AWS	All-wheel steered
BBW	Brake-by-wire
BDC	Bottom dead centre
b–k	Damper–spring
b–m	Damper–mass
b–k–m	Damper–spring–mass
CASE	Computer-aided system engineering
C^2M	Crankshaft and connecting rod (conrod) mechanism
CECM	Coordinate electrical commutation matrixer
CHINT	Charge injection transistor
CHP	Combined heat and power
ChE	Chemo-electrical
Ch-Th	Chemo-thermal
Ch-Th-FM	Chemo-thermo-fluido-mechanical
Ch-Th-F-ME	Chemo-thermo-fluido-mechano-electrical
CMOS	Complementary MOS
COTS	Commercially-off-the-shelf
CPU	Central processing unit
CR	Common rail
CS	Computer system
CUPL	Computer for universal programmable logic
D	Differential
DBW	Drive-by-wire
DC	Direct current
DDPWM	Direct duty-ratio PWM
DG	Distributed generation
DoF	Degrees-of-freedom
DTC	Direct torque control
ECE	External combustion engine
ECL	Emitter-coupled logic
ECM	Electrical commutation matrixer
ECU	Electronic control unit
E^2CMOS	Electrically erasable complementary metal-oxide semiconductor
E^2D	Electrical energy distribution
EFD	Electro-fluido-dynamical
EIC	Electrical integrated circuit
EIM	Electrical integrated matrixer
EMD	Electro-mechano-dynamical
EOF	Electro-osmotic flow

EPD	Electro-plasmo-dynamical
EPROM	Erasable PROM
EMU	Electrical multiple unit
EE	Electro-electrical
ECh	Electro-chemical
EM	Electro-mechanical
E-Th	Electro-thermal
FBW	Fly-by-wire
FCM	Fluidic commutation matrixer
FEM	Finite elements method
FET	Field effect transistor
FFET	Flow field effect transistor
FL	Fuzzy-logic
FM	Fluido-mechanical
FM	Frequency modulation
FOC	Field-oriented control
FPLA	Field programmable logic array
FM	Fluido-mechanical
GCM	Generalised commutation matrixer
GECM	Generic-logic electrical commutation matrixer
GERF	Giant electro-rheological fluid
GML	Generic matrix logic
GTO	Gate-turn-off
HBT	Hetero-junction bipolar transistor
HDI	High-pressure direct injection
HDL	Hardware description language
HE	Hybrid electrical
HEMT	High electron injection transistor
HEV	Hybrid electrical vehicle
HPS	Hybrid power source
HVDC	High-voltage direct current
H&RD	Human and/or robotic-driver
HE	Hydro-electrical
I	Integral
ICE	Internal combustion engine
IEEE	Institution of Electrical and Electronics Engineers
IGBT	Insulated-gate bipolar transistor
IP	Internet protocol
IPM	Interior permanent magnet
IT	Information technology
I/O	Input/output
JFET JUGFET	Junction-gate FET
KFL	Kirchhoff's first law
KSL	Kirchhoff's second law
LAN	Local area network
LED	Light-emitting diode
LM–RM	Linear reciprocating motion to rotary motion
LTQ	Light triggered and quenched
MIMO	Multi-input/multi-output
MCM	Mechano-electrical commutation matrixer

MCT	MOS controlled thyristor
MCU	Microcontroller unit
MEMS	Micro-electro-mechanical system
MESFET	Metal-semiconductor FET
MFD	Magneto-fluido-dynamical
MISFET	Metal-insulator-semiconductor FET
MMC	Man–machine communication
MMD	Magneto-mechano-dynamical
MNOS	Metal-nitride-oxide silicon
MOS	Metal-oxide semiconductor
MOSFET	Metal-oxide semiconductor FET
MPD	Magneto-plasmo-dynamical
MRV	Martian Roving Vehicle
MSI	Medium scale integration
MSI	Mild soft iron
ME	Mechano-electrical
MF	Mechano-fluidic
MM	Mechano-mechanical
MP	Mechano-pneumatic
MV	Mechano-vacuum
N	North
NAEMS	Nano-electro-mechanical system
NF	Neural network and fuzzy logic
NMOS	Non-complementary metal-oxide semiconductor
NMRF	Nano-magneto-rheological fluid
NN	Neural-network
OFET	Organic field effect transistor
OTS	Off-the-shelf
P	Proportional
PBC	Passively based control
PC	Personal computer
PC	Predictive control
PC	Process control
PECM	Programmable-logic electrical commutation matrixer
PDA	Personal digital assistant
PF	Power factor
PHEV	Plug-in hybrid-electrical vehicle
PI	Polarisable interference
PLM	Programmable logic matrixer
PML	Programmable matrix logic
PROM	Programmable read-only memory
PWM	Pulse-width modulation
PM	Pneumo-mechanical
QUIT	Bipolar quantum interference transistor
RB	Reverse blocking
RBSOA	Reverse-biased safe-operating area
RBW	Ride-by-wire
RCGTO	Reverse conducting GTO
RM–LM	Rotary motion to linear reciprocating motion
R&D	Research and development

R-C	Resistor-capacitor
R-L	Resistor-inductor
R-L-C	Resistor-inductor-capacitor
S	South
SBW	Steer-by-wire
SCR	Semiconductor controlled rectifier
SISO	Single-input/single-output
SET	Single electron transistor
SIT	Static induction transistor
SLI	Starting, lighting and ignition
SMES	Superconducting magnetic energy storage
SM&GW	Steered, motorised and/or generatorised wheel
SPWM	Sinusoidal PWM
SR	Semiconductor rectifier
SSI	Standard scale integration
SOP	Sum-of-product
SUPER-HET	Bipolar superconductor hot-electron transistor
SVM	Space vector modulation
SVPWM	Space-vector PWM
TAB	Tape automated bonding
TDC	Top dead centre
TTL	Transistor–transistor logic
UPS	Uninterruptible power supply
UV	Ultra-violet
VAP	Very advanced propulsion
VAT	Very advanced technology
VDU	Video display unit
VLSI	Very large scale integration
VoIP	Voice over Internet protocol
VSC	Voltage-source chopper
VSI	Voltage-source inverter
XBW	X-by-wire
Δ2Y	Delta-to-wye
Π2T	Pi-to-tee
2WD	Two-wheel-drive
2WS	Two-wheel-steered
2D	Two dimensional
3D	Three dimensional
4WA	Four-wheel-absorbed
4WB	Four-wheel-braked
4WD	Four-wheel-driven
4WS	Four-wheel-steered
2×1	Two-by-one
2×2	Two-by-two
3×2	Three-by-two
3×3	Three-by-three
3×5	Three-by-five
5×5	Five-by-five
5×6	Five-by-six

Part I

Towards a unified theory exploiting a dynamical systems approach for physical heterogeneous continuous dynamical hypersystems

Chapter 1

General considerations

CONSILIENCE

Most of the issues that vex humanity daily cannot be solved without integrating knowledge from the natural science with that of the social sciences and humanities. Fluency across the boundaries will provide a clear view of the world as it really is, not as seen through the lens of ideologies and religious dogmas or commanded by myopic response to immediate needs.

E O Wilson

1.1 Dynamical hypersystem definition

Dynamical hypersystems unified theory is a mathematical area that describes the behaviour of heterogeneous and/or homogeneous, complex and/or simple, continuous and/or discrete, dynamical hypersystems, usually by applying differential equations and/or the difference equations of dynamics. When the differential equations of dynamics are applied, the theory is termed 'continuous dynamical hypersystems'. When the difference equations of dynamics are applied, the theory is termed 'discrete dynamical hypersystems'. When the time variable runs over a time set that is discrete over some intervals and continuous over other intervals, or is any arbitrary time set such as a Danish mathematician Georg Cantor (1845–1918) set, one obtains dynamical equations on time scales. Some situations may also be primarily physically and secondarily mathematically modelled by miscellaneous 'operators', such as the differential–difference equations of dynamics. This theory deals with the long-term qualitative behaviour of dynamical hypersystems, and studies the solutions of the dynamical equations of motion of dynamical hypersystems that are primarily mechanical in nature—although this includes both planetary orbits and the behaviour of systems that are electrical, magnetic, fluidic

doi:10.1088/978-0-7503-1350-6ch1

or thermal in nature—and the solutions to partial differential equations that arise in biology. Much of modern research is focused on the study of chaotic dynamical hypersystems. This subject of study is also termed just 'dynamical hypersystems', 'mathematical dynamical hypersystems theory' or 'the mathematical theory of dynamical hypersystems'. The term 'physical heterogeneous continuous dynamical hypersystems', in its most generic form, means the dynamical hypersystems of dynamical systems, hyposystems and components that change over time. The more technical usage, 'physical heterogeneous continuous dynamical hypersystems', refers to a class of mathematical differential equations of dynamics that describe time-based dynamical hypersystems with particular properties.

A 'physical heterogeneous continuous dynamical hypersystem' is complex or simple when it is compiled from many dynamical parts (dynamical systems, hyposystems and components, respectively) that interconnect in intricate manner. This definition has to do with the number and nature of the interconnections. The metric for intricateness is the amount of information contained in the physical heterogeneous continuous dynamical hypersystem.

A physical heterogeneous continuous complex and/or simple dynamical hypersystem (see figure 1.1) presents dynamical complexity when the cause and effect are subtle over time.

Examples: a dramatic variety of effects in the short and long run; dramatically different effects locally and in other dynamical parts (dynamical systems, hyposystems and components, respectively) of the physical heterogeneous continuous dynamical hypersystem; evident interventions cause non-evident consequences.

A physical heterogeneous continuous dynamical hypersystem (figure 1.1) is complex and/or simple when it is composed of a collection of related constituent dynamical parts (dynamical systems, hyposystems and components, respectively), for which the degree and nature of the relationships are imperfectly known.

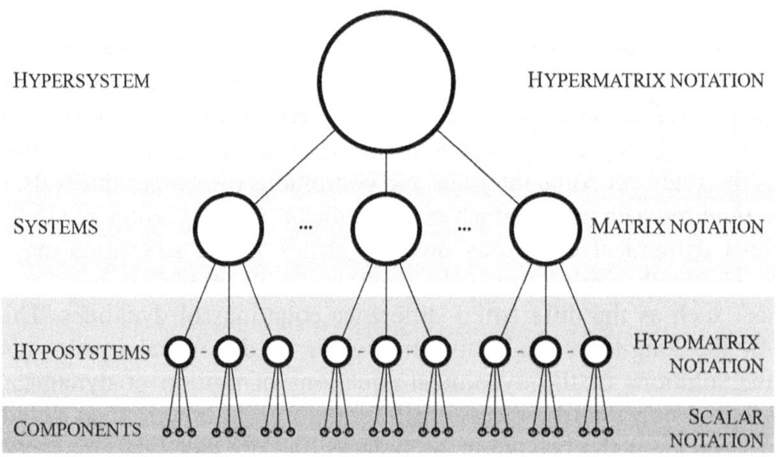

Figure 1.1. Characteristics of physical heterogeneous continuous complex and/or simple dynamical hypersystems.

A physical heterogeneous continuous complex and/or simple dynamical hypersystem has a collection of various dynamical systems, hyposystems and components so connected or related as to perform a particular function not performable by the dynamical systems, hyposystems and components alone.

This necessitates different problem-solving *modi operandi* at various levels of abstraction. Thus, a physical heterogeneous continuous complex and/or simple dynamical hypersystem means a grouping of various dynamical systems, hyposystems and components that operate together for a common purpose.

Scientific dynamical complexity correlates to the behaviour of the macroscopic compilation of parts with the potential to change in time. This is altered from mathematical dynamical complexity, i.e. the number of mathematical operations indispensable to resolve a problem exercised in computer science.

Complexity theory and chaos theory both attempt to reconcile the unpredictability of nonlinear physical heterogeneous continuous complex and/or simple dynamical hypersystems with a sense of underlying order and structure. The implications are: a paradigm where the short-term is predictable but long-term planning is impossible; dramatic unexpected alterations; orderliness can be altered to be more innovative and adaptive.

The Newtonian paradigm is founded on Cartesian reductionism (the machine metaphor) and Cartesian dualism. The body is a physical machine and the mind is separate from the body. The intuitive concept of a machine created from continuous parts that can be reduced to those parts without losing its machine-like character is Cartesian reductionism.

The Newtonian paradigm and the three laws of motion—the general laws of motion—are applied as the basis of the modern scientific method. Dynamics is the centre of the physical idea, which leads to trajectory.

Dynamical complexity is a post-Newtonian paradigm for a unified dynamical systems approach to physical heterogeneous continuous dynamical hypersystems composed of various dynamical constituent parts (dynamical systems, hyposystems and components, respectively).

Dynamical complexity is rooted in the fundamental laws of nature. This opens the way to a systematic study beyond mere analogies and metaphors, a multi-level hierarchical dynamical systems approach with an interdisciplinary dimension.

Dynamical complexity has benefits beyond nonlinear dynamics and statistical physics. By contrast to traditional fundamental science-rooted approaches, dynamical complexity addresses a class phenomenon in which a dynamical hypersystem and an observer may evolve at comparable scales of time and space.

Dynamical complexity comes from the failure of the Newtonian paradigm to be general. Complex and/or simple dynamical hypersystems are disjointed compilations that deal with all of nature, but the dynamical systems approach is invented from complex things and the simple mechanisms are imagined and created by science; experiments involve reducing the dynamical hypersystem to its constituent dynamical parts (dynamical systems, hyposystems and components, respectively) and then studying these parts in a background that is formulated on the subject of dynamics.

How is science pursued? The senses provide information and mental activity makes sense of that sensory information. Arrange a natural dynamical hypersystem into a formal dynamical hypersystem and then manipulate the formal dynamical hypersystem to imitate the causal change in the natural dynamical hypersystem. From the formal dynamical hypersystem, derive an implication that relates to the causal event in the formal dynamical hypersystem and then interpret the formal dynamical hypersystem and verify its ability to represent the causal event in the natural dynamical hypersystem and its constituent dynamical parts (dynamical systems, hyposystems and components, respectively) before studying those parts against a background formulated on the subject of dynamics.

For years, we complied with the Newtonian paradigm as the formal dynamical hypersystem, not even remembering about there being encoding and decoding, and gradually the ontology began to alter so that the Newtonian paradigm in reality replaced or became the physical universe.

As we began to look more deeply into the dynamical systems approach, we came up with aspects that the Newtonian paradigm failed to capture. Then we considered an explanation to be necessary. Dynamical complexity was created. This can be formalised easily. It has an acutely philosophical meaning.

Dynamical complexity is a phrase that expresses that it is laborious to formulate the overall behaviour, even when almost comprehensive information is known about its microscopic dynamical components and their interrelations. A commonly heard spontaneous definition of dynamical complexity is that it is the 'edge of chaos' between order and randomness. A simple set of laws can lead to physical heterogeneous continuous complex and/or simple dynamical hypersystems, correlated to entropy (disorder) and the second law of thermodynamics. That law states that entropy always increases in any closed physical heterogeneous continuous dynamical hypersystem not in equilibrium, and remains constant for a hypersystem which is in equilibrium.

1.2 Conclusions

There are two main modes of physical heterogeneous complex and/or simple dynamical hypersystems: differential equations and difference equations of dynamics (also known as iterated maps). The differential equations of dynamics describe the evolution of physical heterogeneous continuous dynamical hypersystems in continuous time, whereas difference equations (iterated maps) arise in physical heterogeneous discrete dynamical hypersystems where time is discrete. The differential equations of dynamics are applied much more widely in science and engineering, and the author shall therefore concentrate on them. Confining his attention to differential equations of dynamics, the main distinction is between ordinary and partial differential equations of dynamics. His concern here is purely with temporal behaviour, and so he will deal exclusively with ordinary differential equations of dynamics in a holor notation.

In its general conception, a physical heterogeneous continuous complex dynamical hypersystem may be described as a complex of interacting dynamical parts (dynamical systems, hyposystems and components, respectively), together with the relationships among them that permit the identification of a boundary-maintaining entity or process. From this perspective any given physical heterogeneous continuous complex dynamical hypersystem can be described as a collection of dynamical parts (dynamical systems, hyposystems and components, respectively), together with a set of relationships between these dynamical parts. This is the origin of set notation:

$$\text{hypersystem} = (\text{systems, hyposystems, components})$$

with respect to this notation, 'hypersystem (hypermatrix), systems (matrices), hyposystems (hypomatrices), components (scalars)', denoting, respectively, a dynamical hypersystem, a set of dynamical parts distinguished within a dynamical hypersystem and a relation (or, possibly, a set of relations) defined on dynamical parts.

Clearly, the parthood and hypersystemhood properties of dynamical hypersystems reside in dynamical systems, hyposystems and components, respectively. Having defined the concept of dynamical hypersystems we then identify the domain of inquiry of the dynamical systems approach (systems thinking) to be those aspects of dynamical hypersystems that derive from their hypersystemhood properties as opposed to their parthood properties.

A physical heterogeneous continuous complex dynamical hypersystem is non-fragmentable. If it were fragmentable, it would be a machine. Its reduction to dynamical parts (dynamical systems, hyposystems and components, respectively) annihilates important dynamical hypersystem characteristics irreversibly.

A physical heterogeneous continuous complex dynamical hypersystem comprises real dynamical components that are separate from its dynamical parts. There are functional dynamical components defined by the dynamical hypersystem, whose definition relies on the context of the dynamical hypersystem. Outside the dynamical hypersystem, they have no meaning. If removed from the dynamical hypersystem, they lose their original identity.

Physical heterogeneous continuous complex dynamical hypersystems have mathematical models, analytical or synthetic. However, the implements differ. If a synthetic mathematical model can replace an analytical mathematical model, the dynamical hypersystem is fragmentable. There is no 'major mathematical model'. If there were a major mathematical model, all other mathematical models could be created from it and fragmentability would result. Causalities in the dynamical hypersystem are combined when they are dispersed over the parts. The nature of causality necessitates feedbacks and feedforwards (closed loops) that are excluded in the Newtonian paradigm. The important attributes of the physical heterogeneous continuous dynamical hypersystem are beyond algorithmic definition or realisation: a path to negate the well-known thesis that all the mathematical models of computation that have been developed, and all those that may be developed in the future, are equivalent in power. One will never find a more powerful mathematical model.

Dynamical complexity is the feature of a physical heterogeneous continuous dynamical hypersystem that is evident in the inability of any formalism to depict all its properties adequately. It is necessary to find a 'particularly different' mode of interacting with dynamical hypersystems.

Particularly different in the sense that when one creates successful physical models, the formal physical heterogeneous continuous dynamical hypersystems considered necessary to describe each continuous aspect are not derivable from each other.

The Newtonian approach was primarily guided by immediate observations, which is quite natural in such a pioneer's situation. Force, relative motion and momentum formed the cornerstones of Newton's mechanics. Energy, as a more abstract concept, was added to classical mechanics through several steps as the integrated force (Antippa 2000).

Newton's equations of motion were perhaps too excellent for their time, making motion more or less independent of its origin. When combined with Galilean relativity, Newtonian motion found its reference at the observer, an obvious misconception that was inherited by the theory of relativity. Relative velocity is well justified in a kinematical sense as the rate of the change in the distance between an object and the observer. Relative velocity does not, however, relate the quantity of effective mass and kinetic energy stored in a moving object; the build-up of effective mass and kinetic energy is related to the work done in obtaining the motion in the relevant energy frame. Classical ether was implicitly fixed to the observer, which made Newton's laws of motion applicable to any observer *as such*.

The observer-oriented approach claiming the 'same observations' for any observer was reflected in the relativity principle, which was presented to justify the modification of the local ether environment so that observations looked the same for any observer. The result was Lorentz modified inertial frames and space–time modified gravitational frames.

The dynamical systems approach's 'multi-level ether' is built up from the dynamical hypersystem of cascaded energy frames. Motion gained from potential energy in a local frame is paid back in the same frame.

1.3 Summary

The basic quantities of physical heterogeneous complex or simple continuous dynamical hypersystems are defined below.

- **Dynamical quantity:** a physical quantity (position, momentum, phase, electric field…) that depends on time.
- **Dynamical hypersystem:** a set of dynamical quantities that evolves together under a coupled set of equations or maps.
- **Continuous-time versus discrete-time dynamical hypersystems:** this depends on whether time is treated as a continuous or discrete variable:
    * continuous-time dynamical hypersystems are governed by (sets of coupled) **ordinary differential equations** (ODEs), **partial differential equations** (PDEs), integro-differential equations, delay-differential equations of dynamics, etc; e.g. Newton's second law, Maxwell's equations, etc;

* discrete-time dynamical hypersystems are governed by maps (iterate); e.g. $x_{n+1} = F(x_n)$.

- **Linear dynamical hypersystem:** a dynamical hypersystem, whose dynamical quantities respond under the influence of a driving field or driving force *linearly* with respect to that driving field or force. Such continuous dynamical hypersystems are characterised by linear integro-differential equations of dynamics.
- **Nonlinear dynamical hypersystem:** a dynamical hypersystem, whose dynamical quantities respond under the influence of a driving field or driving force *nonlinearly* with respect to that driving field or force. Such continuous dynamical hypersystems are characterised by nonlinear integro-differential equations of dynamics.

Reference

Antippa A F 2000 *Can. J. Phys.* **78** 883–99

Chapter 2

Model definition

THEORY VERSUS MODEL

Insofar as the propositions of mathematics give an account of reality they are not certain; and insofar as they are certain they do not describe reality.
Albert Einstein

A theory has only the alternative of being right or wrong. A model has a third possibility; it may be right but irrelevant.
Manfred Eigen

The model is rational and can be understood by anyone with basic knowledge in physics and mathematics.
Ari Lehto

2.1 Introduction

A basic aim of this textbook is to present a methodology for the formation of physical and mathematical models of physical heterogeneous continuous complex and/or simple dynamical hypersystems. Such models are necessary for the synthesis of these dynamical hypersystems, and help to simplify the computer simulation analytical studies.

In science, models have been known for a very long time. For instance, models (not then so-called) were conceived by the Greek philosopher, Democritus (circa 400 BCE–circa 370 BCE), for the structure of the atom; a model structure of the geocentric universe, the 'Ptolemaic system', was developed by the Greek mathematician, astronomer and geographer, Ptolemy (Claudius Ptolemaeus, circa 90 BCE–circa 168 BCE) in *Almagest* ('The Greatest'); and the model of the heliocentric universe, the 'Copernican system' was developed by the Polish astronomer, Nicolaus Copernicus

doi:10.1088/978-0-7503-1350-6ch2

(1473–1543). Copernicus's famous book, *De Orbium Caelestium Revolutionibus* (1543) was banned by the Roman Catholic Church until 1758.

Ptolemaic astronomy described observations relative to the observer at rest on the Earth. For describing the planetary orbits, a system of epicycles was necessary. Copernican astronomy described planets as objects orbiting the Sun, which meant that the observer was moving with the moving Earth. Such a holistic view created the basis for a practical understanding of the system and the celestial mechanics behind the motion. A German mathematician, astronomer, and astrologer Johannes Kepler (1571–1630) found the mathematics in Copernicus's planetary systems and Newton the physics behind Kepler's equations. Further, an English physicist, and mathematician Issac Newton (1642–1727) formulated the concept of gravitation, and the laws of motion, which were fixed to inertial observers' frames, following the Italian scientist Galileo Galilei's (1564–1648) relativity. A German-born physicist Albert Einstein extended the use of Galilean–Newtonian relativity and observer-oriented inertial frames by redefining the concepts of the coordinate quantities, time and distance. The roots of the energy system-based frames of reference and the zero-energy principle were laid down by a German polymath and philosopher Gottfried Leibniz (1646–1716) when he proposed (in 1694–96) that the build-up of *vis viva* (living force or kinetic energy) is obtained against the release of *vis mortua* (dead force or potential energy) by reflecting the balance of motion and gravitation in spherical closed space; a localised energy object in the dynamical universe framework is like Leibnitz's monad—a perpetual living mirror of the Universe (Suntola 2006).

In his *Baltimore Lectures*, William Thomson (Lord Kelvin, 1824–1907 AD) affirmed that to understand a phenomenon you need to construct its mechanical model. The models are means to and a form of scientific perception and they have found applications in strictly theoretical, analytical studies as well as experimental studies. A notion of the model is in equivalence, which is used as a default here as an unambiguous definition and a generally received classification.

Applying different methods of model formation, a multitude of models of physical heterogeneous continuous dynamical hypersystems may be obtained. The obtained multitude ought to be isomorphic against the functional physical heterogeneous continuous dynamical hypersystem, that is, it ought to exist as an abstract homology between the multitude of models and a modelled physical heterogeneous continuous dynamical hypersystem. Hence, for the same outputs of physical quantities, the dynamical hypersystem and its model ought to be homologically synonymous (in the sense of the selected criterion).

We will take 'identification' in this textbook to mean the formation process for the models of the physical heterogeneous continuous dynamical hypersystems, which aim to estimate their dynamical and static characteristics while using real information on the physical quantities acting on inputs of the physical heterogeneous continuous dynamical hypersystems and their interaction concerning these physical quantities. All the models by convention are divisible into two classes: 'physical models' and 'mathematical models', as shown in figure 2.1.

By a 'physical model' of the physical heterogeneous continuous dynamical hypersystem we mean a model in which the physical processes occurring are

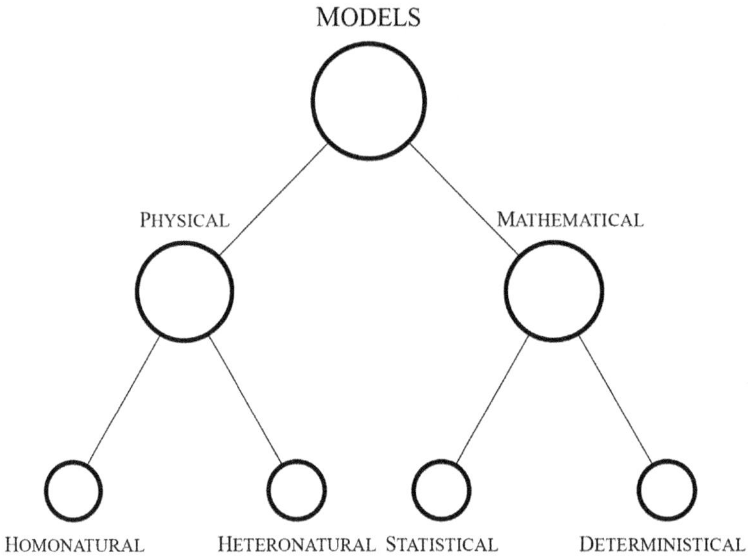

Figure 2.1. Classification of models (Fijalkowski 1987).

dealt with in a way that is equivalent to describing them through mathematical relationships.

Physical models may be divided into two categories (figure 2.1): homonatural mathematical models, which have the same physical nature as the physical heterogeneous continuous dynamical hypersystem, and heteronatural mathematical models, whose physical nature is different from the physical nature of the physical heterogeneous continuous dynamical hypersystem. Such a model turns out, for example, to be a physical model of the physical heterogeneous continuous complex and/or simple dynamical hypersystem, instead of a physical model of the physical heterogeneous continuous dynamical hypersystem's physical commutation matrixer, which is described in chapter 7.

In the 20th century something new emerged that was more adaptable to present-day scientific methodologies—the notion of the 'mathematical model'.

Mathematical models may be divided into two groups (figure 2.1). The first comprises 'deterministic' mathematical models, and the second comprises 'statistical' (random) mathematical models.

A deterministic mathematical model of the physical heterogeneous continuous dynamical hypersystem is a model whose description is possible in the form of the function relationships between inputs and outputs of the physical heterogeneous continuous dynamical hypersystem.

As a rule, a mathematical model is adequate for a physical process, functional and structural dynamical components and superimposed constraints.

A statistical mathematical model of the physical heterogeneous continuous dynamical hypersystem is a model that is formulated while taking into consideration the statistical parameters and the probability expansion function.

2.2 Physical model

Physical models are designated to reflect the physical processes of the physical heterogeneous continuous complex and/or simple dynamical hypersystem. Models of the first kind differ from the physical heterogeneous continuous dynamical hypersystem only by their geometrical dimensions and a range of parameter variations. This is why processes that have a place in the physical heterogeneous continuous dynamical hypersystem and their models only differ quantitatively. This makes it possible to the study the physical nature of the physical heterogeneous continuous dynamical hypersystem, and its transient processes and boundary conditions, using the best computation methods and fundamental theorems.

To estimate the applicability of a formatted physical model that has the same physical nature as the physical heterogeneous continuous dynamical hypersystem, one ought to make a preliminary estimate of the degree of the analogy between the physical processes in the model and those in the physical heterogeneous continuous dynamical hypersystem.

By the formation of physical models, it is possible to use a mathematical description of these physical processes that characterises the physical heterogeneous continuous dynamical hypersystem. The identity of the physical processes occurring in the real physical heterogeneous continuous dynamical hypersystem and its model may be provided by using a probability theory without an aforementioned mathematical description. Physical processes are called 'homothetic' (reciprocal), if the relations of physical quantities that characterise them are constant. The formation process for physical models resolves itself into the selection of model parameters that concern a probability theory and the selection of homological initial and boundary conditions.

The theory of probability is based on the three theorems presented, for example, in the works of Vienikov (1956, 1961). These theorems constitute the mutual relations between the physical quantities which characterise the nature of physical processes. In compliance with the first theorem of probability, in similar physical processes the probability criteria are the same numerically.

To develop physical models that are compatible with the second probability theorem, the number of **degrees of freedom** (DoF) cannot be greater than the number of fundamental units for the measurement system adapted to describe a given physical phenomenon. This means that if a physical process in the physical heterogeneous continuous dynamical hypersystem depends upon q physical quantities, of which b are fundamentals and $d = q - b$ are derivatives, it may be possible to create d probability criteria. In this way, at the formation of the physical models, the physical quantities d should be given going out from the probability criteria. According to the third probability theorem, for physical heterogeneous continuous dynamical hypersystems and physical models in which the conditions of unambiguity are similar, the probability criteria created from the conditions of unambiguity are numerically the same. According to this theorem there should be additional conditions to select a single physical model from all-possible physical models and in this way unambiguously formulate a physical model of the physical heterogeneous continuous dynamical hypersystem. A physical model is 'synchronous' with a physical nature in this case if a physical process runs across in a natural time scale. If a physical model functions in an artificial time scale, then it is 'homogenous' (Pukhov and Khatiashvili 1971).

The ratios between the probability constants are called the 'indicators'; however, the probability criteria are sometimes called the 'invariants'. For the same results, it is allowed to apply an analysis of dimensionality. The formation of physical models on functional and structural dynamical components that differ from the physical nature of these dynamical components is based on the probability of mathematical relationships, which describe physical processes occurring in dynamical systems and 'heteronatural physical models', i.e. with a varying physical nature. So as in 'homonatural physical models', i.e. with the same physical nature, heteronatural physical models are created in this way, and for each real functional and structural dynamical element of the physical heterogeneous continuous dynamical hypersystem there will be a corresponding determined functional and structural dynamical component of the physical model.

2.3 Deterministic mathematical model

Mathematical modelling is a more generalised form of modelling than physical modelling. The approach in mathematical modelling differs somewhat from that in physical modelling. In physical modelling a physical model may have the same physical nature as the heterogeneous continuous dynamical system. In the case where a physical model has been created with different functional and structural dynamical components from those of the real heterogeneous continuous dynamical system, there invariably remains a requirement for maximal identity between the structure and the mathematical relationships' parameters. In mathematical modelling, this requirement is rejected. The real heterogeneous continuous complex and/or simple dynamical hypersystem and its mathematical model ought to be equivalent in the sense of the principle of operation.

For instance, in a mathematical model it is permitted to perform an analytical study by means of a computer, although the processes occurring in the computer may be completely different. Deterministic mathematical models in their essence differ from statistical mathematical models. These differences between the two aforementioned kinds of mathematical models are conditioned by the specifics of different classes of problems, which for their solution require one model's features but not another's. A deterministic mathematical model is one whose behaviour is entirely predictable. The physical heterogeneous dynamical hypersystem is perfectly understood, and then it is possible to predict precisely what will happen.

2.3.1 Synthetic deterministic mathematical model

A synthetic deterministic mathematical model of the physical heterogeneous continuous dynamical hypersystem, relating to dynamical and static states, establishes a set of nonlinear differential equations with partial derivatives, when it imitates in an exact manner the real physical heterogeneous continuous complex and/or simple dynamical hypersystem.

2.3.2 Chaotic deterministic mathematical model

A chaotic deterministic mathematical model is one where the behaviour of the physical heterogeneous continuous complex and/or simple dynamical hypersystem cannot be entirely predicted. The detailed study of chaos is beyond the topic of this textbook; here we only introduce a definition of the chaotic deterministic mathematical model. The most important components of chaotic behaviour are the following.

- Sensitive dependence on initial conditions in a bounded set;
- aperiodicity.

2.4 Statistical mathematical model

A statistical mathematical model has a considerably more formal character than a deterministic mathematical model, because it reflects the physical heterogeneous continuous complex and/or simple dynamical hypersystem, disregarding its concrete physical features.

2.5 Stochastic mathematical model

A stochastic mathematical model is one where the behaviour of the physical heterogeneous continuous complex and/or simple dynamical hypersystem cannot be entirely predicted.

References

Fijalkowski B 1987 *Modele matematyczne wybranych lotniczych i motoryzacyjnych mechano-elektro-termicznych dyskretnych nadsystemów dynamicznych (Mathematical Models of Selected Aviation and Automotive Continuous Dynamical Hyperystems)*, Monografia 53 Politechnika Krakowska imienia Tadeusza Kościuszki, Krakow, 276 p. (In Polish)

Pukhov G E and Khatiashvili C S 1971 *Lokalnyie matematicheskoye modeli sistem upravleniya (Local Mathematical Models of Control Systems)* (Kiiev: Naukova dumka) (In Russian)

Suntola T 2006 The Dynamical Universe: Relativity defines the locally available share of total energy *Physical Interpretations of Relativity Theory, PIRT X* (London: Imperial College)

Vienikov V A 1956 *Fizicheskiye modeli elektricheskikh sistem (Physical models of electrical systems)* (Moscow: GEI) (In Russian)

Vienikov V A 1961 K voprosu o klasifikaciy modeley i metodov modelirovaniya (To question on classification of models and methods of modelling). *Izv. Vuzov, Energetika*, No. 10, (In Russian)

Chapter 3

Dynamical systems approach

Always strive to construct definitions, or at the least a rudimentary image of that which has arisen in your mind; strip it bare and focus on its true nature, considering its entirety through its elements and its elements through its entirety.

Marcus Aurelius

Less emphasis on analysis
More emphasis on synthesis.

Anonymous

3.1 Introduction

The first advance of modern Western science occurred during the Renaissance, a cultural movement that spanned the period roughly from the 14th to the 17th century, beginning in Italy in the late Middle Ages and spreading to the rest of Europe. The prime attributes are 'analytical vision' and 'Cartesian reductionism'. The philosophy of Descartes was not only of prime importance for the advance of science, but it has also had an impressive impact on the Western 'universal' approach to thinking up to the present day. The Eastern 'universal systems thinking', derived from biology and medicine, is more holistic, based on the view that all things and events deduced by the senses are interrelated and connected, and cannot be divided during study; hence it focuses more on the relations between dynamical parts than on the essence of each dynamical part.

The modern universal systems approach or universal systems thinking was founded in the 20th century, but primitive versions emerged several thousand years before. In the ancient period, both Chinese and Greek philosophers created holistic visions of the Universe. At that time, science, philosophy and religion were not separated. These types of holistic visions are intuitive and from a contemporary

doi:10.1088/978-0-7503-1350-6ch3

viewpoint they are founded on primitive systems intuition. As for the science, as Einstein pointed out, 'The only valuable thing is intuition'.

Intuition can be considered as a mode of thinking and also as a kind of knowledge exploitation and creation. Intuition is an ability possessed by everybody. It is strictly related to tacit knowledge. Intuition is a preliminary step when scientists take a universal systems approach to a project. It has been suggested that this kind of intuition be called 'systems intuition' and it has more significance in systematics (systems engineering). Systems intuition or systems intuitive thinking is characterised by both holistic and intuitive vision. The scientific foundation of systems intuition is based on the noetic (cognitive) science initiated in the 20th century by the famous Chinese scientist, Xuesen Qian (Hsue-Shen Tsien), the father of space technology in China. He argued that the limitation of traditional cognitive science is that it focuses largely on analytical (logical) thinking and ignores imaginative thinking, which has the principal role in decisions and innovation. He is also the originator of 'systemology' and a science of thinking.

The neotenic features of systems intuition lie in the concept of the synthesis of imaginative thinking and creative thinking based on experience (Wang 2011).

An idea of the 'dynamical systems approach' in the methodology of the formulation of mathematical models of physical heterogeneous continuous dynamical hypersystems was originated by the author of this book under the influence of a book written by Weinberg (1975, 1979) entitled *An Introduction to General Systems Thinking*, which described different situations and methodological operations appearing in **research and development** (R&D) works step-by-step, showing them against a background of problems for a solution (Fijalkowski 1987).

The dynamical systems approach in the methodology for the formulation of mathematical models of physical heterogeneous continuous dynamical hypersystems facilitates the formation of mathematical models and offers a concept for their analytical study by computer, just using systems thinking, enabling skilful methodological progress.

General systems thinking is an approach to integration that is based on the belief that the dynamical parts of a dynamical hypersystem will act differently when isolated from the dynamical hypersystem's environment or other parts of the dynamical hypersystem. Standing in contrast to positivist and reductionist thinking, systems thinking sets out to view dynamical hypersystems in a holistic manner.

Consistent with systems philosophy, systems thinking concerns understanding a dynamical hypersystem by examining the linkages and interactions between the dynamical parts (dynamical systems, hyposystems and components, respectively) that comprise the whole of the dynamical hyposystem.

When scientists and engineers encounter situations that are complex and messy, systems thinking can help one to understand the situation systemically. This helps scientists and engineers to see the complex dynamical hypersystem—from which they may identify multiple leverage points that can be addressed to support constructive change. It also helps them see the connectivity between the dynamical parts (dynamical systems, hyposystems and components, respectively) in the situation, so as to support joined-up actions.

The never-enough-to-be-regretted Professor Bohdan Walentynowicz (past editor of the Polish *Electrotechnical Review*) in a preface to the first Polish edition of the book written by Weinberg (1979) wrote:

> *Maybe what most accurately indicates the nature of the book written by Weinberg are the words printed on the book jacket of the original. One reads in them about how young workers in science can learn systems thinking in two ways. The first way—the specialist way—depends on mastering the detailed knowledge of a certain field, and then on crossing to generalisations, however at the comprehensible limitation of the field of sight because of the bending domain. The second way—the generalist way—involves learning about thinking, and then thinking in the categories of generalised systems, before the thinking domain, and sometimes it is concurrent with or also integrates specialist methods. This second way of thinking it is not limited to young workers in science, but also applies to mature scholars, engineers, organisers, students and different people.*

The author of this book gives up on defining the notion of the *system*, since this just deals with many authors of prominent books (Wilson 1969, Klir 1979, Weinberg 1979).

However, so the reader of this book can assimilate the notion of the dynamical hypersystem more easily, we will return to the words of Professor Bohdan Walentynowicz (Weinberg 1979), that this is a notion of 'some whole—entering into the composition of the larger wholes, created from parts (the smaller wholes), related in the way a certain structure is assigned, and distinguished with regards to some functions assigned to these wholes'.

The dynamical systems approach in the methodology of formatting mathematical models of physical heterogeneous continuous dynamical hypersystems concerns their analysis and also the projection of their relations, which because of the aim of dynamical systems are interesting.

In getting to know a physical heterogeneous continuous dynamical hypersystem, it is not desirable to examine the whole at once, one should decide on a limitation and investigate it step by step through its dynamical components, its functional and structural dynamical components. It is a matter of choosing what belongs to a given dynamical hypersystem and what does not, and in doing this, a matter of defining what that dynamical hypersystem is.

The physical heterogeneous continuous dynamical system is the thought construction created by the researcher, and so it is—Weinberg (1979) underlines this several times in his book—a certain 'way of seeing the world', for example, the 'physical universe' of the physical heterogeneous continuous complex and/or simple hypersystem.

When considering any functional physical heterogeneous continuous complex and/or simple dynamical hypersystem, encountered as a functional structure, there are a great number of diverse phenomena, which relate to various fields of theoretical and applied sciences.

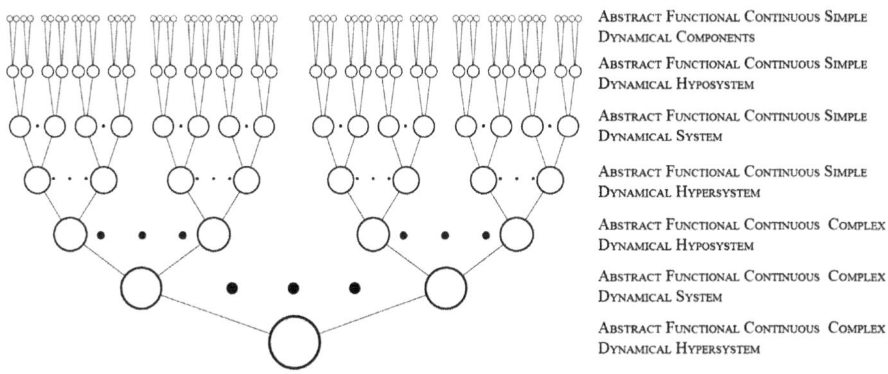

Figure 3.1. A hierarchical structure of a physical model of an abstract functional continuous complex dynamical hypersystem (Fijalkowski 1987).

From a definite point of view, one can perceive similar functional and structural dynamical components in these heterogeneous phenomena (e.g. physical phenomena), which sometimes seem various. To investigate these phenomena from the same point of view or with mutual correlations we introduce in this book an abstract notion that relates to all phenomena in the same degree. This is the notion of the 'abstract functional continuous complex and/or simple dynamical hypersystem' (figure 3.1). However, the principles established for these phenomena were termed the 'dynamical systems approach' in the formulation methodology of the mathematical models for abstract functional continuous complex and/or simple dynamical hypersystems. Thanks to this, in a methodology for the formulation of mathematical models of functional objects (e.g. machine aggregates and executive mechanisms), simplicity was found in the complexity.

Introducing the dynamical systems approach into the methodology for the formation of mathematical models for abstract functional continuous complex and/or simple dynamical hypersystems complies with the idealisation of this textbook—the 'continuous'—letting us describe a dynamical state of a considered complex and/or simple dynamical hypersystem through a finite number of certain physical quantities. The complex and/or simple dynamical hypersystems in which such an idealisation is admissible are termed 'continuous'.

As an example, the abstract functional continuous complex and/or simple dynamical hypersystem (figure 3.1) can be folded, respectively, with the finite number of continuous complex and/or simple dynamical systems, respectively, consisting of the finite number of continuous complex and/or simple dynamical hypersystems, folded with the finite number of continuous complex and/or dynamical hyposystems, or functional and structural dynamical elements, respectively.

The basic requirement of the dynamical systems approach in the methodology for the formation of mathematical models of physical heterogeneous continuous complex and/or simple dynamical hypersystems is the existence of: a researcher studying the functional object (e.g. a machine aggregate or an executive mechanism) and the solved problem, marking the point of the researcher's view of the studied functional object. So now, created by the researcher, the abstract functional continuous complex and/or simple dynamical hypersystem (figure 3.1) is the physical model of the original 'considered' by him from a definite point of view. What applicability does the described hierarchical structure of the physical model of the abstract functional continuous complex dynamical hypersystem have?

Firstly—it contains on every hierarchical level the full set, possible for this level, of the abstract functional continuous complex dynamical systems, complex dynamical hyposystems, simple dynamical hypersystems, simple dynamical systems, simple dynamical hyposystems and simple dynamical components. This means that it can be neither more nor less on the given level by even one abstract functional continuous complex dynamical system, complex dynamical hyposystem, simple dynamical hypersystem, simple dynamical system, simple dynamical hyposystem, or simple dynamical component, and that there can be such kinds that only this physical model simulates. This assures in practice the complexity of the analysis and synthesis, for example, of any functional continuous complex dynamical hypersystem.

Secondly—in the aim of a simulation of a concrete functional object, for example, a physical heterogeneous continuous complex dynamical hypersystem, one selects on every graph branch of the physical model (or one skips certain graph branches), the most distant from the basis of the graph tree (from a continuous complex dynamical hypersystem), such a continuous simple dynamical hypersystem which for researchers still has practical meaning in considerations, for example, of the physical heterogeneous continuous simple dynamical hypersystem. All the chosen (sorted) graph branches of the physical model end on a significant continuous simple dynamical hypersystem and still create the set of elementary functional continuous simple dynamical systems and relationships among them, serving to build, for example, the physical heterogeneous continuous simple dynamical hypersystem. One can build the physical model on this basis, simulating the concrete (e.g. physical) heterogeneous continuous simple dynamical hypersystem, considered from the definite point of vision.

Thirdly—it lets us compare among themselves various concrete (e.g. physical) functional heterogeneous continuous dynamical hypersystems (folded with elementary functional homogeneous continuous simple dynamical systems, simple dynamical hyposystems and simple dynamical components), of the same level by transformation to their functional objects. The aforementioned theses are not only valid for a received division and classification of the physical heterogeneous continuous simple dynamical hypersystems, but also for every different division under the condition that will be kept in the method of formation (introduced in this textbook) for the successive hierarchical levels. The hierarchical structure of the physical model of the abstract heterogeneous continuous simple dynamical hypersphere favours certain proprieties. Namely, when its level (beyond the level making up the complex dynamical hypersystem, that is the object, e.g. the aeroplane, helicopter, automobile, tractor, etc) contains the functional continuous complex dynamical systems as *subjects* acting on the things of their objects, they are on the level directly below and simultaneously the objects of the subjects, and acting

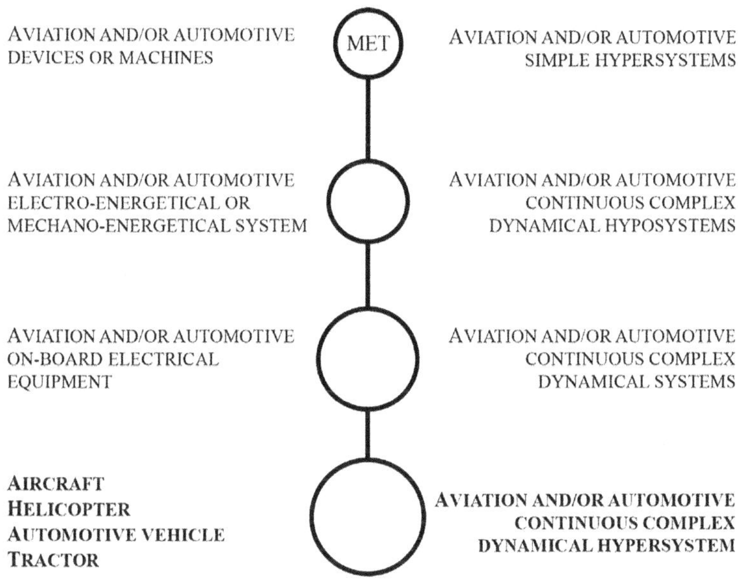

Figure 3.2. A hierarchical structure of a physical model of the aviation or automotive functional heterogeneous continuous complex dynamical hypersystem (Fijalkowski 1987).

of course directly on the higher level. A hierarchical structure on every level, moreover, contains the full set of possible abstract functional homogeneous continuous complex dynamical systems, complex dynamical hyposystems, simple dynamical hypersystems, simple dynamical systems, simple dynamical hyposystems and simple dynamical components created according to the previously described principles. Moreover, this hierarchical structure can at last be extended vertically in an unrestricted way. Therefore it has a theoretical character, because for practical aims the analysis incorporates the set of selected (by the author of this book) holonomic and unholonomic physical heterogeneous continuous complex dynamical hypersystems that are present in this structure (Fijalkowski 1987). It answers the requirements of these definitions for aircraft, helicopters, automobiles and tractor, and one can classify them to a class of phenomena called the physical universe, for instance, of aviation and automotive functional heterogeneous continuous complex dynamical hypersystems (figure 3.2).

In this physical universe, one distinguishes several functional classes of the physical heterogeneous continuous complex dynamical hypersystems, which may be presented in order of growing complexity, although in this book we are only interested in the three initial levels, namely:

- the level of static functional structures incorporating the 'anatomy' of aviation or automotive complex dynamical hypersystems (automobile, tractor, aeroplane and helicopter), i.e. the name of the functional continuous complex dynamical systems and settlement of their mutual positions, that is—among others—the aviation or automotive continuous complex dynamical system, i.e. aviation and automotive electrical equipment;

- the level of functional continuous complex dynamical hyposystems, that is—among others—the aviation or automotive continuous complex dynamical hyposystem, i.e. the aviation or automotive system electro- or mechano-energetic system incorporating the foreseen and repeated influence on itself of concrete simple dynamical hypersystems;
- the level of functional continuous simple dynamical hypersystems, that is—among others—aviation or automotive simple dynamical hypersystems, incorporating elementary continuous simple dynamical systems with definite basic dynamics, that is deterministic, continuous simple dynamical hypersystems, making up aviation or automotive devices and/or machines.

This classification of the complex and/or simple dynamical hypersystem of the level of the higher order contains, respectively, the complex and/or simple dynamical systems of the level of the lower order. Moreover, just this hierarchy of complex and/or simple dynamical hypersystems makes it possible to use the methods applied in the synthesis and analysis of simpler dynamical hypersystems for the formulation of the mathematical models, opinion, building and investigations of more complex and/or simple dynamical hypersystems.

To obtain the picture of the hierarchical function and structure of the concrete physically continuous complex dynamical hypersystem and its functional continuous simple dynamical systems and their functional continuous dynamical simple hypersystems its decomposition was executed. The physical model of the concrete physical heterogeneous (mechano-electro-thermal) continuous complex dynamical hypersystem whose hierarchical structure was built is shown in figure 3.3.

It should be noted that in the process of the successive decomposition of the concrete physical heterogeneous continuous complex dynamical hypersystem we distinguished its functional simple dynamical systems built up from its functional

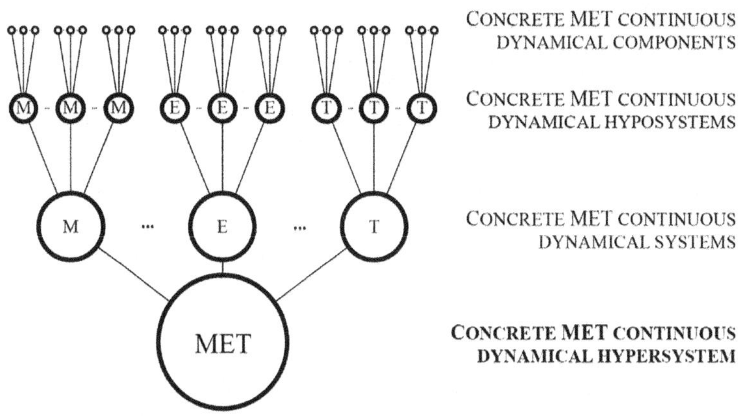

Figure 3.3. A hierarchical structure of the physical model of the physical heterogeneous (mechano-electro-thermal) continuous complex dynamical hypersystem (Fijalkowski 1987).

continuous simple dynamical hyposystems, built-up from its functional continuous simple dynamical components.

Functional heterogeneous continuous complex dynamical hypersystems, simple dynamical systems, simple dynamical hyposystems and simple dynamical component are thus exchangeable notions. Depending on the hierarchical level, prepare the road for the use of the dynamical systems approach in the formulation methodology of the mathematical models for computer analytical studies of both the dynamical and the static processes of the physical heterogeneous continuous complex dynamical hypersystem. A review should be carried out of the classic interpretation and behaviour during a continuous sequence of transformations of the abstract functional continuous simple dynamical hypersystem (figure 3.1).

Attention should be paid nearby to the principal result of the statement about unambiguous solutions of the differential equations of dynamics, because this defines a feature of the half-group called the 'right of causality'. It is shown below that the variation process is a generalisation of this concept.

To illustrate the use of the dynamical systems approach in the formation methodology of mathematical models of functional objects (e.g. machine aggregates and executive mechanisms) in its simplest generalised form, the ways that it leads to mathematical models for the case of selected physical homogeneous continuous simple dynamical hypersystems are shown in later chapters. Finally, to acquaint the reader with the dynamical systems approach in the formulation methodology of mathematical models we present the results of analytical studies of these selected physical heterogeneous continuous dynamical hypersystems.

3.2 The dynamical systems approach to physical heterogeneous continuous dynamical hypersystems

The dynamical systems approach is useful in the development of physical heterogeneous continuous complex and simple dynamical hypersystems.

The premise of the dynamical systems approach is that the development of a physical heterogeneous continuous complex or simple dynamical hypersystem should begin with a thorough understanding of what is intended, including the heterogeneous continuous dynamical hypersystem's dynamical behaviour. If such an understanding can be gained in the early stages of a project, it is possible to eliminate much of the uncertainty now inherent in the process, and to develop physical heterogeneous continuous dynamical hypersystems that are more successful yet less expensive. The understanding of a physical heterogeneous continuous dynamical hypersystem's dynamical behaviour in its environment over time is most often addressed in the design implementation and testing phases of a project, through iterative trial and error.

The dynamical systems approach provides an alternative to this costly process. It allows one to build a usable and comprehensive physical heterogeneous continuous dynamical hypersystem—a physical model that is accurate enough to be relied on and clear enough to be useful. The physical model addresses the usual functional and data flow issues, but also covers the dynamical and behavioural aspects of a physical heterogeneous continuous dynamical hypersystem. This physical model can then be tested with the mathematical model analysis and retrieval tools, which provide extensive mechanisms for inspecting and debugging the specification and retrieving information from it. By testing the physical model systems scientists and engineers can see how the physical heterogeneous continuous dynamical hypersystem as specified would behave if implemented. One can answer detailed 'what if?' questions, following specific scenarios, checking that certain desirable situations will occur in the lifetime of the physical heterogeneous continuous dynamical hypersystem and that other undesirable ones will not.

In this sense, systems scientists and engineers can be said to be playing the role of the eventual user of the physical heterogeneous continuous dynamical hypersystem as well as that of the environment. Moreover, one can also instruct the analytical tools themselves to play the role of the user and the environment. This kind of self-inspection capability has the advantage of testing scenarios that the systems scientists and engineers might not have thought of or might not want to check out individually. Often these scenarios represent unpredictable behaviour.

The physical model must be well structured, intuitive and comprehensive if it is to be useful to the variety of systems scientists and engineers involved, and it must be rigorous and formal if it is to serve as the basis for useful analysis and retrieval. This calls for representing as much of the physical model as possible in visual formalisms—diagrammatical languages with precise meaning.

The systems scientists and engineers model the physical heterogeneous continuous dynamical hypersystem using natural pictorial descriptions that can be understood by physicists, but which also have unambiguous underlying semantics that are meaningful to the computer and can thus be processed by the analytical tools.

The dynamical systems approach meets all of these requirements. It is consistent with existing structured methods, but is more powerful and more flexible.

This textbook explains the philosophy behind the dynamical systems approach. When one says 'approach', one has in mind an entire spectrum of concepts, methods, languages and tools.

The systems scientists and engineers work with a set of computerised tools that support a number of graphical physical commutation matrixers and non-graphical languages, including languages for both physical and mathematical modelling and the theory of holor analysis.

Many of the constituents in the dynamical systems approach are similar to those appearing in other approaches and methodologies for physical heterogeneous continuous dynamical hypersystem design, a fact that should be taken as evidence of their wide appeal. However, there are many aspects of the dynamical systems approach that go far beyond what is currently available elsewhere.

The author has discussed the dynamical systems approach as it applies to all phases of physical heterogeneous continuous dynamical hypersystem development, from physical and mathematical modelling through simulation and analysis to design and implementation.

The dynamical systems approach is based on the preparation and extension of a rigorous physical model, e.g. a physical commutation matrix that captures not only functionality but also dynamical behaviour. These models are constructed using visual formalisms that are clear and intuitive, but precise and rigorous, and thus amenable to computerised analysis.

The dynamical systems approach allows systems scientists and engineers to address the most crucial problem in the successful development of physical heterogeneous continuous dynamical hypersystems: the ability to gain an early understanding of a system's dynamical behaviour and to carry that understanding through the entire development cycle.

The dynamical systems approach will continue to develop as systems scientists and engineers gain the experience needed to advance the state of the art.

The **computer aided systems engineering** (CASE) environment will also continue to evolve: its ability to address future needs is the result of its being based on the firm foundation of comprehensive physical and mathematical modelling through simulation and analysis.

The principal heuristic innovation of the dynamical systems approach is what may be called 'reduction to dynamics' in contrast to 'reduction to dynamical components' as practised in the methodologies of classical science.

3.3 Summary

The dynamical systems approach:

- offers a holistic—from whole to local—approach for describing physical reality;
- produces a comprehensive picture of physical reality;
- produces accurate predictions with straightforward mathematics and clear logic.

The future of the dynamical systems approach is going to be very stimulating over the coming years. Science is developing, changing and expanding—and that is what maintains our enthusiasm.

References

Fijalkowski B 1987 *Modele matematyczne wybranych lotniczych i motoryzacyjnych mechano-elektro-termicznych dyskretnych nadsystemów dynamicznych (Mathematical Models of Selected Aviation and Automotive Continuous Dynamical Hyperystems)*, Monografia 53 Politechnika Krakowska imienia Tadeusza Kościuszki, Krakow, 276 p. (In Polish)

Klir K 1972 *Trends in General System Theory* (New York: Wiley).

Wang Z 2011 Study of System Intuition from Noetic (cognitive) Science Founded by Xuesen Qian. *DUT Presentation,* 2011, In Memorandum of 100 Aniversary of Academician Xuesen Qian. Research Center of Knowledge Wane Sciences and Technology, Dalian University of Technology, 116085 Dalian, China, ss. 1-14

Weinberg G M 1975 *An Introduction to General Systems Thinking* (New York: Wiley) 1975

Weinberg G M 1979 Myslenie systemowe (Systems Thinking). *Wydawnictwo Naukowo-Techniczne*

Wilson W E 1969 *Projektowanie techniczne w ujeciu systemowym (Technical Design in the Systems Approach).* Wydawnictwo Naukowo-Techniczne (In Polish)

Chapter 4

Algorithm for a formulation of the mathematical model

THEORY VERSUS EXPERIMENT

It is the mark of an educated mind to rest satisfied with the degree of precision, which the nature of the subject admits, and not to seek exactness where only approximation is possible.

Aristotle (384 BCE–322 BCE)

Experience is the best teacher.

Marcus Tullius Cicero

Who was it that invented this in-depth analogy, unseeable with the eyes but which the mind can grasp? It is mathematical thought, which holds matter in contempt, attracted to pure form.

Henri Poincaré

While experiment reveals the particular features of every process, the theory permits the establishment of the general laws and thus contributes to the fundamental knowledge of the process.

Rosenthal

Great theoreticians know that hypothesis must be confirmed with experiment.

Albert Einstein

doi:10.1088/978-0-7503-1350-6ch4

The proposal of physical heterogeneous continuous complex or simple dynamical hypersystems is introduced as energy and/or information handling 'physical commutation matrixers', which are identified as physical models. Sets of generalised energy-potential-difference holors and generalised energy-transfer holors and physical homogeneous continuous dynamical systems, hyposystems and components, respectively, are developed. The proposals are then made more precise with reference to mechanical, electrical, fluidic (hydraulic and/or pneumatic), magnetic and thermal homogeneous continuous dynamical systems.

The energetics and informatics restrictions introduced are applied to interconnecting physical homogeneous continuous dynamical systems, hyposystems and components, respectively, and used to identify a selection of interconnectivity constraints and hence obtain mathematical models.

An algorithm for the formulation of the mathematical model of any physical heterogeneous continuous complex or simple dynamical hypersystem comprises not only the steps required for the arrangement of the differential equations of dynamics, but also all those necessary for the identification and verification of the model, namely computer-aided analytical studies, e.g. a computer simulation of the physical heterogeneous continuous complex or simple dynamical hypersystem and the comparison of the results with those of experimental studies. In the general case, the algorithm contains four steps (figure 4.1).

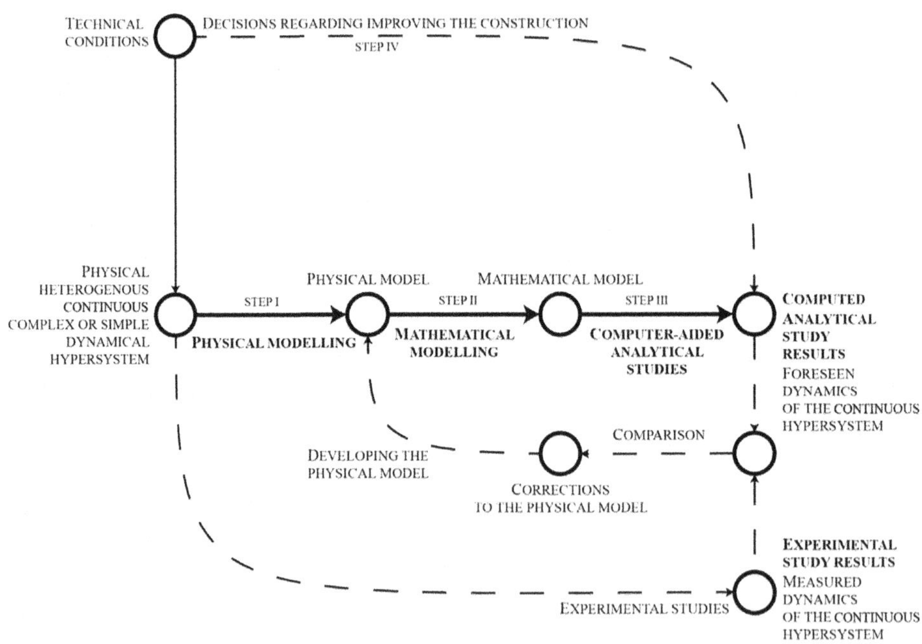

Figure 4.1. An algorithm for the formulation of a mathematical model of the existing and/or designed physical heterogeneous continuous dynamical hypersystem.

- The recognition of the structure, principle of operation and features of the studied physical heterogeneous continuous dynamical hypersystem and the assumptions for simplifying the created physical model. This step of physical modelling has to bring the dynamical and static characteristics of the physical model and the existing (real) physical heterogeneous continuous dynamical hypersystem into sufficient agreement.
- The formulation of the mathematical model comprising a set of differential equations describing the dynamics and statics of the physical model. This step of mathematical modelling does not introduce important new (content-related) information to a description of the physical heterogeneous continuous dynamical hypersystem, and only realises the assumptions of the physical model in mathematical language.
- The analytical studies of the dynamics of the mathematical model. This step consists of the solution of previously arranged differential equations and the design of the dynamical characteristics of the considered mathematical model.
- The comparison of the foreseen dynamics of the physical heterogeneous continuous dynamical hypersystem, which are the results of the analytical studies of the mathematical model with the assumed characteristics of the designed physical heterogeneous continuous dynamical hypersystem, with the results from experimental studies of the already existing (real) physical heterogeneous continuous dynamical hypersystem. This step of verifying the mathematical model in a case of unsatisfactory agreement of the analytical studies' results for the mathematical model with the assumed characteristics has the aim of executing a suitable alteration of the parameters in the physical model, or in the case of the designed physical heterogeneous continuous dynamical hypersystem, the alteration (change) of the physical model, and even the constructional solution (see chapter 3). One should repeat the next step of the algorithm after executing the last step and the settlement of the preliminary project of a construction of the dynamical hypersystem until one has obtained sufficient agreement between the results of the analytical studies of the dynamics and the assumed characteristics of the designed dynamical hypersystem.

In figure 4.1 the four steps of the formulation algorithm for a mathematical model of any existing and/or designed physical heterogeneous continuous dynamical hypersystem are presented by means of a signal-flow graph.

In relation to the obviousness of the subordination the lines of the graph's influence on the respective steps of the algorithm here do not give a precise description of the signal-flow graph, and only direct the attention to stepping out in the feedbacks and feedforwards, realising the verification of the mathematical model with reference to the existing (real) or designed physical heterogeneous continuous dynamical hypersystem.

Chapter 5

Developmental systems approach

There is nothing permanent except change.
Herbalists of Ephesus (circa 525–475 BCE)

Change alone is unchanging.
Heraclitus

Observe constantly that all things take place by change, and accustom thyself to consider that the nature of the Universe loves nothing so much as to change the things which are, and to make new things like them.
Marcus Aurelius

Nothing happens until something moves.
Albert Einstein

To improve is to change; to be perfect is to change often.
Winston Churchill

The world hurries forward—everything changes.
Anonymous

This book attempts to introduce the developmental systems approach into the methodology of physical models of the physical heterogeneous continuous dynamical hypersystems, in addition to the dynamical systems approach. So far this has met with indifference not only from some theorists on the research side of mathematical models, but also some practitioners making experimental studies. In the struggle to improve both the physical model, as well as the (actual or projected) physical heterogenous continuous dynamical hypersystem, the dynamical systems approach still predominates. This leads to some analytical studies researchers falling for the illusion that they are in control of actual physical models. The developmental systems approach turned out not to have been a panaceum for the routine of physical model creation. To break this routine, a considerably stronger impulse is necessary, an

doi:10.1088/978-0-7503-1350-6ch5

impulse that would push researchers to create the macro-conditions needed for computerised analytical studies of mathematical models. If the theorists—systems scientists and engineers—and researchers of mathematical models are interested in their computerised analytical studies as well as their verification (checking) while taking into account the feedback of an algorithm for the formulation of a mathematical model, then sooner or later, they must look towards having a repertoire of methods to improve their physical models. This is an opportunity to recognise the developmental systems approach in the methodology for improving physical models—the idea of improving the physical model, and the comprehensive trouble-shooting processes for dynamical and static problems, which are of interest to researchers. What, therefore, is this idea, and what might be the terms of its usefulness? The basic concept in a glossary of the developmental systems approach in the methodology to create excellent physical models is 'change'. Mathematical models continuously improve due to the effect of the alteration of physical models. Lack of alteration means they are inadequate and unfit. Changes to physical models are therefore in the world of mathematical models, and are important functions of the control of their adequacy. This requires that the mechanisms for responding to the physical model are understood. To be able to control the inadequacy of the mathematical model by means of alterations of the physical model, the theorists—researchers of mathematical model—need diagnostic skills for intrinsic operational states (dynamical and static) or interpretation of the originated observations, based on the resources of knowledge and the capacity to implement the relevant alterations to the physical model. Because recognition of the developmental systems approach in the methodology to improve physical models is a unique strategy for changing them, it requires widespread propagation among mathematical model researchers, who during computerised analytical studies, are working on the principles of skilled incompetence, using a routine to create a physical model and a dullness in systems thinking as a way to reduce a mathematical model's inadequacy. It creates a chance for the researchers of mathematical models to expand their knowledge of the complex solving of dynamical and static problems. It makes it clear that the process of creating a mathematical model is a system of 'learning oneself'. Employing the development systems approach in the methodology for physical model excellence is also a departure from the scheme of the dynamical systems approach. The alteration requires great patience because it is a repetitious process with a multi-step algorithm (see chapter 4). This has disappointed some of the mathematical model researchers, who want to see the immediate effects of great alteration programs on the physical model. At the basis of the diagnosis and intervention that includes the development systems approach in the methodology to improve the physical models lies the mathematical model of 'computer analytical studies in action', based on data derived from feedback (see figure 4.1) between the ideal (calculated in computer analytical studies) and actual (measured in standard experimental studies) operational states (dynamical and static), so there is therefore emptiness, there is a constant flow of alteration data improving the physical model.

The so-called 'external verifier' improvement of a physical model cooperates in a continuous manner with the so-called 'internal verifiers' of the mathematical model—the originators of changes (corrections) or reporters of the problem—and tries to help them to learn about the functional object (the interior of the object and its surroundings) and see the possibilities for the alteration of the physical model, at which they are less sensitive, and address the uncertainty resulting from poor knowledge of physical phenomena, which may include the dynamical systems approach in the methodology to create mathematical models of physical functional objects. The idea of the recognition of the developmental systems approach in the methodology to improve physical models is limited to changes (corrections) to the structure and values of the parameters of a physical model. In this sense, one thinks of alteration of the physical model as a strategy for research intervention. Verifiers also do not forget the difficulties of increasing the physical model, related to the synthesis of the modelled physically continuous dynamical hypersystem, because they cause changes (corrections) in other parts of the physical model of the physical heterogeneous continuous dynamical hypersystem and reciprocal links: the dynamical hypersystem–environment. A comparison of the developmental systems approach in the methodology to create physical models of functional objects and the dynamical systems approach in the methodology to formulate mathematical models of these functional objects shows the sharp differences between them, and these are compiled in table 5.1.

This comparison shows why the dynamical systems approach in the methodology to create the mathematical models of physical objects may not be an alternative to the developmental systems approach, which looks for opportunities to develop the theory of improving not only the physical model, but also a specific existing or designed physical heterogeneous continuous dynamical hypersystem.

The above-mentioned ideas are essentially different intellectual sources and they operate in two different planes: the mathematical model and the original. It is proper to consider whether they can act as a mirror, showing what is important, what is missing in this context and what the next step should be.

Table 5.1. Differences between the developmental systems and dynamical systems approaches (Fijalkowski 1987).

Developmental systems approach	Dynamical systems approach
Alteration	Stability
Concretism	Abstractionism
Empiricism	Naturalism
Emotionalism	Rationalism
Normativism	Descriptivism
Inductionism	Deductionism
Informalism	Formalism
Behaviourism	Cognitivism
Intuitionism	Principlism
Disaggregation	Aggregation

The existing ideas that include the development systems approach in the methodology to improve physical models mostly come from behaviourism. When computer-based analytical studies are performed, it may only be the behaviours of the mathematical models of the function objects that are studied, together with the accompanying impact on various dynamical forces, and these are considered in direct standard experimental studies that just concern the behaviour of a researched functional object.

The researcher is the creator of the changes to the physical model and determines their effective implementation, and therefore they 'get to know each other', which offers the oldest technique for improving physical models (e.g. simulation and analogical modelling), in which a mathematical model is an intellectual source of energy, which is necessary to learn how to maintain computer analytical studies on different qualities of a mathematical model.

The dynamical systems approach in the methodology to create mathematical models of function objects acquires significance for researchers through the processes of perception and valuation. In the dynamical systems approach, researchers will check the structure of the hierarchical physical model and examine what the relation is in this selection of compulsion, using their own knowledge and imagination. They will not think about concepts that are original (the actual function object), but rather about a mathematical model (relationships) to better understand how they define the dynamical systems approach, in which they operate and select a hierarchical structure of the physical model.

The improvement of the physical model has two stages: diagnostics and intervention. The techniques for collecting data at the diagnostic stage include computer analysis and the study of mathematical models and the standard experimental study behaviours of real (existing) or designed physical functional objects in different operational states (dynamic and static). The collected data are classified by the problems, which are solved together with practitioners, i.e. researchers carrying out experimental studies. These diagnoses have an ongoing character (based on the mathematical model 'computerised analytical studies in action'), and therefore are constantly actualised based on data derived from feedback (see figure 4.1). The intervention stage is conceived as a development concept and the implementation of the alteration of the physical model, which can affect the evolution of the alternative behaviours hitherto existing, should be effective and focused on preferring a routine for the physical modelling and invariance. Initially, one tried to alter a physical model using simulation and analogue modelling, but since that yielded small effects, one attempts to make more complex alterations, especially regarding the structures of the hierarchical physical models that condition the mathematical models. Then these hierarchical structures begin to be examined in the broader terms of a functional environment, and are also more generally understood in the meteorological environment.

Why the structure of the hierarchical physical models adapts to the environment, but does not change diametrically, is also reflected on. In view of the above, the attention of researchers shifts from the inside to the outside of the mathematical model. The change to the physical model is the problem. Many of the improvement techniques for physical models therefore stress the methodology for the complex solving of problems. The overall emergent behaviour is difficult to predict, even when the behaviour of the dynamical system, hyposystem and components can be predicted without difficulty. Minor changes to inputs or parameters can cause large changes in behaviour.

Attempts to prevent researchers in mathematical models from 'swinging round each other with their backs' led them to stop disregarding the physical model, not as an exercise in useless conformity, but to resolve issues based on knowledge of logical, dynamic or static processes, and knowledge of the principles of cooperation of a variety of physical heterogeneous continuous dynamical hypersystems. The idea of using the developmental systems approach in the methodology to improve the physical models of physical functional objects was built under the influence of an article by Maslyk (1985): 'Change: the recipe for progress'.

References

Fijalkowski B 1987 *Modele matematyczne wybranych lotniczych i motoryzacyjnych mechano-elektro-termicznych dyskretnych nadsystemów dynamicznych (Mathematical Models of Selected Aviation and Automotive Continuous Dynamical Hyperystems)*, Monografia 53 Politechnika Krakowska imienia Tadeusza Kościuszki, Krakow, 276 p. (In Polish)

Maslyk E 1985 Zmiany: recepta na postep. *Przeglad techniczny*, 1985, No. 6, pp 20–21 (In Polish)

Chapter 6

Synthetic mathematical model of the abstract functional heterogeneous continuous dynamical hypersystem

MATHEMATICAL SUPERIORITY

Mechanics is the paradise of mathematical sciences [&] Mechanical science is most noble and useful beyond all others, since by means of it, all animated bodies that have motion perform all their operations.

Leonardo DaVinci

An intelligence knowing all the forces acting in nature at a given instant, as well as the momentary positions of all things in the Universe, would be able to comprehend in one single formula of the motions of the largest bodies as well as of the lightest atoms in the world, provided that its intellect were sufficiently powerful to subject all data to analysis; to it nothing would be uncertain, the future as well as the past would be present to its eyes.

Pierre Simon Laplace

Mathematics is the most beautiful and most powerful creation of the human spirit.

Stefan Banach

Mathematics is the art of constructing theories that aid in creating an image of reality.

Stanislaw Lojasiewicz

doi:10.1088/978-0-7503-1350-6ch6

6.1 Abstract functional heterogeneous continuous dynamical hypersystem

A synthetic mathematical model of an abstract functional heterogeneous continuous dynamical hypersystem based on the Hamilton–Ostrogradsky variational principle of stationary action initiates an algorithm to formulate a real mathematical model of a physical hypersystem in the dynamical systems approach. This principle may be found in the work of the Irish mathematician, William Rowan Hamilton (1805–65), as published 1830–35. Hamilton assumed that the output dynamical system is 'scleronomic', i.e. it is only subject to stationary constraints, in which the equation of time t does not appear explicitly. In the general case of non-stationary constraints, the principle was formulated and substantiated by the Russian mathematician, Mikhail Vasilevich Ostrogradsky (1801–62), in 1845. In relation to the Hamilton principle, it is sometimes called the 'Hamilton–Ostrogradsky principle'. In the 1760s a Swiss mathematician, Leonhard Euler (1707–83), developed a way of laying out the differential equations of dynamics based on the methods of variational calculus (Euler 1766), while in the 1780s the French mathematician Joseph Louis de Lagrange (1736–1813) developed a stacking of differential equations based on the energetic methods for mechanical homogenous continuous dynamical systems (Lagrange 1788), and the differential equations of both Euler and Lagrange are analogous as they may be applied in the case of a general dynamical hypersystem of many **degrees of freedom** (DoF). In the 1870s the British physicist, James Clark Maxwell (1831–79), developed a way of stacking the differential equations of dynamics that was based on Lagrange's energetic methods for electromechanical continuous dynamical hyposystems that depended on supplementing the Lagrange potential with an analogue of kinetic and potential energy, as well as on finding an analogue of the generalised force (Carvallo 1902, Liennard 1902, Maxwell 1873). By including in the textbook such fundamental concepts of dynamics for the methodology to create a mathematical model of an abstract functional heterogeneous continuous dynamical hypersystem—like the abstract functional continuous dynamical component (figure 3.1)—instead of using the concept of mechanical motion applied by Lagrange, as a material point, for mechanical homogeneous continuous dynamical systems, below we will leave the Euler–Lagrange second-order differential equation of dynamics in the dynamical systems approach and hypermatrix notation. The dynamical systems approach and holor hypermatrix notation take into account the work done by all the generalised forces acting on the abstract functional heterogeneous continuous dynamical hypersystem, regardless of whether or not they are conservative generalised forces.

In addition, the dynamical systems approach and hypermatrix notation have difficulty in distributing the physical variables by applying partial differentiation. In many cases, the structure of the nonlinear abstract functional heterogeneous continuous dynamical hypersystem denies its dynamical components the ability to change their dynamical states freely. A number of conditions and restrictions can hamper their dynamical states. It is said then that constraints are imposed on the dynamics of the dynamical hypersystem. The specific form of these constraints can be very varied, for example, through the combination of two different continuous dynamical hyposystems of **direct current** (DC) or **alternating current** (AC) in the commutator electrical machine by means of an arbitrary physical commutation matrixer (e.g. a mechanical or electrical commutation matrixer)—see chapter 11. If the constraints impose restrictions on the dynamical status of respective parts of the nonlinear abstract functional continuous dynamical hypersystem, they are called 'geometrical constraints'. Moreover, if they impose restrictions on the dynamics of the dynamical hypersystem, i.e. on the values of the generalised velocities of the respective parts of the dynamical hypersystem, then they are called 'kinematical constraints'. However, an inverse affirmation may not be true, i.e. the relationships between the generalised velocities of different respective parts of the dynamical hypersystem, and in particular the limit values of the generalised velocities, may not place restrictions on the dynamical status of the dynamical hypersystem. A convincing example of this is the arbitrary physical commutation matrixer. Relatively recently, it was noted that the existence of constraints does not impose any restrictions on the dynamical state of the nonlinear abstract functional continuous dynamical hypersystem. In his *Mécanique Analytique* (1788) Lagrange did not believe that such constraints could exist, noting that for any mechanical homogeneous continuous dynamical system using ordinary differential equations resulting from the structure of the dynamical system, one may select independent coordinates with independent variations. This oversight became known after many years (Appell 1899a, 1899b, 1893, 1899d, 1903, Bobylev 1892, Beltrami 1895, Tchaplygin 1897a, 1897b, Zhukovsky 1893) when considering rolling solid bodies without slipping after plane or complex surfaces. In 1894, the German physicist, Heinrich Hertz (1857–94), introduced the division of mechanical homogeneous continuous dynamical systems into holonomic and unholonomic ones (Hertz 1894). A nonlinear abstract functional heterogeneous continuous dynamical hypersystem with unholonomic constraints and no integral that does not reduce to the geometrical constraints is called an 'unholonomic physical heterogeneous continuous complex or simple dynamical hypersystem'.

6.2 Independent generalised coordinates

One of the basic concepts of the dynamics of nonlinear abstract functional heterogeneous continuous dynamical hypersystems is the concept of the elementary 'abstract functional homogeneous continuous dynamical system'. This means the dynamical component elementary abstract homogeneous continuous dynamical system whose physical, economic, social, etc structure one can omit when creating a synthetic mathematical model of its dynamics. It is understood that the possibility of such an omission is dependent on the existing conditions of that or another problem. For instance, a planet can be seen as elementary astronomical homogeneous continuous dynamical systems of the sunny homogeneous continuous dynamical hypersystem in the investigation of their movement around the Sun, but one cannot be seen in the investigation of their day and night movement. The dynamical status of elementary abstract functional homogeneous continuous dynamical systems of a nonlinear abstract functional heterogeneous continuous dynamical hypersystem—that is, their position in the generalised multi-dimensional spaces—is determined by a hypermatrix of the independent generalised coordinates

$$\boldsymbol{q} = \left[\|q_1\|^T \|q_2\|^T \|q_3\|^T \cdots \|q_\vartheta\|^T \right]^T.$$

A derivative of the hypermatrix $\boldsymbol{q}$ over time

$$\dot{\boldsymbol{q}} = \frac{\mathrm{d}\boldsymbol{q}}{\mathrm{d}t} = \boldsymbol{v} \tag{6.1}$$

is a hypermatrix of the generalised velocities, and the second derivative

$$\ddot{\boldsymbol{q}} = \frac{\mathrm{d}^2\boldsymbol{q}}{\mathrm{d}t} = \frac{\mathrm{d}\dot{\boldsymbol{q}}}{\mathrm{d}t} = \boldsymbol{a} \tag{6.2}$$

is a hypermatrix of the generalised accelerations of the nonlinear functional heterogeneous continuous dynamical hypersystem.

To determine the dynamical status of the nonlinear functional heterogeneous continuous dynamical hypersystem consisting of N elementary functional homogeneous continuous dynamical systems in multi-dimensional spaces, we should introduce N matrices of the independent generalised coordinates. The total number of matrices of the independent generalised coordinates, which is necessary to introduce the unique identification of the nonlinear functional continuous dynamical hypersystem, is simply the number of DoF. The hypomatrices of these matrices are independent generalised coordinates of elementary functional homogeneous continuous dynamical systems of the nonlinear functional heterogeneous continuous dynamical hypersystem. The mathematical relationship relating the hypermatrix of generalised accelerations with the hypermatrix of the independent generalised coordinates and the hypermatrix of the generalised velocities is called the 'synthetic mathematical model' or the 'differential equation of dynamics' of the nonlinear functional heterogeneous continuous dynamical hypersystem. In relation to the hypermatrix, $\boldsymbol{q}$, it is the second-order differential equation in the hypermatrix notation, and their successive integration allows us, in principle, to determine these hypermatrices, i.e. the trajectories of the dynamics of the nonlinear functional heterogeneous continuous dynamical hypersystem.

6.3 Unconstrained and constrained nonlinear functional heterogeneous continuous dynamical hypersystems—classification of the constraints

The subject covered in this textbook on dynamical analytical studies is the dynamics of the nonlinear abstract functional heterogeneous continuous dynamical hypersystem (consisting of elementary functional homogeneous continuous dynamical systems of this dynamical hypersystem S_υ ($\upsilon = 1, 2, 3, \ldots, N$) relative to a set of matrices of the independent generalised coordinates. On the matrices of the independent generalised coordinates (the generalised positions) and the matrices of the generalised velocities of these elementary homogeneous continuous functional dynamical systems are imposed some restrictions called 'constraints'. A nonlinear functional heterogeneous continuous dynamical hypersystem with such constraints is called an 'unconstrained' one as distinct from 'constrained' dynamical hypersystems, which are not subject to any constraints. The constraints can be expressed in hypermatrix notation by means of an equation[1]

$$\mathscr{F}(\boldsymbol{q}, \dot{\boldsymbol{q}}, t) = 0, \tag{6.3}$$

where after the left side there is a time t and a hypermatrix of the independent generalised coordinates $\boldsymbol{q}$ of all the elementary nonlinear functional homogeneous continuous dynamical systems, S_υ, of the nonlinear functional heterogeneous continuous dynamical hypersystem ($\upsilon = 1, 2, 3, \ldots, N$).

In a particular case, when the hypermatrix of the generalised velocities $\dot{\boldsymbol{q}}$ is not present in the equation of constraints (6.3), the constraints are called 'finite' ones. Analytically, they may be represented with the hypermatrix notation as follows:

$$\mathscr{F}(\boldsymbol{q}, t) = 0. \tag{6.4}$$

In the general case, the constraints (6.3) are called 'differential' ones. In the case where there are finite constraints of the form (6.4), the nonlinear functional heterogeneous continuous dynamical hypersystem may not find itself in an arbitrary dynamical position, or may take an arbitrary position in multi-dimensional space at a given moment t. However, in a case where there are only differential constraints, the dynamical hypersystem may at a given moment t find itself in an arbitrary dynamical state (position) in multi-dimensional space. However, in these dynamical states, the matrices of the generalised velocities of the elementary functional homogeneous continuous dynamical systems may not be arbitrary, because the differential constraints impose the restrictions of the generalised velocities on these matrices. The finite constraints (6.4) are called 'stationary' in the case where in their equation a time t does not appear explicitly. A nonlinear functional heterogeneous continuous dynamical hypersystem that consists of elementary functional homogeneous continuous dynamical systems of this dynamical hypersystem is called a 'holonomic dynamical hypersystem' if the elementary functional heterogeneous continuous dynamical systems are not subject to any integral differential constraints.

For example, the holonomic dynamical hypersystem is every one of the unconstrained nonlinear functional heterogeneous continuous dynamical hypersystems consisting of elementary functional homogeneous continuous dynamical systems, as well as the constrained nonlinear functional heterogeneous continuous dynamical hypersystem with 'finite' and 'differential' constraints, but no integral. In the case of the holonomic nonlinear functional heterogeneous continuous dynamical hypersystem, all constraints can be represented in the form of the finite one. In the case where a nonlinear functional continuous dynamical hypersystem is not

[1] The next $\mathscr{F}(\boldsymbol{q}, \dot{\boldsymbol{q}}, t) = 0$ is an abbreviated function designation $\mathscr{F}\left([\|q_1\|^T \ldots \|q_\vartheta\|^T]^T, [\|\dot{q}_1\|^T \ldots \|\dot{q}_\vartheta\|^T]^T\right)$. An abbreviation of this kind is often used in this book. It is assumed that the functions found in this book are continuous with their derivatives, which are found in the relevant chapters.

subject to any integral differential constraints, it is called an 'unholonomic' dynamical hypersystem. While those with no integral differential constraints are the so-called unholonomic ones, those with integral differential constraints are called 'semi-holonomic'. A nonlinear functional heterogeneous continuous dynamical hypersystem is called a 'scleronomic' one if it is only subject to stationary constraints. If not, the dynamical hypersystem is called a 'rheonomic' one. Nonlinear functional heterogeneous continuous dynamical hypersystems, in which processing takes place in energy generation, conversion, dissipation, or energy storage, divide into:

- the conservative, in which the work of the external generalised forces changes without loss into kinetic or potential energy;
- the dissipative, in which the work of the generalised external forces dissipates into thermal energy.

6.4 Principle of stationary action

The most general law of the dynamics of the conservative nonlinear functional heterogeneous continuous dynamical hypersystem in this textbook is based on the variational principle of stationary action, which originates from the Hamilton–Ostrogradsky principle of least action as defined for conservative dynamical hypersystems after its generalisation to dissipative dynamical hypersystems. In accordance with this principle, each of the nonlinear abstract functional heterogeneous continuous dynamical hypersystems is characterised by some kinetic potential, that is, a function called 'Lagrange's energy function' (or the 'Lagrangian'), which is equal to the surplus of the conservative kinetic co-energy $\mathcal{T}^*$ over the potential energy $\mathcal{U}$ of each of the nonlinear abstract functional heterogeneous continuous dynamical hypersystems, and has been derived, for example, by White and Woodson (1959), in the conventional (Newtonian paradigm) systems approach and scalar notation, but in the dynamical systems approach and hypermatrix notation it is defined as:

$$\mathcal{L}(\boldsymbol{q}, \dot{\boldsymbol{q}}, t) \triangleq \mathcal{T}^*(\boldsymbol{q}, \dot{\boldsymbol{q}}, t) - \mathcal{U}(\boldsymbol{q}, \dot{\boldsymbol{q}}, t), \tag{6.5a}$$

or simply

$$\mathcal{L} \triangleq \mathcal{T}^* - \mathcal{U}. \tag{6.5b}$$

Moreover, a function called 'virtual' or 'prepared' work, which is equal to the surplus of the dissipative kinetic co-energy (dissipation) $\mathcal{T}_R^*$ over the dissipative potential energy $\mathcal{U}_0$ of each of the nonlinear abstract functional heterogeneous continuous dynamical hypersystems, is defined in the dynamical systems approach and hypermatrix notation as:

$$\mathcal{A}(\boldsymbol{q}, \dot{\boldsymbol{q}}, t) \triangleq \mathcal{T}_R^*(\boldsymbol{q}, \dot{\boldsymbol{q}}, t) - \mathcal{U}_F(\boldsymbol{q}, \dot{\boldsymbol{q}}, t), \tag{6.6a}$$

or simply

$$\mathcal{A} \triangleq \mathcal{T}_R^* - \mathcal{U}_F. \tag{6.6b}$$

The variables $\boldsymbol{q}$, $\dot{\boldsymbol{q}}$, t through which the Lagrange energy function and virtual work are expressed are called the 'Lagrange variables'. A set of the values of these variables is characterised by a moment of time and the correspondence to them of the dynamical status of each of the nonlinear abstract functional heterogeneous continuous dynamical hypersystems, i.e. the generalised positions and generalised velocities of the elementary abstract functional heterogeneous continuous complex or simple dynamical systems.

The conservative kinetic co-energy of any of the nonlinear abstract functional heterogeneous continuous dynamical hypersystems is defined in the dynamical systems approach and the hypermatrix notation as:

$$\mathscr{T}^* \triangleq \int_0^{\dot{q}} \boldsymbol{p}^T(\boldsymbol{q}, \dot{\boldsymbol{q}})\mathrm{d}\dot{\boldsymbol{q}} \triangleq \int_0^{\dot{q}} \boldsymbol{p}^T(\boldsymbol{q}, \boldsymbol{v})\mathrm{d}\boldsymbol{v}, \tag{6.7a}$$

where $\boldsymbol{p}(\boldsymbol{q}, \dot{\boldsymbol{q}}) = \boldsymbol{p}(\boldsymbol{q}, \boldsymbol{v})$ is the hypermatrix of conservative generalised momenta and $\boldsymbol{K}(\boldsymbol{q}, \dot{\boldsymbol{q}}) = \boldsymbol{K}(\boldsymbol{q}, \boldsymbol{v}) = \frac{\mathrm{d}\boldsymbol{p}}{\mathrm{d}\dot{\boldsymbol{q}}} = \frac{\mathrm{d}\boldsymbol{p}}{\mathrm{d}\boldsymbol{v}}$ is the hypermatrix of the conservative kinetic differential coefficients (generalised inertia).

The conservative kinetic co-energy of the arbitrary linear abstract functional heterogeneous continuous dynamical hypersystem in the dynamical systems approach and hypermatrix notation is in a homogeneous square form, namely:

$$\mathscr{T}^* \triangleq \frac{1}{2}\boldsymbol{K}(\boldsymbol{q}, \dot{\boldsymbol{q}}) \triangleq \frac{1}{2}\boldsymbol{K}(\boldsymbol{q}, \boldsymbol{v}). \tag{6.7b}$$

The conservative potential energy of the nonlinear abstract functional heterogeneous continuous dynamical hypersystem in the dynamical systems approach and hypermatrix notation is defined as:

$$\mathscr{U} \triangleq \int_0^{q} \boldsymbol{f}^T \mathrm{d}\mathbf{q}, \tag{6.8a}$$

where $\boldsymbol{f}$ is the hypermatrix of conservative generalised forces and $\boldsymbol{P}(\boldsymbol{q}, \dot{\boldsymbol{q}}) = \boldsymbol{P}(\boldsymbol{q}, \boldsymbol{v}) = \frac{\mathrm{d}\boldsymbol{f}}{\mathrm{d}\boldsymbol{q}}$ is the hypermatrix of the conservative potential differential coefficients (generalised stiffnesses).

The dissipative potential energy of any linear functional heterogeneous continuous dynamical hypersystem in the dynamical systems approach and hypermatrix notation is of a homogeneous square form, namely:

$$\mathscr{U} \triangleq \frac{1}{2}\boldsymbol{q}^T\boldsymbol{P}(\boldsymbol{q}, \dot{\boldsymbol{q}})\boldsymbol{q} \triangleq \frac{1}{2}\boldsymbol{q}^T\boldsymbol{P}(\boldsymbol{q}, \boldsymbol{v})\boldsymbol{q}. \tag{6.8b}$$

The conservative kinetic co-energy of any of the nonlinear abstract functional heterogeneous continuous dynamical hypersystems in the dynamical systems approach and hypermatrix notation is defined as:

$$\mathscr{T}^*_R \triangleq \int_0^{t} \mathscr{R}\mathrm{d}t, \tag{6.9}$$

where the integral of the Rayleigh function, i.e. the function of the thermal losses

$$\mathcal{R} \triangleq \mathcal{R}_V + \mathcal{R}_C, \tag{6.10}$$

is a sum of the generalised losses on the viscous friction

$$\mathcal{R}_V \triangleq \int_0^{\dot{q}} \boldsymbol{p}_R(\dot{\boldsymbol{q}})\mathrm{d}\dot{\boldsymbol{q}} \triangleq \int_0^{\boldsymbol{v}} \boldsymbol{p}_R(\boldsymbol{v})\mathrm{d}\boldsymbol{v} \tag{6.11}$$

and the generalised losses on the dry friction

$$\mathcal{R}_C \triangleq \int_0^{\dot{q}} \dot{\boldsymbol{q}}^T(\boldsymbol{\Delta f} \operatorname{sgn} \dot{\boldsymbol{q}})\mathrm{d}\dot{\boldsymbol{q}} \triangleq \int_0^{\boldsymbol{v}} \boldsymbol{v}^T(\boldsymbol{\Delta f} \operatorname{sgn} \boldsymbol{v})\mathrm{dv}, \tag{6.12}$$

where $\boldsymbol{p}_R(\dot{\boldsymbol{q}}) = \boldsymbol{p}_R(\boldsymbol{v})$ is the hypermatrix of the dissipative generalised momenta (dissipation), $\boldsymbol{D}(\dot{\boldsymbol{q}}) = \boldsymbol{D}(\boldsymbol{v}) = \frac{\mathrm{d}\boldsymbol{p}_R}{\mathrm{d}\dot{\boldsymbol{q}}} = \frac{\mathrm{d}\boldsymbol{p}_R}{\mathrm{d}\boldsymbol{v}}$ is the hypermatrix of the dissipative kinetic differential coefficients of the thermal losses caused by the viscous friction and $\boldsymbol{\Delta f}$ is the hypermatrix of the dissipative generalised force drops of the thermal losses caused by the dry friction.

The Rayleigh dissipation function of any of the linear abstract functional heterogeneous continuous dynamical hypersystems is the sum of the generalised losses on the viscous friction, establishing a homogeneous square form

$$\mathcal{R}_V \triangleq \frac{1}{2}\dot{\boldsymbol{q}}^T\boldsymbol{D}(\dot{\boldsymbol{q}})\dot{\boldsymbol{q}} \triangleq \frac{1}{2}\boldsymbol{v}^T\boldsymbol{D}(\boldsymbol{v})\boldsymbol{v}, \tag{6.13}$$

and the generalised losses on the dry friction

$$\mathcal{R}_C \triangleq \frac{1}{2}(\boldsymbol{\Delta f} \operatorname{sgn} \dot{\boldsymbol{q}})\dot{\boldsymbol{q}} \triangleq \frac{1}{2}(\boldsymbol{\Delta f} \operatorname{sgn} \boldsymbol{v})\boldsymbol{v}, \tag{6.14}$$

where $\operatorname{sgn} \dot{\boldsymbol{q}} = [\operatorname{sgn} \|\dot{q}_1\|^T \ldots \operatorname{sgn} \|\dot{q}_8\|^T\,]^T$ and

$$\operatorname{sgn} \dot{\boldsymbol{q}} \triangleq \operatorname{sgn} \boldsymbol{v} \triangleq \begin{cases} +1\ \boldsymbol{v}>0; \\ -1\boldsymbol{v} <0. \end{cases}$$

The dissipative potential energy of any of the nonlinear abstract functional heterogeneous continuous dynamical hypersystems in the dynamical systems approach and hypermatrix notation is assumed as:

$$\mathcal{U}_F \triangleq \int_0^{q} \boldsymbol{f}(t)^T\,\mathrm{d}\boldsymbol{q} = \int_0^{t} \boldsymbol{f}(t)^T\,\dot{\boldsymbol{q}}\mathrm{d}t = \int_0^{t} \boldsymbol{f}(t)^T\,\boldsymbol{v}\,\mathrm{d}t, \tag{6.15}$$

where $\boldsymbol{f}(t)$ is the hypermatrix of the dissipative generalised forces (dissipation) as a function of time.

If a nonlinear abstract functional heterogeneous continuous dynamical hypersystem transmits from the dynamical state described by a hypermatrix of the generalised coordinates

$$\boldsymbol{q} = \boldsymbol{q}^{(1)}$$

occurring in the period $t = t_1$ to the dynamical state described by a hypermatrix of the generalised coordinates

$$\boldsymbol{q} = \boldsymbol{q}^{(2)}$$

occurring in the period $t = t_2$, then this transition is done in such a way that the integral

$$\mathcal{J} \triangleq \int_{t_1}^{t_2} [\mathcal{L}(\boldsymbol{q}, \dot{\boldsymbol{q}}, t) + \mathcal{A}(\boldsymbol{q}, \dot{\boldsymbol{q}}, t)]\mathrm{d}t = \text{minimum}. \tag{6.16}$$

Let $\boldsymbol{q} = \boldsymbol{q}(t)$ be this hypermatrix of generalised coordinates for which the functional $\mathcal{J}$ has the smallest value. This means that $\mathcal{J}$ rose when changing the hypermatrix $\boldsymbol{q}(t)$ on any hypermatrix of the form

$$\boldsymbol{q}(t) + \delta\boldsymbol{q}(t), \tag{6.17}$$

where $\delta\boldsymbol{q}(t)$ is the hypermatrix, in which the matrices and hypomatrices are the elements of the hypermatrix $\boldsymbol{q}(t)$, and are small functions in the entire time period from t_1 to t_2 (called the 'variation' of the hypermatrix $\boldsymbol{q}(t)$). Because at $t = t_1$ and $t = t_2$ the matrices, hypomatrices and elements of the hypermatrix (6.17) should assume one and the same values of the matrices, hypomatrices and components of the hypermatrix $\boldsymbol{q}^{(1)}$ and $\boldsymbol{q}^{(2)}$, then it should be the case that

$$\delta\boldsymbol{q}(t_1) = 0 \tag{6.18a}$$

and

$$\delta\boldsymbol{q}(t_2) = 0. \tag{6.18b}$$

A variation I at a permutation $\boldsymbol{q}$ into $\boldsymbol{q}(t) + \delta\boldsymbol{q}(t)$ may be obtained by the differences

$$\int_{t_1}^{t_2} \mathcal{L}(\boldsymbol{q} + \delta\boldsymbol{q}, \dot{\boldsymbol{q}} + \delta\dot{\boldsymbol{q}}, t)\mathrm{d}t - \int_{t_1}^{t_2} \mathcal{L}(\boldsymbol{q}, \dot{\boldsymbol{q}}, t)\mathrm{d}t$$

and

$$\int_{t_1}^{t_2} \mathcal{A}(\boldsymbol{q} + \delta\boldsymbol{q}, \dot{\boldsymbol{q}} + \delta\dot{\boldsymbol{q}}, t)\mathrm{d}t - \int_{t_1}^{t_2} \mathcal{A}(\boldsymbol{q}, \dot{\boldsymbol{q}}, t)\mathrm{d}t.$$

A distribution of these differences according to $\delta\boldsymbol{q}$ and $\delta\dot{\boldsymbol{q}}$ (in the integral expressions) begins from the first-order terms.

A necessary condition for the integral minimisation of $\mathcal{J}$ is the transformation of all its terms to zero; this is called a 'first variation' (or simply a variation of the integral). In this way, in unison with a symbol of the variational calculus and matrix computation, the Hamilton–Ostrogradsky principle of the stationary action for any nonlinear abstract functional heterogeneous continuous dynamical hypersystem can be written in the form:

$$\mathcal{J} \triangleq \delta \int_{t_1}^{t_2} [\mathcal{L}(\boldsymbol{q}, \dot{\boldsymbol{q}}, t) + \mathcal{A}(\boldsymbol{q}, \dot{\boldsymbol{q}}, t)]\mathrm{d}t \tag{6.19}$$

or as a variation

$$\int_{t_1}^{t_2}\left(\frac{\partial\mathscr{L}}{\partial\boldsymbol{q}}\delta\boldsymbol{q}+\frac{\partial\mathscr{L}}{\partial\dot{\boldsymbol{q}}}\delta\dot{\boldsymbol{q}}+\frac{\partial\mathscr{A}}{\partial\boldsymbol{q}}\delta\boldsymbol{q}+\frac{\partial\mathscr{A}}{\partial\dot{\boldsymbol{q}}}\delta\dot{\boldsymbol{q}}\right)\mathrm{d}t=0.$$

Taking into consideration the fact that $\delta\dot{\boldsymbol{q}}=\frac{\mathrm{d}}{\mathrm{d}t}\delta\boldsymbol{q}$ and integrating the second and fourth terms by parts, one obtains

$$\begin{aligned}\delta\mathscr{J}&=\frac{\partial\mathscr{L}}{\partial\boldsymbol{q}}\delta\boldsymbol{q}\Big|_{t_1}^{t_2}+\int_{t_1}^{t_1}\left(\frac{\partial\mathscr{L}}{\partial\boldsymbol{q}}\delta\boldsymbol{q}+\frac{\mathrm{d}}{\mathrm{d}t}\frac{\partial\mathscr{L}}{\partial\dot{\boldsymbol{q}}}\right)\delta\boldsymbol{q}\mathrm{d}t\\&+\frac{\partial\mathscr{A}}{\partial\boldsymbol{q}}\delta\boldsymbol{q}\Big|_{t_1}^{t_2}+\int_{t_1}^{t_1}\left(\frac{\partial\mathscr{A}}{\partial\boldsymbol{q}}\delta\boldsymbol{q}+\frac{\mathrm{d}}{\mathrm{d}t}\frac{\partial\mathscr{A}}{\partial\dot{\boldsymbol{q}}}\right)\delta\boldsymbol{q}\mathrm{d}t=0.\end{aligned}\tag{6.20}$$

After one considers conditions (6.18), the first and third terms in the above expression disappear. Only the integrals remain, which should be equal to zero for any of the values of the matrices, hypomatrices and elements of the hypermatrix $\delta\boldsymbol{q}(t)$. This is only possible in this case when the under-integral expression will identically be zero. In this way, we obtain the synthetic mathematical model of any nonlinear abstract functional heterogeneous continuous dynamical hypersystem, i.e. a differential equation of dynamics in the dynamical systems approach and hypermatrix notation,

$$\left[\frac{\mathrm{d}}{\mathrm{d}t}\left(\frac{\partial\mathscr{L}}{\partial\dot{\boldsymbol{q}}}\right)-\frac{\partial\mathscr{L}}{\partial\boldsymbol{q}}+\frac{\mathrm{d}}{\mathrm{d}t}\left(\frac{\partial\mathscr{A}}{\partial\dot{\boldsymbol{q}}}\right)-\frac{\partial\mathscr{A}}{\partial\boldsymbol{q}}\right]\delta\boldsymbol{q}=0.\tag{6.21a}$$

A differential equation of dynamics (6.21a) with hypermatrix notation in the dynamical systems approach in the methodology for the formation of mathematical models of abstract functional heterogeneous continuous dynamical hypersystems is unknown in the literature.

The author decided to apply this hypermatrix notation because it is comfortable for the dynamical systems approach in the methodology for the formation of mathematical models. However, this differential equation in the non-matrix, i.e. scalar notation, is well known and in variational calculus it is called 'Euler's differential equation' (1766), while in analytical mechanics it is called 'Lagrange's differential equation' (1788), and it has a somewhat different form (White and Woodson 1959, Szklarski *et al* 1963):

$$\frac{\mathrm{d}}{\mathrm{d}t}\left(\frac{\partial\mathscr{L}}{\partial\dot{q}_i}\right)-\frac{\partial\mathscr{L}}{\partial q_i}+f_{\mathrm{D}}=f_{\mathrm{E}},\tag{6.21b}$$

where q_i, $\dot{q}_i$ are the ith generalised coordinate and velocity, i is the number of DoF, $f_{\mathrm{D}}=\frac{\partial\mathscr{R}}{\partial\dot{q}_i}$ is the generalised dissipation forces, $\mathscr{R}$ is the Rayleigh function and $f_{\mathrm{E}}=f_{\mathrm{E}}(t)$ is the generalised external forces.

If the Lagrange function of the energy and virtual work of an arbitrary nonlinear abstract functional heterogeneous continuous dynamical hypersystem are known, then the Euler–Lagrange second-order differential equation of dynamics in the hypermatrix notation establishes a relationship between the relevant matrices of the hypermatrices of generalised accelerations, velocities and coordinates, that is, it represents a synthetic differential equation of the dynamics, and therefore a synthetic mathematical model of any nonlinear abstract functional heterogeneous continuous dynamical hypersystem.

Substituting expressions (6.5) and (6.6) into the equation of dynamics (6.21a), one obtains a synthetic mathematical model of any nonlinear abstract functional heterogeneous continuous dynamical hypersystem that is more appropriate for computerised analytical studies of the dynamics, namely:

$$\left[\frac{\mathrm{d}}{\mathrm{d}t}\left(\frac{\partial \mathcal{T}^*}{\partial \dot{\boldsymbol{q}}}\right) - \frac{\partial \mathcal{T}^*}{\partial \boldsymbol{q}} + \frac{\partial \mathcal{U}}{\partial \boldsymbol{q}} + \frac{\mathrm{d}}{\mathrm{d}t}\left(\frac{\partial \mathcal{T}^*_R}{\partial \dot{\boldsymbol{q}}}\right) - \boldsymbol{f}^T\right]\delta \boldsymbol{q} = 0, \tag{6.22a}$$

or

$$\left[\frac{\mathrm{d}}{\mathrm{d}t}\left(\frac{\partial \mathcal{T}^*}{\partial \boldsymbol{v}}\right) - \frac{\partial \mathcal{T}^*}{\partial \boldsymbol{q}} + \frac{\partial \mathcal{U}}{\partial \boldsymbol{q}} + \frac{\mathrm{d}}{\mathrm{d}t}\left(\frac{\partial \mathcal{T}^*_R}{\partial \boldsymbol{v}}\right) - \boldsymbol{f}^T\right]\delta \boldsymbol{q} = 0, \tag{6.22b}$$

where

$$\boldsymbol{f}^T = (\mathrm{grad}\,\mathcal{A})^T = \frac{\partial \mathcal{A}}{\partial \boldsymbol{q}} = (\mathrm{grad}\,\mathcal{U}_F)^T = \frac{\partial \mathcal{U}_F}{\partial \boldsymbol{q}}$$

is the hypermatrix of the generalised dissipative input (excitation) forces.

In the above Euler–Lagrange second-order differential equation of the dynamics, one can use hypermatrix notation for the generalised coordinates and velocities and the scalar functions of the conservative kinetic co-energy $\mathcal{T}^*$ and conservative potential energy $\mathcal{U}$ as homogeneous square forms, where the partial derivatives in relation to the hypermatrices of the generalised coordinates and velocities form matrices. The reality of the subsistence of respective matrices of these hypomatrices must be verified in each case. Verification of this will ensure the completeness of the matrices and hypomatrices acting significantly in the abstract functional heterogeneous continuous dynamical hypersystem. This is the advantage of an accepted hypermatrix notation for differential equations of the dynamics of any nonlinear abstract functional heterogeneous continuous dynamical hypersystem and consistent computation in this notation.

A synthetic mathematical model of any nonlinear abstract functional heterogeneous continuous dynamical hypersystem, establishing the Euler–Lagrange second-order differential equation of dynamics in the dynamical systems approach, is written in the hypermatrix form, since the abstract functional heterogeneous continuous dynamical hypersystem is often a set of many functional homogeneous continuous dynamical systems, whose own mathematical models, establishing a set of the Euler–Lagrange differential equations, are written in the matrix form, since each one is often a set of many functional heterogeneous continuous dynamical hyposystems whose mathematical models, forming a subset of the Euler–Lagrange

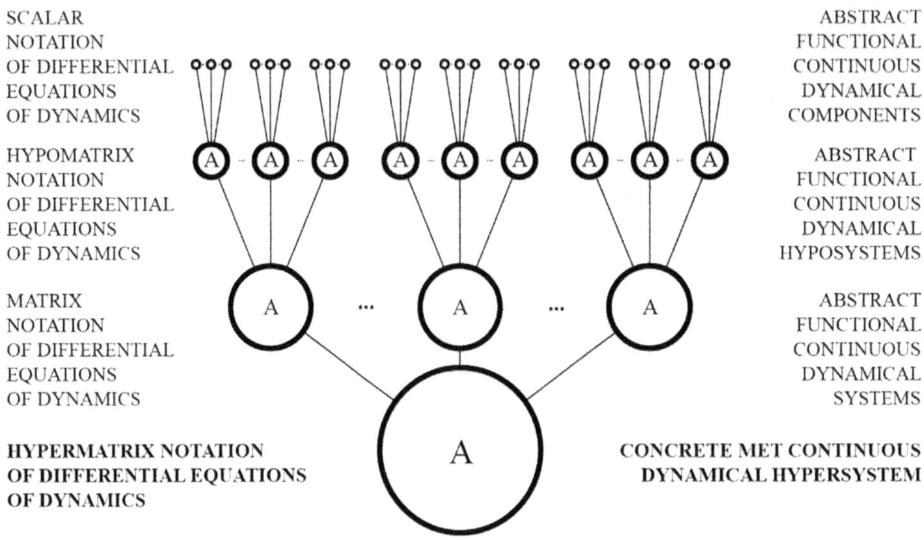

Figure 6.1. A hierarchical structure of a physical model of an abstract functional continuous dynamical hypersystem.

second-order differential equations in the dynamical systems approach, are written in the hypomatrix form, since they are usually a few or more elements of a dynamical hyposystem, which elementary mathematical models, constituting only a single Euler–Lagrange second-order differential equation, do not write in the matrix form of a matrix, but rather in a scalar one (figure 6.1).

Taking into consideration equations (6.7)–(6.15), the Euler–Lagrange second-order differential equations of dynamics (6.22) in the dynamical systems approach can be written in the hypermatrix form as follows:

$$\left[\frac{\mathrm{d}}{\mathrm{d}t}[\boldsymbol{v}^T\boldsymbol{K}(\boldsymbol{q},\,\boldsymbol{v})] - \frac{1}{2}\boldsymbol{v}^T\frac{\partial \boldsymbol{K}(\boldsymbol{q},\,\boldsymbol{v})}{\partial \boldsymbol{q}}\boldsymbol{v} + \boldsymbol{q}^T\boldsymbol{P}(\boldsymbol{q},\,\boldsymbol{v}) + \boldsymbol{v}^T\boldsymbol{D}(\boldsymbol{q},\,\boldsymbol{v}) + \operatorname{sgn}\boldsymbol{v}^{\mathrm{T}}\,\boldsymbol{\Delta f} - \boldsymbol{f}^T\right]\delta\boldsymbol{q} = 0, \tag{6.23a}$$

or simply

$$\left[\frac{\mathrm{d}}{\mathrm{d}t}(\boldsymbol{v}^T\boldsymbol{K}) - \frac{1}{2}\boldsymbol{v}^T\frac{\partial \boldsymbol{K}}{\partial \boldsymbol{q}}\boldsymbol{v} + \boldsymbol{q}^T\boldsymbol{P} + \boldsymbol{v}^T\boldsymbol{D} + \operatorname{sgn}\boldsymbol{v}^T\boldsymbol{\Delta f} - \boldsymbol{f}^T\right]\delta\boldsymbol{q} = 0. \tag{6.23b}$$

Assuming that $\boldsymbol{K}$, $\boldsymbol{P}$ and $\boldsymbol{D}$ are not also the functions of $\boldsymbol{q}$ as $\dot{\boldsymbol{q}} = \boldsymbol{v}$, the Euler–Lagrange second-order differential equations of dynamics (6.23) in the dynamical systems approach can be written in the hypermatrix form as:

$$\left[\frac{\mathrm{d}}{\mathrm{d}t}\boldsymbol{v}^T\boldsymbol{K} + \boldsymbol{q}^T\boldsymbol{P} + \boldsymbol{v}^T\boldsymbol{D} + \operatorname{sgn}\boldsymbol{v}^T\boldsymbol{\Delta f} - \boldsymbol{f}^T\right]\delta\boldsymbol{q} = 0, \tag{6.24a}$$

or

$$\left[\frac{\mathrm{d}}{\mathrm{d}t}\boldsymbol{v}^T\boldsymbol{K} + \int_0^t \boldsymbol{v}^T\mathrm{d}t\boldsymbol{P} + \boldsymbol{v}^T\boldsymbol{D} + \operatorname{sgn}\boldsymbol{v}^T\boldsymbol{\Delta f} - \boldsymbol{f}^T\right]\delta\boldsymbol{q} = 0, \tag{6.24b}$$

$$\left[\boldsymbol{v}^T(s)s\boldsymbol{K} + \boldsymbol{v}^T(s)\frac{\boldsymbol{P}}{s} + \boldsymbol{v}^T(s)\boldsymbol{D} + \operatorname{sgn}\boldsymbol{v}^T(s)\boldsymbol{\Delta f}(s) - \boldsymbol{f}^T(s)\right]\delta\boldsymbol{q}(s) = 0, \tag{6.25a}$$

or

$$\left[\boldsymbol{v}^T(s)\left(s\boldsymbol{K} + \frac{\boldsymbol{P}}{s} + \boldsymbol{D}\right) + \operatorname{sgn}\boldsymbol{v}^T(s)\boldsymbol{\Delta f}(s) - \boldsymbol{f}^T(s)\right]\delta\boldsymbol{q}(s) = 0, \tag{6.25b}$$

where $s = \sigma + j\omega$ is the Laplace operator.

After taking into account the notion of the *impedance* $\boldsymbol{Z}(s)$ and/or the *admittance* $\boldsymbol{Y}(s)$, this yields:

$$\boldsymbol{Z}(s) = \boldsymbol{D} + s\boldsymbol{K} + \frac{\boldsymbol{P}}{s} \tag{6.26a}$$

and

$$\boldsymbol{Y}(s) = \frac{1}{\boldsymbol{Z}(s)} = \frac{1}{\boldsymbol{D}} + \frac{1}{s\boldsymbol{K}} + \frac{s}{\boldsymbol{P}} = \boldsymbol{G} + \frac{\boldsymbol{B}}{s} + s\boldsymbol{S}. \tag{6.26b}$$

The Euler–Lagrange second-order differential equations of dynamics (6.25) in the dynamical systems approach can be written in the hypermatrix form as follows:

$$[\boldsymbol{v}^T(s)\boldsymbol{Z}(s) + \operatorname{sgn}\boldsymbol{v}^T(s)\boldsymbol{\Delta f}(s) - \boldsymbol{f}^T(s)]\,\delta\boldsymbol{q}(s) = 0, \tag{6.27a}$$

or

$$\left[\boldsymbol{v}^T(s)\frac{1}{\boldsymbol{Y}(s)} + \operatorname{sgn}\boldsymbol{v}^T(s)\boldsymbol{\Delta f}(s) - \boldsymbol{f}^T(s)\right]\delta\boldsymbol{q}(s) = 0. \tag{6.27b}$$

The mathematical representation of any measurable physical heterogeneous continuous dynamical hypersystem is greatly facilitated by representing it by a holor[2] that fits the physical concept perfectly.

The term holor is a generalisation of the more familiar mathematical concepts of tensors, vectors, matrices, complex numbers and scalars. Rather than forcing all measurable physical heterogeneous continuous dynamical hypersystems to be imperfectly represented by a few families of mathematical quantities, the study of holors permits the development of a mathematical representation that fits the

[2] Definition from Wikipedia: 'A holor (/ˈhoʊlər/; Greek ὅλος "whole") is a mathematical entity that is made up of one or more *independent quantities* ("merates" as they are called in the theory of holors). Complex numbers, scalars, vectors, matrices, tensors, quaternions, and other hypercomplex numbers are kinds of holors. If proper index conventions are maintained then certain relations of holor algebra are consistent with that of real algebra; i.e. addition and uncontracted multiplication are both commutative and associative.' https://en.wikipedia.org/wiki/Holor.

physical concept perfectly. The set of numbers that specify a holor are termed 'merates'.

The physical quantity holor ought to satisfy two fundamental criteria. The holor must be defined in terms of physically realisable measurements that are significant at physical frequencies. The holor formulation is very simple and modern computer techniques make the design of physical heterogeneous continuous dynamical hypersystems in terms of the generalised external forces holor (termed by the author the 'energy-potential-difference holor', $\boldsymbol{F}_i$) and the generalised velocities holor (termed by the author the 'energy-flow holor', $\boldsymbol{V}^j$) in **two-dimensional** (2D) spaces, and the generalised power holor, $\boldsymbol{P}^K$, in a **three-dimensional** (3D) spaces, a challenge of only moderate difficulty.

The mathematical model, represented by the Euler–Lagrange second-order differential equations of dynamics (6.27) for any physical heterogeneous continuous dynamical hypersystem with energy storage, may originate from holor equations.

For that reason, the resolution of physical heterogeneous continuous dynamical hypersystems for a step-function or sine-wave input-signal holor $\boldsymbol{F}_i$ necessitates the solution of a set of holor equations:

$$\left[\boldsymbol{V}^k\boldsymbol{Z}_{jk} + \operatorname{sgn} \boldsymbol{V}^k\Delta\boldsymbol{F}_{jk} - \boldsymbol{F}_k\right]\delta\boldsymbol{Q}^j = 0, \tag{6.28a}$$

or

$$\left[\boldsymbol{V}^k\frac{\delta_k^i}{\boldsymbol{Y}^{ij}} + \operatorname{sgn} \boldsymbol{V}^k\Delta\boldsymbol{F}_{jk} - \boldsymbol{F}_k\right]\delta\boldsymbol{Q}^j = 0. \tag{6.28b}$$

This appears for the physical and mathematical approaches presented in chapters 7 and 8. One may consider if it is possible to model a physical heterogeneous continuous dynamical hypersystem by algebraic equations from which the holor equations can be created. To deal with this problem one must take the result of the generalised input-signal holor $\boldsymbol{V}^j$ function into account. This nature of the driving holor function in its normal mathematical appearance is not conventionally realised by a physical heterogeneous continuous dynamical hypersystem.

In spite of this, this generalised input signal-holor function has a particular form (sine waveform) that symbolises the conventional generalised driving holor function applied in studies.

After taking into account the notion of the impedance holor, $\boldsymbol{Z}_{ij}$, and/or the admittance holor, $\boldsymbol{Y}^{ji}$, it yields

$$\boldsymbol{Z}_{ij} = \begin{bmatrix} \boldsymbol{Z}_{cc} & -\boldsymbol{Z}_{cs} \\ \boldsymbol{Z}_{sc} & \boldsymbol{Z}_{ss} \end{bmatrix} = \begin{bmatrix} \boldsymbol{D} & -\boldsymbol{X} \\ \boldsymbol{X} & \boldsymbol{D} \end{bmatrix} \tag{6.29a}$$

and

$$\boldsymbol{Y}^{ij} = \frac{\delta_k^i}{\boldsymbol{Z}_{jk}} = \begin{bmatrix} \boldsymbol{Y}^{ss} & \boldsymbol{Y}^{sc} \\ -\boldsymbol{Y}^{cs} & \boldsymbol{Y}^{cc} \end{bmatrix} = \begin{bmatrix} \boldsymbol{G} & \boldsymbol{B} \\ -\boldsymbol{B} & \boldsymbol{G} \end{bmatrix} = \frac{1}{(\boldsymbol{D}\verb|^|2)+(\boldsymbol{X}\verb|^|2)}\begin{bmatrix} \boldsymbol{D} & \boldsymbol{X} \\ -\boldsymbol{X} & \boldsymbol{D} \end{bmatrix}, \tag{6.29b}$$

where the reactance, $\boldsymbol{X}$, and the susceptance, $\boldsymbol{B}$, are as follows:

$$\boldsymbol{X} = \omega \boldsymbol{K} - \frac{1}{\omega}\boldsymbol{P}, \tag{6.30a}$$

$$\boldsymbol{B} = \omega \frac{1}{\boldsymbol{P}} - \frac{1}{\omega \boldsymbol{K}} = \omega \boldsymbol{S} - \frac{1}{\omega}\boldsymbol{H}, \tag{6.30b}$$

and

$$\boldsymbol{Y}^{ij}\boldsymbol{Z}_{jk} = \delta^{i}_{k}. \tag{6.30c}$$

The Euler–Lagrange second-order differential equations of dynamics (6.28) in the dynamical systems approach can be written in the hypermatrix holor form as follows:

$$\begin{aligned} &\left[\boldsymbol{V}^{k}\boldsymbol{Z}_{jk} + \operatorname{sgn} \boldsymbol{V}^{k}\boldsymbol{\Delta F}_{jk} - \boldsymbol{F}_{k}\right]\delta \boldsymbol{Q}^{j} \\ &= \left\{[\boldsymbol{V}^{c} \;\; \boldsymbol{V}^{s}]\begin{bmatrix} \boldsymbol{D} & -\boldsymbol{X} \\ \boldsymbol{X} & \boldsymbol{D} \end{bmatrix} + \operatorname{sgn}[\boldsymbol{V}^{c} \;\; \boldsymbol{V}^{s}]\begin{bmatrix} \boldsymbol{\Delta F}_{c} & 0 \\ 0 & \boldsymbol{\Delta F}_{s} \end{bmatrix} - [\boldsymbol{F}_{c} \;\; \boldsymbol{F}_{s}]\right\}\delta\begin{bmatrix} \boldsymbol{Q}^{c} \\ \boldsymbol{Q}^{s} \end{bmatrix}, \end{aligned} \tag{6.31a}$$

or

$$\begin{aligned} &\left[\boldsymbol{V}^{k}\frac{\delta^{i}_{k}}{\boldsymbol{Y}^{ij}} + \operatorname{sgn} \boldsymbol{V}^{k}\boldsymbol{\Delta F}_{jk} - \boldsymbol{F}_{k}\right]\delta \boldsymbol{Q}^{j} \\ &= \left\{[\boldsymbol{V}^{c} \;\; \boldsymbol{V}^{s}]\frac{1}{(\boldsymbol{G}\wedge 2) + (\boldsymbol{B}\wedge 2)}\begin{bmatrix} \boldsymbol{G} & -\boldsymbol{B} \\ \boldsymbol{B} & \boldsymbol{G} \end{bmatrix}\right. \\ &\quad \left. + \operatorname{sgn}[\boldsymbol{V}^{c} \;\; \boldsymbol{V}^{s}]\begin{bmatrix} \boldsymbol{\Delta F}_{c} & 0 \\ 0 & \boldsymbol{\Delta F}_{s} \end{bmatrix} - [\boldsymbol{F}_{c} \;\; \boldsymbol{F}_{s}]\right\}\delta\begin{bmatrix} \boldsymbol{Q}^{c} \\ \boldsymbol{Q}^{s} \end{bmatrix}, \end{aligned} \tag{6.31b}$$

where

$$\boldsymbol{F}_{i} = (\boldsymbol{F}_{c}, \boldsymbol{F}_{s}) = [\boldsymbol{F}_{c} \;\; \boldsymbol{F}_{s}]^{T} = \begin{bmatrix} \boldsymbol{F}_{c} \\ \boldsymbol{F}_{s} \end{bmatrix} = \begin{bmatrix} \boldsymbol{F}\cos \boldsymbol{\psi}_{F} \\ \boldsymbol{F}\sin \boldsymbol{\psi}_{F} \end{bmatrix} = \boldsymbol{F}\angle\boldsymbol{\psi}_{F}, \tag{6.32a}$$

$$\boldsymbol{V}^{j} = (\boldsymbol{V}^{c}, \boldsymbol{V}^{s}) = [\boldsymbol{V}^{c} \;\; \boldsymbol{V}^{s}]^{T} = \begin{bmatrix} \boldsymbol{V}^{c} \\ \boldsymbol{V}^{s} \end{bmatrix} = \begin{bmatrix} \boldsymbol{V}\cos \boldsymbol{\psi}_{V} \\ \boldsymbol{V}\sin \boldsymbol{\psi}_{V} \end{bmatrix} = \boldsymbol{V}\angle\boldsymbol{\psi}_{V}, \tag{6.32b}$$

and

$$\boldsymbol{F}^{c} = \boldsymbol{F}\cos \boldsymbol{\psi}_{F}, \; \boldsymbol{F}^{s} = \boldsymbol{F}\sin \boldsymbol{\psi}_{F} \tag{6.32c}$$

$$\boldsymbol{V}^{c} = \boldsymbol{V}\cos \boldsymbol{\psi}_{V}, \; \boldsymbol{V}^{s} = \boldsymbol{V}\sin \boldsymbol{\psi}_{V} \tag{6.32d}$$

$$\boldsymbol{\Delta F}_{ij} = \begin{bmatrix} \boldsymbol{\Delta F}_{c} & 0 \\ 0 & \boldsymbol{\Delta F}_{s} \end{bmatrix}. \tag{6.32e}$$

References

Appell P 1893 *Traité de Mécanique Rationnelle* 4 vols (Paris: Gauthier-Villars) (In French)

Appell P 1899a Les mouvements de roulement en dynamique. *Scientia*, No. 4, Paris (In French)

Appell P 1899b Sur les mouvements de roulement, equations du movement analogu.htm es & celles de Lagrange. *C.R.*, T. 129, Paris (In French)

Appell P 1899d Sur une forme gènèrale des èquations de la Dynamique *J. Reine Angew. Math.* Bd. 121, Berlin (In French)

Appell P 1903 Remarques sur les systems non holonomes *J. Math. Pures Appl.* T. 9, pp 27–28 (In French)

Beltrami E 1895 Sulle equazioni diminiche di Lagrange. *Rendiconti dell'Istituto Lombardo*, Ser. II, V. XXVIII, Fasc. XIV, pp. 745–52 (In Italian)

Bobylev D K 1892 O share s giroskopom vnutriy, katyashchemsya po gorizontalnoi ploskosti bez proskalzyvaniya (On a sphere with a gyroscope inside it rolling after horizontal plane without slipping). *Mat, Sb.*, T. 16, Vyp. 3, (In Russian)

Carvallo E 1902 L'Electricité deduite d'experience et remenée an pinxipe des travaux virtuels. *Coll. Scientia*, No. 19, Paris (In French)

Euler L 1766 *Elementa Calculi Variatiorum (The Elements of the Calculus Variations) Originally published in Novi Commentari Academiiae scientium Petropolitanae* **10** 51–93

Hertz H 1894 *Die Prinzipen der Mechanik*. Gesammeite Werke, T. 3, (In German)

Lagrange J L 1788 *Mécanique Analytique*

Lagrange J L 1788 *Mécanique Analytique* (Paris: Chez la Veuve DESAINT, Libraire)

Liennard A 1902 Sur l'application des equations de Lagrange aux phenomènes èlectrodynamik et electromagnetiques. *C.R.*, T. 134, No. 163 (In French)

Maxwell J C 1873 *A Treatise of Electricity and Magnetism* (Oxford: Clarendon)

Szklarski L, Pelczewski W, Kolendowski J, Puchalka T and Komarzewska M 1963 *Dynamika ukladow nieholonomicznych (Dynamics of Electromechanical Systems)* Komitet Elektrotechniki Polskiej Akademii Nauk, Panstwowe Wydawnictwo Naukowe 259 p (In Polish)

Tchaplygin 1897a O dvizheniy tiazhalogo tela vrashcheniya na gorizontalnoy ploskosti (On a motion of the solid body on a horizontal plane). *Trudy otdeleniya fizicheskih nauk obshchva liubiteley iestestvoznaniya*, T. 9, (In Russian)

Tchaplygin 1897b O nekotorom vozmozhnom obobshcheniy teoremy ploshchadey s primenениem k zadache o kataniy sharov (On some possible restriction of a space theory with the application to the exercise on rolling spheres). *Matem. Sb.*, T. 20, (In Russian)

White D C and Woodson H H 1959 *Electromechanical Energy Conversion* (New York: Wiley)

Zhukovsky N E 1893 O giroskopicheskom share D K Bobyleva (On a gyroscopic sphere of D K Bobylev). *Trudy ob-va lyubitieley iestiestvoznaniya, antropologii i etnografii.* T.6, Vyp. 1, (In Russian)

Part II

Towards a unified theory exploiting the theory of holors for physical heterogeneous continuous dynamical hypersystems

IOP Publishing

Mechatronics
Dynamical systems approach and theory of holors
B T Fijalkowski

Chapter 7

Physical heterogeneous continuous dynamical hypersystems as generalised physical commutation matrixers

In most sciences one generation tears down what another has built, and what one has established another undoes. In mathematics alone each generation builds a new story for the old structure.

Hermann Hankel

The fascinating impressiveness of rigorous mathematical analysis, with its atmosphere of precision and elegance, should not blind usto the defects of the premises that condition the whole process

T C Chamberlin

The most excellent practical solutions stem from the most scrupulous theory

Rene Antoine de Reaumur

7.1 Introduction

The majority of high-level engineering activities involve the physical assembly of physical equipment or equipment for which the term 'physical heterogeneous continuous complex or simple dynamical hypersystem' is usually a convenient definition, as it is more difficult to put together a set of physical equipment (systems and hyposystems) or physical parts (components) that create a whole and provide a complete function than it is to give a physical model of their applications as a 'generalised physical commutation matrixer'. The pieces of physical equipment (dynamical components) or the set of physical equipment (dynamical systems and hyposystems) that come together to create a physical heterogeneous continuous dynamical hypersystem are originated on observable facts that students habitually learn in their earliest physics lessons.

doi:10.1088/978-0-7503-1350-6ch7

Physical heterogeneous continuous complex or simple dynamical hypersystems have to be modelled so that a multitude of physical domains are taken into account. This is amplified when micro- or nanouniverse phenomena are studied. Therefore, new approaches and paradigms are needed to simulate these strategically important areas.

Scientists and engineers will face multi-domain modelling and simulation challenges in the future as an everyday occurrence. Currently, many physical heterogeneous continuous dynamical hypersystems already require modelling and simulation from a multi-domain point of view. For instance, the **mechano-electrical/electromechanical** (ME/EM) dynamotor (generator/motor) is a heterogeneous continuous simple dynamical hypersystem requiring both mechanical and electrical modelling.

Since it heats up, thermal modelling is also required. When EM drives are added, the overall system also requires the modelling of electronics, etc. In such physical heterogeneous continuous dynamical hypersystems multi-domain issues become even more natural and evident, as noted by Senturia (1998).

Consider the liquid cooling of electronics, which is becoming a viable option in portable computers and even cellular phones. In addition to modelling the electronics, it is necessary to model the fluidics and the electrostatic and magnetic effects. Radiation must also be taken into account. The task of multi-domain modelling and simulation is formidable.

One clear problem is that although individual micro- and nano-components can be modelled and simulated using **finite element method** (FEM) approaches, there will be more and more systems on a chip. For example, 'lab-on-a-chip' requires not only fluid channel modelling, but also the modelling of micropumps, microvalves and mixers.

FEM mathematical models are not suitable for system studies when the number of dynamical components is in the hundreds or thousands, which will be the case in a nanouniverse. Finding new ways to combine problems in multi-domain mathematical models and simulations for individual nanosystems and nanocomponents is imperative.

The author's intention here is to initiate the application of a dynamical systems approach and holor theories to create firstly a physical model and secondly a mathematical model, and then perform holor algebra analysis for generalised physical heterogeneous continuous complex and/or simple dynamical hypersystems.

The hypothesis of a generalised physical heterogeneous continuous dynamical hypersystem is exceptionally wide. One tries to be more precise by confining the discussion to what this author calls generalised physical commutation matrixers. On the other hand, the term physical commutation matrixer was created to designate the singular nature of a physical heterogeneous continuous dynamical hypersystem in which the generalised physical commutation matrixer's dynamical components correspond to only one physically homogenous continuous dynamical system. Therefore, we will concern ourselves with physical heterogeneous forms of energy, which fall into two main well-known states, namely kinetic energy and potential energy, as shown in table 7.1. For instance, one has to do with electrical, mechanical, fluidic, magnetic or thermal commutation matrixers. The plus point of the dynamical systems approach (systems thinking) is that the *modus operandi* applied for the holor analysis of electrical and magnetic commutation matrixers is very well developed (Fijalkowski 1973, 2011).

Table 7.1. Forms of energy.

Forms of energy All forms of energy fall under two main states			
Kinetic		Potential	
Kinetic energy is energy in motion. It is the motion of waves, electrons, atoms, molecules and objects or substances.		Potential energy is stored energy and the energy of position (gravitational).	
Motion mechanical energy	**The movement of objects or substances from one place to another is motion, e.g. wind and hydropower; water in waterfalls; moving automotive vehicles; and blood travelling through a body.**	**Stored mechanical energy**	**Stored mechanical energy is energy stored in objects by the application of a force. It must push or pull on an object. Compressed springs and stretched rubber bands are examples of gravitational potential energy. When the water is released to spin the turbines, it becomes kinetic energy.**
Electrical energy	**Electrical energy is the movement of electrons, e.g. lighting and electricity. Power lines carry electricity. Electric motors are driven by electromagnetic energy.**	**Gravitational energy**	**Gravitational energy is the energy of place or position; of anything 'up high'. Water in a reservoir behind a hydropower dam is an example of gravitational potential energy. When the water is released to spin the turbines, it becomes kinetic energy**
Thermal energy	**Thermal energy (or heat) is the internal energy in substances; it is the vibration and movement of atoms and molecules within substances. The faster the molecules move, the greater the thermal energy created. Examples include: geothermal energy, viscous and dry friction, and changes in the states of matter.**	**Chemical energy**	**Chemical energy is the energy stored in the bonds of atoms and molecules; it holds molecules together. When bonds are broken, chemical energy is released. Biomass, petroleum, natural gas, propane and coal are examples of stored chemical energy.**
Radiant energy	**Radiant energy is electromagnetic energy that travels in transverse waves. Radiant energy includes visible and laser lights, x-rays, gamma rays and radio waves. Solar energy is an example of radiant energy.**	**Nuclear energy**	**Nuclear energy is the energy stored in the nucleus of an atom; it is the energy that holds the nucleus together. The nucleus of a uranium atom is an example of nuclear energy When the nucleus of an atom splits, nuclear energy is released. Nuclear energy is the most concentrated form of energy. It comprises both fission and fusion.**
Sound energy	**Sound is the movement of energy through substances in longitudinal (compression/rare faction) waves.**		

The same *modus operandi* may be used for problems in mechanical, fluidic and thermal commutation matrixers. Thus, the analogies between problems in dissimilar physical heterogeneous continuous dynamical hypersystems are evident. Consequently, solutions in one physical heterogeneous continuous dynamical system are convertible by comparison to another one.

The physical model of a physical commutation matrixer ought to be qualitative, and the mathematical model of a physical commutation matrixer ought to be quantitative. If it is the latter, there is normally predictability for the experimentation, either to distribute indispensable data or to adapt a physical commutation matrixer to do what it is expected to do.

The generalised dynamical systems approach to presenting a physical situation is normally dealt with firstly as physical modelling and secondly as mathematical modelling. To create a right and proper physical model necessitates a predisposition to examine the physical situation and express it first in qualitative terms.

It is also indispensable, secondly, to think in terms of a mathematical model that can transfer quantitative information regarding the physical situation. To formulate such a mathematical model, on the other hand, it is necessary to have an accurate qualitative depiction of the physically observable facts.

It is just when the qualitative depiction (physical model) is terminated that the quantitative representation (mathematical model) may be completed (Fijalkowski 1987, 2010, 2011).

There is, naturally, a stimulus to accept that when the quantitative representation (mathematical model) has been completed (i.e. the holor algebra formula) the conundrum has truly been resolved. The hazard in this dynamical systems approach is that the physical situations may change. The simple and effective protection against this is to observe the physical situation regularly.

Surely early man observed electrical energy (electricity), whether in the benign form of simple static electricity or in the destructive form of radiant energy (lightning), and wished that one could somehow harness that energy and control it.

At present, we not only know how to generate electrical energy, but how to convert it into useful forms. However, generation and conversion are not enough. We are learning that we do not have an unlimited supply of electrical energy, and nor do we have an unlimited supply of other resources, such as energy supply engineers' time. We need more than just energy, we need *intelligent energy*.

In the generalised sense, **artificial intelligence** (AI) can be considered to be the ability to adapt to one's physical universe, i.e. environment. The interesting (non-trivial) universe tends to be dynamical, i.e. non-static.

As the physical universe changes, intelligent physical heterogeneous continuous complex or simple dynamical hypersystems adapt to the changes. In this adaptation, there is predictably some notion of a 'cost function' that is observed and optimised as a means of controlling the adaptation.

Adaptation can cease when the physical universe impedes change and the cost function is optimised to an acceptable point. Thus, intelligence has three basic elements: the physical universe (environment), the cost function and adaptation.

In terms of energy conversion, the physical universe may include, for example, things like component values: input energy-potential-difference value and quality, output energy-transfer value and temperature. The cost function may include, for example, things like loss or efficiency, ripple, stability, noise, component temperature and pole or zero location.

Adaptation may include, for example, things like changing the relative timing of commutation, adjusting the energy-potential-difference and/or energy-transfer levels, changing the compensation and changing the commutation frequency.

For instance, when the electrical energy supply load increases, an intelligent energy commutator and controller will adjust (adapt to) the dead time between commutation actions to minimise losses (cost function). A non-intelligent electrical energy supply will simply sit and get hot.

One can determine, at least on a qualitative level, the relative AI of energy commutators and controllers within their own species by how they adapt to the physical universe's (environment) changes.

Unregulated commutators and controllers are the least intelligent of the commutators and controllers. Any adaptations to physical universe changes are incidental. As an example, if the load impedance drops, the current only changes according to Ohm's law and is defined by the impedance of the conversion commutator. Feedback has been introduced to electrical energy supplies to increase AI, that is, to improve their ability to respond to changes in the physical universe (environment). If the load impedance drops, the feedback network senses a change in the output voltage and adjusts the operation of the conversion commutator to adapt to this physical universe change. This level of AI is still limited. For example, the feedback is designed for a point or narrow range of operating conditions while assuming given values of the dynamical components. That is, the feedback parameters do not change. The parameters must be selected to give acceptable results over the entire operating range. Because of this constraint, an electrical energy supply designer may spend a considerable amount of time adjusting parameters to satisfy the operational specification over the entire range of operation.

As the years have progressed, so has the AI of energy conversion. Small improvements to the intelligence evolved with the addition of feedforward control, energy-transfer value feedback and temperature correction. For many of these, a cost function can readily be identified and they can be categorised as true adaptations.

In this new era of energy, supplies with embedded macroelectronic commutators (macrocommutators) and microelectronic controllers (microcontrollers) are providing an entirely new level of AI. No longer are static commutators and microcontrollers limited to fixed or narrowly constrained parameters. The parameters can be changed easily from design to design, during operation, and can even be changed intelligently by the static commutator and microcontroller. Numerous cost functions can be tracked (like loss, temperature, noise, etc) and the operation can be adapted based on these cost functions as the physical universe's conditions change. As an example of the AI in the best of today's macrocommutators and microcontrollers, the relative loss during the energy conversion process is monitored as a cost function.

As the physical universe's conditions change and losses are no longer minimal, the energy macrocommutators and microcontrollers adjust or minimise the loss. In the case of less intelligent energy macrocommutators and microcontrollers, as the physical universe's conditions change and impact on efficiency, they simply sit there and run inefficiently, relatively speaking.

Energy conversion is enjoying a new era in terms of intelligent power supply. The new era has come along at the perfect time. The physical universe is changing for energy supply designers; there are fewer designers, more challenging designs that require higher densities and faster time-to-market, greater cost pressures and higher performance targets. These novel intelligent energy macrocommutators and microcontrollers optimise performance so that the designers needs less time to optimise design over all the physical universe's conditions. With their high levels of integration, they provide higher densities and higher performance with an equivalent or even lower cost of ownership.

7.2 The generalised physical commutation matrixer's input- and output-signal holor functions

The aim of this author is to obtain in physical and mathematical form a description of the dynamical behaviour of a physical heterogeneous continuous dynamical hypersystem in terms of some physically significant variables called holors. As the nature of the dynamical hypersystem changes, its holors change. For example, the holors used in electrical homogeneous continuous dynamical systems are voltage holors and current holors; in mechanical homogeneous continuous dynamical systems they are force holors and velocity holors; and in fluidic homogeneous continuous dynamical systems they are pressure holors and volumetric-flow-rate holors. Despite the differences in the physical holors used to characterise physical homogeneous continuous dynamical systems in various physical disciplines, certain fundamental similarities exist, and it is in the analyst's interest to seek out and exploit these similarities in such a manner that the task of physical and mathematical modelling is eased and one's overall insight into the dynamical performance of physical heterogeneous continuous dynamical hypersystems is increased. A suitable unifying conception that may be used for this purpose is energy. A physical heterogeneous continuous complex or simple dynamical hypersystem can be thought of as operating upon a pair of holors whose univalent tensor product is a power holor.

The physical homogeneous continuous dynamical systems, hyposystems and components that make up the physical heterogeneous continuous dynamical hypersystem may thus be thought of as energy commutators and regulators which, depending upon the way they are interconnected, process the energy injected into the physical heterogeneous continuous dynamical hypersystem in a characteristic fashion that is observed as the dynamical hypersystem response.

In the same spirit, the input/output behaviour of a physical heterogeneous continuous dynamical hypersystem follows if energy can be considered as being injected into a dynamical hypersystem via energy ports, with similar energy ports applied to read out the dynamical hypersystem response.

In this book a class of physical heterogeneous continuous complex or simple dynamical hypersystems is identified that a pair of holors has associated with energy conversion.

The generalised forms of these dynamical hypersystem holors are the abstract quantities termed by the author the energy-potential-difference holor and the energy-transfer holor. In this convention, energy-potential-difference holors are referred to as 'transholors' and energy-transfer holors are denoted as 'perholors'.

Elemental dynamical hypersystems' dynamical systems, dynamical hyposystems and dynamical components are then postulated for the fundamental phenomena of energy generation, energy storage and energy dissipation.

The concept of physical heterogeneous continuous dynamical systems as energy static commutators and microcontrollers that interact with inputs and outputs via energy ports is a conceptual physical model that may be represented in terms of what the author calls a physical commutation matrixer, which encompasses a wide range of physical heterogeneous continuous dynamical systems.

In order to develop this notion, however, it is necessary to examine the mechanism of energetic interactions in terms of so-called dynamical hypersystem holors, which determine how and in what sense energy is transmitted.

The generalised physical commutation matrixer's forms for the input- and output-signal holor functions may be obtained from equations (6.25)–(6.27), namely:

$$\boldsymbol{F}_i = \boldsymbol{Z}_{ij}\boldsymbol{V}^j \tag{7.1a}$$

or

$$\boldsymbol{V}^j = \boldsymbol{Y}^{ji}\boldsymbol{F}_i. \tag{7.1b}$$

If the time domain is constrained to $t > 0$, one may obtain the energy-potential-difference-holor (i.e. generalised force) and energy-transfer-holor (i.e. generalised velocity) functions, $\boldsymbol{F}_i$ and $\boldsymbol{V}^j$, respectively, as

$$\boldsymbol{F}_i = (F_c, F_s) = \|F_c \quad F_s\|^T = \begin{Vmatrix} F_c \\ F_s \end{Vmatrix} = \begin{Vmatrix} F\cos\psi_F \\ F\sin\psi_F \end{Vmatrix} = F\angle\psi_F, \tag{7.2a}$$

and

$$\boldsymbol{V}^j = (V^c, V^s) = \|V^c \quad V^s\|^T = \begin{Vmatrix} V^c \\ V^s \end{Vmatrix} = \begin{Vmatrix} V\cos\psi_V \\ V\sin\psi_V \end{Vmatrix} = V\angle\psi_V, \tag{7.2b}$$

where, as a general rule, $\boldsymbol{F}_i$ and $\boldsymbol{V}^j$ are univalent holors with plethos 2, with merates or coordinates (independent components) F_c and F_s as well as V^c and V^s, which may be expressed, respectively, as

$$\left.\begin{aligned} F_c &= F\cos\psi_F \\ F_s &= F\sin\psi_F \end{aligned}\right\}, \tag{7.3a}$$

and

$$\left. \begin{aligned} V^c &= V \cos \psi_V \\ V^s &= V \sin \psi_V \end{aligned} \right\}. \tag{7.3b}$$

The form of the input-signal holors given in equation (7.2) depends on the values F_c and F_s as well as V^c and V^s .

Figures 7.1 and 7.2 show the energy-potential-difference-holor and energy-transfer-holor functions for four possible special combinations of F_c and F_s as well as V^c and V^s.

For $\psi_F = 0$

$$\left. \begin{aligned} F_c &= F \cos \psi_F = F \cos 0 = F \\ F_s &= F \sin \psi_F = F \sin 0 = 0 \end{aligned} \right\}. \tag{7.4a}$$

For $\psi_F = \pi/2$

$$\left. \begin{aligned} F_c &= F \cos \psi_F = F \cos \frac{\pi}{2} = 0 \\ F_s &= F \sin \psi_F = F \sin \frac{\pi}{2} = F \end{aligned} \right\}. \tag{7.4b}$$

Thus, if one can obtain the sinusoidal response of a given physical system to the generalised energy-potential-difference-holor and energy-transfer-holor functions $\boldsymbol{F}_i$ and $\boldsymbol{V}^j$, one may, by the specification of phase angles ψ_F and ψ_V, find the response for all the special combinations. Even if one does not specify the form of ψ_F and ψ_V in the following sinusoidal responses, the examples given in the textbook will examine

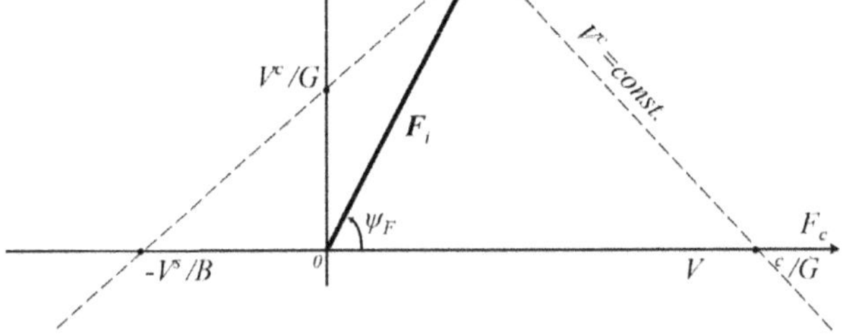

Figure 7.1. A geometrical representation of the energy-potential-difference holor.

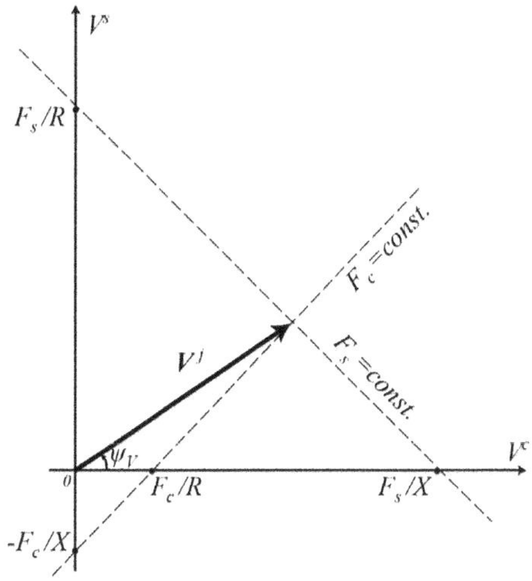

Figure 7.2. A geometrical representation of the energy-transfer holor.

the effect of an unquestionably real quantity. We will continue with the specification of ψ_F and ψ_V as holor quantities until chapter 10, when we will discuss frequency response. An instantaneous value of the step- or sine-wave-input power can be obtained from a product

$$p(t) = f(t)v(t), \tag{7.5}$$

where

$$\left.\begin{aligned} f(t) &= F\sin(\omega t + \psi_F) = \sqrt{2}\,F_{\text{rms}}\sin(\omega t + \psi_F) \\ v(t) &= V_{\max}\sin(\omega t + \psi_V) = \sqrt{2}\,V_{\text{rms}}\sin(\omega t + \psi_V) \end{aligned}\right\}. \tag{7.6}$$

Substituting the energy-potential-difference and energy-transfer instantaneous values from equations (7.6) to equation (7.5), one obtains

$$p(t) = FV\left\{\cos\varphi - \cos\left[2\omega t + \left(\psi_F + \psi_V\right)\right]\right\}, \tag{7.7}$$

where

$$\varphi = \psi_F - \psi_V.$$

To obtain a holor representation of the power one must write equation (7.7) in the form

$$p(t) = FV\left[\cos\varphi - \cos + 2\omega t\cos\left(\psi_F + \psi_V\right) + \sin 2\omega t\sin\left(\psi_F + \psi_V\right)\right]. \tag{7.8}$$

If one designs an independent term by P^p and dependent terms from a doublet frequency by P^q and P^r, then one obtains

$$\left.\begin{aligned} P^p &= FV\cos\lfloor\psi_F-\psi_V\rfloor = FV\cos\varphi = P \\ P^q &= FV\cos(\psi_F+\psi_V) \\ P^r &= FV\sin(\psi_F+\psi_V) \end{aligned}\right\}. \tag{7.9}$$

In this way equation (7.8) can be rewritten in the form

$$p(t) = P^p - P^q\cos 2\omega t + P^r\sin 2\omega t. \tag{7.10}$$

The quantities P_p, P_q and P_r are independent elements (merates) or coordinates of the three-merates monovalent holor of a step- or sine-wave-input power

$$\boldsymbol{P}^{\kappa} = (P^p, P^q, P^r), \tag{7.11}$$

which allows us to eliminate time from equation (7.10).

From the first expression of equations (7.9) it is obtained that an independent element (merate) of the power holor is an active power, which can be measured, for example, in an electrical heterogeneous continuous dynamical system by means of an electrodynamic wattmeter. Residual independent elements (merates) can be also measured or computed with help from physical metrology methods.

In particular circumstances:

a) for an energy-potential-difference-holor phase angle $\psi_F = 0$

$$P^q = FV\cos\varphi = P^p = P;$$
$$P^r = -FV\sin\varphi = -Q; \text{ and}$$

b) for an energy-transfer-holor phase angle $\psi_V = 0$

$$P^q = FV\cos\varphi = P^p = P;$$
$$P^r = FV\sin\varphi = Q.$$

Thus, for the energy-potential-difference-holor phase angle $\psi_F = 0$, or the energy-transfer-holor phase angle $\psi_V = 0$, an independent element (merate) of the holor power P^q is the active power P, however, an independent element (merate) of the power holor P^r is the reactive power Q. To obtain an equation that takes into account a power holor from the energy-potential-difference holor and the energy-transfer holor, one must use an adequate law for the multiplication of these holors.

However, in holor algebra, the conventional tensor product in the form $\boldsymbol{F}_i\boldsymbol{V}^j$ (from Moon and Spencer 1986) is not applied, and in the case of the power holor one uses the particularly useful univalent tensor product

$$\boldsymbol{P}^{\kappa} = \boldsymbol{F}_i \otimes \boldsymbol{V}^j = \gamma^{\kappa i}_{\cdot\cdot j}\boldsymbol{F}_i\boldsymbol{V}^j, \tag{7.12}$$

where $i, j = c, s$; $\kappa = p, q, r$. The univalent tensor product is designed using the symbol $\otimes$, and gives a univalent holor as a result.

From equations (7.12) a power holor

$$\boldsymbol{P}^{\kappa} = \boldsymbol{F}_i \otimes \boldsymbol{V}^j = \gamma^{\kappa i}_{..j} \boldsymbol{F}_i \boldsymbol{V}^j.$$

However, from equation (7.1a) an energy-potential-difference holor

$$\boldsymbol{F}_i = \boldsymbol{Z}_{ij} \boldsymbol{V}^j.$$

Hence, the power holor

$$\boldsymbol{P}^{\kappa} = \boldsymbol{F}_i \otimes \boldsymbol{V}^j = \gamma^{\kappa i}_{..j} \boldsymbol{F}_i \boldsymbol{V}^j = \gamma^{\kappa i}_{..j} \boldsymbol{Z}_{ij} \boldsymbol{V}^k \boldsymbol{V}^j. \tag{7.13}$$

and its independent elements (merates)

$$P^p = \gamma^{pi}_{..j} \boldsymbol{Z}_{ij} \boldsymbol{V}^k \boldsymbol{V}^j, \; P^q = \gamma^{qi}_{..j} \boldsymbol{Z}_{ij} \boldsymbol{V}^k \boldsymbol{V}^j, \; P^r = \gamma^{ri}_{..j} \boldsymbol{Z}_{ij} \boldsymbol{V}^k \boldsymbol{V}^j. \tag{7.14}$$

Taking into consideration equations (7.13) and (7.14), one obtains

$$\left.\begin{aligned} P^p &= R[(V^{c}{}^\wedge 2) + (V^{s}{}^\wedge 2)] \\ P^q &= R[(V^{c}{}^\wedge 2) - (V^{s}{}^\wedge 2)] - 2XV^cV^s \\ P^r &= X[(V^{c}{}^\wedge 2) - (V^{s}{}^\wedge 2)] + 2RV^cV^s \end{aligned}\right\}. \tag{7.15}$$

Substituting equations (7.9) into the first expression from equations (7.15), one obtains

$$P^p = R(V^\wedge 2) = P. \tag{7.16}$$

In this way, one obtains a well-known expression for a sine-wave-input physically continuous dynamical system analysis from which once more one can see that an independent element (merate) of the holor power P^p is a real or active power. Adding the geometrically independent elements (merates) P^q and P^r of the power holor, one derives a well-known expression for a physical effective value of the apparent power S in the form

$$\begin{aligned} \mathrm{sqrt}\left[(P^{q}{}^\wedge 2) + (P^{r}{}^\wedge 2)\right] &= \mathrm{sqrt}\left[(R^\wedge 2) + (X^\wedge 2)\right]\left[(V^{c}{}^\wedge 2) + (V^{s}{}^\wedge 2)\right] \\ &= Z(V^\wedge 2) = FV = S. \end{aligned} \tag{7.17a}$$

Thus, due to the application of holor algebra for a sine-wave-input physically continuous dynamical system analysis, it is possible to have an operation with holor quantities for physically continuous dynamical systems that are analogous to each other as with real quantities for step-input physically continuous dynamical systems. A geometrical representation of the power holor needs 3D space. The coordinates are independent elements (merates) of the power holor P^p, P^q and P^r. The coordinates P^q and P^r are orthogonal to each other, as in equation (7.17*a*). The third axis P^p establishes an angle $\varphi = \psi_V - \psi_i$ with a vector of the apparent power, which a module $S = F\ V$. The power holor $\boldsymbol{P}^{\kappa}$ is hence a vector sum of its independent elements (merates) P^p, P^q and P^r (figure 7.3).

A partial representation of the holor power needs 2D space. If a phase angle of the double-frequency term is small, then a module of the double frequency power can be obtained from a part of the holor diagram shown in figure 7.3, which depicts a well-known holor diagram for the power holor established by independent elements (merates) of the real or active power $P^p = P$ and the reactive power Q. A partial holor diagram of the holor power, however, does not represent the holor power exactly, but a strict representation needs three independent elements (merates) and 3D space. Thus a physical effective value of the apparent power P^{κ} between two points would be:

$$P^{\kappa} = \text{sqrt}\left[\left(P^{p\wedge}2\right) + \left(P^{q\wedge}2\right) + \left(P^{r\wedge}2\right)\right] = S. \tag{7.17b}$$

But what would happen if one moved a physically continuous dynamical system or body from one point to another point in universal space? There would be a change in power between the physically continuous dynamical system or body and the reference point and there would be a rate of change in the power holor $\boldsymbol{P}^{\kappa}$ per unit interval, which can be expressed mathematically as $\omega\boldsymbol{P}^{\kappa}$. Here, a new dimension is introduced; ω, which is an element of time. One cannot deny that there must be time, because one can perceive that what has been is different from what is present and what will be. So one has a past, a present and a future, all of which refer to time.

In the Universe where one exists, time appears to flow equably in one direction, from the past to the future. This is our own timeline. It is, however, also possible that time flows in at least three directions. If this is the case, then time can have its own set of Cartesian coordinates that are similar to but not necessarily the same as those of space. One may assign P^p, P^q and P^r as the coordinates of time. Here one has a universe of at least nine dimensions.

We have always used a notion of the sense of direction. We say that a thing goes forward or backward, or is positive or negative. These senses of direction, therefore, are dimensions in their own right and may be denoted by '+' for forward and '–' for backward.

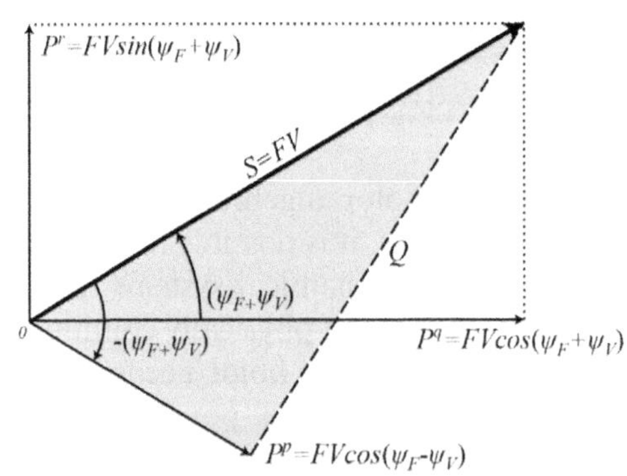

Figure 7.3. A geometrical representation of the power holor.

So far one has only looked at the possible dimensions governing the relation of a local physically continuous dynamical system or body with a reference point in universal space.

Without a concept of relativity, one presumes that rigid bodies are possible, where a rigid body remains unchanged when it is moved from one point to any of the infinite possible locations in universal space. In other words, one assumes that space is independent of its contents. In such a view, our physics is basically Newtonian. The existence of a particular physically continuous dynamical system or body however is also governed by other parameters. Careful observation will give one the idea that all things are in motion. Motion or momentum, therefore, is a property of matter. One could then deduce that nothing exists that does not move. If such is the case, then space could not exist separately from matter, hence all matter has space and all spaces have matter. There is, therefore, no empty space.

7.3 Multi-port, multi-input/multi-output, generalised physical commutation matrixers

In 1820, the marquis de Laplace wrote that 'We ought to regard the present state of the universe as the effect of its antecedent state and as the cause of the state that is to follow. An intelligence knowing all the forces acting in nature at a given instant, as well as the momentary positions of all things in the universe, would be able to comprehend in one single formula the motions of the largest bodies as well as the lightest atoms in the world, provided that its intellect were sufficiently powerful to subject all data to analysis; to it nothing would be uncertain, the future as well as the past would be present to its eyes. The perfection that the human mind has been able to give to astronomy affords but a feeble outline of such an intelligence' (Laplace 1820). This image had a great impact on the dynamical systems approach (systems thinking) in the 1980s. Nowadays, scientists believe that in nature chaos develops in most phenomena and that predicting the future is impracticable in the long term. Controlling chaos is very important from the scientific point of view.

There has been a very rapid revolution in the fields of physics and engineering—especially mechatronics (mechanics, fluidics, pneumatics, photonics and electronics)—because stereotyped ideas in the field of knowledge have been overcome. All progress is stimulated by thinking that is free from dogma and not indiscriminately accepting of existing concepts. It is also well known that perfect practical solutions are born out of ingenious theories. 'There is nothing more practical than a good theory,' said L Boltzmann (1844–1906).

In physics, the concept of matrix-interconnected physical networks, termed physical commutation matrixers (figure 7.4) by the author of this book, was developed from the mathematical matrix notation of the dynamical systems approach (systems thinking) for the formation methodology of physical and mathematical models of mechatronical (e.g. mechanical and/or electrical) heterogeneous continuous dynamical hypersystems with energy-transfer holor conducting horizontal-row and vertical-column conductive collectors (conductors), which are crossed with each other but galvanically separated (insulated) in-between by means of uni- and/or bipolar dissimilar passive or active physical dynamical components and structural devices (e.g. physical valves) (Fijalkowski 1967, 1973, 1982, 1985, 1987, 1995).

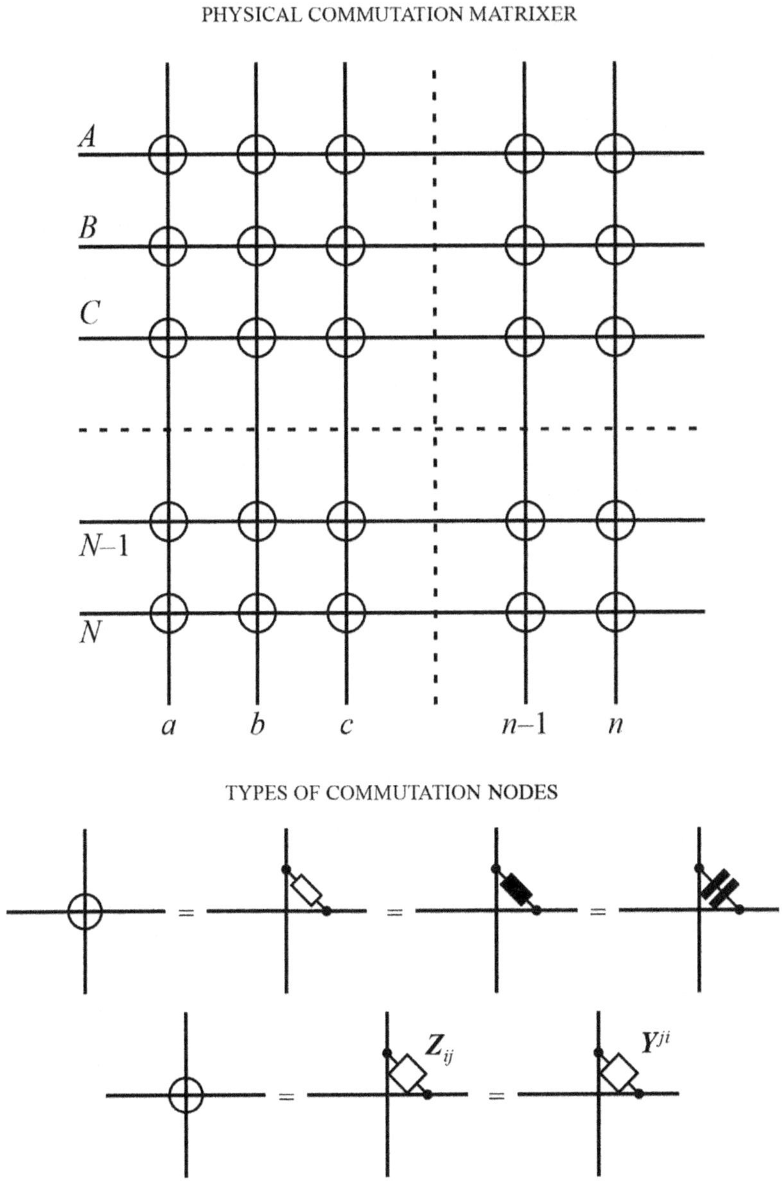

Figure 7.4. A multi-port MIMO generalised physical commutation matrixer (generalized physical model) (Fijalkowski 1967, 1973, 1982, 1985, 1987, 1995).

The author's concept relates to l physical dynamical components that are interconnected in a $(m \times n)$ matrix design with m horizontal-row conductive collectors and n vertical-column conductive collectors, with l, m and n being natural numbers unequal to zero. A first conductive collector end of the m row conductive collectors and n column conductive collectors is connected to an evaluation system and every l physical dynamical component is interconnected between two of the row conductive collectors and the column conductive collectors. The concept of the matrixer arrangement further involves at least one return conductive collector, whose first conductive collector end is connected to the evaluation system and whose second conductive collector end is in contact with the second conductive collector end of one of the row or column conductive collectors (bars or wires). Theoretically and practically, the multi-port, **multi-input/multi output** (MIMO) generalised physical commutation matrixers are the changers of an energy-potential-difference holor (e.g. a generalised force holor); an energy-transfer holor (e.g. a generalised velocity holor); the frequency and number of phases (virtually, static matrix converters, which alter one kind of physical energy with an energy-potential-difference holor, $\boldsymbol{F}_i^P$ (e.g. a voltage holor – $\boldsymbol{U}_i^P$)); an energy-transfer holor, $\boldsymbol{V}_P^j$ (e.g. a current holor, $\boldsymbol{I}_P^j$); the frequency, f_P, and number of phases, m_P, into another kind of physical energy with an energy-potential-difference holor, $\boldsymbol{F}_i^s$, (e.g. a voltage holor, $\boldsymbol{U}_i^s$); an energy-transfer holor, $\boldsymbol{V}_s^j$ (e.g. a current holor, $\boldsymbol{I}_s^j$); and the frequency, f_s, and number of phases, m_s, performing the definite matrix function. The multi-port MIMO generalised physical commutation matrixer concept is an extension of the one- or two-port MIMO generalised physical commutation matrixer concept.

In figure 7.4 there is a generalised physical model (a matrix diagram or matrix map) of the multi-port MIMO generalised physical commutation matrixer realised on unipolar and/or bipolar different passive and/or active physical components and devices (e.g. physical devices). In this physical commutation matrixer, the horizontal collectors (matrix rows) *A, B, C,…, N* are termed primary or input conductive collectors, but the vertical conductive collectors (matrix columns) *a, b, c,…, n* are termed secondary or output conductive collectors. The differentiation of the primary (input) conductive collectors and the secondary (output) conductive collectors in the multi-port MIMO generalised physical commutation matrixer is relative, since the physical energy may be charged as well as discharged (emitted) through the same conductive collectors at any time.

In the general case, the primary conductive collectors are connected with the secondary conductive collectors by means of any uni- and/or bipolar continuous physical dynamical components and structural devices (figure 7.4). Physical energy conversion in any multi-port MIMO generalised physical commutation matrixers takes place in advance, specified or adapted for the given multi-port MIMO generalised physical commutation matrixers' rule algorithms on the basis of the phenomena's *modus operandi*, termed commutation, i.e. an energy-transfer-holor passage (i.e. electric-current-holor passage) switching changeover among preselected conductive collectors of the multi-port MIMO physical commutation matrixer, which connect the terminals of two dissimilar physical heterogeneous continuous dynamical hypersystems—the primary and secondary.

In order to concretise a concept of the physical commutation matrixer it is convenient to use the concept so generalised, which after a reduction in considerations is carried out might formulate specific physical commutation matrixer mathematical models of any physical homogeneous continuous dynamical systems that are applied in practice. This approach permits generalised physical commutation matrixers in a synthetic systems approach on the examination of any specific multi-port MIMO.

Assuming that the excitations of a multi-port MIMO generalised physical commutation matrixer are given in the form of the column hypermatrices or hypervectors of energy sources among the primary (input) and secondary (output) conductive collectors (figure 7.5):

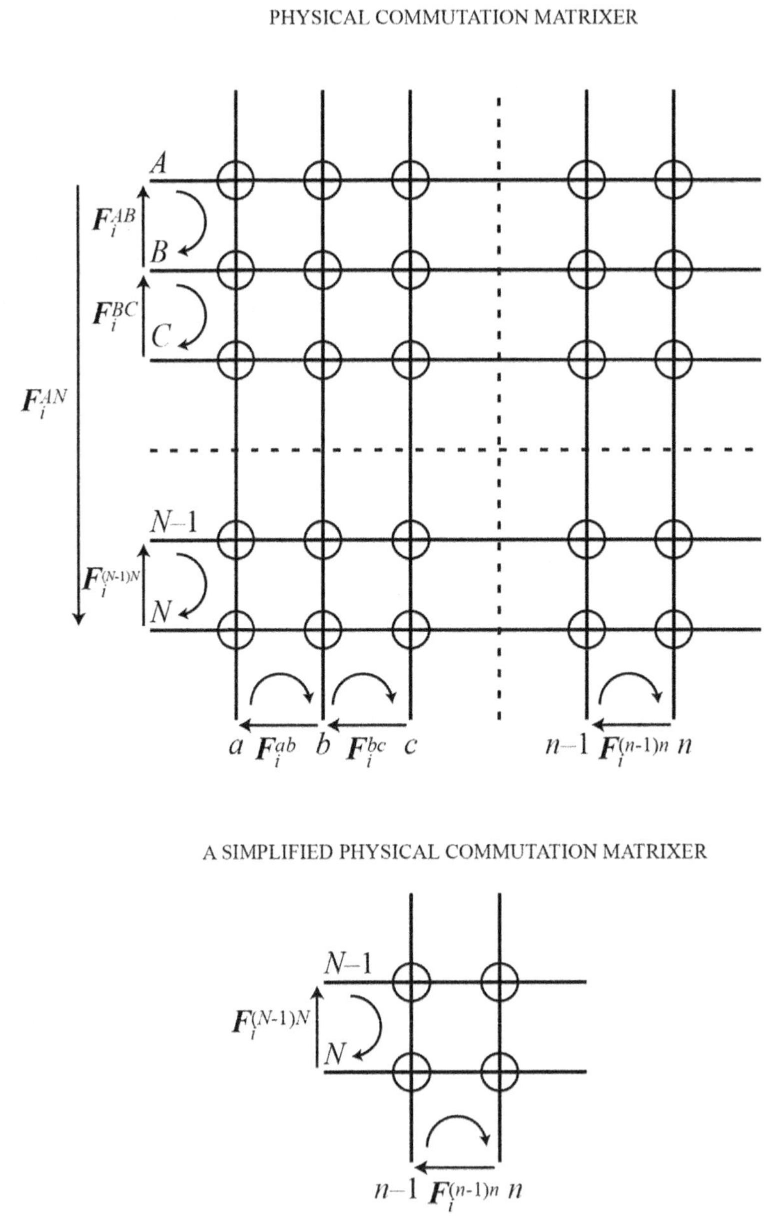

Figure 7.5. A multi-port MIMO generalised physical commutation matrixer with energy-potential-difference holor sources (physical model I) (Fijalkowski 1987).

$$F_i = \begin{Vmatrix} F_i^{P^T} & F_i^{s^T} \end{Vmatrix}^T = \begin{Vmatrix} F_i^{AB^T} & F_i^{BC^T} & F_i^{C(N-1)^T} & \cdots & F_i^{(N-1)N^T} & F_i^{NA^T} \\ F_i^{ab^T} & F_i^{bc^T} & F_i^{c(n-1)^T} & \cdots & F_i^{(n-1)n^T} & F_i^{na^T} \end{Vmatrix}^T \tag{7.18}$$

as the responses of the multi-port MIMO generalised physical commutation matrixer, the column hypermatrices or hypervectors of the outer-mesh energy-transfer holors,

$$V^j = \begin{Vmatrix} V_P^{j^T} V_s^{j^T} \end{Vmatrix}^T = \begin{Vmatrix} V_{AB}^{j^T} & V_{BC}^{j^T} & V_{C(N-1)}^{j^T} & \cdots & V_{(N-1)N}^{j^T} & V_{NA}^{j^T} \\ V_{ab}^{j^T} & V_{bc}^{j^T} & V_{c(n-1)}^{j^T} & \cdots & V_{(n-1)n}^{j^T} & V_{na}^{j^T} \end{Vmatrix}^T, \tag{7.19}$$

a set of the differential equations of dynamics, i.e. the mathematical model of the multi-port MIMO generalised physical commutation matrixer as the physical heterogeneous continuous dynamical hypersystem, may be formulated in the dynamical systems approach and matrix notation as follows:

$$F_i = Z_{ij} V^j, \tag{7.20}$$

where the $i \times j$ hypermatrix Z_{ij} has a dimension of the impedance holor.

To denote the state variables associated with the primary (input) conductive collectors by an index P, and those associated with the secondary (output) collectors by an index s, a set of the differential equations of dynamics, i.e. the mathematical model of the multi-port MIMO generalised physical commutation matrixer, may be expressed as follows:

$$\begin{Vmatrix} F_i^P \\ F_i^s \end{Vmatrix} = \begin{Vmatrix} Z_{ij}^{PP} & Z_{ij}^{Ps} \\ Z_{ij}^{sP} & Z_{ij}^{ss} \end{Vmatrix} \begin{Vmatrix} V_P^j \\ V_s^j \end{Vmatrix}, \tag{7.21a}$$

or in detail

$$\begin{Vmatrix} F_i^{AB} \\ F_i^{BC} \\ F_i^{C(N-1)} \\ \vdots \\ F_i^{NA} \\ F_i^{ab} \\ F_i^{bc} \\ F_i^{c(n-1)} \\ \vdots \\ F_i^{na} \end{Vmatrix} = \begin{Vmatrix} Z_{ij}^{AA} & Z_{ij}^{AB} & Z_{ij}^{AC} & & Z_{ij}^{AN} & Z_{ij}^{Aa} & Z_{ij}^{Ab} & Z_{ij}^{Ac} & & Z_{ij}^{An} \\ Z_{ij}^{BA} & Z_{ij}^{BB} & Z_{ij}^{BC} & \cdots & Z_{ij}^{BN} & Z_{ij}^{Ba} & Z_{ij}^{Bb} & Z_{ij}^{Bc} & \cdots & Z_{ij}^{Bn} \\ Z_{ij}^{CA} & Z_{ij}^{CA} & Z_{ij}^{CC} & & Z_{ij}^{CN} & Z_{ij}^{Ca} & Z_{ij}^{Cb} & Z_{ij}^{Cc} & & Z_{ij}^{Cn} \\ & \vdots & & \ddots & \vdots & & \vdots & & \ddots & \vdots \\ Z_{ij}^{NA} & Z_{ij}^{NB} & Z_{ij}^{NC} & \cdots & Z_{ij}^{NN} & Z_{ij}^{Na} & Z_{ij}^{Nb} & Z_{ij}^{Nc} & \cdots & Z_{ij}^{Nn} \\ Z_{ij}^{aA} & Z_{ij}^{aB} & Z_{ij}^{aC} & & Z_{ij}^{aN} & Z_{ij}^{aa} & Z_{ij}^{ab} & Z_{ij}^{ac} & & Z_{ij}^{an} \\ Z_{ij}^{bA} & Z_{ij}^{bB} & Z_{ij}^{bC} & \cdots & Z_{ij}^{bN} & Z_{ij}^{ba} & Z_{ij}^{bb} & Z_{ij}^{bc} & \cdots & Z_{ij}^{bn} \\ Z_{ij}^{cA} & Z_{ij}^{cB} & Z_{ij}^{cC} & & Z_{ij}^{cN} & Z_{ij}^{ca} & Z_{ij}^{cb} & Z_{ij}^{cc} & & Z_{ij}^{cn} \\ & \vdots & & \ddots & \vdots & & \vdots & & \ddots & \vdots \\ Z_{ij}^{nA} & Z_{ij}^{nB} & Z_{ij}^{nC} & \cdots & Z_{ij}^{cN} & Z_{ij}^{na} & Z_{ij}^{nb} & Z_{ij}^{nc} & \cdots & Z_{ij}^{nn} \end{Vmatrix} \begin{Vmatrix} V_{AB}^j \\ V_{BC}^j \\ V_{C(N-1)}^j \\ \vdots \\ V_{NA}^j \\ V_{ab}^j \\ V_{bc}^j \\ V_{c(n-1)}^j \\ \vdots \\ V_{na}^j \end{Vmatrix}. \tag{7.21b}$$

The coefficients

$$Z_{kl} = \left.\frac{\partial F_{k(k-1)}}{\partial V^{l(l-1)}}\right|_{V^{j(j-1)}} \tag{7.21c}$$

have a dimension of the impedance holor.

To adapt a mathematical model of the multi-port MIMO generalised physical commutation matrixer to the theory of physical heterogeneous continuous dynamical hypersystems demands that one refers to a simplified and very general mathematical model of the multi-port MIMO generalised physical commutation matrixer, derived from the physical model that is shown in figure 7.6.

In order to fulfil the demands of converting one kind of physical energy with the energy-potential-difference holor, F_i^P (e.g. a voltage holor, U_i^P), the energy-transfer holor, V_P^j (e.g. a current holor, I_P^j), the frequency, f_P, and the number of phases, m_P, of a primary physical heterogeneous continuous dynamical hypersystem of the multi-port MIMO generalised physical commutation matrixer into another kind of physical energy with the energy-potential-difference holor, F_i^s (e.g. a voltage holor, U_i^s), the energy-transfer holor, V_s^j (e.g. a current holor, I_s^j), the frequency, f_s, and the number of phases, m_s, of a secondary physically continuous dynamical system of the multi-port generalised physical commutation matrixer—as well as the free passage of physical energy between these dynamical hypersystems—they ought to be connected by means of the multi-port MIMO generalised physical commutation matrixer's uni- and/or bipolar passive and active physical dynamical components and devices.

A physical model of the multi-port MIMO generalised physical commutation matrixer (see figure 7.6) fulfils the aforementioned demands. Uni- and bipolar passive and active physical dynamical components and devices connect both physical heterogeneous continuous dynamical hypersystems of the multi-port MIMO generalised physical commutation matrixer and make free passage of physical energy between them possible.

In the general case, a physical commutation matrixer such as this can connect many physical heterogeneous continuous dynamical hypersystems.

In this book we consider a case where there are only two physical heterogeneous continuous dynamical hypersystems—the primary and the secondary. It is sufficient to understand the multi-port MIMO physical commutation matrixer concept.

Both physical heterogeneous continuous dynamical hypersystems of the multi-port generalised physical matrixer may be considered to be autonomous, i.e. when disconnected from the multi-port MIMO generalised physical commutation matrixer, they have determined potentials on their terminals.

They have energy-potential-difference-holor sources (e.g. voltage-holor sources), energy-transfer-holor sources (e.g. current-holor sources) and physical dynamical components (e.g. inductances, capacitances and resistances), as well as continuous uni- and/or bipolar physical valves.

Assuming that the excitations of a multi-port MIMO generalised physical commutation matrixer are given in the form of the column hypermatrices or

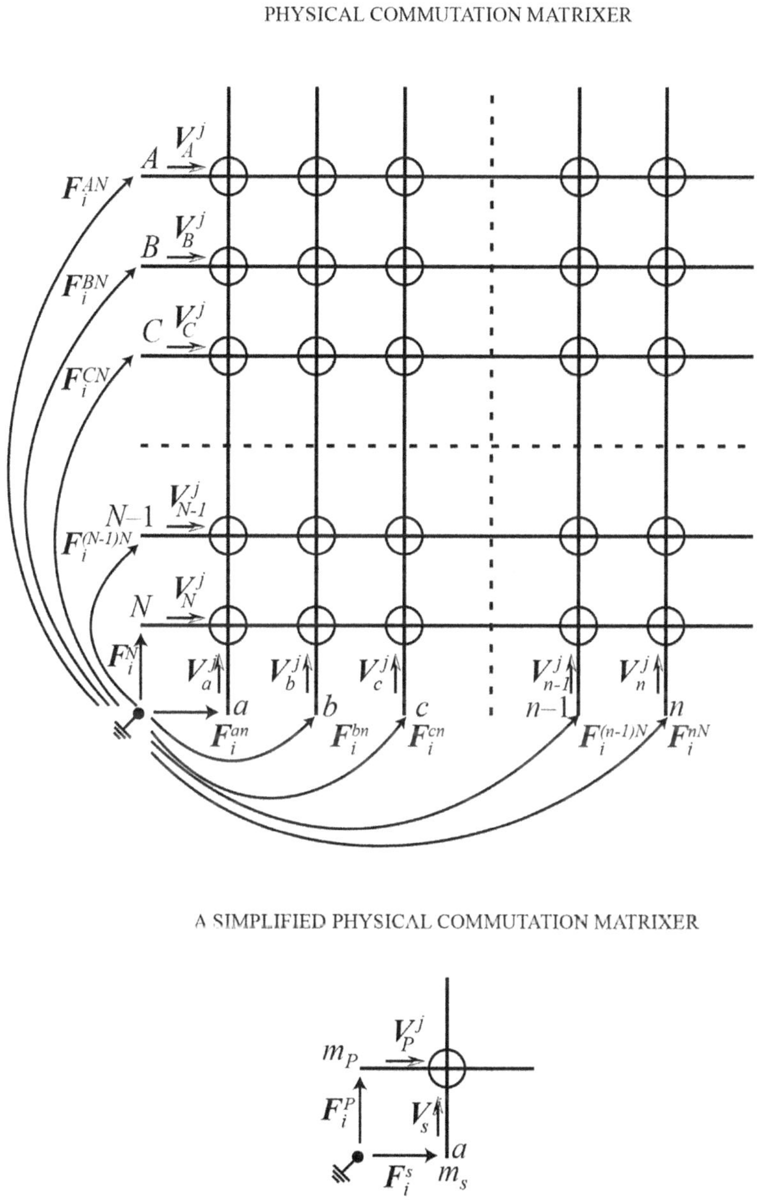

Figure 7.6. A multi-port MIMO generalised physical commutation matrixer with energy-transfer holor sources (physical model II) (Fijalkowski 1987).

hypervectors of source-energy-transfer holors (e.g. current holors) flowing in the primary (input) and secondary (output) conductive collectors (figure 7.6):

$$\boldsymbol{V}^j = \left\| \boldsymbol{V}_P^{j^T} \boldsymbol{V}_s^{j^T} \right\|^T = \left\| \begin{matrix} \boldsymbol{V}_A^{j^T} & \boldsymbol{V}_B^{j^T} & \boldsymbol{V}_C^{j^T} & \cdots & \boldsymbol{V}_N^{j^T} \\ \boldsymbol{V}_a^{j^T} & \boldsymbol{V}_b^{j^T} & \boldsymbol{V}_c^{j^T} & \cdots & \boldsymbol{V}_n^{j^T} \end{matrix} \right\|^T, \tag{7.22}$$

as the responses of the multi-port physical commutation matrixer, the column hypermatrices or the hypervectors of node energy-potential-difference holors (e.g. voltage holors)

$$\boldsymbol{F}_i = \left\| \boldsymbol{F}_i^{P^T} \boldsymbol{F}_i^{s^T} \right\|^T = \left\| \begin{array}{ccccc} \boldsymbol{F}_i^{A^T} & \boldsymbol{F}_i^{B^T} & \boldsymbol{F}_i^{T} & \cdots & \boldsymbol{F}_i^{N^T} \\ \boldsymbol{F}_i^{a^T} & \boldsymbol{F}_i^{b^T} & \boldsymbol{F}_i^{c^T} & \cdots & \boldsymbol{F}_i^{n^T} \end{array} \right\|^T, \tag{7.23}$$

a set of the differential equations of dynamics, i.e. the mathematical model of the multi-port MIMO generalised physical commutation matrixer as the physical heterogeneous continuous dynamical hypersystems, may be formulated in the dynamical systems approach and matrix notation as follows:

$$\boldsymbol{V}^j = \boldsymbol{Y}^{ji} \boldsymbol{F}_i, \tag{7.24}$$

where the $n \times m$ hypermatrix $\boldsymbol{Y}^{ji}$ has a dimension of the admittance holor.

Node energy-potential-difference holors (e.g. voltage holors) consisting of physical elements of the hypermatrix $\boldsymbol{F}_i$ are measured as regards the reference node, which is located on the outside of the multi-port MIMO generalised physical commutation matrixer. For example, in a star connection of the multi-phase generalised physical heterogeneous continuous dynamical hypersystem, the reference node may be a null point of the multi-phase star.

To denote the state variables associated with the primary (input) collectors by an index P, and those associated with the secondary (output) collectors by an index s, the set of differential equations of dynamics, i.e. the mathematical model of the multi-port MIMO generalised physical commutation matrixer, may be expressed as follows:

$$\left\| \begin{array}{c} \boldsymbol{V}_P^j \\ \boldsymbol{V}_s^j \end{array} \right\| = \left\| \begin{array}{cc} \boldsymbol{Y}_{PP}^{ji} & \boldsymbol{Y}_{Ps}^{ji} \\ \boldsymbol{Y}_{sP}^{ji} & \boldsymbol{Y}_{ss}^{ji} \end{array} \right\| \left\| \begin{array}{c} \boldsymbol{F}_i^P \\ \boldsymbol{F}_i^s \end{array} \right\|, \tag{7.25a}$$

or in detail as

$$\left\| \begin{array}{c} \boldsymbol{V}_A^j \\ \boldsymbol{V}_B^j \\ \boldsymbol{V}_C^j \\ \vdots \\ \boldsymbol{V}_N^j \\ \boldsymbol{V}_a^j \\ \boldsymbol{V}_b^j \\ \boldsymbol{V}_c^j \\ \vdots \\ \boldsymbol{V}_n^j \end{array} \right\| = \left\| \begin{array}{cccccccccc} \boldsymbol{Y}_{AA}^{ji} & \boldsymbol{Y}_{AB}^{ji} & \boldsymbol{Y}_{AC}^{ji} & & \boldsymbol{Y}_{AN}^{ji} & \boldsymbol{Y}_{Aa}^{ji} & \boldsymbol{Y}_{Ab}^{ji} & \boldsymbol{Y}_{Ac}^{ji} & & \boldsymbol{Y}_{An}^{ji} \\ \boldsymbol{Y}_{BA}^{ji} & \boldsymbol{Y}_{BB}^{ji} & \boldsymbol{Y}_{BC}^{ji} & \cdots & \boldsymbol{Y}_{BN}^{ji} & \boldsymbol{Y}_{Ba}^{ji} & \boldsymbol{Y}_{Bb}^{ji} & \boldsymbol{Y}_{Bc}^{ji} & \cdots & \boldsymbol{Y}_{Bn}^{ji} \\ \boldsymbol{Y}_{CA}^{ji} & \boldsymbol{Y}_{CA}^{ji} & \boldsymbol{Y}_{CC}^{ji} & & \boldsymbol{Y}_{CN}^{ji} & \boldsymbol{Y}_{Ca}^{ji} & \boldsymbol{Y}_{Cb}^{ji} & \boldsymbol{Y}_{Cc}^{ji} & & \boldsymbol{Y}_{Cn}^{ji} \\ & \vdots & & \ddots & \vdots & & \vdots & & \ddots & \vdots \\ \boldsymbol{Y}_{NA}^{ji} & \boldsymbol{Y}_{NB}^{ji} & \boldsymbol{Y}_{NC}^{ji} & \cdots & \boldsymbol{Y}_{NN}^{ji} & \boldsymbol{Y}_{Na}^{ji} & \boldsymbol{Y}_{Nb}^{ji} & \boldsymbol{Y}_{Nc}^{ji} & \cdots & \boldsymbol{Y}_{Nn}^{ji} \\ \boldsymbol{Y}_{aA}^{ji} & \boldsymbol{Y}_{aB}^{ji} & \boldsymbol{Y}_{aC}^{ji} & & \boldsymbol{Y}_{aN}^{ji} & \boldsymbol{Y}_{aa}^{ji} & \boldsymbol{Y}_{ab}^{ji} & \boldsymbol{Y}_{ac}^{ji} & & \boldsymbol{Y}_{an}^{ji} \\ \boldsymbol{Y}_{bA}^{ji} & \boldsymbol{Y}_{bB}^{ji} & \boldsymbol{Y}_{bC}^{ji} & \cdots & \boldsymbol{Y}_{bN}^{ji} & \boldsymbol{Y}_{ba}^{ji} & \boldsymbol{Y}_{bb}^{ji} & \boldsymbol{Y}_{bc}^{ji} & \cdots & \boldsymbol{Y}_{bn}^{ji} \\ \boldsymbol{Y}_{cA}^{ji} & \boldsymbol{Y}_{cB}^{ji} & \boldsymbol{Y}_{cC}^{ji} & & \boldsymbol{Y}_{cN}^{ji} & \boldsymbol{Y}_{ca}^{ji} & \boldsymbol{Y}_{cb}^{ji} & \boldsymbol{Y}_{cc}^{ji} & & \boldsymbol{Y}_{cn}^{ji} \\ & \vdots & & \ddots & \vdots & & \vdots & & \ddots & \vdots \\ \boldsymbol{Y}_{nA}^{ji} & \boldsymbol{Y}_{aB}^{ji} & \boldsymbol{Y}_{nC}^{ji} & \cdots & \boldsymbol{Y}_{cN}^{ji} & \boldsymbol{Y}_{na}^{ji} & \boldsymbol{Y}_{nb}^{ji} & \boldsymbol{Y}_{nc}^{ji} & \cdots & \boldsymbol{Y}_{nn}^{ji} \end{array} \right\| \left\| \begin{array}{c} \boldsymbol{F}_i^A \\ \boldsymbol{F}_i^B \\ \boldsymbol{F}_i^C \\ \vdots \\ \boldsymbol{F}_i^N \\ \boldsymbol{F}_i^a \\ \boldsymbol{F}_i^b \\ \boldsymbol{F}_i^c \\ \vdots \\ \boldsymbol{F}_i^n \end{array} \right\|. \tag{7.25b}$$

The coefficients

$$\boldsymbol{Y}^{kl} = \left.\frac{\partial \boldsymbol{V}^{k}}{\partial \boldsymbol{F}_{l}}\right|_{\boldsymbol{F}_{i}=0} \qquad (i \neq l), \tag{7.25c}$$

have a dimension of the admittance holor.

As with the mathematical models of physical heterogeneous continuous dynamical hypersystems of generalised physical commutation matrixers shown in figures 7.5 and 7.6, consisting of a set of differential equations of dynamics (7.20) and (7.24), there exists a formulation possibility for the mathematical model of physically continuous dynamical systems of the generalised physical commutation matrixer (figure 7.7) acting as an energy-potential-difference commutator (matrix converter) with an arbitrary number N of primary collectors (inputs), equal to the number m of physical matrix phases and an arbitrary number n of secondary collectors (outputs).

Summing up, a hybrid mathematical model of the multi-port MIMO generalised physical commutation matrixer may be written as follows:

$$\left\|\begin{array}{c}
\boldsymbol{F}_{i}^{AN}\\ \boldsymbol{F}_{i}^{BA}\\ \boldsymbol{F}_{i}^{CB}\\ \vdots\\ \boldsymbol{F}_{i}^{N(N-1)}\\ \boldsymbol{F}_{i}^{an}\\ \boldsymbol{F}_{i}^{ba}\\ \boldsymbol{F}_{i}^{cb}\\ \vdots\\ \boldsymbol{F}_{i}^{n(n-1)}\\ \boldsymbol{V}_{A}^{j}\\ \boldsymbol{V}_{B}^{j}\\ \boldsymbol{V}_{C}^{j}\\ \vdots\\ \boldsymbol{V}_{N}^{j}\,{*}\\ \boldsymbol{V}_{a}^{j}\\ \boldsymbol{V}_{b}^{j}\\ \boldsymbol{V}_{c}^{j}\\ \vdots\\ \boldsymbol{V}_{n}^{j}
\end{array}\right\| = \left\|\begin{array}{cccccccccccccccccccc}
\boldsymbol{Z}_{ij}^{AA} & \boldsymbol{Z}_{ij}^{AB} & \boldsymbol{Z}_{ij}^{AC} & & \boldsymbol{Z}_{ij}^{AN} & \boldsymbol{Z}_{ij}^{Aa} & \boldsymbol{Z}_{ij}^{Ab} & \boldsymbol{Z}_{ij}^{Ac} & & \boldsymbol{Z}_{ij}^{An} & 0 & 0 & 0 & & 0 & 0 & 0 & 0 & & 0\\
\boldsymbol{Z}_{ij}^{BA} & \boldsymbol{Z}_{ij}^{BB} & \boldsymbol{Z}_{ij}^{BC} & \cdots & \boldsymbol{Z}_{ij}^{BN} & \boldsymbol{Z}_{ij}^{Ba} & \boldsymbol{Z}_{ij}^{Bb} & \boldsymbol{Z}_{ij}^{Bc} & \cdots & \boldsymbol{Z}_{ij}^{Bn} & 0 & 0 & 0 & \cdots & 0 & 0 & 0 & 0 & \cdots & 0\\
\boldsymbol{Z}_{ij}^{CA} & \boldsymbol{Z}_{ij}^{CA} & \boldsymbol{Z}_{ij}^{CC} & & \boldsymbol{Z}_{ij}^{CN} & \boldsymbol{Z}_{ij}^{Ca} & \boldsymbol{Z}_{ij}^{Cb} & \boldsymbol{Z}_{ij}^{Cc} & & \boldsymbol{Z}_{ij}^{Cn} & 0 & 0 & 0 & & 0 & 0 & 0 & 0 & & 0\\
 & \vdots & & \ddots & \vdots & & \vdots & & \ddots & \vdots & & \vdots & & \ddots & \vdots & & \vdots & & \ddots & \vdots\\
\boldsymbol{Z}_{ij}^{NA} & \boldsymbol{Z}_{ij}^{NB} & \boldsymbol{Z}_{ij}^{NC} & \cdots & \boldsymbol{Z}_{ij}^{NN} & \boldsymbol{Z}_{ij}^{Na} & \boldsymbol{Z}_{ij}^{Nb} & \boldsymbol{Z}_{ij}^{Nc} & \cdots & \boldsymbol{Z}_{ij}^{Nn} & 0 & 0 & 0 & \cdots & 0 & 0 & 0 & 0 & \cdots & 0\\
\boldsymbol{Z}_{ij}^{aA} & \boldsymbol{Z}_{ij}^{aB} & \boldsymbol{Z}_{ij}^{aC} & & \boldsymbol{Z}_{ij}^{aN} & \boldsymbol{Z}_{ij}^{aa} & \boldsymbol{Z}_{ij}^{ab} & \boldsymbol{Z}_{ij}^{ac} & & \boldsymbol{Z}_{ij}^{an} & 0 & 0 & 0 & & 0 & 0 & 0 & 0 & & 0\\
\boldsymbol{Z}_{ij}^{bA} & \boldsymbol{Z}_{ij}^{bB} & \boldsymbol{Z}_{ij}^{bC} & \cdots & \boldsymbol{Z}_{ij}^{bN} & \boldsymbol{Z}_{ij}^{ba} & \boldsymbol{Z}_{ij}^{bb} & \boldsymbol{Z}_{ij}^{bc} & \cdots & \boldsymbol{Z}_{ij}^{bn} & 0 & 0 & 0 & \cdots & 0 & 0 & 0 & 0 & \cdots & 0\\
\boldsymbol{Z}_{ij}^{cA} & \boldsymbol{Z}_{ij}^{cB} & \boldsymbol{Z}_{ij}^{cC} & & \boldsymbol{Z}_{ij}^{cN} & \boldsymbol{Z}_{ij}^{ca} & \boldsymbol{Z}_{ij}^{cb} & \boldsymbol{Z}_{ij}^{cc} & & \boldsymbol{Z}_{ij}^{cn} & 0 & 0 & 0 & & 0 & 0 & 0 & 0 & & 0\\
 & \vdots & & \ddots & \vdots & & \vdots & & \ddots & \vdots & & \vdots & & \ddots & \vdots & & \vdots & & \ddots & \vdots\\
\boldsymbol{Z}_{ij}^{nA} & \boldsymbol{Z}_{ij}^{nB} & \boldsymbol{Z}_{ij}^{nC} & \cdots & \boldsymbol{Z}_{ij}^{cN} & \boldsymbol{Z}_{ij}^{na} & \boldsymbol{Z}_{ij}^{nb} & \boldsymbol{Z}_{ij}^{nc} & \cdots & \boldsymbol{Z}_{ij}^{nn} & 0 & 0 & 0 & \cdots & 0 & 0 & 0 & 0 & \cdots & 0\\
0 & 0 & 0 & & 0 & 0 & 0 & 0 & & 0 & \boldsymbol{Y}_{AA}^{ji} & \boldsymbol{Y}_{AB}^{ji} & \boldsymbol{Y}_{AC}^{ji} & & \boldsymbol{Y}_{AN}^{ji} & \boldsymbol{Y}_{Aa}^{ji} & \boldsymbol{Y}_{Ab}^{ji} & \boldsymbol{Y}_{Ac}^{ji} & & \boldsymbol{Y}_{An}^{ji}\\
0 & 0 & 0 & \cdots & 0 & 0 & 0 & 0 & \cdots & 0 & \boldsymbol{Y}_{BA}^{ji} & \boldsymbol{Y}_{BB}^{ji} & \boldsymbol{Y}_{BC}^{ji} & \cdots & \boldsymbol{Y}_{BN}^{ji} & \boldsymbol{Y}_{Ba}^{ji} & \boldsymbol{Y}_{Bb}^{ji} & \boldsymbol{Y}_{Bc}^{ji} & \cdots & \boldsymbol{Y}_{Bn}^{ji}\\
0 & 0 & 0 & & 0 & 0 & 0 & 0 & & 0 & \boldsymbol{Y}_{CA}^{ji} & \boldsymbol{Y}_{CA}^{ji} & \boldsymbol{Y}_{CC}^{ji} & & \boldsymbol{Y}_{CN}^{ji} & \boldsymbol{Y}_{Ca}^{ji} & \boldsymbol{Y}_{Cb}^{ji} & \boldsymbol{Y}_{Cc}^{ji} & & \boldsymbol{Y}_{Cn}^{ji}\\
 & \vdots & & \ddots & \vdots & & \vdots & & \ddots & \vdots & & \vdots & & \ddots & \vdots & & \vdots & & \ddots & \vdots\\
0 & 0 & 0 & \cdots & 0 & 0 & 0 & 0 & \cdots & 0 & \boldsymbol{Y}_{NA}^{ji} & \boldsymbol{Y}_{NB}^{ji} & \boldsymbol{Y}_{NC}^{ji} & \cdots & \boldsymbol{Y}_{NN}^{ji} & \boldsymbol{Y}_{Na}^{ji} & \boldsymbol{Y}_{Nb}^{ji} & \boldsymbol{Y}_{Nc}^{ji} & \cdots & \boldsymbol{Y}_{Nn}^{ji}\\
0 & 0 & 0 & & 0 & 0 & 0 & 0 & & 0 & \boldsymbol{Y}_{aA}^{ji} & \boldsymbol{Y}_{aB}^{ji} & \boldsymbol{Y}_{aC}^{ji} & & \boldsymbol{Y}_{aN}^{ji} & \boldsymbol{Y}_{aa}^{ji} & \boldsymbol{Y}_{ab}^{ji} & \boldsymbol{Y}_{ac}^{ji} & & \boldsymbol{Y}_{an}^{ji}\\
0 & 0 & 0 & \cdots & 0 & 0 & 0 & 0 & \cdots & 0 & \boldsymbol{Y}_{bA}^{ji} & \boldsymbol{Y}_{bB}^{ji} & \boldsymbol{Y}_{bC}^{ji} & \cdots & \boldsymbol{Y}_{bN}^{ji} & \boldsymbol{Y}_{ba}^{ji} & \boldsymbol{Y}_{bb}^{ji} & \boldsymbol{Y}_{bc}^{ji} & \cdots & \boldsymbol{Y}_{bn}^{ji}\\
0 & 0 & 0 & & 0 & 0 & 0 & 0 & & 0 & \boldsymbol{Y}_{cA}^{ji} & \boldsymbol{Y}_{cB}^{ji} & \boldsymbol{Y}_{cC}^{ji} & & \boldsymbol{Y}_{cN}^{ji} & \boldsymbol{Y}_{ca}^{ji} & \boldsymbol{Y}_{cb}^{ji} & \boldsymbol{Y}_{cc}^{ji} & & \boldsymbol{Y}_{cn}^{ji}\\
 & \vdots & & \ddots & \vdots & & \vdots & & \ddots & \vdots & & \vdots & & \ddots & \vdots & & \vdots & & \ddots & \vdots\\
0 & 0 & 0 & \cdots & 0 & 0 & 0 & 0 & \cdots & 0 & \boldsymbol{Y}_{nA}^{ji} & \boldsymbol{Y}_{aB}^{ji} & \boldsymbol{Y}_{nC}^{ji} & \cdots & \boldsymbol{Y}_{cN}^{ji} & \boldsymbol{Y}_{na}^{ji} & \boldsymbol{Y}_{nb}^{ji} & \boldsymbol{Y}_{nc}^{ji} & \cdots & \boldsymbol{Y}_{nn}^{ji}
\end{array}\right\| \left\|\begin{array}{c}
\boldsymbol{V}_{AB}^{j}\\ \boldsymbol{V}_{BA}^{j}\\ \boldsymbol{V}_{CB}^{j}\\ \vdots\\ \boldsymbol{V}_{N(N-1)}^{j}\\ \boldsymbol{V}_{an}^{j}\\ \boldsymbol{V}_{ba}^{j}\\ \boldsymbol{V}_{cb}^{j}\\ \vdots\\ \boldsymbol{V}_{n(n-1)}^{j}\\ \boldsymbol{F}_{i}^{A}\\ \boldsymbol{F}_{i}^{B}\\ \boldsymbol{F}_{i}^{C}\\ \vdots\\ \boldsymbol{F}_{i}^{N}\\ \boldsymbol{F}_{i}^{a}\\ \boldsymbol{F}_{i}^{b}\\ \boldsymbol{F}_{i}^{c}\\ \vdots\\ \boldsymbol{F}_{i}^{n}
\end{array}\right\|, \tag{7.26}$$

$$\boldsymbol{Z}_{kl} = \left.\frac{\partial \boldsymbol{F}_{k(k-1)}}{\partial \boldsymbol{V}^{l(l-1)}}\right|_{\boldsymbol{V}^{j(j-1)}} \quad (j \neq l) \text{ and } \boldsymbol{Y}kl = \left.\frac{\partial \boldsymbol{V}^{k}}{\partial \boldsymbol{F}_{l}}\right|_{\boldsymbol{F}_{i}=0} \quad (i \neq l) \tag{7.27}$$

having dimensions of the impedance or admittance holor, respectively.

Assuming that the energy-potential-difference-holor sources applied across the primary collectors (inputs) $A, B, C, \ldots, N$ of a generalised physical commutation matrixer commutator (matrix converter) are given in the form of the energy-potential-difference holors' column matrix or vector (input holors)

$$\boldsymbol{F}_{i}^{P} = \left\| \boldsymbol{F}_{i}^{A^{T}} \;\; \boldsymbol{F}_{i}^{B^{T}} \;\; \boldsymbol{F}_{i}^{C^{T}} \;\; \cdots \;\; \boldsymbol{F}_{i}^{N-1^{T}} \;\; \boldsymbol{F}_{i}^{N^{T}} \right\|^{T}, \tag{7.28}$$

as the primary physical state variables (inputs), and energy-potential-difference-holors' sources, applied across the secondary collectors (outputs) $a, b, c, \ldots, n$ of a generalised physical commutation matrixer commutator (matrix converter), in the form of the energy-potential-difference holors' matrix or vector (output holors)

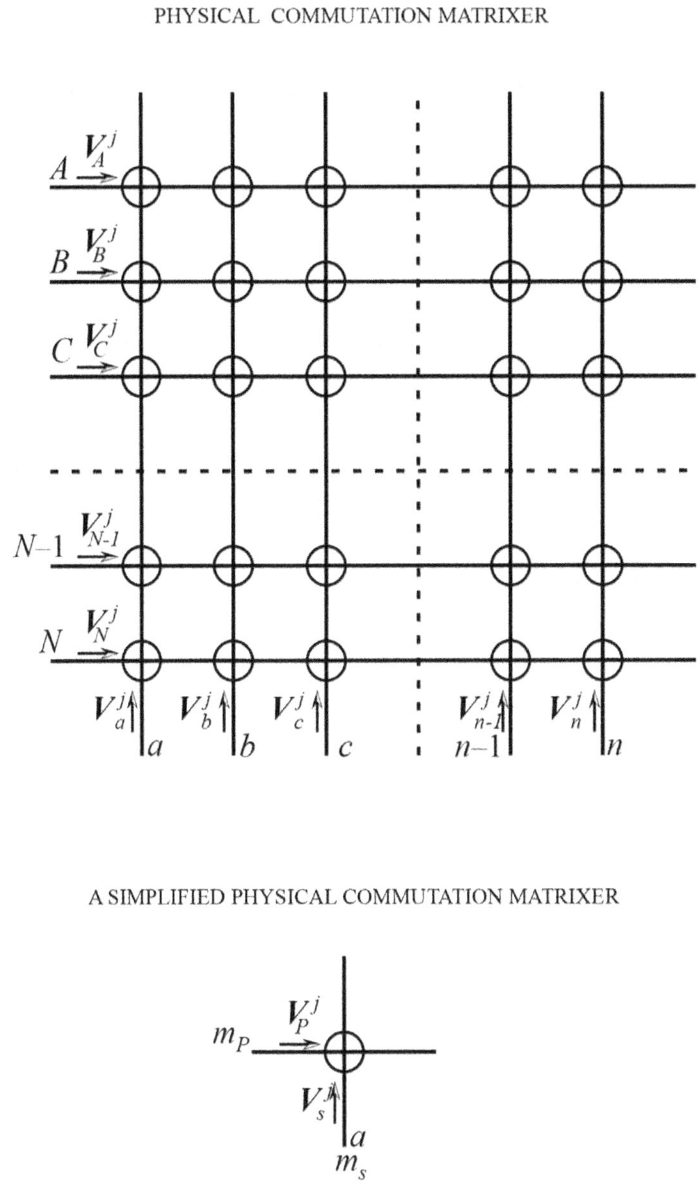

Figure 7.7. Physical model of a generalised physical commutation matrixer commutator (Fijalkowski 1987).

$$\boldsymbol{F}_s^j = \left\| \boldsymbol{F}_a^{j^T} \quad \boldsymbol{F}_b^{j^T} \quad \boldsymbol{F}_c^{j^T} \quad \cdots \quad \boldsymbol{F}_{n-1}^{j^T} \quad \boldsymbol{F}_n^{j^T} \right\|^T, \tag{7.29}$$

as the secondary physical state variables (output holors).

Well, a mathematical model of the physical heterogeneous continuous dynamical hypersystem of a generalised physical commutation matrixer commutator (matrix converter) acting as an energy-potential-difference commutator, making the outputs dependent on the inputs, in the dynamical systems approach and a matrix notation may be written in the form:

$$\boldsymbol{F}_s^j = \boldsymbol{C}_{sP}^{ji}\boldsymbol{F}_i^P, \tag{7.30}$$

or in detail as

$$\begin{Vmatrix} \boldsymbol{F}_a^j \\ \boldsymbol{F}_b^j \\ \boldsymbol{F}_c^j \\ \vdots \\ \boldsymbol{F}_{n-1}^j \\ \boldsymbol{F}_n^j \end{Vmatrix} = \begin{Vmatrix} \boldsymbol{C}_{aA}^{ji} & \boldsymbol{C}_{aB}^{ji} & \boldsymbol{C}_{aC}^{ji} & & \boldsymbol{C}_{a(N-1)}^{ji} & \boldsymbol{C}_{aN}^{ji} \\ \boldsymbol{C}_{bA}^{nm} & \boldsymbol{C}_{bB}^{nm} & \boldsymbol{C}_{bC}^{nm} & \cdots & \boldsymbol{C}_{b(N-1)}^{nm} & \boldsymbol{C}_{bN}^{nm} \\ \boldsymbol{C}_{cA}^{nm} & \boldsymbol{C}_{cB}^{nm} & \boldsymbol{C}_{cC}^{nm} & & \boldsymbol{C}_{c(N-1)}^{nm} & \boldsymbol{C}_{cN}^{nm} \\ & \vdots & & \ddots & \vdots & \\ \boldsymbol{C}_{(n-1)A}^{nm} & \boldsymbol{C}_{(n-1)B}^{nm} & \boldsymbol{C}_{(n-1)C}^{nm} & \cdots & \boldsymbol{C}_{(n-1)(N-1)}^{nm} & \boldsymbol{C}_{(n-1)N}^{nm} \\ \boldsymbol{C}_{nA}^{nm} & \boldsymbol{C}_{nB}^{nm} & \boldsymbol{C}_{nC}^{nm} & & \boldsymbol{C}_{n(N-1)}^{nm} & \boldsymbol{C}_{nN}^{nm} \end{Vmatrix} \begin{Vmatrix} \boldsymbol{F}_i^A \\ \boldsymbol{F}_i^B \\ \boldsymbol{F}_i^C \\ \vdots \\ \boldsymbol{F}_i^{N-1} \\ \boldsymbol{F}_i^N \end{Vmatrix}. \tag{7.31}$$

However, a mathematical model of the physically continuous dynamical system of a generalised physical commutation matrixer commutator (matrix converter) acting as an energy-potential-difference commutator, making the inputs dependent on the outputs, may be written in the form:

$$\|\boldsymbol{F}_i^P\| = \left\|\boldsymbol{C}_{ij}^{Ps}\right\| \|\boldsymbol{F}_s^j\|, \tag{7.32}$$

or in detail as

$$\begin{Vmatrix} \boldsymbol{F}_i^A \\ \boldsymbol{F}_i^B \\ \boldsymbol{F}_i^C \\ \vdots \\ \boldsymbol{F}_i^{N-1} \\ \boldsymbol{F}_i^N \end{Vmatrix} = \begin{Vmatrix} \boldsymbol{C}_{ij}^{Aa} & \boldsymbol{C}_{ij}^{Ab} & \boldsymbol{C}_{ij}^{Ac} & & \boldsymbol{C}_{ij}^{A(n-1)} & \boldsymbol{C}_{ij}^{An} \\ \boldsymbol{C}_{ij}^{Ba} & \boldsymbol{C}_{ij}^{Bb} & \boldsymbol{C}_{ij}^{Bc} & \cdots & \boldsymbol{C}_{IJ}^{B(n-1)} & \boldsymbol{C}_{IJ}^{Bn} \\ \boldsymbol{C}_{IJ}^{Ca} & \boldsymbol{C}_{IJ}^{Cb} & \boldsymbol{C}_{IJ}^{Cc} & & \boldsymbol{C}_{IJ}^{C(n-1)} & \boldsymbol{C}_{IJ}^{Cn} \\ & \vdots & & \ddots & \vdots & \\ \boldsymbol{C}_{IJ}^{(N-1)a} & \boldsymbol{C}_{IJ}^{(N-1)b} & \boldsymbol{C}_{IJ}^{(N-1)c} & \cdots & \boldsymbol{C}_{IJ}^{(N-1)(n-1)} & \boldsymbol{C}_{IJ}^{(N-1)n} \\ \boldsymbol{C}_{IJ}^{Na} & \boldsymbol{C}_{IJ}^{Nb} & \boldsymbol{C}_{IJ}^{Nc} & & \boldsymbol{C}_{IJ}^{N(n-1)} & \boldsymbol{V}_{IJ}^{Nn} \end{Vmatrix} \begin{Vmatrix} \boldsymbol{F}_a^j \\ \boldsymbol{F}_b^j \\ \boldsymbol{F}_c^j \\ \vdots \\ \boldsymbol{F}_{n-1}^j \\ \boldsymbol{F}_n^j \end{Vmatrix}, \tag{7.33}$$

where $\boldsymbol{C}_{kL}^{JI}$ and $\boldsymbol{C}_{ij}^{Kl}$ are the energy-potential-difference coefficients of the commutation nodes kL and Kl, respectively, which may be expressed as follows:

$$\boldsymbol{C}_{kL}^{ji} = \left.\frac{\partial \boldsymbol{F}_k^j}{\partial \boldsymbol{F}_i^L}\right| \boldsymbol{F}_i^I = 0 = \begin{cases} +1 \overset{\Delta}{\to} F_k^j = +F_i^L \\ \;\;0 \overset{\Delta}{\to} F_k^j = 0 \\ -1 \overset{\Delta}{\to} F_k^j = -F_i^L \end{cases}$$

and

$$C_{ij}^{Kl} = \frac{\partial F_i^K}{\partial F_l^j}\Bigg| F_i^j = 0 = \begin{cases} -1 \overset{\Delta}{\rightarrow} F_i^k = -F_i^L \\ 0 \overset{\Delta}{\rightarrow} F_i^k = 0 \\ +1 \overset{\Delta}{\rightarrow} F_i^k = +F_l^j \end{cases}.$$

In the case of the generalized physical commutation matrixer with bipolar (symmetrical) physical valves, the energy-potential-difference coefficients of the commutation nodes $C_{kL}^{ji} = \pm 1$ and $C_{ij}^{Kl} = \mp 1$ represent an ideal physical-valve impedance equal to zero, which is incorporated between the primary collector L and the secondary collector k, or between the secondary collector l and the primary collector K, and correspond to a turn-on state of the physical valves with a sense of the energy-transfer flow direction (polarisation) that is dependent on a sign of these coefficients; however, the energy-potential-difference coefficients of the commutation nodes $C_{kL}^{ji} = 0$ and $C_{ij}^{Kl} = 0$, in which the ideal physical-valve impedance equals infinity, correspond to a turn-off state of the physical valves. Analogously, a mathematical model of the physical heterogeneous continuous dynamical system of a generalised physical commutation matrixer commutator (matrix converter) (figure 7.7), as an energy-transfer commutator with an arbitrary number N of primary collectors (inputs), equals a number m of physical phases and an arbitrary number n of secondary collectors (outputs). Assuming that an energy-transfer holors' sources, flowing in the primary collectors (inputs) $A, B, C, \ldots, N$ of a physical commutation matrixer of the physical commutation matrixer commutator (matrix converter), are given in the form of the energy-transfer holors' column matrix or vector (input holors)

$$V_i^P = \left\| V_i^{A^T} \quad V_i^{B^T} \quad V_i^{C^T} \quad \cdots \quad V_i^{N-1^T} \quad V_i^{N^T} \right\|^T, \tag{7.34}$$

as the primary physical state variables (inputs), and the secondary energy-transfer holors' sources of a generalised physical commutation matrixer of the physical commutator, flowing in the secondary collectors' (outputs) $a, b, c, \ldots, n$ of a physical commutation matrixer, in the form of the secondary energy-transfer holors (output holors)

$$V_s^j = \left\| V_a^{j^T} \quad V_b^{j^T} \quad V_c^{j^T} \quad \cdots \quad V_{n-1}^{j^T} \quad V_n^{j^T} \right\|^T, \tag{7.35}$$

as secondary physical state variables (inputs).

Well, a mathematical model of the physical heterogeneous continuous dynamical hypersystem of a generalised physical commutation matrixer commutator (matrix converter) acting as an energy-transfer commutator, making the outputs dependent on the inputs, in the dynamical systems approach and a matrix notation, may be written in the form:

$$V_s^j = C_{sP}^{ji} V_i^P, \tag{7.36}$$

or in detail as

$$\left\|\begin{matrix} V_a^j \\ V_b^j \\ V_c^j \\ \vdots \\ V_{n-1}^j \\ V_n^j \end{matrix}\right\| = \left\|\begin{matrix} C_{aA}^{ji} & C_{aB}^{ji} & C_{aC}^{ji} & & C_{a(N-1)}^{ji} & C_{aN}^{ji} \\ C_{bA}^{ji} & C_{bB}^{ji} & C_{bC}^{ji} & \cdots & C_{b(N-1)}^{ji} & C_{bN}^{ji} \\ C_{cA}^{ji} & C_{cB}^{ji} & C_{cC}^{ji} & & C_{c(N-1)}^{ji} & C_{cN}^{ji} \\ & \vdots & & \ddots & \vdots & \\ C_{(n-1)A}^{ji} & C_{(n-1)B}^{ji} & C_{(n-1)C}^{ji} & \cdots & C_{(n-1)(N-1)}^{ji} & C_{(n-1)N}^{ji} \\ C_{nA}^{ji} & C_{nB}^{ji} & C_{nC}^{ji} & & C_{n(N-1)}^{ji} & C_{nN}^{ji} \end{matrix}\right\| \left\|\begin{matrix} V_i^A \\ V_i^B \\ V_i^C \\ \vdots \\ V_i^{N-1} \\ V_i^N \end{matrix}\right\| . \tag{7.37}$$

However, a mathematical model of a physically continuous dynamical system of the generalised physical commutation matrixer commutator (the matrix converter shown in figure 7.7), acting as an energy-transfer commutator, making the inputs dependent on the outputs, in the dynamical systems approach and matrix notation may be expressed in the form:

$$V_i^P = C_{ij}^{Ps} V_s^j, \tag{7.38}$$

or in detail as

$$\left\|\begin{matrix} V_i^A \\ V_i^B \\ V_i^C \\ \vdots \\ V_i^{N-1} \\ V_i^N \end{matrix}\right\| = \left\|\begin{matrix} C_{ij}^{Aa} & C_{ij}^{Ab} & C_{ij}^{Ac} & & C_{ij}^{A(n-1)} & C_{ij}^{An} \\ C_{ij}^{Ba} & C^{Bb} & C_{ij}^{Bc} & \cdots & C_{ij}^{B(n-1)} & C_{ij}^{Bn} \\ C_{ij}^{Ca} & C_{ij}^{Cb} & C_{ij}^{Cc} & & C_{ij}^{C(n-1)} & C_{ij}^{Cn} \\ & \vdots & & \ddots & \vdots & \\ C_{ij}^{(N-1)a} & C_{ij}^{(N-1)b} & C_{ij}^{(N-1)c} & \cdots & C_{ij}^{(N-1)(n-1)} & C_{ij}^{(N-1)n} \\ C_{ij}^{Na} & C_{ij}^{Nb} & C_{ij}^{Nc} & & C_{ij}^{N(n-1)} & C_{ij}^{Nn} \end{matrix}\right\| \left\|\begin{matrix} V_a^j \\ V_b^j \\ V_c^j \\ \vdots \\ V_{n-1}^j \\ V_n^j \end{matrix}\right\| , \tag{7.39}$$

where C_{kL}^{ji} and C_{ij}^{Kl} are the energy-transfer coefficients of the commutation nodes kL and Kl, respectively.

To recapitulate, it may be affirmed that an action of a generalised physical commutation matrixer commutator as an energy-transfer commutator (shown in figure 7.7) may be presented by means of its mathematical model, consisting of the equations of unholonomic constraints of an energy-transfer holor in matrix notations (7.38) and (7.39).

For equations of unholonomic constraints of energy-transfer holors in matrix notation formulated in this way, isochronic variations of the generalised coordinates are not already independent of each other and that is why we apply the dynamical systems approach in the methodology for the formulation of mathematical models of physical commutation matrixer commutators (matrix converters) in a matrix notation of the Euler–Lagrange second-order differential equation of dynamics.

7.4 The principal physical commutation matrixer dynamical components

One of our primary objectives is the determination of suitable physical models for physical situations. Specifically, one wants mathematical models based on established natural laws. However, one limits one's consideration to physical situations where the mathematical modelling can be accompanied by linear constant coefficient holor equations.

Mathematical models will only be developed for the most important physical commutation matrixer dynamical components. More complicated physical situations can then be described with combinations of basic physical commutation matrixer dynamical components. It is feasible to depict nearly all of the principal physical situations using single-port (two-terminal) physical commutation matrixer dynamical components, but there are physical situations that necessitate a two-port physical model (with two pairs of terminals).

The common multi-port (two-terminal pairs) physical commutation matrixer dynamical components are described in chapter 8. In spite of this, at this stage one will only consider the single-port (two-terminal) physical commutation matrixer dynamical components.

A general single-port physical commutation matrixer dynamical component is shown schematically in figure 7.8.

The energy-potential-difference holor $\boldsymbol{F}_i$ symbolises a difference in the physical situation on either interface of the physical commutation matrixer dynamical component and the energy-transfer holor $\boldsymbol{V}^j$ symbolises the quantity transmitted through the physical commutation matrixer component.

The value of the single-port depiction lies in the fact that it allows the physical commutation matrixer dynamical components to come together into physical commutation matrixers that can be analysed by a very advanced physical commutation matrixer theory. In this manner, one may apply a universal holor method of analysis for all the physical commutation matrixers.

Normally, there are three basic physical commutation matrixer dynamical components in each of the physical heterogeneous continuous dynamical hypersystems:

- a single energy-dissipation physical commutation matrixer dynamical component in which energy is lost;

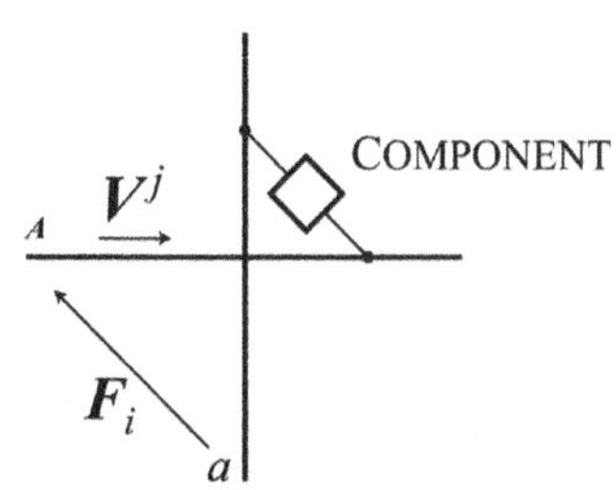

Figure 7.8. Single-port (two-terminal) physical commutation matrixer dynamical component.

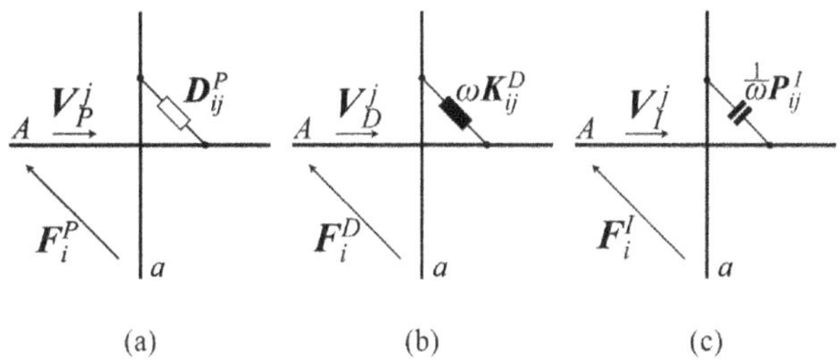

Figure 7.9. Principal physical commutation matrixer dynamical components: (a) energy dissipation; (b) kinetic energy storage; (c) potential energy storage.

- double energy-storage physical commutation matrixer dynamical components in which energy is retained.

Figure 7.9 demonstrates the physical commutation matrixer symbol for each physical commutation matrixer dynamical component performance. The symbol for the energy-dissipation physical commutation matrixer dynamical component (figure 7.9(a)), for instance, symbolises electrical resistance. In spite of this, one applies this symbol to symbolise a physical commutation matrixer dynamical component in any of the physical systems in which the energy-potential-difference holor $\boldsymbol{F}_i$ and the energy-transfer holor $\boldsymbol{V}^j$ are directly proportional to each other.

This physical situation fits into one's observation of the fluidic transmission line. Thus, the fluidic transmission line may be considered as a fluidic resistance.

Correspondingly, the storage physical commutation matrixer dynamical component symbols (figures 7.9(b) and (c)), as a rule, delegate the electrical inductor and capacitor. For one physical solution, one may apply these identical symbols for physical commutation matrixer dynamical components in other physical homogeneous continuous dynamical systems that create the identical relationship between the energy-transfer holor, $\boldsymbol{V}^j$, and the energy-potential-difference holor, $\boldsymbol{F}_i$.

We find, for example, that **mechano-mechanical** (MM) springs produce the same signal holor relation as an inductance and that a mass produces a capacitive-type relation. These physical commutation matrixer dynamical components are considered in detail in chapter 8.

For the state-of-the-art, on the other hand, it will be adequate to examine whether the relations between the energy transfer holor, $\boldsymbol{V}^j$, and the energy-potential-difference holor, F_i, are **proportional** (*P*), **differential** (*D*), or **integral** (*I*). The *D* and *I* physical commutation matrixer dynamical components realise time-domain-dependent operations on the signal holors. These mathematical operations are considered in detail in part II.

7.5 Input- and output-signal holors of physical commutation matrixers

Principally, the physical commutation matrixers that one contends with are composed of specific matrix permutations of single-port physical-commutation-matrixer dynamical components. These matrix permutations when combined together create a more sophisticated mathematical function than is created by a single-port physical commutation matrixer dynamical component. In spite of this, in chapter 10 we discuss mutually coupled physical commutation matrixer dynamical components that have more than one port (two terminals). A conventional physical commutation matrixer with single-port physical commutation matrixer dynamical components is shown in figure 7.10.

The input symbolises a signal holor (either energy-potential-difference or energy-transfer form) that is relevant to the physical commutation matrixer. The input is, for that reason, principally an input-signal holor. The physical commutation matrixer operates on the input-signal holor in relation to the physical commutation matrixer mathematical function and generates an output-signal holor.

The output-signal holor is occasionally termed the sinusoidal response of the physical commutation matrixer. It can be of either the energy-potential-difference or the energy-transfer form.

In conventional applications, a physical commutation matrixer or physical heterogeneous dynamical hypersystem generates particular effects or results, namely those required by the human being operating the assembly of physical equipment or individual equipment.

The normal term used for such effects or results is 'output'. For instance, the output of an automotive vehicle could be its travel velocity. However, in order to obtain an output, there must be some process of giving energy or information (or both) to the physical heterogeneous continuous complex dynamical hypersystem. This is the 'input'. Inputs to an automotive vehicle may include the energy stored in the liquid or gas fuel, the position of the steering wheel, the on/off-road surface and the accelerator- and/or brake-pedal positions. These hypotheses are, as a rule, represented graphically using a physical commutation matrixer of the heterogeneous continuous complex or simple dynamical hypersystem shown in figure 7.11 (Fijalkowski 1987, 2010, 2011).

For an automotive vehicle, if efficiency is of no concern, there is no need to consider the output energy as one of the outputs and nor is there any need to consider the chemical energy of the liquid fuel as one of the inputs. In this case, the

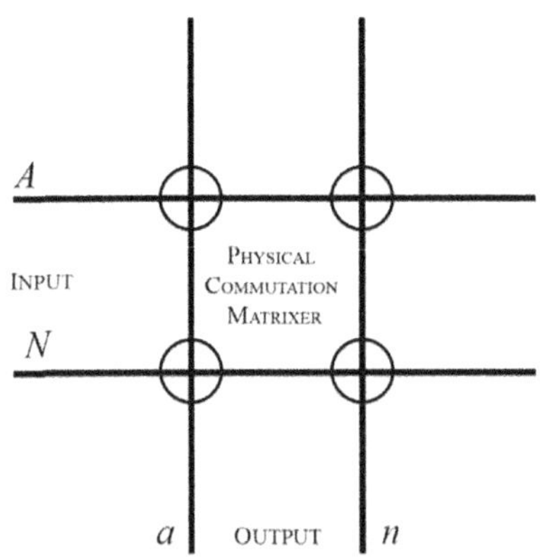

Figure 7.10. Conventional single-port physical commutation matrixer.

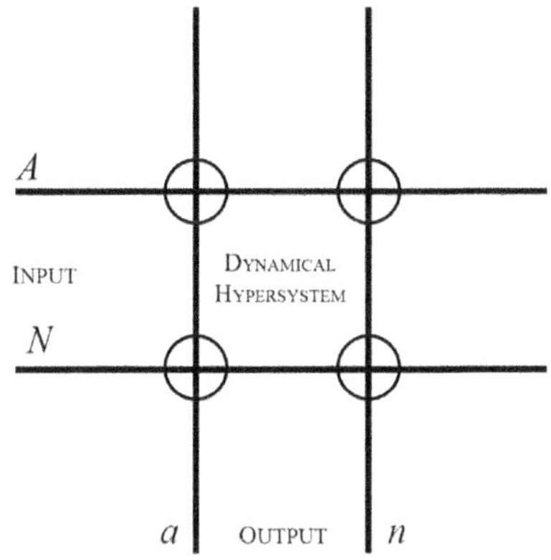

Figure 7.11. Physical commutation matrixer representation of the physical heterogeneous continuous complex or simple dynamical hypersystem.

output torque of the **internal combustion engine** (ICE) or **external combustion engine** (ECE) would be an input to the automotive powertrain system.

The sense of the term 'physical heterogeneous continuous complex or simple dynamical hypersystem' as a result comprises the hypothesis of a boundary within all the dynamical components that create the physical heterogeneous continuous complex or simple dynamical hypersystem.

In this example of the automotive vehicle, it must be noted that it is only the energy and information conversion process that would not be included in the physical system; the mass of the ICE or ECE (both the case and moving parts) would still be included in the physical system.

To some extent, the careful selection of the boundary for a specific problem is an area of expertise, the knowledge of which is one of the qualities of top physical system scientists and engineers.

The terms 'input-signal holor', 'output-signal holor' and 'physical heterogeneous continuous dynamical hypersystem' may as a result give the impression of being somewhat fuzzy.

However, it is elementary to be cognisant of what activates the function (cause) in a physical heterogeneous continuous dynamical hypersystem (input) and what constitutes the effects or results (output).

Normally, there is a minor problem in indicating the inputs and outputs of a physical heterogeneous continuous dynamical hypersystem if it is practical to provide an acceptable qualitative depiction (physical model) of it.

The estimation of the performance of a physical heterogeneous continuous dynamical hypersystem requires the creation of the response (output) to particular expected causes (inputs).

Physical heterogeneous continuous dynamical hypersystems normally traverse a dynamical or transient state in which the holor algebraic values for their outputs vary.

Occasionally, this dynamical state is the decisive point in creating the adequacy of the operation and then it is the dynamical response that ought to be exposed.

In view of the fact that the holor algebraic values of the output vary with time, the easiest method to express the dynamical response is to use a mathematical function of time. Normally, the inputs also vary with time and are expressed by mathematical time functions.

In spite of this, there is a natural thought that the input function can be identified. It is the cause. The output function, by contrast, is the effect, and it relies on both the input and the physical commutation matrixer.

Figure 7.12 displays two conventional input functions, the 'step function input' (figure 7.12(a)) and the 'sine wave input' (figure 7.12(b)).

In a real physical heterogeneous continuous dynamical hypersystem, the input may be recognised *a priori*. For instance, some **human and/or robotic drivers** (H&RD) step down hard on the accelerator pedal, while others drive sensibly. In spite of this, the input functions exposed are in fact trial inputs. They allow one to evaluate the performance of physical heterogeneous continuous dynamical hypersystems and physical commutation matrixers. The output that can be found by computation using the step function input is termed the step response.

The output from a sine-wave input is the sinusoidal response. It is made up of a transient member and a sinusoidal 'steady-state' member.

The sine-wave member of the output has an unchanged frequency, but normally a changed amplitude and phase angle compared to the sine-wave input.

The performance in which the amplitude or effective (root-mean-square, rms) value and phase angle of the outputs for the input sine-wave alter over a selection of various frequencies is termed the frequency response.

In analyses of the physical environment of various conventional responses there should be evidence of good design. As examples, consider the mechanical commutation matrixers of a simple pendulum with a mass at the end of a string, or a bicycle being pedalled along a horizontal road on a calm day, both of which are described in chapter 12 (sections 12.1 and 12.2). Other examples in which the forced and natural responses may readily be identified by a similar argument are the heating and cooling of an oven and the switching on and off of a large fan.

The mathematical model of any physical heterogeneous continuous complex or simple dynamical hypersystem ought to be able to submit solutions that represent the physical situations taken into account above sufficiently well.

The mathematical model of a physical heterogeneous continuous complex or simple dynamical hypersystem is normally created into the set of one holor or more

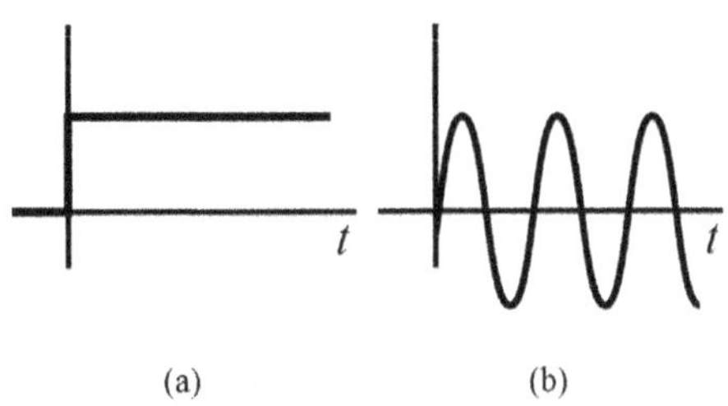

Figure 7.12. Conventional input functions: the step function (a) and the sine wave (b).

scalar equations, the particular expressions of which formulate each particular or principal effect within the physical heterogeneous continuous complex or simple dynamical hypersystem. The solution of a differential equation has two members: the particular integral solution and the complementary solution. The particular integral solution is familiar with the natural response.

Of course, in engineering education it is normal to deal with these solutions using their physical metaphors, i.e. the forced and natural responses. For linear physical heterogeneous continuous dynamical hypersystems (i.e. physical dynamical hypersystems in which each principal physical commutation matrixer dynamical component has the property that its response is directly proportional to the input or its time derivative or time integral), the two component solutions are quite independent of each other. The forced response has, physically and mathematically, a perfectly equivalent structure to the input or driving function. The natural response has a structure that is autonomous of the driving function, even if the magnitude does rely on the input and additionally on the initial stored energy. Alas, nonlinear physical dynamical systems do not have these simplifying properties and the solutions are noticeably more elaborate and out of the scope of this textbook. The solution of the linear holor equations that are normally met in physical commutation matrixers is taken into account in chapters 7 and 8.

7.6 Physical and mathematical modelling methodologies

The principal effects within a physical heterogeneous continuous complex or simple dynamical hypersystem are physical events that may be focused at one point in space or single DoF, but are normally distributed over an area and, as might be anticipated, it is more laborious to express a distributed physical event in detail than it is to depict the physical situation at a point. On the other hand, in various physical events it is evident that taking a limited number of DoFs (points) into account and creating a physical model that reflects the action occurring at these continuous DoFs (points) still offers a sufficiently accurate depiction of the relationships with the remainder of the physical heterogeneous continuous dynamical hypersystem.

This technique of lumping each physical event into an unsophisticated principal physical situation creates a 'lumped' physical model in preference to a 'distributed' physical model. In view of the fact that distributed mathematical models, as a rule, necessitate the solution of sets of partial differential equations and lumped physical models involve ordinary differential equations, there is an unquestionable predilection for using lumped physical models when sensible. Without doubt, there are physical situations in which it is not evident that the circumstances are such that a lumped mathematical model is correct. In such circumstances, it may be indispensable to arrange both lumped and distributed mathematical models and to contrast the solutions. The lumped mathematical model results may be equated with measured sensible records from experiments. Either technique is required for the procedure that is necessary for the formulation of sensible mathematical models.

7.7 Signal holors of physical commutation matrixers

We have mentioned the signal holors of physical commutation matrixers in a few words in section 7.4 and they will be described in chapter 8. They are so essential to physical commutation matrixer analysis that we consider them once more now. Both lumped and distributed physical and mathematical models necessitate the application of mathematical holors to symbolise physical quantities. Despite the fact that the field of study of this textbook is constrained to lumped physical models, the description of physical holors is also relevant to distributed physical models. Principally, there are two sets of physical quantities: those that flow along or through a constrained path, and those that necessitate measurements that must be carried out across a constrained single port or between two terminals. In connecting a measuring instrument across a single port or between two terminals it must be located across the whole thing that is positioned between them, and for that reason such physical quantities are expressed as 'energy-potential-difference holors'.

In connecting a measuring instrument to measure the stream along or through a constrained path it is indispensable to cut the constrained path and include the measuring instrument so that the stream is restrained to flow out through it, and for that reason such physical quantities are expressed as 'energy-transfer holors'. The hypothesis of all physical quantities being physically modelled as either 'energy transfer' or 'energy potential difference' offers a dominant medium for the promotion of analogous physical situations. Normally, these analogies are converted into something understandable only after the mathematical model (in the appearance of a set of equilibrium holor equations) has been completed. In particular, it is possible for a specific mechanical dynamical system, fluidic dynamical system, thermal dynamical system and electrical dynamical system to have equilibrium holor equations, which have the same mathematical model. Having completed a solution for each one of them, the solutions for the others have been successfully determined. The advantage of these analogies can be realised at once if they can be disclosed *a priori* in the physical and mathematical modelling *modus operandi.*

The hypothesis of the energy-transfer and energy-potential-difference holors is the basis for the creation of the analogous physical commutation matrixer diagram, which for practical purposes creates a correlation between the physical model and the perfect mathematical model. This diagram has the configuration of an electrical commutation matrixer.

The major benefit of the analogous physical commutation matrixer diagram is that a diagram with the same conventional configuration physically models each aspect of the physical heterogeneous continuous dynamical hypersystem. For that reason, the procedure for creating a set of equilibrium differential equations for any physical commutation matrixer is an unsophisticated one that is especially easy to submit an application to, and which has the result of presenting a generalised dynamical systems approach (systems thinking) to the analysis of mechanical, electrical, fluidic, magnetic and thermal homogeneous continuous dynamical systems. In the following chapters, the particulars of this generalised dynamical systems approach are worked through.

The energy-transfer and energy-potential-difference holors are recognised for physical heterogeneous continuous dynamical hypersystems concerning single- and/ or multi-port MIMO electrical commutation matrixers, translation mechanical commutation matrixers, rotation mechanical commutation matrixers, fluid flow commutation matrixers and thermal problems' commutation matrixers (chapter 8). As soon as the procedure for realising the analogous physical commutation matrixers has been considered (chapter 9), some particularly important physical situations are taken into account.

Thus, the methodical procedures (chapters 10 to 13) are elaborated with examples considered from any or all these variants of physical heterogeneous continuous dynamical hypersystems so that by the conclusion of the textbook the reader should have the same expertise and self-sufficiency when dealing with problems of physical and mathematical modelling and the theory of holor analysis for any of the physical heterogeneous continuous dynamical hypersystems included in this textbook.

References

Fijalkowski B 1967 Zastosowanie rachunku holorowego w teorii pradow przemiennych (Application of a holor calculus in the theory of alternating currents) *Przeglad elektrotechniczny (Electrotechnics Review)*, Rok XLIII, Zeszyt 2, Warszawa (In Polish)

Fijalkowski B 1973 Kierunki rozwoju napedu elektrycznego lokomotyw (Development trends of the electric drive of locomotives). *Materiały konferencyjne—I Krajowa Konferencja—Pojazdy Szynowe.* Instytut Pojazdow Szynowych Politechniki Krakowskiej, Krakow, pp. 109–128 (In Polish)

Fijalkowski B 1987 *Modele matematyczne wybranych lotniczych i motoryzacyjnych mechanoelektro-termicznych dyskretnych nadsystemów dynamicznych (Mathematical Models of Selected Aviation and Automotive Continuous Dynamical Hyperystems)*, Monografia 53 Politechnika Krakowska imienia Tadeusza Kościuszki, Krakow, 276 p (In Polish)

Fijalkowski B T 1982 Trigistor frequency changer-controlled toothed gearless propulsion systems for electric and hybrid vehicles. *Proc. Europe's International Conference on Electric Road Vehicle Systems DRIVE ELECTRIC (Amsterdam, 25–28 October 1982)* pp 554–567

Fijalkowski B T 1985 On the new concept MMD electrical machines with integral macro-electronic commutators—Development for the future *Proc. First European Conf. on POWER ELECTRONICS AND APPLICATIONS (Brussels, 16–18 October 1985)* vol 2 pp 3.377–3.384

Fijalkowski B T 1995 Emerging and future intelligent aviation and automotive applications of MIMO ASIM static-commutators and ASIC microcontrollers. *Proc. of the NATO Advanced Workshop on Future Trends in Microelectronics: Reflections on the Road to Nanotechnology (Ile de Bendor, France 17–21 July 1995)*. Also a chapter in the book edited by S Luryu, J Xu and A Zaslavsky 1996 *Future Trends in Microelectronics—Reflections on the Road to Nanotechnology* (NATO ASI Series, Series E: Applied Sciences) vol 323 (Dordrecht: Kluwer)

Fijalkowski B T 2010 *Automotive Mechatronics: Operational and Practical Issues.— Vol. 1 (International Series on Intelligent Control, and Automation Science and Engineering—* Vol 47) (Berlin: Springer)

Fijalkowski B T 2011 *Automotive Mechatronics: Operational and Practical Issues.— Vol 1.* (*International Series on Intelligent Control, and Automation Science and Engineering*— Vol 52) (Berlin: Springer)

Laplace P-S 1820 Theorie Analytique des Probabilités (Paris: Mme Ve Courcier)

Moon P H and Spencer D E 1986 *Theory of holors. A generalization of Tensors* (Cambridge: Cambridge University Press)

Senturia S D 1998 CAD challenges for microsensors, micro-actuators, and microsystems *Proc. IEEE* **86** 1611–26

Chapter 8

Physical commutation matrixer dynamical components

A mathematician is a person who can find analogies between theorems; a better mathematician is one who can see analogies between proofs and the best mathematician can notice analogies between theories.

Stefan Banach

One can imagine that the ultimate mathematician is one who can see analogies between analogies.

Stefan Banach

8.1 Introduction

In this textbook, we consider electrical, mechanical, fluidic and thermal physical commutation matrixer dynamical components. A mathematical model of any physical situation requires the use of mathematical variables, for instance, holors, to represent certain physical quantities associated with its behaviour. There are only two principal types of such quantities (chapter 7): those which flow along or through a constrained path (such as water flowing through a pipe), and those which require, whether explicitly or implicitly, measurements that must be made between two points (such as the distance between the points), e.g. generalised coordinates.

When one connects an instrument between two points, it must be placed across everything that lies between the points, and such physical quantities may reasonably be described by energy-potential-difference holors. In measuring the flow along a path, it is necessary to cut the path and insert the instrument so that the flow is constrained to pass through it. Such physical quantities are therefore described by energy-transfer holors. All the physical heterogeneous continuous dynamical hypersystems (consisting of electrical, mechanical, fluidic, magnetic and thermal

doi:10.1088/978-0-7503-1350-6ch8

homogeneous continuous dynamical systems) will have an energy-potential-difference holor and an energy-transfer holor.

The holors are different for each physical heterogeneous continuous dynamical hypersystem. The classification of all physical quantities as either energy-potential-difference holors or energy-transfer holors is the basis for the recognition of analogous physical situations in different physical heterogeneous continuous dynamical hypersystems. This grouping is especially useful if it is done in a manner that ensures that the method of computing the energy for one type of physical heterogeneous continuous dynamical hypersystem, for instance, electrical continuous dynamical systems, is the same as that used for another type of physical heterogeneous continuous dynamical hypersystem, for instance, fluidic dynamical systems. This, however, is not a necessary condition. The essential feature when choosing holors is that the resulting holor equations must have exactly the same mathematical model for all physical heterogeneous continuous dynamical hypersystems.

This book considers electrical, magnetic, mechanical, fluidic, magnetic and thermal homogeneous continuous dynamical systems. In the first four of these, the choice of the signal holors is made so that the univalent tensor product of the energy-transfer and energy-potential-difference holors represents a univalent power.

Thermal continuous dynamical systems are different, however, at the energy-transfer holor itself, the heat flow rate is power.

In choosing the energy-transfer and energy-potential-difference holors it must be kept in mind that the objective in the end is to produce a physical commutation matrixer representation for a physical heterogeneous continuous dynamical hypersystem. This simplifies the problem of describing equilibrium conditions. However, before considering equilibrium in general, it is necessary to identify certain elementary physical situations and derive the elemental holor equations that describe their performance. These correspond to idealised physical situations in which only one physical phenomenon is considered to be present.

One physically models these phenomena with two terminal physical commutation matrixer dynamical components. Figure 8.1 is a schematic representation of the principal single-port or two-terminal physical commutation matrixer dynamical component.

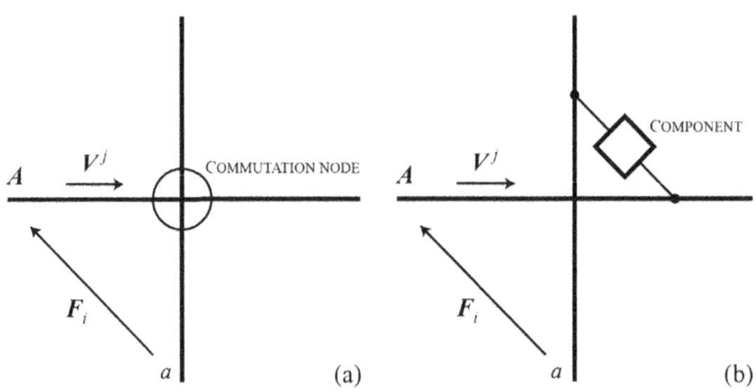

Figure 8.1. Principal single-port or two-terminal physical commutation matrixer dynamical component: (a) commutation node; (b) component.

We designate the point at which the energy transfer enters as terminal A and the point at which the energy transfer leaves as terminal a. There is an energy-potential difference associated with each terminal, which is a measure of the energy in the physical commutation matrixer at that location. The energy-potential-difference holor is the potential difference between terminals A and a. The energy-transfer holor, representing a transmitted physical quantity, passes through the physical dynamical component without change. There are only three principal elemental relations and they are expressed mathematically in holor algebra as:

$$\boldsymbol{F}_i^P = \boldsymbol{D}_{ij}^P \boldsymbol{V}_P^j, \tag{8.1a}$$

$$\boldsymbol{F}_i^D = \omega \boldsymbol{K}_{ij}^D \boldsymbol{V}_D^j, \tag{8.1b}$$

$$\boldsymbol{F}_i^I = \frac{1}{\omega} \boldsymbol{P}_{ij}^I \boldsymbol{V}_I^j, \tag{8.1c}$$

where $\boldsymbol{D}_{ij}^P$, $\boldsymbol{K}_{ij}^D$ and $\boldsymbol{P}_{ij}^I$ are proportionality constants that depend on the physical matrixer's geometrical configuration and material properties.

It should be emphasised that equation (8.1) is merely a general form for physical commutation matrixer dynamical components. The principal physical laws of electrics, mechanics, fluid mechanics and heat transfer can be placed in this format. It is these laws that determine the appropriate signal holors and the specific values of the bivalent holors $\boldsymbol{D}_{ij}^P$, $\boldsymbol{K}_{ij}^D$ and $\boldsymbol{P}_{ij}^I$. In the following sections, we introduce the principal components in mechanical, electrical, fluidic, magnetic and thermal homogeneous continuous dynamical systems.

8.2 Electrical commutation matrixer dynamical components

Electronics (Greek *Εληκτρική*) is the science and technology relating to or associated with electricity, as well as current or voltage for transmitting electrical energy and signals. It has been a subject of physics since at least the early 17th century. The first electrical engineer was probably the British physician, physicist and natural philosopher, William Gilbert (1544–1605), who designed the *versorium* (Latin 'turn around'), which was the first simple electroscope, the first instrument that could detect the presence of static electrically charged objects. He was also the first to describe the evident dissimilarity between magnetism and static electricity and is recognised as having originated the term 'electricity'. More to the point, objects that develop electrification (the act of electrifying, or the state of being charged with electricity) through friction or rubbing, as Gilbert, the originator of the term, applied it, indicate 'dielectrics'. He did not know that, if insulated, any object is one of his 'electrics'. A piece of copper held by a glass handle becomes electrified by friction.

In 1775, a professor at the Royal School in Como, Alessandro Volta (1745–1827), devised the electrophorus, a device that produces a static electrical charge, and by 1800 he had developed the voltaic pile, a forerunner of the **chemo-electrical/electro-chemical** (ChE/ECh) storage battery. However, it was not until the 19th century that research into the subject started to intensify. Notable developments from that period include the work of the Bavarian (German) physicist and mathematician, Georg Ohm (1789–1854), who in 1827 quantified the relationship between electric current and voltage potential difference in a conductor; the work of the British scientist, Michael Faraday (1791–1867), who discovered electromagnetic induction in 1831; and the work of the British mathematical physicist, James Clerk Maxwell (1831–79), who in 1873 published a unified theory of electricity and magnetism in his treatise *Electricity and Magnetism* (Maxwell 1873). During these years, the study of electricity was largely considered to be a subfield of physics. It was not until the late 19th century that universities started to offer degrees in electrical engineering. Darmstadt University of Technology founded the world's first chair and faculty of electrical engineering in 1882. In 1883 Darmstadt University of Technology and Cornell University introduced the world's first courses of study in electrical engineering, and in 1885 University College London founded the UK's first chair of electrical engineering. The University of Missouri subsequently established the USA's first department of electrical engineering in 1886.

The term electronics is used to cover electrical engineering, including the generation, transmission, control and use of all forms of electrical energy. The term electronic engineering is used to include the expanding fields of electronic communications (including computer networks), computers (both hardware and software) and electronic components. These components include microcomputer chips and, increasingly, the optical devices now being used for many applications. In addition, the field of mechatronical control engineering is on the electrical/electronic border and, with its use of computer systems, has a general base. Manufacturing engineering—involving as it does, computer techniques, mechatronical control and electrical engineering—is another important and elementary area within the field of electrics and electronics. Much of the very advanced equipment found in the workplace or home, such as robots and machine tools, domestic appliances, entertainment systems, personal computers and data processing equipment, is made possible by electrical and electronic engineering. This extensive, expanding and exciting subject has interfaces with physics (in the component field), computer science (in software engineering) and mechanics (in manufacture and control).

8.2.1 Electrical-signal holors

The generalised concept of the energy-potential-difference holor, $\boldsymbol{F}_i$, and the energy-transfer holor, $\boldsymbol{V}^j$, as the energy transporting mechanism finds a ready parallel in the voltage holor, $\boldsymbol{V}_i$, and the current holor, $\boldsymbol{I}^j$, of electrical homogeneous continuous dynamical systems. The conceptual similarity of the holor sets leads naturally to the assignment:

the voltage holor, $\boldsymbol{V}_i$, is analogous to the energy-potential-difference holor, $\boldsymbol{F}_i$;
the current holor, $\boldsymbol{I}^j$, is analogous to the energy-transfer holor, $\boldsymbol{V}^j$.

This assignment is compatible with the notions of the energy-potential-difference holor, $\boldsymbol{F}_i$, and the energy-transfer holor, $\boldsymbol{V}^j$, and is used universally. Although the electrical charge holor, $\boldsymbol{Q}^j$, is perhaps the most important electrical quantity from a theoretical viewpoint, the most useful holors for electrical homogeneous continuous dynamical systems are the current holor, $\boldsymbol{I}^j$, and the voltage holor, $\boldsymbol{V}_i$. The current holor, $\boldsymbol{I}^j$, is the rate of the flow of the electrical charge holor, $\boldsymbol{Q}^j$, and is clearly identified as an energy-transfer holor, $\boldsymbol{V}^j$.

The unit of the electrical charge holor, $\boldsymbol{Q}^j$, is the coulomb (C), and the electrical current holor, $\boldsymbol{I}^j$ is the rate of the flow of this charge in coulombs per second (C s^{-1}) or amperes (A).

Thus, in mathematical terms

$$\boldsymbol{I}^j = \omega \boldsymbol{Q}^j.$$

The voltage-potential-difference holor, $\boldsymbol{E}_i^A - \boldsymbol{E}_i^N$, also termed the electrical-potential-difference holor, is the energy-potential-difference holor. The electrical-potential-difference holor is the work done per unit charge in moving the charge between two points at the potentials $\boldsymbol{E}_i^A$ and $\boldsymbol{E}_i^N$. The fact that two points are required in the determination of a voltage-potential-difference holor, or a voltage-potential drop holor, clearly identifies it as the energy-potential-difference holor.

The unit for electrical-potential-difference holor is the volt (V). This is the electrical-potential difference holor when the work involved in moving one coulomb is one joule (J).

Thus the energy holor, or work, $\mathcal{T}$, done in moving a charge holor, $\boldsymbol{Q}^j$, between two points from electrical-potential-difference holors $\boldsymbol{E}_i^A$ to $\boldsymbol{E}_i^N$ is $\mathcal{T}^\times = (\boldsymbol{E}_i^A - \boldsymbol{E}_i^N) \otimes \boldsymbol{Q}^j$. The rate at which this work holor is expended is the power holor, $\boldsymbol{P}^\times$. The power holor is:

$$\boldsymbol{P}^\times = \omega \mathcal{T}^\times = \left(\boldsymbol{E}_i^A - \boldsymbol{E}_i^N\right) \otimes \boldsymbol{I}^j. \tag{8.2}$$

Note that the power holor is the univalent tensor product of the energy-potential-difference holor and the energy-transfer holor. Since the rate of the change of work holor is defined as the power holor, this is the well-known general expression for the power holor, $\boldsymbol{P}^\times$, in an electrical commutation matrixer.

The unit for the power holor is the watt (W), which is one joule per second (J s^{-1}). It also follows that the determination of the energy holor or work holor, $\mathcal{T}^\times$, is simply a matter of integrating equation (8.2):

$$\mathcal{U}^\times = \frac{1}{\omega}\boldsymbol{P}^\times = \frac{1}{\omega}\left[\left(\boldsymbol{E}_i^A - \boldsymbol{E}_i^N\right) \otimes \boldsymbol{I}^j\right] \tag{8.3}$$

Three elementary physical situations are found in electrical commutation matrixers (equation (8.1)) that do not involve an energy source being put into the electrical homogeneous continuous dynamical system. These are commonly known as the ideal resistor, the ideal inductor and the ideal capacitor.

Although real electrical dynamical components only approach the ideal, they can often be physically modelled as a combination of two or more of these ideal electrical dynamical components. All the electrical dynamical components must originate from physically realisable phenomena and obey physical laws.

8.2.2 Electrical-resistance holors

An electrical device in which the terminal voltage holor, V_i, is statically related to the current holor, I^j, flowing through the device is a dissipater of electrical energy. If the device is free from storage and source effects, it is said to have a pure electrical resistance.

The general constitutive relation for a resistance is

$$V_i = \mathcal{F}(I^j). \tag{8.4}$$

The linear form of equation (8.4) was first discovered by Georg Ohm and bears his name. Ohm's law states that

$$V_i = R_{ij} I^j, \tag{8.5}$$

where R_{ij} is termed the resistance holor of the electrical dissipater.

For an electrical homogeneous continuous dynamical system of length l and uniform cross-sectional area S, the resistance is given by

$$R = \rho l / S, \tag{8.6}$$

where ρ is the resistivity of the device material.

The total power dissipated by an electrical resistor is the univalent tensor product of the terminal voltage holor, V_i, and current holor, I^j, which is the sum of the content, G, and the co-content, J, of the electrical resistor

$$P^\chi = V_i \otimes I^j. \tag{8.7}$$

For a linear device, the content, G, and the co-content, J, of the electrical resistor are equal and are given by

$$G = J = \frac{1}{2} R(i\text{^}2). \tag{8.8}$$

The total power dissipated is

$$P^p = \frac{1}{2} R(i\text{^}2) = (v\text{^}2)/R. \tag{8.9}$$

From experience with many metallic electrical commutation matrixer parts, one observes that the voltage-potential-difference holor, $E_m^A - E_m^N$, established in such an electrical commutation matrixer is directly proportional to the current holor, I^j. This observation may be expressed in mathematical form as:

$$E_i^A - E_i^N = R_{ij} I^j = \begin{Vmatrix} E_c^A \\ E_s^A \end{Vmatrix} - \begin{Vmatrix} E_c^N \\ E_s^N \end{Vmatrix} = \begin{Vmatrix} R & 0 \\ 0 & R \end{Vmatrix} \begin{Vmatrix} I^c \\ I^s \end{Vmatrix}. \tag{8.10}$$

This relation is known as the generalised Ohm equation, or Ohm's law.

Note that equation (8.10) is identical in form to the general elemental relation given in equation (8.1*a*), with $R_{ij} = D_{ij}^P$.

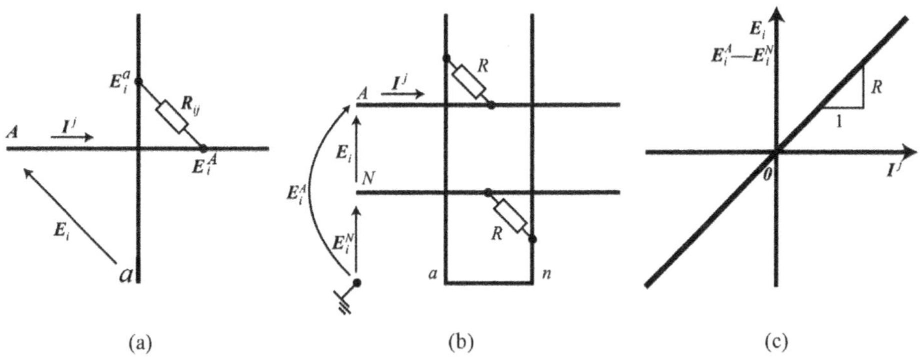

Figure 8.2. Characteristics of ideal resistance.

The physical quantity symbolised by R is termed 'electrical resistance' and is a property of the material and geometry of the electrical commutation matrixer part.

The ideal electrical resistor has the voltage-potential-difference holor/current-holor characteristic shown in figure 8.2.

In this case, the voltage-potential-difference holor is directly proportional to the current holor, as indicated by the line of the constant slope. The slope of this line is the resistance R. The unit of resistance is the ohm (Ω). There are occasions, however, when the reciprocal, G, is more convenient. This is the conductance for which the unit is the siemens (S) or mhos (℧). Note that the ideal resistance dynamical component is bilateral or has two senses of direction. If potential A is greater than potential N, then the energy-potential-difference holor, $\boldsymbol{E}_i^A - \boldsymbol{E}_i^N$, is positive and this defines the positive sense of direction of the current holor. When the potential N is greater than the potential A, the energy-potential-difference holor is negative and the current holor reverses.

In a bilateral resistive dynamical component, a given positive or negative voltage-potential-difference holor causes a current holor of the same magnitude but in the opposite direction. All objects have electrical resistance. However, in electrical commutation matrixers that are used to produce some desired effect on the signal holors, one is dealing with an electrical component that is specially made for the purpose. In general, electrical resistance is obtained from the resistivity and geometry of the wire material.

The resistivity of a wire is expressed as:

$$R = \rho \frac{l}{S}, \ [\Omega], \tag{8.11}$$

where ρ is the resisitivity (Ω m) of the wire material, l is the wire length (m) and S is the wire cross-sectional area (m^2).

The energy dissipated in the resistance can be determined by rearranging equation (8.2) in the form

$$\mathcal{T}^{\chi} = \mathcal{T}^{*\chi} = \frac{1}{\omega} \boldsymbol{P}^{\chi}.$$

When the voltage-potential-difference holor and the current holor are not changing with time, the energy is

$$\mathcal{T} = \mathcal{T}^{*} = R(I^{c\wedge 2})t = \frac{\left[(E_c^{A\wedge 2}) - (E_c^{N\wedge 2})\right]}{R}t. \tag{8.12}$$

Note that the energy increases with time. The resistance is, therefore, an energy-dissipation dynamical component.

8.2.3 Electrical-inductance holor

In the 19th century Joseph Henry and Michael Faraday discovered that the electricity flowing in an electrical homogeneous continuous dynamical system displayed properties that are analogous to the momentum in a mechanical continuous dynamical system. The electricity flowing in a long electrical homogeneous continuous dynamical system displays what Maxwell termed an 'electrokinetic momentum' holor, which is proportional to the current holor, $\boldsymbol{I}^j$, flowing, and constant depending on the physical arrangement of the electrical homogeneous continuous dynamical system.

The quantity of the electrokinetic momentum holor is in time known as the magnetic-flux-linkage holor, $\boldsymbol{\lambda}_i$, and is the total magnetic flux holor, $\boldsymbol{\Phi}_i$, linked by the electrical homogeneous continuous dynamical system.

Similarly, the physical constant, which together with the current holor, $\boldsymbol{L}_{ij}$, determines the flux-linkage holor, $\boldsymbol{\lambda}_i$, is the inductance holor, $\boldsymbol{L}_{ij}$,

$$\boldsymbol{\lambda}_i = \boldsymbol{L}_{ij}\boldsymbol{I}^j. \tag{8.13}$$

The magnetic flux holor, which is linked by the electrical homogeneous continuous dynamical system, determines the voltage holor, $\boldsymbol{V}_i$, developed across the electrical inductor via Faraday's law:

$$\boldsymbol{V}_i = \omega\boldsymbol{\lambda}_i. \tag{8.14}$$

Equation (8.14) is the linear form of the inductive constitutive relation; in general, the relation is nonlinear and may be multi-valued:

$$\boldsymbol{\lambda}_i = \mathcal{F}(\boldsymbol{I}^j). \tag{8.15}$$

For an electrical inductor composed of a coil of N turns wound on a homogeneous core of permeability μ and of uniform cross-sectional area S, the inductance is given by

$$L = \frac{\mu S}{l}(N^{\wedge}2), \tag{8.16}$$

where S is the cross-sectional area of the magnetic homogeneous continuous dynamical system and l is the length of the magnetic homogeneous continuous dynamical system.

For non-magnetic core materials the magnetic permeability, μ, is constant, However, for ferrous-cored inductors the permeability, μ, is a function of the current holor, $\boldsymbol{I}^j$, such that the inductive constitutive relation is nonlinear.

The energy stored in an inductive field is given by the integral

$$\mathcal{T} = \int_0^\lambda i \mathrm{d}\lambda. \tag{8.17}$$

The inductive co-energy is given by the complementary integral

$$\mathcal{T}^* = \int_0^i \lambda \mathrm{d}i. \tag{8.18}$$

If the electrical inductor is linear then the energy, $\mathcal{T}$, and the co-energy, $\mathcal{T}^*$, are equal and are given by

$$\mathcal{T} = \mathcal{T}^* = \frac{1}{2L}(\lambda \hat{} 2) = \frac{1}{2}L(i \hat{} 2). \tag{8.19}$$

The most common form of inductor is a coil of wire wrapped around a core. The core contains a high level of the magnetic-flux holor, $\boldsymbol{\Phi}_i$, due to the current holor, $\boldsymbol{I}^j$, in the conducting wire coil. When the wire coil contains more than one loop, each loop contributes to the magnetic-flux-linkage holor, $\boldsymbol{\lambda}_i^{AN}$, between the two terminals of the coil inductor. When the physical arrangement is such that the magnetic-flux-linkage holor, $\boldsymbol{\lambda}_m^{AN}$, is a single-valued function of the current holor, $\boldsymbol{I}^j$, the dynamical component is termed a pure inductor. This is expressed as:

$$\boldsymbol{\lambda}_i^{AN} = \boldsymbol{L}_{ij}\boldsymbol{I}^j, \tag{8.20}$$

where $\boldsymbol{\lambda}_i^{AN}$ is the magnetical-flux-linkage holor in webers (We) and $\boldsymbol{L}_{ij}$ is the inductance holor in henries (H).

Thus, the pure or ideal inductor has the magnetic-flux-linkage holor, $\boldsymbol{\lambda}_i$/current holor, $\boldsymbol{I}^j$, characteristics shown in figure 8.3.

Faraday's law states that the voltage-potential-difference holor induced in coils by a changing magnetic field is

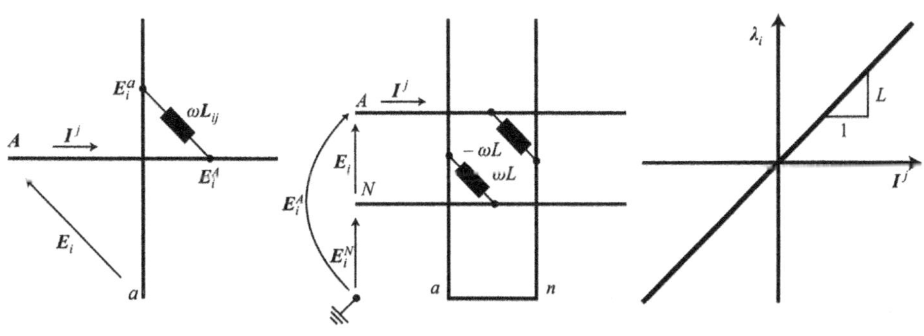

Figure 8.3. Characteristics of ideal inductance.

$$E_i^A - E_i^N = \omega\lambda_i^{Aa}. \tag{8.21}$$

The combination of equations (8.20) and (8.21) yields a relation between the signal holors that is:

$$E_i^A - E_i^N =$$

$$\omega\lambda_i^{Aa} = \omega L_{ij} I^j = \begin{Vmatrix} E_c^A \\ E_s^A \end{Vmatrix} - \begin{Vmatrix} E_c^N \\ E_s^N \end{Vmatrix} = \begin{Vmatrix} 0 & -\omega L \\ \omega L & 0 \end{Vmatrix} \begin{Vmatrix} I^c \\ I^s \end{Vmatrix}. \tag{8.22}$$

Equation (8.22) has the same form as is given in the general elemental relation in equation (8.1*b*). In this case, $K_{ij}^D = L_{ij}$. From the ideal characteristics shown in figure 8.3, one may observe that the ideal inductor is also a bilateral electrical dynamical component.

The magnitude of inductance for a wire wound over a cylindrical core is:

$$L = \frac{4\pi(10)^{-7}\mu S(N)^2}{l}\text{H}, \tag{8.23}$$

where μ is the relative permeability of the core material, S is the area of the core (m^2), l is the length of the core (m) and N is the number of coils. For an air core the relative permeability $\mu = 1$.

When an iron rod is used for the core, the relative permeability may be as high as $(10)^4$, with the actual value depending on the magnetic-flux density.

In actual electrical dynamical components, it is not possible to achieve the ideal characteristics given for the resistances and inductances. In both cases, the dynamical component consists of a length of wire. Thus, the actual inductance always has some resistance and the actual resistance has some inductance. However, we assume throughout this book that the dynamical components are ideal and represent only one phenomenon.

The energy into an inductor is from equations (8.2) and (8.22),

$$\mathcal{T} = L\int_{i_0}^{i} i'di' = \frac{1}{2}L[(i\text{^}2) - (i_0\text{^}2)], \tag{8.24}$$

where i_0 is the initial current.

For the inductor the energy is not directly a function of time. This is because the inductor stores energy rather than dissipating it. The energy stored by the inductor is due to the flow of current and the expression given in equation (8.24) is valid whether the current is constant or not.

The energy stored is not directly dependent on the voltage-potential-difference holor. Indeed, equation (8.24) applies even when the voltage-potential-difference holor is zero. The inductor is in a class of energy-storage dynamical components known as *D*-type (for energy-potential-difference holor) storage.

8.2.4 Electrical-capacitance holor

The electrical inductor stores energy in a magnetic field. The complementary electrical dynamical component, the electrical capacitor, stores energy in an electrical field. Whenever two electrical conductors are at different potentials there is storage of charge upon them. The amount of the electrical-charge holor, $\boldsymbol{Q}^j$, that is accumulated in this way is determined by the voltage-potential-difference holor, $\boldsymbol{V}_i$, across the conductors and the physical properties of the conductors and the medium (dielectric) that separates them.

The constitutive relationship of an electric capacitor therefore has the general form:

$$\boldsymbol{Q}^j = \mathcal{F}(\boldsymbol{V}_i). \tag{8.25}$$

In the linear case this becomes

$$\boldsymbol{Q}^j = \boldsymbol{C}^{ji}\boldsymbol{V}_i, \tag{8.26}$$

where $\boldsymbol{C}^{ji}$ is termed the capacitance holor of the electrical capacitor.

For a pair of parallel plates separated by a distance, d, and with a homogeneous filling of permittivity, ε, the capacitance is given by

$$C = \frac{\varepsilon S}{d}, \tag{8.27}$$

where S is the effective cross-sectional area of the plates.

For almost all the dielectric materials used in electrical capacitors, the permittivity, ε, is a constant such that the capacitive constitutive relation is a constant. The electrical-charge holor, $\boldsymbol{Q}^j$, is related to the current holor, $\boldsymbol{I}^j$, flowing in the capacitor's electrical homogeneous continuous dynamical system by the dynamical relationship:

$$\boldsymbol{I}^j = \omega\boldsymbol{Q}^j. \tag{8.28}$$

The electrical energy stored in an electrical capacitor is given by an integral

$$\mathcal{U} = \int_0^q v \, \mathrm{d}q. \tag{8.29}$$

The electrical co-energy is defined in a complementary integral

$$\mathcal{U}^* = \int_0^v q \, \mathrm{d}v. \tag{8.30}$$

If the electrical capacitor is linear, the capacitive energy, U, and the co-energy, U^*, are equal and are given by

$$\mathcal{U} = \mathcal{U}^* = \frac{1}{2}C(v\hat{}2) = \frac{1}{2C}(q\hat{}2). \tag{8.31}$$

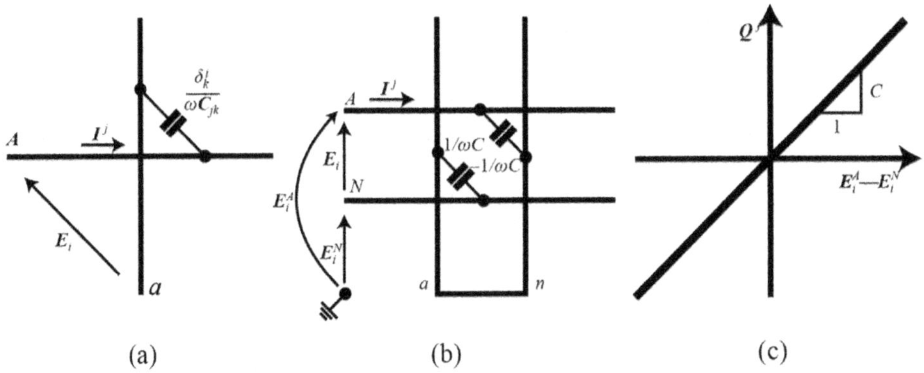

Figure 8.4. Characteristics of ideal capacitance.

Two conductors separated from each other by a dielectric (an insulating material) form a capacitor. In the capacitor, an electrical field is established between the conductors. When the physical arrangement is such that the electrical-charge holor, $\boldsymbol{Q}^j$, on the conductors is a single-valued function of the voltage-potential-difference holor, $\boldsymbol{E}_i^A - \boldsymbol{E}_i^N$, the dynamical component is termed a pure capacitor. This is expressed as:

$$\boldsymbol{Q}^j = \boldsymbol{C}^{ji}\left(\boldsymbol{E}_i^A - \boldsymbol{E}_i^N\right), \tag{8.32}$$

where $\boldsymbol{Q}^j$ is the electrical-charge holor in coulombs and $\boldsymbol{C}^{ji}$ is the capacitance-admittance holor in farads (F).

Thus, the pure or ideal capacitor has the charge-holor/voltage-potential-difference holor characteristics shown in figure 8.4.

In terms of the signal holors, one may replace the electrical charge-holor, $\boldsymbol{Q}^j$, in equation (8.32) by the integral of the current-holor, $\boldsymbol{I}^j$, so that:

$$\boldsymbol{E}_i^A - \boldsymbol{E}_i^N = \frac{\delta_k^i}{\omega \boldsymbol{C}_{jk}}\boldsymbol{I}^j = \begin{Vmatrix} E_c^A \\ E_s^A \end{Vmatrix} - \begin{Vmatrix} E_c^N \\ E_s^N \end{Vmatrix} = \begin{Vmatrix} 0 & \dfrac{1}{\omega C} \\ -\dfrac{1}{\omega C} & 0 \end{Vmatrix} \begin{Vmatrix} I^c \\ I^s \end{Vmatrix}. \tag{8.33}$$

For this type of electrical dynamical component, the governing equation (8.33) has the same mathematical form as the general elemental relation of equation (8.1c), with $\boldsymbol{P}_{jk}^I = \frac{\delta_k^i}{C^{ij}}$. The ideal capacitor is a bilateral dynamical component in the same way as the ideal resistor and ideal inductor.

For a parallel plate capacitor, the magnitude of the capacitance is:

$$C = \frac{\varepsilon S}{d} F, \tag{8.34}$$

where ε is the permeability of the dielectric in farads per meter (F m^{-1}), S is the cross-sectional area (m^2) of the plates and d is the thickness (m) of the dielectric.

The energy associated with a capacitance may be determined from equations (8.2) and (8.33) as:

$$\mathcal{U} \int_{\Delta e_0}^{\Delta e} e' de' = \frac{1}{2} C\left[(\Delta e \wedge 2) - (\Delta e_0 \wedge 2)\right], \tag{8.35}$$

where Δe is the voltage-potential difference which equals $e_1 - e_2$ and Δe_0 is the initial voltage difference.

Since the capacitor stores energy, it is also an energy-storage dynamical component. In this case, energy is stored due to the presence of a voltage difference. The expression for energy given in equation (8.35) is valid whether the voltage-potential difference is constant or not.

The energy stored is not directly dependent on the current. Indeed, equation (8.35) applies even when the current is zero.

The electrical capacitor is in the class of energy storage devices known as *I*-type (for energy-potential-difference holor) storage.

The energy stored by an inductance (equation (8.24)) and by a capacitance (equation (8.33)) may be formulated in a similar way for the analogous storage dynamical components in mechanical and fluidic homogeneous continuous dynamical systems.

The normal symbols for the three electrical commutation matrixer dynamical components are shown in figures 8.2–8.4. The arrow associated with the current holor shows a positive sense of direction for conventional current-holor flow through the dynamical component. The terminal at which the positive current holor enters is positive for each of the dynamical components and polarity indicators must be set up as in the electrical commutation matrixers if the positive signs implied in the elemental holor equations are to apply. At this stage, it can be noted that each of the three electrical dynamical components is significantly different.

One of them, the resistor, dissipates energy, which cannot be recovered by the electrical commutation matrixer. The others store energy, which can be recovered when the electrical commutation matrixer conditions are available. However, in one case, the capacitor (*I*-type storage), the energy is stored by virtue of there being a voltage-potential-difference holor between the two terminals irrespective of whether there is a current holor flow. In the other case, the inductor (*D*-type storage), the opposite statements holds: the energy is stored by virtue of there being a current holor flow through the dynamical component irrespective of whether there is a voltage-potential-difference holor across it.

In this chapter, the electrical commutation matrixer physical models of the storage dynamical components do not include initial storage. That is, i_0, for the inductor, and Δe, for the capacitor, have been assumed to be zero. The analogous mechanical, fluidic and thermal storage dynamical components will also be introduced without initial storage. The physical commutation matrixer models for storage dynamical components with initial storage have been considered in chapter 7.

8.3 Magnetic commutation matrixer dynamical components

Magnetics is the science and technology relating to or associated with a magnetic medium, as well as a magnetic flow (magnetic movement of electrons) for the transmission of magnetic energy and signals. It concerns the production and transferral of magnetic energy by means of various physical phenomena.

One of the great discoveries in the history of physics was made by the British physicist, James Clerk Maxwell (1831–79), in the 19th century. Maxwell found that the two major forms of energy known as electricity and magnetism are not really different from each other, but are instead closely associated. That is, every electrical current has a magnetic field associated with it and every changing magnetic field creates its own electrical current. Maxwell's equations state that magnetic energy and electrical energy are related to one another by nature. Magnets produce potential energy through magnet motion and additionally through the magnetic field emitted. The magnetic moment of a magnet is a quantity that determines the force that the magnet can exert on electrical currents and the torque that a magnetic field will exert on it. A loop of electrical current, a bar magnet, an electron, a molecule and a planet all have magnetic moments. Both the magnetic moment and magnetic field may be considered to be vectors with a magnitude and sense of direction. The sense of direction of the magnetic moment points from the **south** (S) to the **north** (N) pole of a magnet. The magnetic field produced by a magnet is also proportional to its magnetic moment well. More precisely, the term 'magnetic moment' normally refers to a magnetic homogeneous continuous dynamical system's magnetic dipole moment, which produces the first term in the multipole expansion of a general magnetic field.

The dipole dynamical component of an object's magnetic field is symmetric about the direction of its magnetic dipole moment, and decreases as the inverse cube of the distance from the object. The magnetic dipole moment μ can be considered to be a vector quantity with a direction perpendicular to the electrical-current loop in the right-hand-rule direction (see figure 8.5).

Considering magnetic torque as a vector quantity, this can be written as the vector product by

$$\boldsymbol{\tau} = \boldsymbol{\mu} \times \boldsymbol{B}. \tag{8.36}$$

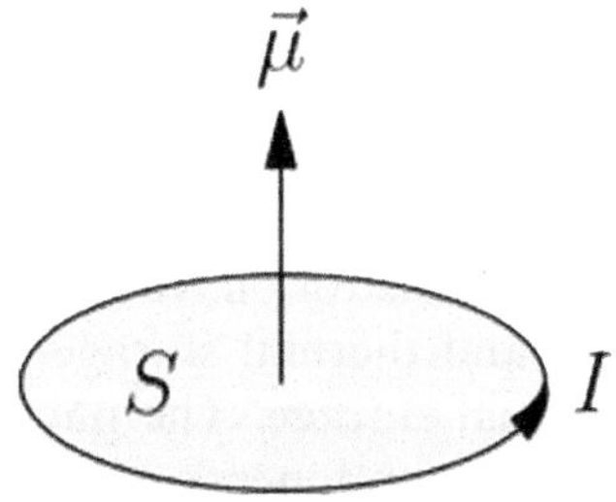

Figure 8.5. Moment μ of a planar current with a magnitude I enclosing an area S.

As seen in figure 8.5, which shows the geometry of an electrical-current loop, this magnetic torque $\boldsymbol{\tau}$ tends to line up the magnetic dipole moment $\boldsymbol{\mu}$ with the external magnetical field $\boldsymbol{B}$, so this represents its lowest energy configuration.

Since this magnetic torque τ acts perpendicularly to the magnetic moment μ, it can cause the magnetic moment μ to precess around the magnetical field B at a characteristic frequency called the 'Larmor frequency'.

A magnetic dipole moment μ in a magnetic field B will have potential energy $\mathcal{U}(\theta)$ that depends upon its orientation with respect to the external magnetic field B. Since magnetic-energy sources are inherently dipole sources that can be visualised as an electrical current I and area S, the energy $\mathcal{U}(\theta)$ is usually expressed in terms of the magnetic dipole moment μ:

$$\mathcal{U}(\theta) = -\mu B, \tag{8.37}$$

where $\mu = IS$ (or $\mu = NIS$ for N electrical-current loops).

The energy $\mathcal{U}(\theta)$ is expressed as a scalar product and implies that the energy is lowest when the magnetic dipole moment μ is aligned with the magnetic field B. The difference in energy $\Delta\mathcal{U}$ between aligned and anti-aligned is

$$\Delta\mathcal{U} = 2\mu B. \tag{8.38}$$

These relationships for a finite electrical-current loop extend to the magnetic dipoles of electron orbits and the intrinsic magnetic moment μ with electron spin. Nuclear magnetic moments are also important.

The potential energy $\mathcal{U}$ of a permanent magnet of magnetic moment μ in a magnetic field B is defined as the action of magnetic force (actually of magnetic torque) on the realignment of the vector of the magnetic dipole moment μ, and is equal to:

$$\mathcal{U} = -\mu B, \tag{8.39}$$

while the magnetic energy stored in an inductor (of inductance L) when electrical current I is passing via it is

$$\mathcal{U} = \frac{1}{2}L(I\verb|^|2). \tag{8.40}$$

This second expression forms the basis for superconducting magnetic energy storage (SMES).

Magnetic energy $\mathcal{U}$ is also stored in an external magnetic field B. The magnetic energy $\mathcal{U}$ per unit volume in a sphere of space of permeability μ containing magnetic field B is:

$$\mathcal{U} = \frac{1}{2}\frac{(B\verb|^|2)}{\mu}. \tag{8.41}$$

In a case where the external magnetic field is non-uniform, there will be a force (proportional to the magnetic field gradient) acting on the magnetic moment itself.

There has been some discussion regarding how to calculate the force acting on a magnetic dipole. There are two expressions for the force acting on a magnetic dipole, depending on whether the physical model used for the dipole is a current loop or two monopoles (analogous to the electrical dipole).

8.3.1 Magnetomotive-force and magnetic-flux holors

In almost all cases, the properties of magnetic homogenous continuous dynamical systems can be physically and mathematically modelled indirectly via their reflected electrical continuous dynamical system properties. Situations arise, however, where the magnetic continuous dynamical system ought to be considered in its own right.

The extensive variable in a magnetic homogeneous continuous dynamical system is the magnetomotive-force holor $\boldsymbol{M}_i$. The intensive variable is associated with the total magnetic flux holor $\boldsymbol{\Phi}^j$.

Hence, according to the energy-potential difference holor $\boldsymbol{F}_i$ and the energy-transfer holor $\boldsymbol{V}^j$ analogy: the magnetomotive-force holor $\boldsymbol{M}_i$ is analogous to the energy-potential difference holor $\boldsymbol{F}_i$, and the magnetomotive-flux holor $\boldsymbol{\Phi}^j$ is analogous to the energy-transfer holor $\boldsymbol{V}^j$.

The storage of energy in a magnetic homogeneous continuous dynamical system is usually physically and mathematically modelled by the reflected electrical characteristics of the inductance holor $\boldsymbol{L}_{ij}$.

In magnetic terms, the property of the inductance holor $\boldsymbol{L}_{ij}$ is a flow storage phenomenon in that it is associated with an accumulation of the magnetomotive-flux holor $\boldsymbol{\Phi}^j$. The complementary form of magnetic energy storage is not normally discernible.

8.3.2 Magnetic-reluctance holor

The magnetising force at any point in a magnetic homogeneous continuous dynamical system is the magnetomotive force M per unit length required at that point to maintain the magnetomotive flux Φ. Lines of magnetomotive flux Φ form closed loops and if H represents the magnetising force then the total magnetomotive force M is the integral of H along the centre flux line, i.e.

$$M = \oint H \, \mathrm{d}l. \tag{8.42}$$

Now consider a closed-iron magnetic homogeneous continuous dynamical system with a cross-sectional area of S (m^2) and a mean length l (m).

The total magnetomotive flux is Φ and is the product of the magnetomotive-flux density B and the cross-sectional area S.

Hence

$$\frac{\Phi}{M} = \frac{BS}{Hl}. \tag{8.43}$$

However, the ratio B/H defines the magnetic permeability μ of the medium supporting the magnetomotive flux Φ, giving

$$M_i = R_{ij}\Phi^j = \begin{Vmatrix} M_c \\ M_s \end{Vmatrix} = \begin{Vmatrix} R & 0 \\ 0 & R \end{Vmatrix} \begin{Vmatrix} \Phi^c \\ \Phi^s \end{Vmatrix}, \tag{8.44}$$

where $R = 1/\mu S$ is the reluctance of the magnetic homogeneous continuous dynamical system and is the magnetic form of energy dissipation.

Magnetic reluctance, or 'magnetic resistance', is a concept used in the analysis of magnetic homogeneous continuous dynamical systems. It is analogous to resistance in an electrical homogeneous continuous dynamical system, but rather than dissipating electrical energy, it stores magnetic energy.

As an electrical field causes an electrical current to follow the path of least resistance, so a magnetic field causes magnetic flux to follow the path of least magnetic reluctance. It is a scalar, extensive quantity, akin to electrical resistance. The units for magnetic reluctance are inverse henries (H^{-1}). The term was coined in 1888 by Oliver Heaviside.

The notion of magnetic resistance was first mentioned by James Joule and the term 'magnetomotive force' was first used by Bosanquet.

The idea for a magnetic flux law, similar to Ohm's law for closed electrical homogeneous continuous dynamical systems, is attributed to H Rowland. Reluctance is usually represented by a cursive capital *R*.

Equation (8.44) is the constitutive relation of a magnetic dissipater. Recalling that the permeability μ is a function of the magnetomotive-flux level, magnetic dissipaters are usually nonlinear in nature with the constitutive relation

$$M = \mathcal{F}(\Phi). \tag{8.45}$$

Figure 8.6 shows a typical constitutive relationship for a magnetic dissipater, together with the magnetic commutation matrixer adopted.

8.3.3 Magnetic-energy-source holors

Fundamentally, contributions to any magnetic homogeneous continuous dynamical system's magnetic moment may come from magnetic-energy sources of two forms: (a) the motion of electrical charges, such as electrical currents, and (b) the intrinsic magnetism of elementary particles, such as the electron.

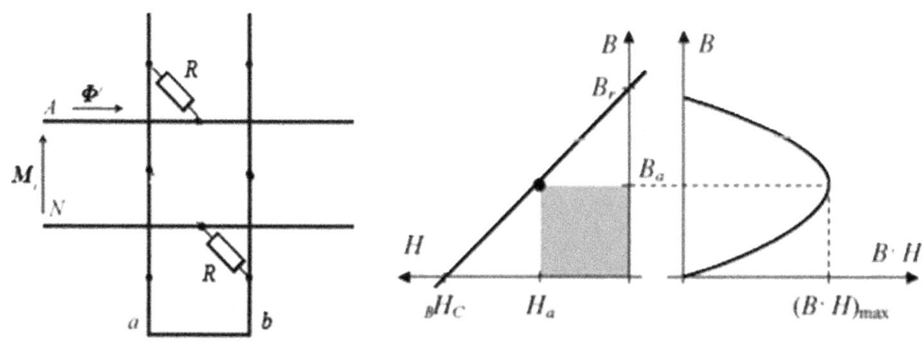

Figure 8.6. Magnetic commutation matrixer and constitutive relation for a magnetic dissipater.

Contributions due to the magnetic-energy sources of the first form can be calculated by knowing the distribution of all the electrical currents (or, alternatively, all the electrical charges and their velocities) inside the magnetic homogeneous continuous dynamical system. On the other hand, the magnitude of each elementary particle's intrinsic magnetic moment is a fixed number, often measured experimentally to a great precision. The sense of direction of the magnetic moment of any elementary particle is entirely determined by the sense of direction of its spin (the minus sign in front of the value above indicates that any electron's magnetic moment is antiparallel to its spin).

The net magnetic moment of any magnetic homogeneous continuous dynamical system is a vector sum of contributions from one or both forms of the sources. For example, the magnetic moment of an atom of hydrogen ^{1}H (the lightest hydrogen isotope, consisting of a proton and an electron) is a vector sum of the following contributions:

- the intrinsic moment of the electron;
- the orbital motion of the electron around the proton;
- the intrinsic moment of the proton.

Similarly, the magnetic moment of a bar magnet is the sum of the intrinsic and orbital magnetic moments of the unpaired electrons of the magnet's material. The most basic form of magnetic-energy-source holor is the permanent magnet, which is a source of the magnetomotive-force holor $\boldsymbol{M}_i$. Every magnet contains two distinct regions, one known as the N pole and the other as the S pole. As with electrical charges, unlike poles attract each other and like poles repel each other.

A study of magnets leads to the introduction of a fascinating hypothesis in energy, the concept of an energy field. An energy field is a sphere in space in which a magnetic, electrical or some other form of force can be considered. The concept of an energy field was some time ago a very intricate one for physicists to understand and accept. How could one object exert a force on another object if the two were not in contact with each other? In time, it became understandable that forces can act at a distance from each other. Electrical charges and magnetic poles seem to exert their forces throughout an energy field along passageways identified as lines of force.

Most magnetic homogeneous continuous dynamical systems display some element of permanent magnetism as either an explicit source or a residual magnetism in the ferrous medium that supports the magnetomotive-flux holor $\boldsymbol{\Phi}^j$.

Another form of magnetic energy source is an electrical inductor (coil) of N turns carrying an electrical current holor $\boldsymbol{I}^j$ coupled to the magnetical homogeneous continuous dynamical system such that the magnetomotive-force holor $\boldsymbol{M}_i$ of the equivalent magnetic energy-source holor is

$$\boldsymbol{M}_i = \boldsymbol{N}_{ij}\boldsymbol{I}^j = \left\| \begin{matrix} M_c \\ M_s \end{matrix} \right\| = \left\| \begin{matrix} N & 0 \\ 0 & N \end{matrix} \right\| \left\| \begin{matrix} I^c \\ I^s \end{matrix} \right\|. \tag{8.46}$$

When electrons move through an inductor wire, a magnetic field B is created around the wire. If the wire carrying the electrons is wound around a metal bar of iron, the iron will become magnetic, and it is then called an 'electromagnet'.

As a result of Maxwell's work, it is often more correct to speak of electromagnetic energy, a form of energy that has both electrical and magnetic dynamical components. Physicists now know that a number of seemingly distinct types of energy are all actually forms of electromagnetic energy. These include x-rays, gamma (Γ) rays, ultraviolet (UV) light, visible light, infrared (IR) radiation, radio waves and microwaves. These forms of electromagnetic energy differ from each other in terms of the wavelength and frequency of the energy wave on which they travel. The waves associated with x-rays, for example, have very short wavelengths and very high frequencies, while the waves associated with microwaves have much longer wavelengths and much lower frequencies.

8.4 Mechanical commutation matrixer dynamical components

Mechanics (Greek *Μηχανική*) is the science and technology relating to or associated with motion, as well as the force or torque for transmission of mechanical energy and signals. In physics it is concerned with the behaviour of physical bodies when exposed to forces or displacements, and the successive effects of the bodies on their environment. The subject has its origins in ancient civilisations. From the Middle Ages on, scientists such as the Polish Renaissance astronomer, Nicolaus Copernicus (1473–1543), the Italian physicist, mathematician, astronomer and philosopher, Galileo Galilei (1564–1642), the German mathematician, astronomer and astrologer, Johannes Kepler (1571–1630), and in particular, the British physicist, mathematician, astronomer, natural philosopher, alchemist and theologian, Isaac Newton (1642–1727), prepared the basis for what is currently identified as classical mechanics.

Mechanics is a subject of classical physics that relates to particles that are either moving with low velocity or are at rest. In addition, it may also be identified as a subject of physics that relates to the motion and force of a particular object. The procedure for determining mechanical commutation matrixer physical models for mechanical dynamical components is similar to that used for electrical commutation matrixer dynamical components. In this case, one applies Newton's laws to mechanical homogeneous continuous dynamical systems and then tries to fit the resulting holor equations into the standard single-port or two-terminal forms. Here again, the holor equations do not always fit the standard forms exactly, and one must use fictitious terminals.

8.4.1 Mechanical-signal holors

The dynamical behaviour of mechanical homogeneous continuous dynamical systems is specified by a set of holor displacements, velocities, forces and torques. An appropriate set of these mechanical-signal holors is sufficient to specify the general motion of a mechanical assembly moving in 3D space. In general, the movement combines the translation and rotation of the mechanical homogeneous continuous dynamical system with respect to some reference framework. In practice, however, a wide variety of situations exist when the motion is either translational or rotational.

Alternatively, the translation and rotation of a mechanical homogeneous continuous dynamical system may be considered as separate but interacting energetic mechanisms.

In any event, it is convenient to distinguish between two kinds of one-port mechanical-commutation matrixer components:

- one-port mechanical-commutation matrixers for translation along a fixed direction coordinate;
- one-port mechanical-commutation matrixers for rotation about a fixed coordinate axis.

The translational motion of a translational mechanical homogeneous continuous dynamical system is graphically represented by a mechanical commutation matrixer of energetically interacting components, where the nature of the interaction is determined by a set of applied and reaction force holors and a corresponding set of component velocity holors. For instance, one thinks of a force holor applied to an ideal mass resulting in an acceleration of the mass such that an equal and opposite reaction force holor is generated. Moreover, the tensor product of the force holor and the resultant velocity holor is the power holor being delivered to the mass. Hence, in the dynamical systems approach (systems thinking), the holors—the force holor and the velocity holor—form a valid energy-potential-difference holor/energy-transfer holor pair. According to the energy-potential-difference holor and the energy-transfer holor notion, the velocity holor is an energy-potential-difference holor, since two points are required to specify the velocity holor of the object, an inertial reference and a point on the object. Similarly, the force holor can be readily visualised as an energy-transfer holor, since it may be measured by reference to one point alone. With this reasoning in mind, the following assignment of holors results. For the mobility analogy:

- the velocity holor is analogous to the energy-potential-difference holor;
- the force holor is analogous to the energy-transfer holor.

Despite the rationale provided by the energy-potential-difference holor and energy-transfer holor concept, this assignment is completely arbitrary.

The dual assignment is equally valid as a way of conceiving and physically and mathematically modelling mechanical homogeneous continuous dynamical systems. In fact, historically the analogy of the force holor with the energy-potential-difference holor preceded the mobility analogy. For this reason, the two analogies are used interchangeably in this book, although the development in this chapter is predominantly carried out in terms of the mobility analogy. For the classical analogy:

- the velocity holor is analogous to the energy-transfer holor;
- the force holor is analogous to the energy-potential-difference holor.

Thus, the key to the mechanical commutation matrixer representation of mechanical homogeneous continuous dynamical systems is the recognition of the force holor as the energy-potential holor for the mechanics of translation and the torque holor as the one for the mechanics of rotation. Here, one notes that it is quite common to speak of force holors being transmitted through certain parts or of torque holors being transmitted along **mechano-mechanical** (MM) or **electro-mechanical** (EM) or **fluido-mechanical** (FM) shafts. This transmission does not require any motion. In the case of rotation, the torque holor is transmitted in the axial sense of direction and is not directly associated with the rotary motion. In addition, there is no physical quantity of fluid or charge being transmitted, and in this sense, the mechanical homogeneous continuous dynamical system is rather different from the electrical homogeneous continuous dynamical systems considered so far. Nevertheless, the concept of a 'flow' of force holor or torque holor is a great help in providing a systematic mechanical commutation matrixer view of mechanical homogeneous continuous dynamical systems.

Consider the modules $\mathcal{A}$, $\mathcal{B}$ and C (figure 8.7(a)) in the equilibrium of a smooth table. A horizontal force holor $\boldsymbol{F}^j$ is applied to module $\mathcal{A}$, as shown.

To demonstrate the transmissibility of the force holor, one draws free-body diagrams of modules A, B, and C in figure 8.7(b) (vertical force holors have been neglected for clarity). Since the modules are in equilibrium, there must be no unbalanced force holor acting on them. The applied force holor $\boldsymbol{F}^j$ acting to the right on module $\mathcal{A}$ must therefore be balanced by a force holor of the same magnitude acting to the left on module $\mathcal{A}$. This latter force holor is applied to module $\mathcal{A}$ by module $\mathcal{B}$.

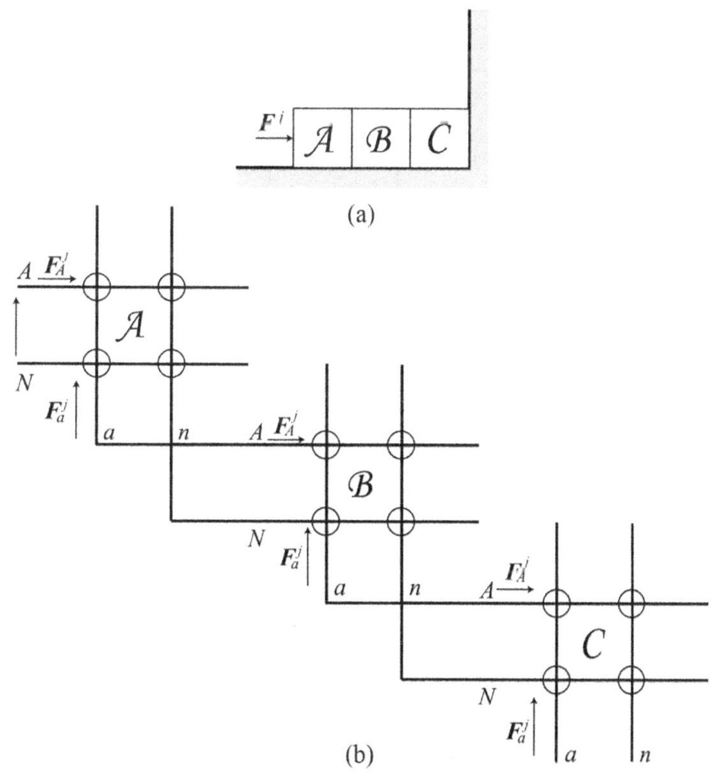

Figure 8.7. Transmission of force holor.

From Newton's third law, every action has an equal and opposite reaction. Since module $\mathcal{B}$ pushes against module $\mathcal{A}$, there is a reaction of module $\mathcal{A}$ against module $\mathcal{B}$. Thus, on the free-body diagram of module B, the force holor from module A acts to the left. To maintain block $\mathcal{B}$ in equilibrium requires a force holor from module $\mathcal{C}$. The process of action and reaction shows that the force holor is transmitted through the modules to the wall.

Rotational motion about a fixed axis of a rotational mechanical homogeneous continuous dynamical system is graphically represented by a mechanical commutation matrixer of energetically interacting dynamical components, where the nature of the interaction is governed by a set of torques and angular velocities. For instance, for the mobility analogy, the torque holor is a moment of the force holor about a point and is consequently an energy-transfer holor in the same way as the force holor is. In a similar fashion, the angular-velocity holor about a fixed axis is readily seen to be an energy-potential-difference holor, thus for rotational motion it is complementary to the mobility analogy. The angular-velocity holor is analogous to the energy-potential-difference holor. The axial torque is analogous to the energy-transfer holor. Thus, in a similar way, one may demonstrate the energy-transfer holor nature of the torque holor. Figure 8.8(a) shows three discs, $\mathcal{A}$, $\mathcal{B}$ and $\mathcal{C}$, connected together by dowel pins.

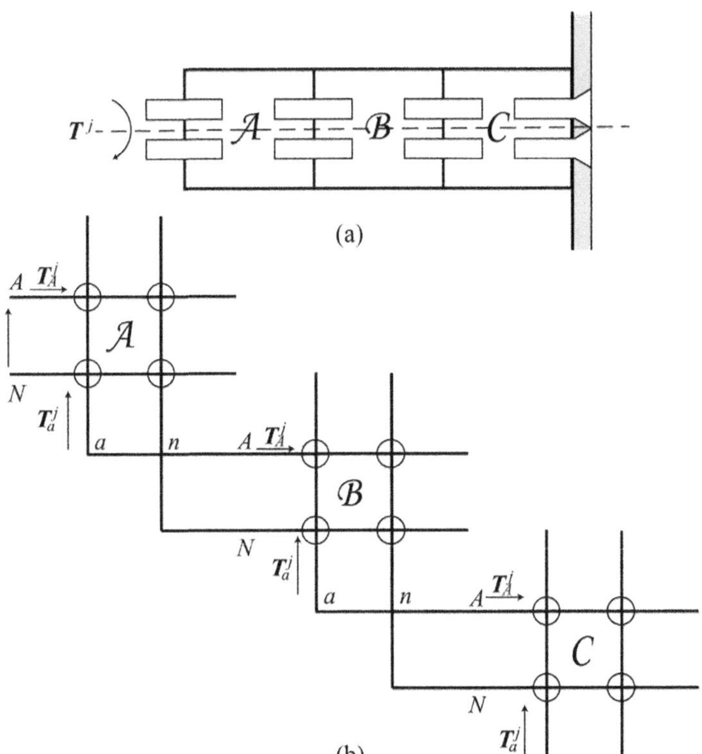

Figure 8.8. Transmission of torque holor.

A torque holor, $\boldsymbol{T}^j$, is applied to disc $\mathcal{A}$. The free-body diagrams of each disc (figure 8.8(b)) show that the torque holor $\boldsymbol{T}^j = \boldsymbol{D}^{ji}\boldsymbol{F}_i$ is transmitted through the discs to the wall. The balancing torque holor, $\boldsymbol{T}^j$, on each disc comes from consecutive applications of the action and reaction principle.

The energy-potential-difference holor in mechanical homogeneous continuous dynamical systems can be a distance holor, a displacement holor, a velocity holor, or an acceleration holor, with the choice being made to ensure that the three elemental physical situations are mathematically modelled by elemental holor equations of the same mathematical form as those in the electrical commutation matrixers, and that the univalent tensor product of the energy-potential-difference holor and the energy-transfer holor is a univalent power holor.

The appropriate energy-potential-difference holor is, therefore, the velocity potential-difference holor $\Delta \boldsymbol{V}_i$ and the dimensional analysis of the univalent tensor product is shown for confirmation

$$\boldsymbol{P}^{\kappa} = \Delta \boldsymbol{V}_i \otimes \boldsymbol{F}^j = \left[\frac{m}{s}\right] \otimes [N] = \left[\frac{J}{s}\right] = [W].$$

Thus, the univalent tensor product of the velocity-potential-difference holor $\Delta \boldsymbol{V}_i$ in m s^{-1} and the force holor $\boldsymbol{F}^j$ in N gives us the power holor directly in W. For rotation, the tensor product of the angular-velocity-potential-difference holor $\Delta\boldsymbol{\Omega}_i$ is in rad s^{-1} and the torque holor $\boldsymbol{T}^j$ is in N · m and W.

8.4.2 Mechanical-resistance or damping holor

A mechanical homogeneous continuous dynamical system that requires a steady force holor $\boldsymbol{F}^j$ to maintain a certain velocity holor $\Delta \boldsymbol{V}_i$ displays dissipative effects. Usually, the dissipation of power occurs because energy is being transformed from kinetic energy to thermal energy by viscous friction. Viscous-force holors $\boldsymbol{F}^j$ have to be overcome whenever neighbouring bodies have a relative-velocity holor $\Delta \boldsymbol{V}_i$. Thus, any arrangement that involves the relative motion of adjacent objects incurs power dissipation. A pure dissipater is one in which the kinetic and potential energy-storage phenomena are absent. Thus, a light, rigid object moving through a viscous fluid or sliding along a rough surface has a constitutive relation that statically relates to the applied force holor $\boldsymbol{F}^j$ and the relative velocity $\Delta \boldsymbol{V}_i$ of the object

$$\boldsymbol{F}^j = \mathcal{F}(\Delta \boldsymbol{V}_i). \tag{8.47}$$

In the linear case the constitutive relation becomes

$$\boldsymbol{F}^j = \boldsymbol{B}^{ji}\Delta \boldsymbol{V}_i. \tag{8.48}$$

The pure dissipater is symbolically represented by the dashpot. This schematically evokes the dashpot devices used to damp the motions of an automotive-vehicle suspension's mechanical homogeneous continuous dynamical systems and electrical meter movements.

The univalent power holor $\boldsymbol{P}^{\kappa}$ absorbed by a dissipator is the univalent tensor product of the velocity holor $\Delta \boldsymbol{V}_i$ and force holor $\boldsymbol{F}^j$, and is obtained from the constitutive relation as the sum of the dissipater content G and the co-content J, namely:

$$\boldsymbol{P}^{\kappa} = \Delta \boldsymbol{V}_i \otimes \boldsymbol{F}^j = \left[\frac{m}{s}\right] \otimes [N] = \left[\frac{J}{s}\right] = [W].$$

For a linear dissipater with the constitutive relation of equation (8.48), the content and co-content are given by the Rayleigh function

$$G = J = \frac{1}{2} b(v\wedge 2). \tag{8.49}$$

The net power absorbed is

$$P^p = b(v\wedge 2) = (F\wedge 2)/. \tag{8.50}$$

Thus, the elemental mechanical situation in which energy is dissipated is that involving friction. Unfortunately, friction effects are varied, although generally three principal forms are recognised:

- static friction;
- Coulomb friction;
- viscous friction.

Static friction is the effect whereby the force holor required to move a stationary object is greater than that required to move the same object when it is moving very slowly. Clearly, this does not conform to the linear relationship, which is analogous to the electrical resistance. Similarly, Coulomb friction, that is the force holor independent of the velocity holor, does not conform to the linear situation in which the force holor is proportional to the velocity holor, and it will be considered in section 8.7. However, there are frictional effects and devices in which the force holor is proportional to the velocity-potential-difference holor, with this being termed viscous friction. These frictional dynamical components are usually termed dampers, although the most common example is probably the shock absorber on an automotive vehicle.

A schematic drawing of a translational damper is shown in figure 8.9(a). The damper has a piston-cylinder arrangement. However, the damping effect is due to a viscous liquid (oil) in the cylinder. The piston and cylinder may move at different velocity holors. Thus, the damper is a single-port or two-terminal device. One terminal is at the piston velocity holor $\boldsymbol{V}_i^A$ and the other terminal is at the cylinder velocity holor $\boldsymbol{V}_i^N$. The shear force holor $\boldsymbol{F}_{\mathrm{T}}^j$ developed in the damper is proportional to the velocity-potential-difference holor $\boldsymbol{V}_i^A - \boldsymbol{V}_i^N$ of the piston and cylinder.

From the free-body diagram of figure 8.9(b), one finds that

$$\boldsymbol{F}_{\mathrm{T}}^j = \boldsymbol{B}_{\mathrm{T}}^{ji}\left(\boldsymbol{V}_i^A - \boldsymbol{V}_i^N\right) = \begin{Vmatrix} F_{\mathrm{T}}^{\mathrm{c}} \\ F_{\mathrm{T}}^{\mathrm{s}} \end{Vmatrix} = \begin{Vmatrix} b & 0 \\ 0 & b \end{Vmatrix} \left(\begin{Vmatrix} V_{\mathrm{c}}^{\mathrm{A}} \\ V_{\mathrm{s}}^{A} \end{Vmatrix} - \begin{Vmatrix} V_{\mathrm{c}}^{\mathrm{N}} \\ V_{\mathrm{s}}^{N} \end{Vmatrix} \right), \tag{8.51}$$

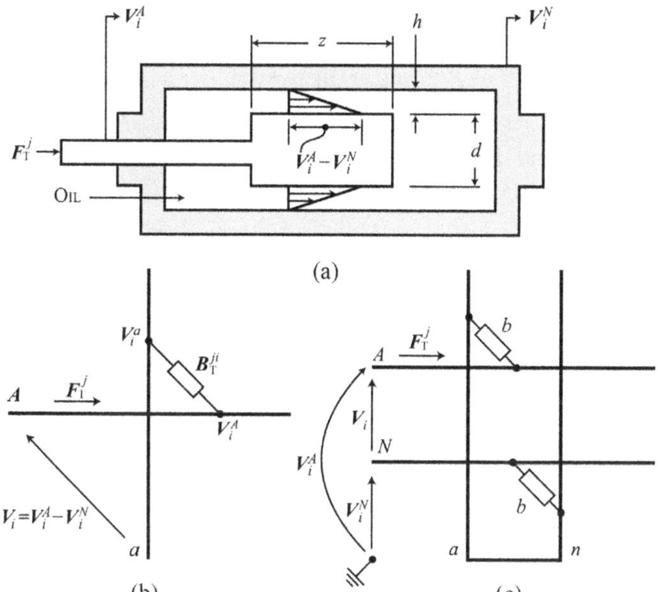

Figure 8.9. Schematic drawing and mechanical commutation matrixers of translational damper.

where $\boldsymbol{B}_T^{ji}$ is the translational damping-admittance holor and b is the translational damping in N·s m^{-1}.

Consideration of the analogous electrical commutation matrixers holors shows that the damping b is therefore analogous to the electrical conductance $G = 1/R$. From equations (8.1a) and (8.51) one may describe damping as:

$$b = G = \frac{1}{R} = \frac{\pi dz}{h}\mu, \tag{8.52}$$

where d is the piston diameter, z is the piston length, h is the shear stress and μ is the viscosity of the oil.

A rotational dissipater of energy display material behaviour determines the device angular-velocity holor $\Delta\boldsymbol{\Omega}_i$ as a function of the device torque holor $\boldsymbol{T}^j$. The dissipater actually converts mechanical energy into thermal energy by the phenomena of viscosity or friction.

Symbolically, a rotational dissipater is represented by a rotational dashpot with a constitutive relation

$$\boldsymbol{T}^j = \mathcal{F}(\Delta\boldsymbol{\Omega}_i), \tag{8.53}$$

where the angular-velocity holor $\Delta\boldsymbol{\Omega}_i$ is measured across the dissipator terminals.

In its linear form the constitutive relation of the rotational dissipater is

$$\boldsymbol{T}^j = \boldsymbol{B}^{ji}\Delta\boldsymbol{\Omega}_i. \tag{8.54}$$

The power dissipated by a rotational dissipater is the sum of the dissipater content G and the co-content J:

$$P = G + J.$$

In the general case, G and J are equal and are given by

$$G = J = \frac{1}{2}B(\omega\wedge 2). \tag{8.55}$$

In rotational mechanical homogeneous continuous dynamical systems, the rotational damping-admittance holor, $\boldsymbol{B}_{R}^{ji}$, is defined by

$$\boldsymbol{T}_{R}^{j} = \boldsymbol{B}_{R}^{ji}\left(\boldsymbol{\Omega}_{i}^{A} - \boldsymbol{\Omega}_{i}^{N}\right) = \begin{Vmatrix} T_{R}^{c} \\ T_{R}^{s} \end{Vmatrix} = \begin{Vmatrix} B & 0 \\ 0 & B \end{Vmatrix}\left(\begin{Vmatrix} \Omega_{c}^{A} \\ \Omega_{s}^{A} \end{Vmatrix} - \begin{Vmatrix} \Omega_{c}^{N} \\ \Omega_{s}^{N} \end{Vmatrix}\right), \tag{8.56}$$

where $\boldsymbol{\Omega}_{i}^{A} - \boldsymbol{\Omega}_{i}^{N}$ is the angular-velocity-potential-difference holor.

The rotational damper has essentially the same geometry as the translational damper. The rotational damping is approximately

$$B = G = \frac{1}{R} = \frac{\pi(d)^{2}z}{2h}\mu. \tag{8.57}$$

The unit of rotational damping is N · m · s.

Figure 8.10 shows the symbols used for damping in translational rotational mechanical homogeneous continuous dynamical systems. The symbols on the left are descriptive indications of the process physical model. As a mechanical commutation-matrixer dynamical component, however, both dampers are represented by the symbol for electrical resistance. Remember, however, that B is analogous to the conductance $G = 1/R$.

In general, frictional effects do not conform exactly to any one of the three physical models of friction. However, other than the further consideration of dry (Coulomb) friction (section 8.7), at this stage, it is quite reasonable to approximate the actual friction characteristics with the nearest equivalent viscous damping. When this is done, the analytical techniques used to predict mechanical homogeneous continuous dynamical system behaviour could then be the same as those used in electrical commutation matrixers.

The power dissipated by a damper is equal to the univalent tensor product of the energy-potential-difference and energy-transfer holors associated with the mechanical commutation matrixer component.

8.4.3 Mechanical inductance holor of the spring

A mechanical homogeneous continuous dynamical system which when subject to a force holor, $\boldsymbol{F}^{j}$, either compresses or elongates without significant acceleration of its elastic dynamical component parts or loss of mechanical energy due to the friction or unrecoverable deformation is a pure translational spring. The energy storage mechanism of a pure spring is the net displacement holor, $\boldsymbol{X}_{i}$, of the spring from its quiescent state. The variable displacement holor is defined as

$$\boldsymbol{X}_{i} = \boldsymbol{X}_{i}^{A} + \Delta\boldsymbol{X}_{i}^{A} - \boldsymbol{X}_{i}^{N} - \Delta\boldsymbol{X}_{i}^{N}. \tag{8.58}$$

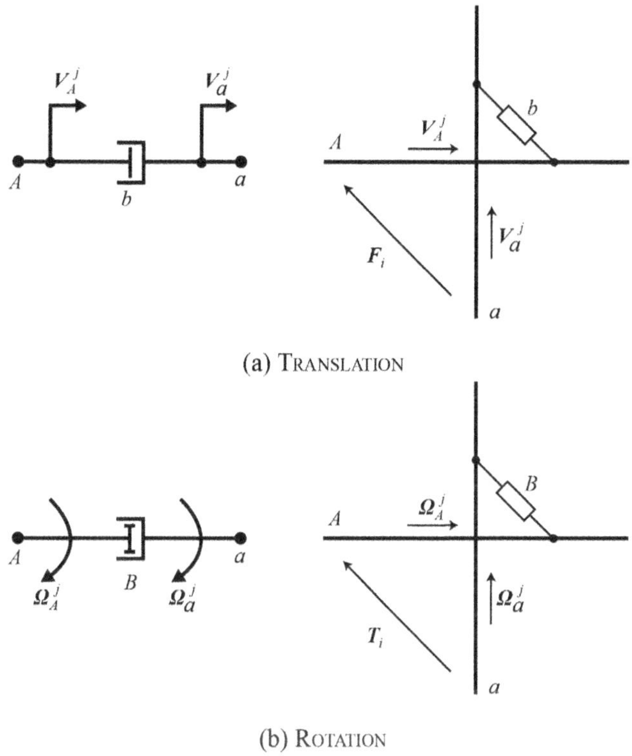

Figure 8.10. Symbols and mechanical commutation matrixers for translational and rotational damping.

The material properties of a spring are specified by a constitutive relation that relates the applied force holor, $\boldsymbol{F}^j$, and the resultant spring deformation holor, $\boldsymbol{X}_i$,

$$\boldsymbol{X}_i = \mathcal{F}(\boldsymbol{F}^j). \tag{8.59}$$

A linear ideal spring obeys Hooke's law and has the constitutive relation

$$\boldsymbol{X}_i = \frac{\delta_k^i}{\boldsymbol{K}_{jk}} \boldsymbol{F}^j, \tag{8.60}$$

where

$$\boldsymbol{K}_{ij} = \begin{Vmatrix} 0 & -k \\ k & 0 \end{Vmatrix}$$

is known as the translational spring stiffness holor, while the inverse multiplied by the Kronecker delta,

$$\frac{\delta_k^i}{\boldsymbol{K}_{jk}} = \frac{1}{(k\wedge 2)} \begin{Vmatrix} 0 & k \\ -k & 0 \end{Vmatrix} = \begin{Vmatrix} 0 & 1/k \\ -1/k & 0 \end{Vmatrix},$$

is commonly known as the 'translational compliance'.

From the above-mentioned argument it is clear that an ideal spring is (according to the mobility analogy) an energy-potential-difference holor, $\boldsymbol{F}^j$, store, since the displacement holor, $\boldsymbol{X}_i$, represents an accumulation of the energy-potential-difference holor, $\boldsymbol{F}^j$. The stored energy in a spring is kinetic energy, $\mathcal{T}$, and both the potential energy, $\mathcal{T}$, and the kinetic co-energy, $\mathcal{T}^*$, can be evaluated from the constitutive relation.

In the linear case, the stored energy and co-energy are given by

$$\mathcal{T} = \mathcal{T}^* = \frac{1}{2}k(f\wedge 2). \tag{8.61}$$

Examples of mechanical homogeneous continuous dynamical systems that can be physically and mathematically modelled as a spring are:

- the leaf spring used in automotive-vehicle suspensions' mechanical homogeneous continuous dynamical systems;
- the pipe U joint common in thermal homogeneous continuous dynamical systems;
- the familiar coil spring used in clock and meter movements.

In fact, in any mechanical homogeneous continuous dynamical system, where compliance effects are very dominant, the power absorbing characteristics can be physically and mathematically modelled as a spring. Another mechanical situation is that of a spring (figure 8.11). The ideal spring (negligible mass) is a direct application of Hooke's law; the extension or compression is directly proportional to the force holor, $\boldsymbol{F}^j_H$, producing it. Normally, this is expressed in the holor form

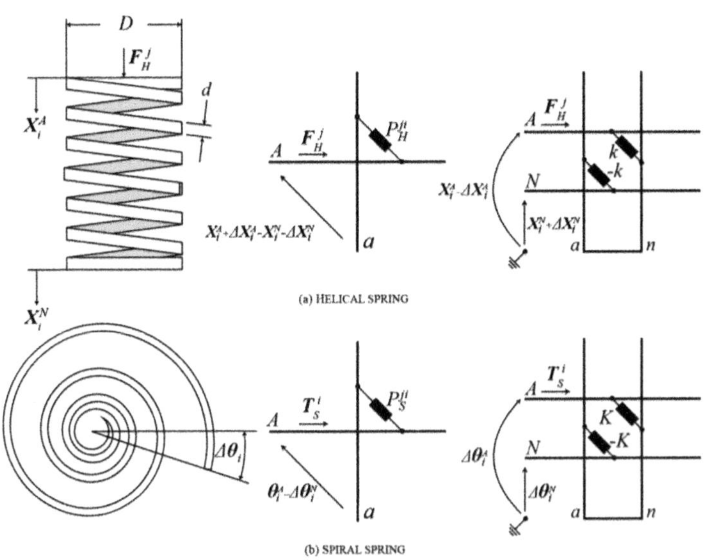

Figure 8.11. Helical and spiral spring configurations with their mechanical commutation matrixers.

$$\boldsymbol{F}_H^j = \boldsymbol{P}_H^{ji}\left(\boldsymbol{X}_i^A + \Delta\boldsymbol{X}_i^A - \boldsymbol{X}_i^N - \Delta\boldsymbol{X}_i^N\right)$$

$$= \left\|\begin{matrix} F_H^c \\ F_H^s \end{matrix}\right\| = \left\|\begin{matrix} 0 & k \\ -k & 0 \end{matrix}\right\| \left(\left\|\begin{matrix} X_c^A + \Delta X_c^A \\ X_s^A + \Delta X_s^A \end{matrix}\right\| - \left\|\begin{matrix} X_c^N + \Delta X_c^N \\ X_s^N + \Delta X_s^N \end{matrix}\right\| \right), \tag{8.62a}$$

where k is the spring stiffness of the translational spring, and $\boldsymbol{X}_i^A$ and $\boldsymbol{X}_i^N$ are the position holors of the ends of the helical spring relative to some fixed reference when the force holor, $\boldsymbol{F}_{\mathrm{H}}^j = 0$. $\Delta\boldsymbol{X}_i^A$ and $\Delta\boldsymbol{X}_i^N$ are the displacement holors of the ends when the force holor, $\boldsymbol{F}_{\mathrm{H}}^n$, is applied.

If one differentiates equation (8.62a) with respect to time (frequency), since $\boldsymbol{X}_i^1$ and $\boldsymbol{X}_i^2$ are constants, one obtains

$$\omega\boldsymbol{F}_H^j = \boldsymbol{P}_H^{ji}\omega\left(\Delta\boldsymbol{X}_i^A - \Delta\boldsymbol{X}_i^N\right) = \boldsymbol{P}_H^{ji}\left(\boldsymbol{V}_i^A - \boldsymbol{V}_i^N\right)$$

$$= \omega\left\|\begin{matrix} F_H^c \\ F_H^s \end{matrix}\right\| = \left\|\begin{matrix} 0 & k \\ -k & 0 \end{matrix}\right\| \omega\left(\left\|\begin{matrix} \Delta X_c^A \\ \Delta X_s^A \end{matrix}\right\| - \left\|\begin{matrix} \Delta X_c^N \\ \Delta X_s^N \end{matrix}\right\| \right) \tag{8.62b}$$

$$= \left\|\begin{matrix} 0 & k \\ -k & 0 \end{matrix}\right\| \left(\left\|\begin{matrix} V_{\mathrm{c}}^A \\ V_{\mathrm{s}}^A \end{matrix}\right\| - \left\|\begin{matrix} V_{\mathrm{c}}^N \\ V_{\mathrm{s}}^N \end{matrix}\right\| \right).$$

This is the same holor form as equation (8.1b). Thus, a spring is the analogous situation to an inductor, although it must be noted that the stiffness, k, is analogous to the reciprocal of inductance, $1/L$.

The principal unit of stiffness is N m^{-1}, which is rather small for most useful springs, and therefore k is more likely to be expressed in kN m^{-1} or MN m^{-1}. For a helical spring, the spring constant may be computed from the formula

$$k = \frac{1}{L} = G\frac{(d)^4}{8N(D)^3}, \tag{8.63}$$

where G is the shear modulus of elasticity, d is the wire diameter, D is the coil diameter and N is the number of coils.

Any mechanical homogeneous continuous dynamical system that is subject to a torsional torque holor, $\boldsymbol{T}^j$, has a resultant angular-displacement holor, $\Delta\boldsymbol{\theta}_i$, measured across the object, and along an axis displays the essential features of a rotational spring. If the object is nominally massless and moves without dissipation, then it is a pure rotational spring. The constitutive relation of a pure torsional spring is

$$\Delta\boldsymbol{\theta}_i = \mathcal{F}(\boldsymbol{T}^j). \tag{8.64}$$

Equation (8.64) states that the angular-displacement holor, $\Delta\boldsymbol{\theta}_i$, measured across the spring is statically determined by the torque holor, $\boldsymbol{T}^j$, applied at the two ends. The angular-displacement holor, $\Delta\boldsymbol{\theta}_i$, is defined in terms of the angular-velocity holor, $\Delta\boldsymbol{\Omega}_i$, by the dynamical relationship

$$\Delta\boldsymbol{\Omega}_i = \omega\Delta\boldsymbol{\theta}_i. \tag{8.65}$$

In equation (8.65), the relative angular velocity, $\Delta\boldsymbol{\Omega}_i$, is defined as positive if the terminal angular velocities, $\Delta\boldsymbol{\Omega}_i^A$ and $\Delta\boldsymbol{\Omega}_i^N$ (themselves measured with respect to an inertial framework), are such that $\Delta\boldsymbol{\Omega}_i^A$ is larger in an anticlockwise sense than $\Delta\boldsymbol{\Omega}_i^N$.

If the constitutive relation of the spring is linear, then

$$\boldsymbol{T}^j = \boldsymbol{K}^{ji}\Delta\boldsymbol{\theta}_i, \tag{8.66}$$

where

$$\boldsymbol{K}^{ji} = \begin{Vmatrix} 0 & -K \\ K & 0 \end{Vmatrix}$$

is known as the torsional spring stiffness holor, while the inverse multiplied by the Kronecker delta,

$$\frac{\delta_k^i}{\boldsymbol{K}^{ij}} = \frac{1}{(K\char`^2)}\begin{Vmatrix} 0 & K \\ -K & 0 \end{Vmatrix} = \begin{Vmatrix} 0 & 1/K \\ -1/K & 0 \end{Vmatrix},$$

is commonly known as the 'torsional compliance'.

According to the mobility analogy, the rotational spring is an energy-potential-difference holor, $\Delta\boldsymbol{\theta}_i$, store with stored potential energy $\mathcal{U}$ and co-energy $\mathcal{U}^*$.

If the store is linear, the potential energy and co-energy are given by

$$\mathcal{U} = \mathcal{U}^* = \frac{1}{2}K(\Delta\theta\char`^2). \tag{8.67}$$

For a torsional spiral spring (figure 8.11(b)), the torque holor, $\boldsymbol{T}_S^j$, is related to the change in the angular-displacement holor, $\Delta\boldsymbol{\theta}_{ii}^S$, as:

$$\boldsymbol{T}_S^j = \boldsymbol{P}_S^{ji}\Delta\boldsymbol{\theta}_i^S = \boldsymbol{P}_S^{ji}(\Delta\boldsymbol{\theta}_i^A - \Delta\boldsymbol{\theta}_i^N)$$

$$= \begin{Vmatrix} T_S^c \\ T_S^s \end{Vmatrix} = \begin{Vmatrix} 0 & K \\ -K & 0 \end{Vmatrix}\left(\begin{Vmatrix} \Delta\theta_c^A \\ \Delta\theta_s^A \end{Vmatrix} - \begin{Vmatrix} \Delta\theta_c^N \\ \Delta\theta_s^N \end{Vmatrix}\right). \tag{8.68}$$

Thus, the spring constant for the torsional spiral spring has the same units as the torque holor (N · m).

Now, deriving equation (8.68) with respect to time (frequency) yields

$$\begin{aligned} \omega\boldsymbol{T}_S^j &= \boldsymbol{P}_S^{ji}\omega\Delta\boldsymbol{\theta}_i^S = \boldsymbol{P}_S^{ji}\omega\left(\Delta\boldsymbol{\theta}_i^A - \Delta\boldsymbol{\theta}_i^N\right) = \boldsymbol{P}_S^{ji}\left(\boldsymbol{\Omega}_i^A - \boldsymbol{\Omega}_i^N\right) \\ &= \omega\begin{Vmatrix} T_S^c \\ T_S^s \end{Vmatrix} = \omega\begin{Vmatrix} 0 & K \\ -K & 0 \end{Vmatrix}\left(\begin{Vmatrix} \Delta\theta_c^A \\ \Delta\theta_s^A \end{Vmatrix} - \begin{Vmatrix} \Delta\theta_c^N \\ \Delta\theta_s^N \end{Vmatrix}\right) \\ &= \begin{Vmatrix} 0 & K \\ -K & 0 \end{Vmatrix}\left(\begin{Vmatrix} \Omega_c^A \\ \Omega_s^A \end{Vmatrix} - \begin{Vmatrix} \Omega_c^N \\ \Omega_s^N \end{Vmatrix}\right), \end{aligned} \tag{8.69}$$

where $\mathbf{\Omega}_1^A$ and $\mathbf{\Omega}_2^N$ are the angular velocity holors of the inner and outer terminals of the spiral spring. Note, also, that the helical spring may also be used in torsion.

The spring constant for a torsional spring with round wire is:

$$K = \frac{1}{L} = E\frac{\pi(d^\wedge 4)}{64l}, \tag{8.70}$$

where E is the modulus of elasticity, d is the wire diameter and l is the total length of the spring.

Figure 8.12 shows the mechanical and mechanical commutation matrixer symbols for springs, i.e. elastic dynamical components. The mechanical commutation matrixer representation for both types is the same as for the electrical inductance and the fluidic inertance. Note, however, that K is analogous to $1/L$.

The energy stored by a spring depends on the energy-transfer holor and the spring constant. Its form is the same as that given for the energy stored in the electrical inductance (equation (8.69)).

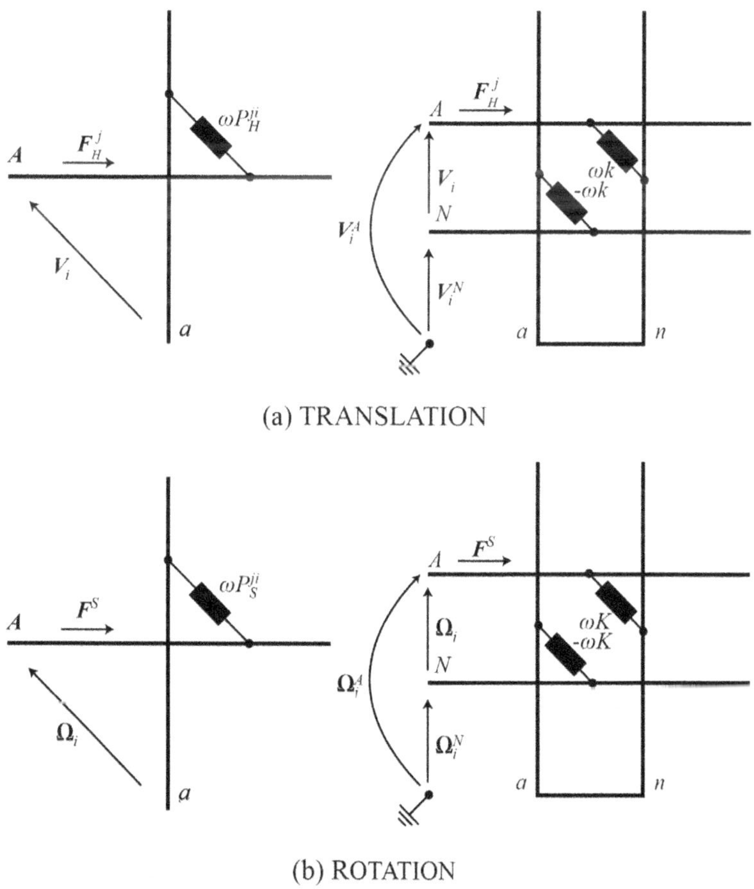

Figure 8.12. Symbols for springs with their mechanical commutation matrixers.

The spring produces a force holor that is proportional to the displacement holor. For some special geometry, a force holor that is proportional to a displacement holor can be obtained without a spring.

Consider the harmonic motion shown in figure 8.13. When a mass is connected to a spring (figure 8.13(a)), the mechanical homogeneous continuous dynamical system will oscillate if the mass is initially displaced and then released.

The spring creates a restoring force holor that is always proportional to the displacement holor. A similar restoring force holor may occur due to the position holor without a spring present. Figure 8.13(b) shows a mass suspended from a string in the form of a pendulum.

Now, if the mass is displaced and released, the mass will oscillate. The position holor of the mass provides the restoring force holor in this case. This force holor can be computed by considering a free-body diagram of the mass (figure 8.14(a)).

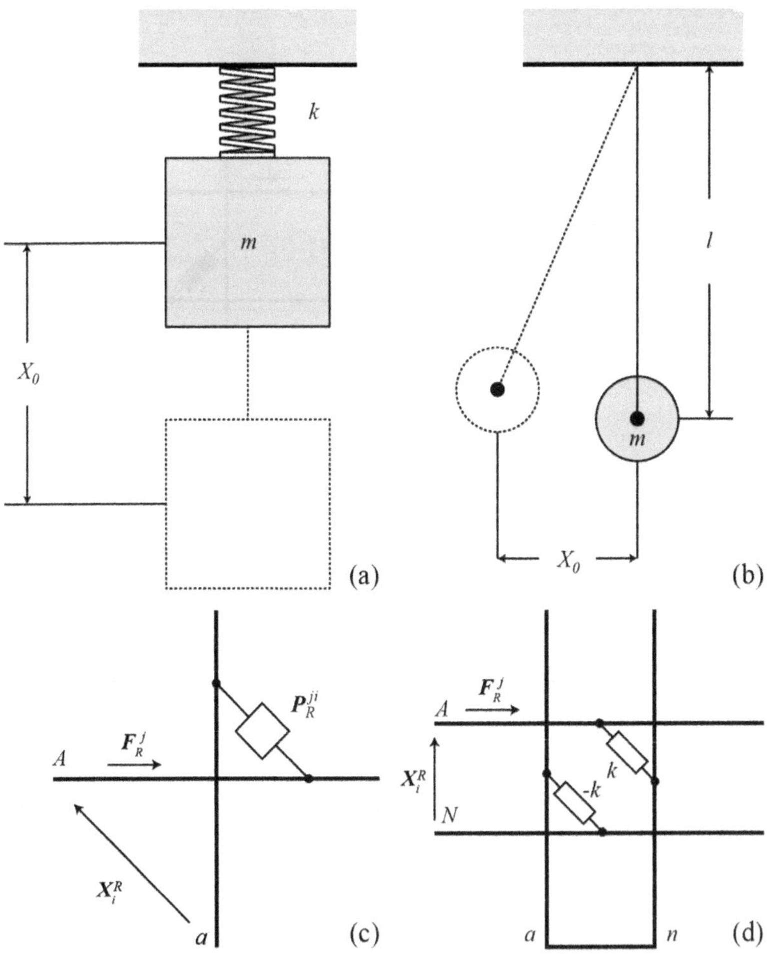

Figure 8.13. Harmonic motion and its mechanical commutation matrixer.

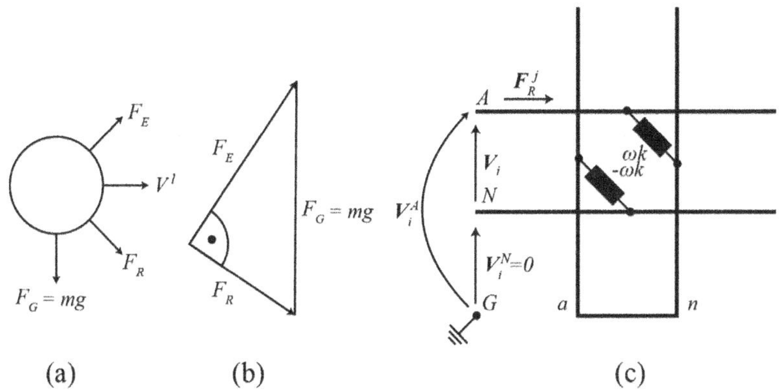

Figure 8.14. Pendulum restoring-force holor and its mechanical commutation matrixer.

From a summation of the force holors (figure 8.14(b)), the relation between the restoring force holor and displacement holor is:

$$\boldsymbol{F}_{\mathrm{R}}^{j} = \boldsymbol{P}_{\mathrm{R}}^{ji}\boldsymbol{X}_{i}^{\mathrm{R}} = \begin{Vmatrix} F_{\mathrm{R}}^{\mathrm{c}} \\ F_{\mathrm{R}}^{\mathrm{s}} \end{Vmatrix} = \begin{Vmatrix} 0 & k \\ -k & 0 \end{Vmatrix} \begin{Vmatrix} X_{\mathrm{c}}^{\mathrm{R}} \\ X_{\mathrm{s}}^{\mathrm{R}} \end{Vmatrix}, \tag{8.71}$$

and in addition, the equivalent spring constant for the pendulum is:

$$k = \frac{1}{L} = \frac{mg}{l},$$

where l is the length of the string in metres (m) and m is the mass in kilograms (kg).

If equation (8.71) is differentiated with respect to time (frequency), the result is:

$$\begin{aligned} \omega\boldsymbol{F}_{\mathrm{R}}^{j} &= \omega\boldsymbol{P}_{\mathrm{R}}^{ji}\boldsymbol{X}_{i}^{\mathrm{R}} = \boldsymbol{P}_{\mathrm{R}}^{ji}\boldsymbol{V}_{i}^{\mathrm{R}} \\ &= \begin{Vmatrix} F_{\mathrm{R}}^{\mathrm{c}} \\ F_{\mathrm{R}}^{\mathrm{s}} \end{Vmatrix} = \omega\begin{Vmatrix} 0 & k \\ -k & 0 \end{Vmatrix} \begin{Vmatrix} X_{\mathrm{c}}^{\mathrm{R}} \\ X_{\mathrm{s}}^{\mathrm{R}} \end{Vmatrix} = \begin{Vmatrix} 0 & k \\ -k & 0 \end{Vmatrix} \begin{Vmatrix} V_{\mathrm{c}}^{\mathrm{R}} \\ V_{\mathrm{s}}^{\mathrm{R}} \end{Vmatrix} \end{aligned} \tag{8.72}$$

The restoring force holor and the velocity holor have the relationship indicated by equation (8.1b), except that again, one has only one terminal.

One must add a fictitious terminal to make this fit into the single-port or two-terminal physical model. In this case, one chooses the ground at the zero velocity holor as the other terminal.

Figure 8.14(c) shows the mechanical commutation matrixer representation for the restoring force holor due to the pendulum position holor. The mechanical dynamical component is shown as 'dashed' because of the assumption of a fictitious terminal.

The equivalent gravitational force for the pendulum is:

$$F^{G} = \mathrm{sqrt}[(F^{R\wedge}2) + (F^{E\wedge}2)]. \tag{8.73}$$

8.4.4 Mechanical-capacitance holor of the mass

A pure translational mass is a mechanical homogeneous continuous dynamical system's rigid object that is moving through a non-dissipative environment. Then, according to Newton's second law, the momentum holor, $\boldsymbol{P}^j$, of the mass is linearly related to the rigid object's velocity holor, $\boldsymbol{V}_i$

$$\boldsymbol{P}^j = \boldsymbol{M}^{ji}\boldsymbol{V}_i, \tag{8.74}$$

where $\boldsymbol{M}^{ji}$ is the Newtonian mass holor of the rigid object.

For the quantity, $\boldsymbol{P}^j$, the momentum holor is defined by

$$\boldsymbol{P}^j = \frac{1}{\omega}\boldsymbol{F}^j \quad \text{or} \quad \boldsymbol{F}^j = \omega\boldsymbol{P}^j, \tag{8.75}$$

where $\boldsymbol{F}^j$ is used to signify the force holor.

The mobility analogy indicates that the quantity momentum holor, $\boldsymbol{P}^j$, is formally analogous to the energy-transfer holor, $\boldsymbol{V}^j$, accumulation, and thus a pure translational mass can be classified as an energy-transfer store. Observe that the velocity holor, $\boldsymbol{V}_i$, is considered to be positive if there is a net movement of terminal N in the sense of direction indicated.

The Newtonian mass, i.e. an inertial dynamical component, has an intrinsically linear constitutive relation and consequently the stored kinetic energy, $\mathcal{T}$, and the kinetic co-energy, $\mathcal{T}^*$, are equal:

$$\mathcal{T} = \mathcal{T}^* = \frac{1}{2}m(v\text{^}2). \tag{8.76}$$

The ideal translational mass moving without resistance is completely described by the dynamical relation given in equation (8.75) or the constitutive or material relation given in equation (8.74). The remarkable thing about the material behaviour of mass as specified by Newton's second law is its linearity.

From Einstein's theory of relativity, it is now established that this is only an approximation that is valid for velocities much lower than the velocity of light.

When this condition does not prevail, the constitutive relation for the momentum, p, of the mass yields

$$p = \frac{mv}{\text{sqrt}[1-(v\text{^}2)/(c\text{^}2)]},$$

where c is the velocity of light.

The elemental mechanical situation of an isolated mass (figure 8.15(a)) can be considered through a simple application of Newton's law:

$$\boldsymbol{F}^j = \omega\boldsymbol{M}^{ji}\boldsymbol{V}_i = \begin{Vmatrix} F^c \\ F^s \end{Vmatrix} = \begin{Vmatrix} 0 & -\omega m \\ \omega m & 0 \end{Vmatrix} \begin{Vmatrix} V_c \\ V_s \end{Vmatrix}, \tag{8.77}$$

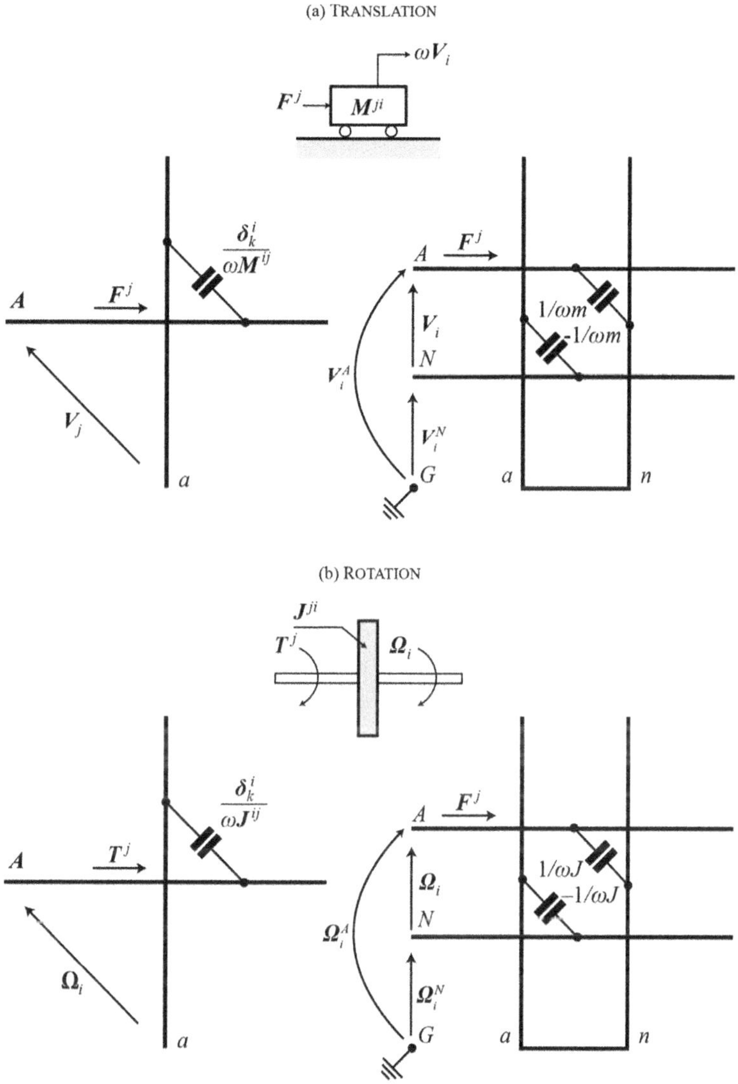

Figure 8.15. Symbols for mass and their mechanical commutation matrixers.

or by integrating with respect to time (frequency):

$$V_i = \frac{\delta_k^i}{\omega M^{ij}} F^j + V_0. \tag{8.78}$$

The initial velocity holor, $V_0 = (0,0)$, equation (8.78), almost fits the mathematical model described in equation (8.1c). The difference is that the mass in not a single-port or two-terminal device. There is only one velocity holor associated with the mass impedance holor. To place this in a mechanical dynamical-component form, one must add the ground terminal at the zero velocity holor to equation (8.78).

Then equation (8.78) becomes:

$$V_i - V_i^G = \frac{\delta_k^i}{\omega M^{ij}} F^j = \begin{Vmatrix} V_c \\ V_s \end{Vmatrix} - \begin{Vmatrix} V_c^G \\ V_s^G \end{Vmatrix} = \begin{Vmatrix} 0 & 1/\omega m \\ -1/\omega m & 0 \end{Vmatrix} \begin{Vmatrix} F^c \\ F^s \end{Vmatrix}, \tag{8.79}$$

where

$$\frac{\delta_k^i}{\omega M^{ij}} = \begin{Vmatrix} 0 & 1/\omega m \\ -1/\omega m & 0 \end{Vmatrix}.$$

For translational mechanical homogeneous continuous dynamical systems, the mass is directly analogous to electrical capacitance. From a mechanical commutation matrixer viewpoint, the mass always represents capacitance to a fictitious ground terminal. A pure rotational mass store is a rigid mechanical homogeneous continuous dynamical system rotating without resistance about a fixed axis. Then, according to Newton's second law, the angular-momentum holor,

$$H^j = J^{ji}\Omega_i, \tag{8.80}$$

where J^{ji} is defined as the moment of inertia of the object about its fixed axis.

The axial torque holor, T^j, applied to the object is defined by the dynamical relation

$$T^j = \omega H^j. \tag{8.81}$$

Note that the energy-transfer accumulation, the angular-momentum holor, H^j, is linearly related through the constitutive relation to the energy-potential difference holor, i.e. the angular-velocity holor, Ω_i. As a result, the stored kinetic energy, $\mathcal{T}$, and the kinetic co-energy, $\mathcal{T}^*$, are equal and are given as

$$\mathcal{T} = \mathcal{T}^* = \frac{1}{2} J(\Omega\text{^}2). \tag{8.82}$$

As with the translational mass, the linear constitutive relation given by equation (8.80) is a direct consequence of neglecting relativistic effects. This is shown in figure 8.15(a). A rotating mass admittance holor (figure 8.15(b)) is also related to its signal holors by a specialisation of Newton's second law.

Thus:

$$T^j = \omega J^{ji}\Omega_i, \tag{8.83}$$

where J^{ji} is the polar-moment-of-inertia admittance holor in $kg \cdot m^2$.

For the rotational mechanical continuous dynamical system, the capacitance equals the polar-moment-of-inertia holor. Figure 8.15 shows the mechanical commutation matrixer representation.

With a fictitious non-rotating ground terminal, equation (8.83) becomes:

$$\boldsymbol{\Omega}_i = \boldsymbol{\Omega}_i^A - \boldsymbol{\Omega}_i^G = \frac{\delta_k^i}{\omega \boldsymbol{J}^{ij}} \boldsymbol{T}^j, \tag{8.84}$$

and moreover, the rotating mass also represents the electrical capacitance to a fictitious ground terminal.

The energy stored by a mass in a translational mechanical homogeneous continuous dynamical system depends on the velocity holor and the mass-admittance holor. For a rotational mechanical homogeneous continuous dynamical system, the energy stored is a function of the angular-velocity holor and the moment-of-inertia admittance holor. The formulation for the energy is the same as that given for the electrical capacitor in equation (8.74).

8.5 Fluidic commutation matrixer components

Fluidics is the science and technology relating to or associated with a compressible medium for transient fluidic energy and signals. It involves using the flows and pressures of fluids in sensing, control and information-processing systems with no moving parts. The physical premises of fluidics are hydraulics and pneumatics, based on the theoretical foundation of fluid dynamics. The term 'fluid' refers to a substance with particles that move and change their relative position easily without a separation of mass and that yields easily to pressure and is capable of flowing, or a substance tending to flow or conform to the outline of its container and that includes a compressible medium (liquids, gases, aerosols and suspensions).

One of the newest technologies, fluidics has in recent years come to compete with mechanical and electrical homogeneous complex and simple dynamical systems.

In the 1930s, a Romanian inventor, aerodynamics pioneer and the builder of an experimental aircraft, Henri Coandă (1886–1972), described what is now known as the Coandă effect, a major contribution to fluidic technology. He observed that as a free jet emerges from a jet nozzle, the stream tends to follow a nearby curved or inclined surface. It also 'attaches' itself to and flows along this surface if the curvature or angle of inclination is not too sharp. Coandă explained this tendency as being caused by the jet stream's entraining (picking up) nearby fluid molecules. When the supply of these molecules is limited by an adjacent surface, a partial vacuum develops between the jet and the surface. If the pressure on the other side of the jet remains constant, the partial vacuum, which is a lower pressure region, will force the jet to bend and attach itself to the wall. Because fluidics is not as rapid as electronics, it is unlikely to compete in fields with ultrahigh-speed requirements. On the other hand, in many applications fluidics is especially advantageous. It is now possible to detect, interlock and power complex operations by using air throughout a system. Controls can be installed by a competent fitter who might not be capable of dealing with electrical homogeneous continuous complex and simple dynamical systems. The elimination of contacted electrical valves (switches) prevents a possible fire hazard. A fluidic valve (i.e. a fluidic diode) is composed solely of a fixed structure, which structure defines an input fluid passage and an output fluid passage, and through these passages fluid normally passes through the input fluid passage into the output fluid passage in a desired or forward sense of direction. The fixed structure comprising fixed fluid-flow control means that it is effective to inhibit the reverse or backward flow of fluid from the output fluid passage through the input fluid passage in a fashion that is analogous to that seen in an electrical valve (i.e. an electrical diode). Although fluidic principles are fairly old, it was not until about 1960 that researchers attempted to use fluidics commercially. The demand for reliable controls in space research stimulated progress.

The 1960s saw the application of fluidics to complex control systems, with the introduction of the fluidic amplifier. A jet of fluid can be deflected by a weaker jet striking it at the side. This provides nonlinear amplification, similar to the controlled electrical valve (i.e. the transistor) used in electrical digital logic. It is mostly used in environments where electrical digital logic would be unreliable, as in fluidic homogeneous continuous complex and simple dynamical systems exposed to high levels of electromagnetic interference or ionising radiation. Nanotechnology considers fluidics to be one of its instruments. In this domain, effects such as fluid–solid and fluid–fluid interface forces are often highly significant. Fluidics has also been used for military applications.

To develop mathematical models for fluidic dynamical components with fluid flow one must apply the holor equations of fluid mechanics to them and try to adapt them to the holor forms given for the principal dynamical components in equation (8.1). In some cases, the fluidic dynamical components analysed by the fluid mechanics holor equations do not match the principal holor forms exactly. When that happens, restrictions are placed on the resulting fluidic dynamical components. One should therefore be aware that the analogy between fluidic dynamical components and electrical dynamical components is not perfect.

8.5.1 Fluidic-signal holors

In a similar way to mechanical homogeneous continuous dynamical systems, fluid motion can be physically and mathematically modelled in terms of elementary source, storage and dissipative mechanisms. However, fluids are generally less dense and more compliant than solids, so the distinction between the massive and compliant parts of fluidic homogeneous continuous dynamical systems cannot generally be made.

Nevertheless, a wide variety of physical homogeneous continuous dynamical systems which use fluids as a working medium are susceptible to the lumped parameter treatment given here.

The kind of fluidic homogeneous continuous dynamical systems dealt with here are the most simple to analyse, consisting of **one-dimensional** (1D) flow in pipe-work, with the energy handling phenomena isolated as specific points.

In contemplating the energy transport mechanism in fluidic homogeneous continuous dynamical systems, one is led to conclude that the fluid-pressure holor, $\boldsymbol{P}_i$, and fluid-flow-rate holor, $\boldsymbol{Q}^j$, form a valid pair of holors with which to associate the generalised energy holor pair:

- the fluid-pressure holor, $\boldsymbol{P}_i$, is analogous to the energy-potential-difference holor, $\boldsymbol{F}_i$;
- the fluid-flow-rate holor, $\boldsymbol{Q}^j$, is analogous to the energy-transfer holor, $\boldsymbol{V}^j$.

Consider a closed pipe in which the fluid pressure is P at a certain cross-section. Pressure is defined as the fluid force per unit area; thus if the cross-sectional area of the pipe is S, the total fluid pressure F is

$$F = \int_S P \, dS. \tag{8.85}$$

If the pressure is constant over the area, the total force is the product of the fluid pressure and the cross-sectional area. The energy associated with fluid flow can now be determined using the relation (cf the potential energy $\mathcal{U}$ of a translational or rotational spring):

$$\mathcal{U} = \int_0^{x_1} F \, dx = \int_0^{x_1} PS \, dx, \tag{8.86}$$

where x is the displacement of fluid along the axis of flow past the reference cross-section, S.

By recalling that the incremental fluid-flow-rate holor, $\delta\boldsymbol{Q}^j$, past the reference cross-section, S, is

$$\delta\boldsymbol{Q}^j = \omega(S\delta\boldsymbol{X}^j), \tag{8.87}$$

the work done holor, $\boldsymbol{\mathcal{U}}^\kappa$, can be written

$$\boldsymbol{\mathcal{U}}^\kappa = \frac{1}{\omega}\boldsymbol{P}^\kappa = \frac{1}{\omega}\left(\boldsymbol{P}_i \otimes \boldsymbol{Q}^j\right). \tag{8.88}$$

This expression means that, in addition to being intuitively valid, the fluid-pressure holor, $\boldsymbol{P}_i$, and the fluid-flow-rate holor, $\boldsymbol{Q}^j$, are true energy holors with a univalent tensor product that equals the univalent fluid-power holor, $\boldsymbol{P}^\kappa$, and the time integral, which is the fluidic energy holor, $\boldsymbol{\mathcal{U}}^\kappa$. There are several fluid dynamical components that store flow energy, with the two most important being the fluid reservoir and the pressurised tank. The quantity that is conserved in the flow of fluid at a junction is the mass-flow holor, $\boldsymbol{M}_i$ (in kg s^{-1}). Strictly speaking, this ought to be the energy-transfer holor. However, it is more convenient for measurement purposes to use the volume-flow holor, $\boldsymbol{Q}^j$ (in m^3 s^{-1}), as the energy-transfer holor. The relation between the mass-flow holor and the volume-flow holor is

$$\boldsymbol{M}_i = \rho_{ij}\boldsymbol{Q}^j,$$

where ρ_{ij} is the fluid-density impedance holor in kg m^{-3}.

Thus, the volume-flow holor will be proportional to the mass-flow holor when the fluid-density impedance holor remains constant through the fluidic commutation matrixer. Under this condition, the volume-flow holor becomes an exact energy-transfer holor. The constant fluid-density condition is approximated very closely when one has a gas medium with low velocity or a liquid medium. One therefore restricts the fluidic dynamical components to liquid flow or incompressible gas flow when one selects the volume-flow holor as the energy-transfer holor. The choice of a suitable energy-potential-difference holor is suggested by the condition that the tensor product of energy-transfer and energy-potential-difference holors should represent the power holor. From this one determines that the units of an appropriate energy-potential holor are N m^{-3} or pascals (Pa). Thus, the pressure-difference holor, $\boldsymbol{P}_i^A - \boldsymbol{P}_i^N$, is clearly indicated as the energy-potential-difference holor in fluidic homogenous continuous dynamical systems.

One interesting result is that it is possible to define the pressure holor in a manner that is directly analogous to the definition of the voltage holor. That is, the pressure-difference holor, $\Delta\boldsymbol{P}_i$, is the work done, $\Delta\mathcal{U}$, in moving the unit volume, ΔV, between two points. This is illustrated by considering the movement of an incremental volume of fluid between two points along a pipe of uniform cross-section, S, as shown in figure 8.16.

The work-done, $\Delta\mathcal{U}$, according to this alternative definition is given by

$$\Delta p = \frac{\Delta\mathcal{U}}{\Delta V}. \tag{8.89}$$

However, the volume may be expressed as $\Delta V = S\Delta X$, and the work done is the net force on the incremental volume multiplied by the distance moved. That is, $\Delta\mathcal{U} = \Delta F\ \Delta X$. The substitution of the work done and volume into equation (8.89) yields:

$$\Delta p = \frac{\Delta\mathcal{U}}{\Delta V} = \frac{\Delta F \Delta X.}{S\Delta X}. = \frac{\Delta F}{S} \tag{8.90}$$

In the limit, this is the force per unit area, which is the normal definition of pressure. Just as there are three elemental physical situations with electrical commutation matrixers, there are three directly analogous situations in problems of the flow of fluids. However, depending on the fluid involved, the linear relationship between energy-transfer and energy-potential-difference holors does not exist over as wide a range of values, and the resulting physical and mathematical models are therefore not quite as accurate. Nevertheless, a significant amount of information can be derived from such models, offering a considerable insight into how fluidic homogeneous continuous dynamical systems behave.

Since the amount of effort involved in obtaining solutions for linear fluidic homogeneous continuous dynamical systems is significantly less than that required

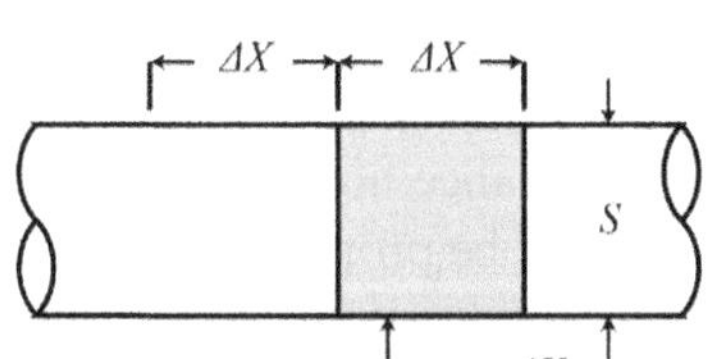

Figure 8.16. Elemental volume within a pipe.

for obtaining them for nonlinear fluidic homogeneous continuous dynamical systems, the linear physical and mathematical model is an attractive tool for determining the general nature of the performance of fluidic homogeneous continuous dynamical systems. It is, of course, necessary to have some appreciation of the conditions under which the linear physical and mathematical model ceases to be accurate, and part of the following discussion is directed towards making this assessment.

8.5.2 Fluidic-resistance holor

Since fluid flow always involves the relative motion of close fluidic homogeneous continuous dynamical systems, there are many fluid devices in which viscous effects can cause the conversion of fluidic kinetic energy to thermal energy and hence dissipate power.

There are, however, two basic mechanisms that fluid dissipaters employ:

- viscous-force holors between the fluid and the retaining pipework;
- viscous-force holors between fluid particles.

The fluid dissipaters that occur because of the former effect are as follows.

(i) Flow through a porous medium: D'Arcy's law for incompressible flow through a porous medium inserted in a fluid line is linear. It relates the relative-pressure holor, $\boldsymbol{P}_i^{AN}$, across the plug to the effective fluid-flow-rate, $\boldsymbol{Q}^j$, through it as

$$\boldsymbol{P}_i^{AN} = \boldsymbol{R}_{ij}^f \boldsymbol{Q}^j, \tag{8.91}$$

where $\boldsymbol{R}_{ij}^f$ is the fluidic-resistance impedance holor.

(ii) Laminar flow through a capillary pipe. For small values of Reynolds numbers, the Hagen–Poiseuille law predicts that power dissipation follows a linear constitutive relation with the fluidic resistance given by

$$R^f = \frac{128\mu l}{\pi(d\text{^}4)}, \tag{8.92}$$

where μ is the fluid viscosity, l is the capillary pipe length and d is the capillary pipe diameter.

The fluid dissipaters that rely upon viscous-force holors between fluid particles are as follows.

(i) Turbulent flow through a long pipe. For incompressible flow at Reynolds numbers in excess of ~2,000, the flow is normally turbulent. In this situation fluid power is dissipated and the dissipater constitutive relation is approximately

$$P^{AN} = aQ\,|Q|^{3/4}, \tag{8.93}$$

where a is a constant dependent upon the fluid properties and pipe geometry.

(ii) Incompressible flow through an orifice. A sudden restriction or change in cross-sectional area over a short length of a pipe constitutes an orifice. Fluid is rapidly accelerated through the orifice with resultant turbulent flow downstream of the restriction. An approximate expression for the constitutive relation of the orifice dissipater is

$$P^{AN} = \frac{\rho}{2(C_d\hat{}2)(S_0\hat{}2)} Q|Q|, \tag{8.94}$$

where ρ is the fluid density, C_d is the orifice coefficient of the discharge, S_0 is the cross-sectional area of the orifice and P^{AN} is the pressure difference measured upstream and downstream of the dissipater.

Formally, the constitutive relation is given by

$$\boldsymbol{P}_i^{AN} = \mathcal{F}(\boldsymbol{Q}^j). \tag{8.95}$$

The power holor, $\boldsymbol{P}^{\kappa}$, dissipated by a fluidic resistance holor, $\boldsymbol{R}_{ij}^{f}$, is

$$\boldsymbol{P}^{\kappa} = \boldsymbol{P}_i^{AN} \otimes \boldsymbol{Q}^j, \tag{8.96a}$$

or only the dissipated power

$$P^p = P^{AN}Q = G + J, \tag{8.96b}$$

where the dissipater content, G, and the co-content, J, are defined in the normal way:

$$G = \int_0^Q P^{AN}\, dQ, \qquad J = \int_0^{P^{AN}} Q\, dP^{AN}. \tag{8.97}$$

For linear dissipaters, the power dissipated is given by

$$P^p = P^{AN}Q = R^f(Q\hat{}2). \tag{8.98}$$

The fluidic resistance physically models the frictional effects that impede the flow of fluid along a pipe. These exist at the interface between the fluid and the pipe, and also between layers of the fluid, which are not moving at the same velocity.

In certain circumstances, the flow of all the particles of fluid is in the sense of direction of the tube, such as the one indicated in figure 8.17, with this form of flow being termed 'laminar'.

The profiles of the fluid velocity holor are shown at three locations in the pipe. At the pipe entrance, the velocity holor profile is uniform. The viscous properties of the fluid cause shear force holors, which act to retard the velocity holor at the boundaries and increase it along the centre-line.

At some distance down the pipe, a stabilised parabolic shape profile is reached. This distance is known as the entrance length, l_e, and it usually varies from about 80 to 120 diameters. Thereafter, the velocity holor profile remains unchanged.

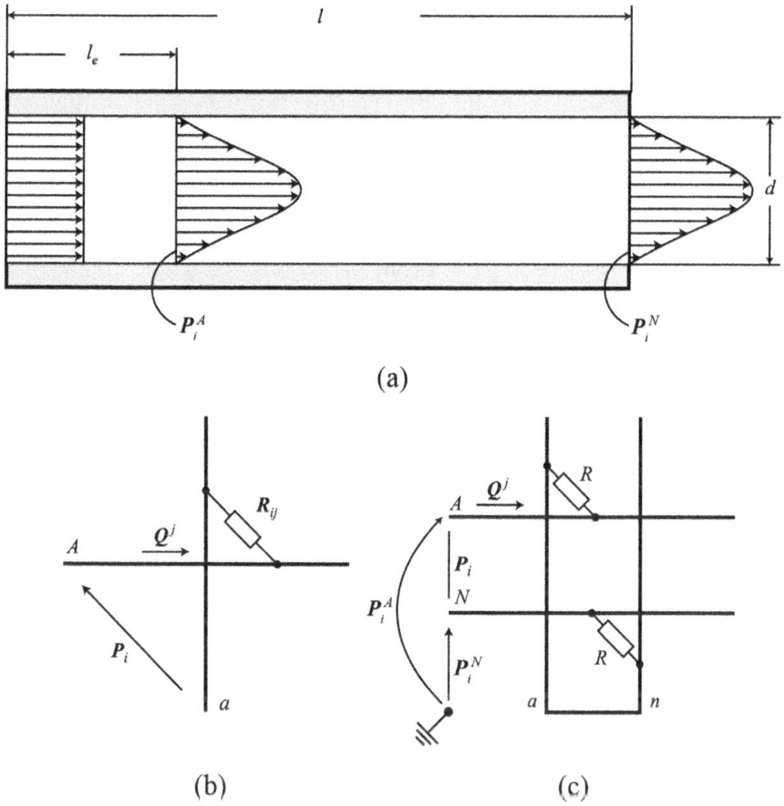

Figure 8.17. Pipe flow with viscous friction and its fluidic commutation matrixer.

In the region where the profile does not change, the steady state momentum equation of the fluid mechanics yields:

$$P_i^A - P_i^N = R_{ij}Q^j = \begin{Vmatrix} P_c^A \\ P_s^A \end{Vmatrix} - \begin{Vmatrix} P_c^N \\ P_s^N \end{Vmatrix} = \begin{Vmatrix} R & 0 \\ 0 & R \end{Vmatrix} \begin{Vmatrix} Q_c^j \\ Q_s^j \end{Vmatrix} \tag{8.99}$$

$$R = \frac{128\mu(l - l_e)}{\pi(d\hat{}4)}, \tag{8.100}$$

where μ is the dynamical viscosity of the fluid (Pa s) and d is the diameter of the pipe (m).

Thus, the laminar flow alone, the pressure-drop holor along a pipe, is proportional to the flow-rate holor and the fluidic resistance is defined in exactly the same manner as the electrical resistance.

The linear portion of the pipe resistance only occurs after the entrance length. Since there is no way to eliminate the nonlinear entrance effects, the usual procedure is to select pipe lengths, l, that are at least four times the entrance length. In this way, the entrance length effects are minimised.

In fact, the fluidic resistance is usually approximated by replacing the distance, $l-l_e$, in equation (8.100) with the length, l, alone. Thus, as a good approximation, when l is much larger than l_e,

$$\boldsymbol{P}_i^A - \boldsymbol{P}_i^N = \boldsymbol{R}_{ij}\boldsymbol{Q}^j = \begin{Vmatrix} P_c^A \\ P_s^A \end{Vmatrix} - \begin{Vmatrix} P_c^N \\ P_s^N \end{Vmatrix} = \begin{Vmatrix} R & 0 \\ 0 & R \end{Vmatrix} \begin{Vmatrix} Q^c \\ Q^s \end{Vmatrix} \tag{8.101}$$

$$R = \frac{128\mu l}{\pi(d\hat{\ }4)}, \tag{8.102}$$

where $\boldsymbol{P}_i^A$ is now the pressure holor at the entrance to the pipe.

One should emphasise that equation (8.101) only holds for laminar flow. When the flow-rate holor increases beyond some critical value the flow breaks up and contains circulating currents termed 'vortices'. This type of flow is termed 'turbulent' and results in an increased rate of fluidic energy dissipation within the fluid. The critical point between laminar and turbulent flow is not precisely known, but may be estimated by means of the Reynolds number ($\mathfrak{R}$), which is a non-dimensional expression involving several properties of the fluid and the geometry of the pipe,

$$\mathfrak{R} = \frac{4\rho q}{\pi\mu d}, \tag{8.103}$$

where q is the velocity of the fluid in m s^{-1}, ρ is the density of the fluid in kg m^{-3}, μ is the dynamical viscosity of the fluid in Pa s and d is the diameter of the pipe in m.

As a rule, if $\mathfrak{R} \leqslant 2 \times 10^3$, the flow is laminar, but it becomes fully turbulent when $\mathfrak{R} = 5 \times 10^3$. No special name has been given to the unit of the fluidic-resistance holor and probably the most convenient unit is N s m^5.

The laminar-flow fluidic resistance given in equation (8.102) is applicable for both liquid and gas flows. A fluidic resistance holor is bilateral if a flow in either sense of direction develops the same magnitude of pressure-difference holor between the end terminals.

In fluidic homogeneous continuous dynamical systems, one may assume at once that the resistance holors are bilateral if only one fluid medium is present. However, in cases where liquid (e.g. water) pipes empty into gas (e.g. air), the resistance holor is not bilateral. For instance, when the resistance is connected between the **mechano-fluidic** (MF) pump and the **fluido-fluidic** (FF) tank, it is bilateral. There is only one medium, the liquid (e.g. water). On the other hand, when the resistance is connected between the liquid MF pump and the gas (e.g. air), the resistance is not bilateral. A reversal of the MF pump will draw in gas (e.g. air).

One may still apply this fluidic commutation matrixer as a physical model, however, if one knows in advance that the flow always remains in the proper unilateral sense of direction from the MF pump to the atmosphere.

The power holor dissipated by a fluidic-resistance holor is equal to the tensor product of the pressure-drop holor and the flow-rate holor associated with the fluidic component, namely:

$$\boldsymbol{P}^{\kappa} = \left(\boldsymbol{P}_i^A - \boldsymbol{P}_i^N\right) \otimes \boldsymbol{Q}^j = \gamma_{..j}^{\kappa i}\left(\boldsymbol{P}_i^A - \boldsymbol{P}_i^N\right)\boldsymbol{Q}^j = \gamma_{..j}^{\kappa i}\boldsymbol{R}_{ij}\boldsymbol{Q}^k\boldsymbol{Q}^j = (P^p, P^q, P^r).$$

However, a pressure-drop holor

$$\boldsymbol{P}_i^A - \boldsymbol{P}_i^N = \boldsymbol{R}_{ij}\boldsymbol{Q}^k.$$

This may also be simpler expressed as

$$P^p = \mathrm{R}\left(q^{\wedge 2}\right),$$

or

$$P^p = \left[(P^A - P^N)^{\wedge}2\right]/\mathrm{R}.$$

8.5.3 Fluidic-inertance holor

The fluid mechanism for storing the energy-potential-difference holor, $\boldsymbol{F}_i$, is the kinetic energy associated with a moving body of fluid. Consider an incompressible fluid flowing without dissipation in a tube and suppose that a plug of fluid of length l can be considered as a rigid mass with a uniform velocity holor, $\boldsymbol{V}^j$.

$$\boldsymbol{F}_i = S\left(\boldsymbol{P}_i^A - \boldsymbol{P}_i^N\right) = Sl\omega\rho_{ij}\boldsymbol{V}^j, \tag{8.104}$$

where $\boldsymbol{P}_i^A - \boldsymbol{P}_i^N$ is the pressure-difference holor across the fluid plug, ρ_{ij} is the fluid-density holor and S is the cross-sectional area of the tube.

The constitutive relation of the fluidic energy-potential-difference holor store can be obtained by manipulating equation (8.104):

$$\boldsymbol{\Gamma}_i^{AN} = \frac{l}{S}\rho_{ij}\boldsymbol{Q}^j, \tag{8.105}$$

where $\boldsymbol{\Gamma}_i^{AN}$ is defined as the fluid momentum holor and is related to the pressure-difference holor, $\boldsymbol{P}_i^{AN}$, across the plug by

$$\boldsymbol{P}_i^{AN} = \boldsymbol{P}_i^A - \boldsymbol{P}_i^N = \omega\boldsymbol{\Gamma}_i^{AN}. \tag{8.106}$$

The material properties of a fluidic energy-potential-difference holor store are defined by its fluidic inertance, L_f, defined from equation (8.105) as

$$L_f = \rho l/S. \tag{8.107}$$

The kinetic energy, $\mathcal{T}$, stored in a fluidic energy-potential-difference holor store is given by

$$\mathcal{T} = \int_0^{\Gamma^{AN}} Q\,\mathrm{d}\Gamma^{AN} = \int_0^{\Gamma^{AN}} \mathcal{F}^{-1}(\Gamma^{AN})\mathrm{d}\Gamma^{AN}. \tag{8.108}$$

The stored kinetic co-energy, $\mathcal{T}*$, is defined by the complementary integral

$$\mathcal{T}* = \int_0^{Q} \Gamma^{AN}\,\mathrm{d}Q = \int_0^{Q} \mathcal{F}(Q)d. \tag{8.109}$$

In the linear case these energy functions are equal and given by

$$\mathcal{T} = \mathcal{T}* = \frac{1}{2}L_f(Q\wedge 2). \tag{8.110}$$

Another physical situation in fluidic homogeneous continuous dynamical systems relates to the inertia effect in the fluid. This is of particular importance with liquids, although it is not necessarily negligible in pneumatic continuous dynamical systems.

Consider a tube such as that indicated in figure 8.18, in which the fluid volume flow-rate holor is $\boldsymbol{Q}^j$ ($\mathrm{m}^3\ \mathrm{s}^{-1}$).

If one restricts one's attention to a section of length l and uniform cross-sectional area S, the mass holor of the fluid is $Sl\rho_{ij}$.

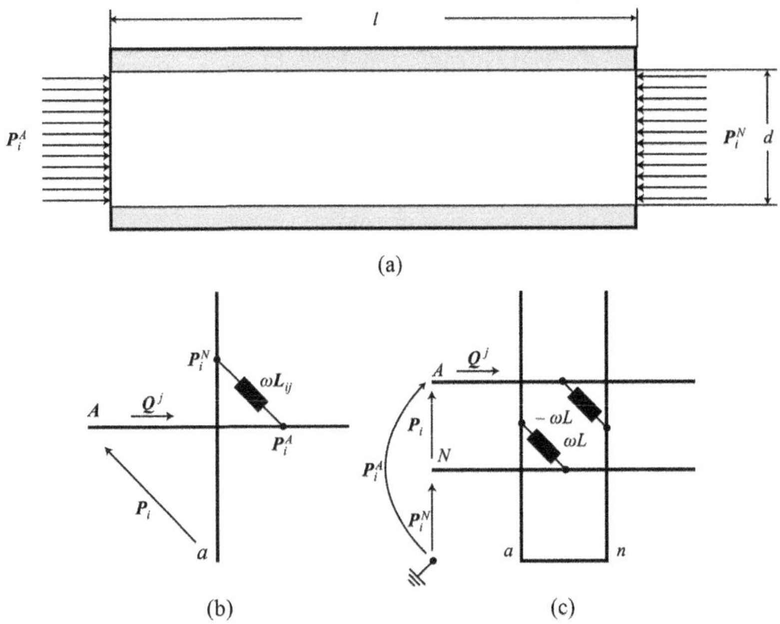

Figure 8.18. Physical model of the fluidic inertance.

The velocity holor of the fluid is $\boldsymbol{Q}^j$

$$\boldsymbol{P}_i^A - \boldsymbol{P}_i^N = \omega \boldsymbol{L}_{ij}\boldsymbol{Q}^j = \begin{Vmatrix} P_c^A \\ P_S^A \end{Vmatrix} - \begin{Vmatrix} P_c^N \\ P_S^N \end{Vmatrix} = \begin{Vmatrix} 0 & -\omega L \\ \omega L & 0 \end{Vmatrix} \begin{Vmatrix} Q^c \\ Q^s \end{Vmatrix}. \tag{8.111}$$

Equation (8.111) has the same form as the general relation given in equation (8.1b). The pressure-difference holor, $\boldsymbol{P}_i^A - \boldsymbol{P}_i^N$, like the energy-potential-difference holor, is directly analogous to the voltage-source holor.

One therefore notes that this equation has the same form as the elemental equation for inductance (equation (8.111)). This effect can therefore be termed 'fluidic inductance', although usually the term 'inertance' is used, since it is the inertia effect that is being physically and mathematically modelled. The fluidic inertance of a pipe is computed from

$$L = \frac{\rho l}{S}. \tag{8.112}$$

The fluidic inertance unit is N s^2 m^{-5}.

The fluidic energy stored by a fluid inertance is dependent upon the flow-rate holor (energy-transfer holor) and inertance is in the same form as given for the electrical inductance holor in equation (8.24). Fluidic inertance has been derived under the assumption of the flow of a frictionless fluid in a pipe. Previously, fluidic resistance was also derived in a pipe for a fluid with friction but no inertia. Both physical dynamical components share the same principal geometry: the pipe. For holor analysis purposes, it is accepted that the friction and inertia effects in the pipe are separate. In any practical case, however, both effects exist simultaneously. Thus, in a real pipe there is both fluidic resistance and fluidic inertance. There is only one flow holor, $\boldsymbol{Q}^j$, passing through the pipe. The complete physical model for a pipe is therefore:

$$\boldsymbol{P}_i^A - \boldsymbol{P}_i^N = \boldsymbol{Z}_{ij}\boldsymbol{Q}^j = \begin{Vmatrix} P_c^A \\ P_S^A \end{Vmatrix} - \begin{Vmatrix} P_c^N \\ P_S^N \end{Vmatrix} = \begin{Vmatrix} R & -\omega L \\ \omega L & R \end{Vmatrix} \begin{Vmatrix} Q^c \\ Q^s \end{Vmatrix} \tag{8.113}$$

The pressure-difference holor, $\boldsymbol{P}_i^A - \boldsymbol{P}_i^N$, is the sum of the pressure-difference holor caused by fluidic resistance and fluidic inertance. This physical model as a fluidic commutation matrixer is shown in figure 8.19.

One uses the same symbols for the fluidic dynamical components that one used previously for the electrical dynamical components. The physical model of a fluid line or fluid pipe consists of a resistance and inertance in series.

The order of the fluidic dynamical components does not matter for the fluid line because there is no definable pressure terminal between the resistance and inertance. This occurs because the effects are not separable in a physical sense, although they were separated analytically. Sometimes one of the effects is much smaller than the other and can be neglected. We indicate in each problem the appropriate physical model for each line fluidic component. However, in general, both effects must be considered.

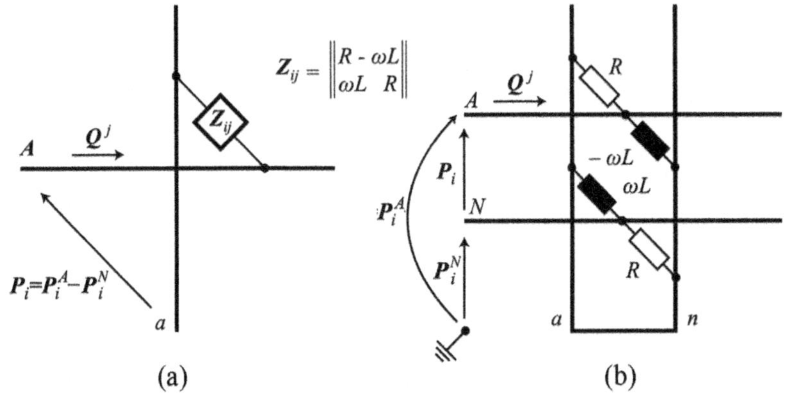

Figure 8.19. Fluidic commutation matrixer as a fluid-line physical model.

8.5.4 Fluidic-capacitance holor

The momentum equation of fluid mechanics was used to define fluidic resistance and inertance. Fluidic capacitance is due to an effect one may observe in the continuity equation of fluid mechanics.

A simple way to demonstrate the formulation of the signal variables in the form of equation (8.1c) is to consider the closed chamber (figure 8.20(a)).

The mass holor, $\boldsymbol{M}_i$, of the fluid in the chamber is equal to the product of the density holor, $\boldsymbol{\rho}_{ij}$, and the volume holor, $\boldsymbol{V}^j$ ($\boldsymbol{M}_i = \boldsymbol{\rho}_{ij}\boldsymbol{V}^j$).

If one takes the time derivative of the mass holor, $\boldsymbol{M}_i$, one obtains an expression for the mass flow, m. In terms of the volume flow, the result is:

$$q = \frac{\mathrm{d}V}{\mathrm{d}t} + \frac{V}{\rho}\frac{\mathrm{d}\rho}{\mathrm{d}t}. \tag{8.114}$$

Equation (8.114) is a form of the continuity equation of fluid mechanics. When the volume of the chamber is fixed, the first term on the right-hand side is equal to zero.

The resulting equation relates the volume-flow holor to the fluid-density holor. However, the principal signal holors are the volume-flow holor and the pressure-difference holor. To find a relation between the signal variables one must specify the connection between pressure and density. This connection depends on the fluid.

For a liquid one may apply Hooke's law to obtain:

$$\frac{1}{\rho}\mathrm{d}\rho = \frac{1}{\beta}\mathrm{d}p, \tag{8.115}$$

where β is the bulk modulus of the liquid in Pa.

When the fluid is a gas, the relation between pressure and density depends on the thermodynamic process. The appropriate results are:

$$\frac{1}{\rho}\mathrm{d}\rho = \frac{1}{p + p_a}\mathrm{d}p \quad \text{(isothermal process)} \tag{8.116a}$$

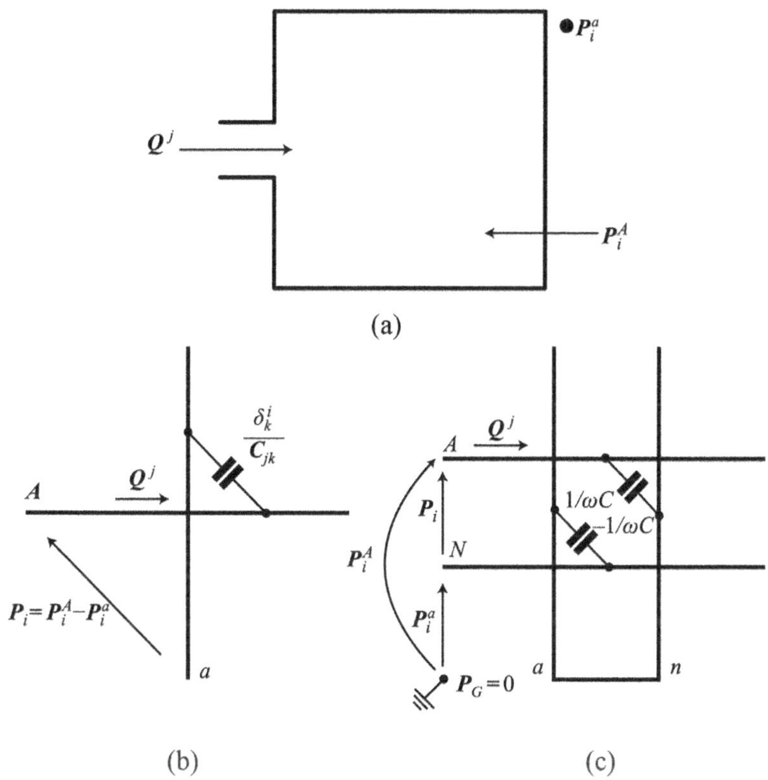

Figure 8.20. Closed-tank fluidic capacitance and its fluidic commutation matrixer.

$$\frac{1}{\rho}\mathrm{d}\rho = \frac{1}{\gamma(p + p_a)}\mathrm{d}p \quad \text{(adiabatic process),} \tag{8.116b}$$

where p_a is the atmospheric pressure and γ is the ratio of the specific heats.

One emphasises the atmospheric pressure here because the absolute pressure is required in equation (8.114) and p is usually the gauge pressure. If equations (8.113) and (8.114) are substituted into equation (8.112) and integration is performed

$$p = \frac{\beta}{V}\int q\,\mathrm{d}t + p(0) \quad \text{(liquid)} \tag{8.117a}$$

$$p = \frac{p + p_a}{V}\int q\,\mathrm{d}t + p(0) \quad \text{(gas, isothermal)} \tag{8.117b}$$

$$p = \frac{\gamma(p + p_a)}{V}\int q\,\mathrm{d}t + p(0) \quad \text{(gas, adiabatic),} \tag{8.117c}$$

where $p(0)$ is the value of the pressure in the chamber at $t = 0$.

If one assumes that $p(0) = 0$, equation (8.112) almost has the form desired in equation (8.1c). The difference between equations (8.114) and (8.117c) is that there is only one pressure in equation (8.116) and a pressure-difference holor is required to fit the standard form. This is a consequence of the fact that the closed tank has only one terminal, which is the internal pressure holor, $\boldsymbol{P}_i$. In effect, one is trying to make a two-terminal fluidic-component physical model out of a physical dynamical component that has only one terminal.

One accomplishes this by assuming a fictitious terminal, $\boldsymbol{P}_i^G$, at zero potential and expressing equation (8.116) as:

$$\boldsymbol{P}_i^A - \boldsymbol{P}_i^G = \frac{\delta_k^i}{\omega \boldsymbol{C}_{jk}} \boldsymbol{Q}^j = \begin{Vmatrix} P_c^A \\ P_s^A \end{Vmatrix} - \begin{Vmatrix} P_c^N \\ P_s^N \end{Vmatrix} = \begin{Vmatrix} 0 & 1/\omega C \\ -1/\omega C & 0 \end{Vmatrix} \begin{Vmatrix} Q^c \\ Q^s \end{Vmatrix}, \tag{8.118}$$

where $C = \frac{V}{\beta}$ for a liquid, $C = \frac{V}{p + p_a}$ for an isothermic gas and $C = \frac{V}{\gamma(p + p_a)}$ for an adiabatic gas.

The unit of fluidic capacitance is $\mathrm{m}^5\ \mathrm{N}^{-1}$.

Thus, the fluidic capacitance for a gas is a function of pressure, p. When the fluidic homogeneous continuous dynamical system operates with small pressure changes, the quantity $p + p_a$ does not change appreciably and the fluidic capacitance is essentially constant. When there are large pressure transients, the fluidic capacitance value is variable. In general, a fluidic component that varies as a function of one of the signal holors is termed a 'nonlinear fluidic component'. Such fluidic dynamical components are beyond the scope of this textbook. For our purposes, all problems relating to gas operation will be restricted to small pressure transients. Then one can treat the fluidic capacitance as a constant.

The fluidic energy stored by a fluidic capacitance is dependent on the pressure-difference holor (energy-potential-difference holor) and has the same form as that given for the electrical capacitance in equation (8.35). The fictitious terminal adopted to complete the two-terminal form is at zero potential. Thus, a closed tank always represents a fluidic capacitance ground. However, since the ground terminal is fictitious, it cannot be located physically. One denotes a fictitious ground terminal by making the fluidic-commutation-matrixer line going to the ground 'dashed'.

A gas-filled tank is bilateral up to a point. It ceases to be bilateral when all the gas has been removed. Figure 8.21(a) shows a fixed volume tank with inlet and outlet flow. The previous reasoning is still applicable. There is only one pressure terminal in the tank and its mathematical model (equation (8.1)) requires two pressure terminals for each fluidic component.

Figure 8.21(b) shows the fluidic commutation matrixer representation for a closed tank capacitor. Note that the liquid filled closed tank is not bilateral. If liquid is removed from the tank, cavitation occurs and one is no longer dealing with a homogeneous medium.

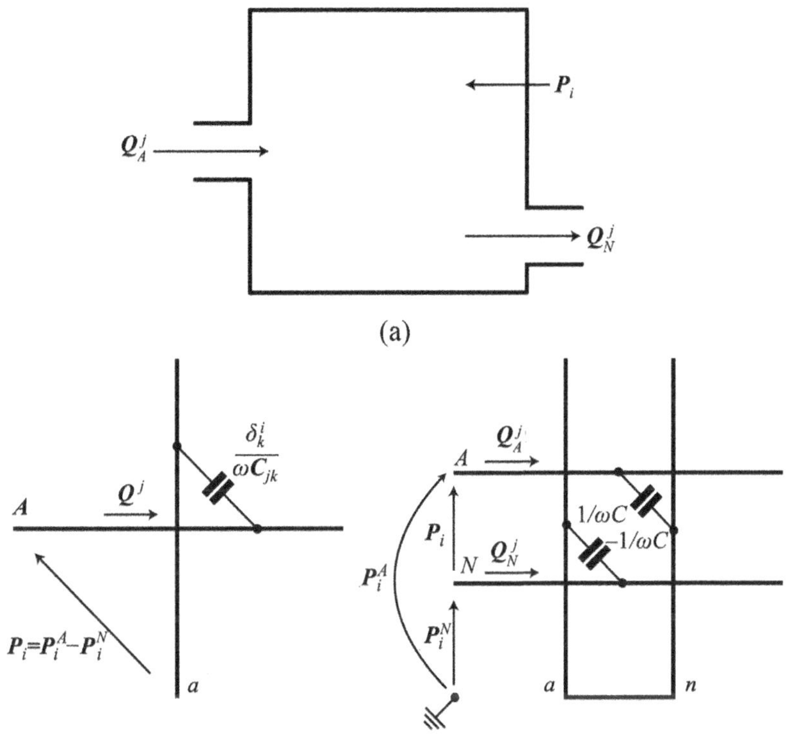

Figure 8.21. Closed-tank fluidic capacitance with inlet and outlet.

Holor analysis of the tank with the inlet- and outlet-flow holors, Q_A^j and Q_N^j, leads to:

$$\begin{aligned} P_i^A - P_i^N &= \frac{\delta_k^i}{\omega C_{jk}}\left(Q_A^j - Q_N^j\right) = \begin{Vmatrix} P_c^A \\ P_s^A \end{Vmatrix} - \begin{Vmatrix} P_c^N \\ P_s^N \end{Vmatrix} \\ &= \begin{Vmatrix} 0 & 1/\omega C \\ -1/\omega C & 0 \end{Vmatrix} \left(\begin{Vmatrix} Q_A^c \\ Q_A^s \end{Vmatrix} - \begin{Vmatrix} Q_N^c \\ Q_N^s \end{Vmatrix} \right) \end{aligned} \tag{8.119}$$

The fluidic commutation matrixer equivalent for this tank is shown in figure 8.21(b). The tank again represents a fluidic capacitance to ground. An open tank or reservoir that is fed from the bottom with working fluid is a store of fluid flow energy. As fluid flows into the reservoir, the potential energy of the fluid mass is increased by the rising fluid level in the tank. The increase in energy manifests itself in an increase in the relative fluid-pressure holor, P_i, measured at an arbitrary fixed point in the tank.

The principle of conservation of material can be applied to obtain

$$\rho_{ij} Q^j = \omega\left(S\rho_{ij} H^j\right), \tag{8.120}$$

where ρ_{ij} is the mass-density impedance holor of the fluid, S is the cross-sectional area of the reservoir and $\boldsymbol{H}^j$ is the fluid-height holor in the reservoir.

This expression can be rearranged in the form:

$$\boldsymbol{V}^j = \boldsymbol{C}^{ji}\boldsymbol{P}_i, \tag{8.121}$$

which is the constitutive relation for the reservoir and relates the fluid-pressure holor, $\boldsymbol{P}_i$, to the fluid-volume holor, $\boldsymbol{V}^j$, in the reservoir.

The fluid-volume holor, $\boldsymbol{V}^j$, is related to the fluid-flow-rate holor, $\boldsymbol{Q}^j$, by the dynamical relation

$$\boldsymbol{Q}^j = \omega \boldsymbol{V}^j = \omega \boldsymbol{C}^{ji}\boldsymbol{P}_i. \tag{8.122}$$

Note that the constitutive relation given by equation (8.121) is linear; this is because incompressible flow and vertical tank sides have been assumed.

The coefficient of a flow store is termed the fluidic capacitance, C; for the open reservoir this is given by

$$C = \frac{S}{\rho g}. \tag{8.123}$$

Consider now the open tank in which a liquid is stored, as indicated in figure 8.22. The previous reasoning is still applicable.

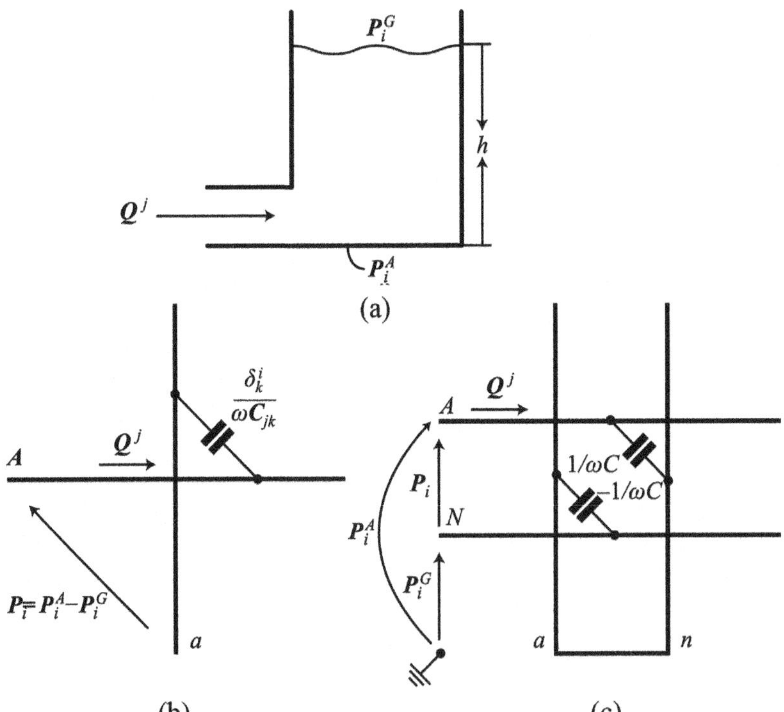

Figure 8.22. Liquid open-tank fluidic capacitance with its fluidic commutation matrixer.

There is only one pressure terminal in the tank and its mathematical model (equation (8.1)) requires two pressure terminals for each fluidic dynamical component.

Holor analysis of the tank with the inlet- and outlet-flow holors, $\boldsymbol{Q}_A^j$ and $\boldsymbol{Q}_N^j$, leads to:

$$\begin{aligned} \boldsymbol{P}_i^A - \boldsymbol{P}_i^G &= \frac{\delta_k^i}{\omega \boldsymbol{C}_{jk}}\left(\boldsymbol{Q}_A^j - \boldsymbol{Q}_N^j\right) = \begin{Vmatrix} P_c^A \\ P_s^A \end{Vmatrix} - \begin{Vmatrix} P_c^N \\ P_s^N \end{Vmatrix} \\ &= \begin{Vmatrix} 0 & 1/\omega C \\ -1/\omega C & 0 \end{Vmatrix} \left(\begin{Vmatrix} Q_A^c \\ Q_A^s \end{Vmatrix} - \begin{Vmatrix} Q_N^c \\ Q_N^s \end{Vmatrix} \right). \end{aligned} \tag{8.124}$$

In this case, the pressure holor, $\boldsymbol{P}_i^A$, is indicated at the bottom of the tank. The pressure is the gravitational force per unit area of the column of liquid above it. If the horizontal area of the tank is S, then the mass holor, $\boldsymbol{M}_i$, of the column of height holor, $\boldsymbol{H}^j$, is

$$\boldsymbol{M}_i = \frac{S}{g}\left(\boldsymbol{P}_i^A - \boldsymbol{P}_i^G\right) = S\rho_{ij}\boldsymbol{H}^j,$$

from which

$$\boldsymbol{P}_i^A - \boldsymbol{P}_i^G = g\rho_{ij}\boldsymbol{H}^j, \tag{8.125}$$

where $g = 9.81 \text{ m s}^{-2}$ is the gravitational constant and $\boldsymbol{P}_i^G$ is the pressure holor above the liquid column.

Before proceeding to identify the fluidic capacitance effect, it may be noted that equation (8.125) provides the justification for using the height of the column of liquid as an alternative measure of pressure, commonly termed 'head'. However, from the discussion at the beginning of this subsection, it should be evident that the product of the head and flow rate cannot possibly result in power as expressed in any acceptable unit.

The resulting confusion with expressions involving fluidic resistance, capacitance and inductance is such that head is not used as an energy-potential-difference holor in the fluidic commutation matrixers analysed in this textbook.

Now, from equation (8.114), when the density does not change but the volume does, $q = \mathrm{d}V/\mathrm{d}t$. In addition, since $V = Sh$, one may express equation (8.125) in terms of the pressure-difference holor and the flow-rate holor as:

$$\boldsymbol{P}_i^A - \boldsymbol{P}_i^G = \frac{\delta_k^i}{\omega \boldsymbol{C}_{jk}}\boldsymbol{Q}^j = \begin{Vmatrix} P_c^A \\ P_s^A \end{Vmatrix} - \begin{Vmatrix} P_c^N \\ P_s^N \end{Vmatrix} = \begin{Vmatrix} 0 & 1/\omega C \\ -1/\omega C & 0 \end{Vmatrix} \begin{Vmatrix} Q^c \\ Q^s \end{Vmatrix} \tag{8.126}$$

Comparison with the standard form in equation (8.1c) shows that the fluidic capacitance of the open tank is

$$C = \frac{S}{\rho g}. \tag{8.127}$$

This fluidic dynamical component is essentially a two-terminal dynamical component without the assumption of a fictitious terminal. One terminal exists at the bottom of the tank and the other is the pressure over the liquid column. Usually, this pressure will be zero gauge pressure and the open tank like the closed tank will represent a fluidic capacitance to ground (figure 8.22(b)). However, this is not necessarily the case. A second fluid flow store is the pressurized tank; these devices store fluid flow energy by compressing the fluid in a sealed chamber. Suppose the tank is filled with fluid of uniform density holor, $\boldsymbol{\rho}_{ij}$, with pressure holor, $\boldsymbol{P}_i$, and absolute temperature, T. The principle of conservation of material may be applied to obtain

$$\boldsymbol{\rho}_{ij}\boldsymbol{Q}^j = \omega\boldsymbol{\rho}_{ij}\boldsymbol{V}_T^j, \tag{8.128}$$

where $\boldsymbol{Q}^j$ is the fluid-flow-rate holor into the tank and $\boldsymbol{V}_T^j$, is the volume holor of the tank, which is assumed constant, i.e. the rigid tank.

The constitutive relation of the pressurised tank now depends upon the nature of the working fluid. More specifically, it depends upon how the mass density holor, $\boldsymbol{\rho}_{ij}$, of the fluid is related to the absolute fluid-pressure holor, $\boldsymbol{P}_i$.

For liquids, a good approximation to this dependence is given by

$$\mathrm{d}\boldsymbol{\rho}_{ij} = \frac{1}{\beta}\boldsymbol{\rho}_{ij}\,\mathrm{d}\boldsymbol{P}_i, \tag{8.129}$$

where β is the fluid bulk modulus of elasticity and is a measure of the 'stiffness' of the fluid.

For gases used in pneumatic homogeneous continuous dynamical systems, the situation is more complex at a given temperature and the pressure holor, $\boldsymbol{P}_i$, is dependent upon the precise specification of the heat transfer that accompanies the detrain upon the gas. Only two specifications are normally considered—isothermal heat transfer and adiabatic heat transfer. These determine two distinct thermodynamic constraints, respectively, the constant temperature processes and the zero heat-exchange processes.

Under these conditions, two distinct bulk modules arise—the isothermal elasticity, k^I, and the adiabatic elasticity, k^A.

For an ideal gas there are:

$$\begin{aligned} &\text{bulk modulus at constant temperature} \Rrightarrow \text{isothermal elasticity, } k^I = p;\\ &\text{bulk modulus at constant heat} \Rrightarrow \text{adiabatic elasticity, } k^A = \gamma p; \end{aligned} \tag{8.130}$$

where $\gamma = C_p/C_v$ is the ratio of the principal specific heats and p is the gaseous pressure.

After manipulating equations (8.128) and (8.129), the constitutive relation of a pressurised tank flow-store holor is

$$\boldsymbol{V}^\kappa = \frac{1}{\beta}\left(\boldsymbol{P}_i \otimes \boldsymbol{V}_T^j\right), \tag{8.131}$$

where β is a constant for most fluids, $\beta = k^I = p$ for isothermal gaseous processes and $\beta = k^A = \gamma p$ for adiabatical gaseous processes.

The total fluid-volume holor, $\boldsymbol{V}^j$ in the pressurised tank is determined from the fluid-flow-rate holor

$$\boldsymbol{Q}^j = \omega \boldsymbol{V}^j. \tag{8.132}$$

The pressurised tank fluid flow store is characterised by the fluidic capacitance, C_f, given by

$$C_f = \frac{V_T}{\beta}. \tag{8.133}$$

The stored flow energy, $\mathcal{U}$, is given by

$$\mathcal{U} = \int_0^V P\,dV = \int_0^V \mathcal{F}^{-1}(V)\,dV. \tag{8.134}$$

Similarly, the flow co-energy $\mathcal{U}^*$ is given by

$$\mathcal{U}^* = \int_0^P \mathcal{F}(P)\,dP \tag{8.135}$$

In the linear case, the two energy functions are equal and given by

$$\mathcal{U} = \mathcal{U}^* = \frac{1}{2} C_f\,(P\wedge 2). \tag{8.136}$$

Consider figure 8.23(a), in which a closed tank is a partially filled tank; the addition of flow, $\boldsymbol{Q}^j$, compresses the gas trapped above the liquid column. As a result, the pressure above the liquid column is no longer constant. The equivalent fluidic commutation matrixer for this circumstance is shown in figure 8.23(b). From the liquid fluidic commutation matrixer viewpoint, there are two fluidic capacitances in series. The resulting fluidic capacitance is reduced. Thus, sealing an open tank will reduce the fluidic capacitance. The liquid open tank fluidic capacitance is only bilateral if there is liquid in the tank. When there are flow reversals, the liquid level in the tank must be checked to ensure that the tank has not emptied. There is still another type of fluidic capacitance. This type features a moving part between two chambers.

Figure 8.24(a) shows a spring-loaded piston separating two chambers at pressure holors, $\boldsymbol{P}_i^A$ and $\boldsymbol{P}_i^N$.

The change in volume depends on the pressure-difference holor and the spring constant, k. If the area of the chamber is S, the volume holor change may be expressed as:

$$\Delta \boldsymbol{V}^j = \boldsymbol{C}^{ji}\left(\boldsymbol{P}_i^A - \boldsymbol{P}_i^N\right). \tag{8.137}$$

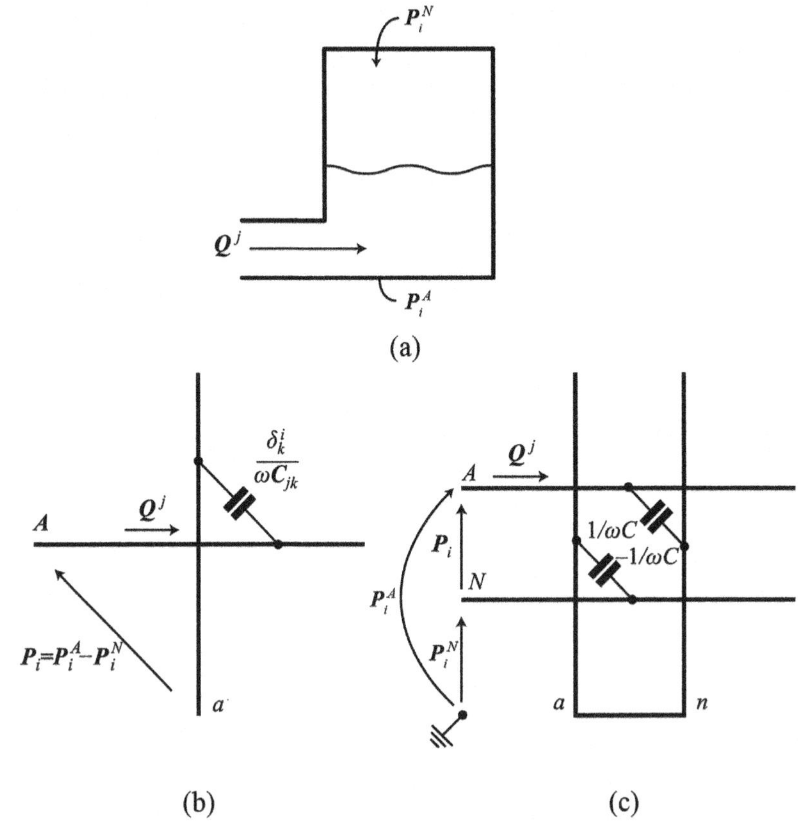

Figure 8.23. Fluidic capacitance for a liquid tank closed at the top.

Moreover, from equation (8.114), for density change, one may relate the pressure-difference holor and the volume-flow holor for this fluidic component as:

$$\boldsymbol{P}_i^A - \boldsymbol{P}_i^G = \frac{\delta_k^i}{\omega \boldsymbol{C}_{jk}} \boldsymbol{Q}^j = \begin{Vmatrix} P_c^A \\ P_s^A \end{Vmatrix} - \begin{Vmatrix} P_c^N \\ P_s^N \end{Vmatrix} = \begin{Vmatrix} 0 & 1/\omega C \\ -1/\omega C & 0 \end{Vmatrix} \begin{Vmatrix} Q^c \\ Q^s \end{Vmatrix} \tag{8.138}$$

Now, once again, comparison with the standard form (equation (8.1c)) shows that this is a two-terminal fluidic capacitance with a capacitance value of

$$C = \frac{S^2}{k}. \tag{8.139}$$

Figure 8.24(b) shows the equivalent fluidic commutation matrixer. The fluidic capacitance computed above is designated C_2. There is a compressibility capacitance associated with each chamber: C_1 and C_3. Most often, the fluidic capacitances C_1 and C_3 have small values. In that case, the spring-loaded piston represents a two-terminal fluidic capacitance.

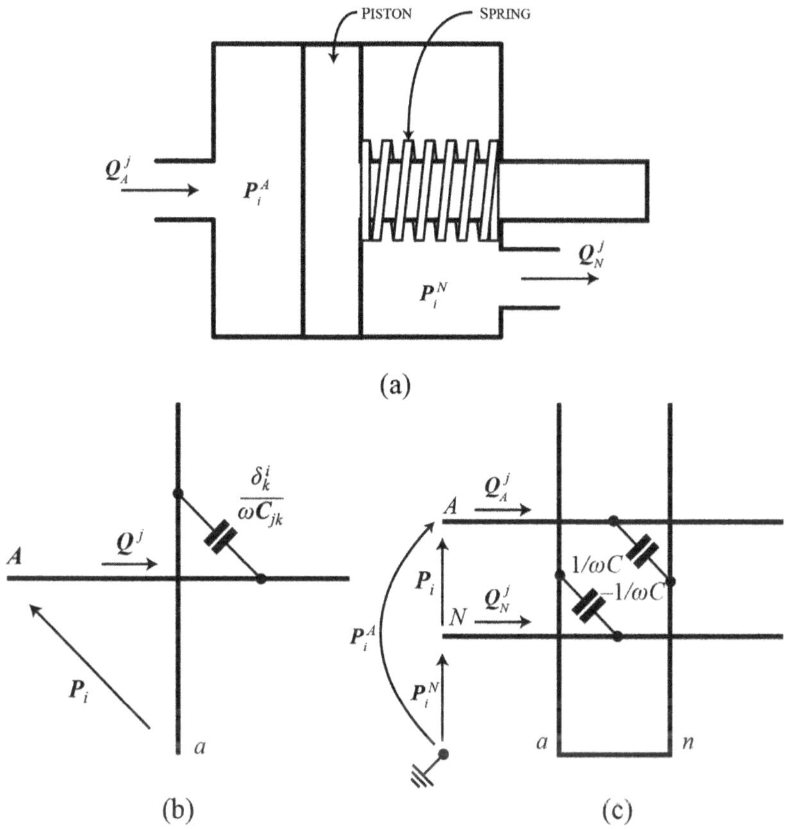

Figure 8.24. Spring-loaded piston-type fluidic capacitance.

8.5.5 Fluidic energy sources

The most frequently encountered source of fluidic energy is the MF pump or MF compressor. In general, these MF machines are neither pure flow nor pressure sources and they are physically and mathematically modelled by a pure source with auxiliary dynamical components to represent imperfections. For a variety of applications, positive-displacement MF pump or MF compressor types can be physically and mathematically modelled as pure flow sources.

If a constant pressure fluidic source is required, a fluid flow store is frequently employed to smooth the pressure variations from a constant-displacement MF pump or MF compressor.

8.6 Thermal commutation matrixer dynamical components

Thermics is the science and technology relating to or associated with a heat (or cold) medium, as well as the use of a heat flow (thermal movements of molecules) or temperature to transmit thermal energy and signals. It concerns producing heat (or cold) and its transfer by means of various physical phenomena, in particular, conduction, convection and radiation.

Thermal phenomena are numerous, complex and often interconnected. Conduction generates thermo-mechanical phenomena, convection thermo-fluidic phenomena and radiation, which brings one back to the atom. There are various sources of thermal energy, from electricity dissipation and radiofrequency through to combustion with thermochemistry and thermofluidics.

The main thermal control dynamical systems are based on the electrical effects of thermoelasticity, which were discovered by a French physicist, Jean Charles Athanasy Peltier (1785–1845), i.e. changes of phase (capillary pumped two-phased fluidic loops) and chemical phenomena (endo- and exothermic effects). Materials do not behave in the same way at very low temperatures (cryogenics) and very high temperatures. Peltier also discovered the calorific effect of electrical current passing through the junction of two different metals, as well as the calorific effect of sunlight. This is now termed the Peltier effect, or the Peltier–Seebeck effect.

The Peltier effect, where electrical current is forced through a junction of two different metals, forms the root of the 12, 24 or 42 V heater/coolers for automotive vehicle use.

Thermics physicists possess exceptional know-how for understanding and mastering all the thermal phenomena that affect basic dynamical components, such as thermal homogeneous continuous dynamical hyposystems, or even complete thermal homogeneous continuous dynamical systems (satellites, aircraft, automobiles, trains, buildings, etc). This knowledge is vital for their design and construction, and for keeping them in action, especially in the case of thermal homogeneous continuous dynamical systems confronted with severe environments (aeronautics, space and defence applications, etc).

Since in the dissipation of energy in all of the physical homogeneous continuous complex and simple dynamical systems considered so far there appears a thermal energy called ‘heat’, it would be logical to extend this physical and mathematical modelling process to thermal commutation matrixers.

The transfer of heat is a complex process, which may occur through a combination of conduction, convection and radiation. Conduction is heat transfer by direct molecular communication. In convection, heat is transported by the movement of fluid masses. Radiation heat transfer, such as the transfer of heat from the Sun to the Earth, operates by wave motion.

We restrict our consideration here to conduction and convection, since these processes fall more naturally into the linear physical commutation matrixer’s physical models (equation (8.1)). In this case, one applies Fourier’s law of heat conduction and Newton’s law of cooling to obtain the appropriate thermal commutation matrixer’s physical models.

8.6.1 Thermal-signal holors

The concept of heat arises from an intuitive judgement of how objects ‘feel’ when touched. The experience of somebody’s tactile sensations is that objects can be ordered in a sequence of 'hotter' or 'colder'. Furthermore, this conceptual property of 'heat' exhibits flow properties. Two bodies, with one hotter than the other, tend to retain this difference if isolated from one another, but if they are brought into close contact they tend to lose any temperature difference. These notions are systematised by the introduction of 'temperature' as a measure of the degree of hotness and the quantity 'heat', which tends to flow from hotter to cooler bodies. A quantitative measure of temperature is obtained by fixing a scale of measurement against some controlled events that are characterised by specific degree of hotness. The temperature scales in common use, Celsius and Fahrenheit, are fixed with respect to the characteristic temperature of water. The amount of heat can be similarly quantified by an interaction experiment between two bodies of holors, $\boldsymbol{T}_i^A$ and $\boldsymbol{T}_i^N$, that are initially at different temperatures. Then, postulating that the amount of heat associated with a body is proportional to the product of mass and temperature, the conservation principle of energy can be invoked to determine the constant of proportionality that characterises the substance.

In this way, the phenomenon of heat can be rationalised as follows. If a mass (admittance holor, $\boldsymbol{m}^{ji}$) of a substance is heated from the temperature holors, $\boldsymbol{T}_i^A$ to $\boldsymbol{T}_i^N$, the amount of the heat holor, $\boldsymbol{H}^j$, required is given by

$$\boldsymbol{H}^j = C_p \boldsymbol{m}^{ji} \left(\boldsymbol{T}_i^A - \boldsymbol{T}_i^N \right), \tag{8.140}$$

where C_p is the specific heat of the substance.

Thus the quantities of the heat-flow-rate holor, $\boldsymbol{\Phi}^j$, and the temperature holor, $\boldsymbol{T}_i$, apparently qualify as an energy-potential-difference holor/energy-transfer holor pair. The heat holor, $\boldsymbol{H}^j$, is readily thought of as an energy-transfer holor, and the temperature holor, $\boldsymbol{T}_i$, is an energy-potential-difference holor, since a temperature datum is required for thermometer calibration. This notion is valid, but should be used with care, since the heat holor, $\boldsymbol{H}^j$, is an energy holor, and so generalisations concerning energy will not carry over.

One may recognise immediately that the heat-flow-rate holor, $\boldsymbol{\Phi}^j$ (e.g. through the walls of buildings), is an energy-transfer holor. However, it is significantly different from the other energy-transfer holors that one has encountered in the electrical, mechanical and fluidic commutation matrixers.

The difference is that the heat-flow holor is the rate at which thermal energy is transferred and has units of power by itself. As a result, the product of the energy-transfer and energy-potential-difference holors is of no particular significance and the energy-potential-difference holor is chosen without any regard to it.

The most convenient energy-potential-difference holor is the temperature-difference holor, $\Delta\boldsymbol{T}_i$, expressed in degrees Celsius (°C). Despite the aforementioned differences, the analogy between the thermal and electrical commutation matrixers is real and useful. The quantity of heat is directly analogous to the quantity of electrical charge and thus, the rate at which electrical charge flows, namely the electrical current. An additional similarity is that temperature is proportional to the energy per unit of heat, whereas voltage is the energy per unit of charge.

8.6.2 Thermal-resistance holor

The flow of heat through a substance is accomplished by thermal resistance and a consequent temperature gradient.

The material relation that governs the energy-transfer holor, $\boldsymbol{\Phi}^j$, of heat by conduction is Fourier's law:

$$\boldsymbol{\Phi}^j = \frac{S}{l}\boldsymbol{\sigma}_C^{ji}\left(\boldsymbol{T}_i^A - \boldsymbol{T}_i^N\right), \tag{8.141}$$

where $\boldsymbol{\sigma}_C^{ji}$ is the thermal-conductivity admittance holor of the material, S is the cross-sectional area of the object, l is the length of the object and $\boldsymbol{T}_i^A - \boldsymbol{T}_i^N$ is the temperature-difference holor across the ends of the object.

If the transport of heat is by convection, the thermal resistance to heat transfer is difficult to analyse. In this instance, the notion of an overall heat transfer coefficient for a specific physical homogeneous continuous dynamical system is employed. The resultant constitutive relation is

$$\boldsymbol{\Phi}^j = S\boldsymbol{C}_H^{ji}\left(\boldsymbol{T}_i^A - \boldsymbol{T}_i^N\right), \tag{8.142}$$

where the heat-coefficient admittance holor, $\boldsymbol{C}_H^{ji}$, is defined in terms of the constitutive relation of a particular object.

The third mechanism for heat transfer is radiation; this is described analytically by the Stefan–Boltzmann law, which describes the power radiated from a black body in terms of its temperature:

$$\Phi = C_R[(T^{A\wedge}4) - (T^{N\wedge}4)], \tag{8.143}$$

where C_R is a constant determined by the geometry of the surfaces radiating and receiving heat.

Like other kinds of physical dissipater components, thermal dissipaters do not lose energy. The heat flow is associated with increase in the entropy of the transporting medium and a net decrease in 'useful' heat energy.

Thermal resistance mathematically models the relationship between the temperature-difference holor and the heat-flow-rate holor for a material of a specific geometry when there is no storage of heat taking place. The absence of storage is merely an approximation, which depends on the properties of the material. We discuss this point more thoroughly at the end of this subsection.

From Fourier's law of heat conduction in the steady state (no storage), one finds that

$$\boldsymbol{\Phi}^j = \boldsymbol{G}^{ji}\left(\boldsymbol{T}_i^A - \boldsymbol{T}_i^N\right) = \frac{\delta_k^i}{\boldsymbol{R}_{jk}}\left(\boldsymbol{T}_i^A - \boldsymbol{T}_i^N\right) \tag{8.144a}$$

$$\delta_k^i = \boldsymbol{R}_{jk}\boldsymbol{G}^{ij} \tag{8.144b}$$

$$G = \sigma_C \frac{S}{l}, \tag{8.144c}$$

where $\boldsymbol{\Phi}^j$ is the heat-flow holor in W, $\boldsymbol{T}_i$ is the temperature holor in °C, $\boldsymbol{G}^{ji}$ is the thermal conductance holor in W °C^{-1}, δ_k^i is the Kronecker delta, $\boldsymbol{R}_{jk}$ is the termal resistance holor in °C W^{-1}, σ_C is the thermal conductivity of the material in W m^{-1} °C $^{-1}$, S is the area through which heat flows in m^2 and l is the length of material in m. Let us clarify thermal resistance by referring to the specific geometry shown in figure 8.25.

Here, a round piece of insulating material has a diameter, d, and a length, l. A collar of perfect insulation that is shown cross-hatched surrounds the piece. The purpose of the collar is to keep the heat flow in the axial direction. Although perfect insulation is not physically realisable, it can be closely approximated. In this case, it is a valuable concept because it helps to define the area through which heat flows clearly. The left face of the piece is at temperature $\boldsymbol{T}_i^A$, the right face is at temperature $\boldsymbol{T}_i^N$, and the ambient temperature is $\boldsymbol{T}_i^a$. For this configuration, equation (8.144a) can be placed in the form of equation (8.1a) as:

$$\boldsymbol{T}_i^A - \boldsymbol{T}_i^N = \boldsymbol{R}_{ij}\boldsymbol{\Phi}^j. \tag{8.145}$$

Thus the thermal resistance of the cylindrical piece with axial conduction heat transfer is:

$$R = \frac{4l}{\pi(d\wedge 2)k}. \tag{8.146}$$

There is heat transferred from the right face of the piece to the ambient air temperature, $\boldsymbol{T}_i^a$. This heat transfer takes place by convection.

From Newton's law of cooling, one may express this heat transfer by:

$$\boldsymbol{\Phi}^j = \boldsymbol{G}^{ji}\left(\boldsymbol{T}_i^A - \boldsymbol{T}_i^N\right) \tag{8.147a}$$

$$G = hS, \tag{8.147b}$$

where h is the surface coefficient (film conductance) in W m^{-2} °C^{-1}.

For the convection heat transfer the thermal resistance has the form:

$$R = \frac{1}{hS} = \frac{4}{\pi(d\wedge 2)h}. \tag{8.148}$$

The surface coefficient, h, depends on the motion of the fluid medium and is given in heat transfer texts for various geometric configurations and surface conditions.

Figure 8.25 shows the thermal resistance physical model as the thermal commutation matrixer for the solid piece and the film. Together they may be physically modelled as two thermal resistances in series.

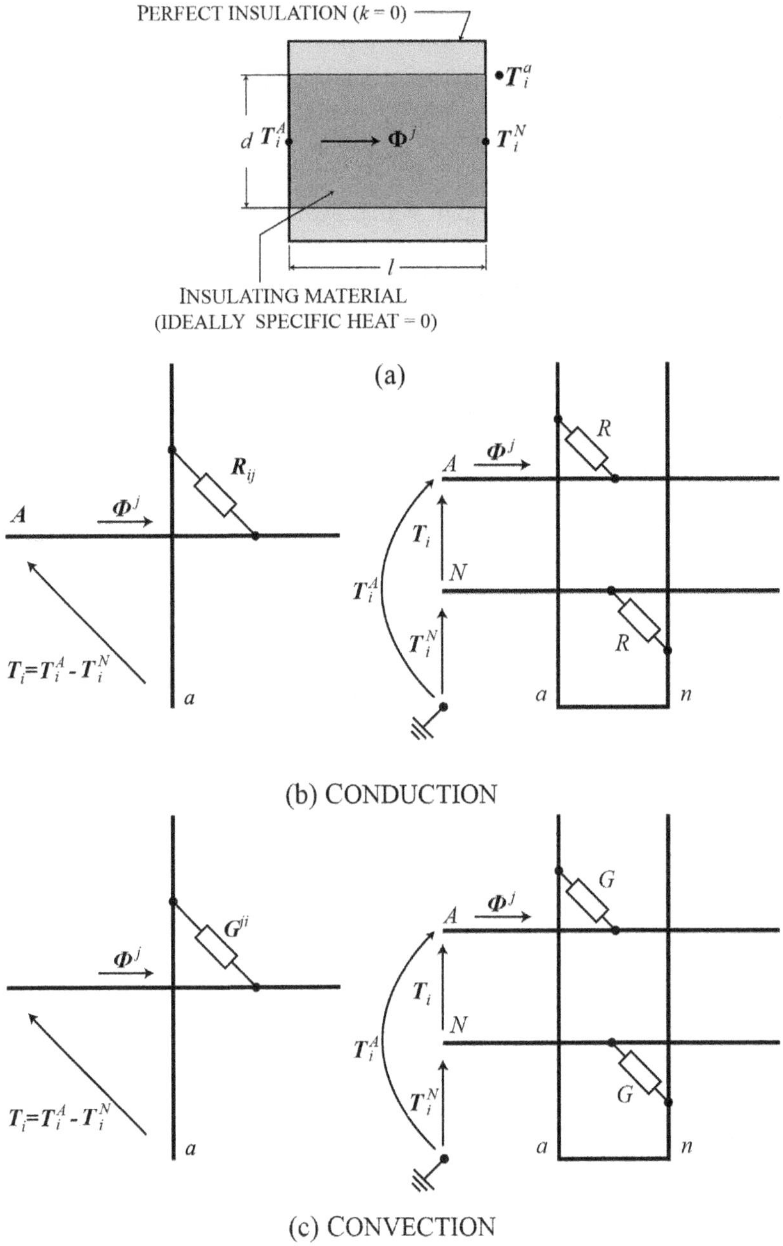

Figure 8.25. Thermal resistance with its thermal commutation matrixers.

8.6.3 Thermal-inertance holor

From the signal holors for thermal commutation matrixers and equation (8.1b), one requires a relation of the type

$$T_i^A - T_i^N = \omega L_{ij} \Phi^j. \tag{8.149}$$

Equation (8.149) would define thermal inertance. However, no physical situation is known to produce such a relation between the temperature-difference holor, $\boldsymbol{T}_i^A - \boldsymbol{T}_i^N$, and the heat flow-rate holor $\boldsymbol{\Phi}^j$. As a result, there is no thermal inertance component analogous to an electrical inductor. There is no thermal component that displays an energy storage mechanism that is complementary to the energy-transfer store.

8.6.4 Thermal-capacitance holor

The ability of a specific material to store heat is a measure of its thermal capacity. The constitutive relation given by equation (8.140) for a thermal-flow store relates in a linear fashion the total heat transferred and the temperature change.

The total heat holor, $\boldsymbol{H}^j$, is related to the heat-flow-rate holor, $\boldsymbol{\Phi}^j$, by the dynamical relation

$$\boldsymbol{\Phi}^j = \omega \boldsymbol{H}^j. \tag{8.150}$$

The total heat energy stored in the energy-transfer store is the time integral of the heat-flow-rate holor:

$$\boldsymbol{H}^j = \frac{1}{\omega}\boldsymbol{\Phi}^j. \tag{8.151}$$

The storage of heat within a mass has long been recognised as a physical process in which thermal energy is regarded almost as a fluid, or as charge, as noted previously.

As a result, the terms 'thermal mass' and 'thermal capacity' appear frequently. The basic equation of heat storage in a mass is:

$$\boldsymbol{\Phi}^j = \omega \boldsymbol{C}^{ji}\boldsymbol{T}_i. \tag{8.152}$$

By comparison with equation (8.1c), one finds that the thermal capacitance, C, is equal to:

$$C = \rho c V, \tag{8.153}$$

where C is the thermal capacitance in W s °C^{-1}, ρ is the mass density in kg m^{-2}, c is the specific heat of the mass in W s kg^{-1} °C^{-1} and V is the volume of the mass in m^3.

The temperature holor, $\boldsymbol{T}_i$, in equation (8.152) refers to the temperature of the mass. For pure heat storage components, the temperature is uniform throughout the mass. Thus, there is only one temperature associated with a mass storage component. To put the mass into the form required for two terminal thermal capacitive components one must assume a frictionless ground terminal at constant temperature holor, $\boldsymbol{T}_i^G$. Then equation (8.152) is written as:

$$\boldsymbol{T}_i^A - \boldsymbol{T}_i^N = \frac{\delta_k^i}{\omega \boldsymbol{C}_{jk}}\boldsymbol{\Phi}^j. \tag{8.154}$$

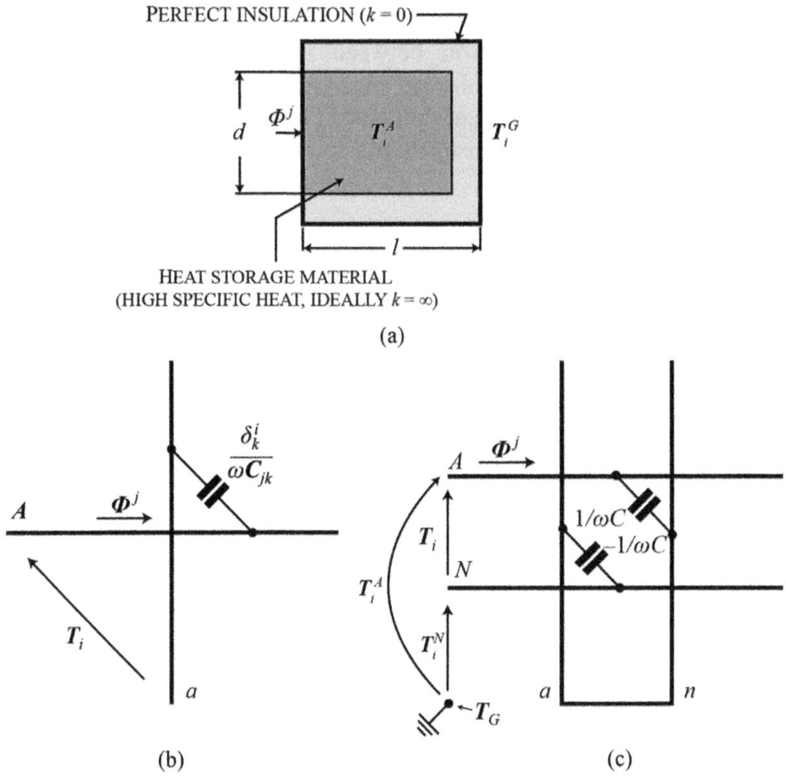

Figure 8.26. Completely insulated thermal capacitance and its thermal commutation matrixer.

Consider the cylindrical mass of the material enclosed by perfect insulation in figure 8.26(a). For pure heat storage, the temperature of the mass is uniform throughout. This means that the materials used for heat storage will be very good heat conductors.

The thermal conductivity of such materials will be high and they will have very low thermal resistance. If heat is applied to the cylindrical mass, the temperature of the mass increases in accordance with equation (8.154). Figure 8.26(b) shows the analogous thermal commutation matrixer's thermal dynamical component.

Once again, the assumption of a fictitious potential terminal leads to a thermal capacitance to ground. All thermal capacitance has this fictitious terminal and, therefore, it always represents thermal capacitance to ground in the thermal-capacitance-matrix sense. Note that because the terminal is fictitious, there is no heat transferred to ground.

If the insulation does not completely surround the heat storage dynamical component, as shown in figure 8.27(a), the thermal capacitance effect is still to ground.

The heat flow that acts to raise the temperature holor, $\boldsymbol{T}_i$, is now $\boldsymbol{\Phi}_A^j - \boldsymbol{\Phi}_N^j$, and the thermal dynamical components may be represented as shown in figure 8.27(b).

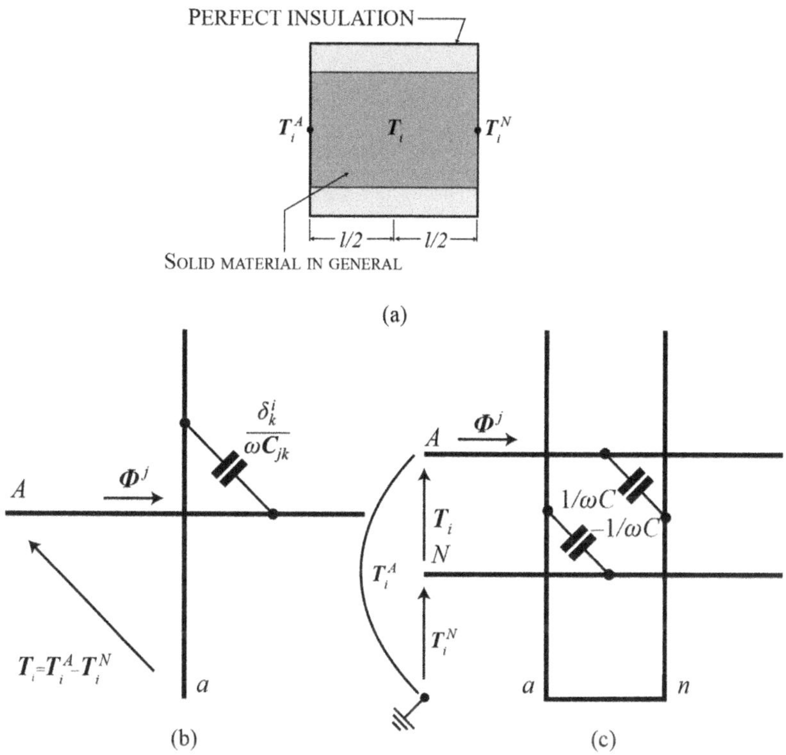

Figure 8.27. Partially insulated thermal capacitance and its thermal commutation matrixer.

8.6.5 General thermal dynamical components

In reality, all thermal dynamical components have both resistance and capacitance. The magnitude of the resistance is inversely proportional to the thermal conductivity of the material.

The magnitude of the capacitance depends on the density-specific heat product. The physical size of the dynamical component is a factor in both the resistance and the capacitance.

Some combinations of material and configuration are predominantly resistive or predominantly capacitive. Such are the physical models already described in section 8.6.

Figure 8.28, however, shows the generalised thermal commutation matrixer physical model for all thermal dynamical components.

The temperature holors $\boldsymbol{T}_i^A$ and $\boldsymbol{T}_i^N$ are the temperatures of the left and right boundaries, and $\boldsymbol{T}_i$ is the mid-plane temperature holor.

The generalised thermal commutation matrixer physical model has half the resistance on each side of the mid-plane and a capacitance to ground from the mid-plane temperature. When the thermal material has a large thermal conductivity, the dynamical component reduces to the thermal commutation matrixer shown in figure 8.27(c). In this case, the resistance is negligibly small and the dynamical component is a capacitance to ground.

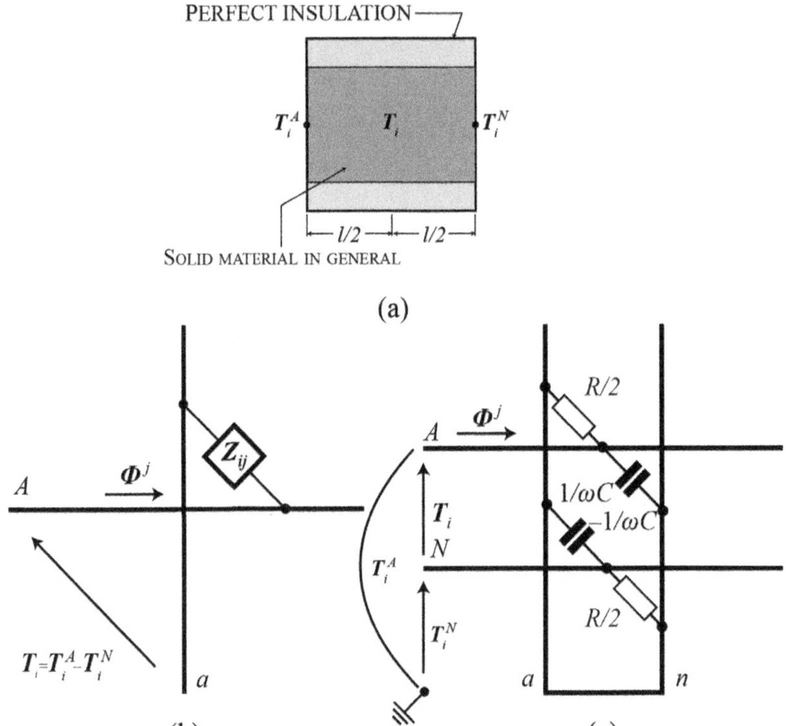

Figure 8.28. General thermal commutation matrixer physical model.

On the other hand, a material with a lower thermal conductivity has resistance. If the specific heat-density product is low, the capacitance will be small and the dynamical component will be purely resistive, as shown in figure 8.25. In the general case, however, both the thermal conductivity and the specific heat product must be considered to physically model a thermal dynamical component.

8.7 Energy sources and loads

The fundamental situations and physical machines and devices discussed so far have implied a source of energy that is quite independent of the physical dynamical component. These dynamical components have been either dissipative or storage physical machines and devices that can only act on the energy while it is within a physical heterogeneous continuous complex and/or simple dynamical hypersystem, and they are therefore called 'passive physical dynamical components'.

The passive physical dynamical components (resistance, inductance and capacitance) are often also called 'physical loads'. By contrast, 'active physical dynamical components' are physical machines and devices that imply situations where energy is supplied to or taken from the physical heterogeneous continuous complex and/or simple dynamical hypersystem. Active physical dynamical components often involve some energy conversion process such as takes place in a **chemo-electrical/electro-chemical** (ChE/ECh) storage battery, **mechano-electrical** (ME) generator, **mechano-fluidic** (MF) pump, **mechano-pneumatic** (MP) compressor, or **chemo-thermal** (Ch-Th) furnace. That is, the source of energy for any physical heterogeneous continuous complex and/or simple dynamical hypersystem is physically modelled by means of active physical dynamical components. Thus, in general, a source supplies energy and a load receives energy. In this section, we discuss various modes of energy sources and loads.

8.7.1 Energy-potential-difference sources

The details of the origin of the energy are often not required; only the form in which it is fed into a physical heterogeneous continuous complex and/or simple dynamical hypersystem is necessary. For example, most ChE/ECh storage batteries are capable of supplying energy to an electrical homogeneous discrete dynamical system with the voltage-difference holor, $\boldsymbol{E}_i$ between its input terminals only changing slightly if the current holor, $\boldsymbol{I}^j$, changes.

The ideal situation would presumably be one in which the voltage-difference holor, $\boldsymbol{E}_i$, is constant, irrespective of the current holor, $\boldsymbol{I}^j$, flowing from the ChE/ECh storage battery. In electrical commutation matrixer analysis, this leads to the idea of 'ideal' voltage source, usually referred to simply as a 'voltage-difference source'.

Pursuing energy-potential-difference holor and energy-transfer holor analogues, one would expect to find ideal pressure-difference sources, velocity sources and temperature sources, which for convenience can be called 'energy-potential-difference sources'.

Figure 8.29(a) shows the static characteristics (the energy-potential-difference holor–energy-flow holor relation) of an ideal energy-potential-difference source.

Such a source can supply energy flow without changing potential. The electrical commutation matrixer representation of an ideal energy-potential-difference source (figure 8.29(b)) is a circle with the polarity indicated by a plus arrow sign inside it. When the polarity is reversed, the device is sometimes termed a 'sink'. However, one may also look on this situation as a negative energy-potential-difference source.

Of course, real energy-potential-difference sources deviate from ideal ones in that a normal pressure MF pump, for example, does not quite maintain constant

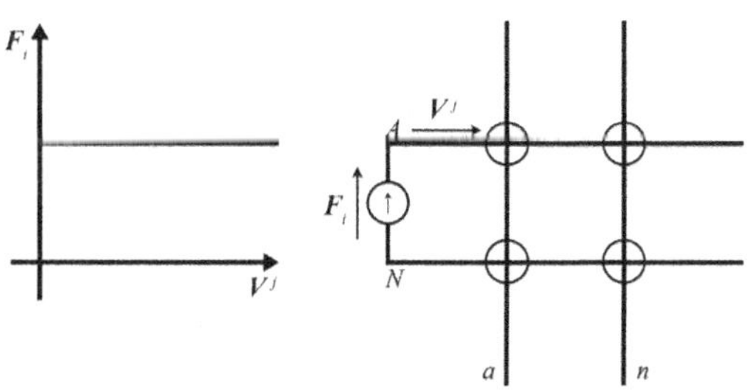

(a) Characteristics (b) Physical commutation matrixer

Figure 8.29. Ideal energy-potential-difference source.

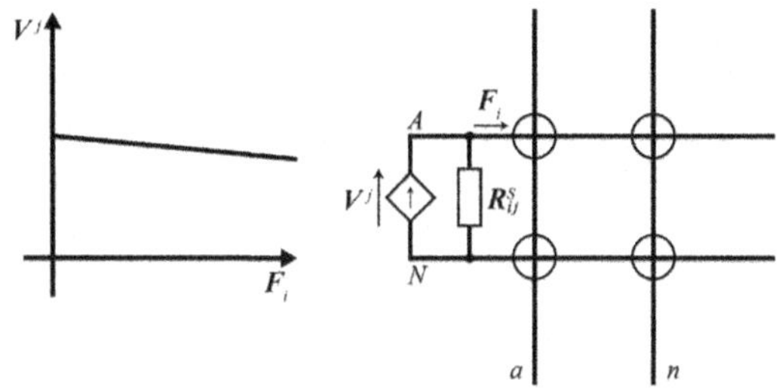

(a) Characteristics (b) Physical commutation matrixer

Figure 8.30. Real energy-potential-difference source.

pressure, irrespective of the flow rate. However, just as a 'real' ChE/ECh storage battery is regarded as a constant voltage with the internal resistance connected in series (physically modelled by a voltage-difference source and series connected resistance), the analogous energy-potential-difference sources can be physically modelled by an energy-potential-difference source and series dissipative dynamical component.

Figure 8.30(a) shows the static characteristics of the real energy-potential-difference source.

The equivalent electrical commutation matrixer representation of the real energy-potential-difference source is shown in figure 8.30(b). The electrical commutation matrixer representation has an ideal energy-potential-difference source and an internal resistance, R^s, in series. It should be noted that for dynamical studies the physical model of an energy-potential-difference source might require at least one storage dynamical component in addition to the energy-potential-difference source and dissipative dynamical component. This should become evident when Thévenin and Norton's theorems are considered in chapter 10.

Situations involving an ideal velocity source are not often encountered in unsophisticated mechanics. The main point is to look to see whether the velocity at a point in the mechanical continuous dynamical system is constrained to some specific value or function of time.

The situation where a body is struck and carried by a much larger mass that is already moving at a constant velocity can be physically modelled by a velocity source, since the velocity of the larger mass is virtually unchanged after contact is made with the smaller mass.

The resulting forces may be sufficiently large to cause damage, and therefore great care is required when subjecting any part of a mechanical homogeneous continuous dynamical system to a suddenly applied constraint of constant velocity. The ideal temperature source is a reasonable physical model of a thermo-statically controlled **electro-thermal** (E-Th) oven, provided that the item to be heated can be inserted without any change in E-Th oven temperature. This is inevitably very difficult to achieve exactly.

Nevertheless, the times involved are such that in many such situations the physical model of the temperature-controlled E-Th oven as an ideal temperature source gives reasonably accurate results.

In general, an ideal energy-potential-difference source is the correct physical model for any situation in which the energy-potential-difference holor is constrained to some specified value or function of time. This is true whether the constraint is produced naturally or whether it is the result of some mechatronical control system. If the end result is a value or functional expression for the energy-potential-difference holor that is independent of the energy-transfer holor, it is appropriate to use an ideal energy-potential-difference source as a physical model. The characteristic holor equation of an energy-potential-difference source for electrical homogeneous continuous dynamical systems may be written as:

$$\boldsymbol{F}_i = \boldsymbol{E}_i - \boldsymbol{R}^s_{ij}\boldsymbol{I}^j, \tag{8.155}$$

where $\boldsymbol{E}_i$ is the voltage-difference holor when there is no current holor, $\boldsymbol{I}^j$.

Note that for the ideal energy-potential-difference source, the internal resistance holor, $\boldsymbol{R}^s_{ij} = [0]$, and the voltage-difference holor, $\boldsymbol{F}_i = \boldsymbol{E}_i$, are independent of the current holor, $\boldsymbol{I}^j$.

8.7.2 Energy-transfer sources

As might be expected, situations arise in which the value of the energy-transfer holor is constant or is some specific function, independent of the energy-potential-difference holor.

Most unsophisticated mechanics are based on the premise that it is possible (at least theoretically) to apply a force holor that is independent of the velocity holor of the point at which it is applied. Such situations can be physically modelled as an ideal force-source holor, usually termed simply a 'force source'.

In electrical homogeneous continuous dynamical systems, there are very few situations where the current holor from a source is naturally constant, but it is quite common for the current holor from a DC power supply to be mechatronically controlled, such that these supplies may be represented by current sources. The ideal current source is one where the current holor is entirely independent of the voltage holor across it.

Figure 8.31(a) shows the static characteristics of an ideal energy-transfer holor or energy-transfer source and its physical commutation matrixer.

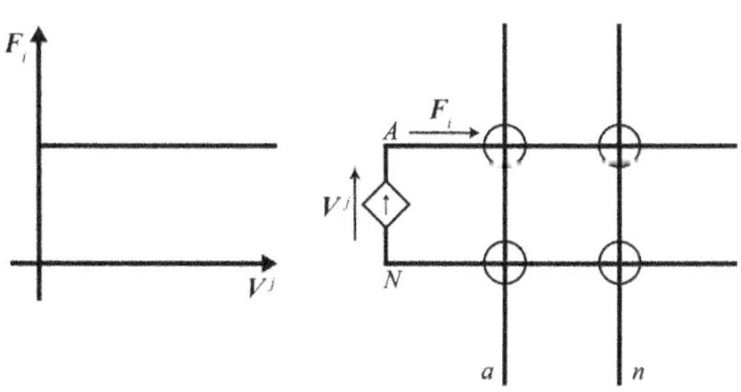

(a) Characteristics (b) Physical commutation matrixer

Figure 8.31. Ideal energy-transfer source and its physical commutation matrixer.

The energy-transfer holor remains constant for all values of the energy-potential-difference holor. The physical commutation matrixer symbol to represent the ideal energy-transfer source is a cornerwise square with an arrow indicating the sense of flow direction along the diagonal line.

At this stage, it may be wise to emphasize the fact that 'source of current' and 'current source' have significantly different meanings. A source of current has absolutely no implications concerning whether the voltage holor or current holor are constant (or neither is). However, a current source unequivocally implies an ideal energy-transfer source, namely one in which the current holor is constant or some function that is independent of the voltage holor.

Similarly, a 'source of force' must not be confused with a 'force source'. For convenience, sources in which the energy-transfer holor is considered are termed 'energy-transfer sources'. In fluidic homogenerous continuous dynamical systems, when this latter term is particularly applicable to MF pumps in which clearly identifiable quantities of liquid are passed through the pump, it is termed 'flow source'. Provided that such MF pumps are driven at constant speed, the rate at which the liquid is delivered is constant.

In the case of thermal continuous dynamical systems, an ordinary E-Th heater without any thermo- control produces heat at a rate that is reasonably independent of the temperature. An ideal flow source static is therefore used to physically model this situation.

In general, an energy-transfer source is used to physically model any situation in which the energy-transfer holor is constrained to some specified value or function of time. This is true regardless of whether the constraint is produced naturally or is the result of some mechatronical control system. If the end result is a value or functional expression for the energy-transfer holor that is independent of the energy-potential-difference holor, it is appropriate to use an ideal energy-transfer source as a physical model for this situation.

In reality, there are no ideal energy-transfer sources. Any real energy-transfer source is affected by the energy-potential-difference holor. Thus, for example, a constant displacement MF pump would slow down if the pressure-difference holor became too large.

Figure 8.32(a). shows the static characteristics of a real energy-transfer source and its physical commutation matrixer.

As a fluidic homogenous continuous dynamical system's component, the real flow source may be considered as an ideal flow source in parallel with an internal resistance, R^s. This is shown in figure 8.32(b).

The equation of the real flow source for fluidic continuous dynamical systems is:

$$V^j = \boldsymbol{Q}^j - \frac{\delta_k^i}{\boldsymbol{R}^s_{jk}}\boldsymbol{P}_i = \boldsymbol{Q}^j - \boldsymbol{G}_s^{ji}\boldsymbol{P}_i, \tag{8.156}$$

where $\boldsymbol{Q}^j$ is the value of the flow-source holor when the pressure-difference holor, $\boldsymbol{P}_i = [0]$.

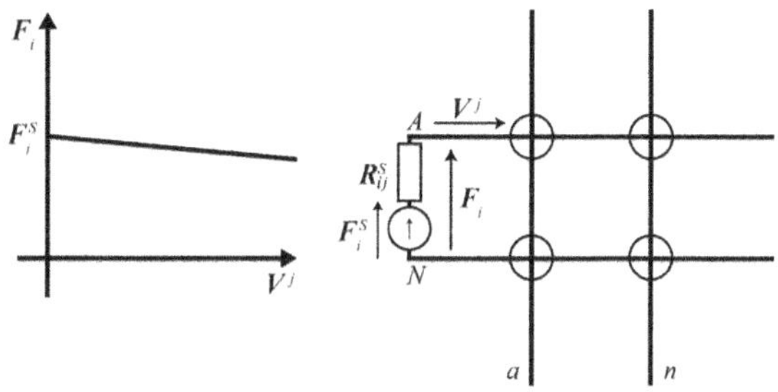

(a) CHARACTERISTICS (b) PHYSICAL COMMUTATION MATRIXER

Figure 8.32. Real energy-transfer source and its physical commutation matrixer.

In this case, an ideal flow source has an infinite value of the internal resistance holor, $\boldsymbol{R}^{s}_{jk}$. Note that under this condition the flow holor does not depend on the pressure-difference holor, $\boldsymbol{P}_i$. The characteristics of real energy-transfer sources and real energy-potential-difference sources are very similar (compare figures 8.30(a) and 8.32(a)).

For a unity value of the internal resistance, equations (8.152) and (8.153) are in the same form and can be identical when $\boldsymbol{E}_i = \boldsymbol{Q}^j$. In this case, one would not designate the dynamical component as an energy-transfer source or an energy-potential-difference source, one would merely designate it as a source.

In general, the active dynamical component can be used to physically model either a source or a sink of energy. If the physical machine or device is acting as a source, positive flow will emerge from the positive terminal for both the energy-potential-difference and the energy-transfer sources. If the physical situation is one that absorbs energy, positive flow will enter at the positive terminal. Although it is not entirely consistent, it is nevertheless convenient to term the active dynamical component a source, whether it is actually acting as a source of energy or a sink.

8.7.3 Loads

Resistance, inductance and capacitance are all load dynamical components. However, resistance is the only load that has a static characteristic. Inductance and capacitance generally have time-dependent signal holors. Figure 8.33(a) shows the characteristics of a resistive load

For a linear resistance, the characteristic is a straight line passing through the origin. The slope of the line is the value of the resistance. Figure 8.33(b) indicates the analogous physical commutation matrixer for the load resistance, R^L.

The relation between the holors for a resistive load can be obtained from equation (8.1a) for this case as:

$$\boldsymbol{F}_i = \boldsymbol{R}^{L}_{ij}\boldsymbol{V}^j. \tag{8.157}$$

The interaction of a source and a load involves some principles of holor theory. Although holor theory is not considered until chapter 10, one may gain some insight now by superimposing the source and load characteristics. For example, suppose that the characteristics of an ideal energy-potential-difference source and a resistive

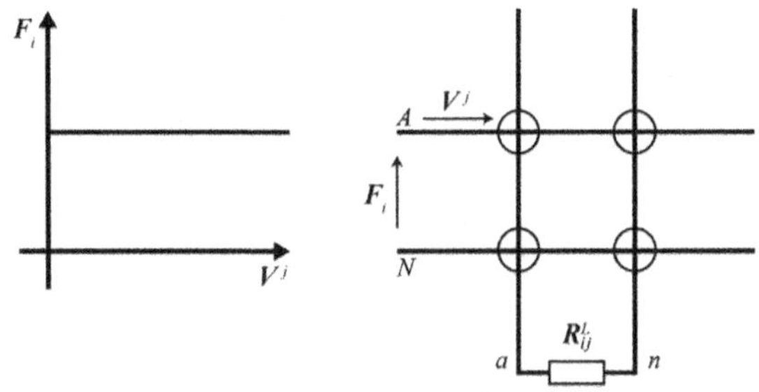

(a) Characteristics (b) Physical commutation matrixer

Figure 8.33. Resistive load and its physical commutation matrixer.

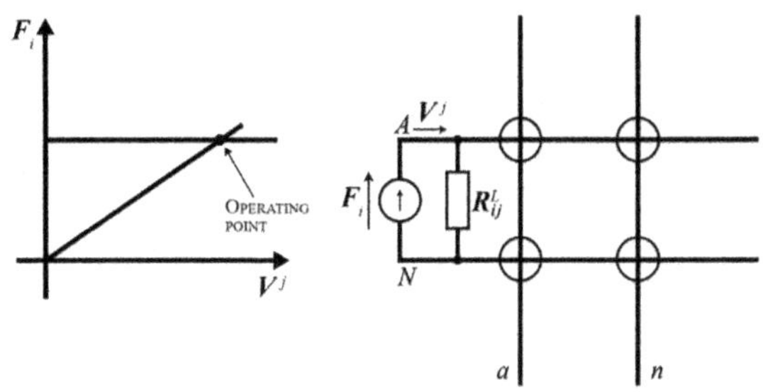

(a) Characteristics (b) Physical commutation matrixer

Figure 8.34. (a) Energy-potential-difference source/resistive load interaction. (b) Physical commutation matrixer connecting an ideal energy-potential-difference source to a resistive load.

load are superimposed (figure 8.34(a)). This is equivalent to the physical commutation matrixer connecting an ideal energy-potential-difference source to a resistive load (figure 8.34(b)).

Let us recall that one may interpret a source as representing the locus of all possible operating states of the energy dynamical component. Similarly, the load represents the locus of all possible values of the signal holors for a resistance.

The physical commutation matrixer configuration is such that both loci must be applicable simultaneously. The only point that permits both loci to coexist is the intersecting point of the characteristics. This point is termed the operating point because it shows the value of the signal holors in the physical commutation matrixer. This is a graphical interpretation of the simultaneous solution of two equations, equations (8.152) and (8.153).

The energy-potential-difference source to load method is especially valuable in linear physical commutation matrixers, where the results are more difficult to obtain analytically.

8.7.4 Coulomb friction

When we considered friction, it was noted that the characteristic of Coulomb friction is that the frictional-force holor is independent of the velocity holor. Accordingly, the appropriate physical model of this situation is a force source (a torque source if rotating) connected such that the flow of the force holor enters at its positive terminal. This physical model is exact, provided that the motion or rotation is restricted to a single sense of motion or rotation direction. If the motion reverses, the polarity of the force source must be reversed, because the frictional-force holor reverses.

Let us examine, for example, the case of a mass holor, $\boldsymbol{M}^{ji}$, moving along a rough horizontal surface with the velocity holor, V_i (figure 8.35).

The normal friction between the mass holor, $\boldsymbol{M}^{ji}$, and the surface holor, $\boldsymbol{N}^j$, is equal to the weight holor, $\boldsymbol{G}^j = g\boldsymbol{M}^{ji}$. The frictional-force holor retarding motion, $\boldsymbol{F}^j$, is $\mu\boldsymbol{N}^j$ (or $\mu\boldsymbol{G}^j$), where μ is the coefficient of the sliding or kinetic friction. Figure 8.36 shows a plot of the velocity-difference holor, $V_i - V_i^G$, to the force holor, $\boldsymbol{F}^j$, characteristics of the friction.

When the velocity-difference holor, $V_i - V_i^G = [0]$ (no motion), the frictional force, $\boldsymbol{F}^j$, is somewhat larger than when the mass is moving. This occurs because the coefficient of the static friction is larger than the coefficient of the kinetic friction.

In reality, the frictional-force holor, $\boldsymbol{F}^j$, decreases slightly with increased velocity-difference holor, $V_i - V_i^G$. Note, however, that the characteristic shown in figure 8.36 is quite similar to the ideal energy-transfer source characteristic shown in figure 8.31(a). As an approximation, therefore, one uses a force source physical model to represent Coulomb friction. If the motion reverses in its sense of direction, the force source is reversed. This is a nonlinear effect that is outside the scope of this text; the

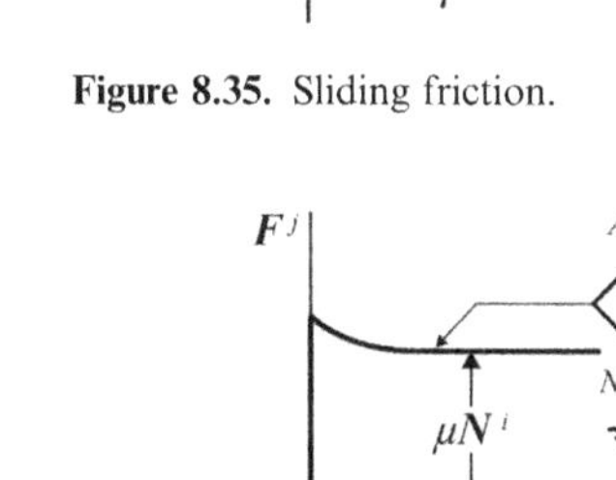

Figure 8.35. Sliding friction.

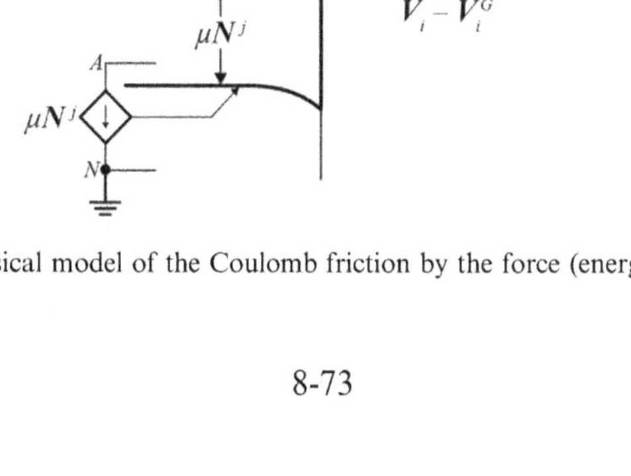

Figure 8.36. Physical model of the Coulomb friction by the force (energy-transfer) source.

analysis of such a situation is usually introduced in texts on mechatronical control systems.

8.7.5 Weight

Another situation involving a constant force holor, $\boldsymbol{F}^j$, which is not necessarily a source of energy, is the weight holor, $\boldsymbol{G}^j$, of a mass holor, $\boldsymbol{M}^{ji}$, subject to vertical motion.

In figure 8.37(a), an applied force holor, $\boldsymbol{F}^j$, is required to balance the weight holor, $\boldsymbol{G}^j = g\boldsymbol{M}^{ji}$, if the mass holor, $\boldsymbol{M}^{ji}$, is to remain stationary, this being the means by which one is aware of the gravitational-force holor, $\boldsymbol{G}^j$, on an object. However, the weight holor of the object is the gravitational force holor, $\boldsymbol{G}^j$, which should be expressed in newtons (N).

Provided that the distances from the centre of mass of the Earth do not vary significantly, the weight holor, $\boldsymbol{G}^j$, is constant and can therefore be physically modelled by means of a force-source holor.

The polarity of the force-source holor will depend on the sense of direction chosen for the positive velocity holor, $\boldsymbol{V}_i$. If the velocity holor, $\boldsymbol{V}_i$, is positive when the motion is upwards, the polarity is that shown in figure 8.37(b).

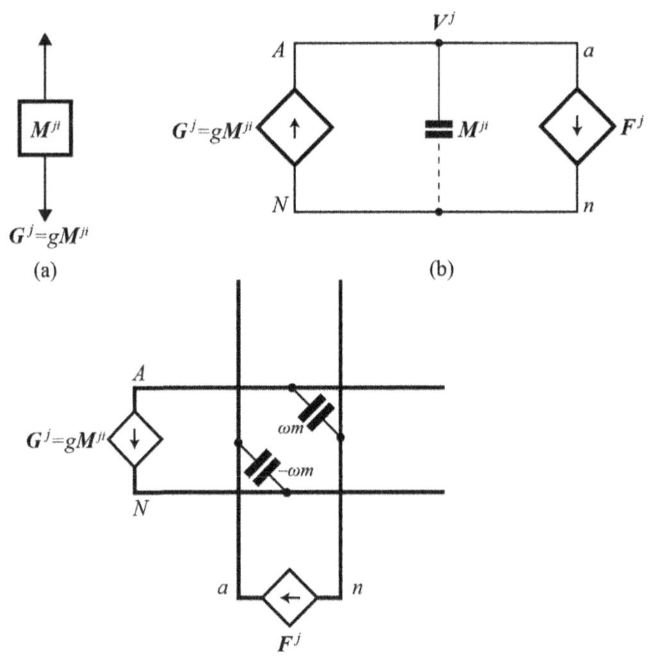

Figure 8.37. Physical modelling of weight.

8.8 Physical dynamical components as mathematical operators

From the fundamental holor equations of all of the passive dynamical components, it can now be noted that the energy-transfer holor is either a multiple, or a time derivative, or a time integral of the energy-potential-difference holor. That is, the voltage-drop holor, V_i, across a resistor, for example, is simply the current holor, I^j, multiplied by a constant, i.e. the resistance holor, R_{ij}. This is true for ideal resistance no matter how the current holor, I^j, varies, and therefore the appearance of a graph of the voltage-drop holor, V_i, against time is identical to that of a graph of the current holor, I^j, against time, except for the scale factor that the value of resistance introduces.

The voltage-difference holor, V_i, across an inductor is the time derivative of the current holor, I^j, flowing through it with the scale factor of the value of inductance holor, L_{ij}, included.

If the current holor, I^j, is constant, the voltage-difference holor, V_i, across an ideal inductance holor, $L_{ij} = [0]$. If the current holor, I^j, is some function of time, the voltage-difference holor, V_i, is then the derivative of that function multiplied by the value of the inductance holor, L_{ij}, as a scale factor.

The voltage-difference holor, V_i, across a capacitor, on the other hand, is the time integral of the current holor, I^j, flowing through it with the scale factor of the capacitance, C_{ik}, included.

If the current holor, I^j, is constant, the voltage-drop holor, V_i, across the capacitor will increase at a uniform rate.

If the current holor, I^j, is some function of time, e.g. sinusoidal, the voltage-drop holor, V_i, is then the integral of that function multiplied by the Kronecker delta, δ^i_k, and divided by the capacitance reactance holor, ωC_{jk}.

The process of integration generally requires evaluation of the constant of integration. However, in this context, the value of the integral is usually known as the instant in time at which the process is considered to start, i.e. $t = 0$ in most cases.

By using the format of a particular integral, the value of the constant of integration is usually obtained automatically by inspection. For example, if one considers a capacitor that is initially charged when a current holor, I^j, is passed through it, the resulting voltage-drop holor, V_i, across the capacitor may be developed from the fundamental holor equation (8.1c):

$$V_i - V_i^0 = \frac{\delta^i_k}{\omega C_{jk}} I^j. \tag{8.158}$$

The second term of the left-hand side represents the voltage-drop holor, V_i^0, resulting from the initial charge on the capacitor.

One should be aware that time integration always adds the integral in an interval to the value of the voltage-drop holor at the beginning of the interval.

8.9 Summary

The process of determining analogous physical situations can now be seen to have two main parts. The first of these involves identifying the analogous energy-potential-difference and energy-transfer holors.

Table 8.1. The analogous energy-potential-difference and energy-transfer holors.

	Energy-potential-difference holor	Energy-transfer holor	Tensor product ⊗
Electrical homogeneous continuous dynamical system	Voltage holor (V)	Current holor (A)	Power holor (W)
Magnetic homogeneous continuous dynamical system	Electromotive-force holor (A turns)	Magnetic-flux holor (Wb)	Power holor (W)
Mechanical homogeneous continuous dynamical system—translation	Velocity holor (m s^{-1})	Force holor (N)	Power holor (W)
Mechanical homogeneous continuous dynamical system—rotation	Angular-velocity holor (rad s^{-1})	Torque holor (N m)	Power holor (W)
Fluidic homogeneous continuous dynamical system	Pressure holor (Pa)	Volume-flow-rate holor (m^3 s^{-1})	Power holor (W)
Thermal homogeneous continuous dynamical system	Temperature-difference holor (°C)	Heat-flow-rate holor (W)	Power holor (W)

Although at first these are selected on the basis that the univalent tensor product will be a univalent power holor, there is no reason not to use the univalent power holor as the energy-transfer holor, if the fundamental holor equations turn out to have the same form, as in the thermal continuous dynamical system. The final choice is shown in table 8.1.

The fundamental physical situations are physically modelled by means of the physical commutation matrixer components. With the physical heterogeneous continuous complex and/or simple dynamical hypersystems having a power holor as the tensor product of the energy-potential-difference and energy-transfer holors (electrical, mechanical and fluidic homogeneous continuous dynamical systems), there are only three physical commutation matrixer dynamical components.

- One dissipates energy, usually as heat with *P*-type energy dissipation, and the other two store energy.
- One of the storage physical dynamical components stores energy when the energy-potential-difference holor exists, whether or not the energy-transfer holor has some value. For convenience, these physical dynamical components can be categorised as *D*-type or kinetic-energy storage physical dynamical components.
- The other physical dynamical component stores energy by virtue of the energy-transfer holor, whether or not the energy-potential-difference holor has some value. For convenience, these physical dynamical components can be categorised as *I*-type or potential-energy storage physical dynamical components.

For thermal homogeneous continuous dynamical systems, there are only two fundamental physical dynamical components: dissipative and kinetic energy storage.

The major forms of fundamental holor equations with impedance holors are:

$$F_i = Z_{ij}^P V^j = D_{ij}^P V^j \quad P\text{-type energy dissipation,} \tag{8.159a}$$

$$F_i = Z_{ij}^D V^j = \omega K_{ij}^D V^j \quad D\text{-type energy storage,} \tag{8.159b}$$

$$F_i = Z_{ij}^I V^j = \frac{1}{\omega} P_{ij}^I V^j \quad \text{I-type energy storage.} \tag{8.159c}$$

The appropriate interpretations of D_{ij}^P, K_{ij}^D and P_{ij}^I are summarised in table 8.2.

The major forms of fundamental holor equations with admittance holors are:

$$V^j = Y_P^{ji} F_i = \frac{\delta_k^i}{D_{jk}^P} F_i \quad P\text{-type energy dissipation,} \tag{8.160a}$$

$$V^j = Y_D^{ji} F_i = \frac{1}{\omega} \frac{\delta_k^i}{K_{jk}^D} F_i \quad I\text{-type energy storage,} \tag{8.160b}$$

$$V^j = Y_I^{ji} F_i = \omega \frac{\delta_k^i}{P_{jk}^I} F_i \quad D\text{-type energy storage.} \tag{8.160c}$$

The appropriate interpretation of $\frac{\delta_k^i}{D_{jk}^P}$, $\frac{\delta_k^i}{K_{jk}^D}$ and $\frac{\delta_k^i}{P_{jk}^I}$ is summarised in table 8.3.

Table 8.2. The analogous impedance physical components.

	P-type D_{ij}^P	D-type K_{ij}^D	I-type P_{ij}^I
Electrical homogeneous continuous dynamical system	R_{ij}	L_{ij}	$\frac{\delta_k^i}{C_{jk}}$
Magnetic homogeneous continuous dynamical system	R_{ij}	—	—
Mechanical homogeneous continuous dynamical system—translation	$\frac{\delta_k^i}{b_{jk}}$	$\frac{\delta_k^i}{k_{jk}}$	$\frac{\delta_k^i}{m_{jk}}$
Mechanical homogeneous continuous dynamical system—rotation	$\frac{\delta_k^i}{B_{jk}}$	$\frac{\delta_k^i}{K_{jk}}$	$\frac{\delta_k^i}{J_{jk}}$
Fluidic homogeneous continuous dynamical system	R_{ij}	L_{ij}	$\frac{\delta_k^i}{C_{jk}}$
Thermal homogeneous continuous dynamical system	R_{ij}	—	$\frac{\delta_k^i}{C_{jk}}$

Table 8.3. The analogous admittance physical components.

	P-type $\frac{\delta^i_k}{D^P_{jk}}$	I-type $\frac{\delta^i_k}{K^D_{jk}}$	D-type $\frac{\delta^i_k}{P^I_{jk}}$
Electrical homogeneous continuous dynamical system	$\boldsymbol{G}^{ij} = \frac{\delta^i_k}{R_{jk}}$	$\frac{\delta^i_k}{L_{jk}}$	$\boldsymbol{C}^{ji}$
Magnetic homogeneous continuous dynamical system	$\boldsymbol{G}^{ij} = \frac{\delta^i_k}{R_{jk}}$	—	—
Mechanical homogeneous continuous dynamical system—translation	$\boldsymbol{b}^{ji}$	$\boldsymbol{k}^{ji}$	$\boldsymbol{m}^{ji}$
Mechanical homogeneous continuous dynamical system—rotation	$\boldsymbol{B}^{ji}$	$\boldsymbol{K}^{ji}$	$\boldsymbol{J}^{ji}$
Fluidic homogeneous continuous dynamical system	$\boldsymbol{G}^{ij} = \frac{\delta^i_k}{R_{jk}}$	$\frac{\delta^i_k}{L_{jk}}$	$\boldsymbol{C}^{ji}$
Thermal homogeneous continuous dynamical system	$\boldsymbol{G}^{ij} = \frac{\delta^i_k}{R_{jk}}$	—	$\boldsymbol{C}^{ji}$

Reference

Maxwell J C 1873 *A Treatise of Electricity and Magnetism* (Oxford: Clarendon)

IOP Publishing

Mechatronics
Dynamical systems approach and theory of holors
B T Fijalkowski

Chapter 9

Principal physical commutation matrixers as physical models

The fascinating impressiveness of rigorous mathematical analysis, with its atmosphere of precision and elegance, should not blind us to the defects of the premises that condition the whole process.

T C Chamberlin

9.1 Introduction

There are two principal procedures to show a physical heterogeneous continuous dynamical hypersystem in a diagram. In one, the particulars of the physical heterogeneous continuous dynamical hypersystem connections are exposed. Diagrams of these features are not easy to put into practice. In the other, the particulars of the connections are neglected. As an alternative, the dynamical components are arrayed in a physical commutation matrixer for maximum intelligibility, and to display how energy-transfer physical holors and energy-potential-difference holors integrate or separate. The physical commutation matrixer diagrams for this depiction ought to be those generally used for analysing the action of a physical heterogeneous continuous dynamical hypersystem. There are conventional procedures for creating equilibrium holor equations by well-defined reference to these physical commutation matrixer diagrams. This eases the formulation of equilibrium holor equations to represent the signal holors.

Physical commutation matrixer diagrams with the same form as electrical-commutation-matrixer diagrams may symbolise the interactions of the essential physical dynamical components in mechanical, fluidic and thermal homogeneous continuous dynamical systems. This creates the same analytical procedures that are accessible for the above-mentioned systems. The effect is that equilibrium holor equations for physical heterogeneous continuous dynamical hypersystems of mechanical, fluidic and thermal dynamical components can be created with the same knowledge and dependability.

9.2 Creation of physical commutation matrixers

The aim of physical commutation matrixer creation is to arrange mechanical, electrical, fluidic and thermal dynamical component configurations into physical commutation matrixer diagrams.

The core of the problem is to detect those points in the physical heterogeneous continuous dynamical hypersystem that may have various values of the energy-potential-difference holor. More to the point, each physical commutation matrixer must have a 'datum' or 'reference' value for the energy-potential-difference holor.

The datum is principally termed the 'ground' of the physical commutation matrixer. The other physical commutation matrixer points are termed 'nodes'. For physical heterogeneous continuous dynamical hypersystems, the number of nodes (excluding a datum) is the same as the number of **degrees of freedom** (DoF).

The procedure for physical-matrixer creation necessitates the detection of the principal effects. In particular, one must detect all the energy-potential-difference holor nodes and then authenticate the adequate physical dynamical component or dynamical components that connect them together. The nodes are then converted into the connections of the physical commutation matrixer branches. Each branch displays one physical dynamical component and relates to one principal physical situation. The procedure is displayed in the examples given in chapters 10–13.

9.3 Principal physical commutation matrixer laws

The objective of physical-commutation-matrixer holor analysis is to derive mathematical relations between output-signal holors and input-signal holors.

The physical commutation matrixer representation of a physical heterogeneous continuous dynamical hypersystem permits one to accomplish this analysis through the application of two principal physical commutation matrixer laws. One of these laws, **Kirchhoff's first law** (KFL), refers to physical conditions at any junctions, and is expressed as:

$$\Sigma V^j = 0. \tag{9.1}$$

For instance, the summation of current holors at a junction of an electrical commutation matrixer is analogous to the summation of force holors in a mechanical commutation matrixer or to the summation of fluid-flow-rate holors in a fluidic commutation matrixer. The fact that the holor-algebraic sum of each of these energy-transfer holors at a junction equals zero is simply an application of the principal ideas of continuity and equilibrium. Since the energy-potential-difference holor in mechanical systems is the force holor, **Kirchhoff's second law** (KSL) can be recognised as being analogous to d'Alembert's principle.

The other physical commutation matrixer law, KSL, refers to physical conditions around a closed path in the physical commutation matrix, and is:

$$\Sigma F_i = 0. \tag{9.2}$$

For instance, the summation of voltage-drop holors in an electrical commutation matrixer is analogous to the summation of velocity-difference holors in mechanical commutation matrixer or the summation of pressure-drop holors in a fluidic commutation matrixer.

The summation of these energy-potential physical holors is sometimes referred to as the compatibility relation.

Although both of Kirchhoff's laws are usually stated (equations (9.1) and (9.2)) as the holor-algebraic sum being zero, it is often simpler to equate one of the physical holors to the sum of the others. For instance, the energy-transfer holor flow into a junction may be set equal to the energy-transfer holor flow out of the junction.

When the principal physical commutation matrixer laws are used in this way it is usually possible to set up the equilibrium holor equations in a form that is suitable for direct mnemonic solution. The generalisation of this procedure for complex physical commutation matrixers is considered in chapter 10. However, for simple series, parallel, series-parallel and parallel-series physical commutation matrixers, one may apply the principal physical commutation matrixer laws without formal procedures.

IOP Publishing

Mechatronics
Dynamical systems approach and theory of holors
B T Fijalkowski

Chapter 10

Generalised impedance and admittance holors of physical commutation matrixers

Always surpass the one you were yesterday.

Yagyu Sekishusai

10.1 Sinusoidal responses to energy-potential-difference holor and energy-transfer holor functions

The holor equations for the generalised physical commutation matrixers were given in equations (8.159) and (8.160), and are

$$\boldsymbol{F}_i = \boldsymbol{Z}_{ij}\boldsymbol{V}^j \tag{10.1}$$

and

$$\boldsymbol{V}^j = \boldsymbol{Y}^{ji}\boldsymbol{F}_i \tag{10.2}$$

where $\boldsymbol{Z}_{ij}$ is the generalised impedance holor for general physical commutation matrixers and $\boldsymbol{Y}^{ji}$ is the generalised admittance holor for general physical commutation matrixers.

10.2 Impedance and admittance of principal physical commutation matrixer dynamical components

It should be possible to obtain a procedure for writing equilibrium holor equations that is homogeneously right and proper for finding the natural response, the forced response, or the complete response.

One may realise this by describing the principal physical commutation matrixer dynamical components in holor-algebraic terms rather than as time derivatives or integrals. The holor-algebraic substitution applied to express the principal physical commutation matrixer dynamical components is termed the 'generalised impedance holor'.

The generalised impedance holor of a physical commutation matrixer dynamical component is defined as the ratio of the amplitude or effective (rms) value of the energy-potential-difference holor to the amplitude or effective (rms) value of the corresponding energy-transfer holor.

The generalised impedance holor, $\boldsymbol{Z}_{ij}$, is a function of the exponent holor, $s = (\sigma, \omega)$, and is expressed mathematically as:

$$\boldsymbol{Z}_{ij} = \frac{\text{effective value of an energy-potential-difference holor}}{\text{effective value of the corresponding energy-transfer holor}}. \quad (10.3)$$

To find the impedance holors of the three principal physical commutation matrixer dynamical components we begin with the elementary relationships given in equation (8.1). The relationships are

$$\boldsymbol{F}_i = \boldsymbol{Z}_{ij}^P \boldsymbol{V}^j = \boldsymbol{D}_{ij}^P \boldsymbol{V}^j \quad P\text{-type energy dissipation}, \quad (10.4a)$$

$$\boldsymbol{F}_i = \boldsymbol{Z}_{ij}^D \boldsymbol{V}^j = \omega \boldsymbol{K}_{ij}^D \boldsymbol{V}^j \quad D\text{-type energy storage}, \quad (10.4b)$$

$$\boldsymbol{F}_i = \boldsymbol{Z}_{ij}^I \boldsymbol{V}^j = \frac{1}{\omega} \boldsymbol{P}_{ij}^I \boldsymbol{V}^j \quad I\text{-type energy storage}. \quad (10.4c)$$

Now we assume that both the energy-potential-difference holor and the energy-transfer holor can be symbolised by holor functions. That is,

$$\boldsymbol{F}_i = (F_c, F_s) == F\angle\psi_F = \|F_c \quad F_s\|^T = \begin{Vmatrix} F_c \\ F_s \end{Vmatrix} = \begin{Vmatrix} F\cos\psi_F \\ F\sin\psi_F \end{Vmatrix} = \boldsymbol{Z}_{ij}\boldsymbol{V}^j, \quad (10.5a)$$

and

$$\boldsymbol{V}^j = (V^c, V^s) = V\angle\psi_V = \|V_c \quad V_s\|^T = \begin{Vmatrix} V^c \\ V^s \end{Vmatrix} = \begin{Vmatrix} V\cos\psi_V \\ V\sin\psi_V \end{Vmatrix} = \boldsymbol{Y}^{ji}\boldsymbol{F}_i. \quad (10.5b)$$

When equations (10.5) are substituted into equations (10.4), we are left with a relationship between the effective (rms) values F and V. If we use the definition of generalised impedance holors as the ratio of F/V, we obtain that:

$$\boldsymbol{Z}_{ij}^P = \boldsymbol{D}_{ij}^P \quad P\text{-type energy dissipation}, \quad (10.6a)$$

$$\boldsymbol{Z}_{ij}^D = \omega \boldsymbol{K}_{ij}^D \quad D\text{-type energy storage}, \quad (10.6b)$$

$$\boldsymbol{Z}_{ij}^I = \frac{1}{\omega} \boldsymbol{P}_{ij}^I \quad I\text{-type energy storage}, \quad (10.6c)$$

where

$$D_{ij}^{P} = \begin{bmatrix} D & 0 \\ 0 & D \end{bmatrix}, \tag{10.7a}$$

$$K_{ij}^{D} = \begin{bmatrix} 0 & -K \\ K & 0 \end{bmatrix}, \tag{10.7b}$$

and

$$P_{ij}^{I} = \begin{bmatrix} 0 & P \\ -P & 0 \end{bmatrix}. \tag{10.7c}$$

Equations (10.6) are therefore the holor-algebraic description of the principal physical commutation matrixer dynamical components that we previously described in equations (10.4) with holor frequency differentials and integrals.

Table 10.1 lists the generalised impedance holors in the various physical homogeneous continuous dynamical systems.

The reciprocal of the generalised impedance holor multiplied by a Kronecker delta is normally a more appropriate procedure to express the principal physical commutation matrixer dynamical components. This is termed the 'generalised admittance holor', $\boldsymbol{Y}^{ji}$, and is defined as

$$\boldsymbol{Y}^{ji} = \frac{\text{effective value of an energy-transfer holor}}{\text{effective value of the corresponding energy-potential-difference holor}}. \tag{10.8}$$

Before proceeding further, it is indispensable to know that conventionally mechanical engineers have used the term 'mechanical impedance holor' with a certain perception of force—*the larger the force necessitated to cause movement, the greater the holor mechanical impedance.* With this assumption, the mechanical impedance holor is

Table 10.1. The analogous impedance physical dynamical components.

	P-type D_{ij}^{P}	D-type ωK_{ij}^{D}	I-type $\frac{1}{\omega}P_{ij}^{I}$
Electrical homogeneous continuous dynamical system	$\boldsymbol{R}_{ij}$	$\omega \boldsymbol{L}_{ij}$	$\frac{1}{\omega}\frac{\delta_k^i}{C_{jk}}$
Magnetic homogeneous continuous dynamical system	$\boldsymbol{R}_{ij}$	—	—
Mechanical homogeneous continuous dynamical system—translation	$\frac{\delta_k^i}{b_{jk}}$	$\omega\frac{\delta_k^i}{k_{jk}}$	$\frac{1}{\omega}\frac{\delta_k^i}{m_{jk}}$
Mechanical homogeneous continuous dynamical system—rotation	$\frac{\delta_k^i}{B_{jk}}$	$\omega\frac{\delta_k^i}{K_{jk}}$	$\frac{1}{\omega}\frac{\delta_k^i}{J_{jk}}$
Fluidic homogeneous continuous dynamical system	$\boldsymbol{R}_{ij}$	$\omega \boldsymbol{L}_{ij}$	$\frac{1}{\omega}\frac{\delta_k^i}{C_{jk}}$
Thermal homogeneous continuous dynamical system	$\boldsymbol{R}_{ij}$	—	$\frac{1}{\omega}\frac{\delta_k^i}{C_{jk}}$

determined by V/F or F/V. Having preferred the theory of the energy-transfer/energy-potential-difference holors as the principal assumption, this textbook uses the electrical commutation matrixer concept of interpreting the generalised impedance holor as the ratio F/V or V/F. In some of the mechatronics literature, the ratio F/V or V/F is termed the 'mobility'.

In spite of this, this text will be understood much more easily if the term 'generalised impedance holor' is used solely to symbolise the ratio of the energy-potential-difference holor to the energy-transfer holor, while the term 'generalised admittance holor' is used solely for the ratio of the energy-transfer holor to the energy-potential-difference holor. This procedure is respected throughout this book.

10.3 Generalised impedance and admittance holors in combination

Generalised impedance and admittance holors can be combined in much the same way as equivalent resistance holors are obtained for series and parallel connections of resistors in electrical dynamical systems. The essential difference is that the energy-storage physical commutation matrixer dynamical components, being terms in ω or $1/\omega$, must have their impedance or admittance holors added algebraically so that in general, an equivalent impedance holor or admittance holor is a polynomial in ω.

The determination of a series or parallel connection must only be made by inspection of the analogous physical matrixer. That is, it is only when the analogous physical commutation matrixer shows unequivocally that the flow of the energy-potential-difference holor is identical for two or more physical commutation matrixer dynamical components that they may be considered as a series physical commutation matrixer.

Similarly, it is only when the analogous physical commutation matrixer shows unequivocally that the energy-potential holor is identical to two or more physical commutation matrixer dynamical components that they may be considered as a parallel physical matrixer.

Figure 10.1(a) shows three impedance holors ($\boldsymbol{Z}_{ij}^1$, $\boldsymbol{Z}_{ij}^2$ and $\boldsymbol{Z}_{ij}^3$) interconnected so that there is a common energy-potential holor. This is, therefore, a series physical commutation matrixer.

Suppose one wishes to represent the physical commutation matrixer using the one with the single equivalent impedance holor ($\boldsymbol{Z}_{ij}^{\mathrm{E}}$) shown in figure 10.1(b). This requires that the applied energy-potential-difference holor ($\boldsymbol{F}_i^{\mathrm{T}}$) produce the same energy-transfer holor in both physical commutation matrixers.

The applied energy-potential-difference holor in figure 10.1(a) may be expressed as:

$$\boldsymbol{F}_i^T = \boldsymbol{F}_i^1 + \boldsymbol{F}_i^2 + \boldsymbol{F}_i^3. \tag{10.9}$$

Alternatively, in terms of impedance holors (equation 10.3), equation (10.9) is:

$$\boldsymbol{F}_i^T = \boldsymbol{Z}_{ij}^1 \boldsymbol{V}^j + \boldsymbol{Z}_{ij}^2 \boldsymbol{V}^j + \boldsymbol{Z}_{ij}^3 \boldsymbol{V}^j = \left(\boldsymbol{Z}_{ij}^1 + \boldsymbol{Z}_{ij}^2 + \boldsymbol{Z}_{ij}^3\right)\boldsymbol{V}^j. \tag{10.10}$$

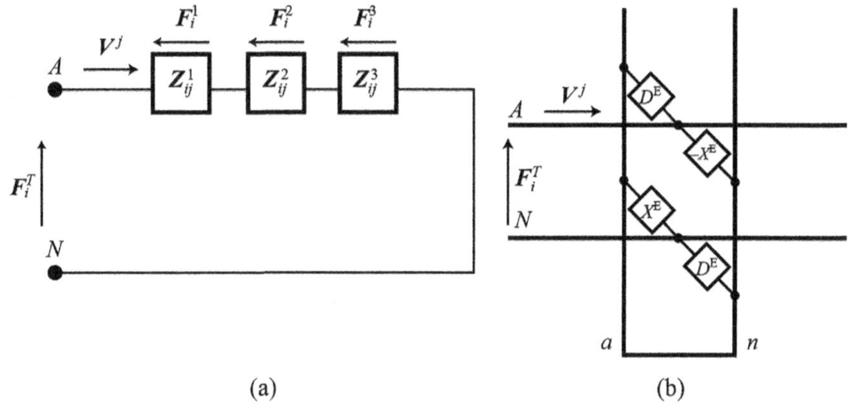

Figure 10.1. Impedance holors in series.

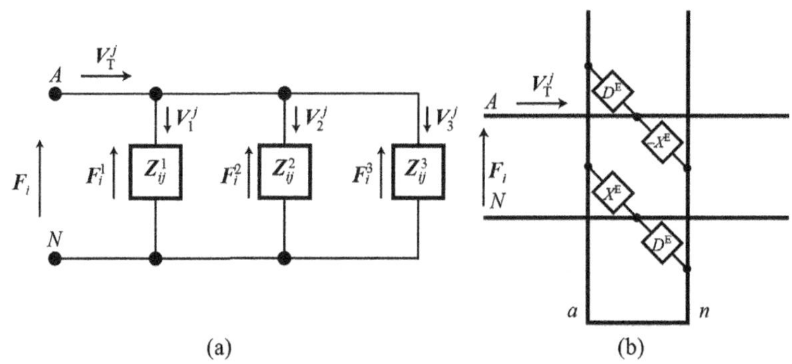

Figure 10.2. Impedance holors in parallel.

The relation between the applied energy-potential difference holor and the energy-transfer holor is in the equivalent physical commutation matrixer (figure 10.1(b)) is:

$$F_i^T = Z_{ij}^E V^j = \begin{bmatrix} F^c \\ F^s \end{bmatrix} = \begin{bmatrix} Z_{cc}^E & -Z_{cs}^E \\ Z_{sc}^E & Z_{ss}^E \end{bmatrix} \begin{bmatrix} V_c \\ V_s \end{bmatrix} = \begin{bmatrix} D^E & -X^E \\ X^E & D^E \end{bmatrix} \begin{bmatrix} V_c \\ V_s \end{bmatrix}. \tag{10.11}$$

A comparison of equations (10.10) and (10.11) shows that the equivalent impedance holor for a series physical commutation matrixer is:

$$Z_{ij}^E = Z_{ij}^1 + Z_{ij}^2 + Z_{ij}^3. \tag{10.12}$$

Let us now consider the equivalent impedance holor for a parallel physical commutation matrixer. Figure 10.2(a) shows three impedance holors (Z_{ij}^1, Z_{ij}^2 and Z_{ij}^3) that have a common energy-potential-difference holor and are therefore in parallel to replace this physical commutation matrixer with a single equivalent impedance holor (Z_{ij}^{E}).

Figure 10.2(b), an energy-transfer physical holor V_T^j, must produce the same energy-potential-difference holor in each physical commutation matrixer.

For the parallel physical matrixer in figure 10.2(a):

$$V_T^j = \left(\frac{\delta_{k(1)}^i}{Z_{jk}^1} + \frac{\delta_{k(2)}^i}{Z_{jk}^2} + \frac{\delta_{k(3)}^i}{Z_{jk}^3} \right) F_i^T. \tag{10.13}$$

For the equivalent physical commutation matrixer shown in figure 10.2(b):

$$\begin{aligned} V_T^j &= \left(\frac{\delta_{k(E)}^i}{Z_{jk}^E} \right) F_i^T = Y_E^{ji} F_i^T = \begin{bmatrix} V_T^s \\ V_T^c \end{bmatrix} = \begin{bmatrix} Y_E^{ss} & Y_E^{sc} \\ Y_E^{cs} & Y_E^{cc} \end{bmatrix} \begin{bmatrix} F_c^T \\ F_s^T \end{bmatrix} \\ &= \begin{bmatrix} G_E & B_E \\ -B_E & G_E \end{bmatrix} \begin{bmatrix} F_c^T \\ F_s^T \end{bmatrix}. \end{aligned} \tag{10.14}$$

From equations (10.13) and (10.14) one may observe that the equivalent admittance holor for a parallel physical commutation matrixer is:

$$\frac{\delta_{k(E)}^i}{Z_{jk}^E} = \frac{\delta_{k(1)}^i}{Z_{jk}^1} + \frac{\delta_{k(2)}^i}{Z_{jk}^2} + \frac{\delta_{k(3)}^i}{Z_{j-k}^3}. \tag{10.15}$$

Figure 10.3(a) shows the admittance holors (Y_1^{ji}, Y_2^{ji} and Y_3^{ji}) in series and figure 10.3(b) shows an equivalent admittance holor Y_E^{ji} representation.

One may determine the relation between the equivalent admittance holor and the individual admittance holors by proceeding directly from the series physical commutation matrixer summation of energy-potential-difference holors given in equation (10.9).

However, since we know that the admittance holor is the multiplied-by-Kronecker-delta reciprocal of the impedance holor, we may use the series physical commutation matrixer result given in equation (10.12) to obtain:

$$\frac{\delta_{k(E)}^i}{Y_E^{ij}} = \frac{\delta_{k(1)}^i}{Y_1^{ij}} + \frac{\delta_{k(2)}^i}{Y_2^{ij}} + \frac{\delta_{k(3)}^i}{Y_3^{ij}}. \tag{10.16}$$

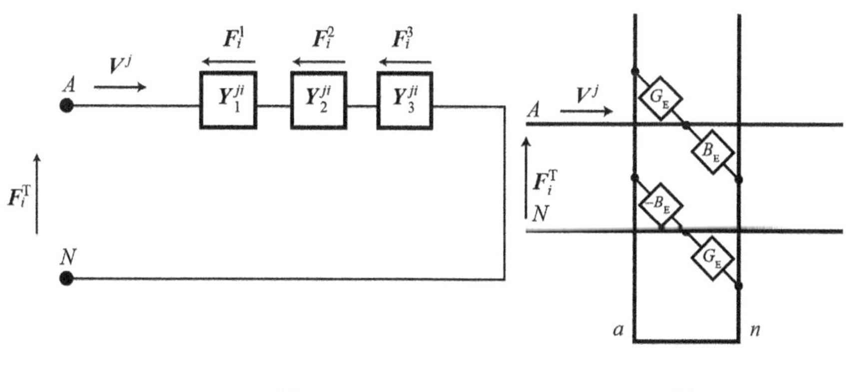

Figure 10.3. Admittance holors in series.

In a similar way, one may find the equivalent admittance holor in parallel. Thus, if one inverts the equivalent impedance holor and the individual impedance holors for the parallel physical commutation matrixer given in equation (10.15), the result is:

$$\boldsymbol{Y}_E^{ji} = \boldsymbol{Y}_1^{ji} + \boldsymbol{Y}_2^{ji} + \boldsymbol{Y}_3^{ji}. \tag{10.17}$$

Admittance holors in parallel are therefore additive, just like impedance holors in series.

Physical commutation matrixer dynamical components may be represented by either impedance holors or admittance holors, whichever is more convenient for a particular physical commutation matrixer. Care must be exercised, however, to indicate clearly whether an impedance holor or an admittance holor is being used.

10.3.1 Driving-point impedance and/or admittance holors

The ratio of an energy-potential-difference holor's effective value to an energy-transfer holor's effective value as observed across the input port of the physical commutation matrixer is termed the 'driving-point' impedance holor ($\boldsymbol{Z}_{ij}^{\mathrm{DP}}$).

One may obtain the driving point impedance holor by combining impedance holors in series and parallel until the entire physical commutation matrixer is represented by a single equivalent impedance holor. The driving-point admittance holor is merely the multiplied-by-Kronecker-delta (reciprocal) of the driving-point impedance holor. The following examples will clarify the procedure.

Figure 10.4(a) shows a physical matrixer with impedance holors, $\boldsymbol{Z}_{ij}^1$, $\boldsymbol{Z}_{ij}^2$, $\boldsymbol{Z}_{ij}^3$ and $\boldsymbol{Z}_{ij}^4$.

One ought to find the driving-point-impedance holor across the port A–N. The procedure is to apply the rules of series and parallel physical commutation matrixers (equations (10.12) and (10.15)) in sequence until only one equivalent impedance holor remains. Accordingly, one begins by combining the series connection of $\boldsymbol{Z}_{ij}^3$ and $\boldsymbol{Z}_{ij}^4$ into an equivalent impedance holor, $\boldsymbol{Z}_{ij}^3 + \boldsymbol{Z}_{ij}^4$. The physical commutation matrixer then appears as shown in figure 10.4(b).

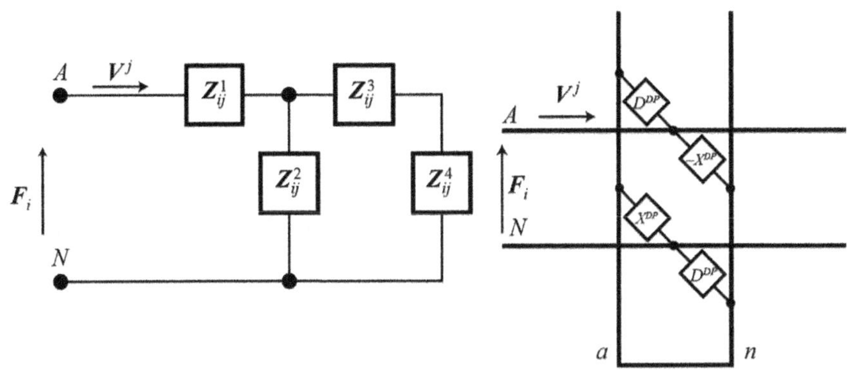

(a) (b)

Figure 10.4. Determination of the driving-point-impedance holor.

Now the parallel combination of $\mathbf{Z}^2_{ij}$ and $\mathbf{Z}^3_{ij} + \mathbf{Z}^4_{ij}$ produces an equivalent impedance holor, according to equation (10.15), of:

$$\frac{\delta^i_{k(E)}}{\mathbf{Z}^E_{jk}} = \frac{\delta^i_{k(2)}}{\mathbf{Z}^2_{jk}} + \frac{\delta^i_{k(3\&4)}}{\mathbf{Z}^3_{jk} + \mathbf{Z}^4_{jk}},$$

or

$$\mathbf{Z}^E_{ij} = \frac{\delta^i_k}{\mathbf{Z}^2_{jk} + \mathbf{Z}^3_{jk} + \mathbf{Z}^4_{jk}} \mathbf{Z}^2_{ij}\left(\mathbf{Z}^3_{ij} + \mathbf{Z}^4_{ij}\right).$$

With this combination, the physical commutation matrixer reduces to the simple series physical commutation matrixer shown in figure 10.4(b).

Finally, the addition of the two impedance holors in series permits us to represent the physical commutation matrixer by a single impedance holor. This is the driving-point impedance holor

$$\begin{aligned}\mathbf{Z}^{DP}_{ij} &= \mathbf{Z}^1_{ij} + \frac{\delta^i_k}{\mathbf{Z}^2_{jk} + \mathbf{Z}^3_{jk} + \mathbf{Z}^4_{jk}} \mathbf{Z}^2_{ij}\left(\mathbf{Z}^3_{ij} + \mathbf{Z}^4_{ij}\right) \\ &= \frac{\delta^i_k}{\mathbf{Z}^2_{jk} + \mathbf{Z}^3_{jk} + \mathbf{Z}^4_{jk}}\left(\mathbf{Z}^1_{ij}\mathbf{Z}^2_{ij} + \mathbf{Z}^1_{ij}\mathbf{Z}^3_{ij} + \mathbf{Z}^1_{ij}\mathbf{Z}^4_{ij} + \mathbf{Z}^2_{ij}\mathbf{Z}^3_{ij} + \mathbf{Z}^2_{ij}\mathbf{Z}^4_{ij}\right).\end{aligned}$$

Figure 10.5(a) shows a physical commutation matrixer in which the dynamical components have been represented by their admittance holors $\mathbf{Y}^j_1$, $\mathbf{Y}^j_2$, $\mathbf{Y}^j_3$ and $\mathbf{Y}^j_4$.

One ought to find the driving-point-admittance holor across the port *A*–*N*.

Thus, one must apply the rules for combining admittance holors (equations (10.16) and (10.17)). The series combinations of $\mathbf{Y}^j_1$, $\mathbf{Y}^j_2$, and $\mathbf{Y}^j_3$, $\mathbf{Y}^j_4$ are shown in figure 10.5(b). This follows directly from equation (10.16). The two resulting

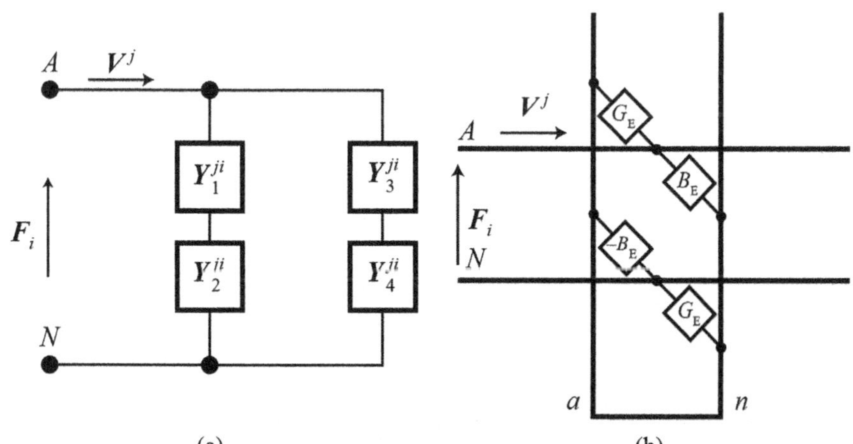

Figure 10.5. Determination of the driving-point-admittance holor.

admittance holors are in parallel and one may be treated by direct addition (equation (10.17)).

The result is:

$$\boldsymbol{Y}_E^{ji} = \frac{\delta_k^i}{\boldsymbol{Y}_1^{ij} + \boldsymbol{Y}_2^{ij}} \boldsymbol{Y}_1^{ji} \boldsymbol{Y}_2^{ji} + \frac{\delta_k^i}{\boldsymbol{Y}_3^{ij} + \boldsymbol{Y}_4^{ij}} \boldsymbol{Y}_3^{ji} \boldsymbol{Y}_4^{ji}$$

$$= \frac{\delta_k^i}{\boldsymbol{Y}_1^{ij}\boldsymbol{Y}_3^{ij} + \boldsymbol{Y}_2^{ij}\boldsymbol{Y}_3^{j} + \boldsymbol{Y}_1^{ij}\boldsymbol{Y}_4^{ij} + \boldsymbol{Y}_2^{ij}\boldsymbol{Y}_4^{ij}} \left(\boldsymbol{Y}_1^{ji}\boldsymbol{Y}_2^{ji}\boldsymbol{Y}_3^{ji} + \boldsymbol{Y}_1^{ji}\boldsymbol{Y}_2^{ji}\boldsymbol{Y}_4^{ji} + \boldsymbol{Y}_1^{ji}\boldsymbol{Y}_3^{ji}\boldsymbol{Y}_4^{ji} + \boldsymbol{Y}_2^{ji}\boldsymbol{Y}_3^{ji}\boldsymbol{Y}_4^{ji}\right).$$

10.4 Physical heterogeneous continuous dynamical hypersystems' transmittance or transfer-function holors

In the previous examples, the ratio of the sinusoidal response to the excitation holor has been either a generalised admittance holor or a generalised impedance holor, since a suitable energy-transfer holor was chosen as the sinusoidal response to an energy-potential-holor excitation, or a suitable energy-potential-difference holor was chosen as the response to an energy-transfer-holor excitation.

In previous chapters, there have been examples where this has not been the case (e.g. the excitation and sinusoidal response are either both energy-potential-difference holors or both energy-transfer holors). It is usual to define a 'physical heterogeneous continuous dynamical hypersystem's transmittance or transfer-function holor', $\boldsymbol{G}_{ij}$, for the ratio of two polynomials in $\boldsymbol{s} = (\sigma, \omega)$.

$$\boldsymbol{G}_{ij} = \frac{b_i\omega^i + b_{i-1}\omega^{i-1} + \cdots + b_0}{a_j\omega^j + a_{j-1}\omega^{j-1} + \cdots + a_0} \tag{10.18a}$$

or

$$\boldsymbol{G}_{ij} = \frac{(\omega - \omega_a)(\omega - \omega_b)(\omega - \omega_c)\cdots(\omega - \omega_i)}{(\omega - \omega_1)(\omega - \omega_2)(\omega - \omega_3)\cdots(\omega - \omega_j)}, \tag{10.18b}$$

where a_j, b_i are the coefficients of the polynomials and ω_a, ω_b, ..., ω_1, ω_2 refer to the roots of the polynomials.

Although, strictly speaking, a physical heterogeneous continuous dynamical hypersystem's transmittance or transfer function is obtained for the forced response only, it is a simple matter to 'recover' the original differential equation by recalling the association between ω and d/dt, and $1/\omega$ and $\int \mathrm{d}t$.

It should also be noted that a physical heterogeneous continuous dynamical hypersystem's transmittance or transfer function implies a single exponential energy source. As a result, the physical heterogeneous continuous dynamical system's transmittance or transfer-function concept is valid only when the energy-storage physical commutation matrixer dynamical components have no initial energy. This is because the initial energy storage is physically modelled with an additional energy source (chapter 8). There are two situations—the energy-potential-difference-holor divider and the energy-transfer-holor divider—that often appear in physical commutation matrixers and are readily treated with physical heterogeneous continuous hypersystem energy-transfer-holor functions.

Figure 10.6 shows the energy-potential-difference-holor divider and the energy-transfer-holor divider physical matrixers.

The energy-potential difference-holor divider physical commutation matrixer (figure 10.6(a)) is basically a series physical commutation matrixer and has only a single energy-transfer holor associated with it.

The physical heterogeneous continuous dynamical hypersystem's transmittance or transfer-function holor that relates the energy-potential-difference-holor output, ($\boldsymbol{F}_i^{\text{out}}$), and the energy-potential-difference-holor input, ($\boldsymbol{F}_i^{\text{in}}$), is:

$$\boldsymbol{G}_{ij} = \frac{\boldsymbol{F}_i^{\text{out}}}{\boldsymbol{F}_i^{\text{in}}} = \frac{\delta_k^i}{\boldsymbol{Z}_{jk}^1 + \boldsymbol{Z}_{jk}^2}\boldsymbol{Z}_{ij}^2 \qquad (10.19a)$$

or

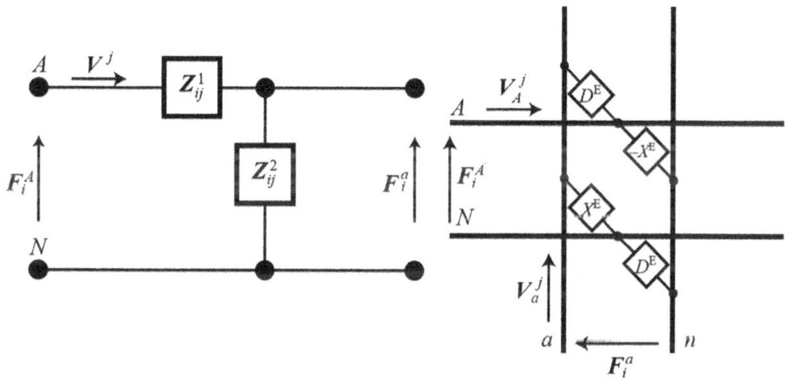

(a) ENERGY POTENTIAL-DIFFERENCE DIVIDER

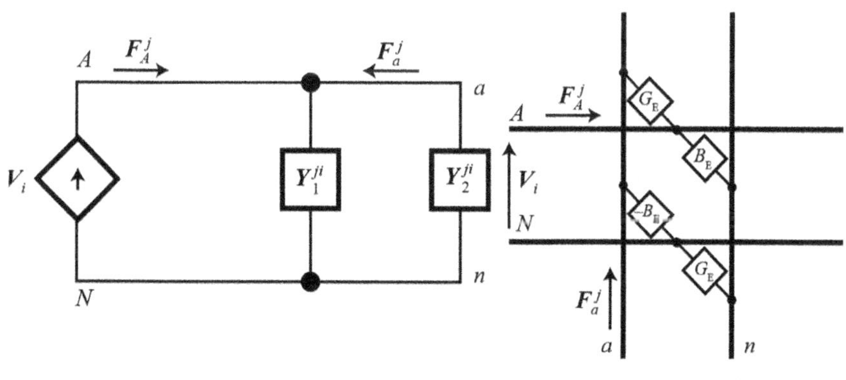

(b) ENERGY TRANSFER DIVIDER

Figure 10.6. Energy-potential-holor divider and energy-transfer-holor divider physical commutation matrixers.

$$G_{ij} = \frac{F_i^{\text{out}}}{F_i^{\text{in}}} = \frac{\delta_k^i}{Y_1^{ij} + Y_2^{ij}} Y_1^{ji}, \tag{10.19b}$$

where equation (10.19b) represents the physical heterogeneous continuous dynamical hypersystem's transmittance or transfer-function holor for the energy-potential-difference-holor divider in terms of admittance holors.

The energy-transfer-holor divider's physical commutation matrixer (figure 10.6(b)) is a parallel physical commutation matrixer. There is only one energy-potential difference-holor across both impedance holors. The physical heterogeneous continuous dynamical hypersystem's energy-transfer holor for the ratio of the energy-transfer-holor output, (V_{out}^j), to the energy-transfer-holor input, (V_{in}^j), is:

$$G_{ij} = \frac{V_{\text{out}}^j}{V_{\text{in}}^j} = \frac{\delta_k^i}{Y_1^{ij} + Y_2^{ij}} Y_2^{ji}, \tag{10.20a}$$

or

$$G_{ij} = \frac{V_{\text{out}}^j}{V_{\text{in}}^j} = \frac{\delta_k^i}{Z_{jk}^1 + Z_{jk}^2} Z_{ij}^1, \tag{10.20b}$$

where equation (10.20b) is the physical heterogeneous continuous dynamical hypersystem's transmittance or transfer-function holor for the energy-transfer-holor divider's physical commutation matrixer in terms of impedance holors.

10.5 General formulation of physical-heterogeneous-continuous-dynamical-hypersystem holor equations

One of the main reasons for seeking a physical commutation matrixer representation of a physical heterogeneous continuous dynamical hypersystem is that it facilitates the procedure of determining the physical heterogeneous continuous dynamical hypersystem equilibrium holor equations.

Simple algorithms exist from which these equations can be mnemonically written, often directly from an inspection of the physical commutation matrixer diagram, which represents a physical model.

In this way, the principal techniques of physical commutation matrixer holor analysis can be applied to physical heterogeneous continuous dynamical hypersystems, other than an electrical homogeneous continuous dynamical system, and the analogous properties of such physical homogeneous continuous dynamical systems become quite evident. Physical commutation matrixer holor analysis is based on the two Kirchhoff laws. These were considered in chapter 9 (equations (9.1) and (9.2)), but are restated here in the form that is most convenient for the physical commutation matrixer holor analysis of more complicated physical commutation matrixers than have been considered so far.

For Kirchhoff's first law:

$$\sum V_{\mathrm{P}}^{j} = \sum V_{\mathrm{S}}^{j}, \tag{10.21a}$$

which means that the sum of all the energy-transfer holors leaving a junction through the passive physical commutation matrixer components connected directly to the junction, V_{P}^{j}, is equal to the sum of all the energy-transfer holors entering the junction from the energy-transfer-holor sources, V_{S}^{j}.

For Kirchhoff's second law:

$$\sum F_{i}^{\mathrm{P}} = \sum F_{i}^{\mathrm{S}}, \tag{10.21b}$$

which states that the sum of the energy-potential difference-holor drops across the passive physical commutation matrixer dynamical components, F_{i}^{P}, around a closed physical commutation matrixer path is equal to the sum of the energy-potential difference holor increases due to the energy-potential difference-holor sources, F_{i}^{S}.

Up to this point, the most complicated physical heterogeneous continuous dynamical hypersystem that has been described in detail (by means of one or two equilibrium holor equations) only has three physical commutation matrixer dynamical components. One is now in a position to develop systematic procedures to obtain sets of equilibrium holor equations from any physical homogeneous continuous dynamical system, no matter how complicated, using its analogous physical commutation matrixer diagram as the common intermediate step.

The physical commutation matrixer diagram is at first simply a graphical representation of the interactions that take place among all the physical commutation matrixer dynamical components of a physical system in qualitative terms, i.e. a physical model.

However, each interaction can be modelled mathematically by means of a simple elemental mathematical relationship and thus the physical commutation matrixer effectively becomes a graphical representation of the equilibrium holor equations.

The problem is, therefore, a matter of finding a method whereby the information contained in the physical commutation matrixer diagram may be qualified quickly and reliably in the form of equilibrium holor equations, which, in general, will be a set of simultaneous holor equations. If one uses the ideas developed in this chapter, one will be able to derive the holor equations more confidently and with less effort.

One therefore implies that all the driving holor functions are sinusoidal, and uses the generalised impedance-holor notation. It should be clearly understood that this does not in any way limit the validity or generality of the resulting holor equations

With some practice, it is a simple matter to write mnemonically the equilibrium holor equations of many physical heterogeneous continuous dynamical hypersystems directly in this form (i.e. 'by inspection' of the physical commutation matrixer), thus reducing the likelihood of error.

We clarify the procedure in the following examples, which are given in chapters 11–13. One may now also formulate the equilibrium holor equations for energy-transfer holors using the mesh method and through inspection of the impedance-holor physical commutation matrixer. We demonstrate the procedures in the examples given in chapter 14, section 14.4.

10.6 Some properties of linear physical heterogeneous continuous dynamical hypersystems

Much of the simplification of the procedure for writing equilibrium holor equations in section 10.5 is due to the basic property of components of a linear physical heterogeneous continuous dynamical hypersystem: holor responses are directly proportional to the driving function (or its derivative or integral). Indeed, this is why the equivalent impedance holor of a series connection is simply the sum of the individual impedance holors.

It should therefore be expected that these basic attributes of the components, which make up a physical heterogeneous continuous dynamical hypersystem, should result in characteristic properties of a linear physical heterogeneous continuous dynamical hypersystem.

Some of these properties are expressed as 'physical commutation matrixer theorems', which greatly simplify the task of eliminating unwanted holors. Others are useful in that they give guidance as to the kinds of measurements that must be taken to obtain a valid mathematical model of the physical heterogeneous continuous dynamical hypersystem.

10.6.1 Two-port single-input/single output physical commutation matrixers

Up to this point, we have frequently focused our attention on the behaviour of a physical commutation matrixer at a specified pair of terminals, i.e. conductive collectors.

Recall that we introduce the Thévenin and Norton equivalent physical commutation matrixer solely to simplify the holor analysis of some physical heterogeneous continuous dynamical systems, to focus our attention on two pairs of terminals, i.e. conductive collectors. One finds this viewpoint convenient in situations where a signal is fed into one pair of terminals, i.e. matrixer-input conductive collectors, and is then extracted at a second pair of terminals, i.e. matrixer-output conductive collectors, after being processed by the physical commutation matrixer. Because the terminal (conductive-collector) pairs represent the points in the physical commutation matrixer where signals are either fed in or extracted, they are referred to as the 'ports' of the physical heterogeneous continuous dynamical hypersystem.

In this section, the author will limit his discussion to physical commutation matrixers in which one has one matrixer-input port and one matrixer-output port. The basic two-port **single-input/single-output** (SISO) physical commutation matrixer is illustrated in figure 10.7.

There are several restrictions on using this physical commutation matrixer. Firstly, no energy can be stored within the physical commutation matrixer. Secondly, there can be no independent energy sources within the physical commutation matrixer. Thirdly, the energy-transfer holor into the port must be equal to the energy-transfer holor out of the port, that is, $\boldsymbol{V}_A^j = \boldsymbol{V}_N^j$ and $\boldsymbol{V}_a^j = \boldsymbol{V}_n^j$. The fourth and

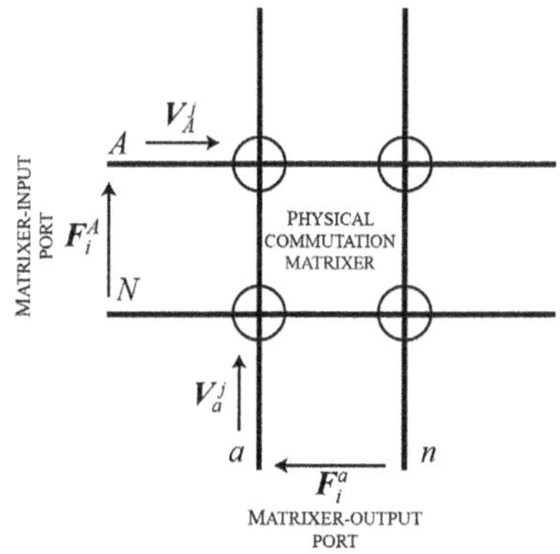

Figure 10.7. The two-port SISO physical commutation matrixer.

final restriction is that all external connections must be made to either the matrixer-input port or the matrixer-output port. No external connections between ports are allowed. In other words, no external connections can be made between the terminals, i.e. the conductive collectors *A–a, A–n, N–a,* or *N–n.* These restrictions simply limit the range of physical commutation matrixer problems to which the two-port SISO formulation is amenable.

The fundamental principle underlying two-port SISO physical commutation matrixer modelling is that only the terminal energy-potential-difference holors and energy-transfer holors, that is V_A^j, F_i^A, V_a^j and F_i^a, are of interest. There is no interest in computing the energy-transfer holors and energy-potential-difference holors that exist inside the physical commutation matrixer. In this section, we formalise the approach by introducing two-port parameters.

Two-port SISO physical commutation matrixer holor equations

In viewing a two-port physical commutation matrixer, we are interested in relating the energy-transfer holor and the energy-potential-difference holor at one port to the energy-transfer holor and energy-potential-difference holor at the second port. The reference polarity of the terminal energy-potential-difference holors and the senses of direction for the references of the energy-transfer holors are shown in figure 10.7. It is worth noting that the references at each port are symmetrical with respect to each other, that is, at each port, the energy-transfer holor is sensed into the upper terminal and each port energy-potential-difference holor is a step up from the lower to the upper terminal. This symmetry makes it easier to generalise the holor analysis of a physical commutation matrixer in terms of two-port SISO black boxes and this is the reason for its nearly universal use in the mechatronics literature. The most general description of the two-port SISO physical commutation matrixer may be carried out in holor calculus. For instance, in the case of purely passive resistive physical commutation matrixers the holor analysis reduces to solving passive resistive physical commutation matrixers. Sinusoidal steady-state problems can be solved by direct analysis in the holor domain, finding the appropriate holor expressions.

In this book, all the equations are written in the holor domain. Thus passive resistive physical commutation matrixers and sinusoidal steady-state solutions simply become special cases. The basic black box in terms of the holor domain variables, i.e. the holors, V_A^j, F_i^A, V_a^j and F_i^a is shown in figure 10.8.

Of the four terminal holors, V_A^j, F_i^A, V_a^j and F_i^a, only two are independent. Thus, for a given physical commutation matrixer, once one specifies two of the holors, one can find the two remaining unknowns. For instance, given F_i^A and F_i^a plus the physical commutation matrixer as the black box, one can determine V_A^j and V_a^j. It follows that one can describe a two-port SISO physical commutation matrixer with just two simultaneous holor equations. Although one needs just two simultaneous holor equations, one has a choice of six different ways in which to combine the four holors.

The six sets of holor equations are:

$$\left.\begin{aligned} F_i^A &= Z_{ij}^{Aa} V_A^j + Z_{ij}^{An} V_a^j \\ F_i^a &= Z_{ij}^{Na} V_A^j + Z_{ij}^{Nn} V_a^j \end{aligned}\right\}, \tag{10.22a}$$

$$\left.\begin{aligned} V_A^j &= Y_{aA}^{ji} F_i^A + Y_{nA}^{ji} F_i^a \\ V_a^j &= Y_{aN}^{ji} F_i^A + Y_{nN}^{ji} F_i^a \end{aligned}\right\}, \tag{10.22b}$$

$$\left.\begin{aligned} F_i^A &= A_{aA}^{ji} F_i^a - A_{ij}^{An} V_a^j \\ V_A^j &= A_{aN}^{ji} F_i^a - A_{ij}^{Nn} V_a^j \end{aligned}\right\}, \tag{10.22c}$$

$$\left.\begin{aligned} F_i^a &= B_{aA}^{ji} F_i^A - B_{ij}^{An} V_A^j \\ V_a^j &= B_{aN}^{ji} F_i^A - B_{ij}^{Nn} V_A^j \end{aligned}\right\}, \tag{10.22d}$$

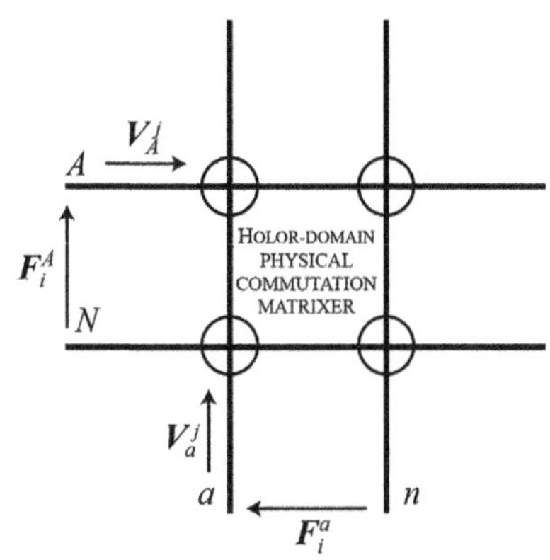

Figure 10.8. The holor-domain two-port SISO physical commutation matrixer.

$$\left.\begin{aligned} F_i^A &= H_{ij}^{Aa} V_A^j + H_{nA}^{ji} F_i^a \\ V_a^j &= H_{ij}^{Na} V_A^j + H_{nN}^{ji} F_i^a \end{aligned}\right\}, \tag{10.22e}$$

$$\left.\begin{aligned} V_A^j &= G_{aA}^{ji} F_i^a + G_{ij}^{An} V_a^j \\ F_i^a &= G_{aN}^{ji} F_i^A + G_{ij}^{Nn} V_a^j \end{aligned}\right\}. \tag{10.22f}$$

These six sets of holor equations can also be thought of as three pairs of mutually inverse relations.

The first set (equations 10.22a) gives the input and output energy-potential-difference holors as functions of input and output energy-transfer holors, whereas the second set (equations 10.22b) gives the inverse relationship, that is, the input and output energy-transfer holors as functions of input and output energy-potential-difference holors. Further study of the remaining four sets will show that equations (10.22c) and (10.22d) are inverse relations, as are equations (10.22e and (10.22f). The coefficients of the energy transfer and/or energy potential difference holors on the right-hand side of equations (10.22a)–(10.22f) are called the 'parameters' of the two-port SISO physical matrixer. Thus, when using equations (10.22a), one refers to the $\boldsymbol{Z}_{ij}$– parameters of the physical matrixer. Similarly, one refers to the $\boldsymbol{Y}^{ij}$– parameters, the $\boldsymbol{A}_{ij}$ and $\boldsymbol{A}^{ji}$– parameters, the $\boldsymbol{B}_{ij}$ and $\boldsymbol{B}^{ji}$– parameters, the $\boldsymbol{H}_{ij}$ and $\boldsymbol{H}^{ji}$– parameters, and the $\boldsymbol{G}_{ij}$ and $\boldsymbol{G}^{ji}$– parameters of the two-port SISO physical commutation matrixer.

The two-port SISO physical commutation matrixer parameters

The parameters for any given physical commutation matrixer can be found by either computations or measurements. The computation, or measurement, to be made is determined directly from the parameter holor equations. For instance, suppose that one's problem is to find the $\boldsymbol{Z}_{ij}$– parameters for a given physical commutation matrixer. It follows directly from equations (10.22*a*.1) that

$$Z_{ij}^{Aa} = \left.\frac{F_i^A}{V_A^j}\right|_{V_a^j=0}, \tag{10.22g}$$

$$Z_{ij}^{An} = \left.\frac{F_i^A}{V_a^j}\right|_{V_A^j=0}, \tag{10.22h}$$

$$Z_{ij}^{Na} = \left.\frac{F_i^a}{V_A^j}\right|_{V_a^j=0}, \tag{10.22i}$$

and

$$Z_{ij}^{Nn} = \left.\frac{F_i^a}{V_a^j}\right|_{V_A^j=0}. \qquad (10.22j)$$

A study of equations (10.22g)–(10.22j) reveals that the four Z_{ij}– parameters can be described as follows:

- Z_{ij}^{Aa} is the impedance holor seen looking into port A when port **a** is open;
- Z_{ij}^{An} is a transfer impedance holor. It is the ratio of the port A energy-potential-difference holor to the port a energy-transfer holor when port A is open;
- Z_{ij}^{Na} is a transfer impedance holor. It is the ratio of the port a energy-potential-difference holor to the port A energy-transfer holor when port a is open;
- Z_{ij}^{Nn} is the impedance holor seen looking into port a when port A is open.

One sees, then, that the impedance parameters can be either computed or measured by first opening port a and determining the ratios F_i^A/V_A^j and F_i^a/V_A^j, and then opening port A and determining the ratios F_i^A/V_a^j and F_i^a/V_a^j. Each parameter is the ratio of an energy-potential-difference holor to an energy-transfer holor and is therefore an impedance holor with a dimension of ohms (Ω).

$$\left.\begin{aligned}
Y_{aA}^{ji} &= \left.\frac{V_A^j}{F_i^A}\right|_{F_i^a=0} \quad [\mho] \\
Y_{nA}^{ji} &= \left.\frac{V_A^j}{F_i^a}\right|_{F_i^A=0} \quad [\mho] \\
Y_{aN}^{ji} &= \left.\frac{V_a^j}{F_i^A}\right|_{F_i^a=0} \quad [\mho] \\
Y_{nN}^{ji} &= \left.\frac{V_a^j}{F_i^a}\right|_{F_i^A=0} \quad [\mho]
\end{aligned}\right\} \qquad (10.22k)$$

$$\left.\begin{aligned}
A_{aA}^{ji} &= \left.\frac{F_i^A}{F_i^a}\right|_{V_a^j} \\
A_{ij}^{An} &= -\left.\frac{F_i^A}{V_a^j}\right|_{F_i^a} \quad [\Omega] \\
A_{aN}^{ji} &= \left.\frac{V_A^j}{F_i^a}\right|_{V_A^j} \quad [\mho] \\
A_{ij}^{Nn} &= -\left.\frac{V_A^j}{V_a^j}\right|_{F_i^A}
\end{aligned}\right\} \qquad (10.22l)$$

$$\left.\begin{aligned}
\boldsymbol{B}_{aA}^{ji} &= \left.\frac{\boldsymbol{F}_i^a}{\boldsymbol{F}_i^A}\right|_{\boldsymbol{V}_A^j=0} \\
\boldsymbol{B}_{ij}^{An} &= -\left.\frac{\boldsymbol{F}_i^a}{\boldsymbol{V}_A^j}\right|_{\boldsymbol{F}_i^A=0} \quad [\Omega] \\
\boldsymbol{B}_{aN}^{ji} &= \left.\frac{\boldsymbol{V}_a^j}{\boldsymbol{F}_i^A}\right|_{\boldsymbol{V}_A^j=0} \quad [\mho] \\
\boldsymbol{B}_{ij}^{Nn} &= -\left.\frac{\boldsymbol{V}_a^j}{\boldsymbol{V}_A^j}\right|_{\boldsymbol{F}_i^A=0}
\end{aligned}\right\} \tag{10.22m}$$

$$\left.\begin{aligned}
\boldsymbol{H}_{ij}^{Aa} &= \left.\frac{\boldsymbol{F}_i^A}{\boldsymbol{V}_A^j}\right|_{\boldsymbol{F}_i^a=0} \quad [\Omega] \\
\boldsymbol{H}_{nA}^{ji} &= \left.\frac{\boldsymbol{F}_i^A}{\boldsymbol{F}_i^a}\right|_{\boldsymbol{V}_A^j=0} \\
\boldsymbol{H}_{ij}^{Na} &= \left.\frac{\boldsymbol{V}_a^j}{\boldsymbol{V}_A^j}\right|_{\boldsymbol{F}_i^a=0} \\
\boldsymbol{H}_{nN}^{ji} &= \left.\frac{\boldsymbol{V}_a^j}{\boldsymbol{F}_i^a}\right|_{\boldsymbol{V}_A^j=0} \quad [\mho]
\end{aligned}\right\} \tag{10.22n}$$

$$\left.\begin{aligned}
\boldsymbol{G}_{aA}^{ji} &= \left.\frac{\boldsymbol{V}_A^j}{\boldsymbol{F}_i^A}\right|_{\boldsymbol{V}_a^j=0} \quad [\mho] \\
\boldsymbol{G}_{ij}^{An} &= \left.\frac{\boldsymbol{V}_A^j}{\boldsymbol{V}_a^j}\right|_{\boldsymbol{F}_i^A=0} \\
\boldsymbol{G}_{aN}^{ji} &= \left.\frac{\boldsymbol{F}_i^a}{\boldsymbol{F}_i^A}\right|_{\boldsymbol{V}_a^j=0} \\
\boldsymbol{G}_{ij}^{Nn} &= \left.\frac{\boldsymbol{F}_i^a}{\boldsymbol{V}_a^j}\right|_{\boldsymbol{F}_i^A=0} \quad [\Omega]
\end{aligned}\right\}. \tag{10.22o}$$

The remaining port parameters are either computed, or measured, using the same thought process that one used to find the $\boldsymbol{Z}_{ij}$– parameters. A given port parameter is obtained by either opening or shortening a port. Furthermore, a given port parameter will be an impedance, an admittance, or a dimensionless ratio. The dimensionless ratio will be either the ratio of two energy-potential-difference holors or the ratio of two energy-transfer holors. These observations are summarised by equations (10.22k)–(10.22o). The two-port SISO physical commutation matrixer parameters are also described in relation to the reciprocal sets of holor equations.

The impedance and admittance parameters are grouped into 'immittance' parameters. The term immittance holor is used to denote a quantity that is either an impedance holor or an admittance holor. The $\boldsymbol{A}_{ij}$ and $\boldsymbol{A}^{ji}$ or $\boldsymbol{B}_{ij}$ and $\boldsymbol{B}^{ji}$ parameters are called the 'transmission' parameters because they describe the energy-potential-difference holor and energy-transfer holor at one end of the two-port in terms of the energy-potential-difference holor and energy-transfer holor at the other end.

Thus the $\boldsymbol{A}_{ij}$ and $\boldsymbol{A}^{ji}$ or $\boldsymbol{B}_{ij}$ and $\boldsymbol{B}^{ji}$ parameters give a measure of how the energy-potential-difference holor and energy-transfer holor are transmitted through the physical commutation matrixer. The immittance and transmission parameters are characterised as being the natural choices for relating the port holors. That is, they relate either energy-potential-difference holors to energy-transfer holors or matrixer-input holors to matrixer-output holors.

The $\boldsymbol{H}_{ij}$ and $\boldsymbol{H}^{ji}$ and $\boldsymbol{G}_{ij}$ and $\boldsymbol{G}^{ji}$– parameters, on the other hand, relate cross-holors, that is, a matrixer-input energy-potential-difference holor and matrixer-output energy-transfer holor to a matrixer-output energy-potential-difference holor and matrixer-input energy-transfer holor. Therefore, the $\boldsymbol{H}_{ij}$ and $\boldsymbol{H}^{ji}$ and the $\boldsymbol{G}_{ij}$ and $\boldsymbol{G}^{ji}$– parameters are called 'hybrid' parameters.

Analysis of the terminated two-port SISO physical commutation matrixer

In the typical application of a two-port physical commutation matrixer, it is driven at port $\boldsymbol{A}$ and loaded at port $\boldsymbol{a}$. The holor domain physical commutation matrixer diagram for the typically terminated two-port physical commutation matrixer is shown in figure 10.9, where $\boldsymbol{Z}_{ij}^{G}$ represents the internal impedance holor of the internal energy-potential-difference source holor, $\boldsymbol{F}_{i}^{G}$, and $\boldsymbol{Z}_{ij}^{L}$ is the load impedance holor.

The analysis of the physical commutation matrixer in figure 10.9 involves expressing the terminal energy-transfer holors and energy-potential-difference holors as functions of the two-port SISO physical commutation matrix parameters, $\boldsymbol{F}_{i}^{G}$, $\boldsymbol{Z}_{ij}^{G}$ and $\boldsymbol{Z}_{ij}^{L}$.

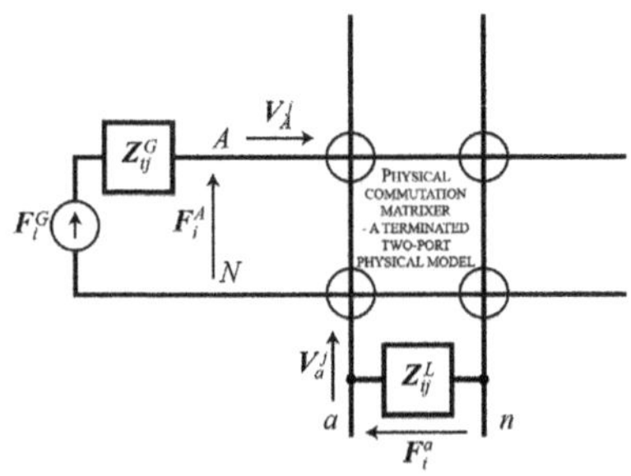

Figure 10.9. A terminated two-port SISO physical model of the physical commutation matrixer.

Six characteristics of the terminated two-port SISO physical commutation matrixer define its terminal behaviour:

- the matrixer-input impedance holor, $Z_{ij}^{A} = F_i^A / V_A^j$, or admittance holor, $Y_A^{ji} = V_A^j / F_i^A$;
- the matrixer-input energy-transfer holor V_a^j;
- the Thévenin energy-potential-difference and impedance holors, $V_{A(\mathrm{TH})}^{j}$, $Z_{ij}^{A(\mathrm{TH})}$, with respect to port $\boldsymbol{a}$;
- the energy-transfer gain holor, V_a^j / V_A^j;
- the energy-potential-difference gain holor, F_i^a / F_i^A;
- the energy-potential-difference gain holor, F_i^a / F_i^G.

To illustrate how these six characteristics are derived, we develop the expressions using the Z_{ij}– parameters to mathematically model the two-port SISO physical commutation matrixer portion. The derivation of the desired expressions involves the algebraic holor manipulation of any one of the two-port SISO physical commutation matrixer holor equations, along with the two constraint holor equations imposed by the terminations.

If one uses the Z_{ij}– parameter holor equations, the four-holor equations that describe the physical commutation matrixer in figure 10.9 are:

$$F_i^A = Z_{ij}^{Aa} V_A^j + Z_{ij}^{An} V_a^j, \tag{10.23}$$

$$F_i^a = Z_{ij}^{Na} V_A^j + Z_{ij}^{Nn} V_a^j, \tag{10.24}$$

$$F_i^A = F_i^G - Z_{ij}^G V_A^j, \tag{10.25}$$

$$F_i^a = -Z_{ij}^L V_a^j. \tag{10.26}$$

Expressions (10.23) and (10.24) describe the constraints imposed by the terminations.

Interconnected two-port SISO physical matrixers

In the design of large physical heterogeneous continuous complex dynamical hypersystems, it is usually easier to synthesise the complex dynamical hypersystem by first designing simple dynamical systems and/or dynamical hyposystems of them. The physical heterogeneous continuous complex dynamical hypersystem is then fabricated by interconnecting these simpler, easier-to-design, smaller physical commutation matrixer units. If the physical heterogeneous continuous simple dynamical systems and/or dynamical hyposystems are physically modelled by two-port physical commutation matrixers, then the synthesis of the physical heterogeneous continuous complex dynamical hypersystem involves the analysis of interconnected two-port SISO physical matrixers.

There are five fundamental ways of interconnecting two-port SISO physical commutation matrixers: (a) cascade, (b) series, (c) parallel, (d) series-parallel and

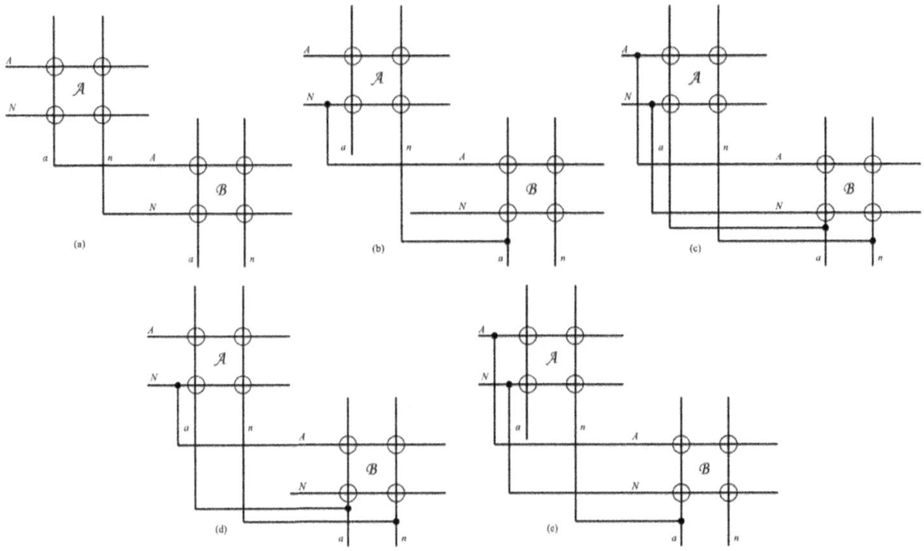

Figure 10.10. The five basic interconnections of two-port SISO physical commutation matrixers: (a) cascade, (b) series, (c) parallel, (d) series-parallel and (e) parallel-series.

(e) parallel-series interconnections. These five basic interconnections are illustrated in figure 10.10.

We only analyse and illustrate the cascade connection in this section. However, in passing, one can obtain the parameters that describe the interconnected physical commutation matrixers by the simple addition of the parameters of the individual physical commutation matrixers.

In particular, the $\boldsymbol{Z}_{ij}$– parameters describe the series connection, the $\boldsymbol{Y}^{ji}$– parameters describe the parallel connection, the $\boldsymbol{H}_{ij}$ and $\boldsymbol{H}^{ji}$– parameters describe the series-parallel connection, and the $\boldsymbol{G}_{ij}$ and $\boldsymbol{G}^{ji}$– parameters describe the parallel-series connection.

The interested reader can find a detailed discussion of these four interconnections, e.g. for electrical commutation matrixers, in the literature (Ruston and Bordogna 1966).

The cascade connection is important because of its frequent occurrence in the physical modelling of physical heterogeneous continuous complex dynamical hypersystems.

Unlike the other four basic interconnections, there are no restrictions on using the parameters of the individual two-port SISO physical commutation matrixers to obtain the parameters of the interconnected physical commutation matrixers.

The $\boldsymbol{A}_{ij}$ and $\boldsymbol{A}^{ji}$– parameters are best suited for describing the cascade connections using the physical commutation matrixer shown in figure 10.11, where one has used a single prime to denote the $\boldsymbol{A}_{ij}$ and $\boldsymbol{A}^{ji}$– parameters of the first physical commutation matrixer and a double prime to denote the $\boldsymbol{A}_{ij}$ and $\boldsymbol{A}^{ji}$– parameters of the second physical commutation matrixer.

The matrixer-output energy-potential-difference holor and the energy-transfer holor of the first physical commutation matrixer are labelled $\boldsymbol{F}_a^I$ and $\boldsymbol{V}_I^a$, and the matrixer-input energy-potential-difference holor and the energy-transfer holor of the

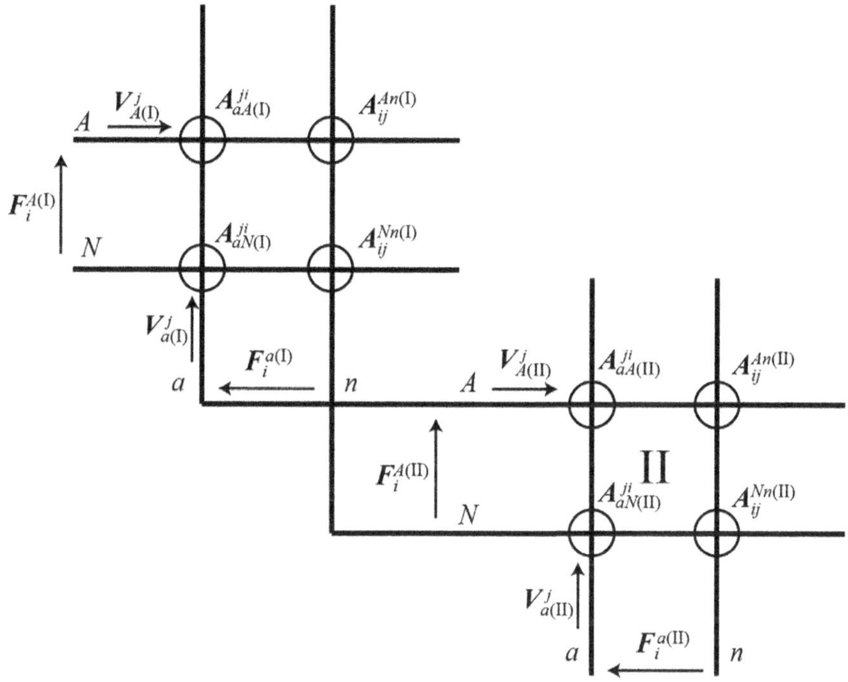

Figure 10.11. Cascade connection of two-port SISO physical commutation matrixers.

second physical commutation matrixer are labelled $F_i^{A(I)}$ and $V^j_{A(I)}$. The problem is to derive the A_{ij} and A^{ji}– parameter holor equations that relate $F_i^{a(I)}$ and $V^j_{a(I)}$ to $F_i^{A(I)}$ and $V^j_{A(I)}$. That is, one seeks the pair of holor equations

$$F_i^A = A^{ji}_{aA} F_i^a - A_{ij}^{An} V^j_a, \tag{10.27}$$

$$V^j_A = A^{ji}_{aN} F_i^a - A_{ij}^{Nn} V^j_a, \tag{10.28}$$

where the A_{ij} and A^{ji}– parameters are given explicitly in terms of the A_{ij} and A^{ji}– parameters of the individual physical commutation matrixers.

One begins the derivation by noting from figure 10.11 that

$$F_i^A = A^{ji}_{aA(I)} F_i^{a(I)} - A_{ij}^{An(I)} V^j_{a(I)}, \tag{10.29}$$

and

$$V^j_A = A^{ji}_{aA(I)} F_i^{a(I)} - A_{ij}^{Nn(I)} V^j_{a(I)}. \tag{10.30}$$

Next, one observes that the interconnection means that $F_i^{a(I)} = F_i^{A(I)}$ and $V^j_{a(I)} = V^j_{A(I)}$.

When these constraints are substituted into equations (10.29) and (10.30), one obtains

$$F_i^A = A_{aA(I)}^{ji} F_i^{A(I)} - A_{ij}^{An(I)} V_{A(I)}^j, \tag{10.31}$$

and

$$V_A^j = A_{aA(I)}^{ji} F_i^{A(I)} - A_{ij}^{Nn(I)} V_{A(I)}^j. \tag{10.32}$$

The energy-potential-difference holor, $F_i^{A(I)}$, and the energy-transfer holor, $V_{A(I)}^j$, through the A_{ij} and A^{ji}– parameters of the second physical commutation matrixer, thus

$$F_i^{A(I)} = A_{aA(II)}^{ji} F_i^a - A_{ij}^{An(II)} V_a^j \tag{10.33}$$

and

$$V_{A(I)}^j = A_{aA(II)}^{ji} F_i^a - A_{ij}^{Nn(II)} V_a^j. \tag{10.34}$$

One can substitute equations (10.33) and (10.34) into equations (10.31) and (10.32) to generate the sought-after relationships between F_i^A, V_A^j and F_i^a, V_a^j.

One obtains

$$F_i^A = \left(A_{aA(I)}^{ji} A_{aA(II)}^{ji} + A_{ij}^{An(I)} A_{aA(II)}^{ji}\right) F_i^{\,a} - \left(A_{aA(I)}^{ji} A_{ij}^{An(II)} + A_{ij}^{An(I)} A_{ij}^{An(II)}\right) V_a^j \tag{10.35}$$

and

$$V_A^j = \left(A_{aA(I)}^{ji} A_{aA(II)}^{ji} + A_{ij}^{Nn(I)} A_{aA(II)}^{ji}\right) F_i^a - \left(A_{aA(I)}^{ji} A_{ij}^{Nn(II)} + A_{ij}^{Nn(I)} A_{ij}^{Nn(II)}\right) V_a^j. \tag{10.36}$$

By comparing equations (10.35) and (10.36) with equations (10.27) and (10.28), one obtains the desired expressions for the A_{ij} and A^{ji}– parameters of the interconnected physical commutation matrixers, namely,

$$A_{aA}^{ji} = A_{aA(I)}^{ji} A_{aA(II)}^{ji} + A_{ij}^{An(I)} A_{aA(II)}^{ji}, \tag{10.37}$$

$$A_{ij}^{An} = A_{aA(I)}^{ji} A_{ij}^{An(II)} + A_{ij}^{An(I)} A_{ij}^{Nn(II)}, \tag{10.38}$$

$$A_{aN}^{ji} = A_{aA(I)}^{ji} A_{aA(II)}^{ji} + A_{ij}^{Nn(I)} A_{aA(II)}^{ji}, \tag{10.39}$$

$$A_{ij}^{Nn} = A_{aA(I)}^{ji} A_{ij}^{Nn(II)} + A_{ij}^{Nn(I)} A_{ij}^{Nn(II)}. \tag{10.40}$$

One can use equations (10.27)–(10.40) to analyse the cascade connection of the two physical commutation matrixers. If more than two physical commutation matrixer units are connected in cascade, the A_{ij} and A^{ji}—parameters of the equivalent two-port physical commutation matrixer can be found by successively reducing the original set of two-port SISO physical commutation matrixers one pair a time.

Summary

Physical models of two-port SISO physical commutation matrixers are useful when it is convenient to describe the performance of the physical commutation matrixer in terms of its behaviour at two pairs of terminals. Each pair of terminals is referred to as a port.

The port where the signal is fed into the physically continuous dynamical system is called the 'matrixer input' and the port where the signal is extracted is called 'matrixer output'.

The physical model of the two-port SISO physical commutation matrixer is restricted to physical homogeneous continuous dynamical systems where: (i) there are no independent energy sources in the internal physical commutation matrixer between the ports; (ii) there is no energy stored in the physical commutation matrixer; and (iii) the passive components (e.g. *b–k–m* for a mechanical commutation matrixer or *R–L–C* for an electrical commutation matrixer) are linear and time-invariant. It is also understood that no external connections are made between the matrixer-input and matrixer-output ports.

The physical model of the two-port SISO physical commutation matrixer is described in terms of four holors: the matrixer-input—energy-potential-difference holor (F_i^A) and energy-transfer holor (V_A^j), and the matrixer-output—energy-potential-difference holor (F_i^a) and energy-transfer holor (V_a^j).

The mathematical model of the two-port SISO physical commutation matrixer is formulated in the holor domain so that the physical commutation matrixers and holor domain are treated as special cases, that is, $\omega^j = 0$ for the translational-motion physical commutation matrixers (e.g. for DC electrical commutation matrixers) and $\omega > 0$ for sinusoidal steady-state rotary-motion physical commutation matrixers (e.g. for AC electrical commutation matrixers). Only two of the four holors are independent, which means that there are six possible ways to relate the four holors.

The six sets of holor equations that group the holors are:

(i) the energy-potential-difference holors as functions of the energy-transfer holors (Z_{ij}– parameters);

(ii) the energy-transfer holors as functions of the energy-potential-difference holors (Y^{ji}– parameters);

(iii) the matrixer-input port holors as functions of the matrixer-output port holors (A_{ij} and A^{ji}– parameters);

(iv) the matrixer-output port holors as functions of the matrixer-input port holors (B_{ij} and B^{ji}– parameters);

(v) the matrixer-input energy-potential-difference holor and the matrixer-output energy-transfer holor as functions of the matrixer output energy-potential-difference holor and the matrixer-input energy-transfer holor (H_{ij} and H^{ji}– parameters);

(vi) the matrixer-output energy-potential-difference holor and the matrixer-input energy-transfer holor as functions of the matrixer-input energy-potential-difference holor and the matrixer-output energy-transfer holor ($\boldsymbol{G}_{ij}$ and $\boldsymbol{G}^{ji}$– parameters).

These six sets of holor equations can be grouped into three sets of reciprocal pairs and identified by the parameters as:

(i) the immittance parameters ($\boldsymbol{Z}_{ij}$, $\boldsymbol{Y}^{ji}$);
(ii) the transmission parameters ($\boldsymbol{A}_{ij}$ and $\boldsymbol{A}^{ji}$, $\boldsymbol{B}_{ij}$ and $\boldsymbol{B}^{ji}$);
(iii) the hybrid parameters ($\boldsymbol{H}_{ij}$ and $\boldsymbol{H}^{ji}$, $\boldsymbol{G}_{ij}$ and $\boldsymbol{G}^{ji}$).

Because all six sets of holor equations involve the same four holors, all the parameters are interrelated. Thus if one knows one set of parameters, one can compute the remaining five sets.

If the two-port SISO physical commutation matrixer satisfies the principle of reciprocity, it is referred to as a reciprocal two-port SISO physical commutation matrixer. For reciprocal physical commutation matrixers, special relationships exist among the port parameters. In particular, $\boldsymbol{Z}_{ij}^{An} = \boldsymbol{Z}_{ij}^{Na}$, $\boldsymbol{Y}_{aA}^{ji} = \boldsymbol{Y}_{aN}^{ji}$, $\Delta_A = \Delta_a = 1$, $\boldsymbol{H}_{ij}^{Na} = -\boldsymbol{H}_{nA}^{ji}$ and $\boldsymbol{G}_{ij}^{An} = -\boldsymbol{G}_{aN}^{ji}$.

If a reciprocal two-port SISO physical commutation matrixer is symmetrical, the matrixer-input and matrixer-output ports can be interchanged without distributing the values of the port holors. For symmetrical two-port SISO physical commutation matrixers, '$\boldsymbol{Z}_{ij}^{Aa} = \boldsymbol{Z}_{ij}^{Nn}$, $\boldsymbol{Y}_{nA}^{ji} = \boldsymbol{Y}_{aN}^{ji}$, $\boldsymbol{A}_{ij}^{Na} = \boldsymbol{A}_{aA}^{ji}$, $\boldsymbol{B}_{ij}^{Nn} = \boldsymbol{B}_{aA}^{ji}$ and $\Delta_A = \Delta_a = 1$

10.6.2 Physical commutation matrixer having three independent nodes

Consider a set of node holor equations, the basic form of which we have seen is:

$$\boldsymbol{V}^j = \boldsymbol{Y}^{ji}\boldsymbol{F}_i. \tag{10.41}$$

In the case of a physical commutation matrixer with three independent nodes, the details are:

$$\begin{bmatrix} \boldsymbol{V}_a^j \\ \boldsymbol{V}_b^j \\ \boldsymbol{V}_c^j \end{bmatrix} = \begin{bmatrix} \boldsymbol{Y}_{aA}^{ji} & \boldsymbol{Y}_{aB}^{ji} & \boldsymbol{Y}_{aC}^{ji} \\ \boldsymbol{Y}_{bA}^{ji} & \boldsymbol{Y}_{bB}^{ji} & \boldsymbol{Y}_{bC}^{ji} \\ \boldsymbol{Y}_{cA}^{ji} & \boldsymbol{Y}_{cB}^{ji} & \boldsymbol{Y}_{cC}^{ji} \end{bmatrix} \begin{bmatrix} \boldsymbol{F}_i^A \\ \boldsymbol{F}_i^B \\ \boldsymbol{F}_i^C \end{bmatrix}. \tag{10.42}$$

The solution has the basic form:

$$\boldsymbol{F}_i = \frac{\delta_k^j}{\boldsymbol{Y}^{ji}}\boldsymbol{V}^j = z_{ik}\boldsymbol{V}^i, \tag{10.43}$$

where

$$z_{IJ} = \begin{bmatrix} z_{ij}^{Aa} & z_{ij}^{Ab} & z_{ij}^{Ac} \\ z_{ij}^{Ba} & z_{ij}^{Bb} & z_{ij}^{Bc} \\ z_{ij}^{Ca} & z_{ij}^{Cb} & z_{ij}^{Cc} \end{bmatrix} = \frac{1}{\boldsymbol{\Delta}} \begin{bmatrix} \boldsymbol{\Delta}_{Aa} & -\boldsymbol{\Delta}_{Ba} & \boldsymbol{\Delta}_{Ca} \\ -\boldsymbol{\Delta}_{Ab} & \boldsymbol{\Delta}_{Bb} & -\boldsymbol{\Delta}_{Cb} \\ \boldsymbol{\Delta}_{Ca} & -\boldsymbol{\Delta}_{Cb} & \boldsymbol{\Delta}_{Cc} \end{bmatrix},$$

where $\boldsymbol{\Delta}$ is the determinant of the admittance holor, $\boldsymbol{Y}^{ji}$, and $\boldsymbol{\Delta}_{jk}$ is the co-factor of the admittance holor, $\boldsymbol{y}^{jk}$.

The terms in the inverse physical-holor matrix have some physical significance, in that they are driving point and transfer impedance holors. That is, the term z_{Jj} is the impedance holor between the node j and ground when the energy sources connected to all the other nodes are suppressed (i.e. set to zero), which can be recognised as the driving point impedance holor discussed in section 10.3.

The term z_{ik} is the ratio of the energy-potential difference holor between the node i and ground to the energy-transfer holor entering the node k when all the other energy sources are suppressed, and is usually termed the transfer impedance holor. It is therefore theoretically possible to take appropriate measurements on a physical heterogeneous continuous dynamical hypersystem to yield values for the entire array of the driving point and transfer impedance holors so that the impedance-holor matrix, z_{ij}, may be measured.

By inverting it, the impedance-holor matrix, $\boldsymbol{Y}^{ji}$, of the corresponding node holor equations can be obtained. Reversal of the procedure to obtain the node holor equations 'by inspection' of the physical commutation matrixer will then produce a mathematical model of the physical heterogeneous continuous dynamical hypersystem.

In practice, it is rarely necessary to carry out this inversion process because the form of the mathematical model can be determined if one knows the number of holors required. However, this does indicate the form of the measurements that must be taken on a physical heterogeneous continuous dynamical hypersystem, including the constraints that must be applied.

It should be evident that if one considers a set of mesh holor equations with the form

$$\boldsymbol{F}_i = \boldsymbol{Z}_{ij} \boldsymbol{V}^j, \tag{10.44}$$

the solution is

$$\boldsymbol{V}^j = \frac{\left[\delta_k^i\right]}{\boldsymbol{Z}_{jk}} \boldsymbol{F}_i = \boldsymbol{y}^{ji} \boldsymbol{F}_i, \tag{10.45}$$

where $\boldsymbol{y}^{ji}$ is the array of driving point and transfer admittance holors.

Note that there is no simple relation between the $\boldsymbol{y}^{ij}$ (i.e. the driving point and transfer admittance holors) and $\boldsymbol{Y}^{ji}$ (i.e. the admittance holor array of the node holor equations). Nor is there any simple relation between the driving point and transfer impedance holor matrix, z_{ik}, and the mesh equation impedance holor matrix, $\boldsymbol{Z}_{ij}$.

10.6.3 Superposition theorem

Returning to equation (10.43), one can write out the solution for $\boldsymbol{F}_i^A$ as:

$$\boldsymbol{F}_i^A = z_{ij}^{Aa}\boldsymbol{V}_a^j + z_{ij}^{Ab}\boldsymbol{V}_b^j + z_{ij}^{Ac}\boldsymbol{V}_c^j. \tag{10.46}$$

One notes that for this physical heterogeneous continuous dynamical hypersystem with three energy sources the solution has three components, each of which is the value of $\boldsymbol{F}_i^A$, while the other energy sources are set zero.

The superposition theorem uses the property of a linear physical heterogeneous continuous dynamical hypersystem to obtain the total sinusoidal-response holor as the sum of the response holors for each driving function holor or energy source holor taken one at a time while all the other energy sources are suppressed. As an example, a physical commutation matrixer consisting of two impedance holors is driven by an energy-potential-difference holor and an energy-transfer-source holor, as shown in figure 10.12(a).

Find the energy-potential-difference across, $\boldsymbol{Z}_{ij}^2$. The procedure is simply to obtain the holors of $\boldsymbol{F}_i^a$ when only one energy source is acting at a time, and to add those solutions. In this case, there are two energy sources, so the problem is to find $\boldsymbol{F}_i^a$ in each of figures 10.12(b) and 10.12(c) and add them.

From figure 10.12(b), the first component of $\boldsymbol{F}_i^a$ can be found directly by noting that this is an energy-potential-difference holor divider,

$$\boldsymbol{F}_i^a = \frac{\delta_k^i}{\boldsymbol{Z}_{jk}^1 + \boldsymbol{Z}_{jk}^2}\boldsymbol{Z}_{ij}^2\boldsymbol{F}_i^A.$$

From figure 10.12(c), the second component of $\boldsymbol{F}_i^a$ is found by noting that this is the energy-potential-difference holor across the parallel combination of $\boldsymbol{Z}_{jk}^1$ and $\boldsymbol{Z}_{jk}^2$.

$$\boldsymbol{F}_i^a = \frac{\delta_k^i}{\boldsymbol{Z}_{jk}^1 + \boldsymbol{Z}_{jk}^2}\boldsymbol{Z}_{ij}^2\boldsymbol{Z}_{ij}^1\boldsymbol{V}_a^j.$$

The complete response is the sum of these two expressions:

$$\boldsymbol{F}_i^a = \frac{\delta_k^i}{\boldsymbol{Z}_{jk}^1 + \boldsymbol{Z}_{jk}^2}\boldsymbol{Z}_{ij}^2\boldsymbol{F}_i^A + \frac{\delta_k^i}{\boldsymbol{Z}_{jk}^1 + \boldsymbol{Z}_{jk}^2}\boldsymbol{Z}_{ij}^2\boldsymbol{Z}_{ij}^1\boldsymbol{V}_a^j = \frac{\delta_k^i}{\boldsymbol{Z}_{jk}^1 + \boldsymbol{Z}_{jk}^2}\boldsymbol{Z}_{ij}^2\left(\boldsymbol{F}_i^A + \boldsymbol{Z}_{ij}^1\boldsymbol{V}_a^j\right).$$

Some physical commutation matrixers require more than one energy-potential-difference source holor. For instance, certain types of physical amplifiers require a positive and a negative energy-potential-difference source holor for proper operation, as well as an alternating (e.g. sinusoidal) signal energy source.

In this section, a general method is presented for evaluating series-parallel physical commutation matrixers having multiple energy-potential difference source

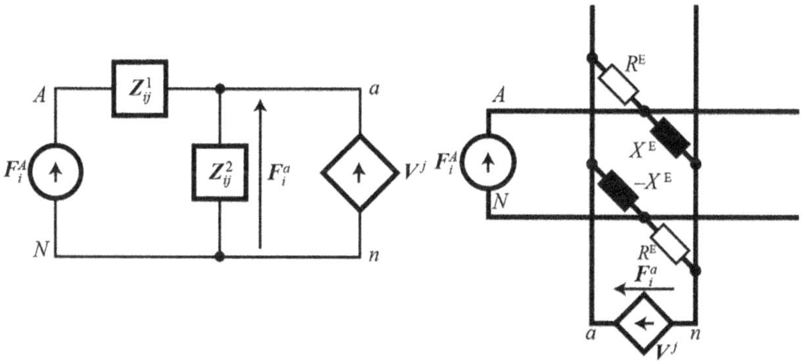

(a) COMPLETE PHYSICAL COMMUTATION MATRIXER

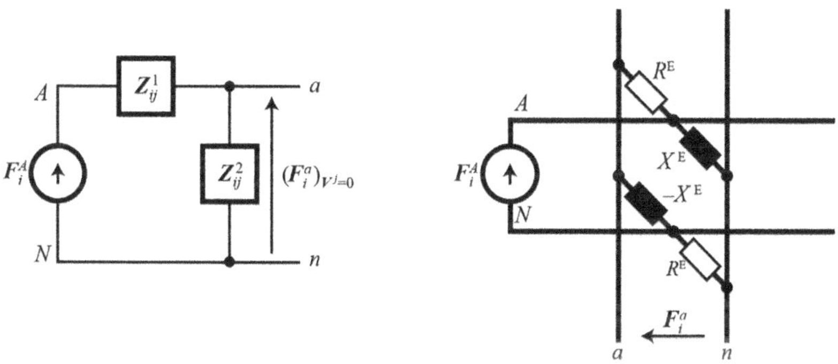

(b) $V^j\!=\!0$

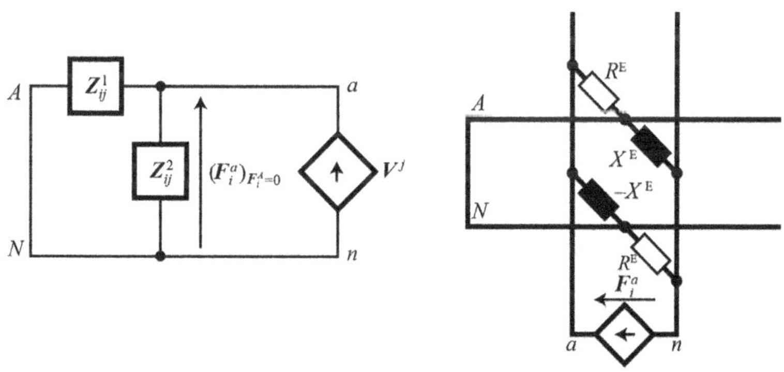

(c) $F_i^A = 0$

Figure 10.12. Superposition with its physical commutation matrixers.

holors. The method is illustrated with three-energy-source physical commutation matrixers, but it can be extended to physical commutation matrixers with any number of energy sources, where the basic principle will be the same.

The superposition theorem is a way to determine energy-transfer holors and energy-potential-difference holors in a linear physical commutation matrixer that has multiple energy sources by taking one energy source at a time. The other energy sources are replaced by their internal impedance holors.

Recall that the ideal energy-potential difference source holor has a zero internal impedance holor. All energy-potential difference source holors will be treated as ideal in order to simplify the coverage. A general statement of the superposition theorem is:

The energy-transfer holor in any given branch of a multiple source linear physical commutation matrixer can be found by determining the energy-transfer holors in that particular branch, produced by each energy source acting alone, with all the other energy sources replaced by their internal impedances.

The total energy-transfer holor in the branch is the holor algebraic sum of the individual energy-transfer source holors in that branch. The steps for applying the superposition theorem are as follows.

(i) Take one energy-potential difference source holor at a time and replace each of the other energy-potential-difference source holors with a short, which represents the zero internal impedance holor.

(ii) Determine the particular energy-transfer-holor or energy-potential-difference holor that one wants to find, just as if there were only one energy-potential-difference source holor in the physical commutation matrixer. This is a component of the total energy-transfer holor or energy-potential-difference holor that one is looking for.

(iii) Take the next energy-potential difference source holor in the physical commutation matrixer and repeat steps (i) and (ii) for each energy source.

(iv) To find the actual energy-transfer-holor in a given branch, add or subtract the component energy-transfer holors to or from each individual energy source.

(v) If the energy-transfer holors are in the same sense of direction, add them, and if they are in the opposite sense of direction, subtract them, with the sense of direction of the resulting energy-transfer holor being the same as that of the largest of the individual energy-transfer holors. Once the energy-transfer holor has been found, the energy-potential difference holor can be determined by a generalised holor equation.

An example of the dynamical systems approach to superposition is illustrated in figure 10.13 for a series-parallel physical commutation matrixer with two energy-potential-difference source holors. Study the steps in this figure.

10.6.4 Reciprocity theorem

We previously noted that when all of the dynamical components in a physical system or physical commutation matrixer are bilateral, the terms $\boldsymbol{Y}^{jk}$ and $\boldsymbol{Y}^{kj}$ are equal in the node holor equations, and $\boldsymbol{Z}_{jk}$ and $\boldsymbol{Z}_{kj}$ are equal in the mesh holor equations, and therefore the admittance-holor and impedance-holor matrices $[\boldsymbol{Y}^{ji}]$ and $[\boldsymbol{Z}_{ij}]$ are

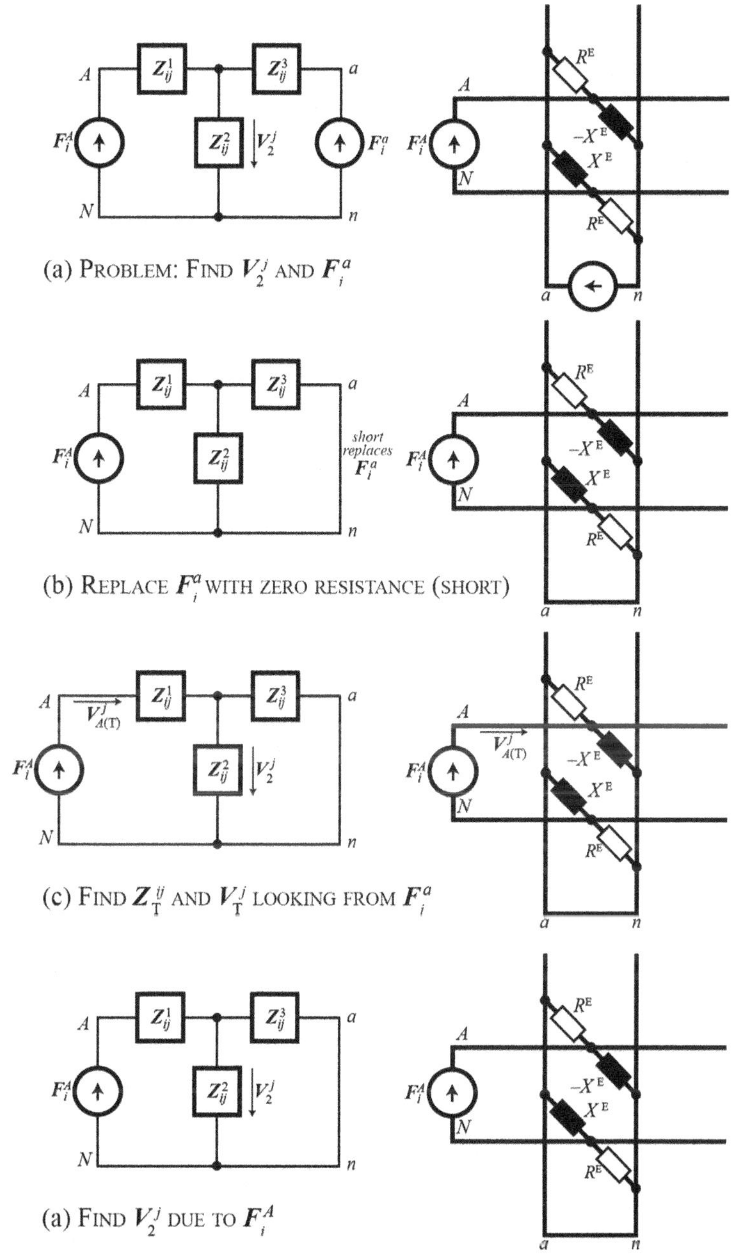

Figure 10.13. Illustration of the superposition theorem.

symmetrical. As a result, their inverses, $[z_{ij}]$ and $[y^{ji}]$, are also symmetrical. Although this is of little value as an aid to analyse a physical commutation matrixer, a corollary of the reciprocity theorem is of some value in physical modelling.

If a set of measurements on a physical system shows that any pair of transfer impedance holors or admittance holors is not reciprocal (i.e. $z_{ik} \neq z_{ki}$ or $y^{jk} \neq y^{kj}$)

the physical heterogeneous continuous dynamical hypersystem cannot be physically modelled using only bilateral dynamical components and independent energy sources such as those being considered in this textbook.

10.6.5 Thévenin's theorem

Thévenin's theorem is a very convenient method of obtaining a simple physical commutation matrixer that is equivalent to a section of a complete physical commutation matrixer. To some extent, it is an extension of the physical modelling of real energy sources as an ideal energy source associated with an impedance holor. The section of the physical commutation matrixer to be physically modelled or simplified will always have at least one energy source and a port (a pair of terminals) at which it is connected to the remainder of the complete physical commutation matrixer. This is shown symbolically in figure 10.14(left).

The Thévenin theorem states simply that the physical commutation matrixer shown in figure 10.14(right) will produce exactly the same energy-transfer holor/energy-potential-difference holor relationship at port terminals *a* and *n* provided that:

- $\boldsymbol{F}_i^{\mathrm{TH}}$ is the energy-potential-difference holor between *a* and *n* when the energy-transfer holor $\boldsymbol{V}^j = (0, 0)$, i.e. the open-output physical commutation matrixer energy-potential-difference holor $\boldsymbol{F}_i$;
- $\boldsymbol{Z}_{ij}^{\mathrm{TH}}$ is the bivalent impedance holor between port terminals *a* and *n* when all the energy sources within the section of the physical commutation matrixer are suppressed, i.e. the driving-point impedance holor.

Thévenin's theorem provides a method for simplifying a physical commutation matrixer to a standard equivalent form. In many cases, this theorem can be used to simplify the holor analysis of series-parallel physical commutation matrixers.

The Thévenin equivalent form of any two-terminal impedance-holor physical commutation matrixer of an equivalent energy-potential-difference-holor source ($\boldsymbol{F}_i^{\mathrm{TH}}$) and an equivalent impedance holor ($\boldsymbol{Z}_{ij}^{\mathrm{TH}}$) is arranged as shown in figure 10.14.

The values of the equivalent energy-potential-difference holor and the impedance holor depend on the values in the original physical matrixer. Any two-terminal

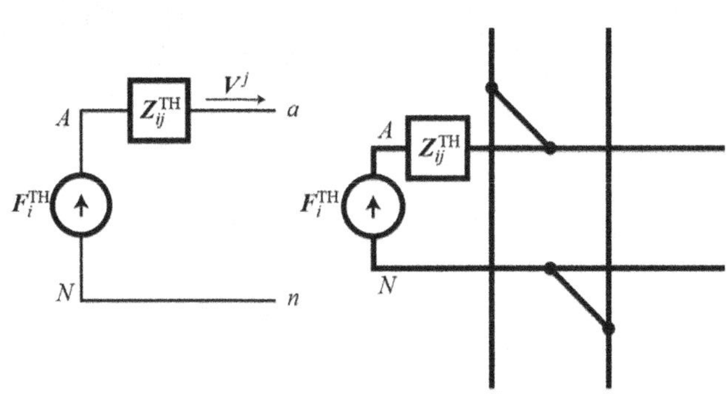

Figure 10.14. The general form of a Thévenin equivalent physical commutation matrixer is an energy-potential difference source holor in series with an impedance holor.

impedance holor can be simplified to a Thévenin equivalent, regardless of its complexity.

Thévenin's equivalent energy-potential-difference holor and equivalent impedance holor

The equivalent energy-potential-difference holor, $\boldsymbol{F}_i^{\mathrm{TH}}$, is one part of the complete Thévenin equivalent physical commutation matrixer. The other part is an equivalent impedance holor, $\boldsymbol{Z}_{ij}^{\mathrm{TH}}$.

The Thévenin equivalent energy-potential-difference holor ($\boldsymbol{F}_i^{\mathrm{TH}}$) is the open-output (no-load) holor between two specified output terminals in a physical commutation matrixer. Any dynamical component connected between these two terminals effectively 'sees' $\boldsymbol{F}_i^{\mathrm{TH}}$ in series with $\boldsymbol{Z}_{ij}^{\mathrm{TH}}$, as defined by Thévenin's theorem.

The Thévenin equivalent impedance holor ($\boldsymbol{Z}_{ij}^{\mathrm{TH}}$) is the total impedance holor appearing between two specified output terminals in a physical commutation matrixer with all of the energy sources replaced by their internal impedance holors (which for ideal energy-potential-difference source holors are zero).

What equivalency means in Thévenin's theorem

Although a Thévenin equivalent physical commutation matrixer is not of the same form as the original physical commutation matrixer, it acts the same in terms of the output energy-potential-difference holor and the energy-transfer holor. This condition is sometimes known as 'terminal equivalency', because both physical commutation matrixers look the same from the 'viewpoint' of the two output terminals.

The Thévenin equivalent of a physical commutation matrixer

To find the Thévenin equivalent of any physical commutation matrixer, one must determine the equivalent energy-potential-difference holor, $\boldsymbol{F}_i^{\mathrm{TH}}$, and the equivalent impedance holor, $\boldsymbol{Z}_{ij}^{\mathrm{TH}}$. For instance, the Thévenin equivalent for the physical commutation matrixer between output terminals *a* and *b* is developed in figure 10.15.

In figure 10.15(a), the energy-potential-difference holor across the designated terminals *a* and *n* is the Thévenin equivalent energy-potential-difference holor. In this particular physical commutation matrixer, the energy-potential-difference holor from *a* to *n* is the same as the energy-potential-difference holor across $\boldsymbol{Z}_{ij}^2$, because there is no energy-transfer holor through $\boldsymbol{Z}_{ij}^3$ and, therefore, no energy-potential holor drop across it. $\boldsymbol{F}_i^{\mathrm{TH}}$ is expressed as follows for this particular example:

$$\boldsymbol{F}_i^{\mathrm{TH}} = \frac{\delta_k^i}{\boldsymbol{Z}_{jk}^1 + \boldsymbol{Z}_{jk}^2}\boldsymbol{Z}_{ij}^2\boldsymbol{F}_i^{\mathrm{S}}.$$

In figure 10.15(b), the impedance holor between terminals *a* and *n* with the energy source replaced by a short to represent its zero internal impedance holor is the Thévenin equivalent impedance holor. In this particular physical commutation

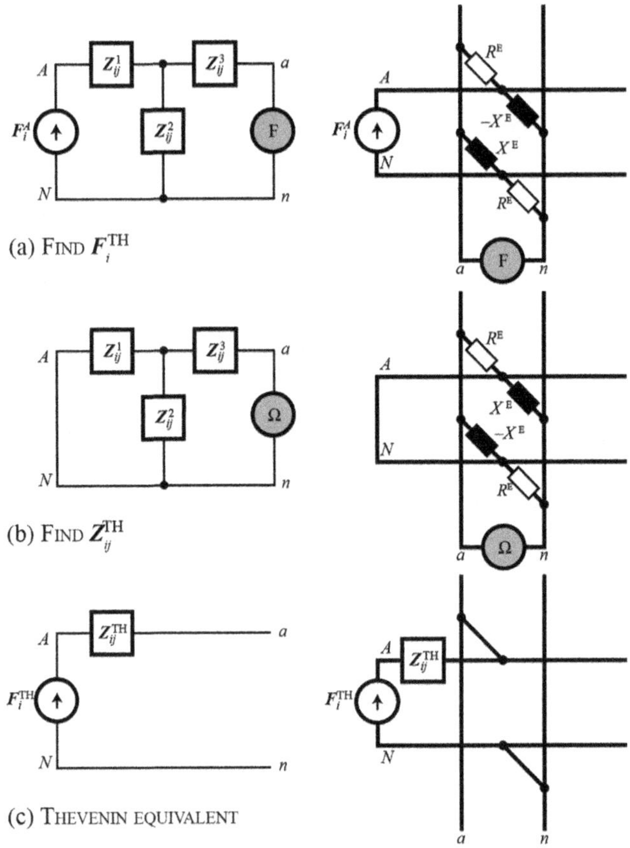

Figure 10.15. The simplification of a physical commutation matrixer by Thèvenin's theorem.

matrixer, the impedance holor from a to b is Z_{ij}^3 in series with the parallel combination of Z_{ij}^1 and ij.

Therefore, Z_{ij}^{TH} is expressed as follows:

$$Z_{ij}^{TH} = Z_{ij}^3 + \frac{\delta_k^i}{Z_{jk}^1 + Z_{jk}^2} Z_{ij}^1 Z_{ij}^2.$$

The Thévenin equivalent physical commutation matrixer is shown in figure 10.15(c).

Thévenin equivalency depends on the viewpoint

The Thévenin equivalent for any physical commutation matrixer depends on the location of the two output terminals from between which the physical commutation matrixer is 'viewed'. Any given physical commutation matrixer can have more than one Thévenin equivalent, depending on how the output terminals are designated.

In figure 10.16(a), when F_i^{TH} is viewed from between terminals a and c, it is the energy-potential-difference holor across $Z_{ij}^2 + Z_{ij}^3$ and can be expressed using the energy-potential-difference-holor divider formula as

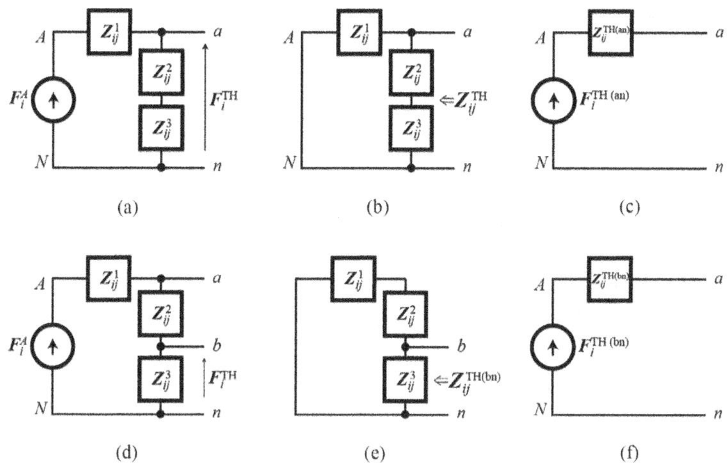

Figure 10.16. A physical commutation matrixer Thévenised from two different sets of terminals: (a), (b) and (c) illustrate one set of terminals and (d) (e), and (f) illustrate another set of terminals. (The $\boldsymbol{F}_i^{\mathrm{TH}}$ and $\boldsymbol{Z}_{ij}^{\mathrm{TH}}$ holors are different for each case.)

$$\boldsymbol{F}_{i(ac)}^{\mathrm{TH}} = \frac{\delta_k^i}{\boldsymbol{Z}_{jk}^1 + \boldsymbol{Z}_{jk}^2 + \boldsymbol{Z}_{jk}^3}\left(\boldsymbol{Z}_{ij}^2 + \boldsymbol{Z}_{ij}^3\right)\boldsymbol{F}_i^{\mathrm{S}}.$$

Also, as shown in figure 10.16(b), the impedance holor between the terminals a and c is $\boldsymbol{Z}_{ij}^2 + \boldsymbol{Z}_{ij}^3$ in parallel with $\boldsymbol{Z}_{ij}^1$ (the energy source is replaced by a short) and can be expressed as

$$\boldsymbol{Z}_{ij(ac)}^{\mathrm{TH}} = \boldsymbol{Z}_{ij}^1 \| \left(\boldsymbol{Z}_{ij}^2 + \boldsymbol{Z}_{ij}^3\right) = \frac{\delta_k^i}{\boldsymbol{Z}_{jk}^1 + \boldsymbol{Z}_{jk}^2 + \boldsymbol{Z}_{jk}^3}\boldsymbol{Z}_{ij}^1\left(\boldsymbol{Z}_{ij}^2 + \boldsymbol{Z}_{ij}^3\right).$$

The resulting Thévenin equivalent physical commutation matrixer is shown in figure 10.16(c). When it is viewed from between terminals b and c, as indicated in figure 10.16(d), $\boldsymbol{F}_{i(bc)}^{\mathrm{TH}}$ is the energy-potential-difference holor across $\boldsymbol{Z}_{ij}^3$ and can be expressed as

$$\boldsymbol{F}_{i(bc)}^{\mathrm{TH}} = \frac{\delta_k^i}{\boldsymbol{Z}_{jk}^1 + \boldsymbol{Z}_{jk}^2 + \boldsymbol{Z}_{jk}^3}\boldsymbol{Z}_{ij}^3\boldsymbol{F}_i^{\mathrm{S}}.$$

As shown in figure 10.16(c), the impedance holor between terminals b and c in parallel with the series combination of $\boldsymbol{Z}_{ij}^1 + \boldsymbol{Z}_{ij}^2$:

$$\boldsymbol{Z}_{ij(bc)}^{\mathrm{TH}} = \boldsymbol{Z}_{ij}^3 \| \left(\boldsymbol{Z}_{ij}^1 + \boldsymbol{Z}_{ij}^2\right) = \frac{\delta_k^i}{\boldsymbol{Z}_{jk}^1 + \boldsymbol{Z}_{jk}^2 + \boldsymbol{Z}_{jk}^3}\boldsymbol{Z}_{ij}^3\left(\boldsymbol{Z}_{ij}^1 + \boldsymbol{Z}_{ij}^2\right).$$

Thévenising a bridge physical commutation matrixer

The usefulness of Thévenin's theorem can be illustrated when it is applied, for example, to a bridge physical commutation matrixer. For instance, consider the case

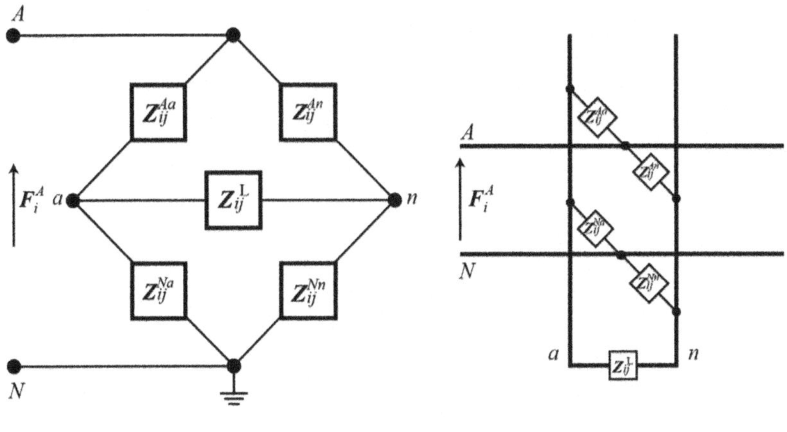

(a) (b)

Figure 10.17. A bridge physical commutation matrixer with a load impeder connected between the output terminals is not a straightforward series-parallel physical commutation matrixer.

where a load impeder is connected to the output terminals of a bridge, as shown in figure 10.17.

The bridge physical commutation matrixer is very difficult to analyse because it is not a straightforward series-parallel arrangement when an impeder, such as the very small impeder of a galvanometer, is connected between the output terminals a and n. There are no impeders that are in series or in parallel with another impeder. Using Thévenin's theorem, one can simplify the bridge physical commutation matrixer to an equivalent physical commutation matrixer viewed from the load impeder, as shown step-by-step in figure 10.18. Study the steps in the figure carefully. Once the equivalent physical commutation matrixer for the bridge is found, the energy-potential-difference holor and the energy-transfer holor for any value of load impeder can be determined easily using a generalised Ohm holor equation.

Summary of Thévenin's theorem

Remember, the Thévenin equivalent physical commutation matrixer for any impedance-holor physical commutation matrixer is *always* an equivalent energy-potential-difference source holor in series with an equivalent impedance holor, regardless of the original physical commutation matrixer that it replaces.

The significance of a Thévenin theorem is that the equivalent physical commutation matrixer can replace the original physical commutation matrixer as far as any external load is concerned. Any load impeder connected between the terminals of a Thévenin equivalent physical matrixer will have the same energy-transfer holor through it and the same energy-potential-difference holor across it as if it were connected to the terminals of the original physical commutation matrixer.

The steps for applying Thévenin's theorem are as follows.

(i) Open the two terminals (remove only load) between which one wants to find the Thévenin equivalent physical commutation matrixer.

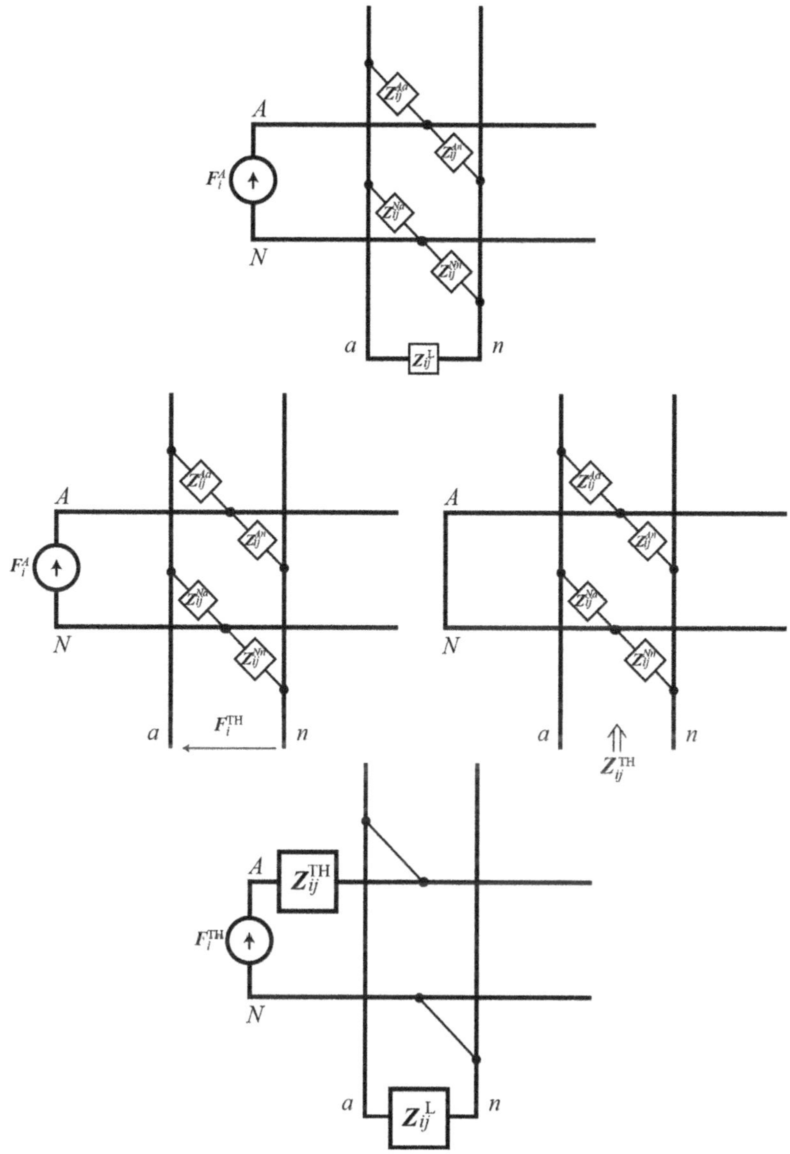

Figure 10.18. Simplifying a bridge physical commutation matrixer with Thévenin's theorem.

(ii) Determine the energy-potential-difference holor (F_i^{TH}) across the two open terminals.
(iii) Determine the impedance holor (Z_{ij}^{TH}) between the two terminals with all the energy sources replaced by their internal impedance holors. An ideal energy-potential-difference-source holor is replaced by a short.
(iv) Connect F_i^{TH} and Z_{ij}^{TH} in series to produce the complete Thévenin equivalent for the original physical commutation matrixer.

(v) Replace the load removed in step (i) across the terminals of the Thévenin equivalent physical matrixer. The load energy-transfer holor and the load energy-potential-difference holor can now be computed using only Ohm's law, and they have the same value as the load energy-transfer holor and the load energy-potential holor in the original physical commutation matrixer.

10.6.6 Norton's theorem

Like Thévenin's theorem, Norton's theorem provides a method for reducing a more complex physical commutation matrixer to a simpler form. The basic difference is that Norton's theorem gives an equivalent energy-transfer-source holor in parallel with an equivalent impedance holor.

The form of Norton's equivalent physical commutation matrixer is shown in figure 10.19. Regardless of how complex the original physical commutation matrixer is, it can always be reduced to this equivalent form.

The equivalent energy-transfer-source holor is designated V_N^j, and the equivalent impedance holor is designated Z_{ij}^N.

To apply Norton's theorem, one must know how to find the two quantities V_N^j and Z_{ij}^N. Once one knows them for a given physical commutation matrixer, one simply connects them in parallel to obtain the complete Norton physical commutation matrixer.

Norton's equivalent energy-transfer holor

As stated, V_N^j is one part of the complete Norton equivalent physical commutation matrixer; Z_{ij}^N is the other part. V_N^j is defined as the short-circuit energy-transfer holor between two points in a physical commutation matrixer. Any dynamical component connected between these two points effectively 'sees' an energy-transfer-holor source of value V_N^j in parallel with Z_{ij}^N. To illustrate, suppose that a holor-impedance physical commutation matrixer of some kind has an impeder connected between two points in the physical commutation matrixer, as shown in figure 10.20(a).

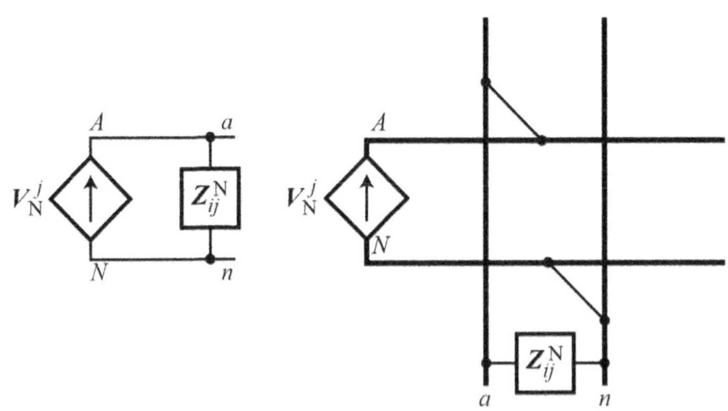

Figure 10.19. The form of Norton's equivalent physical commutation matrixer.

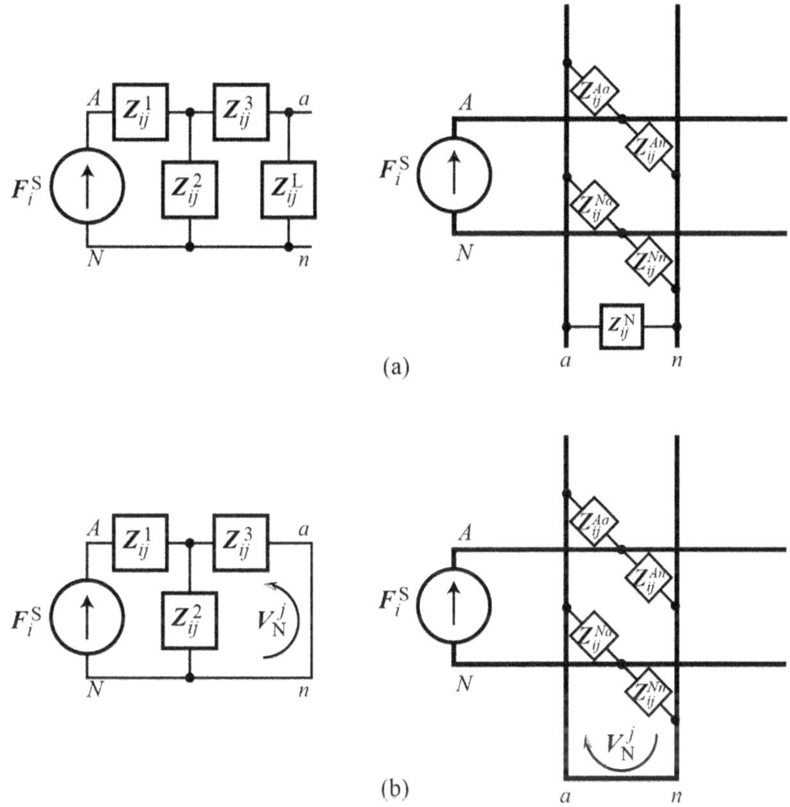

Figure 10.20. Determining the Norton equivalent energy-transfer holor, V_N^j.

One wishes to find the Norton physical commutation matrixer that is equivalent to the one shown to be *'seen'* by the load impedance holor, Z_{ij}^{L}. To find V_{N}^j, compute the energy-transfer holor between points a and b with these two points shorted, as shown in figure 10.20(b).

To determine V_{N}^j for the physical commutation matrixer in figure 10.20, short terminals a and b, as shown in figure 10.20(b). V_{N}^j is the energy-transfer holor through the short and is computed as follows.

First, the total impedance holor seen by the energy-potential-difference source holor is

$$Z_{ij}^{\mathrm{T}} = Z_{ij}^1 + \frac{\delta_k^l}{Z_{jk}^2 + Z_{jk}^3} Z_{ij}^2 Z_{ij}^3.$$

The total energy-transfer holor from the energy-potential-difference-source holor is

$$V_{\mathrm{T}}^j = \frac{\delta_k^i}{Z_{jk}^{\mathrm{T}}} F_i^{\mathrm{S}} = Y_{\mathrm{T}}^{ji} F_i^{\mathrm{S}}.$$

Now apply the energy-transfer-holor divider formula to find V_{N}^{j} (the energy-transfer holor through the short),

$$V_{\mathrm{N}}^{j} = \frac{\delta_{k}^{i}}{Z_{jk}^{2} + Z_{jk}^{3}} Z_{ij}^{2} V_{\mathrm{T}}^{j}.$$

This is the value for the equivalent Norton energy-transfer-holor source.

Norton's equivalent impedance holor

One defines a Norton equivalent impedance holor Z_{ij}^{N} in the same way as Z_{ij}^{TH}: it is the total impedance holor appearing between two terminals in a given physical commutation matrixer with all the energy sources replaced by their internal impedance holors.

We find Z_{ij}^{N} for the physical commutation matrixer in figure 10.20(b). First, we reduce F_{i}^{S} to zero by shorting it as shown in figure 10.20(b).

Looking in at terminals *a* and *b*, one can see that the parallel combination of Z_{ij}^{1} and Z_{ij}^{2} is in series with Z_{ij}^{3}.

Thus,

$$Z_{ij}^{\mathrm{N}} = Z_{ij}^{3} + \frac{1}{2} Z_{ij}^{1}.$$

Norton's theorem is a dual of Thévenin's theorem in that the equivalence for a section is expressed in terms of an energy-transfer-holor source and a shunting admittance holor. This is indicated in figure 10.19, where the equivalent is provided.

- V_{N}^{j} is the energy-transfer holor flow when the energy-potential-difference holor between *a* and *n* is zero, i.e. the short-circuit energy-transfer holor.
- Y_{N}^{ji} is the admittance holor between terminals *a* and *b* when all the energy sources within the section of the physical commutation matrixer are suppressed. i.e. the driving-point admittance holor.

The Norton equivalent can be obtained by series-parallel combinations, by the solution of mesh holor equations, or by a transformation of the Thévenin equivalent.

In all the equivalent physical commutation matrixers, one has considered it is important to note that some transformation is lost when one is working with the equivalent. The equivalence is valid only at the terminals we have designated *a* and *b*.

In effect, these equivalences are convenient methods to eliminate unwanted physical holors, thereby simplifying some of the mathematical manipulation, which is required. We have shown above how to find the two equivalent components of a Norton equivalent physical commutation matrixer, V_{N}^{j} and Z_{ij}^{N} or Y_{N}^{ji}. Keep in mind that these physical holors can be found for any linear physical heterogeneous continuous dynamical hypersystem. Once these are known, they must be connected in parallel to form the Norton equivalent physical commutation matrixer, as illustrated in figure 10.20

Summary of Norton's theorem
Any load impeder connected between the terminals of a Norton equivalent physical commutation matrixer will have the same energy-transfer holor through it and the same energy-potential-difference holor across it that it would if it were connected to the terminals of the original physical commutation matrixer.

The steps for applying Norton's theorem are as follows.

(i) Short the two terminals between which one wants to find the Norton equivalent physical commutation matrixer.
(ii) Determine the energy-transfer holor (V_N^j) through the shorted terminals.
(iii) Determine the impedance holor (Z_{ij}^N) or the admittance holor (Y_N^{ji}) between the two terminals (opened) with all energy-potential-difference-holor sources shorted and all the energy-transfer-holor sources opened ($Z_{ij}^N = Z_{ij}^{TH}$).
(iv) Connect V_N^j and Z_{ij}^N or Y_N^{ji} in parallel to produce the complete Norton equivalent for the original physical matrixer. This is Norton's equivalent physical commutation matrixer via the energy source conversion method.

10.6.7 Millman's theorem

Millman's theorem permits any number of parallel energy-potential-difference source holors to be reduced to a single equivalent energy-potential-difference source holor. It simplifies finding the energy-potential-difference holor across, or the energy-transfer holor through, a load impeder.

Millman's theorem gives the same results as Thévenin's theorem for the special case of parallel energy-potential-difference source holors. A conversion via Millman's theorem is illustrated in figure 10.21.

Millman equivalent energy-potential holor and equivalent impedance holor

Millman's theorem gives us a formula for computing the equivalent energy-potential-difference holor, F_i^{EQ}. To find F_i^{EQ}, convert each of the parallel energy-potential-difference-holor sources into an energy-transfer-holor source, as shown in figure 10.22.

In figure 10.22(b), the total energy-transfer holor from the parallel energy-transfer source holors is

$$V_T^j = V_1^j + V_2^j + V_3^j + \cdots + V_X^j.$$

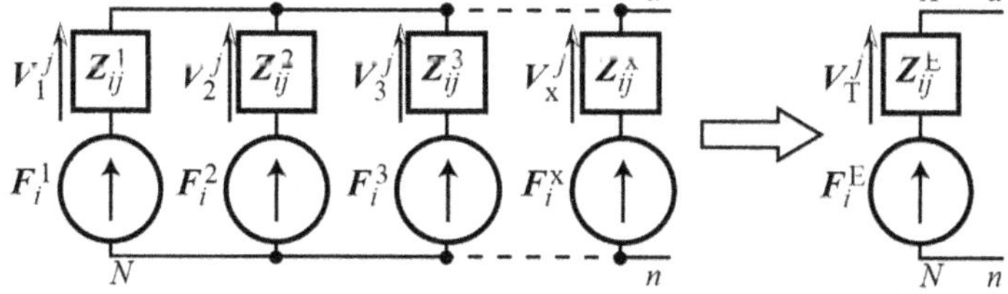

Figure 10.21. Reduction of parallel energy-potential source holors to a single equivalent energy-potential source holor.

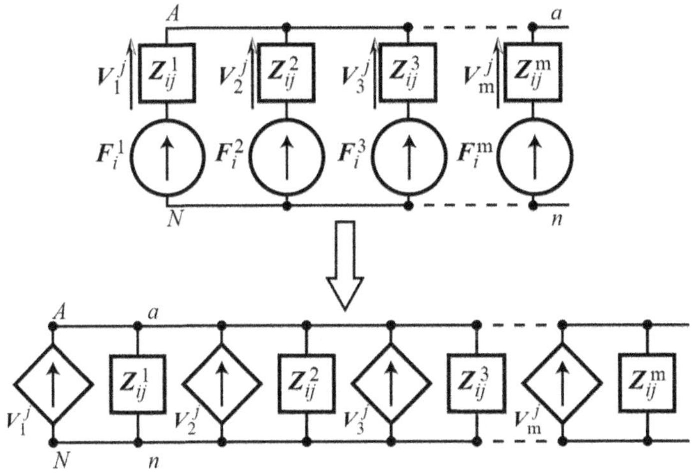

Figure 10.22. Parallel energy-potential source holors converted to energy-transfer source holors.

The total admittance holor between terminals a and b is

$$\boldsymbol{Y}_T^{ji} = \boldsymbol{Y}_1^{ji} + \boldsymbol{Y}_2^{ji} + \boldsymbol{Y}_3^{ji} + \cdots + \boldsymbol{Y}_X^{ji},$$

where $\boldsymbol{Y}_{\mathrm{T}}^{ji} = \delta_k^i / \boldsymbol{Z}_{jk}^{\mathrm{T}}$, $\boldsymbol{Y}_1^{ji} = \delta_k^i / \boldsymbol{Z}_{jk}^1$, and so on.

Remember, the energy-transfer sources are effectively open. Therefore, by Millman's theorem, the equivalent impedance holor, $\boldsymbol{Z}_{ij}^{\mathrm{EQ}}$, is the total impedance holor, $\boldsymbol{Z}_{ij}^{\mathrm{T}}$.

$$\boldsymbol{Z}_{ij}^{EQ} = \frac{\delta_k^i}{\boldsymbol{Y}_T^{ij}} = \frac{\delta_k^i}{\left(\delta_k^i / \boldsymbol{Z}_{jk}^1\right) + \left(\delta_k^i / \boldsymbol{Z}_{jk}^2\right) + \left(\delta_k^i / \boldsymbol{Z}_{jk}^3\right) + \cdots + \left(\delta_k^i / \boldsymbol{Z}_{jk}^X\right)}.$$

By Millman's theorem, the equivalent energy-potential-difference holor is

$$\boldsymbol{F}_i^{\mathrm{EQ}} = \boldsymbol{Z}_{ij}^{\mathrm{EQ}} \boldsymbol{V}_{\mathrm{T}}^j,$$

where $\boldsymbol{V}_{\mathrm{T}}^j$ is expressed as follows:

$$\begin{aligned} \boldsymbol{V}_T^j &= \boldsymbol{V}_1^j + \boldsymbol{V}_2^j + \boldsymbol{V}_3^j + \cdots + \boldsymbol{V}_X^j \\ &= \frac{\delta_k^i}{\boldsymbol{Z}_{jk}^1} \boldsymbol{F}_i^1 + \frac{\delta_k^i}{\boldsymbol{Z}_{jk}^2} \boldsymbol{F}_i^2 + \frac{\delta_k^i}{\boldsymbol{Z}_{jk}^3} \boldsymbol{V}_i^3 + \cdots + \frac{\delta_k^i}{\boldsymbol{Z}_{jk}^X} \boldsymbol{V}_i^X \\ &= \boldsymbol{Y}_1^{ji} \boldsymbol{F}_i^1 + \boldsymbol{Y}_2^{ji} \boldsymbol{F}_i^2 + \boldsymbol{Y}_3^{ji} \boldsymbol{F}_i^3 + \cdots + \boldsymbol{Y}_X^{ji} \boldsymbol{F}_i^X. \end{aligned}$$

The following is the formula for the equivalent energy-potential-difference holor:

$$\boldsymbol{F}_i^{\mathrm{EQ}} = \frac{\delta_k^i}{\boldsymbol{Y}_1^{ij} + \boldsymbol{Y}_2^{ij} + \boldsymbol{Y}_3^{ij} + \cdots + \boldsymbol{Y}_{\mathrm{X}}^{ij}} \left(\boldsymbol{Y}_1^{ji} \boldsymbol{F}_i^1 + \boldsymbol{Y}_2^{ji} \boldsymbol{F}_i^2 + \boldsymbol{Y}_3^{ji} \boldsymbol{F}_i^3 + \cdots + \boldsymbol{Y}_{\mathrm{X}}^{ji} \boldsymbol{F}_i^{\mathrm{X}} \right).$$

The equivalent energy-potential-difference source holor has a polarity such that the total energy-transfer holor through a load impedance holor will be in the same sense of direction as in the original physical commutation matrixer.

Reference

Ruston H and Bordogna J 1966 *Electric Networks: Functions, Filters, Analysis* (New York: McGraw-Hill)

Chapter 11

Electrical homogeneous continuous dynamical systems as DC and AC electrical commutation matrixers

How would people on the Earth do,
Thrown among the endless stars
Without light and warmth,
Electric beams and magnetic forces?
How would it be? It's easy to guess:
A gloomy time of chaos and cold would come.
So, sunbeams be praised!
Let's raise a glass to our Sun!

Adam Mickiewicz

11.1 Introduction

Direct current (DC) is a continuous flow of electricity in one sense of direction through a wire or conductor. DC is created by generators, such as fuel or photovoltaic cells; by static electricity and lightning; by ChE/ECh storage batteries; and by DC commutatorless homopolar generators and MCM or ECM AC–DC commutator generators.

It flows from a high to a low potential; for example, in a ChE/ECh storage battery, it flows from a positive to a negative pole. Any device that relies on ChE/ECh storage batteries—e.g. a flashlight, portable CD player, or laptop computer—operates on DC. When represented graphically, DC voltage appears as a straight line, usually flat (see figure 11.1—'A direct current').

Alternating current (AC) is electricity that changes its sense of direction at regular intervals. It builds up to a maximum voltage in one sense of direction and decreases

doi:10.1088/978-0-7503-1350-6ch11

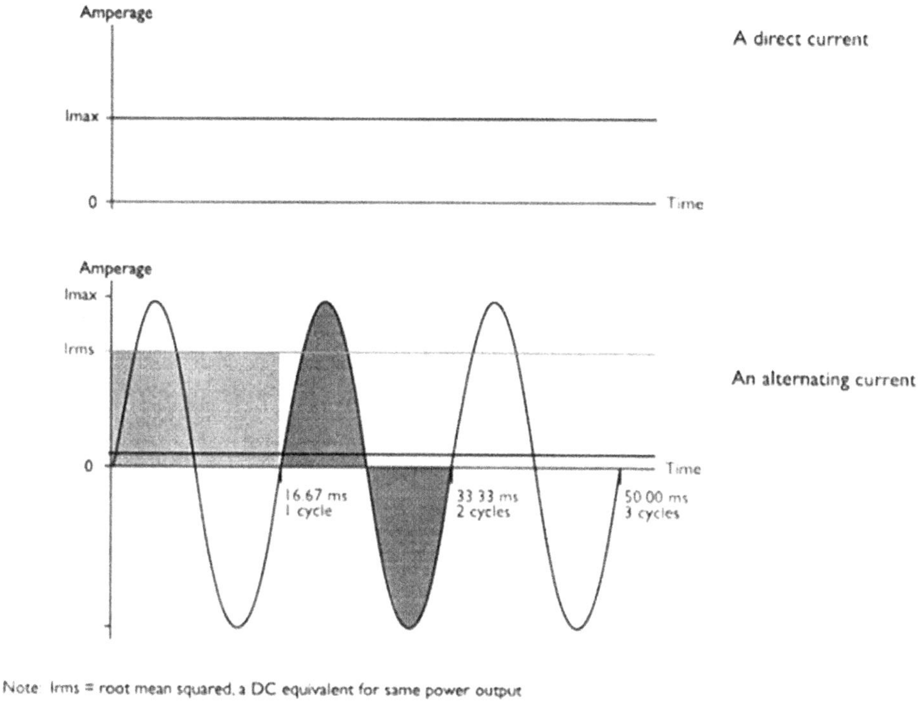

Figure 11.1. DC versus AC (EPRI 2006).

to zero; then builds up to a maximum in the opposite sense of direction; and then returns to zero once more (see figure 11.1—'An alternating current'). This complete sequence, or cycle, repeats, and the rate at which it repeats is called the 'frequency' of the current. In Europe, the AC energy provided to a home outlet has a frequency of 50 cps, and in the USA the frequency is 60 cps. This is expressed as 50 or 60 Hz, with hertz (Hz) being a unit that is equal to one cycle per second (cps).

The steady-state behaviour laws of DC electrical homogeneous continuous dynamical systems that are energised by DC energy sources are an important area of study for several reasons.

First, the DC energy generation, **high voltage direct current** (HVDC) transmission, distribution and consumption of electrical energy occurs under essentially DC steady-state conditions.

Second, it is possible to predict the behaviour of DC electrical homogeneous continuous dynamical systems to AC sinusoidal energy sources if the DC behaviour is understood.

Third, it is often easier to carry out the design of DC electrical homogeneous continuous dynamical systems based on steady-state DC behaviour. Thus, design specifications are spelled out in terms of a desired steady-state DC response.

The DC electrical homogeneous continuous dynamical system is designed to meet these DC characteristics. If these specifications are satisfied, the designer knows that the

DC electrical homogeneous continuous dynamical system will respond satisfactorily to DC inputs.

In view of the fact that one has been able to analyse relatively simple resistive DC electrical homogeneous continuous dynamical systems through the intelligent application of Kirchhoff's laws in combination with Ohm's law, this approach can also be used for all AC electrical homogeneous continuous dynamical systems, but as the AC electrical homogeneous continuous dynamical systems become structurally more complicated and involve more and more structural elements, one will soon find this direct method becoming quite cumbersome.

One purpose is to introduce two powerful techniques of DC and AC electrical homogeneous continuous dynamical system holor algebra steady-state analysis that facilitate the holor algebra analysis of sophisticated AC homogeneous continuous dynamical system's structures:

- the node-voltage holor method;
- the mesh-current holor method.

In addition to these general holor methods of holor algebra and steady-state analysis, we will also discuss some additional techniques for simplifying DC and AC electrical homogeneous continuous dynamical systems. We have already seen how series-parallel reductions and **pi-to-tee** (Π2T) or **delta-to-wye** (Δ2Y) transformations can be used to simplify a given structure. We now add energy-source transformation and Thévenin–Norton equivalent DC and AC electrical homogeneous continuous dynamical systems to our list of simplification holor techniques.

Before we begin our discussion of the node-voltage and mesh-current holor methods of DC and AC electrical homogeneous continuous dynamical system steady-state analysis, let us take a breather and think about the foundations that have been laid to this point.

Firstly, we introduce a voltage holor and a current holor as the parameters of the impedance holor. These three types of basic structural components of DC and AC electrical homogeneous continuous dynamical systems were chosen to start our discussion of DC and AC electrical homogeneous continuous dynamical system steady-state analysis because all the basic analytical techniques can be explored using interconnections of these structural components.

Thus, we introduce DC and AC electrical homogeneous continuous dynamical system steady-state analysis through the direct application of Ohm's law and Kirchhoff's laws. Ohm's law is critical because it describes the relationship between the current and the conductors of an impeder, and Kirchhoff's laws are important because they describe the constraints imposed on currents and voltages due to the interconnections of the basic structural components.

One is now ready to introduce the node-voltage and mesh-current holor methods of DC and AC electrical homogeneous continuous dynamical system steady-state analysis. Keep in mind that these two holor methods are of interest because they give us two systematic holor methods for describing DC and AC electrical homogeneous continuous dynamical systems with the minimum number of simultaneous equations.

In order to discuss the more elegant holor methods of DC and AC electrical homogeneous continuous dynamical system analysis, one must define a few basic terms that will enable one to give a clear, concise description of the important features of a given DC and AC electrical homogeneous continuous dynamical system.

Thus far, all of the DC and AC electrical homogeneous continuous dynamical systems have been 'planar' DC and AC electrical homopolar continuous dynamical systems, that is, dynamical systems that can be drawn on a plane such that no branches cross over each other. A DC and an AC electrical homogeneous continuous dynamical system that can be drawn with branches crossing over each other is still considered 'planar' if it can be redrawn with no crossover branches.

The node-voltage holor method can be applied to both planar and nonplanar DC and AC electrical homogeneous continuous dynamical systems, whereas in the nonplanar case the mesh-current holor method may be replaced by a technique known as the loop-current holor method.

Although we will not discuss the loop-current holor method, we wish to mention the fact that once one understands the mesh-current holor method, the transition to the loop-current holor method is not difficult. One has already defined what is meant by an ideal DC and AC electrical homogeneous continuous dynamical system's basic structural components. And when the basic DC and AC electrical homogeneous continuous dynamical system components are interconnected to form a DC or an AC electrical homogeneous continuous dynamical system, the resulting interconnections are described in terms of nodes, paths, branches, loops and meshes. One defined both a node and a closed path, or loop.

One purpose here is to review those definitions and then expand one's vocabulary to include the terms path, branch and mesh.

- A 'node' is a point in a DC and an AC electrical homogeneous continuous dynamical system where two or more of its structural elements join.
- A 'path' is formed whenever a set of a DC and AC electrical homogeneous continuous dynamical system's adjoining basic structural components is traced, in order, without passing through a connecting node more than once.
- A 'closed path', or 'loop', is created by starting at a selected node and then tracing through a set of a DC and AC electrical homogeneous continuous dynamical system's connected structural elements in such a manner that one returns to the original starting node without passing through any intermediate node more than once.
- A 'branch' is a path that connects two nodes.
- A 'mesh' is a special type of loop, that is, it does not contain any other loops within it.

The greatest advantage of both the node-voltage and mesh-current holor methods is that they reduce the number of simultaneous equations that must be manipulated. They also require the analyst to be quite systematic in organising and writing the required simultaneous equations. A first possible approach is to compare the number of simultaneous equations required for each holor method and then select the one that requires the fewest. A second is to analyse the location of energy sources (voltage and/or current sources) within the DC and AC electrical homogeneous continuous dynamical system structure.

Voltage sources may require extra effort in formulating holor node-voltage equations, whereas current sources may require extra effort in formulating mesh-current equations.

Another point to consider when choosing between the two holor methods is what information about the DC or AC electrical homogeneous continuous dynamical system being analysed is of primary interest. In other words, a complete solution of a DC and an AC electrical homogeneous continuous dynamical system may not be needed, and therefore the particular piece of information that is of interest may influence what method is used.

Even though the node-voltage and mesh-current holor methods are powerful techniques for solving DC and AC electrical homogeneous continuous dynamical systems, one is still interested in holor methods that can be used to simplify DC and AC electrical homogeneous continuous dynamical systems.

One begins to expand our list of simplifying holor techniques with energy source transformations. An 'energy source transformation' allows one to replace a voltage source in series with a resistor or impeder by a current source in parallel with the same resistor or impeder, or vice versa.

If one's interest in a DC and AC electrical homogeneous continuous dynamical system's behaviour is focused on a pair of conductors, the Thévenin and Norton theorem equivalent DC and AC electrical homogeneous continuous dynamical systems, which one is about introduce, are also extremely valuable aids in holor algebra analysis. However, it ought to be noted that Thévenin and Norton theorem equivalent DC and AC electrical homogeneous continuous dynamical systems can be used to represent any electrical homogeneous continuous dynamical system made up of linear structural components. This equivalence will hold for all possible values of load resistor or impedance. The Thévenin theorem resistor or impedance is the ratio of the open-circuit voltage holor to the short-circuit current holor.

The Norton theorem equivalent DC and AC electrical continuous dynamical system consists of an independent current source in parallel with the Norton theorem equivalent resistor or impedance. It can be derived from the Thévenin theorem equivalent DC and AC electrical homogeneous continuous dynamical system simply by making an energy source transformation. Thus the Norton current holor equals the short-circuit current holor at the conductive conductors of interest and the Norton resistor or impedance holor is identical to the Thévenin resistor or impedance holor.

The most distinguishing characteristic of a linear DC and AC electrical homogeneous continuous dynamical system is the principle of *'superposition'*, which states that whenever a linear DC and AC electrical homogeneous continuous dynamical system is excited, or driven by more than one independent energy source, one can find the total response by finding the response of each independent energy source separately and then summing the individual responses.

Since one is dealing with DC and AC electrical homogeneous continuous dynamical systems made up of interconnected linear structural components, one can apply the principle of superposition directly to the holor algebra analysis of such DC and AC electrical homogeneous continuous dynamical systems when they are driven by more than one independent energy source.

Summing up, the purpose of this textbook has been to introduce extremely useful holor algebra techniques of DC and AC electrical homogeneous continuous dynamical system analysis. The node-voltage and mesh-current holor techniques are important because they enable us to minimise the number of simultaneous equations needed to describe a DC and an AC electrical homogeneous continuous dynamical system. These two techniques are also valuable because they force us to take a systematic holor approach to writing DC and AC electrical homogeneous continuous dynamical system equations.

Energy source transformation and Thévenin–Norton equivalent DC and AC electrical homogeneous continuous dynamical systems—two additional methods of simplifying DC and AC electrical homogeneous continuous dynamical systems—are useful when the performance of a DC and an AC electrical homogeneous continuous dynamical system focuses on its behaviour at a specific pair of conductors.

The principle of superposition is an important concept when two or more energy sources are present in the DC or AC electrical homogeneous continuous dynamical system, because it enables one to isolate the effect of each Thévenin or Norton source on the total response of the DC or AC electrical homogeneous continuous dynamical system.

It is also important to keep in mind that although we have used the impedance holor in AC electrical homogeneous continuous dynamical systems to illustrate the various analytical techniques, the techniques themselves can also be applied to DC electrical homogeneous continuous dynamical systems. Thus, in this chapter, we begin discussing the holor algebra analysis of linear, lumped-parameter, single- and/or polyphase AC electrical homogeneous continuous dynamical systems that contain an impedance holor with inductances and capacitances as well as resistances.

11.2 Electrical homogeneous continuous dynamical systems' holor algebra, steady-state DC analysis

11.2.1 DC holor algebra, steady-state analysis

The simplest DC electrical homogeneous continuous dynamical system (figure 11.2) consists of an electrical energy source (e.g. a ChE/ECh storage battery, a DC homopolar generator, or a DC heteropolar generator, i.e. an **electrical commutation matrixer** (ECM) AC–DC commutator generator), a receiver (or simply load) of an electrical energy (e.g. an electrical lamp, electrical heating device, DC homopolar motor, or DC heteropolar motor, i.e. an ECM DC–AC commutator motor), and four conductive collectors connecting the receiver and energy-source single port (two terminals). The energy source transforms chemical, mechanical, thermal or other

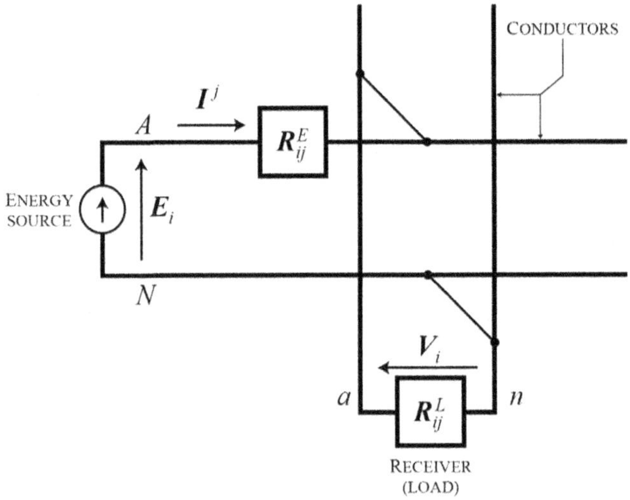

Figure 11.2. Simple DC electrical commutation matrixer.

energy into electromagnetic energy, which, for brevity, is often simply called 'electrical energy'; the electrical energy receiver (load), by contrast, transforms the electrical energy to other forms of energy: mechanical, thermal, radiant and so on. The energy source, together with the conductors and receiver (load) connected to it, forms a closed loop, along which an uninterrupted stream of electrical charges (an electrical current) flows. This is called a DC 'electrical commutation matrixer'. With respect to the energy source, the conductive collectors and receiver (load) form the so-called 'external circuit'.

The direct cause of the flow of electrical current in an electrical commutation matrixer is the self- or mutual-induced voltage source of the power source. Conventionally, the sense of the current direction in the DC electrical commutation matrixer is taken to coincide with the sense of the induced voltage (in figure 11.2 the senses of these directions are shown by arrows). The terminal of the voltage source through which the current 'leaves' for the external circuit is called the 'positive' (+) one, and the terminal through which the current 'enters' the voltage source from the external circuit is called the *'negative'* (−) one. The sense of the current direction is thus 'from to' in the external circuit and 'from to' inside the voltage source. Both the external circuit and the voltage source possess resistance, the value of which depends upon the material, shape and dimensions of the conductive collectors comprising these parts of the electrical matrixer. The magnitude or intensity of the current (or, simply, the current) in the electrical circuit depends on the magnitude of the induced voltage and the resistance of the whole DC electrical commutation matrixer. The relationship between these three quantities is expressed by Ohm's law, which plays a very important role in electrical engineering, being the basis of many practical calculations. According to this law, the current holor $\boldsymbol{I}^j$ in an electrical commutation matrixer is directly proportional to the voltage-source holor $\boldsymbol{E}_i$ and 'inversely' proportional to the resistance holor $\boldsymbol{R}_{ij}$ (or proportional to the conductance holor $\boldsymbol{G}^{ji}$) of the whole DC electrical commutation matrixer:

$$\boldsymbol{I}^j = \frac{\delta^i_k}{\boldsymbol{R}_{jk}} \boldsymbol{E}_i = \boldsymbol{G}^{ji} \boldsymbol{E}_i, \tag{11.1}$$

where the conductance holor $\boldsymbol{G}^{ji}$ matrix is the inverse of the resistance holor $\boldsymbol{R}_{jk}$ matrix:

$$\boldsymbol{G}^{ij} \boldsymbol{R}_{jk} = \delta^i_k,$$

and moreover, the current holor $\boldsymbol{I}^j$, the voltage-source holor $\boldsymbol{E}_i$, the conductance holor $\boldsymbol{G}^{ji}$ and the resistance holor $\boldsymbol{R}_{ij}$ matrices can be written directly from relationships:

$$\boldsymbol{I}^j = (I^c, 0) = \begin{bmatrix} I^c \\ 0 \end{bmatrix};$$

$$\boldsymbol{E}_i = (E^c, 0) = \begin{bmatrix} E^c \\ 0 \end{bmatrix};$$

$$\boldsymbol{R}_{ij} = \begin{bmatrix} R & 0 \\ 0 & R \end{bmatrix};$$

$$\boldsymbol{G}^{ji} = \begin{bmatrix} G & 0 \\ 0 & G \end{bmatrix}.$$

Ohm's law can be applied to the DC electrical commutation matrixer as a whole and, separately, to any portion of the DC electrical commutation matrixer outside the energy source. In this case, the induced voltage source holor $\boldsymbol{E}_i$ in equation (11.1) must be replaced by the voltage holor across the portion $\boldsymbol{V}_i$ (i.e. the difference in potential between the ends of the portion), while $\boldsymbol{R}_{ij}$ is the resistance holor of the given portion and not of the whole DC electrical commutation matrixer. Thus,

$$\boldsymbol{I}^j = \frac{\delta^i_k}{\boldsymbol{R}_{jk}} \boldsymbol{V}_i = \boldsymbol{G}^{ji} \boldsymbol{V}_i, \tag{11.2}$$

i.e. the current holor in the given portion of the DC electrical commutation matrixer is directly proportional to the voltage across it, and inversely proportional to the resistance holor $\boldsymbol{R}_{ij}$, or proportional to the conductance holor $\boldsymbol{G}^{ji}$ of the portion.

If the internal resistance holor of an energy source is significant, then the voltage holor (or potential-difference holor) $\boldsymbol{V}_i$ across the port terminals of the induced voltage-source holor $\boldsymbol{E}_i$ is less than its induced-voltage holor $\boldsymbol{E}_i$ by the 'internal voltage-drop' holor, which, according to Ohm's law, is equal to the product of the current holor $\boldsymbol{I}^j$ multiplied by the 'internal resistance' holor $\boldsymbol{R}^E_{ij}$ of the induced voltage-source holor $\boldsymbol{E}_i$, i.e.

$$\boldsymbol{V}_i = \boldsymbol{E}_i - \boldsymbol{R}^E_{ij} \boldsymbol{I}^j. \tag{11.3}$$

Often the internal resistance holor $\boldsymbol{R}^E_{ij}$ of the induced voltage-source holor $\boldsymbol{E}_i$ is relatively low, and the last term on the right-hand side of equation (11.3) may be disregarded. In this case, the voltage holor $\boldsymbol{V}_i$ of the energy source is approximately equal to the induced voltage-source holor $\boldsymbol{E}_i$.

When the external circuit is open, i.e. under so-called 'no-load conditions' of the induced voltage source, $\boldsymbol{I}^j = (0,0)$, and, irrespective of the magnitude $\boldsymbol{R}_{ij}^E$, equation (11.3) gives

$$\boldsymbol{V}_i = \boldsymbol{E}_i. \tag{11.4}$$

If the resistance holor $\boldsymbol{R}_{ij}$ of the external circuit ('external resistance holor') becomes low compared with the internal resistance holor $\boldsymbol{R}_{ij}^E$ of the induced voltage source, then one may say that $\boldsymbol{V}_i = (0,0)$ and that the energy source is short-circuited, in which case, according to equation (11.3), the current holor

$$\boldsymbol{I}^j = \frac{\delta_k^i}{\boldsymbol{R}_{jk}^E}\boldsymbol{E}_i = \boldsymbol{G}_E^{ji}\boldsymbol{E}_i \tag{11.5}$$

can reach a very high value, since $\boldsymbol{R}_{ij}^E$ is usually many times less than the resistance of the loads. In certain computations, it is more convenient to use the reciprocal multiplied by the Kronecker delta of the resistance holor $\boldsymbol{R}_{jk}^E$, i.e. the conductance holor $\boldsymbol{G}_E^{ji}$.

11.2.2 Node-voltage holor method

The direct use of Kirchhoff's laws for determining the currents in complicated DC electrical commutation matrixers demands the simultaneous solution of a considerable number of equations, making the computations extremely time-consuming. However, there are a number of methods, based on the same Kirchhoff laws that obviate the solving of a set of equations or reduce the number of equations and, hence, significantly cut the computations.

One of these is the 'node-voltage method', which may be applied when the DC electrical commutation matrixer contains only two nodes (figure 11.3), where the internal conductance (or resistance) holors of the energy sources are assumed to be included in the conductances (or inductances) of the branches, or may be reduced to

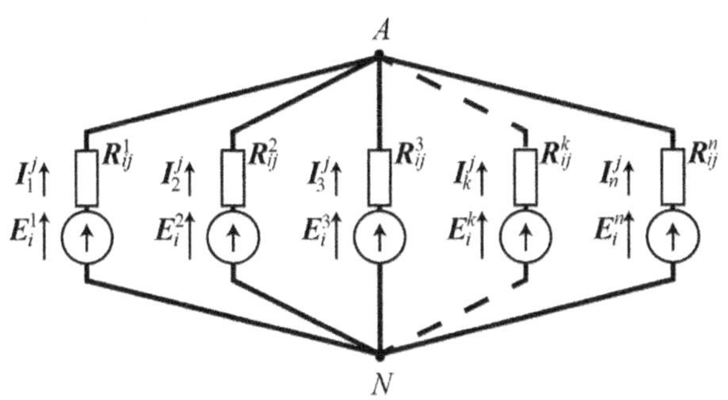

Figure 11.3. DC electrical commutation matrixer with two nodes.

such a form by converting deltas into wyes or vice versa. The number of branches N connecting the two nodes may be arbitrary.

The voltage-source holor $\boldsymbol{V}_i^{AN}$ between the nodes (the 'node-voltage holor') may be considered as the difference between the induced voltage-source holor $\boldsymbol{E}_i^E$ of any (jth) branch (which also includes the internal voltage-drop holor $\boldsymbol{R}_{ij}^E \boldsymbol{I}^j$ in the resistance holor $\boldsymbol{R}_{ij}^E$ of that branch (which also includes the internal resistance holor of the energy source)), i.e.

$$\boldsymbol{V}_i^{AN} = \boldsymbol{E}_i^E - \boldsymbol{R}_{ij}^E \boldsymbol{I}^j$$

if positive senses of directions from node N to node A are chosen for all current holors. Whence the current holor in the jth branch is

$$\boldsymbol{I}^j = \frac{\delta_k^i}{\boldsymbol{R}_{jk}^E}\left(\boldsymbol{E}_i^E - \boldsymbol{V}_i^{AN}\right) = \boldsymbol{G}_E^{ji}\left(\boldsymbol{E}_i^E - \boldsymbol{V}_i^{AN}\right). \tag{11.6}$$

But, according to **Kirchhoff's first law** (KFL), the sum of the currents meeting at node A must be equal to zero, i.e.

$$\frac{\delta_k^i}{\boldsymbol{R}_{jk}^E}\left(\boldsymbol{E}_i^E - \boldsymbol{V}_i^{AN}\right) = \boldsymbol{G}_E^{ji}\left(\boldsymbol{E}_i^E - \boldsymbol{V}_i^{AN}\right) = 0,$$

or

$$\frac{\delta_k^i}{\boldsymbol{R}_{jk}^E}\boldsymbol{E}_i^E = \frac{\delta_k^i}{\boldsymbol{R}_{jk}^E}\boldsymbol{V}_i^{AN}; \quad \boldsymbol{G}_E^{ji}\boldsymbol{E}_i^E = \boldsymbol{G}_E^{ji}\boldsymbol{V}_i^{AN},$$

whence

$$\boldsymbol{V}_i^{AN} = \begin{bmatrix} V_c^{AN} \\ 0 \end{bmatrix} = \frac{\boldsymbol{G}_E^{ji}\boldsymbol{E}_i^E}{\boldsymbol{G}_E^{ji}} = V_c^{AN} = \frac{G_1E_1 + G_2E_2 + \cdots + G_NE_N}{G_1 + G_2 + \cdots + G_N}. \tag{11.7}$$

Equation (11.7) gives us the node voltage; it is then easy to find the required current from equation (11.6). If the induced voltage of one of the branches has a reverse sense of direction to that shown in figure 11.3 or is missing altogether, then it must be included with a minus sign or omitted entirely in equations (11.6) and (11.7), respectively. If in one of the branches $\boldsymbol{E}_i^E < \boldsymbol{V}_i^{AN}$, then, according to equation (11.6), the current holor $\boldsymbol{I}^j$ will be negative, i.e. sensed in direction from node A to node N. In this case, it is more convenient to write equation (11.6) in the form

$$\boldsymbol{I}^j = \frac{\delta_k^i}{\boldsymbol{R}_{jk}^E}\left(\boldsymbol{V}_i^{AN} - \boldsymbol{E}_i^E\right) = \boldsymbol{G}_E^{ji}\left(\boldsymbol{V}_i^{AN} - \boldsymbol{E}_i^E\right)$$

in order to obtain the absolute magnitude of the required current.

11.2.3 Loop-current holor method

A quite different application occurs in electrical commutation matrixer theory (Staff 1940, Kron 1959, Bewley 1961). If a DC electrical commutation matrixer with several energy sources has more than two nodes, the currents in it may be found by the 'method of loop-currents'. This is one of the most universal methods for solving DC electrical commutation matrixers.

In a number of cases, a DC electrical commutation matrixer can be considered as consisting of a set of adjoining loops (two in figure 11.4), each of which forms a polygon made up of several branches of the DC electrical commutation matrixer (without any diagonals).

Some branches of the DC electrical commutation matrixer are common to two adjacent loops, while others form an external circuit where each branch only occurs in one loop.

The required branch current holor $\boldsymbol{I}^J$ of the DC electrical commutation matrixer must satisfy a set of holor equations based on KFL. If a certain current holor $\boldsymbol{I}^J$, known as the 'loop' holor, is assigned to each loop in an arbitrary sense of direction, and the actual current in the common branch of two adjacent loops is regarded as the algebraic sum of the corresponding loop currents, then KFL will be automatically satisfied, since all the loop currents in one branch flow towards a node, and in the other branch they flow away from the node.

When determining the loop currents it is only necessary that the actual currents should satisfy **Kirchhoff's second law** (KSL). The DC electrical commutation matrixer shown in figure 11.4 consists of, e.g. two loops and consequently has two loop-current holors: $\boldsymbol{I}_A^j$ and $`\boldsymbol{I}_a^j$. The current in any external branch will be equal in magnitude to the corresponding loop-current holor (as the only current in that branch) and the sign will be dependent on whether its sense of direction is the same as that of the loop-current holor or the opposite. If for all j loops one writes holor equations based on KSL, the result will be a set of j equations with j unknown loop-current holors. The solutions of these equations will first give all the loop-current holors and then all the actual-current holors. For instance, figure 11.4 shows a two-loop DC electrical commutation matrixer.

The applied voltage holors are $\boldsymbol{V}_i^A$, and $\boldsymbol{V}_i^a$ and the resulting current holors are $\boldsymbol{I}_A^j$ and $`\boldsymbol{I}_a^j$. The holors are related by

$$\boldsymbol{V}_i = \boldsymbol{R}_{ij}\boldsymbol{I}^J, \tag{11.8}$$

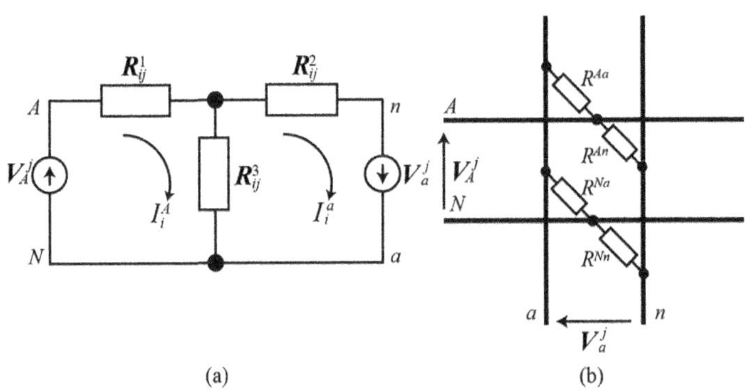

Figure 11.4. Two-loop DC electrical commutation matrixer.

where the voltage holor $\boldsymbol{V}_i$, the current holor $\boldsymbol{I}^J$ and the resistance holor $\boldsymbol{R}_{ij}$ in matrix notation are

$$\boldsymbol{V}_i = \begin{bmatrix} V_i^A \\ V_i^a \end{bmatrix} = \begin{bmatrix} \begin{Vmatrix} V_c^A \\ 0 \end{Vmatrix} \\ \begin{Vmatrix} V_c^a \\ 0 \end{Vmatrix} \end{bmatrix} = \begin{bmatrix} V^A \\ V^a \end{bmatrix};$$

$$\boldsymbol{I}^J = \begin{bmatrix} \boldsymbol{I}_A^j \\ \boldsymbol{I}_a^j. \end{bmatrix} = \begin{bmatrix} \begin{Vmatrix} I_A^c \\ 0 \end{Vmatrix} \\ \begin{Vmatrix} I_a^c \\ 0 \end{Vmatrix} \end{bmatrix} = \begin{bmatrix} I_A \\ I_a. \end{bmatrix};$$

$$\boldsymbol{R}_{ij} = \begin{bmatrix} R_{ij}^{Aa} & R_{ij}^{An} \\ R_{ij}^{Na} & R_{ij}^{Nn} \end{bmatrix} = \begin{bmatrix} R^{Aa} & R^{An} \\ R^{Na} & R^{Nn} \end{bmatrix} = \begin{bmatrix} R_1 + R_3 & -R_3 \\ -R_3 & R_2 + R_3 \end{bmatrix}.$$

Thus, the loop linear matrix equation is

$$\begin{bmatrix} V^A \\ V^a \end{bmatrix} = \begin{bmatrix} R^{Aa} & R^{An} \\ R^{Na} & R^{Nn} \end{bmatrix} \begin{bmatrix} I_A \\ I_a \end{bmatrix} = \begin{bmatrix} R_1 + R_3 & -R_3 \\ -R_3 & R_2 + R_3 \end{bmatrix} \begin{bmatrix} I_A \\ I_a \end{bmatrix},$$

or the loop linear algebraic equations are

$$V^A = R^{Aa} I_A + R^{An} I_a = (R_1 + R_3) I_A - R_3 I_a;$$
$$V^a = R^{Na} I_A + R^{Nn} I_a = -R_3 I_A + (R_2 + R_3) I_a.$$

A DC electrical commutation matrixer of any number of loops may also be expressed by equation (11.8), although, of course, the matrix $\boldsymbol{R}_{ij}$ will be different for each DC electrical commutation matrixer. In the usual problem of the holor analysis of an N-loop DC electrical commutation matrixer (Moon and Spencer 1986), $\boldsymbol{R}_{ij}$ are given and the current holors $\boldsymbol{I}^k$ are to be determined. The result can be written directly from equation (11.8):

$$\boldsymbol{I}^k = \boldsymbol{G}_E^{ki} \boldsymbol{V}_i, \qquad (11.9)$$

where the conductance holor $\boldsymbol{G}^{ji}$ is the inverse of the resistance holor $\boldsymbol{R}_{ik}$ matrix:

$$\boldsymbol{G}^{ki} \boldsymbol{R}_{ij} = \delta_j^k.$$

Evidently, the customary problem of DC electrical commutation matrixer holor analysis does not involve any question of coordinate transformation (Moon and Spencer 1986), so $\boldsymbol{V}_i$, $\boldsymbol{I}^J$ and, $\boldsymbol{R}_{ij}$ are merely holors

$$
\begin{aligned}
&V_i = (V^A,\ V^B, \ldots, V^N);\\
&I^J = (I_a,\ I_b, \ldots, I_n);\\
&R_{ij} = \begin{bmatrix} R^{Aa} & R^{Ab} & \ldots & R^{An} \\ R^{Ba} & R^{Bb} & \ldots & R^{Bn} \\ \vdots & \vdots & \ddots & \vdots \\ R^{Na} & R^{Nb} & \ldots & R^{Nn} \end{bmatrix}.
\end{aligned}
$$

For a DC electrical commutation matrixer consisting of N loops, one uses equation (11.8), namely:

$$V_i = R_{ij} I^J,$$

or

$$
\begin{bmatrix} V^A \\ V^B \\ \vdots \\ V^N \end{bmatrix} = \begin{bmatrix} R^{Aa} & R^{Ab} & \ldots & R^{An} \\ R^{Ba} & R^{Bb} & \ldots & R^{Bn} \\ \vdots & \vdots & \ddots & \vdots \\ R^{Na} & R^{Nb} & \ldots & R^{Nn} \end{bmatrix} \begin{bmatrix} I_a \\ I_b \\ \vdots \\ I_n \end{bmatrix}.
$$

Thus, for an N-loop DC electrical commutation matrixer, one obtains the following set of linear algebraic loop equations:

$$
\begin{aligned}
V^A &= R^{Aa} I_a + R^{Ab} I_b + \cdots + R^{An} I_n;\\
V^B &= R^{Ba} I_a + R^{Bb} I_b + \cdots + R^{Bn} I_n\\
&\ldots\ldots\ldots\ldots\ldots\ldots\ldots\ldots\ldots\ldots\ldots;\\
V^N &= R^{Na} I_a + R^{Nb} I_b + \cdots + R^{Nn} I_n.
\end{aligned}
$$

The solution of which, for an arbitrary current I_n, has the form

$$I_n = \frac{\Delta^N}{\Delta}, \tag{11.10}$$

where the determinant is

$$\Delta = \begin{vmatrix} R^{Aa} & R^{Ab} & \ldots & R^{An} \\ R^{Ba} & R^{Bb} & \ldots & R^{Bn} \\ \vdots & \vdots & \ddots & \vdots \\ R^{Na} & R^{Nb} & \ldots & R^{Nn} \end{vmatrix}. \tag{11.11}$$

Moreover, the determinant Δ^N is obtained from Δ by substituting the applied voltages $V^A, V^B, \ldots, V^N$ for the resistances of the Nth column.

The determinants Δ and Δ^N may be computed by rules that may be found in reference books on mathematics. Equation (11.10) serves to determine any loop current. The extension to AC electrical commutation matrixers is obvious. In AC electrical commutation matrixer holor synthesis, however, the equation of transformation may arise.

11.2.4 Superposition method

The determinant Δ^N in equation (7.26) can be given in the form:

$$\Delta^N = \Delta^{Na}E^A + \Delta^{Nb}E^B + \cdots + \Delta^{Nm}E^M + \cdots + \Delta^{Nn}E^N$$

by expanding it in terms of components of the Nth column.

Here, the algebraic complements Δ^{Nm} are obtained from the determinant if one strikes out the Nth column and the mth row and multiplies the determinants thus obtained by $(1)^{N+m}$.

Substituting this expression for Δ^N into equation (11.10), one obtains

$$I_N = \frac{\Delta^{Na}}{\Delta}E^A + \frac{\Delta^{Nb}}{\Delta}E^B + \frac{\Delta^{Nc}}{\Delta}E^C + \cdots + \frac{\Delta^{Nn}}{\Delta}E^N. \quad (11.12)$$

Every term on the right-hand side of this expression represents the current in the loop, provided applied voltage is present in one loop and absent in the other loops.

Thus, equation (11.12) shows the current in any loop and, hence, in any branch it can be considered as the superimposition of the partial currents arising in that branch due to each applied voltage in the loop, and, consequently, to each applied voltage in the branch separately. This is known as the 'superposition principle', and represents a special case of the principle of the independent action of forces, which is widely used in physics.

The superposition principle can serve as the basis for a method of solving electrical commutation matrixers—the 'superposition method'.

In a DC electrical commutation matrixer of many energy sources, determinations are successively made of the partial currents produced by the action of each applied voltage (the other applied voltages each time being put equal to zero).

Next, the partial currents produced by separate applied voltages in the same branches are algebraically added (superimposed). In this way, one finds the resultant currents produced by the combined action of all the applied voltages. For instance, in the DC electrical commutation matrixer in figure 11.5(a), which contains two energy sources, the applied induced voltage E_i^a is initially eliminated (but the internal resistance R_E^a of the energy source is retained).

This makes it possible to consider the DC electrical commutation matrixer as one series parallel connected matrixer component and to then simply determine all the partial currents $I_{1(I)}$, $I_{2(I)}$ and $I_{3(I)}$ caused by the action of the applied induced voltage E_i^A alone (figure 11.5(b)).

Then in a similar manner one determines the partial currents $I_{1(II)}$, $I_{2(II)}$ and $I_{3(II)}$ produced by the single applied induced voltage E_i^a (figure 11.5(c)).

Finally, determinants produced by the independent actions of the applied induced voltages are superimposed, and the resultant currents in the DC electrical commutation matrixer (which contains two energy sources) are determined:

$$I_1 = I_{1(I)} - I_{1(II)}; \quad I_2 = I_{2(I)} - I_{2(II)}; \quad I_3 = I_{3(I)} - I_{3(II)}.$$

However, the superposition method has a serious shortcoming: the partial currents, which flow in different senses of directions in a given branch, are often several times

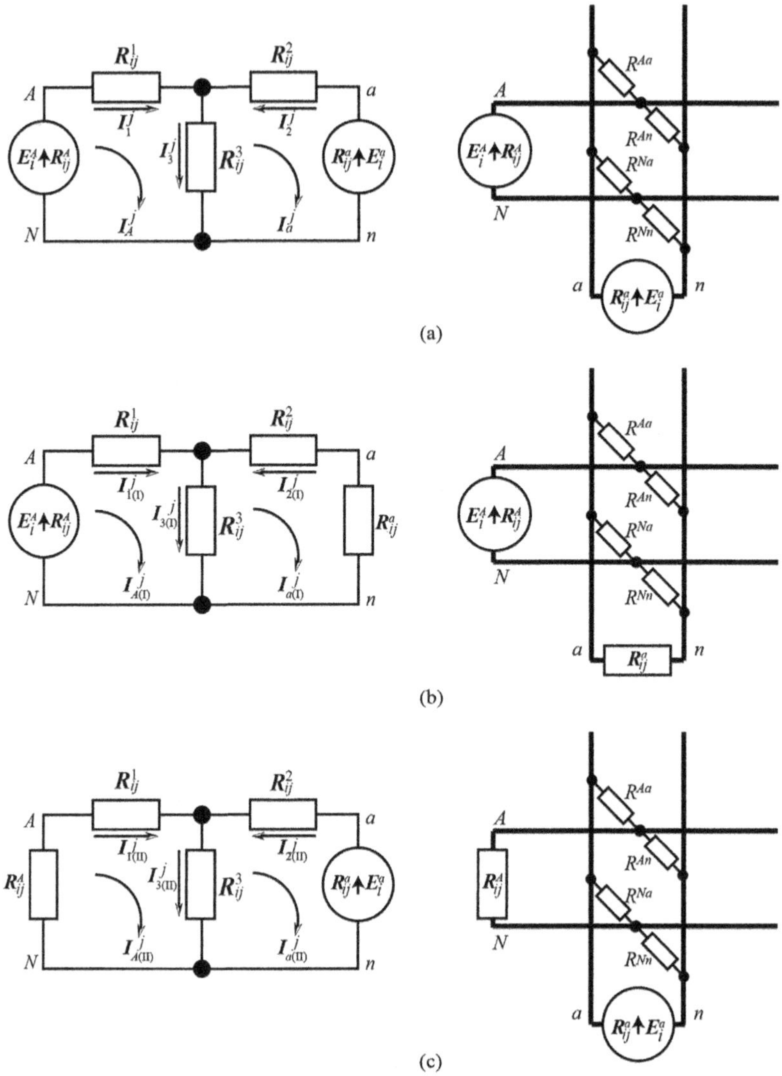

Figure 11.5. Solution by superposition method.

greater than the resultant current. Because of this, a small error (percentage-wise) in computing the partial currents leads to an impermissibly great error in the magnitude of the resultant current. For this reason, the superposition-holor method is used comparatively rarely.

11.2.5 The DC energy source

A DC 'energy source' produces electrical energy. It may act as a 'voltage source' (independent or dependent) that produces a voltage constant with time, or as a 'current source' (independent or dependent) that produces a current that is also constant with time. We use the constant function throughout in the discussion below.

11.2.6 The generation of DC energy

A current or voltage is termed 'direct' if it does not periodically change its sense of direction and magnitude with time. DC energy is obtained from AC–DC static or rotary converters (AC–DC rectifiers) or an MCM AC–DC commutator generator, or an AC commutatorless alternator with an ECM AC–DC commutator acting as the AC–DC rectifier.

An MCM AC–DC commutator generator is an electrical machine that converts mechanical energy into electrical energy. For its operation, an MCM AC–DC commutator generator depends upon electromagnetic induction. In an MCM AC–DC commutator generator, the armature *1* (figure 11.6), which carries a winding, is driven by a prime mover in the magnetic field of the **north** (N) and **south** (S) poles of electromagnet *2*. The AC induced voltage generated in the armature winding is collected by means of an MCM AC–DC commutator *3* and brushes, and delivered into an external circuit. The current delivered into the external circuit is DC because of the action of the MCM AC–DC commutator.

The prime mover may be a steam engine, a steam turbine, a gas turbine, an external or internal combustion engine, a windmill, an AC commutatorless synchronous motor or an MCM or ECM DC–AC commutator synchronous or induction motor.

In 1889, the noted US inventor Thomas Alva Edison (1843–1931) wrote in *Scientific American* that 'My personal desire would be to prohibit entirely the use of alternating currents. They are as unnecessary as they are dangerous, I can therefore see no justification for the introduction of a system which has no element of permanency and every element of danger to life and property.' Edison's 19th-century electrical energy distribution system relied on DC energy generation, delivery and use. This pioneering system, however, turned out to be impractical and uneconomical, largely because in the 19th century DC energy generation was limited

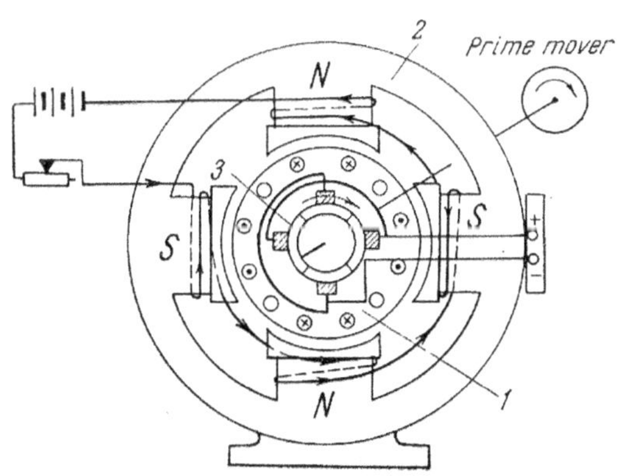

Figure 11.6. Diagram of an MCM AC–DC commutator generator (Kuznetsov 1960).

to a relatively low voltage potential and DC energy could not be transmitted above a small amount for more than a kilometre or mile.

Edison's electrical energy plants had to be local affairs, sited near the load, or the load had to be brought close to the DC commutatorless homopolar generator or MCM AC–DC commutator heteropolar generator.

AC energy distribution was far superior for the needs of a robust electrical infrastructure. Unlike DC electrical energy, the voltage of AC could be stepped up with relatively simple EE transformer devices for distance transmission and subsequently stepped down for delivery to appliances and equipment in the home or factory. As a consequence, the Hungarian-born US electrical engineer and inventor Nikola Tesla's (1856–1943) invention of a relatively simple polyphase AC commutatorless hyposynchronous (induction) motor meant end users needed AC, which could be generated at large central electrical energy plants for high-voltage bulk delivery over long distances. (See section 11.3.3 for more on the attributes of AC energy and DC energy, and why AC energy originally prevailed.)

Despite a vigorous campaign against the adoption of alternating current, Edison could not overcome the shortcomings of his DC energy system. The AC energy system won out, and today utilities generate, transmit and deliver electricity in the form of AC energy.

Despite the fact that **high-voltage direct current** (HVDC) is now a viable means of long-distance electrical energy transmission and is used in above a hundred applications worldwide (Woodford 1998, Barthold 2006), no one is advocating a wholesale change of the infrastructure from AC–DC, as this would be wildly impractical.

Although in Edison's time, DC was impractical for transmission beyond a distance of a kilometre or mile, today HVDC can transmit bulk energy over very long distances and also enables the interconnection of incompatible power grids. Electrical valves that can convert high voltage from AC–DC and back again were needed for HVDC to work, so such conversion was enabled with the development of high-voltage ECM AC–DC/DC–AC commutators (static converters) with unilateral, mercury-arc electrical valves capable of handling high voltages.

This technology was first deployed in 1954 by the noted Swedish electrical engineers Uno Lamm and Gunnar Engström (ASEA), for a transmission system between the island of Gotland and the Swedish mainland. The system was rated at 20 MW and 100 kV and transmitted power over 91 km (57 miles). Development in the 1960s of high-voltage unipolar semiconductor-controlled electrical valves, called 'thyristors', helped boost transmission capacity, increasing the cost-effectiveness of HVDC. Advances in the technology continue apace, with refinements in microprocessor control and other developments enhancing performance.

Over 100 HVDC systems, including many in Europe, Africa, Asia, Asia–Pacific, and North and South America, are now in service around the world. The largest, in Brazil, is rated at 600 kV. According to ABB, a key supplier of HVDC systems, newer designs have expanded the power range, extending the economic power range of HVDC systems from 90 MW to 1 GW (EPRI 2006).

HVDC can be the most economical option in some cases. Depending on the distances involved, an overhead DC transmission line and towers may be less expensive per unit of length than an overhead AC line. Although AC–DC/DC–AC converter stations are more costly than AC line terminating stations, in some cases transmission lengths as low as 640 km (400 miles) are sufficient to make the DC energy system more economical. The 'break even' distance needed for economical deployment of HVDC may be less when submarine or underground DC cable transmission is involved. In these cases, AC cable transmission can be no longer than about 48 km (30 miles), but distances of 288 km (180 miles) have been achieved with HVDC cables, and systems covering over 1.122 km (700 miles) are in the planning stages (EPRI 2006).

HVDC cables may make it easier and more economical to site high-voltage lines. Although electrical energy demand in certain areas may increase, installing high-voltage lines, or electrical energy plants near to load centres, typically meets with public opposition. Using compact HVDC underground cables as city infeeds from remote areas may be a means to overcome the political problems associated with building such new systems.

Moreover, in some circumstances, by upgrading or replacing existing AC transmission lines with HVDC ones, the electrical energy transfer capability of existing rights of way may be substantially increased.

Greater electrical energy flow control may be afforded to marketers and system operators. Various restructuring and deregulation schemes have forced the electrical energy system to operate in ways that it was not designed to. The electrical energy system was, in effect, designed to operate as a private toll road, and is now expected to operate as an open-access highway, handling millions of transactions daily (Galvin Electricity Initiative 2005).

AC networks do not easily accommodate the desired electrical energy flow control for today's system, so power marketers and system operators may require the electrical energy flow control capability provided by HVDC transmission. HVDC may be the only feasible means of interconnecting electrical energy networks. HVDC can interconnect two separate electrical energy networks so that electrical energy can be exchanged between them. Two AC energy systems can be connected by installing an AC–DC/DC–AC converter station in each system, with an interconnecting superconductive DC link between them, making it possible to transfer electrical energy even though the AC energy systems so connected remains asynchronous. For instance, an AC energy system may not be synchronized to neighbouring networks, even though the physical distance between them is quite small. This is the case in Japan, which has asynchronous networks: half the country is served by a 60 Hz network, and the other half by a 50 Hz system. It is physically impossible to connect the two together by direct AC methods—only DC works. HVDC can be a barrier to cascading blackouts. Perhaps one of the most vivid examples of a benefit of HVDC is related to its effect on grid reliability. HVDC lines cannot be overloaded and energy flow can be controlled for grid stability. This was illustrated during the 'Northeast blackout' of 2003, when the Quebec/Canada grid was unaffected, since it is interconnected to the neighbouring system with HVDC.

As a result, more HVDC interconnections between asynchronous networks such as those in the east and west of the USA and in Texas have been advocated. Additional HVDC links are being considered, most notably the **Trans-America Generation Grid** (TAGG), a connection of the eastern and western networks (EPRI 2006).

However, a new debate is arising over AC versus DC: should DC energy delivery systems displace or augment the AC energy distribution system in buildings or other small, distributed applications?

Edison's original vision for a system that has DC generation, energy delivery and end-use loads may come to fruition—at least for some types of installations. Facilities such as data centres, campus-like groups of buildings, or building hyposystems may find a compelling value proposition in using DC energy. Several converging factors have spurred the recent interest in DC energy delivery. One of the most important is that an increasing number of microprocessor-based electronic devices use DC energy internally, converted inside the device from standard AC supply. Another factor is that new distributed resources such as solar **photovoltaic cells** (PVC) and **fuel cells** (FC) produce DC energy, while ChE/ECh storage batteries and other technologies store it. So why not have a DC energy distribution system as well? Why not eliminate the equipment that converts DC–AC for electrical energy distribution and then back again to DC at the appliance?

Advocates point to the greater efficiency and reliability of a DC energy delivery system. Eliminating the need for multiple conversions could potentially prevent energy losses of up to 35%. Less waste heat and a less complicated conversion system could also potentially translate into lower maintenance requirements, longer-lived system components and lower operating costs. In a larger context, the deployment of DC energy delivery systems as part of AC–DC 'hybrid' buildings—or as a DC energy micro-grid 'island' that can operate independently of the bulk electrical energy grid—could enhance the reliability and security of the electrical energy system.

Benefits and drivers of DC energy delivery systems

Several potential benefits are driving new-found interest in DC energy delivery systems.

- *Increasingly, equipment operates on DC, requiring conversion from AC electrical energy sources.* All microprocessors require DC and many devices operate internally on DC energy since it can be regulated precisely for sensitive components. Building electrical systems are fed with AC that is converted to DC at every fluorescent ballast, and then there are computer systems and other electronic devices. As one specialist electronics manufacturer put it, 'DC is the blood of electronics' (Savage 2003). AC–DC conversions within these devices waste power. The electrical energy supplies that convert high-voltage AC power into the low-voltage DC energy needed by the electronic equipment used in commercial buildings and data centres typically operate at roughly 65–75% efficiency, meaning that 25–35% of all the electrical energy consumed is wasted.

About half the losses are from AC–DC conversions, the rest from stepping down DC voltage in DC–DC conversions. Simply getting rid of the losses from AC–DC conversion could reduce energy losses by about 10–20%. Likewise, an increasing number of portable gadgets such as cell phones and **personal digital assistants** (PDAs) require an AC–DC adapter, which also results in electrical energy losses during conversion. Considered in aggregate, the millions of AC–DC conversions necessitated for the operation of electronics extract a huge electrical energy loss penalty.

- *Distributed generation systems produce DC energy*. Many distributed generation electrical energy sources, such as PVCs and FCs, and advanced energy storage systems (ChE/ECh storage batteries, flywheels and ultra capacitors), produce electrical energy in the form of DC energy. Other devices can also be suited to DC output, such as Ch-Th-FM microturbines and PM wind turbines. Even **hybrid electric vehicles** (HEV) could serve as MCM AC–DC commutator generators in emergencies with the right equipment to connect them to the electrical energy system. The electrical energy losses entailed in converting DC–AC energy for distribution could be eliminated with DC energy delivery, enhancing efficiency and reliability and system cost-effectiveness. For instance, EPRI Solutions estimated that the total lifecycle cost of PVC electrical energy for certain DC applications could be reduced by more than 25% compared to using a conventional DC–AC approach—assuming that the specific end-use applications are carefully selected (EPRI PEAC Corp 2002, Barker *et al* 2003, Crudele *et al* 2004). The costs of new distributed generation sources such as PVCs are still high, so the optimisation of designs with DC energy delivery may help spur adoption and efficient operation.
- *Storage devices such as ChE/ECh storage batteries, flywheels and capacitors store and deliver DC energy*. This again helps to avoid unnecessary conversions between AC and DC.
- *DC electrical energy could help power hybrid electric vehicles, transit buses, and commercial fleets*. **Plug-in hybrid electric vehicles** (PHEV) can travel greater distances on electricity than today's hybrids since they have larger ChE/ECh storage batteries. These batteries store DC energy, so charging them with electricity from solar PVCs and other distributed electrical energy sources could reduce reliance on gasoline, enhancing security and emergency preparedness.
- *DC energy delivery could potentially enhance electrical energy efficiency in data centres, a pressing need*. One of the most promising potential applications of DC energy delivery is in data centres, which have densely packed racks of servers that use DC electrical energy. In such centres, AC is converted to DC at the **uninterruptible power supply** (UPS) to facilitate storage, and then converted back again to push out to the servers, before being converted one more time to DC at each individual server. These conversions waste power and generate considerable heat, which must be removed by air-conditioning systems, resulting in high electricity costs. For instance, in a 10–15 MW data centre, as much as 2–3 MW may be lost because of electrical energy conversions (EPRI 2006). As these centres install

ever more dense configurations of server racks, DC energy delivery systems may be a means to reduce their rocketing electrical energy needs.

- *Improved ECM DC–AC commutators, acting as DC–AC inverters and macro-electronics, allow DC energy to be converted easily and efficiently to AC energy and to different voltage levels.* Component improvements enable greater efficiency than in the past, and improve the economics of hybrid AC–DC systems. Although improved macro-electronics also enhance AC-only electrical energy systems, such enabling technology makes the DC energy delivery option feasible as well.
- *The evolution of central electrical energy architecture in computers and other equipment simplifies DC energy delivery systems.* At present, delivering DC energy to a computer requires input at multiple voltages to satisfy the electrical energy requirements of various internal components (RAM, processors, etc). The development of central electrical energy architecture, now underway, will enable the input of one standardised DC voltage at the port, streamlining delivery system design.
- *DC energy delivery may enhance micro-grid system integration, operation and performance.* A number of attributes make DC energy delivery appealing for use in micro-grids. With DC energy distribution, solid-state commutation (switching) can quickly interrupt faults, making for better reliability and power quality. If tied into the AC transmission system, a DC energy micro-grid makes it easy to avoid back-feeding surplus generation and fault contributions into the bulk utility system (by the use of an ECM AC–DC commutator acting as an AC–DC rectifier that only allows unilateral (one-way) electrical energy flow). In addition, in a low-voltage DC energy system, such as would be suitable for a home or group of homes, a line of a given voltage rating can transmit much more DC energy than AC energy.

  Of course, while DC electrical commutation matrixers are widely used in energy-consuming devices and appliances, DC energy delivery systems are not commonplace, and therefore face the obstacles any new system design or technology must overcome. For any of the benefits outlined above to be realised, testing, development and demonstration are needed to determine the true potential and market readiness of DC energy delivery, as outlined in (EPRI 2006).

High-density computer environments, utilising AC energy supply blade server technology, are presenting major challenges to operators and owners. Typical facility infrastructures cannot accommodate the added heat loads and reliability requirements of such installations (Baldwin and Geary 2006).

Servers equipped with DC energy supplies, instead of AC energy supplies, operate with 30–40% less heat, reduce electrical energy consumption up to 30%, increase server reliability, offer flexibility to installations and decrease maintenance requirements. Utilising current **off-the-shelf** (OTS) components, it is possible to install and operate a high-reliability DC energy distribution system for a DC powered server environment—without ChE/ECh storage batteries or complicated AC energy system drawbacks.

High-reliability DC energy distribution systems could be the next-generation electrical energy systems for server environments. The benefits of DC energy systems are as follows (Baldwin and Geary 2006).

- Lower component count equals higher system efficiencies, greater reliability, lower maintenance costs and lower total costs.
- They are modular and flexible: systems can grow with load requirements.
- They are environmentally flexible: system front-end components can be located in non-conditioned spaces or on the raised floor near the load.
- Flywheel energy storage eliminates the need for ChE/ECh storage batteries.
- Universal starline busway provides a double end feed feature to permit redundant DC energy sources at critical loads.
- No downstream static or transfer switches are required, as voltage-matched DC systems can inherently be coupled together.
- DC energy distribution eliminates harmonics.
- There is simplified grounding—positive grounding or negative grounding.
- A **very advanced technology** (VAP) maintenance management system contributes to higher system reliability and cost savings by managing equipment documentation effectively.
- System configuration is a unique application and partnership of existing and proven **commercially off-the-shelf** (COTS) equipment.
- This DC energy system has no requirement for an **uninterruptible power system** (UPS) in order to provide high system reliability with utility power outage ride through.
- DC energy distribution eliminates power factor concerns.
- The heat load at server racks is reduced by 20–40%.
- Electrical energy consumption is reduced by up to 30%.
- Electrical energy reliability is increased by as much as 27%.

Another **information technology** (IT) application that may lend itself to DC power delivery is supply for **power over the Ethernet** (PoE). According to the Institution of Electrical and Electronics Engineers (IEEE), PoE technology allows **internet protocol** (IP) telephones, wireless **local area network** (LAN) access points and other appliances to receive power as well as data over LAN cabling without needing or modifying the existing ethernet infrastructure.

The LAN cables carry both data and power, just as traditional telephone lines carry both voice and power on the same lines. PoE can supply power to computer phones (also known as **voice over internet protocol** (VoIP)), as well as other devices, such as web cameras, electronic badge readers and even electric guitars and other musical instruments. The voltage for PoE is 48 V, with about 12 to 15 W of available power at the device. A DC–DC converter transforms the 48 V to the lower voltages needed for electronics. The power infrastructure must be able to support the increased power requirements of PoE.

In Cisco Systems (2006) it is stated that the overall power budget for both the switch and powered devices is 5–6 kW. The ultimate source of the power may be AC, but as with the increased power requirements of data centres, facility managers must consider how best to deliver this energy, taking into consideration component costs, efficiency and cooling.

Cisco Systems (2006) noted that its technology can support DC power delivery for the network that is suitable 'for installations where highly available power delivery is critical and where an investment in DC power infrastructure is considered to be a business benefit.'

11.3 Electrical homogeneous continuous dynamical systems holor algebra, steady-state AC analysis

11.3.1 Sinusoidal AC holor algebra, steady-state analysis

The steady-state behaviour laws of AC electrical homogeneous continuous dynamical systems energised by AC sinusoidal energy sources are an important area of study for several reasons. First, the generation, transmission, distribution and consumption of electrical energy occur under essentially AC sinusoidal steady-state conditions. Second, it is possible to predict the behaviour of AC electrical homogeneous continuous dynamical systems to non-sinusoidal energy sources if the sinusoidal behaviour is understood. Third, it is often easier to carry out the design of AC electrical homogeneous continuous dynamical systems based on steady-state sinusoidal behaviour. Thus, the design specifications are spelled out in terms of a desired steady-state sinusoidal response. The AC electrical homogeneous continuous dynamical system is designed to meet these sinusoidal characteristics. If these specifications are satisfied, the designer knows that the AC electrical homogeneous continuous dynamical system will respond satisfactorily to non-sinusoidal inputs.

11.3.2 The sinusoidal AC energy source

A sinusoidal AC electrical energy source produces electrical energy. It may act as a sinusoidal voltage source (independent or dependent) that produces a voltage that varies sinusoidally with time, or as a sinusoidal current source (independent or dependent) that produces a current that also varies sinusoidally with time. One can express a sinusoidally varying function using either the sine function or the cosine function. There is no clear-cut choice for either function. Although either works equally well in sinusoidal steady-state analysis, it is important to recognise that both functional forms cannot be used simultaneously. We will use the sine function throughout in the discussion below.

11.3.3 The generation of AC

A current or voltage is termed 'alternating' if it periodically changes its sense of direction and magnitude. The complete period of a current variation is usually termed simply the period T of the AC, and the number of such periods per second

$$f = T^{-1} \tag{11.13}$$

is termed the frequency. The unit of frequency is s^{-1}, cycles per second (cps), and it is sometimes, especially in Europe, termed hertz (Hz).

In a more restricted sense, AC is a periodically varying current, the average value of which, over a period, is zero. The early electrical energy systems developed by Thomas Edison generated and delivered DC. However, DC energy systems had many limitations; most notably, the power typically could not be practically transmitted beyond a distance of a little more than a kilometre or approximately a mile. Moreover, because changing the voltage of DC was extremely inefficient, the delivery of DC energy in Edison's time meant that separate electric lines had to be installed to supply power to appliances and equipment of different voltages, an economically and physically impractical approach. Another limitation was that DC current incurred considerable power losses. Edison's concept for electrification of the USA—which included royalties from his patents on DC energy systems—was to deploy relatively small-scale individual DC energy plants to serve small areas, such as Pearl Street Station, which powered a part of New York City's financial district (EPRI 2006).

By contrast, the US inventor George Westinghouse's (1846–1914) polyphase AC energy system—invented by Nikola Tesla and used with EE transformers developed by the US electrician William Stanley, Jr (1858–1916), who also worked for Westinghouse—proved to be far superior technically and economically. The voltage of AC could be stepped up or decreased to enable long-distance power transmission and distribution to end-use equipment.

Thomas Alva Edison fought vociferously against the use of AC-based systems, which he claimed would be dangerous because of the high voltage at which power would need to be transmitted over long distances. Edison even went so far as to demonstrate the danger of AC by using it to electrocute a Coney Island elephant named *Topsy*, who had killed three men. He also electrocuted numerous cats and dogs procured from neighbourhood boys. However, despite proving that AC could be an effective means of electrocuting these hapless creatures, its superiority to DC for transmission and distribution was compelling. AC energy can be produced by large ME generators, and the AC voltage can be stepped up or down for transmission and delivery. The distance limitation of DC energy and the difficulties of changing DC voltages proved critical factors in abandoning DC energy systems in favour of those based on AC.

With DC energy systems, electrical energy had to be generated close to where it was used. This resulted in problematical reliability and economics. If the local plant failed, the entire system was down. And because initial DC energy systems were devoted to lighting loads and the systems only generated electrical energy at times of high usage, the cost of energy was high—often more than $1 (US dollar) per kWh when adjusted for inflation to present dollars (2005)—compared to an average cost for residential electricity today of 8.58¢ (US cents, $0.0858) per kWh (Energy Market Profiles database 2006).

Engineers wanted to interconnect systems to improve reliability and overcome the economic limitations of DC energy systems. If one area's power were out because of a problem at the ME generator, then the adjacent town would be available to pick up the load. In addition, by interconnecting isolated systems, a greater diversity of load was obtained, which would improve load factor and enable more economical operation of the generation plants.

Another major driver was the desire to make use of **hydro-electrical** (HE) energy sources located far from urban load centres, which made long-distance transmission—and therefore AC—essential. By using EE transformers, the AC voltage can be stepped up to high levels so that electricity can be distributed over long distances at low currents, and hence with low losses. EE transformers that could efficiently adjust voltage levels in different parts of the system and help minimise the inherent energy losses associated with long-distance distribution were a critical enabling technology that led to today's AC-dominated energy distribution system. EE transformers do not work with DC energy. William Stanley of the Westinghouse Company (IEEE HISTORY CENTER 2006) first demonstrated effective EE transformers in 1886. Stanley first demonstrated the potential of EE transformers to enable AC energy transmission at Main Street in Great Barrington, Massachusetts, USA.

Similar AC energy systems that used EE transformers eventually replaced Thomas Edison's DC energy transmission systems. Stanley's installation in Great Barrington was the first such system to include all of the basic features of large electrical energy systems, as they still exist more than one hundred years later.

Other factors led to the preference for AC energy transmission instead of DC energy delivery—most notably a desire for large-area grids relying on centralised electrical energy plant, such as HE dams. Having a transmission and distribution system that could provide HE to cities or to remotely located industries was also an economic imperative. Such development relied not only on EE transformers, but also on the development of polyphase AC commutatorless synchronous generators.

Up to the end of the 1870s, AC was not used. Soon after, AC became widely used in Europe, especially in France. The first AC energy generation plant was built at Ames, Colorado, USA. Electricity produced here in the spring of 1891 was transmitted 4 km (2.6 miles) over rugged and at times inaccessible terrain to provide electrical energy to operate the EM motor-driven mill at the Gold King mine. This pioneering demonstration of the practical value of transmitting electrical energy was a significant precedent in the USA for much larger plants at Niagara Falls (in 1895) and elsewhere. Electricity at Ames was generated at 3 kV, 133 Hz, single-phase AC, by a 74 kW (100 hp) Westinghouse AC commutatorless synchronous alternator (EPRI 2006).

In 1891, the eminent Polish electrician, Mikhail Dolivo-Dobrovolsky (1862–1919), was the genius who designed the first long-distance three-phase AC energy transmission in the world. The electrical energy transmission was from Lauffen am Neckar waterfall (near Heidelberg, Germany), where there was a **hydro-mechanical** (HM) turbine with a three-phase AC commutatorless synchronous generator and a step-up EE transformer, to Frankfurt am Main, the site of the 1891 International Electro-Technical Exhibition. Here, there was a step-down three-phase EE transformer that powered a three-phase AC commutatorless hyposynchronous (induction) EM motor, which drove a **mechano-hydraulic** (MH) pump (see figure 11.7).

In this three-phase AC energy transmission system, the three-phase EE transformers, the three-phase AC commutatorless hyposynchronous (squirrel-cage) EM motor and the system itself were all invented by Dolivo-Dobrovolsky. Here, 149 kW (200 hp) electrical energy was transmitted over a distance of 175 km (109 miles) at a

Figure 11.7. The three-phase AC commutatorless synchronous generator installed at Lauffen am Neckar, Germany. Credit: Historisches Museum, Frankfurt, Germany.

line voltage of 15 kV and a frequency of 30–40 Hz. The electrical energy transmission efficiency was 75% (Hoseman *et al* 1999). This was the first long-distance transmission of electrical energy for utility. This event also confirmed the first efficient long-distance transmission system that used a three-phase AC dynamical system.

The confirmation of the effectiveness of a three-phase AC energy transmission system was a sign of the end of the 'war of currents'. This episode confirmed the dominance of Dobrovolsky's three-phase AC energy transmission system over other systems. Three-phase AC energy transmission systems would prove over the next 20 years to be better practice for transmitting electrical energy than Edison's DC and Tesla's two-phase AC energy transmission systems.

In the generation, transmission and distribution of electrical energy, AC has two basic advantages over DC.

- The possibility (through using EE transformers) of increasing and reducing voltage simply and economically; this is especially important for long-distance electrical energy transmission.
- Simple design and, hence, cheaper electrical machines.

At present, AC of various frequencies is applied in diverse fields of electrical engineering. The highest frequencies are used in radio and television: from 100 kHz (long waves) up to 20 MHz (short waves), and then even up to several GHz (centimetric waves). A wide frequency band (from 500 Hz to 50 MHz) is used in various electrical furnaces.

One finds frequencies of the order of hundreds and thousands of cycles per second in telephone networks. Frequencies of 150–2000 Hz are used in powering high-speed AC EM motors. The lowest frequencies 50 or 60 Hz are used for lighting and for powering EM motors with rotation speeds not exceeding 3000 or 3600 revolutions per minute (rpm), respectively. Thus, electric railways use 15 or $16^2/_3$–50, or 25–60 Hz AC. However, such low frequencies cannot be used for lighting purposes, since at 15, $16^2/_3$, or 25 Hz, the light intensity varies with the current variation, producing a noticeable effect on the eye at frequencies below 40 cps.

On the other hand, as will be shown later, the voltage losses in AC energy transmission increase with increasing frequency. For this reason, a standard frequency of 50 Hz has been established for all power stations in Europe. Electric power stations in the USA operate at frequencies of 25 and 60 Hz.

At present, electrical-valve oscillators usually obtain high-frequency alternating currents. On the other hand, rotary or linear AC commutatorless ME generators produce industrial 50 or 60 Hz AC. Alternating currents of higher frequency (from 500 Hz to 10 kHz) are obtained from specially designed AC commutatorless generators and/or electrical-valve DC–AC, AC–DC–AC, or AC–AC commutators (converters).

The basis of AC generation is the law of electromagnetic induction. For instance, rotary AC commutatorless generators are ordinarily designed to produce the AC in the armature phase windings of the stationary part of the electric machine (the 'stator'). These phase windings are fitted into 'slots' into the inner surface of the stator (figure 11.8), and the separate leads are interconnected in such a way that the energy-source voltages produced in them combine.

The moving part of the AC generator (the 'rotor') is a flywheel with **north** (N) and **south** (S) magnetic pole pieces arranged alternately around the rim. The magnetic-field excitation windings for these magnetic poles are connected to two conductive commutators' 'slip rings', mounted on the shaft but insulated from it. Carbon brushes against the conductive commutators' slip rings supply DC to the magnetic-field windings. The DC is obtained from AC–DC converters (rectifiers) or a separate MCM AC–DC commutator generator, termed a 'DC exciter', often mounted on the same shaft as the AC commutatorless alternator. Use is also made of AC generators with **interior permanent magnets** (IPM), which, of course, do not have any rotor magnetic-field windings.

One turn of a two-pole AC commutatorless generator's rotor corresponds to one cycle of an AC energy-source voltage in the stator armature phase windings, and if the rotor makes n revolutions per second (rps), the frequency of the induced energy-source voltage is

$$f = n. \tag{11.14}$$

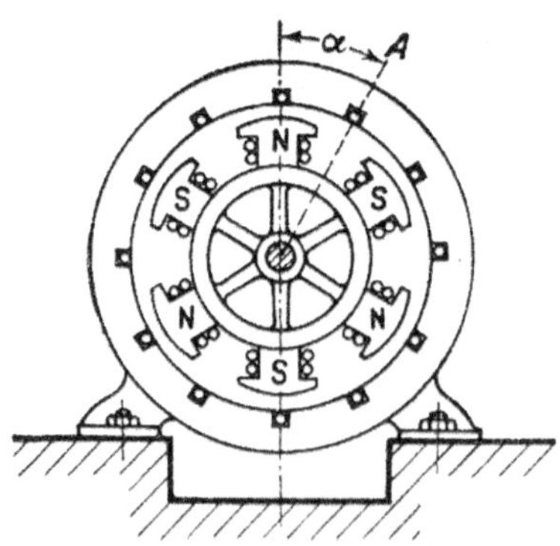

Figure 11.8. Schematic cross-section of a six-pole AC generator (Kasatkin and Perekalin 1970).

Since the angular velocity of the AC ME generator is, in this case,

$$\omega = 2\pi n,$$

a simple relationship exists between it and the frequency of the induced energy-source voltage:

$$\omega = 2\pi f = 2\pi / T. \quad (11.15)$$

For an AC ME generator with p pairs of magnetic poles,

$$f = pn. \quad (11.16)$$

Independently of the AC generator construction, the value ω determined from equation (11.15) is termed the 'angular frequency' of the AC. Thus, the angular frequency is the angular velocity of a two-pole AC generator, the energy-source voltage of which has a frequency f. The unit of angular frequency is s^{-1}.

The standard 50 or 60 Hz frequency corresponds to an angular frequency

$$\omega = 2\pi f = 2\pi 50 = 314 \text{ s}^{-1},$$

or

$$\omega = 2\pi f = 2\pi 60 = 377 \text{ s}^{-1}.$$

11.3.4 Sinusoidal alternating current

The alternating energy-source voltage of a commutatorless AC generator and the AC can have highly diversified variation waveforms, which in large measure depend on constructional factors, for instance, the shape of the rotor magnetic-pole pieces.

The most important periodic waveforms are the sine or cosine waveforms, which are the only periodic functions with derivatives like themselves. The technical significance of this fact is that the shape of the waveforms of voltages and currents (figure 11.9) is the same in all the components of an AC electrical continuous dynamical system, i.e. sinusoidal or harmonic.

The analytical expressions for the sinusoidal functions of instantaneous values of the voltages and currents are

$$\left.\begin{aligned} v(t) &= V_{\max} \sin(\omega t + \psi_v) = \sqrt{2}\, V_{\text{rms}} \sin(\omega t + \psi_v) \\ i(t) &= I_{\max} \sin(\omega t + \psi_i) = \sqrt{2}\, I_{\text{rms}} \sin(\omega t + \psi_i) \end{aligned}\right\}, \quad (11.17)$$

where the small letters $v(t)$ and $i(t)$ denote the instantaneous values of voltage and current, i.e. their values at an arbitrary time t, the capital letters $V_{\max}$ and $I_{\max}$ denote the corresponding amplitudes, i.e. the maximum values, $V_{rms} = V$ and $I_{rms} = I$ denote the corresponding effective values, and ψ_v and ψ_i denote the corresponding phase angles, depending on the values of the voltage and current at the instant chosen as the time origin ($t = 0$).

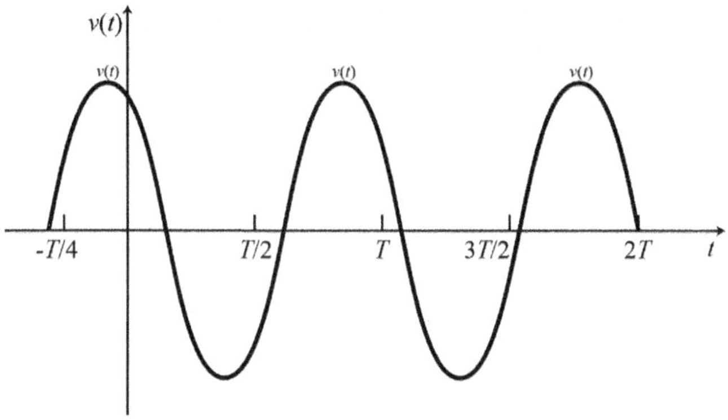

Figure 11.9. Sinusoidal waveform of voltage or current.

If one takes the initial time to be when the current passes through its maximum, $\sin(\omega t + \psi_i)$ at $t = 0$ must be equal to unity:

$$\sin(\omega t + \psi_i) = 1 \text{ and } \psi_i = \pi/2.$$

The various stages of the processes described by the sinusoidal functions, namely, growth, decay, passage through zero or a maximum, and so on, are termed phases. The phase of a process at any time t is defined by the angle $\omega t + \psi$ of the sine and is termed the 'phase angle'.

The angle ψ that determines the phase at a given initial instant ($t = 0$) is termed the 'initial phase', and the difference between the initial phases ψ_v and ψ_i of the two functions is termed the 'phase difference' or 'phase shift'.

11.3.5 Sinusoidal voltage and current waveforms—holor representation

In an AC electrical homogeneous continuous dynamical system one usually deals with sinusoidal voltage and current waveforms that are all of the same frequency, but that may be shifted from each other in time. We start by discussing sinusoidal voltage and current waveforms. Very simple sinusoidal voltage and current instantaneous values from using the trigonometric identity are (Fijalkowski 1967):

$$\left.\begin{aligned} v(t) &= \sqrt{2}\,U \sin(\omega t + \psi_v) = \sqrt{2}\,V[\sin(\omega t \cos\psi_v + \cos\omega t \sin\psi_v)] \\ i(t) &= \sqrt{2}\,I \sin(\omega t + \psi_i) = \sqrt{2}\,I[\sin(\omega t \cos\psi_i + \cos\omega t \sin\psi_i)] \end{aligned}\right\}, \tag{11.18}$$

in which ψ_v is the voltage phase angle and ψ_i is the current phase angle.

Now, one must be able to convert equations (11.18) of the forms

$$\left.\begin{aligned} v(t) &= \sqrt{2}[V_s \sin\omega t + V_c \cos\omega t] \\ i(t) &= \sqrt{2}[I^s \sin\omega t + I^c \cos\omega t] \end{aligned}\right\}. \tag{11.19}$$

Then, comparing equations (11.19) with equations (11.18), one obtains

$$V_c = V\cos\psi_v,\ V_s = V\sin\psi_v. \tag{11.20}$$

The quantities V_c and V_s are independent elements (merates) of an effective value of the voltage holor

$$\boldsymbol{V}_i = (V_c, V_s). \tag{11.21}$$

The above-written holor fully represents a sinusoidal voltage.

Analogically, a sinusoidal current can be represented by an effective value of the current holor

$$\boldsymbol{I}^j = (I^c, I^s), \tag{11.22}$$

where

$$I^c = I\cos\psi_i,\ I^s = I\sin\psi_i. \tag{11.23}$$

Thus, the sinusoidal voltage holor $\boldsymbol{V}_i$ and/or the current holor $\boldsymbol{I}^j$ are defined as mathematical entities or 'wholes' built from two independent elements (merates) V_c and V_s, as well as I^c and I^s, which carry the effective values and phase angle information of sinusoidal functions to the doublet (dipole) functions of two independent quantities that cannot be added directly.

Due to the fact that independent elements of the voltage holor as well as the current holor depend solely on the effective values and phase angles of these quantities, then the voltage holor and the current holor can be represented in the same way as in complex numbers, namely:

$$\boldsymbol{V}_i = V\angle\psi_v \tag{11.24}$$

and

$$\boldsymbol{I}^i = I\angle\psi_i. \tag{11.25}$$

As in equations (11.21)–(11.25), a holor quantity will be represented throughout the text by a boldface letter.

In studying equations (11.24) and (11.25), we observe that the shorthand abbreviation $\angle$ in the angle notation of the phase angles belongs to holors that carry the effective values and phase angles of the given sinusoidal functions. We use this notation extensively in the material that follows. These holors are by definition the 'holor representation', or 'holor transform', of the given sinusoidal functions. Thus, the holor transform transfers the sinusoidal function from the time domain to the holor domain. Thus far, we have emphasized moving from the sinusoidal function to its holor transform. However, it should be apparent that one can also reverse the process. That is, given the holors, one can write the expressions for the sinusoidal functions.

Thus, if one is given that $\boldsymbol{V}_i = V\angle\psi_v$ and $\boldsymbol{I}^j = I\angle\psi_i$, the expressions are $v(t) = \sqrt{2}\,U\sin(\omega t + \psi_v)$ and $i(t) = \sqrt{2}\,I\sin(\omega t + \psi_i)$, since one has decided to use the sine function for all sinusoids.

Observe that one cannot deduce the value of ω from the holor. The holors only carry the magnitude of the effective values and the phase angle information. The step of going from the holor transform to the time-domain expression is referred to as finding the inverse holor transform and is formalised by the equation

$$\left.\begin{aligned} \mathbb{H}^{-1}\{V\angle\psi_u\} &= \sqrt{2}\{V\angle\omega t + \psi_v\} \\ \mathbb{H}^{-1}\{I\angle\psi_i\} &= \sqrt{2}\{I\angle\omega t + \psi_i\} \end{aligned}\right\}, \tag{11.26}$$

where the notations $\mathbb{H}^{-1}\{V\angle\psi_v\}$ and $\mathbb{H}^{-1}\{I\angle\psi_i\}$ are read as the inverse transforms of $V\angle\psi_u$ and $I\angle\psi_i$.

Equation (11.26) tells us that to find the inverse holor transform one multiplies the holors by $\sqrt{2}\{V\angle\omega t\}$ and $\sqrt{2}\{I\angle\omega t\}$ and then extracts the second merates V_s and I^s of the products, respectively. The holor transform is so useful in electric system analysis because it reduces the problem of finding the maximum amplitude and phase angle of the steady-state sinusoidal response to the algebra of holors. In a linear electrical continuous dynamical system driven by sinusoidal energy sources, the steady-state response is also sinusoidal.

The holor transform is also useful in electric system analysis because it applies directly to the sum of sinusoidal functions. Since electrical continuous dynamical system analysis involves summing voltages and currents, the importance of this observation is obvious.

The holor representation of the sum is the sum of the holors of the individual independent elements (merates). The superiority of the holor approach to adding sinusoidal functions should be apparent. One should note that the holor approach requires the ability to move back and forth between the polar and rectangular forms.

11.3.6 The passive AC electrical-homogenous-continuous-dynamical-system components in the holor domain

The systematic application of the holor transform in AC electrical homogeneous continuous dynamical system analysis requires two steps.

First, one must establish the relationship between the current holor and the voltage holor of the passive AC electrical homogeneous continuous dynamical system's components.

Second, one must develop the holor-domain version of Kirchhoff's laws, which we discuss in the next section. In this section, we establish the relationship between the current holor and the voltage holor at the conductive collectors of the resistor, inductor and capacitor. We begin with the resistor. We use the passive sign convention in all the derivations. It follows directly from Ohm's law that if the current in a resistor varies sinusoidally with time, that is, if $i(t) = \sqrt{2}I\sin(\omega t + \psi_i)$, then the voltage at the conductive collectors of the resistor (figure 11.10) will be

$$i(t) = R\sqrt{2}I\sin(\omega t + \psi_i), \tag{11.27}$$

where $I_{\max} = \sqrt{2}I$ is the maximum amplitude of the current in amperes (A) and ψ_i is the phase angle of the current.

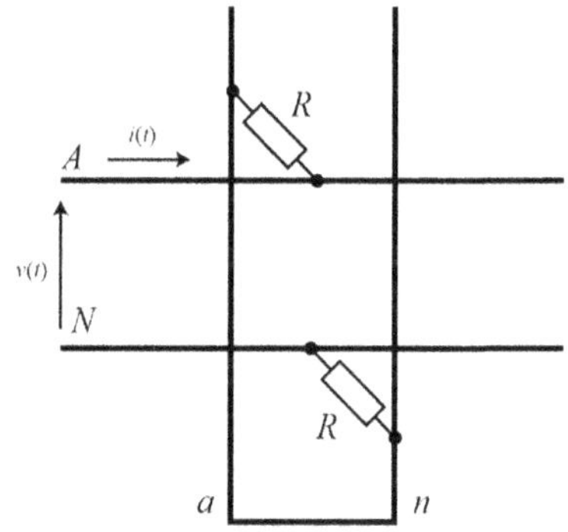

Figure 11.10. A resistive AC electrical-homogeneous-continuous-dynamical-system's dynamical component carrying a sinusoidal AC.

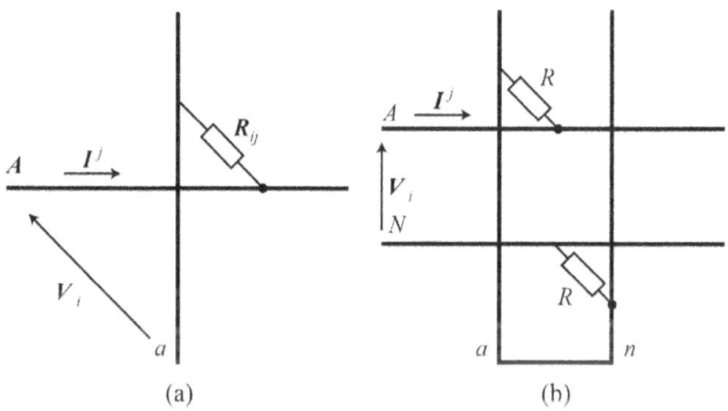

Figure 11.11. The holor domain equivalent AC electrical homogeneous continuous dynamical system for a resistor.

The holor transform of this voltage holor is

$$V_i = \boldsymbol{R}_{ij} I \angle\psi_i. \tag{11.28}$$

However, $I\angle\psi_i$ is the holor representation of the sinusoidal current holor; thus, one can write equation (11.28) as

$$V_i = \boldsymbol{R}_{ij}\boldsymbol{I}^j = \begin{bmatrix} R & 0 \\ 0 & R \end{bmatrix} \begin{Vmatrix} I^c \\ I^s \end{Vmatrix}, \tag{11.29}$$

which tells one that the voltage holor at the conductive collectors of a resistor is simply the resistance holor $\boldsymbol{R}_{ij}$ times the current holor.

An AC electrical homogeneous continuous dynamical system diagram for a resistor in the holor domain is shown in figure 11.11.

This phase relationship is shown in figure 11.12.

The signals are said to be in time phase because they both reach corresponding values on their respective waveforms at the same time (for example, they are at their positive maxima at the same instant).

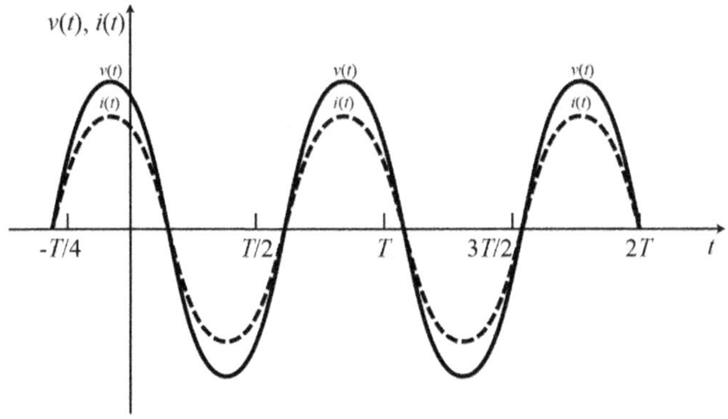

Figure 11.12. A resistive AC electrical-homogeneous-continuous-dynamical-system's dynamical component carrying a sinusoidal AC.

The relationship between the current and voltage at the conductive connectors of the inductor is derived by assuming a sinusoidal current and using $L\mathrm{d}i/\mathrm{d}t$ to establish the corresponding voltage.

Thus, given $i(t) = \sqrt{2}I\sin(\omega t + \psi_i)$, one sees that the expression for the voltage is

$$v(t) = L\frac{\mathrm{d}i(t)}{\mathrm{d}t} = \omega L\sqrt{2}I\cos(\omega t + \psi_i). \tag{11.30}$$

Now one rewrites equation (11.30) using the sine function:

$$v(t) = L\frac{\mathrm{d}i(t)}{\mathrm{d}t} = \omega L\sqrt{2}I\sin(\omega t + \psi_i + \pi/2). \tag{11.31}$$

The holor representation of the voltage holor given by equation (11.31) is

$$V_i = X_{ij}^L I^j = \begin{bmatrix} 0 & -\omega L \\ \omega L & 0 \end{bmatrix} I\angle\psi_i + \pi/2 = \begin{bmatrix} 0 & -\omega L \\ \omega L & 0 \end{bmatrix} I^j. \tag{11.32}$$

From equation (11.32), one sees that the voltage holor at the conductive conductors of an inductor equals the inductive-reactance holor X_{ij}^L times the current holor.

The holor domain equivalent AC electrical homogeneous continuous dynamical system is shown in figure 11.13. The phase relationship between the current holor and the voltage holor at the conductive conductors of an inductor is contained in equation (11.32), from which we can see that the voltage and current will be out of phase by exactly $\pi/2$.

In particular, the voltage will lead the current by $\pi/2$ or, what is equivalent, the current will lag the voltage by $\pi/2$. This concept of 'voltage leading current' or 'current lagging voltage' is made clear by the plot shown in figure 11.14, where one notes that as time increases from zero, the voltage comes to a particular instantaneous value on the sine function ahead of the current.

For instance, it reaches its negative peak exactly $\pi/2$ before the current reaches its negative peak.

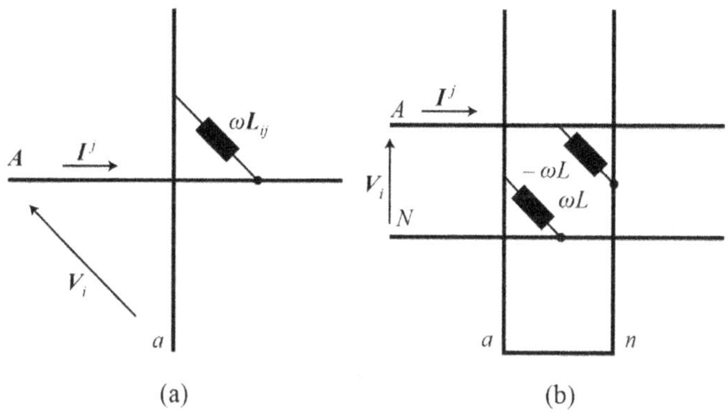

Figure 11.13. The holor domain equivalent AC electrical homogeneous continuous dynamical system for an inductor.

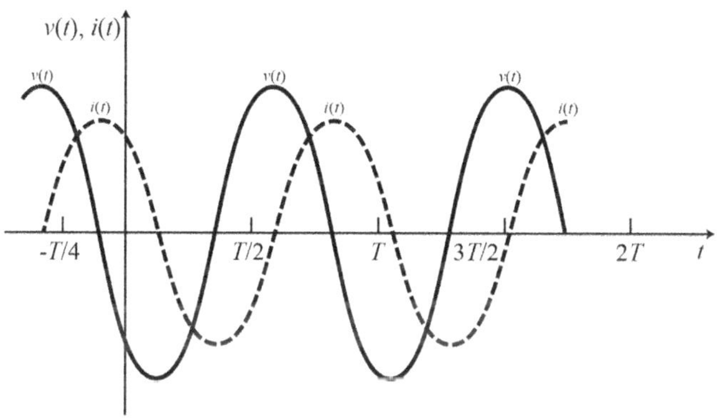

Figure 11.14. An inductive AC electrical-homogeneous-continuous-dynamical-system's dynamical component carrying a sinusoidal AC.

The same observation can be made with respect to the 'zero-going positive' crossing or the positive peak. Thus, one says that the voltage holor 'leads' the current holor by $\pi/2$ or, what is equivalent, the current holor 'lags' behind the holor voltage by $\pi/2$. The phase shift can also be expressed as a period T in seconds (s).

A phase shift of $\pi/2$ corresponds to a quarter of a period, hence the voltage holor leads the holor current by $T/4$, or $1/4f$ s.

The relationship between the current and voltage at the conductive conductors of a capacitor follows directly from the derivation of equation (11.32).

That is, if one notes that for a capacitor,

$$i(t) = C\frac{\mathrm{d}v(t)}{\mathrm{d}t}, \tag{11.33}$$

and then assume

$$v(t) = \sqrt{2}\,V\sin(\omega t + \psi_u),$$

it follows that

$$\boldsymbol{I}^j = \boldsymbol{Y}_C^{ji}\boldsymbol{V}_i = \frac{\delta_k^i}{\boldsymbol{Z}_{jk}^C}\boldsymbol{V}_i = \begin{Vmatrix} I^c \\ I^s \end{Vmatrix} = \begin{bmatrix} 0 & -\omega C \\ \omega C & 0 \end{bmatrix}\begin{Vmatrix} V_c \\ V_s \end{Vmatrix}. \tag{11.34}$$

Now, if one writes equation (11.34) for the voltage holor as a function of the current holor, one obtains

$$\boldsymbol{V}_i = \boldsymbol{X}_{ij}^C\boldsymbol{I}^j = \begin{Vmatrix} V_c \\ V_s \end{Vmatrix} = \begin{bmatrix} 0 & \dfrac{1}{\omega C} \\ -\dfrac{1}{\omega C} & 0 \end{bmatrix}\begin{Vmatrix} I^c \\ I^s \end{Vmatrix}. \tag{11.35}$$

From equation (11.35), one sees that the equivalent AC electrical continuous dynamical system for the capacitor in the holor domain is as shown in figure 11.15.

The voltage holor across the conductive conductors of a capacitor will lag behind the capacitor current holor by exactly $\pi/2$.

One can easily see this by rewriting equation (11.35) as

$$\boldsymbol{V}_i = \boldsymbol{X}_{ij}^C\boldsymbol{I}^j = \begin{Vmatrix} V_c \\ V_s \end{Vmatrix} = \begin{bmatrix} 0 & \dfrac{1}{\omega C} \\ -\dfrac{1}{\omega C} & 0 \end{bmatrix}\begin{Vmatrix} I^c \\ I^s \end{Vmatrix}, \tag{11.36}$$

where $\boldsymbol{I}^j = \| I^c \quad I^s \|^T = I\angle\psi_i - \pi/2$.

The alternative way to express the phase relationship contained in equation (11.36) is to say that the current holor leads the voltage holor by $\pi/2$.

The phase relationship between the current holor and voltage holor at the conductive conductors of a capacitor is shown in figure 11.16.

We conclude our discussion of the passive AC electrical-homogeneous-continuous-dynamical-system's dynamical components in the holor domain with a very important observation. When one compares equations (11.29), (11.32) and (11.35), one notes that they are all of the form

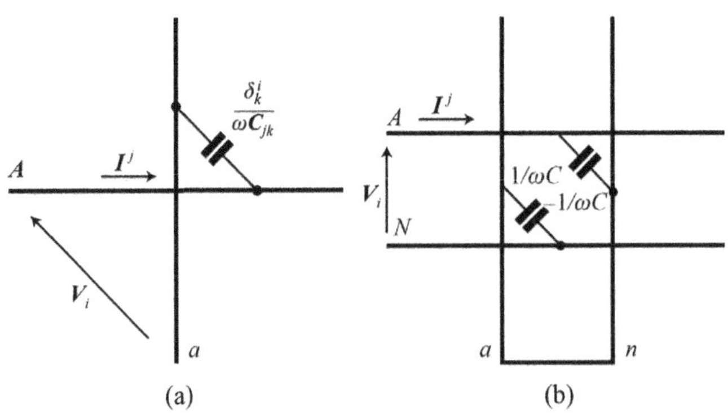

Figure 11.15. The holor domain equivalent AC electrical continuous dynamical system for a capacitor.

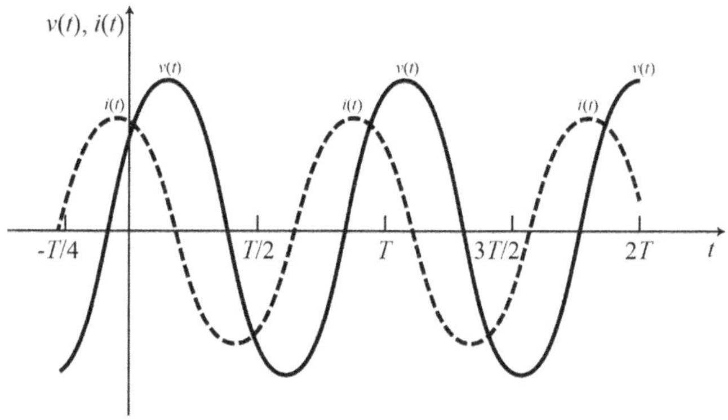

Figure 11.16. A capacitive AC electrical-homogeneous-continuous-dynamical-system's dynamical component carrying a sinusoidal AC.

$$V_i = Z_{ij} I^j, \tag{11.37}$$

where Z_{ij} represents the 'impedance holor matrix' of the AC electrical-continuous-dynamical system components.

Thus, the impedance of a resistor is R, the impedance of an inductor is ωL and the impedance of a capacitor is $1/\omega C$. In all cases, the impedance holor is measured in ohms (Ω). The concept of impedance is very powerful in sinusoidal steady-state analysis and we will have much more to say about the usefulness in subsequent sections.

The term 'inductive reactance' is used to describe the product ωL. The reactance of an inductor is ωL (Ω), whereas the impedance of an inductor is ωL (Ω).

The term 'capacitive reactance' is used to describe the quantity $-1/\omega C$. The reactance of the capacitor is $1/\omega C$ (Ω), whereas the impedance of a capacitor is $-1/\omega C$ (Ω). Here, we will make one final comment about the impedance holor. Note that although the impedance holor is a bivalent holor, it is a matrix Z_{ij}.

Finally, we remind the reader that if the reference sense of direction for the current holor in a passive AC electrical-homogeneous-continuous-dynamical-system's dynamical component is in the sense of direction of the voltage holor increase across the dynamical component, a minus sign must be inserted into the equation that relates the holor voltage to the current holor.

11.3.7 An impedance-holor representation

Applying KSL to the R–L–C AC electrical homogeneous continuous dynamical system shown in figure 11.17 gives one

$$v(t) = Ri(t) + L\frac{\mathrm{d}i(t)}{\mathrm{d}t} + \frac{1}{C}\int_0^t i(t)\mathrm{d}t. \tag{11.38}$$

Substituting to equation (11.38) an instantaneous value of the current from equations (11.18), one has

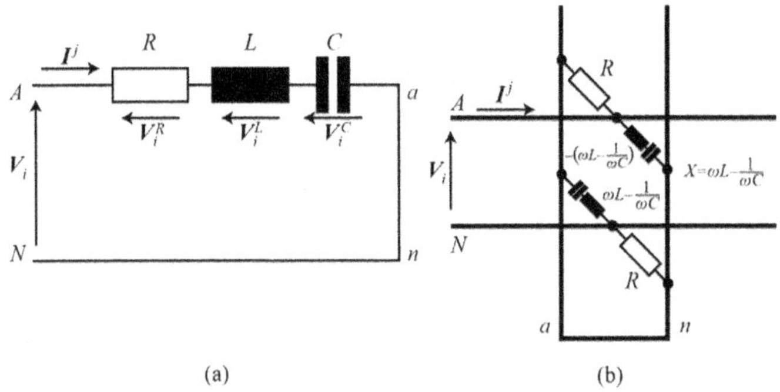

Figure 11.17. Holor equivalent ECM of the *R–L–C* AC electrical homogeneous continuous dynamical system.

$$v(t) = \sqrt{2}I\left[(R\cos\psi_i - X\sin\psi_i)\sin\omega t + (R\sin\psi_i + X\cos\psi_i)\cos\omega t\right], \qquad (11.39)$$

where the passive reactance is given as

$$X = \omega L - 1/\omega C. \qquad (11.40)$$

Substituting to equation (11.39) an instantaneous value of the voltage from equations (11.18), one has

$$\begin{gathered}(V\cos\psi_v - RI\cos\psi_i + XI\sin\psi_i)\sin\omega t \\ +(V\sin\psi_v - RI\sin\psi_i - XI\cos\psi_i)\cos\omega t = 0.\end{gathered} \qquad (11.41)$$

But if $\sin\omega t$ and $\cos\omega t$ are the independent linear functions, one has

$$V\cos\psi_v = RI\cos\psi_i - XI\sin\psi_i \qquad (11.42)$$

$$V\sin\psi_v = RI\sin\psi_i + XI\cos\psi_i. \qquad (11.43)$$

Comparing equations (11.21) and (11.24) with equations (11.42) and (11.43), one obtains the relationship between the independent elements (merates) of the voltage holor V_i and independent elements (merates) of the current holor $\boldsymbol{I}^j$, namely:

$$V_c = RI^c - XI^s,\; V_s = XI^c + RI^s. \qquad (11.44)$$

Assuming that

$$Z_{cc} = R,\; Z_{cs} = -X,\; Z_{sc} = X,\; Z_{ss} = R,$$

equation (11.44) can be rewritten as

$$V_c = Z_{cc}I^c - Z_{cs}I^s,\; V_s = Z_{sc}I^c + Z_{ss}I^s. \qquad (11.45)$$

Thus, the relationship between the current holor $\boldsymbol{I}^j$ and the voltage holor can be described by an expression

$$V_i = \boldsymbol{Z}_{ij}\boldsymbol{I}^j, \qquad (11.46)$$

which defines a new bivalent holor $\boldsymbol{Z}_{ij}$, termed the bivalent impedance holor.

The impedance holor written in a general form establishes a matrix:

$$\boldsymbol{Z}_{ij} = \begin{bmatrix} Z_{cc} & -Z_{cs} \\ Z_{sc} & Z_{ss} \end{bmatrix}. \tag{11.47}$$

In particular circumstances, the following applies.

a) For a series R–L–C AC electrical homogenous continuous dynamical system, the impedance holor $\boldsymbol{Z}_{ij}$ is well known in the (Fijalkowski 1967) matrix form of the impedance holor matrix $[Z]$, namely:

$$\boldsymbol{Z}_{ij} = \begin{bmatrix} R & -X \\ X & R \end{bmatrix} = \begin{bmatrix} R & 1/\omega C + \omega L \\ \omega L - 1/\omega C & R \end{bmatrix}, \tag{11.48}$$

where $X = \omega L - 1/\omega C$.

b) For an ideal active resistance R, the passive reactance $X = 0$, hence

$$\boldsymbol{Z}_{ij} = \begin{bmatrix} R & 0 \\ 0 & R \end{bmatrix} = R\begin{bmatrix} 1 & 0 \\ o & 1 \end{bmatrix} = R1.$$

c) For a series R–L AC electrical homogeneous continuous dynamical system, the passive reactance $X = \omega L$, hence

$$\boldsymbol{Z}_{ij} = \begin{bmatrix} R & -\omega L \\ \omega L & R \end{bmatrix}.$$

d) For a series R–C AC electrical homogeneous continuous dynamical system, the passive reactance $X = 1/\omega C$, hence

$$\boldsymbol{Z}_{ij} = \begin{bmatrix} R & 1/\omega C \\ -1/\omega C & R \end{bmatrix}.$$

Applying the holor algebra laws that govern the bivalent holors, one also obtains the impedance holors for other very complicated AC electrical homogeneous continuous dynamical systems.

11.3.8 An admittance-holor representation

Equation (11.46) for the current holor can be also rewritten in the form

$$\boldsymbol{I}^j = \boldsymbol{Y}^{ji} V_i, \tag{11.49}$$

where the admittance holor $\boldsymbol{Y}^{ji}$ is the inverse of impedance holor:

$$\boldsymbol{Y}^{ij} \boldsymbol{Z}_{jk} = \delta^i_k, \tag{11.50}$$

where δ^j_k is a Kronecker delta.

For instance, for a series R–L–C AC electrical homogeneous continuous dynamical system,

$$Y^{ji} = \begin{bmatrix} Y^{ss} & Y^{sc} \\ -Y^{cs} & Y^{cc} \end{bmatrix} = \begin{bmatrix} G & B \\ -B & G \end{bmatrix} = \frac{1}{(R\text{^}2) + (X\text{^}2)} \begin{bmatrix} R & X \\ -X & R \end{bmatrix}. \tag{11.51}$$

11.3.9 AC power holor representation

An instantaneous value of the AC power can be obtained from a product

$$p(t) = u(t)i(t). \tag{11.52}$$

Substituting to equation (11.52) instantaneous values of the voltage and current from equations (11.18), one obtains

$$p(t) = VI\left\{\cos\varphi - \cos\left[2\omega t + (\psi_v + \psi_i)\right]\right\}, \tag{11.53}$$

where

$$\varphi = \psi_v - \psi_i.$$

To obtain a holor representation of the AC power one must write equation (11.53) in the form

$$p(t) = UI\left[\cos\varphi - \cos + 2\omega t\cos(\psi_v + \psi_i) + \sin 2\omega t\sin(\psi_u + \psi_i)\right]. \tag{11.54}$$

If one designs an independent term by P^p and dependent terms from a doublet frequency by P^q and P^r, then one obtains

$$\left.\begin{aligned} P^p &= UI\cos\lfloor\psi_v - \psi_i\rfloor = UI\cos\varphi = P \\ P^q &= UI\cos(\psi_v + \psi_i) \\ P^r &= UI\sin(\psi_v + \psi_i) \end{aligned}\right\}. \tag{11.55}$$

In this way, equation (11.54) can be rewritten in the form

$$p(t) = P^p - P^q\cos 2\omega t + P^r\sin 2\omega t. \tag{11.56}$$

The quantities P^p, P^q and P^r are independent elements (merates) or coordinates of the three-merate monovalent holor of an AC power

$$\boldsymbol{P}^\kappa = \left(P^p, P^q, P^r\right), \tag{11.57}$$

which allows us to eliminate time from equation (11.56).

From the first expression of equations (11.55) it results that an independent element (merate) of the power holor is an AC active power, which can be measured by means of an electrodynamic wattmeter.

Residual independent elements (merates) can also be measured or computed with the aid of electrical metrology methods.

In particular circumstances, the following applies.

a) For a voltage-holor phase angle $\psi_v = 0$

$$P^q = VI\cos\varphi = P^p = P;$$

$$P^r = -VI\sin\varphi = -Q.$$

b) For a current-holor phase angle $\psi_i = 0$

$$P^q = VI\cos\varphi = P^p = P;$$

$$P^r = VI\sin\varphi = Q.$$

Thus, for the voltage-holor phase angle ψ_v=0 or the current-holor phase angle $\psi_i = 0$, an independent element(merate) of the holor power P^q is the active AC power P, however, an independent element (merate) of the power holor P^r is the reactive AC power Q.

To obtain an equation that takes a power holor from the voltage holor and the current holor into account one must use an adequate law for the multi-application of these holors. However, in holor algebra, the conventional tensor product in the form $\boldsymbol{V}_i\boldsymbol{I}^j$ (due to Moon and Spencer (1986)) is not for application in the case of the power holor, where one uses the particularly useful univalent tensor product

$$\boldsymbol{P}^\kappa = \boldsymbol{V}_i \otimes \boldsymbol{I}^j = \gamma^{\kappa i}_{..j}\boldsymbol{V}_i\boldsymbol{I}^j, \tag{11.58}$$

where $i, j = c, s$ and $\kappa = p, q, r$. The univalent tensor product is designed using a symbol $\otimes$, giving a univalent holor as a result.

Hence, applying a holor representation of the power in the AC electrical homogeneous continuous dynamical system analysis, one obtains a full analogy with the DC electrical current analysis.

Independent elements (merates) of the power holor can be obtained from equation (11.55), namely:

$$P^p = \gamma^{pi}_{..j}\boldsymbol{V}_i\boldsymbol{I}^j, \quad P^q = \gamma^{qi}_{..j}\boldsymbol{V}_i\boldsymbol{I}^j, \quad P^r = \gamma^{ri}_{..j}\boldsymbol{V}_i\boldsymbol{I}^j, \tag{11.59}$$

where

$$\gamma^{pi}_j = \begin{bmatrix} \gamma^{pc}_{..c} & \gamma^{pc}_{..s} \\ \gamma^{ps}_{..c} & \gamma^{ps}_{..s} \end{bmatrix} = \begin{bmatrix} 1 & 0 \\ 0 & 1 \end{bmatrix}, \quad \gamma^{qi}_j = \begin{bmatrix} \gamma^{qc}_{..c} & \gamma^{qc}_{..s} \\ \gamma^{qs}_{..c} & \gamma^{qc}_{..s} \end{bmatrix} = \begin{bmatrix} 1 & 0 \\ 0 & -1 \end{bmatrix},$$

$$\gamma^{ri}_j = \begin{bmatrix} \gamma^{rc}_{..c} & \gamma^{rc}_{..s} \\ \gamma^{rs}_{..c} & \gamma^{rc}_{..s} \end{bmatrix} = \begin{bmatrix} 1 & 0 \\ 0 & 1 \end{bmatrix}.$$

Applying a convention of holor summation, one obtains

$$\left.\begin{aligned} P^p &= \gamma^{pc}_{..c}V_cI^c + \gamma^{pc}_{..s}V_cI^s + \gamma^{ps}_{..c}V_sI^c + \gamma^{ps}_{..s}V_sI^s \\ P^q &= \gamma^{qc}_{..c}V_cI^c + \gamma^{qc}_{..s}V_cI^s + \gamma^{qs}_{..c}V_sI^c + \gamma^{qs}_{..s}V_sI^s \\ P^r &= \gamma^{rc}_{..c}V_cI^c + \gamma^{rc}_{..s}V_cI^s + \gamma^{rs}_{..c}V_sI^c + \gamma^{rs}_{..s}V_sI^s \end{aligned}\right\}. \tag{11.60}$$

Substituting independent elements (merates) of the holor voltage from equations (11.20) and independent elements (merates) of the holor current from equations (11.23) to equations (11.60), one obtains

$$\left.\begin{aligned} P^p &= \gamma^{pc}_{..c}\cos\psi_v\cos\psi_i + \gamma^{pc}_{..s}\sin\psi_v\sin\psi_i \\ &\quad + \gamma^{ps}_{..c}\sin\psi_v\cos\psi_i + \gamma^{ps}_{..s}\sin\psi_v\sin\psi_i \\ P^q &= \gamma^{qc}_{..c}\cos\psi_v\cos\psi_i + \gamma^{qc}_{..s}\sin\psi_v\sin\psi_i \\ &\quad + \gamma^{qs}_{..c}\sin\psi_v\cos\psi_i + \gamma^{qs}_{..s}\sin\psi_v\sin\psi_i \\ P^r &= \gamma^{rc}_{..c}\cos\psi_v\cos\psi_i + \gamma^{rc}_{..s}\sin\psi_v\sin\psi_i \\ &\quad + \gamma^{rs}_{..c}\sin\psi_v\cos\psi_i + \gamma^{rs}_{..s}\sin\psi_v\sin\psi_i \end{aligned}\right\}. \tag{11.61}$$

Comparing equations (11.61) with equations (11.55), one obtains

$$\gamma^{pc}_{..c} = \gamma^{ps}_{..s} = 1, \quad \gamma^{pc}_{..s} = \gamma^{ps}_{..c} = 0,$$

$$\gamma^{qc}_{..c} = 1, \quad \gamma^{qs}_{..s} = -1, \quad \gamma^{qc}_{..s} = \gamma^{qs}_{c} = 0,$$

$$\gamma^{rc}_{..c} = \gamma^{rs}_{..s} = 0, \quad \gamma^{rc}_{..s} = \gamma^{rs}_{..c} = 1.$$

The aforementioned independent elements (merates) of a trivalent holor γ can also be rewritten in the form of the following matrices:

$$\gamma^{pi}_{..j} = \begin{bmatrix} 1 & 0 \\ 0 & 1 \end{bmatrix}, \quad \gamma^{qi}_{..j} = \begin{bmatrix} 1 & 0 \\ 0 & -1 \end{bmatrix}, \quad \gamma^{ri}_{..j} = \begin{bmatrix} 0 & 1 \\ 1 & 0 \end{bmatrix}. \tag{11.62}$$

Taking into account equations (11.62), one can express equations (11.60) as follows:

$$P^p = V_c I^c + V_s I^s, \quad P^q = V_c I^c - V_s I^s, \quad P^r = V_c I^s + V_s I^c. \tag{11.63}$$

Routinely, to compute an AC power, an analogous expression to that for computing a DC power is needed in the form

$$P = R(I\hat{\ }2).$$

From equations (11.55), a power holor

$$\boldsymbol{P}^{\kappa} = \boldsymbol{V}_i \otimes \boldsymbol{I}^j = \gamma^{\kappa i}_{..j}\boldsymbol{V}_i\boldsymbol{I}^j.$$

However, from equation (11.37) a voltage holor

$$\boldsymbol{V}_i = \boldsymbol{Z}_{ij}\boldsymbol{I}^j.$$

Hence, the power holor

$$\boldsymbol{P}^{\kappa} = \boldsymbol{V}_i \otimes \boldsymbol{I}^j = \gamma^{\kappa i}_{..j}\boldsymbol{V}_i\boldsymbol{I}^j = \gamma^{\kappa i}_{..j}\boldsymbol{Z}_{ij}\boldsymbol{I}^k\boldsymbol{I}^j, \tag{11.64}$$

and its independent elements (merates)

$$P^p = \gamma^{pi}_{..j}\boldsymbol{Z}_{ij}\boldsymbol{I}^k\boldsymbol{I}^j, \quad P^q = \gamma^{qi}_{..j}\boldsymbol{Z}_{ij}\boldsymbol{I}^k\boldsymbol{I}^j, \quad P^r = \gamma^{ri}_{..j}\boldsymbol{Z}_{ij}\boldsymbol{I}^k\boldsymbol{I}^j. \tag{11.65}$$

Applying a convention of holor summation, one obtains

$$\left.\begin{aligned} P^p &= \gamma^{pm}_{..n} Z_{cc}(I^c{}^\wedge 2) + \gamma^{pm}_{..n} Z_{cc} I^c I^s + \gamma^{pm}_{..n} Z_{cs} I^s I^c + \gamma^{pm}_{..n} Z_{cs}(I^s{}^\wedge 2) \\ &\quad + \gamma^{pm}_{..n} Z_{sc}(I^c{}^\wedge 2) + \gamma^{pm}_{..n} Z_{sc} I^c I^s + \gamma^{pm}_{..n} Z_{ss} I^s I^c + \gamma^{pm}_{..n} Z_{ss}(I^s{}^\wedge 2) \\ P^q &= \gamma^{qm}_{..n} Z_{cc}(I^c{}^\wedge 2) + \gamma^{qm}_{..n} Z_{cc} I^c I^s + \gamma^{qm}_{..n} Z_{cs} I^s I^c + \gamma^{qm}_{..n} Z_{cs}(I^s{}^\wedge 2) \\ &\quad + \gamma^{qm}_{..n} Z_{sc}(I^c{}^\wedge 2) + \gamma^{qm}_{..n} Z_{sc} I^c I^s + \gamma^{qm}_{..n} Z_{ss} I^s I^c + \gamma^{qm}_{..n} Z_{ss}(I^s{}^\wedge 2) \\ P^r &= \gamma^{rm}_{..n} Z_{cc}(I^c{}^\wedge 2) + \gamma^{rm}_{..n} Z_{cc} I^c I^s + \gamma^{rm}_{..n} Z_{cs} I^s I^c + \gamma^{rm}_{..n} Z_{cs}(I^s{}^\wedge 2) \\ &\quad + \gamma^{rm}_{..n} Z_{sc}(I^c{}^\wedge 2) + \gamma^{rm}_{..n} Z_{sc} I^c I^s + \gamma^{rm}_{..n} Z_{ss} I^s I^c + \gamma^{rm}_{..n} Z_{ss}(I^s{}^\wedge 2) \end{aligned}\right\}. \tag{11.66}$$

Taking into consideration equations (11.48) and (11.61), one obtains

$$\left.\begin{aligned} P^p &= R[(I^c{}^\wedge 2) + (I^s{}^\wedge 2)] \\ P^q &= R[(I^c{}^\wedge 2) - (I^s{}^\wedge 2)] - 2XI^c I^s \\ P^r &= X[(I^c{}^\wedge 2) - (I^s{}^\wedge 2)] + 2RI^c I^s \end{aligned}\right\}. \tag{11.67}$$

Substituting equations (11.66) into the first expression from equations (11.67), one obtains

$$P^p = R(I{}^\wedge 2) = P. \tag{11.68}$$

In this way, one obtains a well-known expression for an AC electrical continuous dynamical system analysis, from which we can again see that an independent element (merate) of the holor power P^p is an AC real or active power.

Adding geometrically independent elements (merates) P^q and P^r of the power holor, one derives a well-known expression for an AC effective value of the apparent power S in the form

$$\begin{aligned} \text{sqrt}\Big[(P^q{}^\wedge 2) + (P^r{}^\wedge 2)\Big] &= \text{sqrt}\Big[(R{}^\wedge 2) + (X{}^\wedge 2)\Big]\Big[(I^c{}^\wedge 2) + (I^s{}^\wedge 2)\Big] \\ &= Z(I^{\wedge 2}) = VI = S. \end{aligned} \tag{11.69a}$$

Thus, due to an application of the holor algebra for a sinusoidal AC electrical continuous dynamical system analysis, an operation is possible with holor quantities for the AC electric continuous dynamical systems that are analogous with real quantities for the DC electrical continuous dynamical systems.

11.3.10 A geometrical interpretation of the voltage, current and power holors

The voltage, current and power holors introduced can be represented geometrically in 2D or 3D spaces. If one assumes that independent elements (merates) of the current holors I^c and I^s are rectangular coordinates in 2D space (figure 11.18), then the first equation of equations (11.44) can be regarded as an equation of the straight line V_c = const, whose slope is described by $\boldsymbol{Z}_{ij}$. The second straight line V_s = const results from the second equation of equations (11.44).

An impedance holor $\boldsymbol{Z}_{ij}$ can be represented using a pair of orthogonal straight lines (see figure 11.18). The current holor stretches out from the beginning of the coordinate system to cut each of these straight lines. In particular circumstances, for instance, for an ideal active resistance, these orthogonal straight lines run across vertically and horizontally.

The voltage holor may also be represented geometrically in a similar way to the current holor, as shown in figure 11.19.

A geometrical representation of the power holor needs 3D space. The coordinates are independent elements (merates) of the power holor P^p, P^q and P^r. The coordinates P^q and P^r are orthogonal to each other with regard to equation (11.69). The third axis P^p establishes an angle $\varphi = \psi_V - \psi_i$ with a vector of the apparent power, with a module $S = V\,I$.

The power holor $\boldsymbol{P}^\kappa$ is hence a vector sum of its independent elements (merates) P^p, P^q and P^r (figure 11.20).

A partial representation of the holor power needs 2D space. If a phase angle of the double-frequency term is small, then a module of the double-frequency power can be obtained from a part of the holor diagram shown in figure 11.20, which depicts a well-known holor diagram for a power holor established by independent elements (merates) of the real or active power $P^p = P$ and the reactive power Q.

A partial holor diagram of the holor power, however, does not represent the holor power exactly, and a strict representation needs three independent elements (merates) and 3D space. Thus a physically effective value of the apparent power P^κ between two terminals would be:

$$P^\kappa = \mathrm{sqrt}[(P^{p\wedge}2) + (P^{q\wedge}2) + (P^{r\wedge}2)] = S. \qquad (11.69b)$$

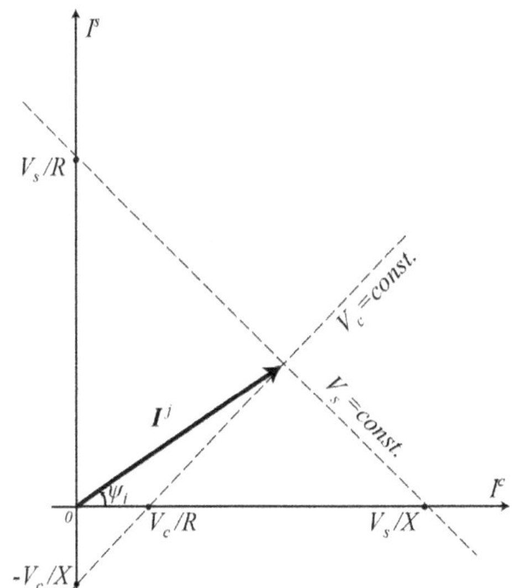

Figure 11.18. A geometrical representation of the current holor.

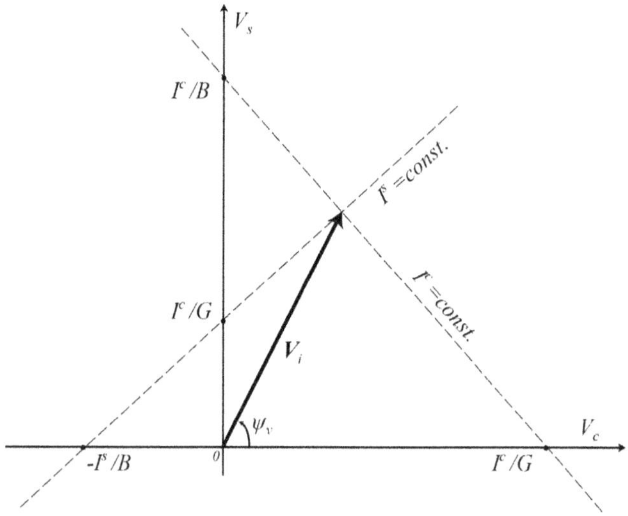

Figure 11.19. A geometrical representation of the voltage holor.

11.3.11 Conclusion

Holor algebra, like complex number algebra, allows one to eliminate time from equations describing a physical action of the sinusoidal physical variables, as well as a geometrical interpretation of the fundamental quantities of the alternating current as real quantities in 2D and 3D spaces.

Holor algebra allows one to isolate fundamental electrical quantities into individual terms, such as the voltage holor, current holor, impedance holor, admittance holor and power holor. It seems that a basic value of holor algebra lies not only in computing processes, but also in the theoretical domain, as it allows the sinusoidal AC fundamental quantities to be represented in a way that is logical and approved by traditional mathematics.

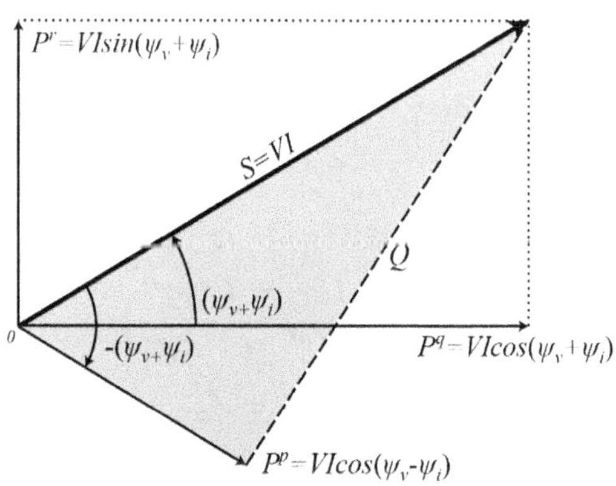

Figure 11.20. A geometrical representation of the power holor.

11.4 The single-phase AC electrical homogeneous continuous dynamical system as the single-phase AC electrical commutation matrixer

11.4.1 Introduction

Almost all AC electrical homogeneous continuous dynamical systems operate with sinusoidal alternating currents and voltages. Fortunately, AC electrical homogeneous continuous dynamical system analysis can be essentially the same as the analysis presented for DC electrical homogeneous continuous dynamical systems. It is only necessary to develop some understanding of current, voltage, impedance and power holors (and of the holor algebra by which they are computed) to be able to apply the same DC electrical homogeneous continuous dynamical system knowledge to them.

Before starting, it might be helpful to look at a standard and simple single-phase AC electrical homogeneous continuous dynamical system to see how much like DC it really is, and to see what new ideas have to be developed to apply DC techniques to AC. Consider the single-phase AC electrical homogeneous continuous dynamical system in figure 11.21(a) and the similar DC electrical homogeneous continuous dynamical system in figure 11.21(b).

Figure 11.21(a) shows a series with an AC energy-source (voltage source) and three passive components in series. In addition to the resistor R there are two other components. C is a capacitor and L is an inductor.

Figure 11.21(b) shows a similar DC electrical homogeneous continuous dynamical system with the same configuration. All three passive components here are resistors. From DC electrical homogeneous continuous dynamical system analysis, it is possible to come to the solution for it (figure 11.21(b)) as follows:

$$I = \frac{1}{R} V = \frac{1}{R_1 + R_3 + R_3} V,$$

where

$$R = R_1 + R_3 + R_3.$$

The solution for a single-phase AC electrical homogeneous continuous dynamical system (figure 11.21(a)) is almost the same:

$$\boldsymbol{I}^j = \frac{\delta_k^i}{\boldsymbol{Z}_{jk}} \boldsymbol{V}_i = \frac{\delta_k^i}{\boldsymbol{Z}_{jk}^1 + \boldsymbol{Z}_{jk}^2 + \boldsymbol{Z}_{jk}^3} \boldsymbol{V}_i,$$

where

$$\boldsymbol{I}^j = \|I^c \; I^s\|^T = I \angle \psi_i;$$
$$\boldsymbol{V}_i = \|V_c \; V_s\|^T = V \angle \psi_u;$$
$$\boldsymbol{Z}_{ij} = \boldsymbol{Z}_{ij}^1 + \boldsymbol{Z}_{ij}^2 + \boldsymbol{Z}_{ij}^3.$$

Here the boldface letters $\boldsymbol{V}_i$ and $\boldsymbol{I}^j$ stand for AC voltage holors and current holors. As might be guessed, the AC voltage holor produces an AC current holor. The

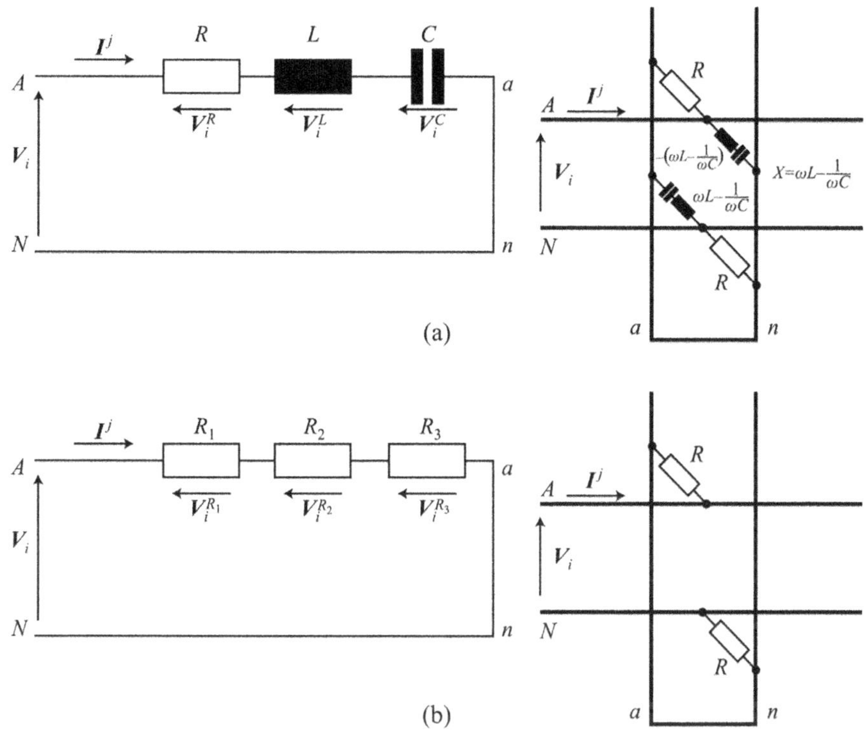

Figure 11.21. Single-phase AC electrical homogeneous continuous dynamical system and a similar DC electrical homogenous continuous dynamical system.

angles ψ_i and ψ_u describe the phase angle associated with AC current holors and voltage holors, respectively, as explained in section 11.3.

The AC impedance holor ($\boldsymbol{Z}^1_{jk}$) of a component is a more general form of DC resistance (R), namely:

$$\boldsymbol{Z}^1_{ij} = \begin{bmatrix} R & 0 \\ 0 & R \end{bmatrix} = R\begin{bmatrix} 1 & 0 \\ 0 & 1 \end{bmatrix}.$$

Alternatively, to put it another way, in addition to R, single-phase AC electrical homogeneous continuous dynamical systems have two other forms of impedance holor ($\boldsymbol{Z}^2_{ij}$ and $\boldsymbol{Z}^3_{ij}$), which distinguish between the resistive kind of impedance holor and that produced with inductors L or capacitors C, namely:

$$\boldsymbol{Z}^2_{ij} = \begin{bmatrix} 0 & -\omega L \\ \omega L & 0 \end{bmatrix}$$

and

$$\boldsymbol{Z}^3_{ij} = *\begin{bmatrix} 0 & 1/\omega C \\ -1/\omega C & 0 \end{bmatrix}.$$

11.4.2 Series and/or parallel transformations

The rules for combining impedance holors in series and/or parallel are the same as those for resistors. The only difference is that combining impedance holors involves the algebraic manipulation of hypernumbers, i.e. holors.

Impedance holors in series can be combined into a single impedance holor $\boldsymbol{Z}_{ij}$ by simply adding the individual impedance holors.

Thus, the series single-phase AC electrical homogeneous continuous dynamical system's equivalent impedance holor $\boldsymbol{Z}_{ij}$ is

$$\boldsymbol{Z}_{ij} = \begin{bmatrix} R & 1/\omega C - \omega L \\ \omega L - 1/\omega C & R \end{bmatrix} = \begin{bmatrix} R & -X \\ X & R \end{bmatrix},$$

where $X = \omega L - 1/\omega C$.

Put another way, the problem is defined in general terms by the series AC single-phase electrical homogeneous continuous dynamical system in figure 11.22, where the impedance holors $\boldsymbol{Z}_{ij}^1, \boldsymbol{Z}_{ij}^2, \ldots, \boldsymbol{Z}_{ij}^s$ are connected in series between the conductors A, N.

The impedance holors are in series because they carry the same current holor $\boldsymbol{I}^j$. From equation ($\boldsymbol{V}_I = \boldsymbol{Z}_{IJ}\boldsymbol{I}^j$) one knows that the voltage-drop holor across each impedance holor will be

$$\boldsymbol{Z}_{ij}^1\boldsymbol{I}^j,\ \boldsymbol{Z}_{ij}^2\boldsymbol{I}^j,\ \ldots,\ \boldsymbol{Z}_{ij}^s\boldsymbol{I}^j$$

and from **Kirchhoff's second law** (KSL), one has

$$\boldsymbol{V}_i^{AN} = \boldsymbol{Z}_{ij}^1\boldsymbol{I}^j + \boldsymbol{Z}_{ij}^2\boldsymbol{I}^j + \cdots + \boldsymbol{Z}_{ij}^s\boldsymbol{I}^j = \left(\boldsymbol{Z}_{ij}^1 + \boldsymbol{Z}_{ij}^2 + \cdots + \boldsymbol{Z}_{ij}^s\right)\boldsymbol{I}^j. \tag{11.70}$$

The equivalent impedance holor between the conductive conductors A, N is

$$\boldsymbol{Z}_{ij}^{AN} = \frac{\boldsymbol{V}_i^{AB}}{\boldsymbol{I}^j} = \boldsymbol{Z}_{ij}^1 + \boldsymbol{Z}_{ij}^2 + \cdots + \boldsymbol{Z}_{ij}^s. \tag{11.71}$$

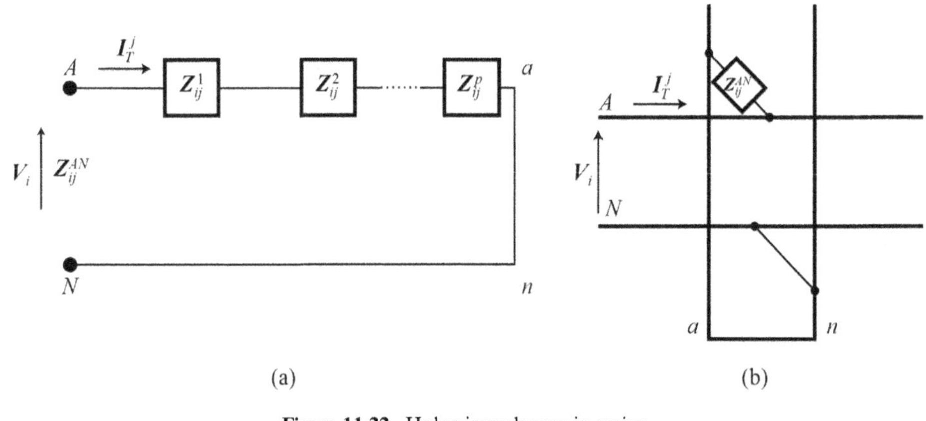

Figure 11.22. Holor impedances in series.

The impedance holor connection in parallel for the AC electrical homogeneous continuous dynamical system can be reduced to a single equivalent impedance holor by multiplying by a Kronecker delta relationship

$$\frac{\delta^i_{k(AN)}}{Z^{AN}_{jk}} = \frac{\delta^i_{k(1)}}{Z^1_{jk}} + \frac{\delta^i_{k(2)}}{Z^2_{jk}} + \cdots + \frac{\delta^i_{k(p)}}{Z^p_{jk}}, \tag{11.72a}$$

or the impedance holor connection in parallel for the AC electrical homogeneous continuous dynamical system can be reduced to a single equivalent holor admittance

$$Y^{ji}_{AN} = Y^{ji}_1 + Y^{ji}_3 + \cdots + Y^{ji}_p. \tag{11.72b}$$

The parallel connection of holor impedances is illustrated in figure 11.23. Note that the impedance holors are in parallel because they have the same holor voltage across their conductors *A, N*.

The derivation of equation (11.72) follows directly from figure 11.23. One simply combines **Kirchhoff's first law** (KFL) with the holor-domain version of Ohm's law, that is

$$I^j = I^j_1 + I^j_2 + \cdots + I^j_p,$$

or

$$\frac{\delta^i_{k(AN)}}{Z^{AN}_{jk}} U^{AN}_i = \frac{\delta^i_{k(1)}}{Z^1_{jk}} U^{AN}_i + \frac{\delta^i_{k(2)}}{Z^2_{jk}} U^{AN}_i + \cdots + \frac{\delta^i_{k(p)}}{Z^p_{jk}} U^{AN}_i, \tag{11.73a}$$

or

$$Y^{ji}_{AN} U^{AN}_i = Y^{ji}_1 U^{AN}_i + Y^{ji}_2 U^{AN}_i + \cdots + Y^{ji}_p U^{AN}_i. \tag{11.73b}$$

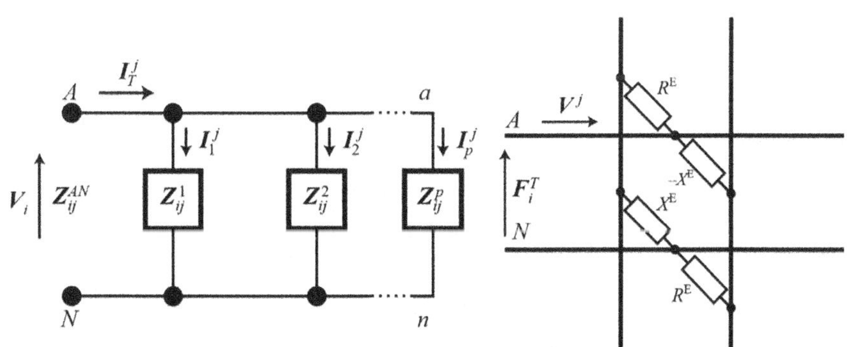

Figure 11.23. Holor impedances in parallel.

Once the common voltage holor term is cancelled out of equation (11.37b) one sees that one has equation (11.72).

Equation (11.72) can also be expressed in terms of an admittance holor, which is defined as the reciprocal of the holor impedances multiplied by a Kronecker delta $\boldsymbol{\delta}_k^i$ and is denoted by the boldface letter $\boldsymbol{Y}_{AN}^{ji}$. Thus

$$\boldsymbol{Y}_{AN}^{ji} = \begin{bmatrix} Y_{AN}^{ss} & Y_{AN}^{sc} \\ Y_{AN}^{cs} & Y_{AN}^{cc} \end{bmatrix} = \begin{bmatrix} G & B \\ -B & G \end{bmatrix} = \frac{\delta_{k(AN)}^i}{\boldsymbol{Z}_{jk}^{AN}} = \frac{1}{(R\hat{}2) + (X\hat{}2)} \begin{bmatrix} R & X \\ -X & R \end{bmatrix}, \quad (11.74)$$

where

$$X = \omega L - 1/\omega C.$$

Like conductance G, the admittance holor $\boldsymbol{Y}_{AN}^{ji}$ is measured in siemens (S) and commonly expressed in mhos (℧).

When one uses equation (11.74) in equation (11.72), one has

$$\boldsymbol{Y}_{AN}^{ji} = \boldsymbol{Y}_1^{ji} + \boldsymbol{Y}_2^{ji} + \cdots + \boldsymbol{Y}_p^{ji}. \quad (11.75)$$

One also notes from equation (11.72) that for the exceptional case of just two impedance holors in parallel,

$$\boldsymbol{Z}_{ij}^{AN} = \frac{\delta_k^i}{\boldsymbol{Z}_{jk}^1 + \boldsymbol{Z}_{jk}^2} \boldsymbol{Z}_{ij}^1 \boldsymbol{Z}_{ij}^2. \quad (11.76)$$

It is also worth noting the holor admittance of each of the ideal passive single-phase AC electrical homogeneous continuous dynamical system's components. In particular, the admittance holor of an individual resistor is

$$\boldsymbol{Y}_R^{ji} = \begin{bmatrix} Y_R^{ss} & Y_R^{sc} \\ Y_R^{cs} & Y_R^{cc} \end{bmatrix} = \begin{bmatrix} G & 0 \\ 0 & G \end{bmatrix} = \frac{\delta_{k(R)}^i}{\boldsymbol{Z}_{jk}^R} = \frac{1}{(R\hat{}2)} \begin{bmatrix} R & 0 \\ 0 & R \end{bmatrix} = \begin{bmatrix} \frac{1}{R} & 0 \\ 0 & \frac{1}{R} \end{bmatrix} [℧].$$

We have already noted in our discussion of resistive single-phase AC electrical homogeneous continuous dynamical systems that the reciprocal of the resistance holor multiplied by a Kronecker delta δ_k^i is termed the 'conductance' holor.

The admittance holor of the individual inductor is

$$\boldsymbol{Y}_L^{ji} = \begin{bmatrix} Y_L^{ss} & Y_L^{sc} \\ Y_L^{cs} & Y_L^{cc} \end{bmatrix} = \begin{bmatrix} 0 & B \\ -B & 0 \end{bmatrix} = \frac{\delta_{k(L)}^i}{\boldsymbol{Z}_{jk}^L} = \frac{1}{(\omega\hat{}2)(L\hat{}2)} \begin{bmatrix} 0 & \omega L \\ -\omega L & 0 \end{bmatrix} = \begin{bmatrix} 0 & \frac{1}{\omega L} \\ -\frac{1}{\omega L} & 0 \end{bmatrix} [℧].$$

The quantity above is referred to as the 'inductive-susceptance' holor of the inductor.

The admittance holor of the individual capacitor is

$$Y_C^{ji} = \begin{bmatrix} Y_C^{ss} & Y_C^{sc} \\ Y_C^{cs} & Y_C^{cc} \end{bmatrix} = \begin{bmatrix} 0 & B \\ -B & 0 \end{bmatrix} = \frac{\delta^i_{k(C)}}{Z^C_{jk}} = \frac{1}{\left(\frac{1}{\omega}^\wedge 2\right)\left(\frac{1}{C}^\wedge 2\right)} \begin{bmatrix} 0 & -\frac{1}{\omega C} \\ \frac{1}{\omega C} & 0 \end{bmatrix}$$

$$= \begin{bmatrix} 0 & -\omega C \\ \omega C & 0 \end{bmatrix} [\mho].$$

The quantity above is referred to as the 'capacitive-susceptance' holor of the capacitor.

11.4.3 Delta-to-wye or pi-to-tee equivalent

The AC electrical homogeneous continuous dynamical system's bridge configuration in figure 11.24 introduces an intersection of holor impedances that warrants further discussion.

An interesting thing about this AC electrical homogeneous dynamical system is that the interconnected impeders cannot be reduced to a single equivalent holor impedance across the conductive conductors of the energy source if one is restricted to the simple series or parallel equivalent AC electrical homogeneous continuous dynamical systems that were introduced earlier in this chapter.

The interconnected impeders can be reduced to a single equivalent impeder by means of a delta-to-wye (Δ2Y) or pi-to-tee (Π2T) equivalent AC electrical continuous dynamical system. The Δ and Y structures occur in a variety of single- and/or polyphase AC electrical homogeneous continuous dynamical systems and therefore one will still find Δ2Y transformation a helpful tool in AC electrical continuous dynamical system analysis.

The Z_{ij}^{Aa}, Z_{ij}^{An}, and Z_{ij}^{an} or Z_{ij}^{Na}, Z_{ij}^{Nn} and Z_{ij}^{an} in the AC electrical homogeneous continuous dynamical system shown in figure 11.24 are referred to as a delta (Δ) interconnection because the interconnection looks like the Greek letter Δ. It is also referred to as a pi (Π) interconnection because the Δ can be shaped into a Π without

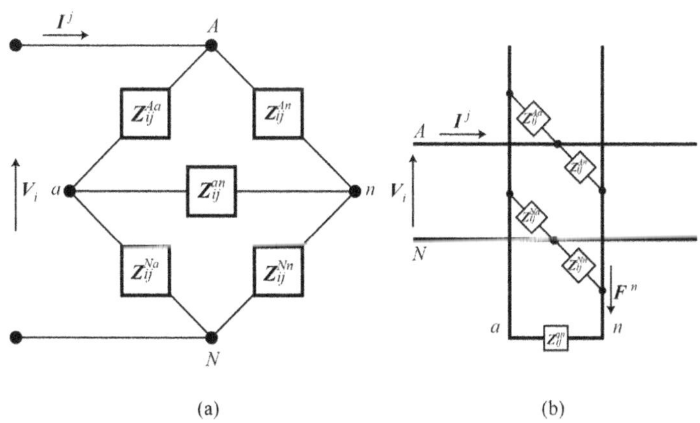

Figure 11.24. The AC electrical homogeneous continuous dynamical system's bridge configuration.

disturbing the electrical equivalence of the two configurations. The electrical equivalence between the Δ and the Π interconnections is apparent from figure 11.25.

The Δ2Y equivalent AC electrical homogeneous continuous dynamical system means that the Δ configuration can be replaced by a Y configuration, such that the conductive conductors' behaviour for the two configurations will be identical.

The Δ2Y transformation is illustrated in figure 11.26, where the two AC electrical homogeneous continuous dynamical systems shown are equivalent at the conductive conductors *a, b* and *c*, provided that

$$\boldsymbol{Z}_{ij}^{an} = \frac{\delta_k^i}{\boldsymbol{Z}_{jk}^{bc} + \boldsymbol{Z}_{jk}^{ca} + \boldsymbol{Z}_{jk}^{ab}} \boldsymbol{Z}_{ij}^{ca} \boldsymbol{Z}_{ij}^{ab}; \tag{11.77}$$

$$\boldsymbol{Z}_{ij}^{bn} = \frac{\delta_k^i}{\boldsymbol{Z}_{jk}^{bc} + \boldsymbol{Z}_{jk}^{ca} + \boldsymbol{Z}_{jk}^{ab}} \boldsymbol{Z}_{ij}^{ab} \boldsymbol{Z}_{ij}^{bc}; \tag{11.78}$$

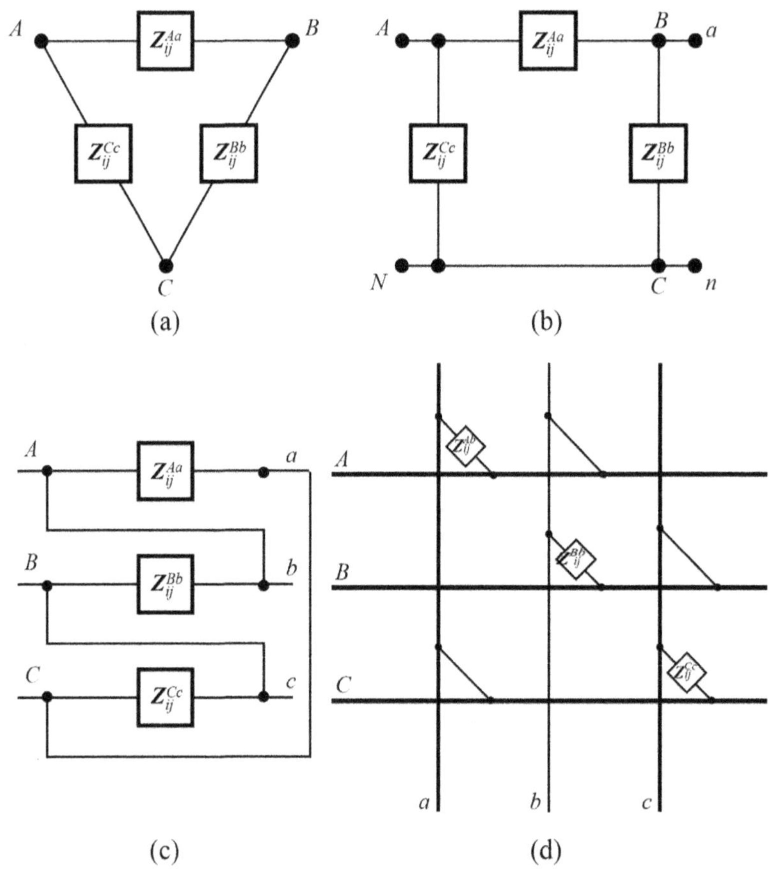

Figure 11.25. A Δ configuration viewed as a Π configuration.

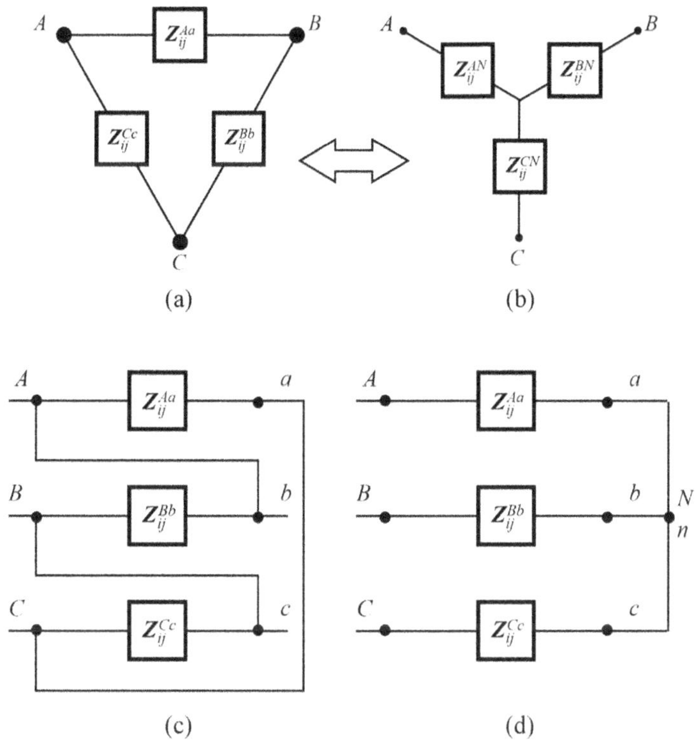

Figure 11.26. The Δ2Y transformation.

$$\boldsymbol{Z}_{ij}^{cn} = \frac{\delta_k^i}{\boldsymbol{Z}_{jk}^{bc} + \boldsymbol{Z}_{jk}^{ca} + \boldsymbol{Z}_{jk}^{ab}} \boldsymbol{Z}_{ij}^{bc} \boldsymbol{Z}_{ij}^{ca}. \tag{11.79}$$

The Y configuration is also referred to as the tee (T) configuration, because the Y structure can be replaced by the T structure without disturbing the electrical equivalence of the two structures.

Equations (11.77)–(11.79) give the Y-connected impeders as functions of the Δ-connected impeders. It is also possible to reverse the Δ2Y or Π2T transformation. That is, one can start with the Y structure and replace it with an equivalent Δ structure.

The electrical equivalence of the Y and T configurations is apparent from figure 11.27.

The expressions for the three Δ-impeder holors, $\boldsymbol{Z}_{ij}^{bc}$, $\boldsymbol{Z}_{ij}^{ca}$ and $\boldsymbol{Z}_{ij}^{ab}$, as functions of the three Y-impeder holors $\boldsymbol{Z}_{ij}^{an}$, $\boldsymbol{Z}_{ij}^{bn}$ and $\boldsymbol{Z}_{ij}^{cn}$ are

$$\boldsymbol{Z}_{ij}^{ab} = \frac{\delta_k^i}{\boldsymbol{Z}_{jk}^{cn}} \left(\boldsymbol{Z}_{ij}^{an} \boldsymbol{Z}_{ij}^{bn} + \boldsymbol{Z}_{ij}^{bn} \boldsymbol{Z}_{ij}^{cn} + \boldsymbol{Z}_{ij}^{cn} \boldsymbol{Z}_{ij}^{an} \right); \tag{11.80}$$

$$\boldsymbol{Z}_{ij}^{bc} = \frac{\delta_k^i}{\boldsymbol{Z}_{jk}^{an}} \left(\boldsymbol{Z}_{ij}^{an} \boldsymbol{Z}_{ij}^{bn} + \boldsymbol{Z}_{ij}^{bn} \boldsymbol{Z}_{ij}^{cn} + \boldsymbol{Z}_{ij}^{cn} \boldsymbol{Z}_{ij}^{an} \right); \tag{11.81}$$

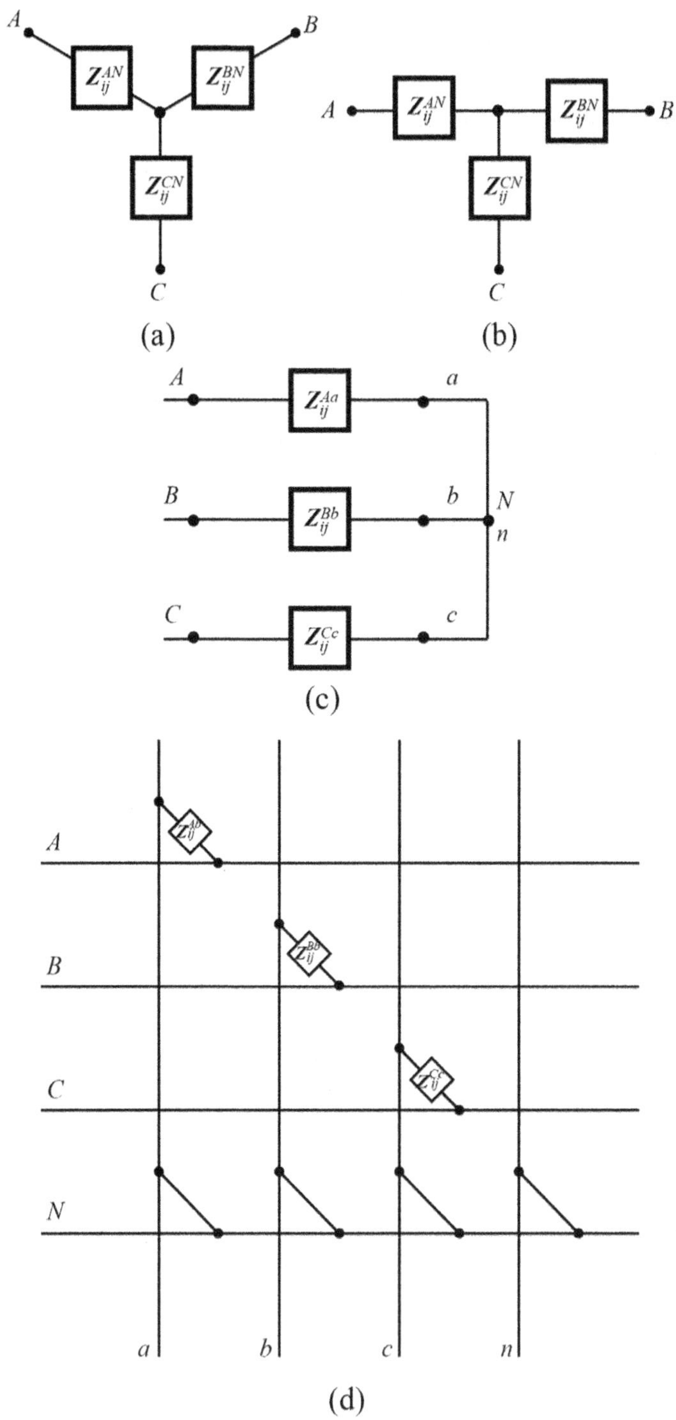

Figure 11.27. A Y configuration viewed as a T configuration.

$$Z_{ij}^{ca} = \frac{\delta_k^i}{Z_{jk}^{bn}}\left(Z_{ij}^{an}Z_{ij}^{bn} + Z_{ij}^{bn}Z_{ij}^{cn} + Z_{ij}^{cn}Z_{ij}^{an}\right). \tag{11.82}$$

The derivation of equations (11.77)–(11.82) can be accomplished by noting that the two AC electrical systems are by definition equivalent with respect to their conductive collectors' behaviour. Thus if each AC electrical homogeneous continuous dynamical system is placed in a black box one cannot tell by external measurements whether the black box contains a set of Δ-connected impeders or a set of Y-connected impeders. This can only be true if the impedance holor between the corresponding conductive conductor pairs is the same for each black box. For instance, one must see the same impedance holor between conductive conductors a and b whether one uses the Δ-connected set or the Y-connected set. It follows directly that

$$Z_{ij}^{ab} = \frac{\delta_k^i}{Z_{jk}^{bc} + Z_{jk}^{ca} + Z_{jk}^{ab}} Z_{ij}^{ab}\left(Z_{ij}^{bc} + Z_{ij}^{ca}\right) = Z_{ij}^{an} + Z_{ij}^{bn}; \tag{11.83}$$

$$Z_{ij}^{bc} = \frac{\delta_k^i}{Z_{jk}^{bc} + Z_{jk}^{ca} + Z_{jk}^{ab}} Z_{ij}^{bc}\left(Z_{ij}^{ca} + Z_{ij}^{ab}\right) = Z_{ij}^{bn} + Z_{ij}^{cn}; \tag{11.84}$$

$$Z_{ij}^{ca} = \frac{\delta_k^i}{Z_{jk}^{bc} + Z_{jk}^{ca} + Z_{jk}^{ab}} Z_{ij}^{ca}\left(Z_{ij}^{ab} + Z_{ij}^{bc}\right) = Z_{ij}^{an} + Z_{ij}^{cn}. \tag{11.85}$$

One can obtain equations (11.77)–(11.82) by straightforward holor algebraic manipulation of equations (11.83)–(11.85).

In addition to single-phase AC electrical homogeneous continuous dynamical systems, other polyphase ones are also used for special purposes. For instance, a 'conventional two-phase' AC electrical homogeneous continuous dynamical system consists of two energy-source voltage holors. The minimum phase difference between these voltage holors is π.

An 'unconventional two-phase' AC electrical homogeneous continuous dynamical system does not follow the preceding ideas. It consists of two energy-source voltage holors whose phase difference is π/2.

The 'five-phase' electrical homogeneous continuous dynamical system consists of five energy-source voltage holors. The minimum phase difference between them is 2π/5, etc.

11.5 The polyphase AC electrical homogeneous continuous dynamical system as the polyphase AC electrical commutation matrixer

11.5.1 Introduction

In addition to single-phase AC electrical homogeneous continuous dynamical systems, other polyphase AC electrical homogeneous dynamical systems are also used for special purposes. For instance, a three-phase AC electrical homogeneous continuous dynamical system consists of three voltage-source holors. The minimum phase difference between these voltage-source holors is $2\pi/3$. Similarly, a five-phase electrical homogenous continuous dynamical system consists of five voltage-source holors. The minimum phase difference between them is $2\pi/5$. The unconventional two-phase AC electrical homogeneous continuous dynamical system does not follow the preceding ideas. It consists of two voltage-source holors whose phase difference is $\pi/2$. Note that this is not the conventional two-phase one discussed below in section 11.6. 'Polyphase AC electrical homogeneous continuous dynamical system' is the name given to an assemblage of AC electrical homogenous continuous dynamical systems in which the voltage-source holors have the same frequency but are shifted in phase.

The entirety of voltage-source holors, load voltage holors and current holors in a polyphase AC electrical homogeneous continuous dynamical system are also termed a 'polyphase AC electrical homogenous continuous dynamical system's holors'. The separate parts comprising a polyphase AC electrical homogeneous continuous dynamical system are termed the 'phases'. Note: do not confuse them with the phase angles of the sinusoidal waveforms of the voltage and/or current phases, which are likewise termed 'phase' for short.

A polyphase AC electrical homogeneous continuous dynamical system is termed a 'two-, three-, five-phase…' one, depending on the number of phases. If the voltage-source holors, load-voltage holors, or current holors forming a polyphase AC electrical homogeneous continuous dynamical system have identical amplitudes in all phases of the AC electrical homogeneous continuous dynamical system and are shifted in phase relative to one another by equal angles, then the AC electrical continuous dynamical system is termed 'balanced' or 'symmetrical' (figures 11.28 and 11.29).

The letters *A, B, C* denote the separate phases of a polyphase AC electrical homogeneous continuous dynamical system, etc. It is usual to have the sequence or order of the phases such that the voltage-source holor in each successive phase lags the voltage-source holor of the preceding phase, i.e. the voltage-source holor of phase *B* lags behind the voltage-source holor of phase *A*, the voltage-source holor of phase *C* lags behind the voltage-source holor of phase *B*, and so on. Such a phase sequence is termed a 'positive sequence'. However, there are occasions when several voltage-source holors of the AC electrical homogeneous continuous dynamical

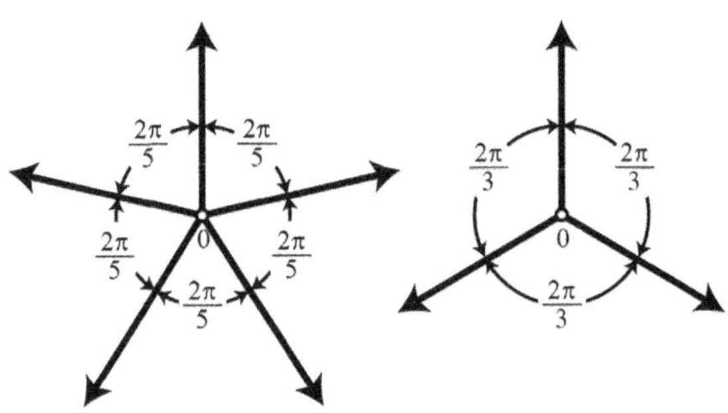

Figure 11.28. Balanced voltage-source holors of five- and three-phase AC electrical homogeneous continuous dynamical systems.

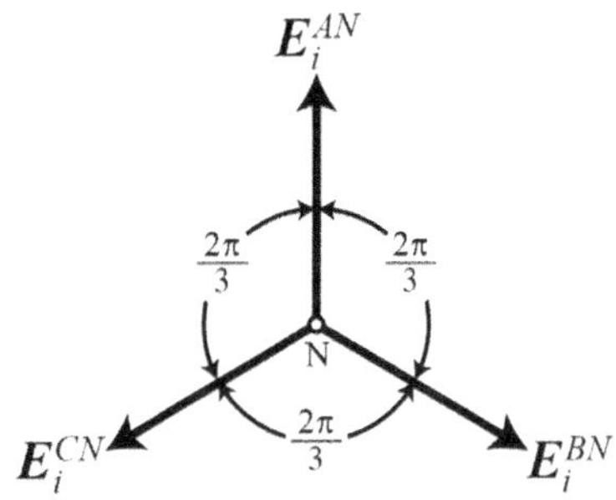

Figure 11.29. Balanced voltage-source holors of a three-phase AC electrical homogeneous continuous dynamical system.

systems act in the same polyphase AC electrical homogeneous continuous dynamical system and not all of them have the same sequence.

An AC electrical homogeneous continuous dynamical system in which the voltage-source holor in each successive phase leads the preceding voltage-source holor is termed a 'negative sequence' AC electrical homogeneous continuous dynamical system. For instance, the phase sequence determines a sense of the rotation direction of polyphase EM motors.

11.6 The two-phase AC electrical homogeneous continuous dynamical system as the two-phase AC electrical commutation matrixer

11.6.1 Introduction

By far the most widespread system is the conventional two-phase AC electrical homogeneous continuous dynamical one, since, of all the AC electrical homogeneous continuous dynamical systems, it is the most economical for electrical energy transmission. Electrical energy supply to households, commercial establishments and other small consumers is usually distributed as single phase, but with two available voltages—220 V or 110 V, or 440 V or 220 V. The energy source in a conventional two-phase electrical homogeneous continuous dynamical system is a single-phase ME generator, which supplies the primary winding of a small (often only 15–25 kVA) EE transformer with centre-tapped secondary winding. Figure 11.30(a) shows how the two-phase voltages are derived from the centre-tapped winding of an EE transformer for two-phase AC electrical homogeneous continuous dynamical systems. The advantages of this two-phase AC electrical homogeneous continuous dynamical supply system is that small loads (such as electric lamps and hand tools) can be accommodated with more safety 220 V or 110 V, while larger loads (particularly heavy electric heating devices and machines) can have the advantage of higher voltage and lower current. Furthermore, if the 220 V or 110 V loads are reasonably balanced between the two phases, some of the advantages of lower current are retained even for them. For safety reasons, the centre (neutral) conductor is grounded, as shown in figure 11.30(a).

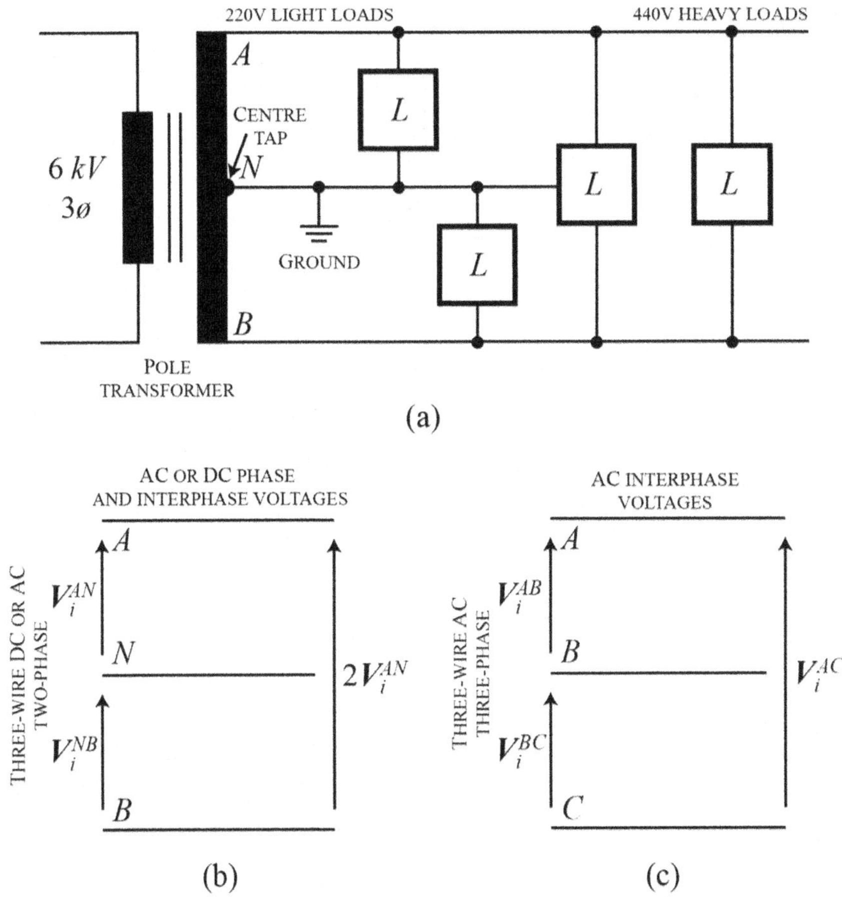

Figure 11.30. Two-phase AC electrical homogeneous continuous dynamical systems and three-phase AC electrical homogeneous continuous dynamical systems are easily distinguished by taking the voltmeter between three pairs of conductors.

A further advantage is that no part of the two-phase AC electrical homogeneous continuous dynamical system is more than 220 V or 110 V above ground, even for 440 V or 220 V appliances. It is easy to determine whether a conventional two-phase AC electrical homogeneous continuous dynamical system is connected according to this scheme (figure 11.30(b)). Of the three voltage readings between the conductors, two will be equal and the third will be twice as great.

The two-voltage, three-conductor conventional two-phase AC electrical homogeneous continuous dynamical system is also used infrequently with the two-phase DC electrical homogeneous continuous dynamical system. The phase voltage-source waveforms of a conventional balanced two-phase AC electrical homogeneous continuous dynamical system are shifted relative to each other by equal phase angles of π (figure 11.31).

Accordingly, the instantaneous values of the phase voltage-source waveforms are expressed by the following formulae:

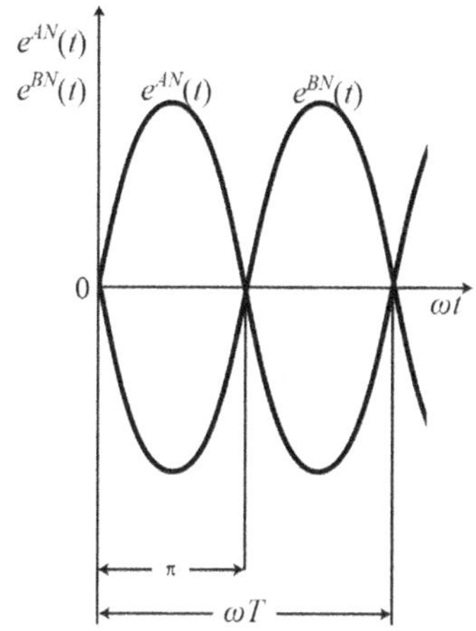

Figure 11.31. The phase voltage-source waveforms of a conventional balanced two-phase AC electrical homogeneous continuous dynamical system.

$$\left.\begin{aligned} e_{AN}(t) &= E_{\max}\sin\omega t = \sqrt{2}E_{\text{rms}}\sin\omega t \\ e_{BN}(t) &= E_{\max}\sin(\omega t+\pi) = \sqrt{2}E_{\text{rms}}\sin(\omega t+\pi) \end{aligned}\right\}, \tag{11.86}$$

if $t = 0$ corresponds to the instant when the voltage of phase A passes through zero and becomes positive. The last formula in equations (11.86) can also be written in the form

$$e_{BN}(t) = E_{\max}\sin(\omega t-\pi) = \sqrt{2}E_{\text{rms}}\sin(\omega t-\pi). \tag{11.87}$$

In the holor notation of the phase-voltage holor, the real axis is usually made to coincide with the energy-source-voltage holor of phase A.

The effective values ($E_{\text{rms}} = E_\phi$) of the conventional balanced two-phase AC electrical homogeneous continuous dynamical system are represented by the holors

$$\left.\begin{aligned} \boldsymbol{E}_{ij}^{AN} &= E_\phi\angle 0 \\ \boldsymbol{E}_{ij}^{BN} &= E_\phi\angle\pi \end{aligned}\right\}. \tag{11.88}$$

Figure 11.32 shows the waveforms of the instantaneous energy-source voltages for both phases; it is easy to see that at any instant the algebraic sum of the two energy-source voltages is zero.

This result can also be established analytically with the aid of the relations in equations (11.86), but it is easier to use the holor diagram (figure 11.33) for this purpose. Indeed, combining the voltage holor $\boldsymbol{E}_{ij}^{BN}$, one finds that, due to the symmetry, the voltage holor $\boldsymbol{E}_{ij}^{BN}$ is directed vertically downwards, i.e. it is opposite to the holor voltage $\boldsymbol{E}_{ij}^{AN}$.

Forming π angle with the voltage holor $\boldsymbol{E}_{ij}^{BN}$, this is equal in magnitude to the voltage holor $\boldsymbol{E}_{ij}^{AN}$. Therefore,

$$\boldsymbol{E}_i^{AN} + \boldsymbol{E}_i^{BN} = \boldsymbol{E}_i^{AN} + \left(-\boldsymbol{E}_i^{AN}\right) = 0, \tag{11.89}$$

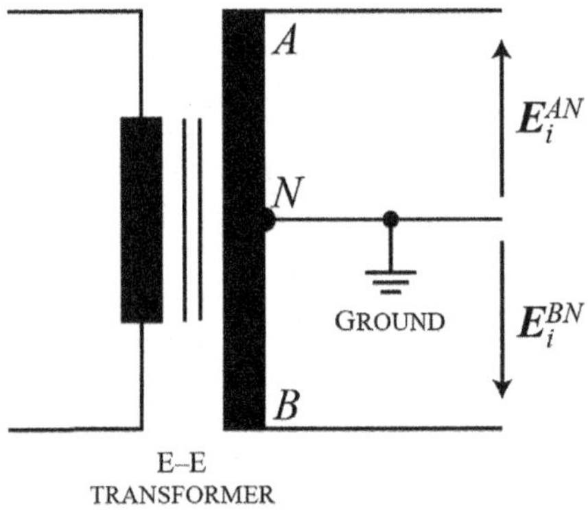

Figure 11.32. Voltage-source holors of the two-phase EE transformer's secondary winding.

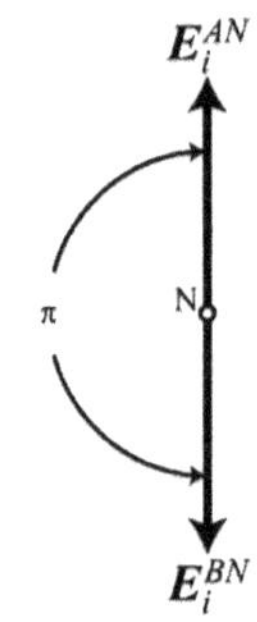

Figure 11.33. Holor diagram of the voltage-source holors of the conventional balanced two-phase AC electrical homogeneous continuous dynamical system.

and since the sum of the instantaneous values $e_{AN}(t) + e_{BN}(t)$ represents the sum of the $\boldsymbol{E}_i^{AN}$ and $\boldsymbol{E}_i^{BN}$ holor projections multiplied by $\sqrt{2}$, or the projections of their sum on the vertical, one has, at any instant,

$$e_{AN}(t) + e_{BN}(t) = 0. \tag{11.90}$$

It is obvious that the positive senses of the directions of the energy-source voltages and currents must be chosen the same for all phases, for instance, away from the reader into the plane of the image shown in figure 11.33.

Let us decide then to consider the 'beginning' of each phase as the point where the positively directed current comes out of the EE transformer's secondary winding into the external AC electrical homogenous continuous dynamical system, and the 'end' of the phase as the point where the positively directed current leaves the external circuit and enters the EE transformer's secondary winding. Thus, the energy-source voltages in the EE transformer secondary phases are positive when they act from the ends of phases (*N*) to their beginning (*A* and *B*). In AC electrical homogenous continuous dynamical system diagrams, a two-phase EE transformer's secondary winding is ordinarily depicted in the form of two secondary windings *NA* and *NB* (figure 11.32). If the load phases connected to the beginnings and ends of each phase have identical (in magnitude and waveform) holor impedances $\boldsymbol{Z}_{ij}^{AN}$ and

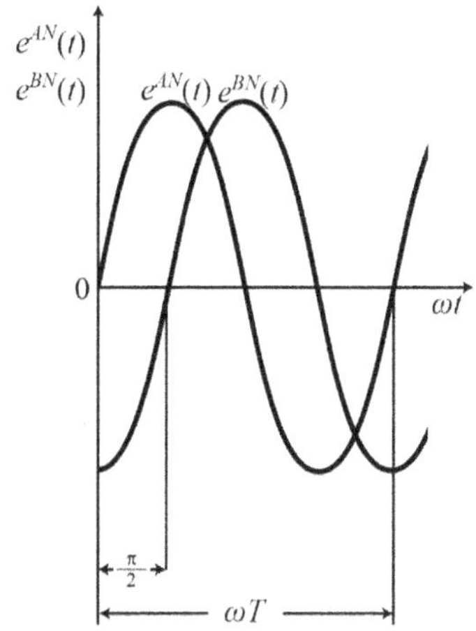

Figure 11.34. The phase energy-source voltage waveforms of an unconventional balanced two-phase AC electrical homogeneous continuous dynamical system.

$\boldsymbol{Z}_{ij}^{BN}$, the current holors $\boldsymbol{I}_{AN}^{j}$ and $\boldsymbol{I}_{BN}^{j}$ will likewise be identical in magnitude, and shifted in phase relative to the corresponding holor energy-source voltage holors by equal angles φ. They will also be shifted in phase relative to each other by π, i.e. the mutual distribution of their waveforms will not differ from that of the waveforms of the voltage-source holors in figure 11.32, and, hence, the holors of the effective values of the holor currents will be connected by the relation

$$\boldsymbol{I}_{AN}^{j} + \boldsymbol{I}_{BN}^{j} = 0, \tag{11.91}$$

and the instantaneous values of the currents by

$$i_{AN}(t) + i_{BN}(t) = 0. \tag{11.92}$$

Such a load for a two-phase AC electrical homogeneous continuous dynamical system is termed a 'balanced' or 'symmetrical' load. A two-phase electrical homogeneous continuous dynamical system, where two conductors to its own load phase independently connect each EE transformer's secondary winding phase, is termed an 'independent' two-phase AC electrical homogeneous continuous dynamical system. Thus, an independent two-phase AC electrical homogeneous continuous dynamical system demands the use of six conductors to transmit electrical energy from the EE transformer to the loads.

The phase energy-source voltages of an unconventional balanced two-phase AC electrical homogeneous continuous dynamical system are shifted relative to each other by equal phase angles of $\pi/2$ (figure 11.34).

Accordingly, the instantaneous values of the phase energy-source voltages are expressed by the following formulae:

$$\left.\begin{aligned} e_{AN}(t) &= E_{\max} \sin \omega t = \sqrt{2}\, E_{\text{rms}} \sin \omega t \\ e_{BN}(t) &= E_{\max} \sin(\omega t + \pi/2) = \sqrt{2}\, E_{\text{rms}} \sin(\omega t + \pi/2) \end{aligned}\right\}, \tag{11.93}$$

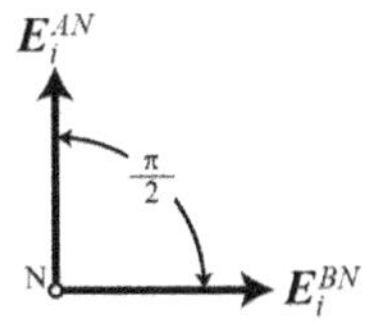

Figure 11.35. Holor diagram of the voltage-source holors of the unconventional balanced two-phase AC electrical homogeneous continuous dynamical system.

$$\left.\begin{aligned} E_i^{AN} &= E_\phi \angle 0 \\ E_i^{BN} &= E_\phi \angle \pi/2 \end{aligned}\right\}. \tag{11.95}$$

if $t = 0$ corresponds to the instant when the voltage of phase A passes through zero and becomes positive.

The last formula in equations (11.94) can also be written in the form

$$e_{BN}(t) = E_{\max}\sin\left(\omega t - \frac{\pi}{2}\right) = \sqrt{2}E_{\text{rms}}\sin\left(\omega t - \frac{\pi}{2}\right). \tag{11.94}$$

In the holor notation of the phase-voltage holor, the real axis is usually made to coincide with the energy-source voltage holor of phase A. The effective values ($E_{\text{rms}} = E_\phi$) of the unconventional balanced two-phase AC electrical homogeneous continuous dynamical system are represented by the holors (figure 11.35).

It is most interesting that one advantage of the Δ–Δ or Δ–Y three-phase AC electrical homogeneous continuous dynamical system in industry is the possibility of using the 'open delta' (i.e. unconventional unbalanced two-phase AC electrical homogeneous continuous dynamical system) in the case where one energy-source phase has been lost. In figure 11.36(a), the three-phase energy sources are shown as three EE transformer secondaries (secondary phase windings)—the common industrial situation in the USA.

If one transformer is burned out or faulty, as shown in figure 11.36(b), it can simply be left out and the AC electrical homogeneous continuous dynamical system will still be capable of supplying a reduced amount of three-phase electrical energy. In this situation, the Δ2Δ or Δ2Y three-phase AC electrical homogeneous continuous dynamical system operates as an unconventional unbalanced two-phase AC electrical homogeneous continuous dynamical system.

11.7 The three-phase AC electrical homogeneous continuous dynamical system as the three-phase AC electrical commutation matrixer

11.7.1 Introduction

The generation, transmission, distribution and utilisation of large quantities of electrical energy can be achieved using polyphase AC electrical homogeneous continuous dynamical systems. The most common is the three-phase AC electrical homogeneous

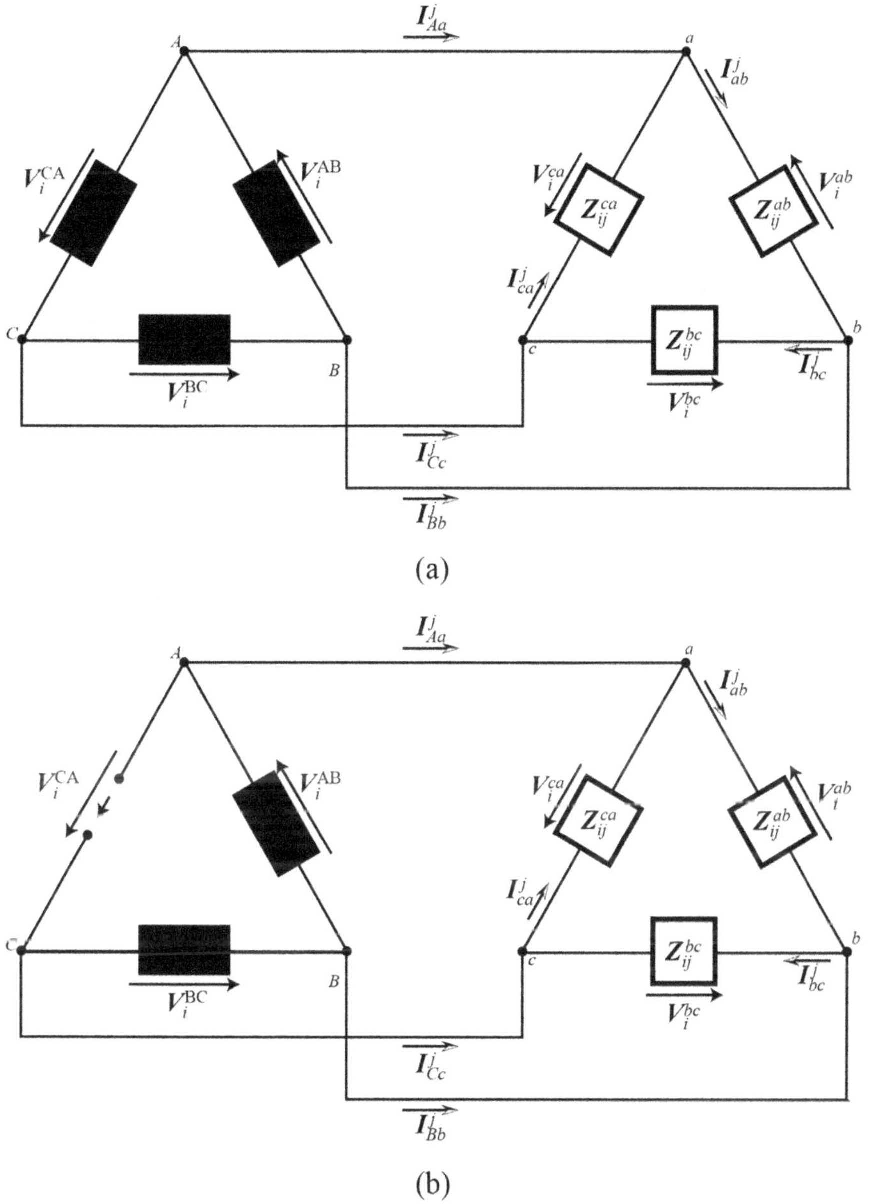

Figure 11.36. The Δ2Δ or Δ2Y three-phase AC electrical homogeneous continuous dynamical system.

continuous dynamical system, since, of all the AC electrical homogeneous continuous dynamical systems, it is most economical for electrical energy transmission.

The basic structure of a three-phase AC electrical homogenous continuous dynamical system consists of primary energy sources connected to loads via EE transformers and transmission lines. One can reduce the problem to the analysis of an AC electrical homogeneous continuous dynamical system consisting of a primary

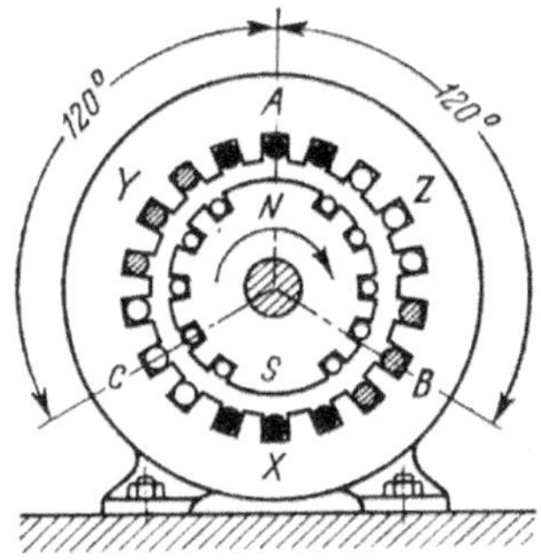

Figure 11.37. Cross-section of the two-pole three-phase AC ME generator (Kasatkin and Perekalin 1970).

voltage source connected to a load via a transmission line. The omission of the EE transformer as a structural element in the AC electrical homogeneous continuous dynamical system simplifies the discussion without jeopardising a basic understanding of the computations involved.

The primary energy source in a three-phase AC electrical homogenous continuous dynamical system is a three-phase AC generator, which differs from a single-phase AC generator in that on the inner surface of the stator (figure 11.37) it has three separate armature-phase windings ('phases'), the axes of which (*A, B,* and *C*) form angles of $2\pi/3$ between each other (if the rotor of the AC generator has one pair of magnetic poles).

Each phase consists of two groups of conductors, the energy-source voltages of which are opposite in direction at any given instant. Hence, the groups *A, B* and *C,* shifted relative to each other by $2\pi/3$, will be termed the 'positive sides' or 'beginnings' of the phases, and the groups diametrically opposite to them, *X, Y* and *Z,* will be termed the 'negative sides' or 'ends' of the phases *A, B* and *C.*

All the conductors of the positive and negative sides of each phase are connected in series to each other (on the end surfaces of the stator), so that their energy-source voltages are added together, forming the phase energy-source voltages. Note: by the phase axis is understood the direction of the radius that passes through the centre of the positive portion of the phase winding.

11.7.2 Three-phase energy sources

As mentioned above, three-phase energy sources consist of AC generators that have three separate armature-phase windings distributed around the periphery of the stator. Each armature-phase winding comprises one phase of the AC generator.

The rotor of the ME generator is an electromagnet with magnetic-field winding, driven at a synchronous speed by a prime mover such as an external combustion engine (i.e. a steam or gas turbine) or an internal combustion engine (i.e. a diesel engine).

As the electromagnet is rotated past the three armature-phase windings, a sinusoidal AC armature-phase voltage is induced in each armature-phase winding.

The armature-phase windings are designed so that the sinusoidal AC armature-phase voltages induced in them are equal in amplitude and out of phase with each other by exactly $2\pi/3$.

Since the armature-phase windings are stationary with respect to the rotating electromagnet, the frequency of the armature-phase voltages induced in each armature-phase winding is the same.

Normally, the impedances of each armature-phase winding on a three-phase AC generator are very small compared to the other impedances in the AC electrical homogenous continuous dynamical system.

Therefore, to an approximation, each armature-phase winding can be physically modelled in an AC electrical homogeneous continuous dynamical system by an ideal sinusoidal energy source.

There are two ways of interconnecting the separate armature-phase windings to form a three-phase energy source. The armature-phase windings can be connected together in either a triangle or delta (Δ) or a three-ray star or wye (Y) configuration.

The delta and wye (Y) connections are shown in figure 11.38, where ideal energy sources are used to physically model the armature-phase windings of the three-phase AC generator.

The common terminal in the Y-connected energy source, labelled n in figure 11.38(a), is known as the 'neutral terminal' of the three-phase energy source. The neutral terminal may or may not be available for external connections.

If the impedances of each armature-phase winding are not negligible, the three-phase energy source is physically modelled by placing the armature-phase winding impedance in series with an ideal sinusoidal energy source.

In view of the fact that all armature windings on the AC generator are of the same structure, the armature-phase winding impedances are assumed to be identical. The armature-phase winding impedance of three-phase AC generators is inductive.

A physical model of a three-phase energy source, including the armature-phase winding impedance, is shown in figure 11.39, in which R_w is the winding resistance and X_w is the inductive reactance of the winding.

The fact that a three-phase energy source can be either Δ-connected or Y-connected means that the basic AC electrical homogeneous continuous dynamical system in figure 11.40 can take four different configurations, since the three-phase loads can also be either Δ- or Y-connected.

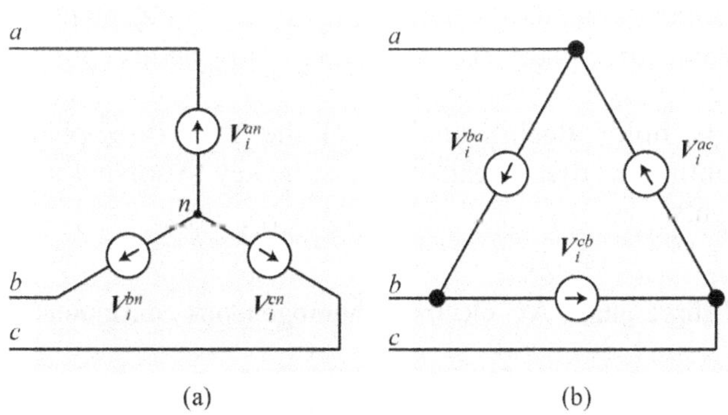

Figure 11.38. The two basic connections of an ideal three-phase energy source. (a) Δ-connected energy source. (b) Y-connected energy source.

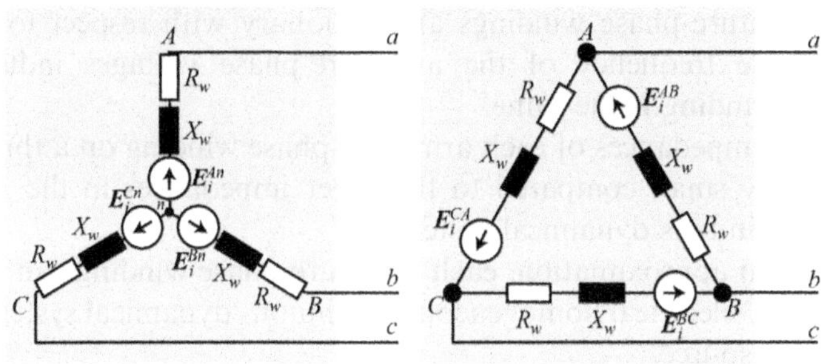

Figure 11.39. A physical model of a three-phase energy source with armature-phase winding impedance. (a) Δ-connected energy source. (b) Y-connected energy source.

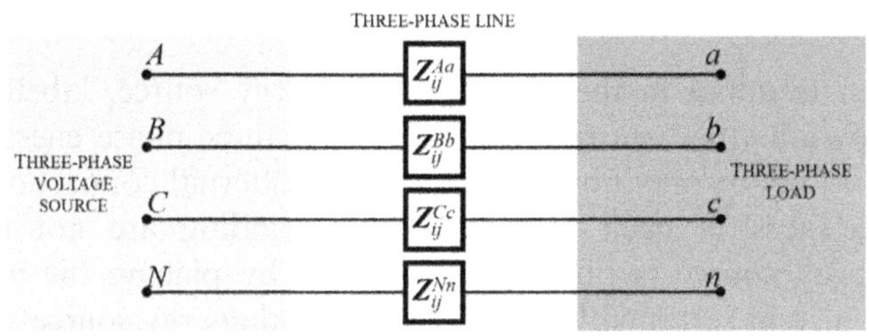

Figure 11.40. Basic AC electrical homogeneous continuous dynamical system.

The four possible arrangements are:

- a Δ-connected energy source and a Δ-connected load;
- a Δ-connected energy source and a Y-connected load;
- a Y-connected energy source and a Y-connected load;
- a Y-connected energy source and a Δ-connected load.

We begins our holor algebra analysis with the third arrangement. After holor analysing the Y–Y three-phase AC electrical homogeneous continuous dynamical system, we show how for balanced three-phase AC electrical homogeneous continuous dynamical systems the remaining three arrangements can be reduced to a Y–Y equivalent three-phase AC electrical homogeneous continuous dynamical system.

In other words, holor algebra analysis of the Y–Y three-phase AC electrical homogeneous continuous dynamical system is the key to solving all balanced three-phase arrangements.

11.7.3 Balanced three-phase AC electrical homogeneous continuous dynamical systems

The phase or line-to-neutral voltage-source holors of a balanced three-phase AC electrical homogeneous continuous dynamical system are shifted relative to each other by equal phase angles of $2\pi/3$ (figure 11.41).

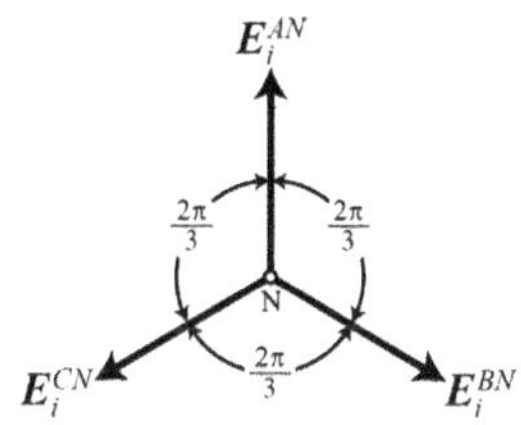

Figure 11.41. Phase or line-to-neutral voltage-source holors of a balanced three-phase AC electrical homogeneous continuous dynamical system.

In discussing three-phase AC electrical homogeneous continuous dynamical systems, it is standard practice to refer to the three phases as *A, B* and *C*. Furthermore, the *A* phase is almost always used as the reference phase. The three voltages that comprise the three-phase set are referred to as the *A*-phase energy-source voltage, the *B*-phase energy-source voltage and the *C*-phase energy-source voltage. Since the phase or line-to-neutral energy-source voltages are out of phase by exactly $2\pi/3$, two possible phase relationships can exist between the *A*-phase energy-source voltage and the *B*-phase and *C*-phase energy-source voltages. One possibility is for the *B*-phase energy-source voltage to lead the *A*-phase energy-source voltage by $2\pi/3$, in which case the *C*-phase energy-source voltage must lead the *A*-phase energy-source voltage by $2\pi/3$. This phase relationship is known as the *ABC* or 'positive phase sequence'. The only other possibility is for the *B*-phase energy-source voltage to lead the *B*-phase energy-source voltage by $2\pi/3$, in which case the *C*-phase energy-source voltage must lag the *A*-phase energy-source voltage by $2\pi/3$. This phase relationship is known as the *ACB* or 'negative phase sequence'. Accordingly, the instantaneous values of the phase or line-to-neural energy-source voltages for negative phase sequences are expressed by the following formulae:

$$\left.\begin{aligned} e_{AN}(t) &= E_{\max} \sin \omega t = \sqrt{2}\,E_{\text{rms}} \sin \omega t \\ e_{BN}(t) &= E_{\max}\sin(\omega t + 2\pi/3) = \sqrt{2}\,E_{\text{rms}}\sin(\omega t + 2\pi/3) \\ e_{CN}(t) &= E_{\max}\sin(\omega t + 4\pi/3) = \sqrt{2}\,E_{\text{rms}}\sin(\omega t + 4\pi/3) \end{aligned}\right\}, \tag{11.96}$$

corresponding to the instant when the voltage of phase *A* passes through zero and becomes positive.

The last formula in equations (11.96) can also be written in the form

$$e_{CN}(t) = E_{\max}\sin(\omega t - 2\pi/3) = \sqrt{2}\,E_{\text{rms}}\sin(\omega t - 2\pi/3). \tag{11.97}$$

In the holor notation of the phase or line-to-neutral energy-source voltage, the real axis is usually made to coincide with the energy-source voltage of phase *A*. In holor notation, the two possible sets of effective values ($E_{\text{rms}} = E_{\phi}$) of the phase or line-to-neutral energy-source voltages of the balanced three-phase AC electrical homogeneous continuous dynamical system are represented by the following holors.

a) For negative phase sequences:

$$\left.\begin{aligned} \boldsymbol{E}_i^{AN} &= E_\phi \angle 0 \\ \boldsymbol{E}_i^{BN} &= E_\phi \angle \frac{2\pi}{3} \\ \boldsymbol{E}_i^{CN} &= E_\phi \angle \frac{4\pi}{3} = E_\phi \angle -\frac{2\pi}{3} \end{aligned}\right\}. \tag{11.98a}$$

b) For positive phase sequences:

$$\left.\begin{aligned} \boldsymbol{E}_i^{AN} &= E_\phi \angle 0 \\ \boldsymbol{E}_i^{BN} &= E_\phi \angle - 2\pi/3 \\ \boldsymbol{E}_i^{CN} &= E_\phi \angle - 4\pi/3 = E_\phi \angle + 2\pi/3 \end{aligned}\right\}. \tag{11.98b}$$

The phase sequence of the energy-source voltage holors given in equations (11.98a) is the *ACB*, or negative, sequence. The phase sequence of the energy-source voltage holors given in equations (11.98b) is the *ABC*, or positive, sequence. The holor diagram representations of the voltage-holor sets given by equations (11.98a) and (11.98b) are shown in figure 11.42, from which one can determine the phase sequence by noting the order of the subscripts as one moves clockwise around the diagram.

The fact that a three-phase AC electrical homogeneous continuous dynamical system can have one of two possible phase sequences is an important characteristic that must be taken into consideration whenever two separate polyphase AC electrical homogeneous continuous dynamical systems are operated in parallel.

The two polyphase AC electrical homogeneous continuous dynamical systems can only operate in parallel if they have the same phase sequence. Another important

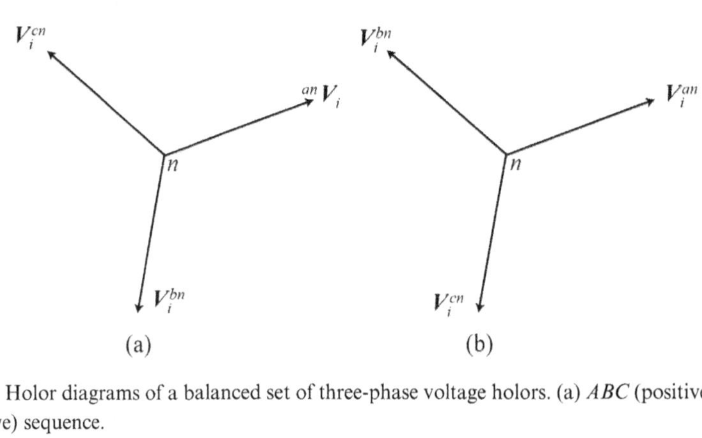

Figure 11.42. Holor diagrams of a balanced set of three-phase voltage holors. (a) *ABC* (positive) sequence. (b) *ACB* (negative) sequence.

characteristic of a set of balanced three-phase line-to-neutral energy-source voltage holors is that the sum of the line-to-neutral voltage-source holors adds to zero.

Thus, using either equations (11.98a) or equations (11.98b), one has

$$\boldsymbol{E}_i^{AN} + \boldsymbol{E}_i^{BN} + \boldsymbol{E}_i^{CN} = 0. \tag{11.99}$$

The fact that the sum of the line-to-neutral energy-source voltage holors adds to zero also means that the sum of the instantaneous values of the line-to-neutral energy-source is zero, that is,

$$e_{AN}(t) + e_{BN}(t) + e_{CN}(t) = 0. \tag{11.100}$$

Another noteworthy observation is that if one knows the phase sequence and one voltage holor in the set, one knows the entire set. Thus if one has a balanced three-phase AC electrical homogeneous continuous system, one can focus on determining the voltage holor (or current holor) in one phase, because once one knows one phase quantity, one automatically knows the corresponding quantity in the other two phases.

Figure 11.43 shows the waveforms of the instantaneous line-to-neutral energy-source voltages for all three phases; it is easy to see that at any instant the algebraic sum of the three line-to-neutral energy-source voltages is zero.

This result can also be established analytically with the aid of the relations in equations (11.100), but it is easier to use the holor diagram (figure 11.44) for this purpose. Indeed, combining the holors $\boldsymbol{E}_i^{BN}$ and $\boldsymbol{E}_i^{CN}$, one finds that, due to the symmetry, the holor $\boldsymbol{E}_i^{BN} + \boldsymbol{E}_i^{CN}$ is sensed vertically downwards, i.e. it is opposite to the holor $\boldsymbol{E}_i^{AN}$.

Forming $\pi/3$ angles with the $\boldsymbol{E}_i^{BN}$ and $\boldsymbol{E}_i^{CN}$ holors (or *NB* and *BD*), this result becomes one of the sides of the equilateral triangle *NBD* and, consequently, is equal in magnitude to the holor $\boldsymbol{E}_A$.

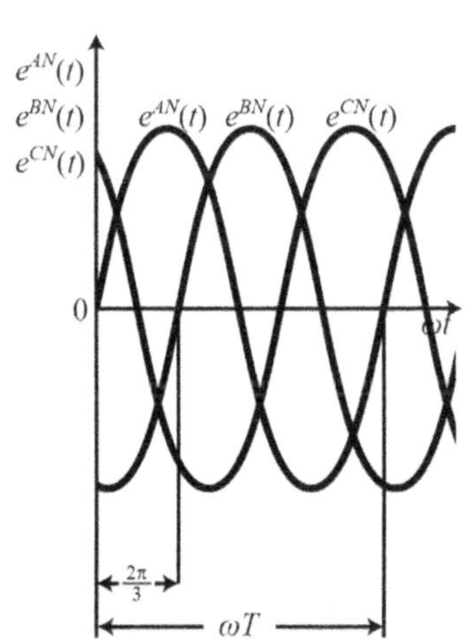

Figure 11.43. Energy-source voltages of the three-phase ME generator.

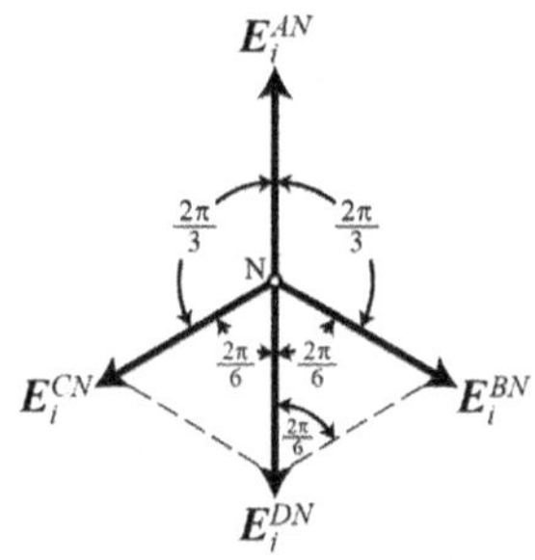

Figure 11.44. The sum of the voltage-source holors.

Therefore,

$$E_i^{AN} + E_i^{BN} + E_i^{CN} = E_i^{AN} + \left(-E_i^{AN}\right) = 0, \tag{11.101}$$

and since the sum of the instantaneous values $e_{AN}(t) + e_{BN}(t) + e_{CN}(t)$ represents the sum of the E_i^{AN}, E_i^{BN} and E_i^{CN} holor projections multiplied by $\sqrt{2}$, or the projections of their sum on the vertical, one has, at any instant,

$$e_{AN}(t) + e_{BN}(t) + e_{CN}(t) = 0. \tag{11.102}$$

It is obvious that the positive senses of the directions of the energy-source line-to-neutral voltages and currents must be chosen the same for all phases, for instance, away from the reader into the plane of the image shown in figure 11.44.

Let us decide, then, to consider the beginning of each phase to be the point where the positively directed current comes out of the ME generator's stator winding into the external AC electrical homogenous continuous dynamical system, and let us consider the end of the phase to be the point where the positively directed current leaves the external AC electrical homogeneous continuous dynamical system and enters the ME generator's stator winding. Thus, the energy-source voltages in the ME generator phases are positive when they act from the ends of phases (*X, Y* and *Z*) to their beginnings (*A, B* and *C*). In AC electrical homogeneous continuous dynamical system diagrams, a three-phase ME generator is ordinarily depicted in the form of three stator windings *XA, YB* and *ZC* (figure 11.45).

If the loads connected to the beginnings and ends of each phase have identical (in magnitude and waveform) impedance holors Z_{ij}^{AN}, Z_{ij}^{BN}, and Z_{ij}^{CN}, the current holors $I_{Aa,}^{j}$, I_{Bb}^{j} and I_{Cc}^{j} will likewise be identical in magnitude, and shifted in phase relative to the corresponding voltage-source holors by equal angles φ (figure 11.46).

They will also be shifted in phase relative to each other by $2\pi/3$, i.e. the mutual distribution of their waveforms will not differ from that of the waveforms of the line-to-neutral voltage-source holors in figure 11.43, and hence the values of the effective values of the line-current holors will be connected by the relation

$$I_{Aa}^{j} + I_{Bb}^{j} + I_{Cc}^{j} = 0 \tag{11.103}$$

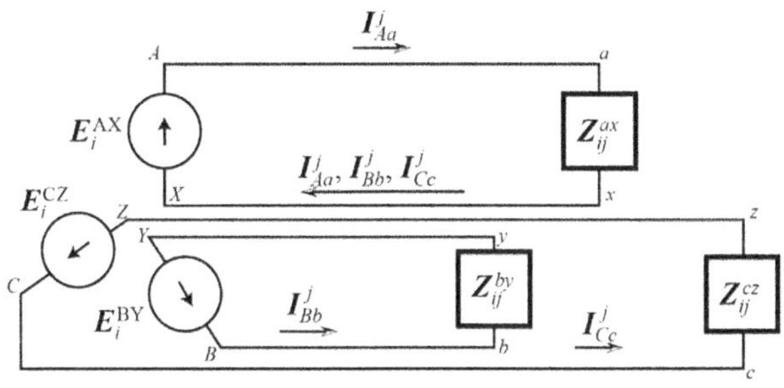

Figure 11.45. Schematic diagram of a three-phase generator feeding three loads.

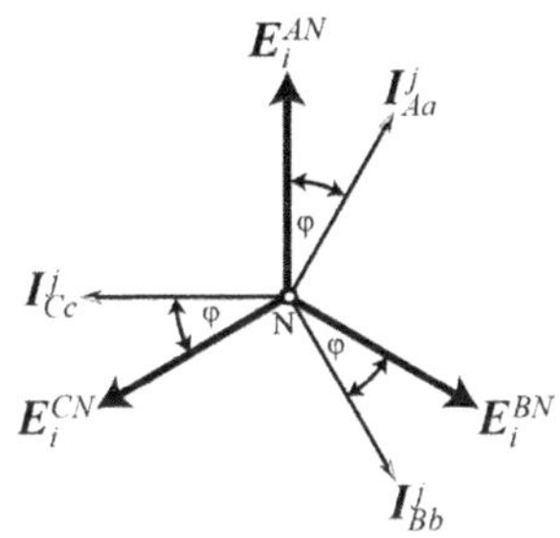

Figure 11.46. Holor diagram of balanced three-phase AC electrical homogeneous continuous dynamical system.

and the instantaneous values of the line currents will be connected by

$$i_{Aa}(t) + i_{Bb}(t) + i_{Cc}(t) = 0, \tag{11.104}$$

Such a load for a three-phase electrical homogeneous continuous dynamical system is termed a 'balanced' or 'symmetrical load'.

A three-phase electrical homogeneous continuous dynamical system where each ME generator phase is independently connected by two conductors to its own load is termed an 'independent' three-phase AC electrical homogeneous continuous dynamical system. Thus, an independent three-phase AC electrical homogeneous continuous dynamical system demands the use of six conductors to transmit electrical energy from the ME generator to the loads.

11.7.4 Wye connection of the AC electrical homogenous continuous dynamical system

The number of conductors can be reduced if, instead of three separate return conductors, one common conductor is used (figure 11.47), and one joins to it all the phase ends of the ME generator or EE transformer secondaries and load phases (let us agree that the positive sense of the current direction in the load phases is from the beginning of the phase to its end). The operating conditions of the load phases are in no way changed, since the voltage between the beginning and end of each phase, as in an independent three-phase AC electrical homogeneous continuous dynamical system, is equal to the voltage of the corresponding phase of the ME generator or EE transformer secondaries (neglecting the impedances of the connecting

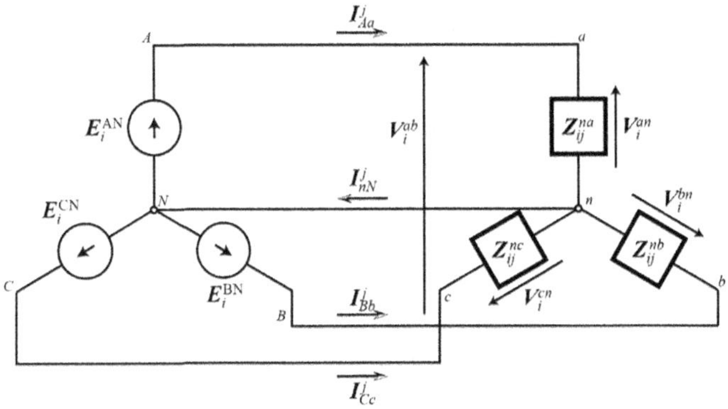

Figure 11.47. A three-arm star or wye (Y) connection of the three-phase AC electrical homogeneous continuous dynamical system.

conductors). In this way, a four-conductor Y-connected three-phase AC electrical homogeneous continuous dynamical system is obtained.

In circuit diagrams, it is customary to represent the phases of the ME generator or EE transformer secondaries and load phases in the form of a three-arm star, or more often, a wye (Y), placing the load phases parallel to the corresponding ME generator phases or EE transformer secondaries (figure 11.47).

The common point where all the phase terminals of the ME generator (or EE transformer secondaries) and load phases are joined is termed the 'neutral point' or 'neutral', and the common return conductor joining the neutral points of the ME generator (or EE transformer secondaries) and load is termed the 'neutral conductor'.

The current in this conductor, by KFL, is equal to the sum of the current holors I^j_{Aa}, I^j_{Bb} and I^j_{Cc}:

$$I^j_{Nn} = I^j_{Aa} + I^j_{Bb'} + I^j_{Cc}. \tag{11.105}$$

For balanced load phases when $I^j_{Aa} + I^j_{Bb'} + I^j_{Cc} = 0$, there will be no current in the neutral conductor (hence the term 'neutral'), and the neutral conductor becomes unnecessary. By omitting the neutral conductor, one obtains a three-wire Y-connected three-phase AC electrical homogeneous continuous dynamical system. Such an AC electrical homogeneous continuous dynamical system can only function normally when the load is distributed equally among the phases. Otherwise, the phase or line-to-neutral effective (rms) values of the load-voltage holors will no longer be identical and equal to the ME generator (or EE transformer secondaries) phase or line-to-neutral effective (rms) values of the energy-source voltage holors.

The effective (rms) values of the voltages between the beginnings of the ME generator (or EE transformer secondaries) or load phases and the neutral point are designated V^{an}_{Φ}, V^{bn}_{Φ} and V^{cn}_{Φ} and are termed the 'phase' or 'line-to-neutral load voltages'.

In three-phase AC electrical homogeneous continuous dynamical system analysis, the phase or line-to-neutral load voltages are often given instead of the phase or line-to-neutral energy-source voltages E^{AN}_{Φ}, E^{BN}_{Φ} and E^{CN}_{Φ}, and they are equal to the phase or line-to-neutral energy-source voltages when the ME generator or EE transformer winding impedances can be neglected as compared to the load phase impedances.

The effective (rms) values of the voltages between the phase beginnings *a* and *b*, *b* and *c*, and *c* and *a* are designated V_Φ^{ab}, V_Φ^{bc} and V_Φ^{ca} and are termed the 'line' or 'line-to-line load voltages'. If the neutral point of the ME AC generator or EE AC transformer is chosen as a point of zero potential, the potentials of points *a, b* and *c* will be equal to the corresponding phase or line-to-neutral load voltages V_Φ^{an}, V_Φ^{bn} and V_Φ^{cn}.

The line or line-to-line load voltage holors, regarded as differences of potential, will be related to the phase or line-to-neutral load-voltage holors by the expressions

$$\boldsymbol{V}_i^{ab} = \boldsymbol{V}_i^{an} - \boldsymbol{V}_i^{bn}, \qquad \boldsymbol{V}_i^{bc} = \boldsymbol{V}_i^{bn} - \boldsymbol{V}_i^{cn}, \qquad \boldsymbol{V}_i^{ca} = \boldsymbol{V}_i^{cn} - \boldsymbol{V}_i^{an}. \tag{11.106}$$

This subtraction is performed on the holor diagram (figure 11.48) by holor addition of the holors $\boldsymbol{V}_i^{an}$, $\boldsymbol{V}_i^{bn}$ and $\boldsymbol{V}_i^{cn}$ to the holors $-\boldsymbol{V}_i^{bn}$, $-\boldsymbol{V}_i^{cn}$ and $-\boldsymbol{V}_i^{an}$, respectively, which are equal and opposite to the holors $\boldsymbol{V}_i^{bn}$, $\boldsymbol{V}_i^{cn}$ and $\boldsymbol{V}_i^{an}$.

It can be seen from the holor diagram that the magnitude and direction of the line-voltage holors can be obtained more simply (by joining by a straight line) from the ends of the holors of the corresponding phase-load-voltage holors; the line-voltage holor is directed from the end of the subtrahend holor to the end of the minuend holor. The three line-voltage holors are equal in magnitude and shifted in phase (relative to each other) by $2\pi/3$, in the same way as the phase-voltage holors. Here, the line-voltage holor three-ray star (Y) is rotated forward relative to the phase-voltage three-ray star or wye (Y), through an angle of $\pi/6$. The relationship between the absolute value U_λ of any line voltage and the absolute value V_Φ of any phase voltage can be found by considering the isosceles triangle *nab* (or *nbc* or *nca*) with a vertex angle of $2\pi/3$.

Dropping a perpendicular from the vertex of the obtuse angle to the opposite side, for example *ab*, one has

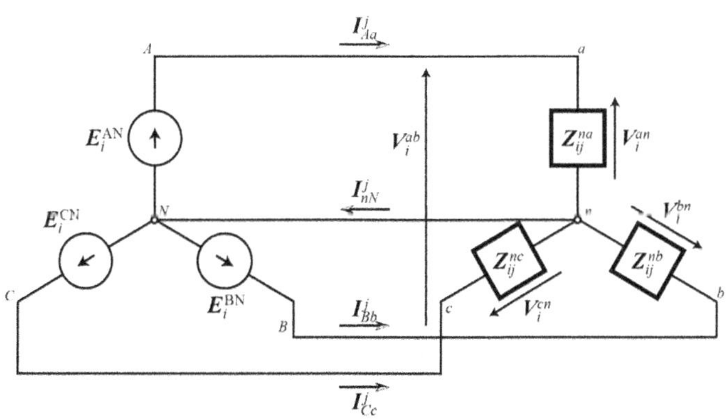

Figure 11.48. Phase or line-to-neutral and line or line-to-line load-voltage holors.

$$\frac{V_{\lambda}^{ab}}{2} = V_{\Phi}^{an} \cos\frac{\pi}{6} = V_{\Phi}^{an}\frac{\sqrt{3}}{2}, \tag{11.107}$$

whence

$$V_{\lambda} = \sqrt{3}\, V_{\Phi}. \tag{11.108}$$

This important relation refers equally to a three-phase AC electrical homogeneous continuous dynamical system with a neutral conductor and one without one.

11.7.5 Delta connection of the AC electrical homogeneous continuous dynamical system

If three load impedance holors $\boldsymbol{Z}_{ij}^{ab}$, $\boldsymbol{Z}_{ij}^{bc}$ and $\boldsymbol{Z}_{ij}^{ca}$ are connected directly between the conductors of a three-conductor line from the terminals of a three-phase ME generator, without using a neutral conductor, the result is a triangle, or more often a delta (Δ), connection (figure 11.49).

Here, one need not distinguish between the phase and line or line-to-line load voltage holors, since the voltage holor between the beginning and end of each load phase is at the same time the line or line-to-line load-voltage holor. But there is a difference between the phase load-current holors $\boldsymbol{I}_{ab}^{j}$, $\boldsymbol{I}_{bc}^{j}$ and $\boldsymbol{I}_{ca}^{j}$, whose positive senses of directions are chosen from a to b, b to c, and c to a, and the line-load current holors, which, as before, are considered positive in the senses of directions from the ME generator or EE transformer to the load phases.

From KFL,

$$\boldsymbol{I}_{Aa}^{j} = \boldsymbol{I}_{ab}^{j} - \boldsymbol{I}_{ca}^{j}, \quad \boldsymbol{I}_{Bb}^{j} = \boldsymbol{I}_{bc}^{j} - \boldsymbol{I}_{ab}^{j}, \quad \boldsymbol{I}_{Cc}^{j} = \boldsymbol{I}_{ca}^{j} - \boldsymbol{I}_{bc}^{j}, \tag{11.109}$$

that is, any one of the line-current holors is equal to the holor difference in the current holors of two phases directly connected to the given conductor of the line;

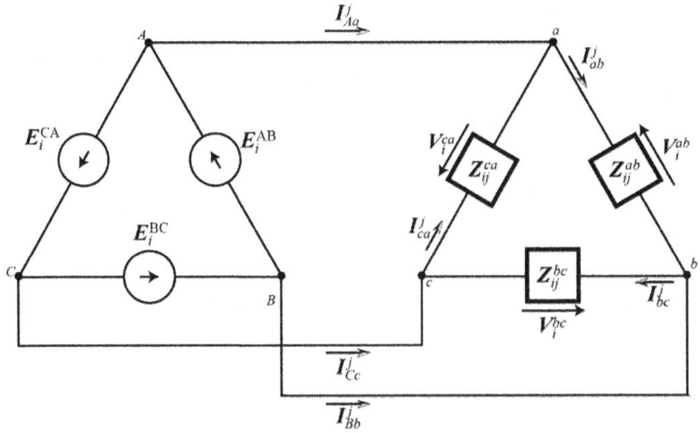

Figure 11.49. A triangle or delta (Δ)-connected three-phase AC electrical homogeneous continuous dynamical system.

the minuend is the phase-current holor directed from the conductor, and the subtrahend is the phase-current holor directed to the line.

From the relationships (11.109) it follows that, for any values of the phase currents, the holor sum of the line-current holors is

$$\boldsymbol{I}^{j}_{Aa} + \boldsymbol{I}^{j}_{Bb} + \boldsymbol{I}^{j}_{Cc} = 0. \tag{11.110}$$

For balanced phases (that is, for load phase impedance holors $\boldsymbol{Z}^{ab}_{ij}$, $\boldsymbol{Z}^{bc}_{ij}$ and $\boldsymbol{Z}^{ca}_{ij}$ identical in magnitude and character), when the phase or line-to-neutral current holors $\boldsymbol{I}^{j}_{ab}$, $\boldsymbol{I}^{j}_{bc}$ and $\boldsymbol{I}^{j}_{ca}$ are equal in magnitude and differ in phase by the same angles relative to the corresponding line or line-to-line load voltage holors $\boldsymbol{V}^{ab}_{i}$, $\boldsymbol{V}^{bc}_{i}$ and $\boldsymbol{V}^{ca}_{i}$ (figure 11.50), forming a symmetrical three-arm wye (Y), the line-current holors $\boldsymbol{I}^{j}_{Aa}$, $\boldsymbol{I}^{j}_{Bb}$ and $\boldsymbol{I}^{j}_{Cc}$ also become equal and symmetrically distributed. Here, the wye (Y) line-current holor is rotated backwards relative to the wye (Y) phase-current holor through a $\pi/6$ angle.

Each of the line-current holors thus represents the base of an isosceles triangle with a $2\pi/3$ vertex angle, and for this reason the same relationship exists between the line- and phase-current holors for a symmetrical load as between the line voltage and phase voltage in a wye (Y) connection, namely,

$$I_{\lambda} = \sqrt{3}\, I_{\Phi}. \tag{11.111}$$

This important relation makes it rather easy to compute the line currents in a symmetrical load. If the impedances of conductors joining delta (Δ)-connected load phases to the energy source or to any apparatus connected in these conductors must be taken into consideration, one should transform the impedance delta into an equivalent wye (Y), which, in the case of a symmetrical delta, is also symmetrical.

The impedance of each line then becomes connected in series with the impedance of each phase of the wye (Y), and it may be replaced by overall impedance. In this case, each energy-source phase must be directly connected between two conductors of the three-conductor line. If, in this case, phase AB of energy source ($\boldsymbol{E}^{AB}_{i}$) is connected between conductors A and B (figure 11.51), phase ($\boldsymbol{E}^{BC}_{i}$) between conductors B and C, and phase ($\boldsymbol{E}^{CA}_{i}$) between conductors C and A, and in such a way that the energy-source voltages have the positive senses of directions shown in

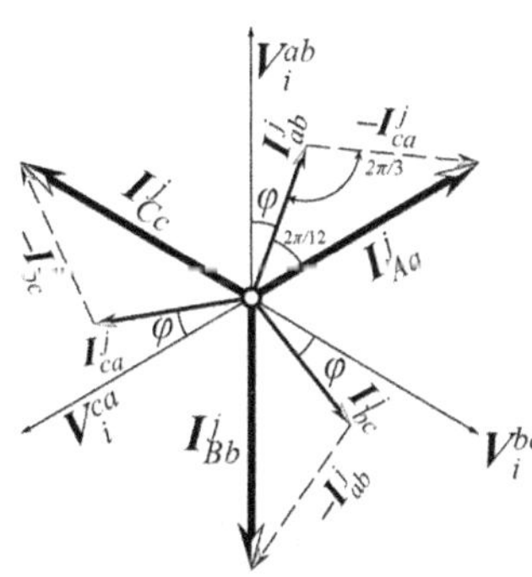

Figure 11.50. Line-current holors under balanced loading of phases.

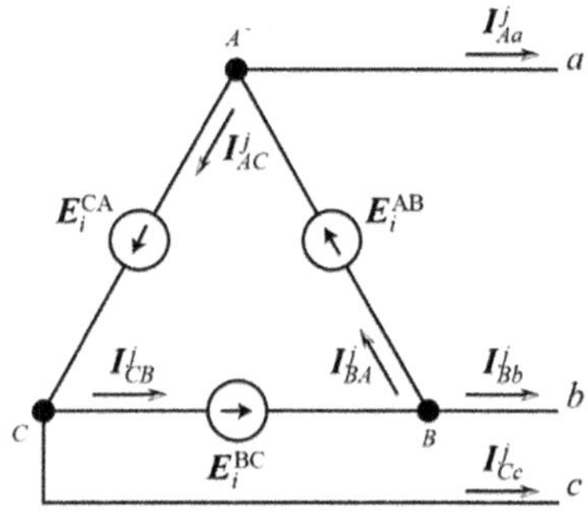

Figure 11.51. Triangle or delta (Δ)-connected three-phase ME generator.

figure 11.51, then the following relationships will hold (neglecting the winding impedances):

$$V_i^{ab} = E_i^{AB}, \quad V_i^{bc} = E_i^{BC}, \quad V_i^{ca} = E_i^{CA},$$

and consequently, the usual line-voltage sequence, in which V_i^{bc} lags V_i^{ab} by $2\pi/3$ and V_i^{ca} lags V_i^{bc} by $2\pi/3$, is retained.

Thus, the three energy-source phases must be connected in series with each other, that is, the end of phase *AB* must be connected with the beginning of phase *BC*, and the end of phase *CA* with the beginning of phase *AB*.

At first sight it may seem that the series connection of the three energy-source phases in a closed delta (Δ) is equivalent to a joint short circuit, as the case would be, for instance, in such a connection of DC energy sources.

However, since in a three-phase AC electrical continuous dynamical system $e(t)^{AB} + e(t)^{BC} + e(t)^{CA} = 0$ at any instant, not only is there no short circuit in the given case, but, in the absence of a load, the current in the closed delta (Δ) of the energy-source is zero. However, this only holds true when the sum of the phase energy-source voltages is zero. But often the phase energy-source voltages are not strictly sinusoidal waveforms and do not fully satisfy the relationship. Therefore, delta (Δ)-connected three-phase ME generators are rarely encountered. On the other hand, delta (Δ)-connected EE transformers are just as common as wye (Y)-connected ones.

In the case of a symmetrical (balanced) load when the line-current holors I^j_{Aa}, I^j_{Bb}, and I^j_{Cc} are equal in magnitude and shifted in phase relative to each other by 2/3, any one of the energy-source phase currents is

$$I_\phi = \frac{I_\lambda}{\sqrt{3}},$$

as was also the case in a balanced three-phase AC electrical homogeneous continuous dynamical system for the phase currents of the loads, and it follows that if the phases of both the energy source and the load are delta (Δ) connected, then, in a balanced three-phase AC electrical homogeneous continuous dynamical system, the currents in the corresponding energy-source and load phases will be equal in magnitude and phase.

11.7.6 Unbalanced wye-connected load of the AC electrical homogeneous continuous dynamical system

Turning to a holor algebra analysis of unbalanced three-phase loads, let us first consider the case of a wye (Y)-connected energy source and load where the line conductors, including the neutral, possess a finite (and not negligibly small) impedance.

Incidentally, the impedances of the phase conductors connected in series with the load phases may be grouped with the impedances of these phases, and then one only needs to take the neutral-conductor impedance into account.

Let us choose the neutral of the energy source as zero potential and then aim to compute the mode of operation of the three-phase AC electrical homogeneous continuous dynamical system (figure 11.52), if one knows the energy-source phase-voltage holors $\boldsymbol{E}_i^{AN}$, $\boldsymbol{E}_i^{BN}$ and $\boldsymbol{E}_i^{CN}$, that is, the potentials at points *A, B* and *C,* or *a, b* and *c,* the load-phase impedance holors $\boldsymbol{Z}_{ij}^{an}$, Z_{ij}^{bn} and Z_{ij}^{cn}, and the impedance holor $\boldsymbol{Z}_{ij}^{Nn}$ of the neutral conductor.

Since the given AC electrical homogeneous continuous dynamical system has two nodes (the energy source and load neutrals), the problem is easily solved using the node-voltage holor method.

One first determines the voltage holor between the neutral points, which is termed displacement of the neutral, that is, the potential of the neutral point *n* of the load, from the formula

$$\boldsymbol{V}_i^{Nn} = \frac{\delta_k^i}{\dfrac{\delta_k^i}{\boldsymbol{Z}_{jk}^{an}} + \dfrac{\delta_k^i}{\boldsymbol{Z}_{jk}^{bn}} + \dfrac{\delta_k^i}{\boldsymbol{Z}_{jk}^{cn}} + \dfrac{\delta_k^i}{\boldsymbol{Z}_{jk}^{Nn}}} \left(\frac{\delta_k^i}{\boldsymbol{Z}_{jk}^{an}} \boldsymbol{E}_i^{AN} + \frac{\delta_k^j}{\boldsymbol{Z}_{jk}^{bn}} \boldsymbol{E}_i^{BN} \frac{\delta_k^i}{\boldsymbol{Z}_{jk}^{cn}} \boldsymbol{E}_i^{CN} \right). \qquad (11.112)$$

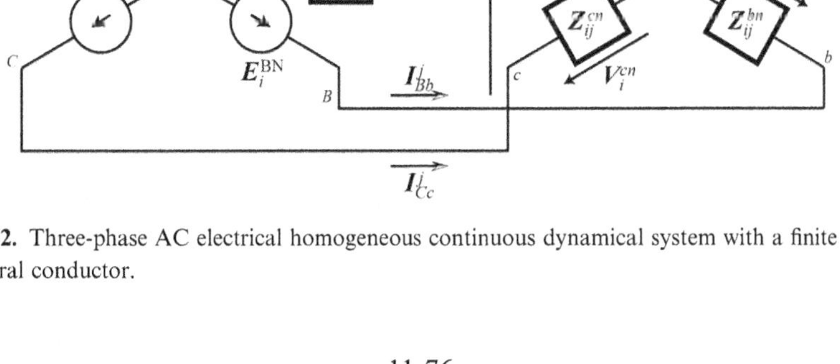

Figure 11.52. Three-phase AC electrical homogeneous continuous dynamical system with a finite impedance of the neutral conductor.

Replacing all the $\frac{\delta_k^i}{Z_{jk}}$ s by the corresponding admittance holors Y^{ji}, one writes the formula in a more concise form:

$$V_i^{Nn} = \frac{\delta_k^i}{Y_{an}^{ij} + Y_{bn}^{ij} + Y_{cn}^{ij} + Y_{Nn}^{ij}} \left(Y_{an}^{ij} E_i^{AN} + Y_{bn}^{ji} E_i^{BN} + Y_{cn}^{ij} E_i^{CN} \right). \tag{11.113}$$

If there is no neutral conductor, that is, if $Z_{ik}^{Nn} = \infty$, then $Y_{Nn}^{ji} = 0$, and the last term in the denominator is removed. If the energy source is given a symmetrical energy-source voltage-holor three-phase AC electrical homogeneous continuous dynamical system

$$\left. \begin{aligned} E_i^{AN} &= \left(E_c^{AN}, E_s^{AN}\right) = E_\phi \angle 0, \\ E_i^{BN} &= \left(E_c^{BN}, E_s^{BN}\right) = E_\phi \angle -2\pi/3, \\ E_i^{CN} &= \left(E_c^{CN}, E_s^{CN}\right) = E_\phi \angle -4\pi/3 = E_\phi \angle 2\pi/3, \end{aligned} \right\} \tag{11.114}$$

and the current holors

$$I_{Aa}^j = Y_{an}^{ji} E_i^{AN}, \quad I_{Bb}^j = Y_{bn}^{ji} E_i^{BN}, \quad I_{Cc}^j = Y_{cn}^{ji} E_i^{CN}, \tag{11.115}$$

then by substituting expressions (11.115) into formula (11.104), one obtains a formula more suitable for computation

$$V_i^{Nn} = \frac{\delta_k^j}{Y_{an}^{ji} + Y_{bn}^{ji} + Y_{cn}^{ji} + Y_{Nn}^{ji}} \left(I_{Aa}^j + I_{Bb}^j + I_{Cc}^j \right). \tag{11.116}$$

For a balanced load, when $Y_{an}^{ji} = Y_{bn}^{ji} = Y_{cn}^{ji} = Y^{ji}$, $U_i^{Nn} = 0$, since $E_i^{AN} + E_i^{BN} + E_i^{CN} = 0$ (as in figure 11.52).

When the neutral displacement V_m^{Nn} has been found, it becomes easy to determine the phase-voltage holors of the load:

$$V_i^{an} = E_i^{AN} - V_i^{Nn}, \quad V_i^{bn} = E_i^{BN} - V_i^{Nn}, \quad V_i^{cn} = E_i^{CN} - V_i^{Nn}. \tag{11.117}$$

The current holor in the neutral conductor may be found as the holor sum of the line- or phase-current holors,

$$I_{Nn}^j = I_{Aa}^j + I_{Bb}^j + I_{Cc}^j,$$

or from the formula

$$I_{Nn}^j = Y_{Nn}^{ji} V_i^{Nn}.$$

If the impedance holor of the neutral conductor can be ignored, $Z_{ik}^{Nn} = 0$, then $Y_{Nn}^{ji} = \infty$ and, from equation (11.103), $V_i^{Nn} = 0$, that is, the neutral point potentials of the energy source and load are the same. Consequently, the current holors may be computed directly from formula (11.115).

11.7.7 Unbalanced delta-connected load of the AC electrical homogeneous continuous dynamical system

Let us consider the problem of an unbalanced delta (Δ)-connected load, the vertices of which are connected to the energy source through impedance holors Z_{ik}^{Aa}, Z_{ik}^{Bb} and Z_{ik}^{Cc}, which may be identical or different (figure 11.53).

This problem is simply solved by the delta (Δ) to wye (Y) transformation according to the familiar formulae:

$$Z_{ik}^{an} = \frac{\delta_k^j}{Z_{ik}^{bc} + Z_{ik}^{ca} + Z_{ik}^{ab}} Z_{ij}^{ca} Z_{ij}^{ab};$$

$$Z_{ik}^{bn} = \frac{\delta_k^j}{Z_{ik}^{bc} + Z_{ik}^{ca} + Z_{ik}^{ab}} Z_{ij}^{ab} Z_{ij}^{bc};$$

$$Z_{ik}^{cn} = \frac{\delta_k^j}{Z_{ik}^{bc} + Z_{ik}^{ca} + Z_{ik}^{ab}} Z_{ij}^{bc} Z_{ij}^{ca}.$$

Knowing the admittance holors of each phase,

$$Y_{Aa}^{ji} = \frac{\delta_k^j}{Z_{ik}^{Aa} + Z_{ik}^{an}}, \quad Y_{Bb}^{ji} = \frac{\delta_k^j}{Z_{ik}^{Bb} + Z_{ik}^{bn}}, \quad Y_{Cc}^{ji} = \frac{\delta_k^j}{Z_{ik}^{Cc} + Z_{ik}^{cn}},$$

it becomes easy (from equation (11.112)) to find the neutral displacement V_i^{Nn} by putting $Y_{NN}^{ji} = 0$ and then compute the line-current holors:

$$I_{Aa}^j = Y_{Aa}^{ji}\left(E_i^{AN} - V_i^{Nn}\right),$$

$$I_{Bb}^j = Y_{Bb}^{ji}\left(E_i^{BN} - V_i^{Nn}\right),$$

$$I_{Cc}^j = Y_{Cc}^{ji}\left(E_i^{CN} - V_i^{Nn}\right),$$

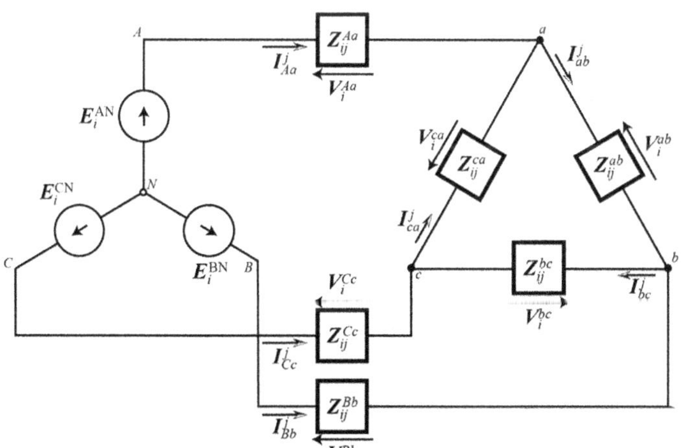

Figure 11.53. Unbalanced delta (Δ)-connected load of a three-phase AC electrical homogeneous continuous dynamical system.

which are, at the same time, the phase-current holors of the equivalent wye (Y)-connected load. Then, the phase-voltage holors of this equivalent wye (Y)-connected load will be

$$V_i^{an} = Z_{ij}^{an} I_{Aa}^j, \quad V_i^{bn} = Z_{ij}^{bn} I_{Bb}^j, \quad V_i^{cn} = Z_{ij}^{cn} I_{Cc}^j,$$

and its line-voltage holors, which are also the delta (Δ) line-voltage holors, will be

$$V_i^{ab} = V_i^{an} - V_i^{bn}, \quad V_i^{bc} = V_i^{bn} - V_i^{cn}, \quad V_i^{ca} = V_i^{cn} - V_i^{an}.$$

Finally, the phase-current holors of the load are

$$I_{ab}^j = \frac{\delta_k^j}{Z_{ik}^{ab}} V_i^{ab}, \quad I_{bc}^j = \frac{\delta_k^j}{Z_{ik}^{bc}} V_i^{bc}, \quad I_{ca}^j = \frac{\delta_k^j}{Z_{ik}^{ca}} V_i^{ca}.$$

11.7.8 Power holor computations in balanced three-phase AC electrical homogeneous continuous dynamical systems

An instantaneous value of the AC phase power can be obtained from a product

$$p_\phi(t) = v_\phi(t) i_\phi(t). \tag{11.118}$$

Substituting to equation (11.118) instantaneous values of the phase voltage and phase current from equations (11.18), one obtains

$$p_\phi(t) = V_\phi I^\phi \left\{ \cos\varphi_\phi - \cos\left[2\omega t + \left(\psi_v^\phi - \psi_i^\phi\right)\right]\right\}, \tag{11.119}$$

where

$$\varphi_\phi = \psi_v^\phi - \psi_i^\phi.$$

To obtain a holor representation of the phase power one must write equation (11.117) in the form

$$p_\phi(t) = V_\phi I^\phi \left[\cos\varphi_\phi - \cos 2\omega t \cos\left(\psi_v^\phi + \psi_i^\phi\right) + \sin 2\omega t \sin\left(\psi_u^\phi + \psi_i^\phi\right)\right]. \tag{11.120}$$

In this way, equation (11.118) can be rewritten in the form

$$p_\phi(t) = P_\phi^p - P_\phi^q \cos 2\omega t + P_\phi^r \sin 2\omega t. \tag{11.121}$$

The quantities P^p, P^q and P_ϕ^r are independent elements (merates) or coordinates of the three-merate monovalent holor of an AC phase-power holor,

$$P_\phi^\kappa = \left(P_\phi^p, P_\phi^q, P_\phi^r\right), \tag{11.122}$$

and this allows time to be eliminated from equation (11.123).

If one designed an independent term by P_{ϕ}^{p} and dependent terms from a doublet frequency by P_{ϕ}^{q} and P_{ϕ}^{r}, then one would obtain for the balanced load the following expressions for the phase-power holor independent elements (merates):

$$\left.\begin{aligned} P_{\phi}^{p} &= V_{\phi} I^{\phi} \cos\left(\psi_{u}^{\phi} - \psi_{i}^{\phi}\right) = V_{\phi} I^{\phi} \cos \varphi_{\phi} = P_{\phi} \\ P_{\phi}^{q} &= V_{\phi} I^{\phi} \cos\left(\psi_{u}^{\phi} - \psi_{i}^{\phi}\right) \\ P_{\phi}^{r} &= V_{\phi} I^{\phi} \sin\left(\psi_{u}^{\phi} - \psi_{i}^{\phi}\right) \end{aligned}\right\}. \tag{11.123}$$

From the first expression of equations (11.121) it results that an independent element (merate) of the power holor is an AC phase average or active power, which can be measured by means of an electrodynamic wattmeter.

Residual independent elements (merates) can be also measured or computed with the aid of electrical metrology methods. In particular circumstances, the following apply.

a) For a voltage phase angle $\psi_{v} = 0$

$$P_{\phi}^{q} = U_{\phi} I^{\phi} \cos \varphi_{\phi} = P_{\phi}^{0} = P_{\phi};$$

$$P_{\phi}^{r} = -U_{\phi} I^{\phi} \sin \varphi_{\phi} = -Q_{\phi}.$$

b) For a current phase angle $\psi_{i} = 0$

$$P_{\phi}^{q} = U_{\phi} I^{\phi} \cos \varphi_{\phi} = P^{0} = P_{\phi};$$

$$P_{\phi}^{r} = U_{\phi} I^{\phi} \sin \varphi_{\phi} = Q_{\phi}.$$

Thus, for the voltage phase angle $\psi_{V}^{\phi} = 0$ or the current phase angle $\psi_{i}^{\phi} = 0$, an independent element (merate) of the power holor P_{ϕ}^{q} is the AC average or active power P_{ϕ}; however, an independent element (merate) of the power holor P_{ϕ}^{r} is the AC reactive power Q_{ϕ}.

A phase-power holor can be also postulated by special tensor multiplying the phase-voltage holor and phase-current holor together so that

$$\boldsymbol{P}_{\boldsymbol{\phi}}^{\kappa} = \boldsymbol{V}_{i}^{\phi} \otimes \boldsymbol{I}_{\boldsymbol{\phi}}^{j} = \gamma_{..j}^{\kappa i} \boldsymbol{V}_{i}^{\phi} \boldsymbol{I}_{\boldsymbol{\phi}}^{j} = \gamma_{..j}^{\kappa i} \boldsymbol{Z}_{ij}^{\phi} \boldsymbol{I}_{\boldsymbol{\phi}}^{k} \boldsymbol{I}_{\boldsymbol{\phi}}^{j} = \left(P_{\phi}^{p},\ P_{\phi}^{q},\ P_{\phi}^{r}\right). \tag{11.124}$$

However, a voltage holor

$$\boldsymbol{V}_{i}^{\phi} = \boldsymbol{Z}_{ij}^{\phi} \boldsymbol{I}_{\boldsymbol{\phi}}^{k}.$$

Thus the power holor associated with any one phase of Y-connected load can also be computed using the holor algebra techniques introduced in chapter 11 or from equation (11.122); thus

$$\boldsymbol{P}_{\boldsymbol{\phi}}^{\kappa} = \left(P_{\phi}^{p},\ P_{\phi}^{q},\ P_{\phi}^{r}\right). \tag{11.125}$$

Thus far, holor algebra analysis of balanced three-phase AC electrical continuous dynamical systems has been limited to the determination of voltage holors and current holors in a given AC electrical homogeneous continuous dynamical system.

One begins by discussing the average or active power delivered to a balanced Y-connected load. A balanced Y-connected load, along with the pertinent voltage holors and current holors, is shown in figure 11.54.

The average or active power associated with any one phase can be computed using the holor algebra techniques introduced in chapter 11.

Using equation (11.123) as a starting point, one finds that one can express the average or active power associated with the A phase of the load as

$$P_A^p = \left|V_{ij}^{an}\right|\left|I_{Aa}^j\right|\cos\left(\psi_v^{an} - \psi_i^{Aa}\right), \tag{11.126}$$

where ψ_v^{an} and ψ_i^{Aa} denote the phase angles of U_{ij}^{an} and I_{Aa}^j, respectively.

Using the notation introduced in equation (11.124), one finds that the powers associated with the B and C phases are

$$P_B^p = \left|V_i^{bn}\right|\left|I_{Bb}^j\right|\cos\left(\psi_v^{bn} - \psi_i^{Bb}\right) \tag{11.127}$$

and

$$P_C^p = \left|V_i^{cn}\right|\left|I_{Cc}^j\right|\cos\left(\psi_v^{cn} - \psi_i^{Cc}\right). \tag{11.128}$$

In equations (11.124)–(11.126), all currents and voltages are written in terms of the effective (rms) values of the sinusoidal function that they represent.

In a balanced three-phase AC electrical homogeneous continuous dynamical system, the effective value of each line-to-neutral voltage is the same, as is the effective value of the phase current. The argument of the cosine functions is also the same for all three phases. To emphasise these observations, one introduces the following notation to facilitate further discussion of power computations in balanced three-phase AC electrical homogeneous continuous dynamical systems.

$$V_\phi = |V_i^{an}| = |V_i^{bn}| = |V_i^{cn}|, \tag{11.129}$$

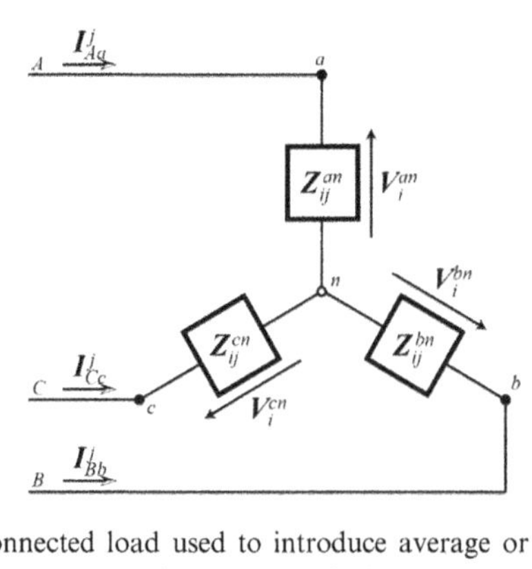

Figure 11.54. A balanced Y-connected load used to introduce average or active power computations in a three-phase AC electrical homogeneous continuous dynamical system.

$$I^{\phi} = |\boldsymbol{I}_{Aa}^{j}| = |\boldsymbol{I}_{Bb}^{j}| = |\boldsymbol{I}_{Cc}^{j}|, \tag{11.130}$$

and

$$\varphi_{\phi} = \psi_{v}^{an} - \psi_{i}^{Aa} = \psi_{v}^{bn} - \psi_{i}^{Bb} = \psi_{v}^{cn} - \psi_{i}^{Cc}. \tag{11.131}$$

One also notes from the above observations that for a balanced three-phase AC electrical homogeneous continuous dynamical system the power delivered to each phase of the load is the same. Thus

$$P_{A}^{p} = P_{B}^{p} = P_{C}^{p} = P_{\phi}^{p} = V_{\phi} I^{\phi} \cos \varphi_{\phi}, \tag{11.132}$$

where P_{ϕ}^{p} stands for the average or active power per phase.

Power is a scalar quantity, and so the three-phase total, average or active power is simply the average power in three phases. For this reason, the total average or active power delivered to the balanced Y-connected load is simply three times the power per phase. Thus

$$P_{T}^{p} = 3P_{\phi}^{p} = 3V_{\phi} I^{\phi} \cos \varphi_{\phi}. \tag{11.133}$$

But for a delta,

$$V_{\lambda} = V_{\phi} \text{ and } I^{\lambda} = \sqrt{3}\, I^{\phi}$$

and for a wye (Y)

$$I^{\lambda} = I^{\phi} \text{ and } V_{\lambda} = \sqrt{3}\, V_{\phi}.$$

Hence, for either it is also desirable to express the total power in terms of the effective value of the line voltage and the effective value of the line current.

If one lets V represent the effective value of the line voltage and I^{λ} represent the effective value of the line current, then one can modify equation (11.134) as follows:

$$P_{T}^{p} = 3\left(\frac{V_{\phi}}{\sqrt{3}}\right) I^{\lambda} = \sqrt{3}\, V_{\lambda} I^{\lambda} \cos \varphi_{\phi}. \tag{11.134}$$

In deriving equation (11.132), one uses the fact that for a balanced Y-connected load, the effective value of the phase voltage is the effective value of the line voltage divided by $\sqrt{3}$, and the effective value of the line current is equal to the effective value of the phase current. In using equation (11.132) to compute the total power delivered to the load, it is important to remember that φ_{ϕ} is the phase angle between the phase voltage and the phase current. Thus equation (11.132) is the basic average or active power formula for a balanced three-phase AC electrical continuous dynamical system in either wye (Y) or delta (Δ). Note that the term power factor of a three-phase AC electrical continuous dynamical system, $\cos\varphi_{\phi}$, is the 'power factor of the phase'. Thus, the term 'power factor' has no meaning for a badly unbalanced three-phase AC electrical continuous dynamical system.

If the load is Δ-connected, the computation of holor power is basically the same as for a Y-connected load. A Δ-connected load, along with the pertinent holor voltages and holor currents, is shown in figure 11.55.

From figure 11.55 it follows that the phase average or active power associated with each phase is

$$P_A^p = \left|V_i^{ab}\right|\left|I_{ab}^j\right|\cos\left(\psi_v^{ab} - \psi_i^{ab}\right), \tag{11.135}$$

where ψ_v^{ab} and ψ_i^{ab} denote the phase angles of V_{ij}^{ab} and I_{ab}^j, respectively.

Using the notation introduced in equation (11.133), one finds that powers associated with the B and C phases are

$$P_B^p = \left|V_i^{bc}\right|\left|I_{bc}^j\right|\cos\left(\psi_v^{bc} - \psi_i^{bc}\right) \tag{11.136}$$

and

$$P_C^p = \left|V_i^{ca}\right|\left|I_{ca}^j\right|\cos\left(\psi_v^{ca} - \psi_i^{ca}\right).$$

For balanced Δ-connected load,

$$V_\phi = |V_i^{ab}| = |V| = |V_i^{ca}|, \tag{11.138}$$

$$I^\phi = |I_{ab}^j| = |I_{bc}^j| = |I_{ca}^j|, \tag{11.139}$$

and

$$\varphi_\phi = \psi_v^{ab} - \psi_i^{ab} = \psi_v^{bc} - \psi_i^{bc} = \psi_v^{ca} - \psi_i^{ca}. \tag{11.140}$$

One also notes from the above observations that for a balanced three-phase AC electrical continuous dynamical system the power delivered to each phase of the load is the same. Thus

$$P_A^p = P_B^p = P_C^p = P_\phi^p = U_\phi I^\phi \cos\varphi_\phi. \tag{11.141}$$

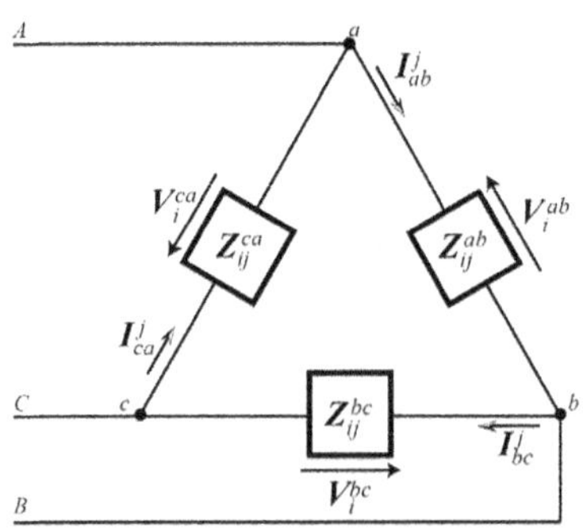

Figure 11.55. A Δ-connected load used to discuss power computations.

It is worth noting that equation (11.129) is the same as equation (11.130). This is equivalent to saying 'in a balanced load the average or active power per phase is equal to the product of the effective (rms) value of the phase voltage, the effective (rms) value of the phase current and the cosine of the angle between the phase voltage and phase current'.

The total power delivered to a balanced Δ-connected load is

$$P_T^p = 3P_\phi^p = 3V_\phi I^\phi \cos\varphi_\phi = 3V_\lambda\left(\frac{I^\lambda}{\sqrt{3}}\right)\cos\varphi_\phi = \sqrt{3}\,V_\lambda I^\lambda \cos\varphi_\phi. \tag{11.142}$$

Note that equation (11.140) is the same as equation (11.132).

For the phase real or average power P_ϕ^p a third factor, $\cos\varphi_\phi$, that is, the **power factor** (PF) enters the equation, so that

$$P_\phi^p = V_\phi I^\phi \cos\varphi_\phi,$$

where φ is the power factor angle between V_ϕ and I^ϕ. Note that these equations yield results in watts (W).

11.7.9 Delta three-phase AC electrical homogeneous continuous dynamical systems

Electrical energy is universally produced, transmitted and distributed via three-phase AC electrical homogeneous continuous dynamical systems with three or four conductors. Such an arrangement is simply three single-phase (two collectors) AC electrical homogeneous continuous dynamical systems combined. The three single-phase AC electrical homogeneous continuous dynamical systems are constrained in a certain relationship with one another.

Figure 11.56 illustrates the development of a delta (Δ), three-phase AC electrical homogeneous continuous dynamical system.

In the three single-phase AC electrical homogeneous continuous dynamical systems (figure 11.56(a)), the energy-source phase voltage-source holors $\boldsymbol{E}_i^{AN}$, $\boldsymbol{E}_i^{BN}$

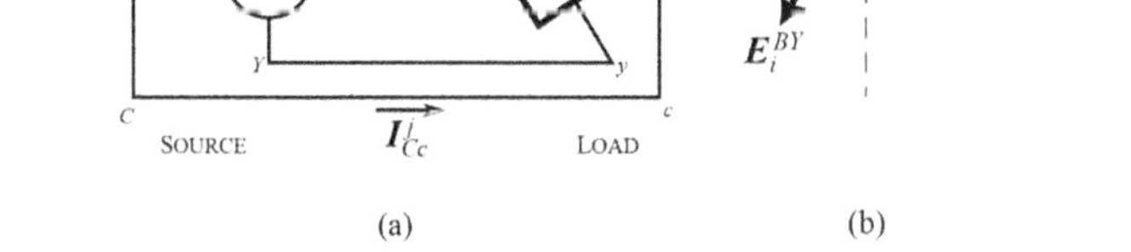

Figure 11.56. The three single-phase AC homogeneous continuous dynamical systems can be combined into a Δ three-phase AC homogeneous continuous dynamical system in (a). The voltage-source holors must be related in phase, as sketched. $\boldsymbol{E}_i^{BY}$ and $\boldsymbol{E}_i^{CZ}$ may be interchanged.

and $\boldsymbol{E}_{\mathrm{i}}^{CN}$ are equal in amplitude and 2π/3 apart in the *n* phase, as shown in the holor diagram (figure 11.56(b)).

If each energy-source phase voltage holor is equal in amplitude and if each of their load impedance holors are equal, the phase-current holors $\boldsymbol{I}_{AN}^{j}$, $\boldsymbol{I}_{BN}^{j}$ and $\boldsymbol{I}_{CN}^{j}$ will be also equal. The phase-current holors will be along the corresponding voltage holors and will consequently also be 2π/3 apart.

Suppose now that in the interest of saving conductors, the six conductors are combined into three, as shown in figure 11.57(a).

Figure 11.57(b) shows that to keep the load voltage holors uniform under unbalanced load, the wye (Y) three-phase AC electrical homogeneous continuous dynamical system requires a fourth neutral conductor. None of the three load current holors (or their corresponding current holors in the three energy sources,

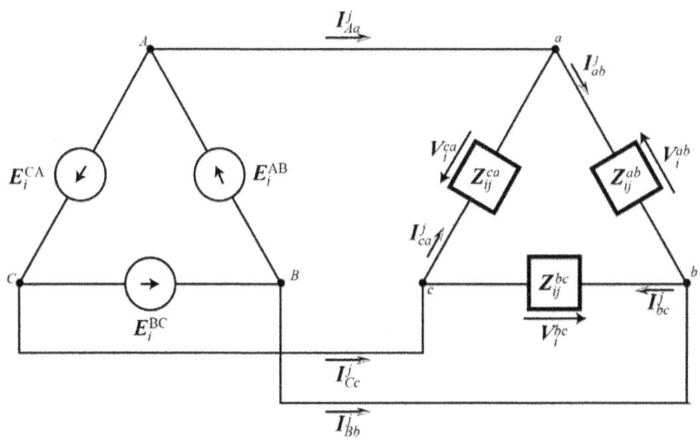

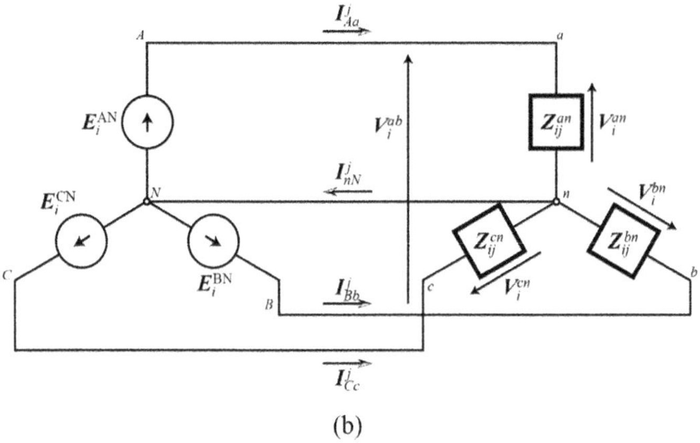

Figure 11.57. Delta (a) and wye (b) three-phase AC electrical homogeneous continuous dynamical systems. Note the carefully chosen symmetrical notation (signs and arrows) for the voltage and current holors.

which in power plants are usually the secondary windings of EE transformers) will be affected. The energy sources and loads should act as before.

It might be objected that the three energy sources $\boldsymbol{E}_i^{AB}$, $\boldsymbol{E}_i^{BC}$ and $\boldsymbol{E}_i^{CA}$ have been short-circuited on each other. However, adding three equal voltages that are $2\pi/3$ apart around the energy source loop on the left side of the circuit makes a zero sum. In practice, when these AC electrical homogeneous continuous dynamical systems are set up, great care must be exercised to ensure that one holor voltage is not connected backwards, destroying the zero sum effect. In the three-phase AC electrical homogeneous continuous dynamical system it will be noted that the 'line' current holors, labelled $\boldsymbol{I}_{Aa}^j$, $\boldsymbol{I}_{Bb}^j$ and $\boldsymbol{I}_{Cc}^j$, are each the difference of two phase current holors, for instance $\boldsymbol{I}_{Aa}^j = \boldsymbol{I}_{AN}^j - \boldsymbol{I}_{CN}^j$.

However, a carefully drawn holor diagram will show that these holors ($\boldsymbol{I}_{Aa}^j$, and $-\boldsymbol{I}_{CN}^j$), when combined, are $\pi/3$ apart, so that with three equal loads

$$\boldsymbol{I}_\lambda^j = \sqrt{3}\,\boldsymbol{I}_\phi^j. \tag{11.143}$$

Moreover, in any Δ three-phase AC electrical homogeneous continuous dynamical system,

$$V_i^\phi = V_i^\lambda. \tag{11.144}$$

Note that $\sqrt{3} = 2\cos\pi/3 = 1.732$.

Three-phase AC electrical homogeneous continuous dynamical systems have two principal advantages. Each conductor carries 1.732 times as much current as any of the six conductors of figure 11.56(a). If the same size conductors were used, the $R(I\wedge 2)$ losses would be three times as great as for any of the six conductors. However, there are only three conductors. Increasing their cross-sectional area by only 50% would reduce the losses to the original case. Thus 25% of the copper is saved. This saving of copper in AC three-phase transmission lines and electrical machines is significant.

The other advantage is even more important: electrical energy flow is smoothed. A single-phase AC electrical homogeneous continuous dynamical system with a frequency of 50 or 60 Hz power pulses 100 or 120 times a second, respectively.

Thus, all single-phase electrical machines must be constructed to withstand not only peak torques and stresses, but also a continual 100 or 120 Hz vibration.

It can be easily shown that in three-phase electrical machines the power flow is completely uniform. Electrical machines can be smaller for the same ratings; vibration problems are reduced. Three-phase AC electrical homogeneous continuous dynamical systems with equal phase loads, as shown in figure 11.57, are termed 'balanced'. Phase loads in balanced three-phase AC electrical homogeneous continuous dynamical systems must have the same in phase angle as well as magnitude. Equation (11.143) is only true for balanced delta (Δ) three-phase AC electrical homogeneous continuous dynamical systems.

11.7.10 Wye three-phase AC electrical homogeneous continuous dynamical systems

There is a second way to connect three-phase AC electrical homogeneous continuous dynamical systems; it is termed wye (Y) and illustrated in figure 11.57(b).

The central node to which all energy sources are connected (and referred) is termed the 'neutral' or N point. It is often referred to as 'ground', and is frequently actually grounded for safety reasons. Similarly, there is a neutral point in the load connection. Here the phase-current holors ($\boldsymbol{I}^j_{AN}$, $\boldsymbol{I}^j_{BN}$ and $\boldsymbol{I}^j_{CN}$) are equal to the line-current holors ($\boldsymbol{I}^j_{AB}$, $\boldsymbol{I}^j_{BC}$ and $\boldsymbol{I}^j_{CA}$). However, the line voltage holors (sometimes termed 'line-to-line' voltage holors) $\boldsymbol{E}_i^{AB}$, $\boldsymbol{E}_i^{BC}$ and $\boldsymbol{E}_{\mathrm{i}}^{CA}$ are $\sqrt{3}$ times the phase voltage holors $\boldsymbol{E}_i^{AN}$, $\boldsymbol{E}_i^{BN}$ and $\boldsymbol{E}_{\mathrm{i}}^{CN}$. If this Y three-phase AC electrical homogeneous continuous dynamical system were to have only three conductors, there would be some interesting problems with unbalanced loads.

The holor voltage drops on the load phases could adjust themselves differently for each phase (to enable KFL to be satisfied). To overcome these unbalance problems a fourth conductive conductor is usually run in Y three-phase AC electrical homogeneous continuous dynamical systems (see figure 11.57(b)).

This neutral conductive conductor carries the current holor unbalances and keeps the voltage-drop holor magnitudes equal for all load phases. Since the Y three-phase AC electrical homogeneous continuous dynamical system is designed to be quite well balanced, the neutral conductive conductor's current holor will be small, and there is no need to use a large conductor for this purpose. Moreover, since the neutral conductive conductor is run at ground potential, it has no need for expensive and bulky insulation.

Single-phase loads on Y three-phase AC electrical homogeneous continuous dynamical systems are generally taken from one hot conductor and the neutral.

The term 'hot conductor' refers to one of the three-phase conductors as opposed to the ground or neutral one.

Thus for four-conductor Y three-phase AC electrical homogeneous continuous dynamical systems, whether balanced or not, and for balanced three-conductor Y three-phase AC electrical homogeneous continuous dynamical systems,

$$\boldsymbol{V}_i^{\phi} = \sqrt{3}\,\boldsymbol{V}_i^{l} \tag{11.145a}$$

and

$$\boldsymbol{I}_{\lambda}^{j} = \boldsymbol{I}_{\phi}^{j}. \tag{11.145b}$$

While Y three-phase AC electrical homogeneous continuous dynamical systems have the disadvantage of requiring this fourth (usually smaller) conductive conductor, they have several advantages over Δ three-phase AC electrical homogeneous continuous dynamical systems. There is a natural ground in the Y three-phase AC electrical homogeneous continuous dynamical system, which is often quite helpful for single-phase loads. For a given phase voltage holor, transmission is at higher voltage holors, resulting in smaller line-current holors and further copper savings.

Older industrial Δ three-phase AC electrical homogeneous continuous dynamical systems can have their capacity increased by almost 75% by adding a fourth conductor, raising the distribution voltage holor and reconnecting load in wye (Y).

The Y and Δ three-phase AC electrical homogeneous continuous dynamical systems of figure 11.57 can be generalised as follows. All three-phase AC electrical homogeneous continuous dynamical systems comprise three holor voltages that are equal in magnitude but 2π/3 apart. Their current holors depend, however, on the nature of load.

If the three load impedance holors are equal in magnitude and phase angle, the three phase-current holors will be equal in magnitude and the three holor line currents will be equal in magnitude. This is termed a balanced three-phase AC electrical homogeneous continuous dynamical system.

Most such three-phase AC electrical homogeneous continuous dynamical systems can usually be analysed one phase at a time, or, at worst, by mesh-current holor methods. There are many single-phase loads—lights are a prime example. There are two possible orders of 'phase sequence'.

Standing at the point marked x in figure 11.58, one sees the three holors (2π/3 apart) rotating counter-clockwise. First $\boldsymbol{V}_i^{an}$ passes one, then $\boldsymbol{V}_i^{bn}$, then $\boldsymbol{V}_i^{cn}$.

The phase sequence seen is a, b, c. If $\boldsymbol{V}_i^{bn}$ and $\boldsymbol{V}_i^{cn}$ were unchanged in position, the sequence would be a, c, b. These two sequences are often termed 'positive' and 'negative', respectively.

11.7.11 Holor algebra analysis of the wye–wye three-phase AC electrical homogenous continuous dynamical system

One begins holor algebra analysis of the Y2Y three-phase AC electrical homogeneous continuous dynamical system by assuming that the system is not a balanced (symmetrical) three-phase AC electrical homogeneous continuous dynamical system.

One does this so that one can show for a three-phase AC electrical homogeneous continuous dynamical system what one means by a balanced system, and what the consequences of being balanced are in terms of the holor algebra analysis of the Y2Y three-phase AC electrical homogeneous continuous dynamical system.

The general Y2Y three-phase AC electrical homogeneous continuous dynamical system is illustrated in figure 11.59, where one has included a fourth conductor that connects the energy source neutral N to the load neutral n. The fourth conductor is only possible in the Y2Y arrangement.

We also mention that for convenience in drawing the holor diagram, one has transformed the Y-connections into 'tipped over tees'.

In figure 11.59, $\boldsymbol{Z}_{ik}^{AN}$, $\boldsymbol{Z}_{ik}^{BN}$ and $\boldsymbol{Z}_{ik}^{CN}$ represent the internal impedance holors associated with each phase winding of the energy source; $\boldsymbol{Z}_{ik}^{Aa}$, $\boldsymbol{Z}_{ik}^{Bb}$ and $\boldsymbol{Z}_{ik}^{Cc}$ represent the impedance holor associated with each phase conductor of the transmission line connecting the energy source to the load; $\boldsymbol{Z}_{ik}^{Nn}$ is the impedance holor of the neutral conductor that connects the energy-source neutral to the load neutral; and $\boldsymbol{Z}_{ik}^{an}$, $\boldsymbol{Z}_{ik}^{bn}$ and $\boldsymbol{Z}_{ik}^{cn}$ represent the impedance holors of each phase of the load.

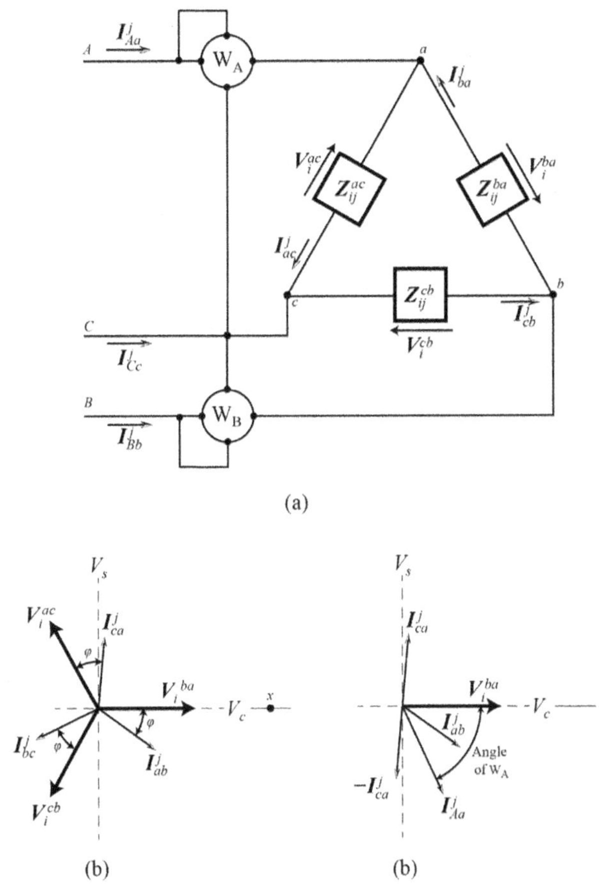

Figure 11.58. The $m - 1$ wattmeter method—two wattmeters for three lines. The sum of the wattmeter indications is the total active power in the three loads.

The general Y2Y three-phase AC electrical homogeneous continuous dynamical system in figure 11.59 can be described by a single node-voltage holor equation. Using the energy-source neutral as the reference node and letting $\boldsymbol{E}_i^{Nn}$ denote the node voltage holor between the nodes N and n, one finds that the node-voltage holor equation is

$$\begin{aligned}\frac{\delta_k^i}{\boldsymbol{Z}_{jk}^{Nn}}\boldsymbol{E}_i^{Nn} &= \frac{\delta_k^i}{\boldsymbol{Z}_{jk}^{an} + \boldsymbol{Z}_{jk}^{Aa} + \boldsymbol{Z}_{jk}^{AN}}\left(\boldsymbol{E}_i^{Nn} - \boldsymbol{E}_j^{AN}\right) \\ &+ \frac{\delta_k^i}{\boldsymbol{Z}_{jk}^{bn} + \boldsymbol{Z}_{jk}^{Bb} + \boldsymbol{Z}_{jk}^{BN}}\left(\boldsymbol{E}_i^{Nn} - \boldsymbol{E}_i^{BN}\right) \\ &+ \frac{\delta_k^i}{\boldsymbol{Z}_{jk}^{cn} + \boldsymbol{Z}_{jk}^{Cc} + \boldsymbol{Z}_{jk}^{CN}}\left(\boldsymbol{E}_i^{Nn} - \boldsymbol{E}_{-}^{CN}\right) = 0.\end{aligned} \tag{11.146}$$

Before pursuing equation (11.142) any further, let us pause to observe that the AC electrical homogeneous continuous dynamical system holor analysis techniques

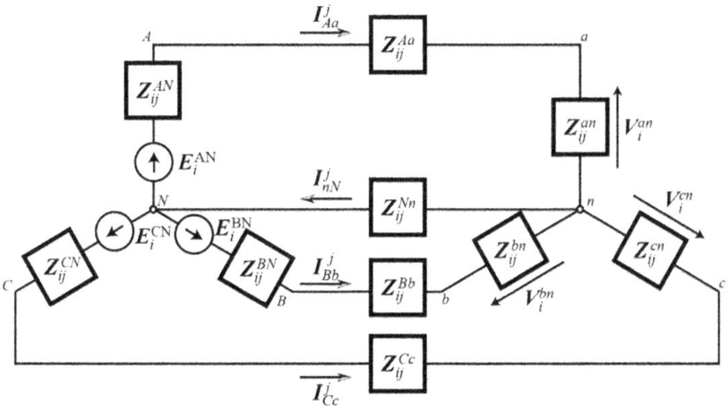

Figure 11.59. A Y2Y three-phase AC electrical homogeneous continuous dynamical system.

discussed in the earlier chapters are directly applicable to three-phase AC electrical homogeneous continuous dynamical systems. Thus, it is not necessary to introduce new holor analytical techniques in order to analyse three-phase AC electrical homogeneous continuous dynamical systems.

However, as we see in the remainder of this chapter, if a three-phase AC electrical homogenous continuous dynamical system is balanced, one can take some significant holor analytical shortcuts to predict the behaviour of the system.

The general Y2Y three-phase AC electrical homogeneous continuous dynamical system in figure 11.59 is a balanced one if it satisfies all of the following four criteria:

- $\boldsymbol{E}_i^{AN}$, $\boldsymbol{E}_i^{BN}$ and $\boldsymbol{E}_i^{CN}$ form a set of balanced three-phase energy-source winding voltage holors;
- $\boldsymbol{Z}_{jk}^{AN} = \boldsymbol{Z}_{jk}^{BN} = \boldsymbol{Z}_{jk}^{CN}$ form a set of balanced three-phase internal energy source winding impedance holors;
- $\boldsymbol{Z}_{jk}^{Aa} = \boldsymbol{Z}_{jk}^{Bb} = \boldsymbol{Z}_{jk}^{Cc}$ form a set of balanced three-phase transmission line impedance holors;
- $\boldsymbol{Z}_{jk}^{an} = \boldsymbol{Z}_{jk}^{bn} = \boldsymbol{Z}_{jk}^{cn}$ form a set of balanced three-phase load-phase impedance holors.

There is no restriction on the neutral-conductor impedance holor ($\boldsymbol{Z}_{ik}^{Nn}$), its value has no effect on whether the Y2Y three-phase AC electrical homogeneous continuous dynamical system is balanced.

If the Y2Y three-phase AC electrical homogeneous continuous dynamical system is balanced, equation (11.146) tells one that $\boldsymbol{U}_i^{Nn}$ must be zero.

To see that, let

$$\boldsymbol{Z}_{jk}^{\phi} = \boldsymbol{Z}_{jk}^{an} + \boldsymbol{Z}_{jk}^{Aa} + \boldsymbol{Z}_{jk}^{AN} \tag{11.147}$$

and then rewrite equation (11.142) as

$$\left(\frac{\delta_k^i}{\boldsymbol{Z}_{jk}^{Nn}} + 3\frac{\delta_k^i}{\boldsymbol{Z}_{jk}^{\phi}}\right)\boldsymbol{U}_i^{Nn} = \frac{\delta_k^i}{\boldsymbol{Z}_{jk}^{\phi}}\left(\boldsymbol{E}_i^{AN} + \boldsymbol{E}_i^{AN} + \boldsymbol{E}_i^{AN}\right). \tag{11.148}$$

The right-hand side of equation (11.144) is zero because by hypothesis the numerator is a set of balanced three-phase voltage holors and $\boldsymbol{Z}_{ik}^{\phi}$ is not zero. The only value of $\boldsymbol{U}_i^{Nn}$ that satisfies equation (11.144) is zero. Therefore, for a Y2Y balanced three-phase AC electrical homogeneous continuous dynamical system,

$$\boldsymbol{V}_i^{Nn} = 0. \tag{11.149}$$

Equation (11.145) is an extremely important result. If $\boldsymbol{V}_i^{Nn}$ is zero, there is no difference in potential between the energy-source neutral N and the load neutral n, and consequently the current holor in the neutral conductor is zero. These observations tell us that one can either remove the neutral conductor from a balanced Y2Y configuration ($\boldsymbol{I}_{Nn}^{j} = 0$) or replace it by a perfect short circuit between the nodes N and n ($\boldsymbol{V}_i^{Nn} = 0$).

One finds both equivalents convenient to use when physically modelling balanced three-phase AC electrical homogeneous continuous dynamical systems.

Now let us turn our attention to what effect balanced conditions have on the three transmission-line-current holors. It follows directly from figure 11.59 that when the three-phase AC electrical homogeneous continuous dynamical system is balanced, the three transmission-line-current holors will be

$$\boldsymbol{I}_{Aa}^{j} = \frac{\delta_k^i}{\boldsymbol{Z}_{jk}^{an} + \boldsymbol{Z}_{jk}^{Aa} + \boldsymbol{Z}_{jk}^{AN}}\left(\boldsymbol{E}_i^{AN} - \boldsymbol{E}_i^{Nn}\right) = \frac{\delta_k^i}{\boldsymbol{Z}_{jk}^{\phi}}\boldsymbol{E}_i^{AN} = \boldsymbol{Y}_{\phi}^{ji}\boldsymbol{E}_i^{AN}; \tag{11.150}$$

$$\boldsymbol{I}_{Bb}^{j} = \frac{\delta_k^i}{\boldsymbol{Z}_{jk}^{an} + \boldsymbol{Z}_{jk}^{Aa} + \boldsymbol{Z}_{jk}^{AN}}\left(\boldsymbol{E}_i^{BN} - \boldsymbol{E}_i^{Nn}\right) = \frac{\delta_k^i}{\boldsymbol{Z}_{jk}^{\phi}}\boldsymbol{E}_i^{BN} = \boldsymbol{Y}_{\phi}^{ji}\boldsymbol{E}_i^{BN}; \tag{11.151}$$

and

$$\boldsymbol{I}_{Cc}^{j} = \frac{\delta_k^i}{\boldsymbol{Z}_{jk}^{an} + \boldsymbol{Z}_{jk}^{Aa} + \boldsymbol{Z}_{jk}^{AN}}\left(\boldsymbol{E}_i^{CN} - \boldsymbol{E}_i^{Nn}\right) = \frac{\delta_k^i}{\boldsymbol{Z}_{jk}^{\phi}}\boldsymbol{E}_i^{BN} = \boldsymbol{Y}_{\phi}^{ji}\boldsymbol{E}_i^{BN}, \tag{11.152}$$

from which one sees that in a balanced, Y2Y three-phase AC electrical homogeneous continuous dynamical system the three transmission-line-current holors form a balanced set of three phase-current holors. Thus, a current holor in each transmission-line will be equal in amplitude and frequency and will be exactly $2\pi/3$ out of phase with the other two transmission-line-current holors. This tells us that if one computes the current holor $\boldsymbol{I}_{Aa}^{j}$, one can write down the transmission-line-current holors $\boldsymbol{I}_{Bb}^{j}$ and $\boldsymbol{I}_{Cc}^{j}$ without further computation. One implies by this statement that the phase sequence is known.

One can use equation (11.146) to construct an equivalent single-phase AC electrical continuous dynamical system of the balanced Y2Y three-phase AC electrical homogeneous continuous dynamical system.

It follows from equation (11.146) that the current holor in the A-phase conductor line is simply the voltage-source holor generated in the A-phase armature winding of the AC EM generator divided by the total impedance holor in the A phase of the Y2Y three-phase AC electrical homogeneous continuous dynamical system.

Thus, equation (11.146) describes the simple single-phase AC electrical homogeneous continuous dynamical system shown in figure 11.60, where a perfect short circuit has replaced the neutral conductor. Here, we offer a word of caution.

The current holor in the neutral conductor of figure 11.60 is not the current holor in the neutral conductor of a balanced three-phase AC electrical homogeneous continuous dynamical system.

The current holor in the neutral conductor is

$$I^j_{Nn} = I^j_{Aa} + I^j_{Bb} + I^j_{Cc}, \tag{11.153}$$

whereas the current holor of the neutral conductor in figure 11.60 is I^j_{Aa}.

Thus, the AC electrical homogeneous continuous dynamical system in figure 11.59 gives the correct value of the transmission-line-current holor, but only the A-phase component of the neutral current holor.

Whenever the equivalent single-phase AC electrical homogeneous continuous dynamical system in figure 11.60 is applicable, the transmission-line-current holors form a balanced three-phase set and the right-hand side of equation (11.149) sums to zero.

Once one knows the transmission-line-current holors in the AC electrical homogeneous continuous dynamical system in figure 11.59, it is a relatively simple task to compute any voltage holors that are of interest. In particular, one is interested in the relationship between the line-to-line voltage holors and the line-to-neutral voltage holors. One will establish this relationship at the load terminals. The observations one makes will also apply at the energy-source terminals. The line-to-line voltage holors at the terminals of the load in terms of the line-to-neutral voltage holors at the three-phase load are

$$V^{ab}_i = V^{an}_i - V^{bn}_i; \tag{11.154}$$

$$V^{bc}_i = V^{bn}_i - V^{cn}_i; \tag{11.155}$$

and

$$V^{ca}_i = V^{cn}_i - V^{an}_i. \tag{11.156}$$

The double-subscript notation in the voltage holor equations indicates that the voltage holor is a drop from the first subscript to the second subscript.

The relationships given by equations (11.150)–(11.152) are shown in figure 11.61. Because one is interested in the balanced state, one has to omit the neutral conductor

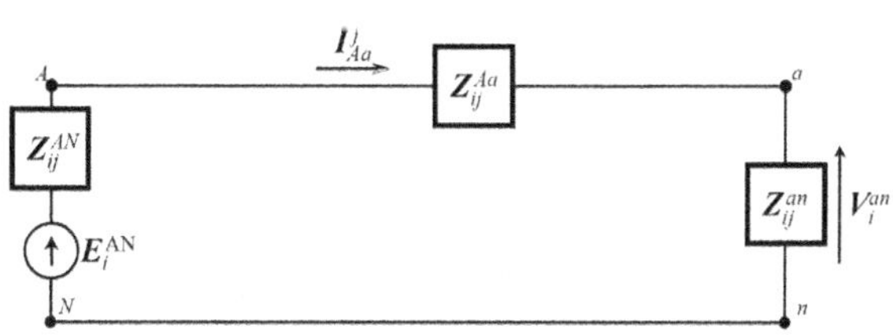

Figure 11.60. An equivalent single-phase AC electrical homogeneous continuous dynamical system.

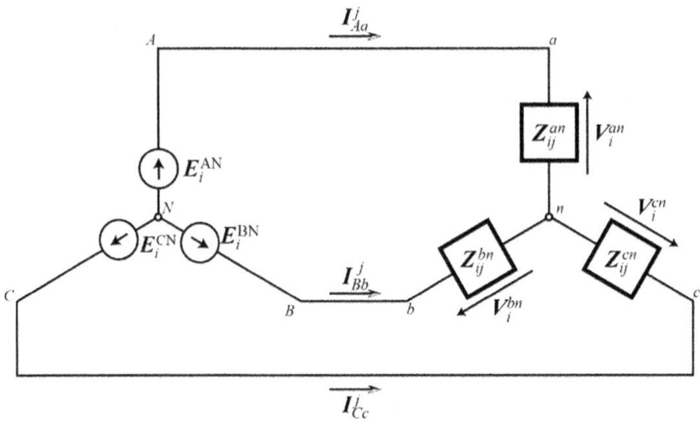

Figure 11.61. Line-to-line and line-to-neutral voltage holors.

from figure 11.61. To show the relationship between the line-to-line voltage holors and the line-to-neutral voltage holors, one assumes a negative, or *acb*, sequence.

One arbitrarily chooses the line-to-neutral voltage holor as the reference. Thus,

$$V_i^{an} = V_\phi \angle 0; \tag{11.157}$$

$$V_i^{bn} = V_\phi \angle 2\pi/3; \tag{11.158}$$

and

$$V_i^{cn} = V_\phi \angle 4\pi/3 = V_\phi \angle - 2\pi/3, \tag{11.159}$$

where U_Φ represents an effective value of the line-to-neutral voltage holor.

When one substitutes equations (11.153)–(11.155) into equations (11.150)–(11.152), respectively, one obtains

$$V_i^{ab} = V_i^{an} - V_i^{bn} = V_\phi \angle 0 - V_\phi \angle \frac{2\pi}{3} = \sqrt{3}\, V_\phi \angle \frac{11\pi}{6}.; \tag{11.160}$$

$$V_i^{bc} = V_i^{bn} - V_i^{cn} = V_\phi \angle \frac{2\pi}{3} - V_\phi \angle \frac{4\pi}{3} = \sqrt{3}\, V_\phi \angle \frac{\pi}{2}; \tag{11.161}$$

and

$$V_i^{ca} = V_i^{cn} - V_i^{an} = V_\phi \angle 4\pi/3 - V_\phi \angle 0 = \sqrt{3}\, V_\phi \angle 7\pi/6. \tag{11.162}$$

From equations (11.156)–(11.158), one sees that:

- the effective value of the line-to-line voltage holor is $\sqrt{3}$ times the effective value of the line-to-neutral voltage holor;
- the line-to-line voltage holors form a balanced three-phase set of voltage holors;
- the set of line-to-line voltage holors lags the set of line-to-neutral voltage holors by $\pi/6$.

These observations are summarised in the holor diagram of figure 11.62.

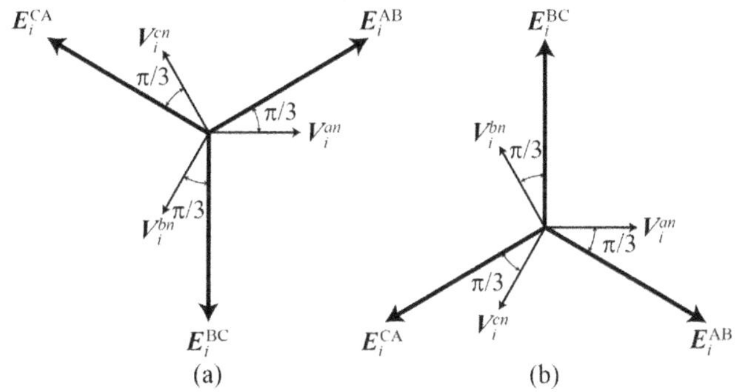

Figure 11.62. Holor diagrams showing the relationship between line-to-line voltage holors and line-to-neutral voltage holors in a balanced Y–Y three-phase AC electrical homogeneous continuous dynamical system: (a) *acb* sequence and (b) *abc* sequence.

One now observes that if the line-to-neutral voltage holor is known at some point in the balanced Y2Y three-phase AC electrical homogeneous continuous dynamical system, the line-to-line voltage holors at the same point are also known, and vice versa. In addition, one must make some additional comments on terminology. In the Y2Y three-phase AC electrical homogeneous continuous dynamical system, the line-to-neutral voltage holor is also termed the 'phase voltage holor'. For concision, the line-to-line voltage holor is also termed the 'line voltage holor'.

The 'phase-current holor' is defined as the current holor in each phase of the load or, at the energy-source end of the AC electrical homogeneous continuous dynamical system, the current holor in each phase of the AC EM generator.

The 'line-current holor' is defined as the current holor of each phase of the transmission line. For the Y–Y arrangement, the phase-current holor and line-current holor are identical. Because three-phase AC electrical homogeneous continuous dynamical systems are designed to handle large quantities of electrical energy, all voltage-holor and current-holor specifications and computations are given in terms of effective values or **root-mean-square** (rms) values. Finally, the Greek letter phi (Φ) is usually used in the literature to denote a per-phase quantity.

11.7.12 Holor algebra analysis of the wye-delta three-phase AC electrical homogeneous continuous dynamical system

If the load in a three-phase AC electrical homogeneous continuous dynamical system is connected in a **delta** (Δ), it can be transformed into a **wye** (Y) by using the Δ2Y transformation discussed in section 11.7. When the load is balanced, the impedance holor of each leg of the Y-load is one-third of the impedance holor of the Δ load. Thus

$$\boldsymbol{Z}_{ij}^{Y} = \frac{1}{3}\boldsymbol{Z}_{ij}^{\Delta}, \tag{11.163}$$

which follows directly from equations (11.166)–(11.168).

Once the Δ load has been replaced by its Y equivalent, the Y-energy-source, Δ-load, three-phase AC electrical homogeneous continuous dynamical system can be physically modelled by the equivalent single-phase AC electrical homogeneous continuous dynamical system in figure 11.60.

After one has the equivalent single-phase AC electrical homogeneous continuous dynamical system to compute the line or line-to-line-current holors, one can find the current holor in each leg of the original Δ load by simply dividing the line or line-to-line-current holors by $\sqrt{3}$ and shifting them through π/6.

This relationship between line or line-to-line-current holors and phase or line-to-neutral current holors in the Δ load can be derived using the AC electrical homogeneous continuous dynamical system in figure 11.63.

When a load or energy source is connected in a delta (Δ), the current in each leg of the delta is the phase current and the voltage across each leg is the phase voltage. One can see from figure 11.63 that in the Δ configuration, the phase voltage is identical to the line voltage. To see the relationship between the phase-current holors and the line-current holors, one will assume a negative phase sequence and let I_ϕ represent an effective value of the phase current. It follows, then, that

$$\boldsymbol{I}^j_{AB} = I_\phi \angle 0; \tag{11.164}$$

$$\boldsymbol{I}^j_{BC} = I_\phi \angle 2\pi/3; \tag{11.165}$$

and

$$\boldsymbol{I}^j_{CA} = I_\phi \angle 4\pi/3 = I_\phi \angle -2\pi/3. \tag{11.166}$$

In writing these equations, one has arbitrarily selected $\boldsymbol{I}^n_{AB}$ as the reference current holor. One can write the line-current holors in terms of the phase-current holors by direct application of KFL. Thus

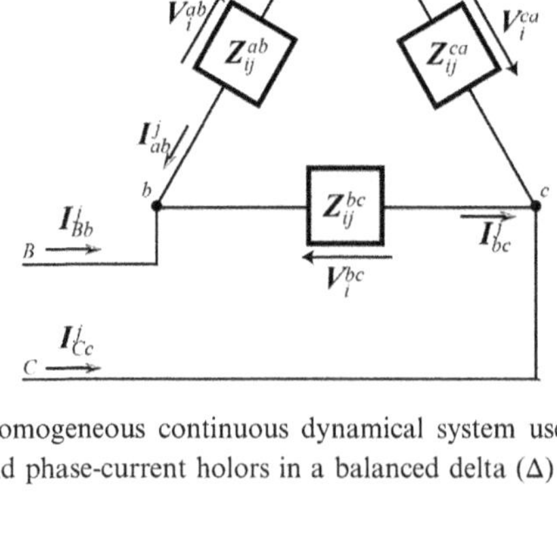

Figure 11.63. AC electrical homogeneous continuous dynamical system used to establish the relationship between line-current holors and phase-current holors in a balanced delta (Δ) load.

$$I^j_{Aa} = I^j_{AB} - I^j_{CA} = I_\phi\angle 0 - I_\phi\angle 4\pi/3 = \sqrt{3}I_\phi\angle\pi/6; \qquad (11.167)$$

$$I^j_{Bb} = I^j_{BC} - I^j_{AB} = I_\phi\angle 2\pi/3 - I_\phi\angle 0 = \sqrt{3}I_\phi\angle 7\pi/6; \qquad (11.168)$$

and

$$I^j_{Cc} = I^j_{CA} - I^j_{BC} = I_\phi\angle 4\pi/3 - I_\phi\angle 2\pi/3 = \sqrt{3}I_\phi\angle 11\pi/6. \qquad (11.169)$$

When one compares equations (11.163)–(11.165) with equations (11.160)–(11.162), one sees that an effective value of the line-current holors is $\sqrt{3}$ times an effective value of the phase-current holors, and the set of line-current holors 'leads' the set of phase-current holors by π/6. One will leave it to the reader to verify that for a positive phase sequence, the line-current holors are $\sqrt{3}$ times larger than the phase-current holors and 'lag' the phase-current holors by π/6. The relationship between the line-current holors and the phase-current holors of a Δ-connected load is summarised in figure 11.64.

11.7.13 Holor algebra analysis of the delta–wye three-phase AC electrical homogenerous continuous dynamical system

In the Δ2Y three-phase AC electrical homogeneous continuous dynamical system, the energy source is Δ connected and the load is Y connected. One can obtain the equivalent single-phase AC electrical homogeneous continuous dynamical system by replacing the balanced Δ-connected energy source by a Y equivalent. One can obtain the Y equivalent of the energy source by dividing the internal phase voltage holors of the Δ energy source by $\sqrt{3}$ and shifting this set of three-phase voltage holors by π/6 if the phase sequence is negative and by – π/6 if the phase sequence is positive. The internal impedance holor of the Y equivalent is one-third of the internal impedance holor of the Δ energy source.

The Y-equivalent AC electrical homogeneous continuous dynamical system of a positive-sequence Δ-connected energy source is illustrated in figure 11.65.

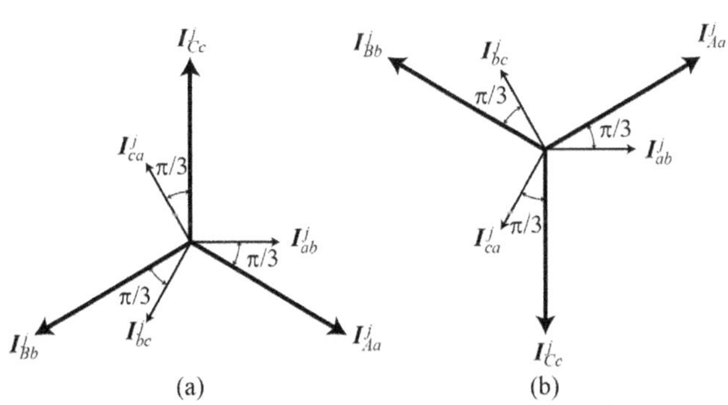

Figure 11.64. Holor diagrams showing the relationship between line-current holors and phase-current holors in a delta-connected load. (a) Negative sequence. (b) Positive sequence.

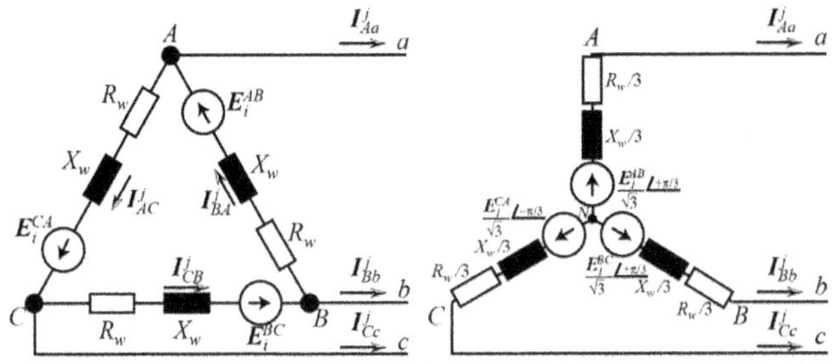

Figure 11.65. The Y equivalent of a balanced three-phase Δ-connected energy source of a positive phase sequence. (a) Δ energy source and (b) Y equivalent.

For a positive phase sequence, the set of Δ energy-source current holors (I^j_{BA}, I^j_{CB} and I^j_{AC} in figure 11.65) leads the set of line or line-to-line-current holors I^j_{Aa}, I^j_{Bb} and I^j_{Cc} by π/6. An effective value of the phase or line-to-neutral currents is $1/\sqrt{3}$ times an effective value of the line or line-to-line currents. For a negative phase sequence, the phase or line-to-neutral currents in the energy source lag the line or line-to-line currents. To show that the Y energy source of figure 11.65(b) is equivalent to the Δ energy source of figure 11.65(a), it is only necessary to show that the two AC electrical homogeneous continuous dynamical systems produce the same terminal conditions for any balanced 'external' connections applied to the terminals *A, B* and *C.* The two test conditions that are easiest to prove are open circuit and short circuit. For open-circuit conditions, the three line-current holors are zero and the two AC electrical homogeneous continuous dynamical systems are equivalent if they deliver the same voltage holors between the terminals *A, B* and *C.* For an external short circuit connecting the terminals *A, B* and *C*, the line or line-to-line voltage holors are zero and the two AC electrical homogeneous continuous dynamical systems are equivalent if they deliver the same line or line-to-line-current holors.

11.7.14 Holor algebra analysis of the delta–delta three-phase AC electrical homogeneous continuous dynamical system

In the Δ–Δ three-phase AC electrical homogeneous continuous dynamical system, both the energy source and the load are Δ connected. The equivalent single-phase AC electrical homogeneous continuous dynamical system of balanced Δ2Δ three-phase AC electrical continuous dynamical system is obtained by replacing both the energy source and the load with their Y equivalents. As before, the Y-equivalent AC electrical homogeneous continuous dynamical system is used to solve for line or line-to-line-current holors and phase line-to-neural voltage holors. Once one knows the line or line-to-line-current holors, one can find the phase or line-to-neutral current holors in both the load and the energy source using the holor techniques described in sections 11.6 and 11.7. The phase or line-to-neutral voltage holors can be converted to line or line-to-line voltage holors, as described in section 11.5.

11.7.15 Delta-to-wye transformations

The rules of combining impedance holors for making **delta-to-wye** ($\Delta 2Y$) transformations are the same as those for resistor holors. The only difference is that combining holor impedances involves the algebraic manipulation of hypernumbers, i.e. holors. The $\Delta 2Y$ discussed in section 11.7 concerning resistive AC electrical homogeneous continuous dynamical systems also applies to impedance holors. The Δ-connected impedance holors along with the Y-equivalent AC electrical homogeneous continuous dynamical system are defined in figure 11.66.

The Y-impedance holors as functions of the Δ-impedance holors are given by the following:

$$\boldsymbol{Z}_{ij}^{an} = \frac{\delta_k^i}{\boldsymbol{Z}_{jk}^{bc} + \boldsymbol{Z}_{jk}^{ca} + \boldsymbol{Z}_{jk}^{ab}} \boldsymbol{Z}_{ij}^{ca} \boldsymbol{Z}_{ij}^{ab}; \qquad (11.170)$$

$$\boldsymbol{Z}_{ij}^{bn} = \frac{\delta_k^i}{\boldsymbol{Z}_{jk}^{bc} + \boldsymbol{Z}_{jk}^{ca} + \boldsymbol{Z}_{jk}^{ab}} \boldsymbol{Z}_{ij}^{ab} \boldsymbol{Z}_{ij}^{bc}; \qquad (11.171)$$

$$\boldsymbol{Z}_{ij}^{cn} = \frac{\delta_k^i}{\boldsymbol{Z}_{jk}^{bc} + \boldsymbol{Z}_{jk}^{ca} + \boldsymbol{Z}_{jk}^{ab}} \boldsymbol{Z}_{ij}^{bc} \boldsymbol{Z}_{ij}^{ca}. \qquad (11.172)$$

It is also possible to reverse the $\Delta 2Y$ transformation. That is, one can start with the Y structure and replace it with an equivalent Δ structure.

The Δ-impedance holors as functions of the Y-impedance holors are given by the following:

$$\boldsymbol{Z}_{ij}^{ab} = \frac{\delta_k^i}{\boldsymbol{Z}_{jk}^{cn}} \left(\boldsymbol{Z}_{ij}^{an} \boldsymbol{Z}_{ij}^{bn} + \boldsymbol{Z}_{ij}^{bn} \boldsymbol{Z}_{ij}^{cn} + \boldsymbol{Z}_{ij}^{cn} \boldsymbol{Z}_{ij}^{an} \right); \qquad (11.173)$$

$$\boldsymbol{Z}_{ij}^{bc} = \frac{\delta_k^i}{\boldsymbol{Z}_{jk}^{an}} \left(\boldsymbol{Z}_{ij}^{an} \boldsymbol{Z}_{ij}^{bn} + \boldsymbol{Z}_{ij}^{bn} \boldsymbol{Z}_{ij}^{cn} + \boldsymbol{Z}_{ij}^{cn} \boldsymbol{Z}_{ij}^{an} \right); \qquad (11.174)$$

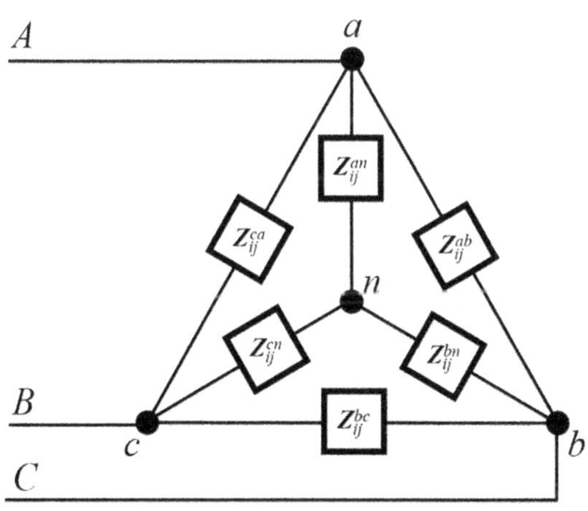

Figure 11.66. The delta-wye ($\Delta 2Y$) transformation.

$$Z_{ij}^{ca} = \frac{\delta_k^i}{Z_{jk}^{bn}}\left(Z_{ij}^{an}Z_{ij}^{bn} + Z_{ij}^{bn}Z_{ij}^{cn} + Z_{ij}^{cn}Z_{ij}^{an}\right). \tag{11.175}$$

The derivation of equations (11.166)–(11.168) or equations (11.169)–(11.171) follows the same thought process as was used to derive the corresponding equations for pure resistive AC electrical homogeneous continuous dynamical systems. In fact, if one compares equations (11.77)–(11.79) with equations (11.166)–(11.168), and equations (11.80)–(11.82) with equations (11.169)–(11.171), one can see that the resistance holor $\boldsymbol{R}_{ij}$ has been replaced by the impedance holor $\boldsymbol{Z}_{ij}$. The usefulness of the Δ-Y transformation in holor-domain AC electrical homogeneous continuous dynamical system analysis will be illustrated in some examples below.

11.7.16 Repetitive example I

In section 11.4, one discussed the simple electrical energy distribution system illustrated in figure 11.67. This is termed a 'single-phase energy distribution system'. It utilises a single energy source, a single-phase ME generator, or an EE transformer secondary.

We shall now discuss an electrical energy distribution system that utilises three energy sources, or a three-phase ME AC generator, or EE AC transformer secondaries that differ in phase from each other by 2π/3. This is termed a 'three-phase AC electrical homogeneous continuous dynamical system'.

A physical model for a three-phase ME generator or EE transformer secondaries is shown in figure 11.68.

For the most part, three-phase EM generators and EE transformers are constructed as one single ME generator or EE transformer. However, figure 11.68 is an accurate physical model for an ideal three-phase ME generator or EE transformer secondaries.

In figure 11.68 one shows the transmission line leaving the ME generator or EE transformer secondaries. The conductors *A–a, B–b* and *C–c* are termed the lines, while the conductor *N–n* is termed 'neutral'. One can talk of line or line-to-line energy-source voltages and phase or line-to-neutral energy-source voltages. Let us compute these holor values.

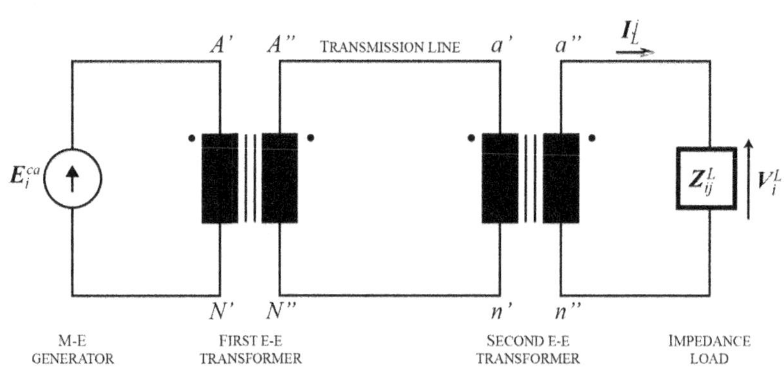

Figure 11.67. A simplified electrical energy distribution system.

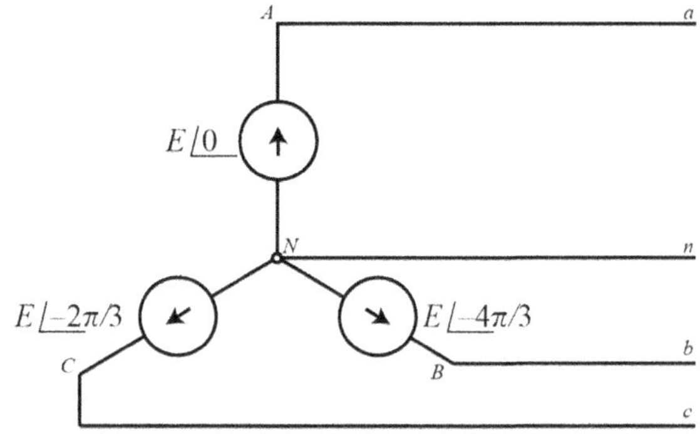

Figure 11.68. A physical model of a three-phase ME generator or EE transformer secondaries.

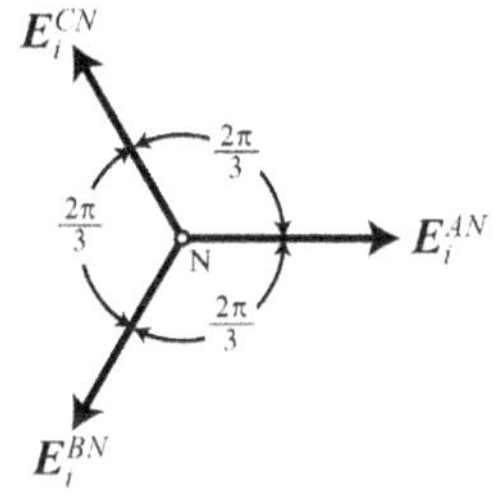

Figure 11.69. A holor representation of the holor phase or line-to-neutral energy-source voltages.

The phase or line-to-neutral energy-source voltage holors for a positive phase sequence are

$$\left.\begin{aligned} \boldsymbol{E}_i^{AN} &= E_\phi \angle 0 \\ \boldsymbol{E}_i^{BN} &= E_\phi \angle - 2\pi/3 \\ \boldsymbol{E}_i^{CN} &= E_\phi \angle - 4\pi/3 = E_\phi \angle + 2\pi/3 \end{aligned}\right\}. \tag{11.176}$$

A holor diagram representing these voltage holors is shown in figure 11.69. Note that three voltage holors differ in phase by $2\pi/3$.

Now let us compute the line or line-to-line energy-source voltage holors.

$$\boldsymbol{E}_i^{AB} = \boldsymbol{E}_i^{AN} - \boldsymbol{E}_i^{BN} = E_\varnothing - E_\varnothing \angle - 2\pi/3. \tag{11.177}$$

Proceeding similarly, one has

$$\left.\begin{aligned} \boldsymbol{E}_i^{AB} &= \sqrt{3}\, E_\varnothing \angle \pi/6 \\ \boldsymbol{E}_i^{BC} &= \sqrt{3}\, E_\varnothing \angle \pi/6 - 2\pi/3 \\ \boldsymbol{E}_i^{CA} &= \sqrt{3}\, E_\varnothing \angle \pi/6 - 4\pi/3 \end{aligned}\right\}. \tag{11.178}$$

The line or line-to-line energy-source voltage holors are also three-phase voltage holors. That is, they are equal and their phase angles differ by $2\pi/3$.

The line or line-to-line energy-source voltage holors are displaced by $\pi/6$ from the phase or line-to-neutral energy-source voltage holors. The line or line-to-line energy-source voltage holors are $\sqrt{3}$ times as large as the phase or line-to-neutral energy-source voltage holors.

Let us now analyse the three-phase AC electrical homogeneous continuous dynamical system of figure 11.70. This represents a three-phase AC generator or EE AC transformer secondaries connected to a three-phase load. The internal energy-source impedance holors $\boldsymbol{Z}_{ij}^{AN}$, $\boldsymbol{Z}_{ij}^{BN}$ and $\boldsymbol{Z}_{ij}^{CN}$ are included to physically model the three-phase AC generator losses or the losses of the EE AC transformer secondaries.

The three-phase load consists of impedance holors $\boldsymbol{Z}_{ij}^{an}$, $\boldsymbol{Z}_{ij}^{bn}$ and $\boldsymbol{Z}_{ij}^{cn}$, and this is termed a Y-connected three-phase load. Note that its configuration resembles the letter Y (which is upside down in this case).

The impedance holors $\boldsymbol{Z}_{ij}^{Aa}$, $\boldsymbol{Z}_{ij}^{Bb}$, $\boldsymbol{Z}_{mn}^{Cc}$ and $\boldsymbol{Z}_{ij}^{Nn}$ are included to physically model the transmission-line losses.

If one applies mesh-current holor analysis to this three-phase AC electrical homogeneous continuous dynamical system, one obtains

$$\left.\begin{aligned}
\boldsymbol{E}_i^{AN} &= E_\phi\angle 0 = \left(\boldsymbol{Z}_{ij}^{AN} + \boldsymbol{Z}_{ij}^{Aa} + \boldsymbol{Z}_{ij}^{an} + \boldsymbol{Z}_{ij}^{Nn}\right)\boldsymbol{I}_{Aa}^{j} + \boldsymbol{Z}_{ij}^{Nn}\boldsymbol{I}_{Bb}^{j} + \boldsymbol{Z}_{ij}^{Nn}\boldsymbol{I}_{Cc}^{j}\\
\boldsymbol{E}_i^{BN} &= E_\phi\angle -\frac{2\pi}{3} = \boldsymbol{Z}_{ij}^{Nn}\boldsymbol{I}_{Aa}^{j} + \left(\boldsymbol{Z}_{ij}^{BN} + \boldsymbol{Z}_{ij}^{Bb} + \boldsymbol{Z}_{ij}^{bn} + \boldsymbol{Z}_{ij}^{Nn}\right)\boldsymbol{I}_{Bb}^{j} + \boldsymbol{Z}_{ij}^{Nn}\boldsymbol{I}_{Cc}^{j}\\
\boldsymbol{E}_i^{CN} &= E_\phi\angle -\frac{4\pi}{3} = \boldsymbol{Z}_{ij}^{Nn}\boldsymbol{I}_{Aa}^{j} + \boldsymbol{Z}_{ij}^{Nn}\boldsymbol{I}_{Bb}^{j} + \left(\boldsymbol{Z}_{ij}^{CN} + \boldsymbol{Z}_{ij}^{Cc} + \boldsymbol{Z}_{ij}^{cn} + \boldsymbol{Z}_{ij}^{Nn}\right)\boldsymbol{I}_{Cc}^{j}
\end{aligned}\right\}. \tag{11.179a}$$

Equations (11.179a) can also be written in matrix notation as follows:

$$\begin{bmatrix}\boldsymbol{E}_i^{AN}\\ \boldsymbol{E}_i^{BN}\\ \boldsymbol{E}_i^{CN}\end{bmatrix} = \begin{bmatrix}\boldsymbol{Z}_{ij}^{AN} + \boldsymbol{Z}_{ij}^{Aa} + \boldsymbol{Z}_{ij}^{an} + \boldsymbol{Z}_{ij}^{Nn} & \boldsymbol{Z}_{ij}^{Nn} & \boldsymbol{Z}_{ij}^{Nn}\\ \boldsymbol{Z}_{ij}^{Nn} & \boldsymbol{Z}_{ij}^{BN} + \boldsymbol{Z}_{ij}^{Bb} + \boldsymbol{Z}_{ij}^{bn} + \boldsymbol{Z}_{ij}^{Nn} & \boldsymbol{Z}_{ij}^{Nn}\\ \boldsymbol{Z}_{ij}^{Nn} & \boldsymbol{Z}_{ij}^{Nn} & \boldsymbol{Z}_{ij}^{CN} + \boldsymbol{Z}_{ij}^{Cc} + \boldsymbol{Z}_{ij}^{cn} + \boldsymbol{Z}_{ij}^{Nn}\end{bmatrix}\begin{bmatrix}\boldsymbol{I}_{Aa}^{j}\\ \boldsymbol{I}_{Bb}^{j}\\ \boldsymbol{I}_{Cc}^{j}\end{bmatrix}, \tag{11.179b}$$

or

$$\begin{bmatrix}\boldsymbol{E}_i^{AN}\\ \boldsymbol{E}_i^{BN}\\ \boldsymbol{E}_i^{CN}\end{bmatrix} = \begin{bmatrix}\boldsymbol{Z}_{ij}^{\phi AA} & \boldsymbol{Z}_{ij}^{Nn} & \boldsymbol{Z}_{ij}^{Nn}\\ \boldsymbol{Z}_{ij}^{Nn} & \boldsymbol{Z}_{ij}^{\phi BB} & \boldsymbol{Z}_{ij}^{Nn}\\ \boldsymbol{Z}_{ij}^{Nn} & \boldsymbol{Z}_{ij}^{Nn} & \boldsymbol{Z}_{ij}^{\phi CC}\end{bmatrix}\begin{bmatrix}\boldsymbol{I}_{Aa}^{j}\\ \boldsymbol{I}_{Bb}^{j}\\ \boldsymbol{I}_{Cc}^{j}\end{bmatrix}, \tag{11.179c}$$

where

$$\boldsymbol{Z}_{ij}^{AN} + \boldsymbol{Z}_{ij}^{Aa} + \boldsymbol{Z}_{ij}^{an} + \boldsymbol{Z}_{ij}^{Nn} = \boldsymbol{Z}_{ij}^{\phi AA},$$

$$\boldsymbol{Z}_{ij}^{BN} + \boldsymbol{Z}_{ij}^{Bb} + \boldsymbol{Z}_{ij}^{bn} + \boldsymbol{Z}_{ij}^{Nn} = \boldsymbol{Z}_{ij}^{\phi BB},$$

and

$$\boldsymbol{Z}_{ij}^{CN} + \boldsymbol{Z}_{ij}^{Cc} + \boldsymbol{Z}_{ij}^{cn} + \boldsymbol{Z}_{ij}^{Nn} = \boldsymbol{Z}_{ij}^{\phi CC}.$$

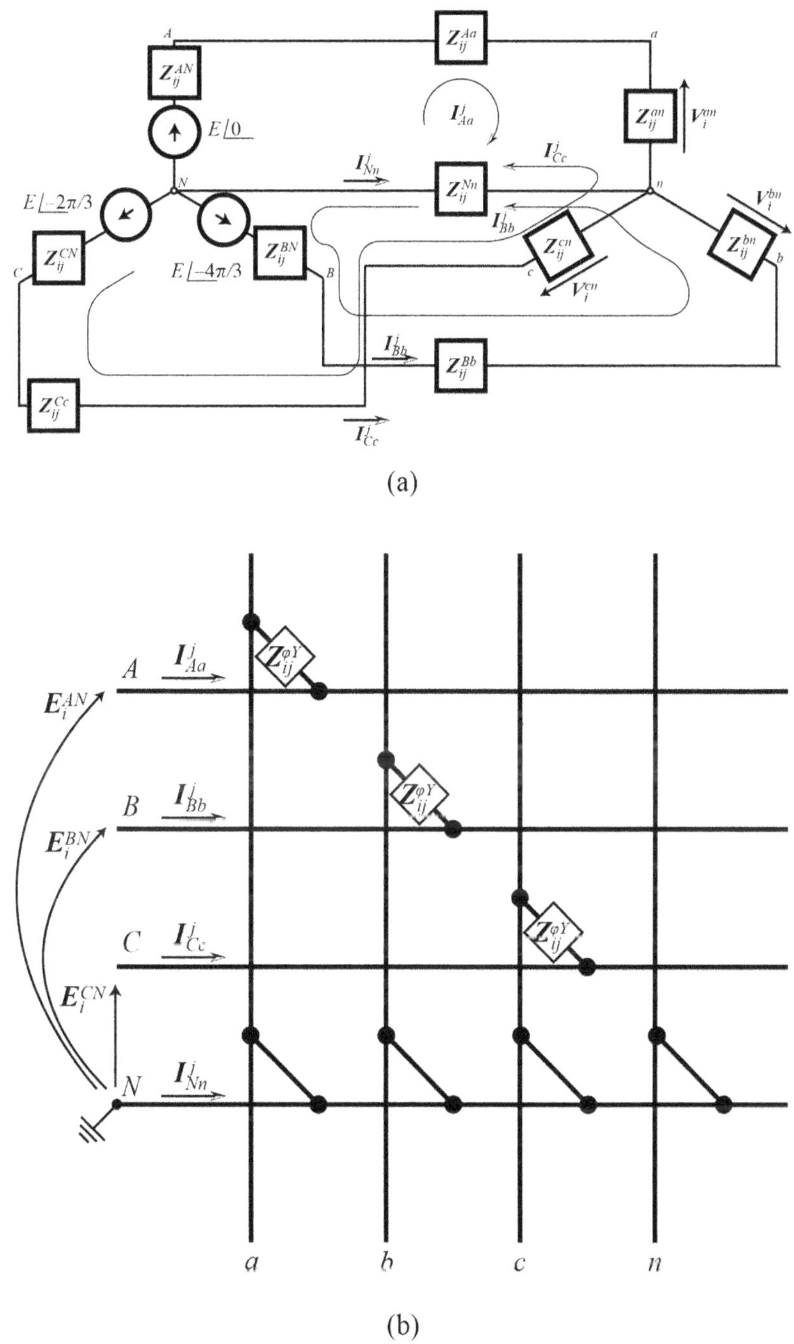

Figure 11.70. A three-phase AC electrical homogeneous continuous dynamical system with a wye (Y)-connected three-phase load.

These can be solved in the same way as any of the sets of simultaneous holor equations that we have considered. An especially interesting case occurs when the three-phase AC electrical homogeneous continuous dynamical system is balanced.

In this case, one assumes that

$$\boldsymbol{Z}_{ij}^{an} = \boldsymbol{Z}_{ij}^{bn} = \boldsymbol{Z}_{ij}^{cn} = \boldsymbol{Z}_{ij}^{\varphi Y}. \tag{11.180}$$

In addition, one assumes that the three internal energy-source impedance holors and line impedance holors, as well as the neutral impedance holor, respectively, are equal.

$$\boldsymbol{Z}_{ij}^{AN} = \boldsymbol{Z}_{ij}^{BN} = \boldsymbol{Z}_{ij}^{CN} = \boldsymbol{Z}_{ij}^{\gamma}. \tag{11.181a}$$

$$\boldsymbol{Z}_{ij}^{Aa} = \boldsymbol{Z}_{ij}^{Bb} = \boldsymbol{Z}_{ij}^{Cc} = \boldsymbol{Z}_{ij}^{\lambda}, \text{ and } \boldsymbol{Z}_{ij}^{Nn} = \boldsymbol{Z}_{ij}^{\nu}. \tag{11.181b}$$

Equations (11.179a) for the Y–Y balanced three-phase AC electrical homogeneous continuous dynamical system now become

$$\left.\begin{aligned}
\boldsymbol{E}_i^{AN} &= E_\phi\angle 0 = \left(\boldsymbol{Z}_{ij}^{\gamma} + \boldsymbol{Z}_{ij}^{\lambda} + \boldsymbol{Z}_{ij}^{\varphi Y} + \boldsymbol{Z}_{ij}^{\nu}\right)\boldsymbol{I}_{Aa}^{j} + \boldsymbol{Z}_{ij}^{\nu}\boldsymbol{I}_{Bb}^{j} + \boldsymbol{Z}_{ij}^{\nu}\boldsymbol{I}_{Cc}^{j} \\
\boldsymbol{E}_i^{BN} &= E_\phi\angle -\frac{2\pi}{3} = \boldsymbol{Z}_{ij}^{\nu}\boldsymbol{I}_{Aa}^{j} + \left(\boldsymbol{Z}_{ij}^{\gamma} + \boldsymbol{Z}_{ij}^{\lambda} + \boldsymbol{Z}_{ij}^{\varphi Y} + \boldsymbol{Z}_{ij}^{\nu}\right)\boldsymbol{I}_{Bb}^{j} + \boldsymbol{Z}_{ij}^{\nu}\boldsymbol{I}_{Cc}^{j} \\
\boldsymbol{E}_i^{CN} &= E_\phi\angle -\frac{4\pi}{3} = \boldsymbol{Z}_{ij}^{\nu}\boldsymbol{I}_{Aa}^{j} + \boldsymbol{Z}_{ij}^{\nu}\boldsymbol{I}_{Bb}^{j} + \left(\boldsymbol{Z}_{ij}^{\gamma} + \boldsymbol{Z}_{ij}^{\lambda} + \boldsymbol{Z}_{ij}^{\varphi Y} + \boldsymbol{Z}_{ij}^{\nu}\right)\boldsymbol{I}_{Cc}^{j}
\end{aligned}\right\}. \tag{11.182a}$$

Equations (11.182a) can also be written in matrix notation as follows:

$$\begin{bmatrix} \boldsymbol{E}_i^{AN} \\ \boldsymbol{E}_i^{BN} \\ \boldsymbol{E}_i^{CN} \end{bmatrix} = \begin{bmatrix} \boldsymbol{Z}_{ij}^{\gamma} + \boldsymbol{Z}_{ij}^{\lambda} + \boldsymbol{Z}_{ij}^{\varphi Y} + \boldsymbol{Z}_{ij}^{\nu} & \boldsymbol{Z}_{ij}^{\nu} & \boldsymbol{Z}_{ij}^{\nu} \\ \boldsymbol{Z}_{ij}^{\nu} & \boldsymbol{Z}_{ij}^{\gamma} + \boldsymbol{Z}_{ij}^{\lambda} + \boldsymbol{Z}_{ij}^{\varphi Y} + \boldsymbol{Z}_{ij}^{\nu} & \boldsymbol{Z}_{ij}^{\nu} \\ \boldsymbol{Z}_{ij}^{\nu} & \boldsymbol{Z}_{ij}^{\nu} & \boldsymbol{Z}_{ij}^{\gamma} + \boldsymbol{Z}_{ij}^{\lambda} + \boldsymbol{Z}_{ij}^{\varphi Y} + \boldsymbol{Z}_{ij}^{\nu} \end{bmatrix} \begin{bmatrix} \boldsymbol{I}_{Aa}^{j} \\ \boldsymbol{I}_{Bb}^{j} \\ \boldsymbol{I}_{Cc}^{j} \end{bmatrix}, \tag{11.182b}$$

or

$$\begin{bmatrix} \boldsymbol{E}_i^{AN} \\ \boldsymbol{E}_i^{BN} \\ \boldsymbol{E}_i^{CN} \end{bmatrix} = \begin{bmatrix} \boldsymbol{Z}_{ij}^{\phi Y} & \boldsymbol{Z}_{ij}^{\nu} & \boldsymbol{Z}_{ij}^{\nu} \\ \boldsymbol{Z}_{ij}^{\nu} & \boldsymbol{Z}_{ij}^{\phi Y} & \boldsymbol{Z}_{ij}^{\nu} \\ \boldsymbol{Z}_{ij}^{\nu} & \boldsymbol{Z}_{ij}^{\nu} & \boldsymbol{Z}_{ij}^{\phi Y} \end{bmatrix} \begin{bmatrix} \boldsymbol{I}_{Aa}^{j} \\ \boldsymbol{I}_{Bb}^{j} \\ \boldsymbol{I}_{Cc}^{j} \end{bmatrix}, \tag{11.182c}$$

where

$$\boldsymbol{Z}_{ij}^{\gamma} + \boldsymbol{Z}_{ij}^{\lambda} + \boldsymbol{Z}_{ij}^{\varphi} + \boldsymbol{Z}_{ij}^{\nu} = \boldsymbol{Z}_{ij}^{\phi Y}.$$

If one assumes that $\boldsymbol{Z}_{ij}^{\gamma} = 0$, because the energy-source winding impedance holors are small and may be neglected for the Y–Y balanced three-phase AC electrical homogeneous continuous dynamical system, equation (11.182b) can be written in matrix notation as follows:

$$\begin{bmatrix} \boldsymbol{E}_i^{AN} \\ \boldsymbol{E}_i^{BN} \\ \boldsymbol{E}_i^{CN} \end{bmatrix} = \begin{bmatrix} \boldsymbol{Z}_{ij}^{\lambda} + \boldsymbol{Z}_{ij}^{\varphi Y} + \boldsymbol{Z}_{ij}^{\nu} & \boldsymbol{Z}_{ij}^{\nu} & \boldsymbol{Z}_{ij}^{\nu} \\ \boldsymbol{Z}_{ij}^{\nu} & \boldsymbol{Z}_{ij}^{\lambda} + \boldsymbol{Z}_{ij}^{\varphi Y} + \boldsymbol{Z}_{ij}^{\nu} & \boldsymbol{Z}_{ij}^{\nu} \\ \boldsymbol{Z}_{ij}^{\nu} & \boldsymbol{Z}_{ij}^{\nu} & \boldsymbol{Z}_{ij}^{\lambda} + \boldsymbol{Z}_{ij}^{\varphi Y} + \boldsymbol{Z}_{ij}^{\nu} \end{bmatrix} \begin{bmatrix} \boldsymbol{I}_{Aa}^{j} \\ \boldsymbol{I}_{Bb}^{j} \\ \boldsymbol{I}_{Cc}^{j} \end{bmatrix}. \quad (11.182d)$$

If one assumes that $\boldsymbol{Z}_{ij}^{\lambda} = 0$, because the transmission-line impedance holors are neglected for the short transmission lines for the Y–Y balanced three-phase AC electrical homogeneous continuous dynamical system, equation (11.182d) can be written in matrix notation as follows:

$$\begin{bmatrix} \boldsymbol{E}_i^{AN} \\ \boldsymbol{E}_i^{BN} \\ \boldsymbol{E}_i^{CN} \end{bmatrix} = \begin{bmatrix} \boldsymbol{Z}_{ij}^{\varphi Y} + \boldsymbol{Z}_{ij}^{\nu} & \boldsymbol{Z}_{ij}^{\nu} & \boldsymbol{Z}_{ij}^{\nu} \\ \boldsymbol{Z}_{ij}^{\nu} & \boldsymbol{Z}_{ij}^{\varphi Y} + \boldsymbol{Z}_{ij}^{\nu} & \boldsymbol{Z}_{ij}^{\nu} \\ \boldsymbol{Z}_{ij}^{\nu} & \boldsymbol{Z}_{ij}^{\nu} & \boldsymbol{Z}_{ij}^{\varphi Y} + \boldsymbol{Z}_{ij}^{\nu} \end{bmatrix} \begin{bmatrix} \boldsymbol{I}_{Aa}^{j} \\ \boldsymbol{I}_{Bb}^{j} \\ \boldsymbol{I}_{Cc}^{j} \end{bmatrix}. \quad (11.182e)$$

If one assumes that $\boldsymbol{Z}_{ij}^{\nu} = 0$, because the neutral-conductor impedance holor does not exist for the Y–Y balanced three-phase AC electrical homogeneous continuous dynamical system, equation (11.182e) can be written in matrix notation as follows:

$$\begin{bmatrix} \boldsymbol{E}_i^{AN} \\ \boldsymbol{E}_i^{BN} \\ \boldsymbol{E}_i^{CN} \end{bmatrix} = \begin{bmatrix} \boldsymbol{Z}_{ij}^{\varphi Y} & 0 & 0 \\ 0 & \boldsymbol{Z}_{ij}^{\varphi Y} & 0 \\ 0 & 0 & \boldsymbol{Z}_{ij}^{\varphi Y} \end{bmatrix} \begin{bmatrix} \boldsymbol{I}_{Aa}^{j} \\ \boldsymbol{I}_{Bb}^{j} \\ \boldsymbol{I}_{Cc}^{j} \end{bmatrix}. \quad (11.182f)$$

Now add the three equations (11.183)

$$\begin{aligned} \boldsymbol{E}_i^{AN} + \boldsymbol{E}_i^{BN} + \boldsymbol{E}_i^{CN} &= E_{\phi}(\angle 0 + \angle - 2\pi/3 + \angle - 4\pi/3) \\ &= \left(\boldsymbol{Z}_{ij}^{\gamma} + \boldsymbol{Z}_{ij}^{\lambda} + \boldsymbol{Z}_{ij}^{\varphi Y} + 3\boldsymbol{Z}_{ij}^{\nu}\right)\left(\boldsymbol{I}_{Aa}^{j} + \boldsymbol{I}_{Bb}^{j} + \boldsymbol{I}_{Cc}^{j}\right). \end{aligned} \quad (11.183)$$

Noting that

$$1(\angle 0 + \angle - 2\pi/3 + \angle - 4\pi/3) = 0, \quad (11.184)$$

one has

$$\boldsymbol{I}_{Aa}^{j} + \boldsymbol{I}_{Bb}^{j} + \boldsymbol{I}_{Cc}^{j} = 0. \quad (11.185)$$

But,

$$\boldsymbol{I}_{Nn}^{j} = -\left(\boldsymbol{I}_{Aa}^{j} + \boldsymbol{I}_{Bb}^{j} + \boldsymbol{I}_{Cc}^{j}\right).$$

Hence

$$\boldsymbol{I}_{Nn}^{j} = 0. \quad (11.186)$$

Thus, in a Y–Y balanced three-phase AC homogeneous electrical continuous dynamical system, the current holor in the neutral conductor is zero. It is very advantageous if the current in a transmission line is zero. Then there is no energy loss in that conductor, and nor is there any voltage drop across it.

Let one contrast three-phase energy transmission with single-phase energy transmission. Suppose that one has three ME generators or three EE transformer secondaries and three loads.

If one connects them as three isolated single-phase AC electrical homogeneous continuous dynamical systems, and three loads, then one needs six sets of conductors. That is, each transmission line consists of a pair of conductors (see figure 11.67).

If each conductor of the transmission line has a resistance R_λ, and the current in each line has an effective value I_{rms}, the energy dissipated will be $6R_\lambda(I_{\text{rms}}\hat{}2)$.

Now let us contrast this with the balanced three-phase case. Only four conductors are used, and in one of them the current is zero. Thus the energy dissipated by all transmission lines is $3R_\lambda.(I_{\text{rms}}\hat{}2)$. Thus, the energy loss is halved. Not only is there a saving in energy, but there is also a saving in conductors, since three-phase transmission requires four conductors rather than six.

Since there is no current in the neutral conductor, it seems as though it could be omitted. If the load were always balanced, one would indeed not need the neutral conductor. However, if the load is not exactly balanced, the neutral current will not be zero.

Consider that figure 11.70 represents an energy distribution system; $\boldsymbol{Z}_{ij}^{an}$, $\boldsymbol{Z}_{ij}^{bn}$ and $\boldsymbol{Z}_{ij}^{cn}$ could represent three different neighbourhoods. Statistically, they would be equal loads, but actually they would differ slightly. In this case, the neutral current would be much smaller than the line current, but would not be zero. For this reason, the neutral conductor is always present in a three-phase energy distribution system.

We have not shown any EE transformers in the three-phase energy-distribution system model of figure 11.70. However, the ideas presented in figure 11.67 are incorporated into the three-phase energy distribution system.

Let us continue with holor analysis of figure 11.70. Assume that the three-phase electrical homogeneous continuous dynamical system is balanced. Since $\boldsymbol{I}_{Nn}^{j} = 0$, the voltage drop across $\boldsymbol{Z}_{ij}^{Nn} = \boldsymbol{Z}_{ij}^{\nu} = 0$.

Then, from equation (11.183), one has

$$\left.\begin{aligned} \boldsymbol{I}_{Aa}^{j} &= \frac{\delta_k^i}{\boldsymbol{Z}_{jk}^{\gamma} + \boldsymbol{Z}_{jk}^{\lambda} + \boldsymbol{Z}_{jk}^{\varphi}}\boldsymbol{E}_i^{AN} = \frac{\delta_k^i}{\boldsymbol{Z}_{jk}^{\phi}}\boldsymbol{E}_i^{AN} = \boldsymbol{Y}_{\phi}^{ji}\boldsymbol{E}_i^{AN} \\ \boldsymbol{I}_{Bb}^{j} &= \frac{\delta_k^i}{\boldsymbol{Z}_{jk}^{\gamma} + \boldsymbol{Z}_{jk}^{\lambda} + \boldsymbol{Z}_{jk}^{\varphi}}\boldsymbol{E}_i^{BN} = \frac{\delta_k^i}{\boldsymbol{Z}_{jk}^{\phi}}\boldsymbol{E}_i^{BN} = \boldsymbol{Y}_{\phi}^{ji}\boldsymbol{E}_i^{BN} \\ \boldsymbol{I}_{Cc}^{j} &= \frac{\delta_k^i}{\boldsymbol{Z}_{jk}^{\gamma} + \boldsymbol{Z}_{jk}^{\lambda} + \boldsymbol{Z}_{jk}^{\varphi}}\boldsymbol{E}_m^{CN} = \frac{\delta_k^i}{\boldsymbol{Z}_{jk}^{\phi}}\boldsymbol{E}_i^{CN} = \boldsymbol{Y}_{\phi}^{ji}\boldsymbol{E}_i^{CN} \end{aligned}\right\}, \tag{11.187}$$

where

$$Z^{\gamma}_{jk} + Z^{\lambda}_{jk} + Z^{\varphi}_{jk} = Z^{\phi}_{jk},$$

and

$$\frac{\delta^i_k}{Z^{\gamma}_{jk} + Z^{\lambda}_{jk} + Z^{\varphi}_{jk}} = \frac{\delta^i_k}{Z^{\phi}_{jk}} = Y^{ji}_{\phi}.$$

Applying loop-current holor analysis, one has

$$\left.\begin{aligned} E_i^{AB} &= E_\phi\left(\angle 0 - \angle -\frac{2\pi}{3}\right) = \left(Z_{ij}^{AN} + Z_{ij}^{Aa} + Z_{ij}^{Bb} + Z_{ij}^{ab}\right) I_{Aa}^j - Z_{ij}^{Bb} I_{Bb}^j - Z_{ij}^{Aa} I_{Cc}^j \\ E_i^{BC} &= E_\phi\left(\angle -\frac{2\pi}{3} - \angle -\frac{4\pi}{3}\right) = -Z_{ij}^{Bb} I_{Aa}^j + \left(Z_{ij}^{BN} + Z_{ij}^{Bb} + Z_{ij}^{bc} + Z_{ij}^{Cc}\right) I_{Bb}^j - Z_{ij}^{Cc} I_{Cc}^j \\ E_i^{CA} &= E_\phi\left(\angle -\frac{4\pi}{3} - \angle 0\right) = -Z_{ij}^{Aa} I_{Aa}^j - Z_{ij}^{Cc} I_{Bb}^j + \left(Z_{ij}^{CN} + Z_{ij}^{Aa} + Z_{ij}^{Cc} + Z_{ij}^{ca}\right) I_{Cc}^j \end{aligned}\right\}. \quad (11.188a)$$

Equations (11.188a) can also be written in matrix notation as follows:

$$\begin{bmatrix} E_i^{AB} \\ E_i^{BC} \\ E_i^{CA} \end{bmatrix} = \begin{bmatrix} Z_{ij}^{AN} + Z_{ij}^{Aa} + Z_{ij}^{Bb} + Z_{ij}^{ab} & -Z_{ij}^{Bb} & -Z_{ij}^{Aa} \\ -Z_{ij}^{Bb} & Z_{ij}^{BN} + Z_{ij}^{Bb} + Z_{ij}^{bc} + Z_{ij}^{Cc} & -Z_{ij}^{Cc} \\ -Z_{ij}^{Aa} & -Z_{ij}^{Cc} & Z_{ij}^{CN} + Z_{ij}^{Aa} + Z_{ij}^{Cc} + Z_{ij}^{ca} \end{bmatrix} \begin{bmatrix} I_{Aa}^j \\ I_{Bb}^j \\ I_{Cc}^j \end{bmatrix}, \quad (11.188b)$$

or

$$\begin{bmatrix} E_i^{AB} \\ E_i^{BC} \\ E_i^{CA} \end{bmatrix} = \begin{bmatrix} Z_{ij}^{\phi\Delta} & -Z_{ij}^{Bb} & -Z_{ij}^{Aa} \\ -Z_{ij}^{Bb} & Z_{ij}^{\phi\Delta} & -Z_{ij}^{Cc} \\ -Z_{ij}^{Aa} & -Z_{ij}^{Cc} & Z_{ij}^{\phi\Delta} \end{bmatrix} \begin{bmatrix} I_{Aa}^j \\ I_{Bb}^j \\ I_{Cc}^j \end{bmatrix}, \quad (11.188c)$$

where

$$Z_{ij}^{AN} + Z_{ij}^{Aa} + Z_{ij}^{Bb} + Z_{ij}^{ab} = Z_{ij}^{\phi\Delta},$$

$$Z_{ij}^{BN} + Z_{ij}^{Bb} + Z_{ij}^{bc} + Z_{ij}^{Cc} = Z_{ij}^{\phi BC},$$

and

$$Z_{ij}^{CN} + Z_{ij}^{Aa} + Z_{ij}^{Cc} + Z_{ij}^{ca} = Z_{ij}^{\phi CA}.$$

The procedures discussed for solving simultaneous equations can be used here. An especially interesting case occurs when the three-phase AC electrical homogeneous continuous dynamical system is balanced.

11.7.17 Repetitive example II

Figure 11.71 shows a three-phase AC homogenous electrical continuous dynamical system with a triangle or delta (Δ)-connected three-phase load. The procedures

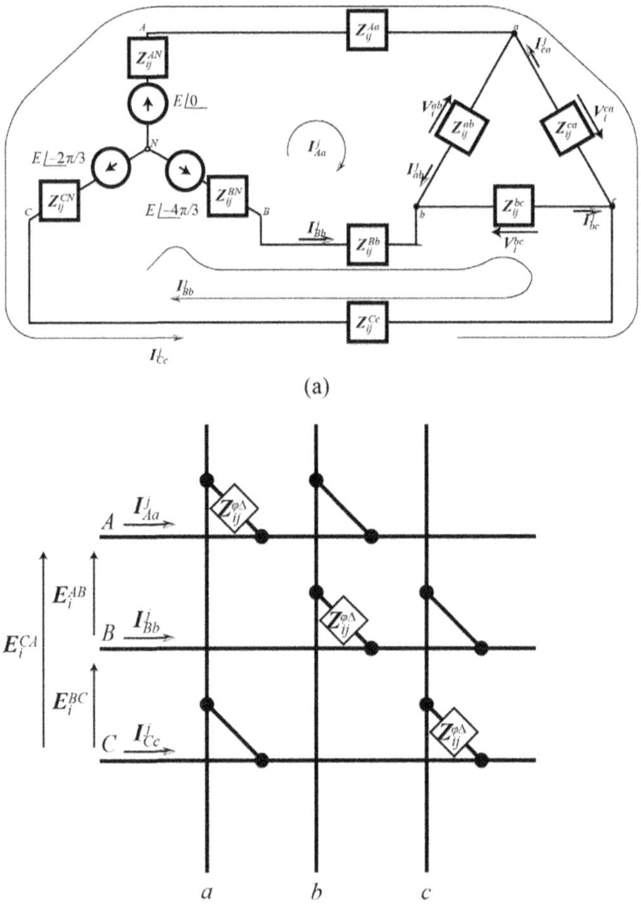

(a)

(b)

Figure 11.71. A three-phase AC electrical homogeneous continuous dynamical system with a delta (Δ)-connected three-phase load.

discussed for solving simultaneous equations can be used here. An especially interesting case occurs when the three-phase AC electrical homogeneous continuous dynamical system is balanced.

In this case, one assumes that

$$\boldsymbol{Z}_{ij}^{ab} = \boldsymbol{Z}_{ij}^{bc} = \boldsymbol{Z}_{ij}^{ca} = \boldsymbol{Z}_{ij}^{\varphi\Delta}. \tag{11.189}$$

In addition, one assumes that the three internal energy-source impedance holors and line impedance holors are equal.

$$\boldsymbol{Z}_{ij}^{AN} = \boldsymbol{Z}_{ij}^{BN} = \boldsymbol{Z}_{ij}^{CN} = \boldsymbol{Z}_{ij}^{\gamma}, \tag{11.189a}$$

$$Z_{ij}^{Aa} = Z_{ij}^{Bb} = Z_{ij}^{Cc} = Z_{ij}^{\lambda}. \tag{11.189b}$$

Equations (11.188a) for the Y2Δ balanced three-phase AC electrical homogeneous continuous dynamical system now become

$$\left.\begin{aligned} E_i^{AB} &= E_\phi\left(\angle 0 - \angle -\frac{2\pi}{3}\right) = \left(Z_{ij}^{\gamma} + Z_{ij}^{\lambda} + Z_{ij}^{\lambda} + Z_{ij}^{\varphi\Delta}\right)I_{Aa}^{j} - Z_{ij}^{\lambda}I_{Bb}^{j} - Z_{ij}^{\lambda}I_{Cc}^{j} \\ E_i^{BC} &= E_\phi\left(\angle -\frac{2\pi}{3} - \angle -\frac{4\pi}{3}\right) = -Z_{ij}^{\lambda}I_{Aa}^{j} + \left(Z_{ij}^{\gamma} + Z_{ij}^{\lambda} + Z_{ij}^{\varphi\Delta} + Z_{ij}^{\lambda}\right)I_{Bb}^{j} - Z_{ij}^{\lambda}I_{Cc}^{j} \\ E_i^{CA} &= E_\phi\left(\angle -\frac{4\pi}{3} - \angle 0\right) = -Z_{ij}^{\lambda}I_{Aa}^{j} - Z_{ij}^{\lambda}I_{Bb}^{j} + \left(Z_{ij}^{\gamma} + Z_{ij}^{\lambda} + Z_{ij}^{\lambda} + Z_{ij}^{\varphi\Delta}\right)I_{Cc}^{j} \end{aligned}\right\} \tag{11.190a}$$

$$\begin{bmatrix} E_i^{AB} \\ E_i^{BC} \\ E_i^{CA} \end{bmatrix} = \begin{bmatrix} Z_{ij}^{\gamma} + Z_{ij}^{\lambda} + Z_{ij}^{\lambda} + Z_{ij}^{\varphi\Delta} & -Z_{ij}^{\lambda} & -Z_{ij}^{\lambda} \\ -Z_{ij}^{\lambda} & Z_{ij}^{\gamma} + Z_{ij}^{\lambda} + Z_{ij}^{\varphi\Delta} + Z_{ij}^{\lambda} & -Z_{ij}^{\lambda} \\ -Z_{ij}^{\lambda} & -Z_{ij}^{\lambda} & Z_{ij}^{\gamma} + Z_{ij}^{\lambda} + Z_{ij}^{\lambda} + Z_{ij}^{\varphi\Delta} \end{bmatrix} \begin{bmatrix} I_{Aa}^{j} \\ I_{Bb}^{j} \\ I_{Cc}^{j} \end{bmatrix}. \tag{11.190b}$$

The neutral is not connected. The load is called a delta (Δ) configuration because of its resemblance to the Greek letter delta.

If one assumes that $Z_{ij}^{\gamma} = 0$, and $Z_{ij}^{\lambda} = 0$, because the energy-source winding impedance holors and the transmission-line impedance holors, respectively, are small and may be neglected for the Y2Δ balanced three-phase AC electrical homogeneous continuous dynamical system, equation (11.190b) can be written in matrix notation as follows:

$$\begin{bmatrix} E_i^{AB} \\ E_i^{BC} \\ E_i^{CA} \end{bmatrix} = \begin{bmatrix} Z_{ij}^{\varphi\Delta} & 0 & 0 \\ 0 & Z_{ij}^{\varphi\Delta} & 0 \\ 0 & 0 & Z_{ij}^{\varphi\Delta} \end{bmatrix}. \tag{11.190c}$$

Let us now consider the case of the Y2Δ balanced three-phase AC electrical homogeneous continuous dynamical system. Note that equations (11.185) and (11.186) apply.

Substituting these equations and adding equation (11.184), one has

$$0 = Z_{ij}^{\varphi\Delta}(I_{Aa}^{j} + I_{Bb}^{j} + I_{Cc}^{j}). \tag{11.191}$$

Hence,

$$I_{Aa}^{j} + I_{Bb}^{j} + I_{Cc}^{j} = 0. \tag{11.110}$$

In fact, it can be shown that if the three-phase load is balanced, the load current holors will be equal in magnitude and will differ in phase by 2π/3.

11.8 Parallel-series resistance-inductance-capacitance AC electrical commutation matrixer

One will find the node-to-datum holor equations in matrixer form for the parallel-series **resistance-inductance-capacitance** (*R–L–C*) AC electrical commutation matrixer shown in figure 11.72(a). One will also express the effective (rms) value of $\boldsymbol{E}_i^2$ in terms of the electrical commutation matrixer components and input sources. Figure 11.72(b) shows the admittance-holor diagram representation of the AC electrical commutation matrixer. There are a total of four nodes or terminals. However, two of them (the ground terminal and the voltage-source terminal holor, $\boldsymbol{E}_i$) are already defined.

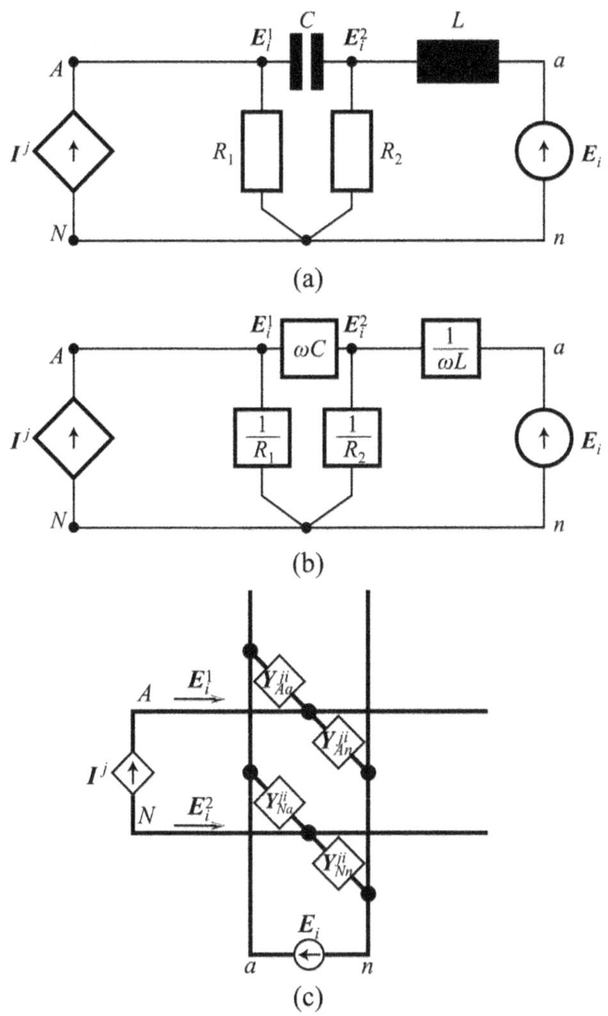

Figure 11.72. AC electrical commutation matrixer.

Thus, there are only two independent nodes and the admittance-holor matrix will be of the form:

$$\begin{matrix}\text{Node1} \Longrightarrow \\ \text{Node2} \Longrightarrow\end{matrix} \begin{bmatrix} \boldsymbol{Y}^{ji}_{Aa} & \boldsymbol{Y}^{ji}_{An} \\ \boldsymbol{Y}^{ji}_{Na} & \boldsymbol{Y}^{ji}_{Nn} \end{bmatrix} \begin{bmatrix} \boldsymbol{E}^1_i \\ \boldsymbol{E}^2_i \end{bmatrix} = \begin{bmatrix} \boldsymbol{I}^j \\ \left\| \begin{matrix} 0 & \frac{1}{\omega L} \\ -\frac{1}{\omega L} & 0 \end{matrix} \right\| \boldsymbol{E}_i \end{bmatrix}.$$

First, let us consider the flow into the nodes from the inputs. The flow into node 1 from the input voltage-source holors is merely the effective value of the current-holor source, $\boldsymbol{I}^j$. However, in this case the flow into node 2 depends on the voltage-source holor, $\boldsymbol{E}_i$ and the inductive admittance holor, $\frac{\delta^i_k}{\omega \boldsymbol{L}_{jk}}$. This is a flow holor, $\frac{\delta^i_k}{\omega \boldsymbol{L}_{jk}} \boldsymbol{E}_i$. As a result, the known column-vector flow-holor is as indicated above.

The matrix admittance-holor component $\boldsymbol{Y}^{ji}_{11}$ is the parallel combination of $1/R_1$ and ωC. This is the condition that exists when all terminals except $\boldsymbol{E}^1_i$ are grounded. Thus

$$\boldsymbol{Y}^{ji}_{Aa} = \left\| \begin{matrix} 1/R_1 & \omega C \\ -\omega C & 1/R_1 \end{matrix} \right\|.$$

To find the matrix admittance-holor element $\boldsymbol{Y}^{jii}_{Nn}$ one reverses the procedure. Now all the terminals except $\boldsymbol{E}^2_i$ are grounded.

The admittance holor from node 2 in this case is the parallel combination of ωC, $1/R_2$ and $1/\omega L$ or

$$\boldsymbol{Y}^{ji}_{Nn} = \left\| \begin{matrix} 1/R_2 & 1/\omega L + \omega C \\ -1/\omega L - \omega C & 1/R_2 \end{matrix} \right\|.$$

In this case, there is only one common term between $\boldsymbol{Y}^{ji}_{11}$ and $\boldsymbol{Y}^{ji}_{12}$, and it is ωC, so that:

$$\boldsymbol{Y}^{ji}_{An} = \boldsymbol{Y}^{ji}_{Na} = \left\| \begin{matrix} 0 & -\omega C \\ \omega C & 0 \end{matrix} \right\|.$$

Moreover, the complete hypermatrix representation is:

$$\begin{bmatrix} \left\| \begin{matrix} \frac{1}{R_1} & \omega C \\ -\omega C & \frac{1}{R_1} \end{matrix} \right\| & \left\| \begin{matrix} 0 & -\omega C \\ \omega C & 0 \end{matrix} \right\| \\ \left\| \begin{matrix} 0 & -\omega C \\ \omega C & 0 \end{matrix} \right\| & \left\| \begin{matrix} \frac{1}{R_2} & \frac{1}{\omega L} - \omega C \\ -\frac{1}{\omega L} + \omega C & \frac{1}{R_2} \end{matrix} \right\| \end{bmatrix} \begin{bmatrix} \left\| \begin{matrix} E^1_c \\ E^1_s \end{matrix} \right\| \\ \left\| \begin{matrix} E^2_c \\ E^2_s \end{matrix} \right\| \end{bmatrix} = \begin{bmatrix} \left\| \begin{matrix} I^c \\ I^s \end{matrix} \right\| \\ \left\| \begin{matrix} 0 & 1/\omega L \\ -1/\omega L & 0 \end{matrix} \right\| \left\| \begin{matrix} E_c \\ E_s \end{matrix} \right\| \end{bmatrix}.$$

The effective value of $\boldsymbol{E}_i^2$ in terms of determinant is:

$$E_i^2 = \begin{Vmatrix} E_c^2 \\ E_s^2 \end{Vmatrix} = \frac{\begin{vmatrix} \begin{Vmatrix} 1/R_1 & \omega C \\ -\omega C & 1/R_1 \end{Vmatrix} & \begin{Vmatrix} I^c \\ I^s \end{Vmatrix} \\ \begin{Vmatrix} 0 & -\omega C \\ \omega C & 0 \end{Vmatrix} & \begin{Vmatrix} 0 & 1/\omega L \\ -1/\omega L & 0 \end{Vmatrix}\begin{Vmatrix} E_c \\ E_s \end{Vmatrix} \end{vmatrix}}{\begin{vmatrix} \begin{Vmatrix} 1/R_1 & \omega C \\ -\omega C & 1/R_1 \end{Vmatrix} & \begin{Vmatrix} 0 & -\omega C \\ \omega C & 0 \end{Vmatrix} \\ \begin{Vmatrix} 0 & -\omega C \\ \omega C & 0 \end{Vmatrix} & \begin{Vmatrix} 1/R_2 & 1/\omega L + \omega C \\ -1/\omega L - \omega C & 1/R_2 \end{Vmatrix} \end{vmatrix}}$$

and, evaluating the determinants,

$$E_i^2 = \begin{Vmatrix} E_c^2 \\ E_s^2 \end{Vmatrix}$$

$$= \frac{\begin{Vmatrix} 1/R_1 & \omega C \\ -\omega C & 1/R_1 \end{Vmatrix}\begin{Vmatrix} 0 & 1/\omega L \\ -1/\omega L & 0 \end{Vmatrix}\begin{Vmatrix} E_c \\ E_s \end{Vmatrix} - \begin{Vmatrix} 0 & -\omega C \\ \omega C & 0 \end{Vmatrix}\begin{Vmatrix} I^c \\ I^s \end{Vmatrix}}{\begin{Vmatrix} 1/R_1 & \omega C \\ -\omega C & 1/R_1 \end{Vmatrix}\begin{Vmatrix} 1/R_2 & 1/\omega L + \omega C \\ -1/\omega L - \omega C & 1/R_2 \end{Vmatrix} - \begin{Vmatrix} 0 & -\omega C \\ \omega C & 0 \end{Vmatrix}\begin{Vmatrix} 0 & -\omega C \\ \omega C & 0 \end{Vmatrix}}.$$

11.9 Application of Thévenin's theorem

We may now obtain the electrical commutation-matrixer transfer function $\boldsymbol{E}_i^{\mathrm{L}}/\boldsymbol{E}_k$ for the electrical commutation matrixer shown in figure 11.73(a). The procedure is to obtain the Thévenin equivalent of the section of the original electrical commutation matrixer reproduced in figure 11.73(b). Since there is no current-holor flow through $\boldsymbol{Z}_{ij}^3$, the open output electrical-matrixer's voltage holor is obtained directly from the voltage-holor divider relation:

$$\boldsymbol{E}_i^{\mathrm{TH}} = \frac{\boldsymbol{\delta}_k^i}{\boldsymbol{Z}_{jk}^1 + \boldsymbol{Z}_{jk}^3}\boldsymbol{Z}_{ij}^2\boldsymbol{E}_i.$$

The equivalent impedance holor is the driving-point impedance holor between port terminals a and b when the voltage-source holor is suppressed and this is readily seen to be the equivalent of $\boldsymbol{Z}_{ij}^3$ in series with the parallel combination of $\boldsymbol{Z}_{ij}^1$ and $\boldsymbol{Z}_{ij}^3$.

$$\boldsymbol{Z}_{ij}^{\mathrm{TH}} = \boldsymbol{Z}_{ij}^1 + \frac{\delta_k^i}{\boldsymbol{Z}_{jk}^1 + \boldsymbol{Z}_{jk}^2}\boldsymbol{Z}_{ij}^1\boldsymbol{Z}_{ij}^2 = \frac{\delta_k^i}{\boldsymbol{Z}_{jk}^1 + \boldsymbol{Z}_{jk}^2}\left(\boldsymbol{Z}_{ij}^1\boldsymbol{Z}_{ij}^2 + \boldsymbol{Z}_{ij}^2\boldsymbol{Z}_{ij}^3 + \boldsymbol{Z}_{ij}^3\boldsymbol{Z}_{ij}^1\right).$$

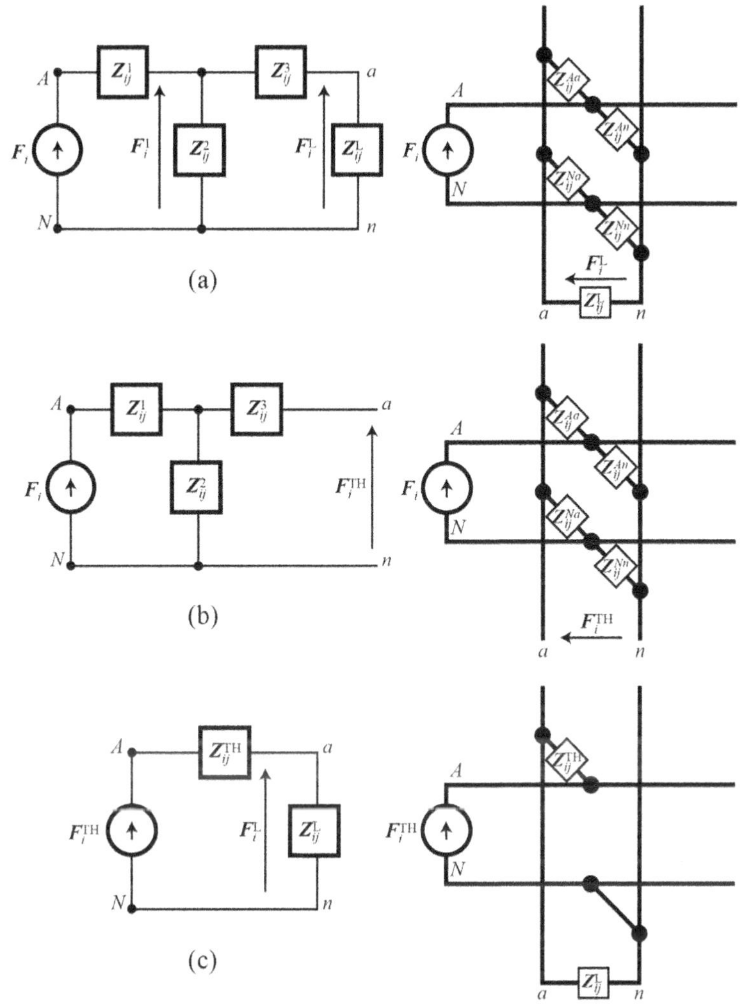

Figure 11.73. Thévenin equivalent electrical commutation matrixer.

Using these values in figure 11.73(c), one obtains:

$$\frac{\boldsymbol{E}_i^{\mathrm{L}}}{\boldsymbol{E}_k^{\mathrm{TH}}} = \frac{\delta_k^i}{\boldsymbol{Z}_{jk}^{\mathrm{TH}} + \boldsymbol{Z}_{jk}^{\mathrm{L}}} \boldsymbol{Z}_{ij}^{\mathrm{L}},$$

and substituting for $\boldsymbol{E}_i^{\mathrm{TH}}$ and $\boldsymbol{Z}_{ij}^{\mathrm{TH}}$ gives

$$\begin{aligned} \frac{\boldsymbol{E}_i^{\mathrm{L}}}{\boldsymbol{E}_k} &= \frac{\delta_k^i}{\boldsymbol{Z}_{jk}^1 + \boldsymbol{Z}_{jk}^2} \boldsymbol{Z}_{ij}^2 . \frac{\delta_k^i}{\boldsymbol{Z}_{jk}^{\mathrm{TH}} + \boldsymbol{Z}_{jk}^{\mathrm{L}}} \boldsymbol{Z}_{ij}^{\mathrm{L}} \\ &= \frac{\delta_k^i}{\boldsymbol{Z}_{jk}^1 \boldsymbol{Z}_{jk}^2 + \boldsymbol{Z}_{jk}^2 \boldsymbol{Z}_{jk}^3 + \boldsymbol{Z}_{jk}^3 \boldsymbol{Z}_{jk}^1 + \boldsymbol{Z}_{jk}^{\mathrm{L}} \left(\boldsymbol{Z}_{jk}^1 + \boldsymbol{Z}_{jk}^2 \right)} \boldsymbol{Z}_{ij}^2 \boldsymbol{Z}_{ij}^{\mathrm{L}} . \end{aligned}$$

This electrical commutation matrixer is sufficiently simple to check this answer by obtaining the solution to the node holor equations of the entire electrical commutation matrixer.

$$\begin{bmatrix} \frac{\delta_k^i}{Z_{jk}^1}+\frac{\delta_k^i}{Z_{jk}^2}+\frac{\delta_k^i}{Z_{jk}^3} & -\frac{\delta_k^i}{Z_{jk}^3} \\ -\frac{\delta_k^j}{Z_{jk}^3} & \frac{\delta_k^j}{Z_{jk}^3}+\frac{\delta_k^j}{Z_{jk}^L} \end{bmatrix} \begin{bmatrix} E_i^1 \\ E_i^L \end{bmatrix} = \begin{bmatrix} \frac{\delta_k^i}{Z_{jk}^1} E_i \\ 0 \end{bmatrix}.$$

The solution is

$$E_i^L = \frac{\delta_k^i}{\left(\frac{\delta_k^i}{Z_{jk}^1}+\frac{\delta_k^i}{Z_{jk}^2}+\frac{\delta_k^i}{Z_{jk}^3}\right)\left(\frac{\delta_k^i}{Z_{jk}^3}+\frac{\delta_k^i}{Z_{jk}^L}\right)-\left(\frac{\delta_k^i}{Z_{jk}^3}\right)\left(\frac{\delta_k^i}{Z_{jk}^3}\right)} \frac{\delta_k^i}{Z_{jk}^3}\frac{\delta_k^i}{Z_{jk}^1} E_i$$

$$= \frac{\delta_k^i}{\frac{\delta_k^i}{Z_{jk}^L}\left(Z_{jk}^1 Z_{jk}^2 + Z_{jk}^2 Z_{jk}^3 + Z_{jk}^3 Z_{jk}^1\right) + Z_{jk}^1 + Z_{jk}^2} Z_{ij}^2 E_i$$

$$= \frac{\delta_k^i}{Z_{jk}^1 Z_{jk}^2 + Z_{jk}^2 Z_{jk}^3 + Z_{jk}^3 Z_{jk}^1 + Z_{jk}^L\left(Z_{jk}^1 + Z_{jk}^2\right)} Z_{ij}^2 Z_{ij}^L E_i.$$

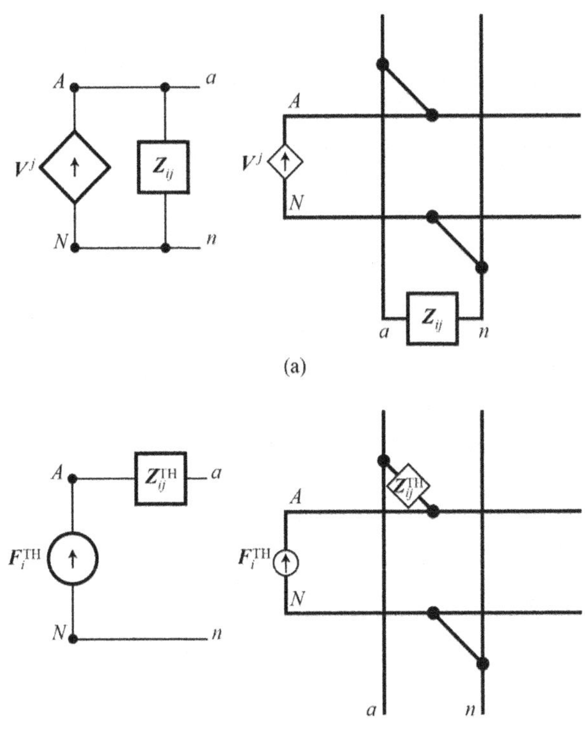

Figure 11.74. Application of Thévenin's theorem. (a) Original electrical commutation matrixer. (b) Thévenin equivalent electrical commutation matrixer.

In some cases the process of obtaining the Thévenin equivalent by the simple procedure used in this example of the electrical commutation matrixer becomes complicated and it is simpler to obtain the open-output voltage holor and driving-point impedance holor directly from a set of node holor equations.

A simple application of Thévenin's theorem that is often very useful is that of transforming an energy-transfer-source holor and the impedance holor connected across its terminals (figure 11.74(a)) to an equivalent series connected energy-

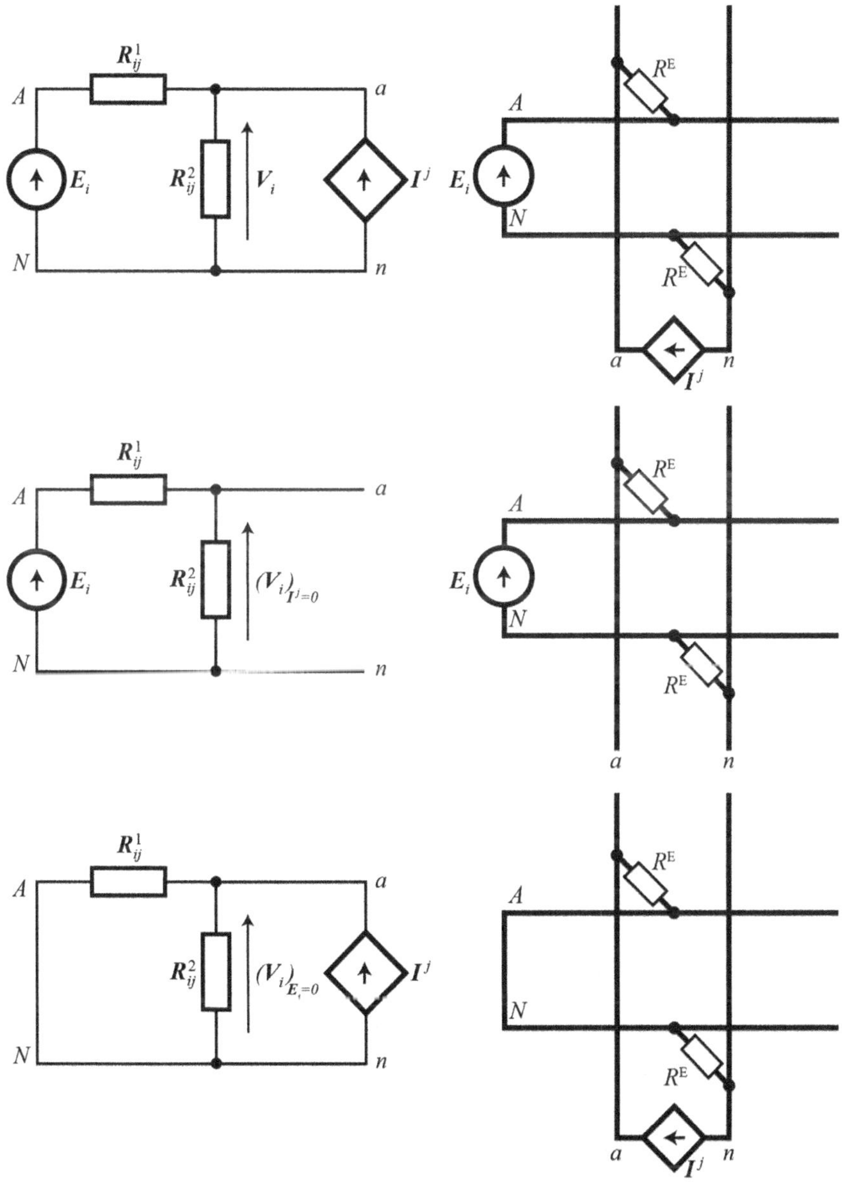

Figure 11.75. Superposition theorem.

potential-source holor and impedance holor (figure 11.74(b)). The electrical commutation matrixers are equivalent provided that $\boldsymbol{E}_i^{\mathrm{TH}} = \boldsymbol{Z}_{ij}\boldsymbol{I}^j$ and $\boldsymbol{Z}_{ij}^{\mathrm{TH}} = \boldsymbol{Z}_{ij}$.

11.10 Application of superposition theorem

An electrical commutation matrixer consisting of two resistors is driven by a voltage-source holor and a current-source holor, as shown in figure 11.75(a).

One will find the voltage holor across the resistor R_2. The procedure is simply to obtain the values of $\boldsymbol{E}_i^2$ when only one energy source is acting at a time and then add these solutions.

In this case, there are two energy sources, so the problem is to find $\boldsymbol{E}_i^2$ in each of the electrical commutation matrixers shown in figures 11.75(b) and (c) and add them. From figure 11.75(b), the first component is a voltage divider holor

$$\boldsymbol{E}_i^2 = \begin{bmatrix} E_c^2 \\ E_s^2 \end{bmatrix} = \frac{1}{\begin{bmatrix} R_1 & 0 \\ 0 & R_1 \end{bmatrix} + \begin{bmatrix} R_2 & 0 \\ 0 & R_2 \end{bmatrix}} \begin{bmatrix} R_2 & 0 \\ 0 & R_2 \end{bmatrix} \begin{bmatrix} E_c \\ E_s \end{bmatrix} = \frac{R_2}{R_1 + R_2}\boldsymbol{E}_i.$$

From figure 11.75(c), the second component of $\boldsymbol{E}_i^2$ is found by noting that this is the voltage holor across the parallel combination of R_1 and R_2

$$\boldsymbol{E}_i^2 = \begin{bmatrix} E_c^2 \\ E_s^2 \end{bmatrix} = \frac{1}{\begin{bmatrix} R_1 & 0 \\ 0 & R_1 \end{bmatrix} + \begin{bmatrix} R_2 & 0 \\ 0 & R_2 \end{bmatrix}} \begin{bmatrix} R_1 & 0 \\ 0 & R_1 \end{bmatrix} \begin{bmatrix} R_2 & 0 \\ 0 & R_2 \end{bmatrix} \begin{bmatrix} I^c \\ I^s \end{bmatrix} = \frac{R_1 R_2}{R_1 + R_2}\boldsymbol{I}^j.$$

The complete sinusoidal response is the sum of these two expressions:

$$\boldsymbol{E}_i^2 = \frac{R_2}{R_1 + R_2}\boldsymbol{E}_i + \frac{R_1 R_2}{R_1 + R_2}\boldsymbol{I}^j = \frac{R_2}{R_1 + R_2}\left(\boldsymbol{E}_i + R_1\boldsymbol{I}^j\right).$$

References

Baldwin M and Geary D 2006 High reliability DC power distribution system (HRDCPDS) for critical environments (DCUPS). A patent pending system presentation, Baldwin Technologies Inc., Environmental Support Solutions, www.baldwintec.com (46 slides)

Barthold L 2006 Current-Modulated HVDC Transmission: An Efficient Way to Convert AC Lines to DC *PowerPoint* presentation by Ram Adapa, PhD, EPRI

Bewley L V 1961 *Tensor Analysis of Electric Circuits and Machines* (New York: Ronald Press)

Cisco Systems 2006 Power over Ethernet: Cisco Inline Power and IEEE 802.3af White Paper, available for download at http://cisco.com

Crudele F D, Key T and Mansoor A 2004 Commercial and Industrial Applications Getting Ready for Direct-Current Power Distribution *IEEE Power Systems Conf. and Exposition* (New York)

Energy Market Profiles database 2006 Energy Use Strategic Service Online Tool *EPRI Solutions* (Primen), www.primen.com.

EPRI 2006 DC Power Production, Delivery and Utilization. *An EPRI White Paper*, June 2007. www.netpower.se/documents/EPRI_DCpower_WhitePaper_June_2006.pdf

EPRI PEAC Corp 2002 Development and Demonstration of DC Photovoltaic Applications: Task 1 Project Results *Draft report EPRI Solutions*

Fijalkowski B 1967 Zastosowanie rachunku holorowego w teorii pradow przemiennych (Application of a holor calculus in the theory of alternating currents) *Przeglad elektrotechniczny (Electrotechnics Review)* (Warszawa: Rok XLIII, Zeszyt 2). (In Polish)

Galvin Electricity Initiative 2005 Transforming Electricity Service Reliability and Value for the 21st Century, available for download at www.galvinelectricity.org.

Hoseman G, Krolikowski L, Raatz E, Shapilov E D, Turek-Kwiatkowska L and Vieselovsky O N 1999 Zhizn i delyatelnost Mikhaila Dolivo-Dobrovolskogo—vydaiushchegosia inzhenera elektrika evropeiskogo masshtaba *Proc. of the 4th Int. Conf. on Unconventional Electromechanical and Electrical Systems* (St Petersburg) vol. 1 pp 59–70 (In Russian)

IEEE History Center 2006 *History Center*. Rutgers-The State University, Contact history@ieee.org. Material available at the IEEE website, www.ieee.org/organizations/ history_center

Kasatkin A and Perekalin M 1970 *Basic Electrical Engineering* (Moscow: Mir Publishers)

Kron G 1959 *Tensors for Circuits* (New York: Dover)

Kuznetsov M 1960 *Fundamentals of Electrical Engineering* (Moscow: Mir Publishers)

Moon P H and Spencer D E 1986 *Theory of holors. A generalization of Tensors* (Cambridge: Cambridge University Press) (Digitally printed, 2005)

Savage P 2003 What would Thomas Edison do? *Energy Pulse,* September 2003

Staff of the Department of Electrical Engineering at MIT 1940 *Electric circuits* (New York: Wiley) ch. 8

Woodford D 1998 *HDVC Transmission* (Manitoba: HVDC Research Centre) www.hvdc.ca

Chapter 12

Mechanical homogenous continuous dynamical systems as mechanical commutation matrixers

Mechanics is the paradise of mathematical sciences … Mechanical science is most noble and useful beyond all others, since by means of it, all animated bodies that have motion perform all their operations.

Leonardo DaVinci

12.1 Simple plane-motion mechanical homogenous continuous dynamical system

Consider a simple plane-motion mechanism in txerms of a parameter holor. In this case, it is possible to express the coordinates x and y of all the significant points of the mechanism by means of simple analytical expressions containing a single parameter holor. It may be advantageous in such a case to determine the absolute velocity and the absolute acceleration of the various points of the mechanism directly, since the components of the velocity and acceleration of a given point may be obtained by differentiating the coordinates x and y of that point.

Let one consider the rod-length holor $\boldsymbol{L}_i$, whose extremities slide, respectively, in a horizontal and a vertical track (figure 12.1).

The coordinates x_A and y_B of the extremities of the rod AB may be expressed in terms of the angle θ the rod AB forms with the vertical:

$$\begin{aligned}\boldsymbol{L}_i = (l_c, l_s) = \|l_c \quad l_s\|^T = \left\|\begin{matrix} l_c \\ l_s \end{matrix}\right\| \\ = (y_B, x_A) = \|y_B \quad x_A\|^T = \left\|\begin{matrix} y_B \\ x_A \end{matrix}\right\| = \left\|\begin{matrix} l\cos\theta \\ l\sin\theta \end{matrix}\right\| = l\angle\theta.\end{aligned} \tag{12.1}$$

doi:10.1088/978-0-7503-1350-6ch12

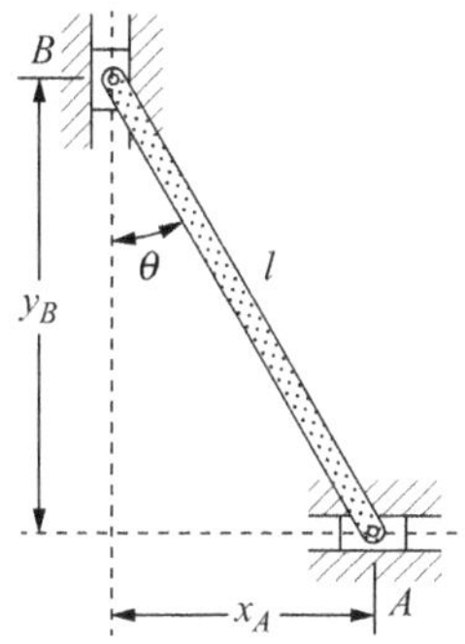

Figure 12.1. Plane motion (adapted from Beer and Johnston 1984).

Differentiating equations (12.1) twice with respect to t, one writes

$$v_B = \dot{y}_B = -\dot{\theta} l \sin\theta$$

$$a_B = \ddot{y}_B = -(\dot{\theta}\text{^}2) l \cos\theta - \ddot{\theta} l \sin\theta$$

$$v_A = \dot{x}_A = \dot{\theta} l \cos\theta$$

$$a_A = \ddot{x}_A = -(\dot{\theta}\text{^}2) l \sin\theta + \ddot{\theta} l \cos\theta.$$

Recalling that $\dot{\theta} = \omega$ and $\ddot{\theta} = \alpha$, one obtains

$$v_B = -\omega l \sin\theta \qquad v_A = \dot{x}_A = \omega l \cos\theta \tag{12.2}$$

$$a_B = -(\omega\text{^}2) l \cos\theta - \alpha l \sin\theta \qquad a_A = -(\omega\text{^}2) l \sin\theta + \alpha l \cos\theta. \tag{12.3}$$

One notes that a positive sign for v_A or a_A indicates that the velocity v_A or the acceleration a_A has a sense of direction to the right; a positive sign for v_B or a_B has a sense of direction upward. Equation (12.2) may be used, for example, to determine v_B and ω when v_A and θ are known. Substituting for ω in equation (12.3), one may then determine a_B and α if a_A is known.

12.2 Simple pendulum mechanical homogeneous continuous dynamical system (approximate solution)

Most of the vibrations encountered in mechanical homogeneous continuous dynamical systems may be represented by a simple harmonic motion. Many others, although of a different type, may be approximated by a simple harmonic motion, provided that their amplitude remains small.

Consider a simple pendulum mechanical homogeneous continuous system, consisting of a bob of mass m attached to the end of a string or cord of length l, which may oscillate on a vertical plane (figure 12.2(a)).

If a constant force is applied periodically, the pendulum can be made to swing with constant amplitude. If the force is no longer applied, the pendulum will

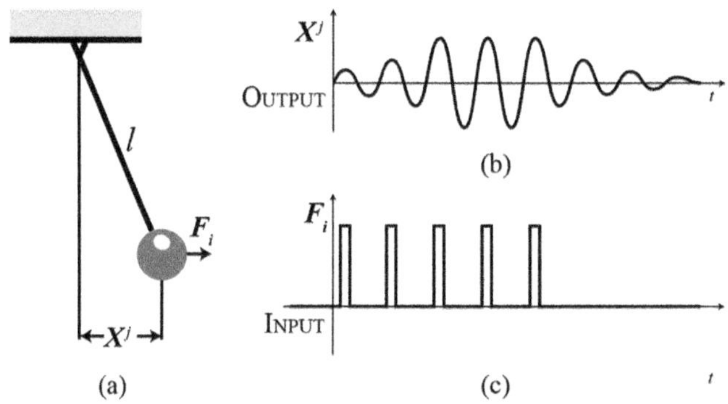

Figure 12.2. Response of pendulum.

continue to swing, although its amplitude will decrease due to air resistance (figure 12.2(b)).

At a given time, the string forms an angle θ with the vertical. The force holors acting on the bob are its weight holor $\boldsymbol{G}_i$ and the force holor $\boldsymbol{P}_i$ exerted by the string.

Resolving the holor $\boldsymbol{R}_i = \boldsymbol{M}_{ij}\boldsymbol{A}^j$ into tangential and normal components with $\boldsymbol{R}_i^T = \boldsymbol{M}_{ij}\boldsymbol{A}_T^j$, the sense of direction is to the right, i.e. to the sense of direction corresponding to increasing values of an angle θ, and observing that $\boldsymbol{A}_T^j = l\boldsymbol{A}^j = l\ddot{\theta}$, one writes

$$-G\sin\theta = ml\ddot{\theta}.$$

Noting that $G = mg$, and dividing through by ml, one obtains

$$\ddot{\theta} + \frac{g}{l}\sin\theta = 0. \tag{12.4}$$

For oscillations of small amplitude, one may replace $\sin\theta$ by θ, expressed in radians, and write

$$\ddot{\theta} + \frac{g}{l}\theta = 0. \tag{12.5}$$

Equation (12.5) shows that the equation obtained is that of a simple harmonic motion, and that the circular frequency ω of the oscillations is equal to sqrt (g/l).

Taking into consideration the fact that the period $T = 2\pi/\omega$, one obtains the period of the small oscillations of a simple pendulum of length l:

$$T = \frac{2\pi}{\omega} = 2\pi\ \mathrm{sqrt}\left(\frac{l}{g}\right). \tag{12.6}$$

The main point to note is that there is motion (output) even when there is no applied force (input). The form this motion takes is a property of the pendulum and is termed the 'natural response' or 'force-free response'.

The force is applied periodically to take advantage of this fundamental property of the mechanical homogeneous continuous dynamical system, the difference between

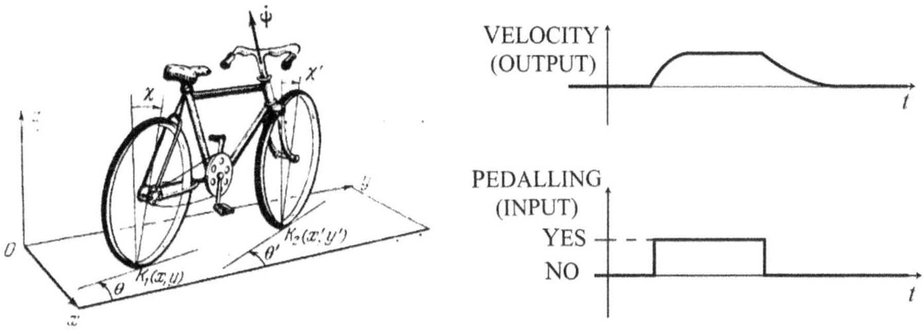

Figure 12.3. Response of bicycle (Neimark and Fufaiev 1972).

the natural response and the actual response being the component of the response directly attributable to the applied force. It is usually termed the 'forced response'.

12.3 Bicycle mechanical homogeneous continuous dynamical system

An example in which the forced response is more easily seen is that of a bicycle being pedalled along a horizontal road on a calm day. If the rider applies a constant force to the pedals, the velocity will increase until a steady value is reached (figure 12.3).

This velocity is directly related to the applied force and is the forced response of the mechanical homogeneous continuous dynamical system. After the rider stops pedalling, the bicycle still moves, although at a decreasing speed. Again, there is a response without any stimulus. This is the natural response of the mechanical homogeneous continuous dynamical hypersystem, which can also be identified as being the difference between the forced response and the actual response during acceleration.

12.4 Damper–spring–mass mechanical homogeneous continuous dynamical system I

Take into account the **damper–spring–mass** (*b–k–m*) mechanical homogeneous continuous dynamical system (I) depicted in figure 12.4(a), in which a human being submits a force holor $\boldsymbol{F}_A^j$ to one end of the damper. The initial action is to detect all the components of the mechanical homogeneous continuous dynamical system, which are constrained to move at the same values of velocity holors, and to signify the sense of direction in which these velocity holors will be considered as positive. This action is shown in figure 12.4(b). All the velocity holors are taken to be positive in the same sense of direction.

This eases the creation of equilibrium holor equations and formulates results that are considerably easier to deduce.

For the mechanical homogeneous continuous dynamical system under consideration, three velocity holors are not effectively the same. These are the velocity holor at the force holor end of the damper ($\boldsymbol{V}_i^1$), the velocity holor at the point between damper and spring ($\boldsymbol{V}_i^2$) and the velocity holor of the mass ($\boldsymbol{V}_i^3$).

For this reason, the equivalent mechanical commutation matrixer diagram will have three nodes or junctions other than the one indicating the datum velocity holor

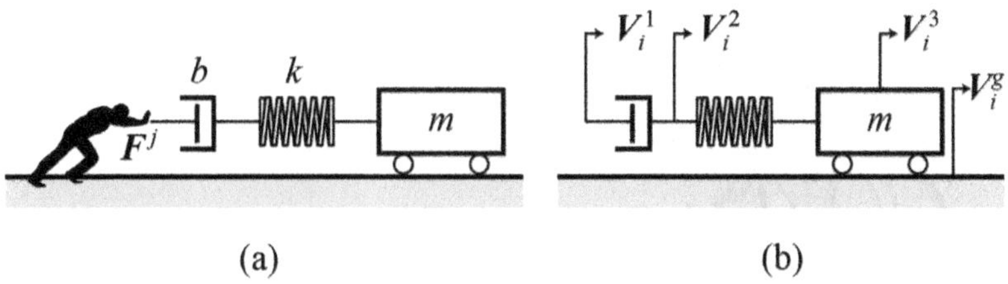

Figure 12.4. Damper–spring–mass mechanical homogeneous continuous dynamical system (I).

(V_i^g). These junctions, depicted in figure 12.5(a), are the configuration on which the mechanical commutation matrixer components are incorporated. The junctions symbolise the physical holors V_i^1, V_i^2, V_i^3 and V_i^g. For simplicity and intelligibility, the ground terminal, V_g may be prolonged along a line at the same potential. The passive components are positioned in the mechanical commutation matrixer diagram by detecting the velocity difference (energy potential) that is necessitated for each principal holor equation. For instance, the damping, b, physically models an effect that is determined by the difference between V_i^1 and V_i^2.

Thus, the damping component is connected between junctions 1 and 2. Damping is mechanical resistance and is a dissipative component. One represents it in a physical commutation matrixer diagram by an electrical resistance symbol. The spring stiffness, k, also physically models an effect that depends on velocity difference. In this case, the velocities at the ends of the spring are V_i^2 and V_i^3. The spring component is therefore connected between these two nodes. The spring is analogous to the electrical inductance and has the same mechanical commutation matrixer symbol.

Now the physical holor equation that describes the motion of the mass (equation (3.48)) shows that it is only dependent on one velocity. In this case, the velocity holor is V_i^3. Since all mechanical commutation matrixer components require two terminals, one adds a terminal that does not change the physical relations. This terminal is the ground junction at zero velocity holor. When a fictitious terminal is used, the mechanical commutation matrixer diagram contains a dashed line to show this. The mechanical mass contains a dashed line to show this. The mechanical mass is analogous to an electrical capacitance with one terminal always connected to ground.

Figure 12.5(b) shows the mechanical-matrixer diagram for the damper–mass–spring (b–m–k) mechanical continuous dynamical system. The driving force holor F_A^j exerted by the human being is physically modelled by an energy-transfer-source holor connected between junctions V_i^1 and V_i^g. Since the force is applied in the sense of direction of the positive velocity holor, the flow must be shown as entering junction V_i^1. Note that the force holor representation is complete because it also indicates the necessary reaction force holor on the datum in the opposite sense of direction (away from junction V_i^g).

The electrical analogue of this particular mechanical commutation-matrix is a series R–L–C electrical commutation matrixer with a current-holor source input. However, it must be emphasised that other **damper–mass–spring** (b–m–k) mechanical homogeneous continuous dynamical systems in which the components appear in

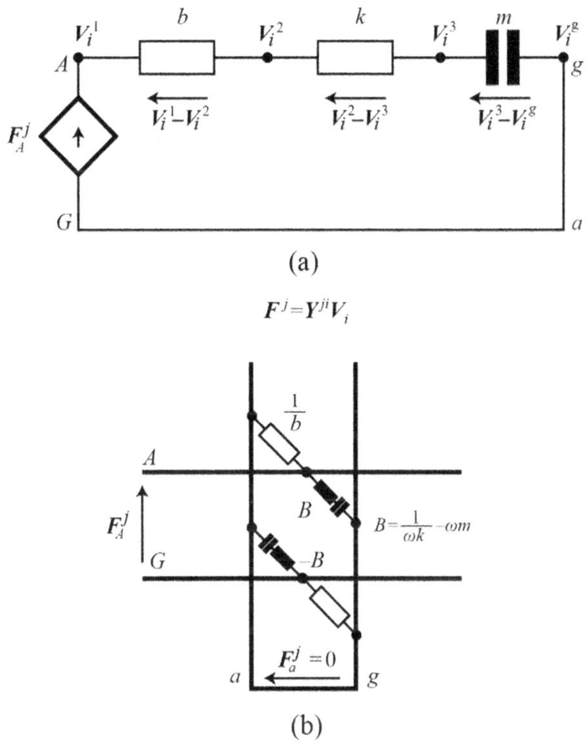

Figure 12.5. Damper–mass–spring (b–m–k) mechanical homogenous continuous dynamical system (I) and its mechanical commutation matrixer.

tandem do not necessarily result in a series mechanical commutation matrixer. This point is illustrated in the following example.

12.5 Damper–spring–mass mechanical homogeneous continuous dynamical system (II)

Figure 12.6(a) shows a **damper–spring–mass** (b–k–m) mechanical homogeneous continuous dynamical system (II) in which the force is applied to the mass. Here again, the mechanical homogeneous continuous dynamical system has the components in tandem. However, in this case, the order is different. With this mechanical commutation matrixer, there are only two velocity holors that can be identified as independent.

These are the velocity holor of the mass (designated as V_i^1) and the velocity holor of the point between the spring k and damper b (designated as V_i^2).

Once again, the reference terminal is the velocity holor of the ground (V_i^{G}). The mass m represents a capacitance between the V_i^1 and V_i^{G}. The spring is equivalent to an inductance between V_i^1 and V_i^2, and the damper b is a resistance that goes from V_i^2 to V_i^{G}. The resulting mechanical commutation matrixer is shown in figure 12.6(b). In this case, the passive components do not form a series mechanical commutation

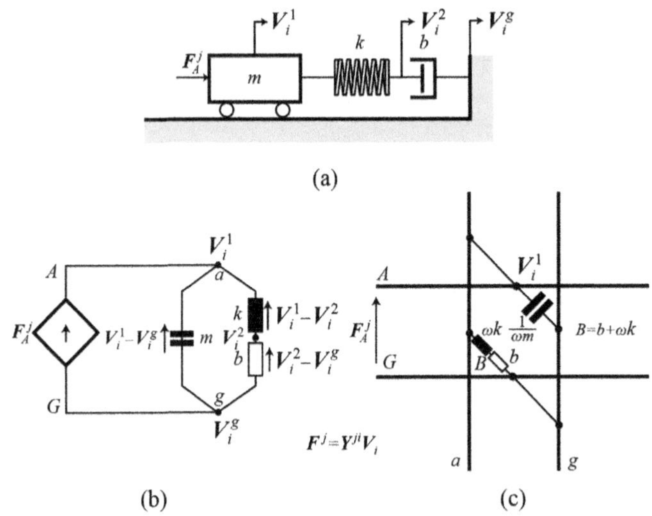

Figure 12.6. Damper–spring–mass (b–k–m) mechanical homogeneous continuous dynamical system (II) and its mechanical commutation matrixer.

matrixer. Thus, it is important not to jump to conclusions about series or parallel equivalents based solely on damper–mass–spring sketches.

The analogous mechanical commutation matrixer representation can only be obtained by determining the terminals and connecting them with the appropriate component. Before proceeding to more complicated examples with more than one energy source, it will be useful to discuss the polarity of energy sources further. Polarity concepts are especially pertinent when the energy sources are not all acting in the same sense of direction.

Consider the isolated mass shown in figure 12.7, in which several possible physical situations are indicated. In all cases, the sense of direction selected for the velocity holor (V_i) is conceptually positive. Then the application of the fundamental holor equation

$$F^j = \begin{bmatrix} 0 & \omega m \\ -\omega m & 0 \end{bmatrix} V_i$$

requires that the force holor F^j is positive from the terminal V_i through the component to the ground terminal. Thus, the physical situations presented in figures 12.7(a) and (b) are identical in the sense that the force holor F^j applied to the mass m acts in the same sense of direction as the assumed positive sense of direction for the velocity holor V_i. In other words, the force holor F^j tends to increase the velocity holor V_i.

On the other hand, the force holor F^j in figure 12.7(c) is directed to the left, while the assumed positive velocity holor V_i's sense of direction is to the right. As a result, the corresponding mechanical commutation matrixer representation shows a force-source holor that is directed away from the terminal. When two force holors F_A^j and F_a^j act on a mass m, the mechanical commutation matrixer has two energy sources and therefore contains three branches, which meet at a junction. In the case

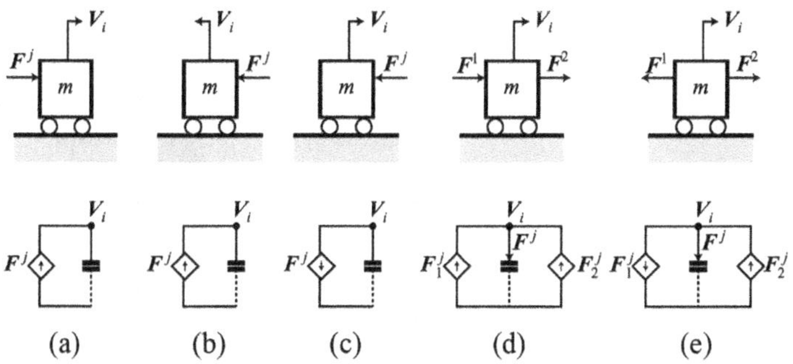

Figure 12.7. Polarity of energy sources.

of two force holors F_A^j and F_a^j in the same sense of direction (figure 12.7(d)), the previous discussion indicates that each energy source should have its positive flow towards the junction.

Of course, this is because both force holors F_A^j and F_a^j act to increase the velocity holor V_i. The continuity of flow requires that the total force holor passing through the mass is the sum of the flows entering the junction. That is,

$$F^j = F_A^j + F_a^j.$$

In conventional mechanics terms, the sum of the unbalanced force holors F_A^j and F_a^j determines the acceleration of the mass m. The mechanical commutation matrixer approach, therefore, is merely another way of recognising and developing principal physical relations.

Suppose now that the force holors F_A^j and F_a^j are acting in opposite senses of direction, as in figure 12.7(e).

The force-source holor (F_a^j) in the sense of direction of the positive velocity holor (V_i) is physically modelled by an energy-transfer holor (F_a) directed towards the terminal (F_a^j). This is the force-source holor (F_a^j) that tends to increase the velocity holor V_i.

The opposite force holor F_A^j is physically modelled by an energy-transfer-source holor directed away from the terminal and has an energy-transfer holor at the terminal that yields

$$F^j = F_a^j - F_A^j.$$

Both physical situations (figures 12.7(d) and (e)) are in exact agreement with physical reality. The junction of the mechanical commutation matrixer representation corresponds to the free-body diagram of mechanics. However, for complicated mechanical commutation matrixers, several free-body diagrams may be required. An advantage of the mechanical commutation matrixer representation is that it is completely described by a single diagram.

For physical situations that are physically modelled by velocity-source holors, the procedure is simply to observe whether the energy source is in a positive or negative sense of direction. If it is positive, the positive terminal of the velocity-source holor is connected to the junction. When a velocity-source holor is directed opposite to the assumed positive

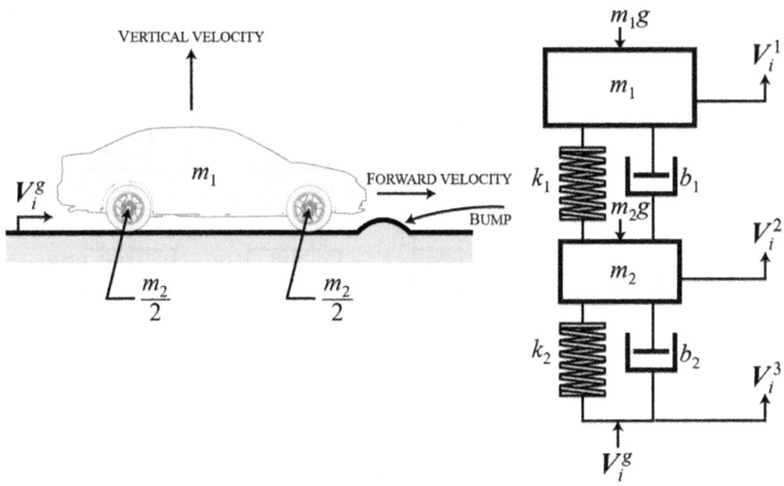

Figure 12.8. Automotive vehicle's suspension damper–spring–mass (*b–k–m*) mechanical homogeneous continuous dynamical system.

velocity-holor sense of direction, the negative terminal of the energy source is connected to the junction. These points are illustrated in the following examples.

12.6 Automotive vehicle's suspension mechanical commutation matrixer

Figure 12.8(a) shows an automotive vehicle approaching a bump in the on/off road surface. The suspension mechatronical control system of the automotive vehicle supports the vehicle body by means of springs (k_1) and shock absorbers (b_1).

The wheel-tyres exhibit a spring effect (k_2) that is close to the ideal spring, but the hysteresis normally associated with rubber pneumatic springs is physically modelled by an equivalent damper (b_2).

The constraint at the point of contact between the wheel-tyre and the on/off road surface is that the position of this point is defined.

If the automotive vehicle is driven at constant forward velocity along the on/off road surface, the vertical velocity holor constraint is, therefore, appropriately physically modelled by a velocity-source holor (V_i^S).

In addition, there are force-source holors to physically model the mass of the automotive vehicle and the mass of the wheels. These considerations lead to the ***damper–spring–mass*** (*b–k–m*) configuration shown in figure 12.8(b).

To obtain the mechanical commutation matrixer diagram one must first determine the number of independent velocity holors associated with the mechanical commutation matrixer. Each mass may move independently and this leads to terminals representing V_i^1 and V_i^2. The point of contact with the ground is V_i^3. The velocity holor of the ground is V_i^G.

Thus, there are three terminals plus a ground on the mechanical commutation matrixer diagram (figure 12.9). The connection of components between the terminals is obtained by referring to the image in figure 12.8(b).

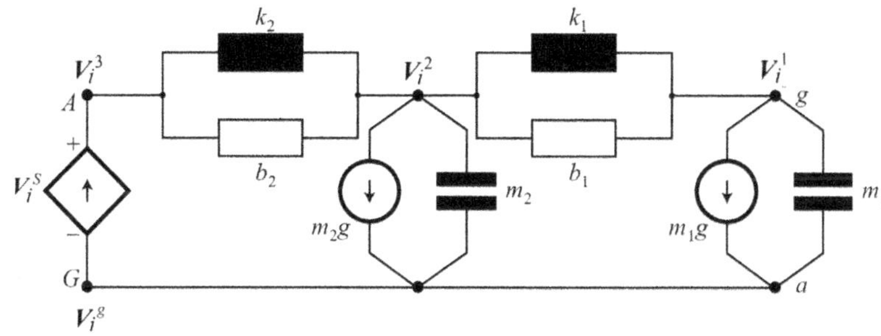

Figure 12.9. Mechanical homogeneous continuous dynamical system.

The velocity-source holor V_i^S acts in the sense of direction of the positive velocity holor. As a result, the positive terminal of this energy source is connected to V_i^3. This means that $V_i^S = V_i^3 - V_i^G$.

Had the reference-velocity holor sense of direction been selected as down, the negative terminal of V_S would have been attached to V_i^3.

In that case, $V_i^S = V_i^G - V_i^3$. The force-source holors due to the weight of the automotive vehicle (m_1g) and the weight of the wheels (m_2g) are directed away from their respective terminals (V_i^1 and V_i^2). Obviously, if the reference velocity holor was down, these sources would be pointed toward the terminals.

12.7 Determination of the analogous impedance holor of mechanical commutation matrixers

Since the physical commutation matrixer diagram is the basis on which the generalised dynamical systems approach rests, it is appropriate to review the process by which the physical commutation matrixer was obtained in previous chapters.

Let us consider the ***damper–spring–mass*** (*b–k–m*) mechanical homogeneous continuous dynamical system in figure 12.10(a).

The first step is to identify all the parts, which are constrained to move at the same velocity and show the sense of direction chosen to be positive. Number these points, since they will form the junctions or nodes of the mechanical-commutation matrixer diagram. In this case, there are three (3) independent velocity holors, V_i^1, V_i^2, V_i^3, and a reference velocity holor, V_i^g.

The holor equations are usually easier to write and the results easier to interpret if the positive holor velocity is taken in the same sense of direction at each part of the mechanical homogeneous continuous dynamical system. Now draw the 'matrix skeleton' of the mechanical commutation matrixer, i.e. the three (3) nodes corresponding to the three (3) parts of the mechanical continuous dynamical system at which the velocity holors are different, and the datum or reference node. In figure 12.10(b), these are indicated as V_i^1, V_i^2, V_i^3 and V_i^g.

The impedance-holor blocks corresponding to the passive mechanical commutation matrixer components are added to the velocity-difference holor, which is required in the force-velocity relation of the mechanical-commutation-matrixer

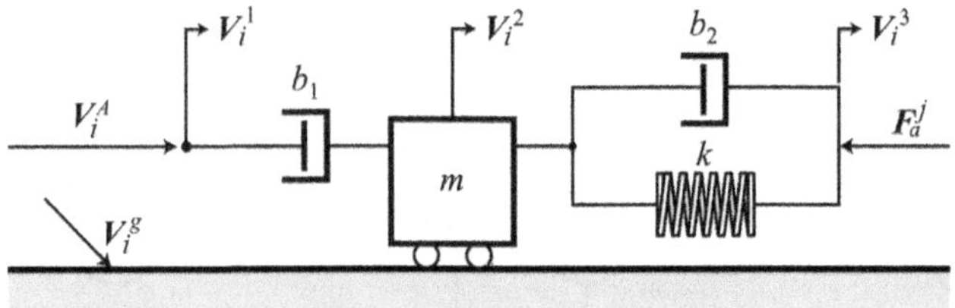

(a) MECHANICAL ARRANGEMENT

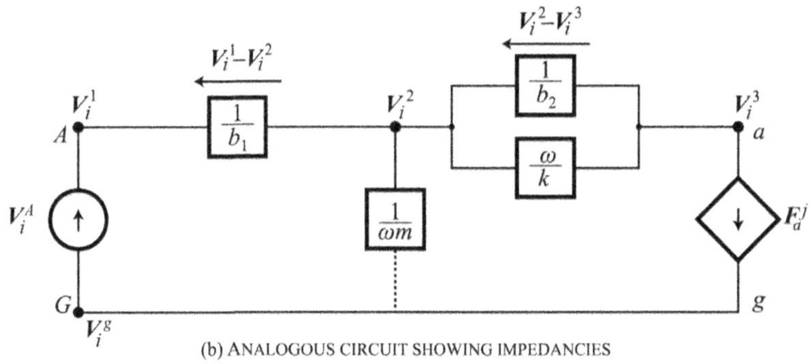

(b) ANALOGOUS CIRCUIT SHOWING IMPEDANCIES

Figure 12.10. Determination of the analogous mechanical circuit, showing impedances.

component. The damper, b_1, physically models an effect that is dependent on the relative velocity holor between nodes 1 and 2. Therefore, the impedance-holor block $(1/b_1)$ is inserted between these nodes. The damper, b_2, and the spring, k, both depend on the relative velocity holor between nodes 2 and 3. Thus, the damper block $(1/b_2)$ is connected in parallel with the spring impedance-holor block (ω/k) between nodes 2 and 3. The mass, m, on the other hand, physically models an effect that is dependent on the absolute velocity holor of node 2. To make this a two-terminal mechanical commutation matrixer component, one uses the ground as a fictitious terminal and connects the mass impedance holor block $(1/\omega m)$ between $\boldsymbol{V}_i^2$ and $\boldsymbol{V}_i^g$. Now the branches with the energy sources can be added. The velocity-source holor at node 1 constrains node 1 to move to the right at a velocity holor $\boldsymbol{V}_i$ relative to the datum. Therefore, on the mechanical commutation matrixer diagram this is represented by a velocity-source holor connected between node 1 and the datum. It is positive at node 1 because the velocity-source holor constrains node 1 to move in the positive sense of direction.

The force-source holor, $\boldsymbol{F}^j$, acting at node 3 would have moved node 3 in the negative sense of direction if it had been acting separately. Thus, the 'transfer of force' into node 3 must also be negative and the arrow is fixed away from node 3.

12.8 Damper–spring–mass mechanical homogeneous continuous dynamical system (III)

The **damper–spring–mass** (b–k–m) mechanical homogeneous continuous dynamical system (III) shown in figure 12.11(a) consists of three dampers, one spring and two masses. There are two known force-holor inputs ($\boldsymbol{F}_a^j$ and $\boldsymbol{F}_c^j$). One must present the node-to-datum holor equations in hypermatrix form for this mechanical homogeneous continuous dynamical system. Figure 12.11(b) shows the mechanical-commutation-matrixer representation of the *b–m–k* mechanical homogeneous continuous dynamical system. Since there are three unknown velocity holors, the required hypermatrix must be of the form:

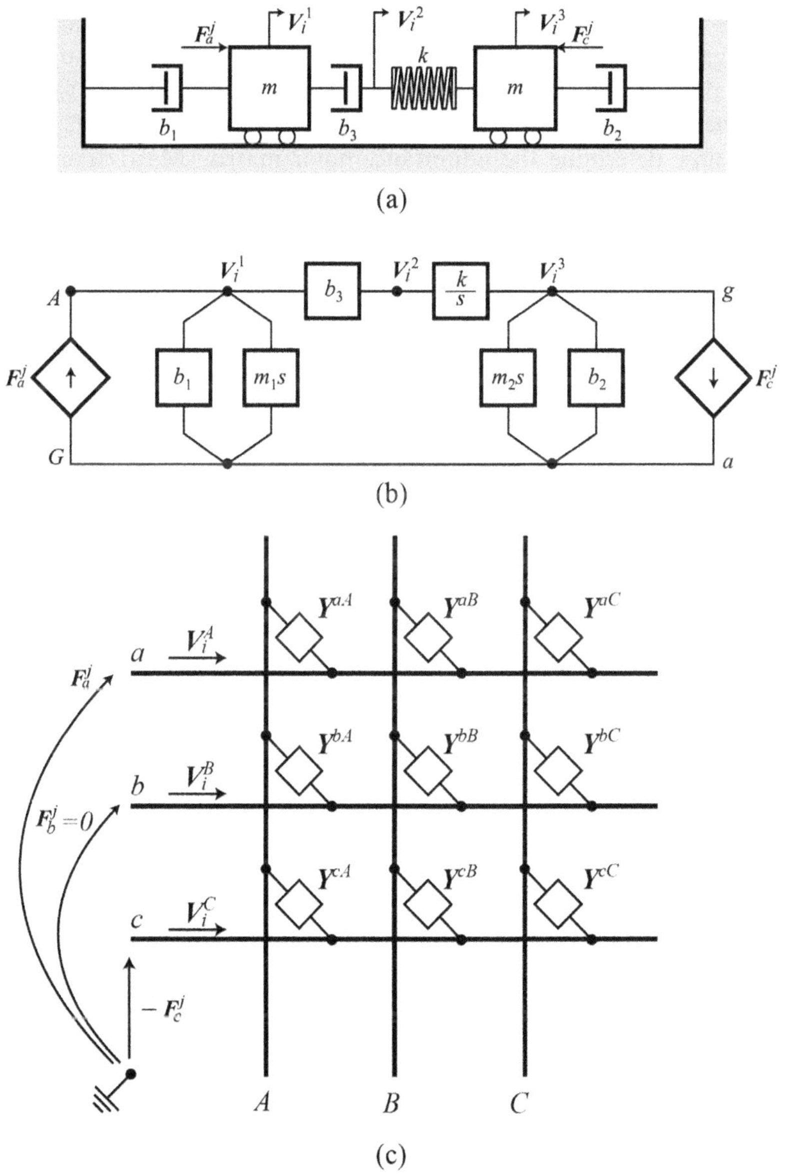

Figure 12.11. Damper–spring–mass (*b–k–m*) mechanical homogeneous continuous dynamical system (III) and its mechanical commutation matrixer.

$$\begin{array}{l}\text{node1} \Longrightarrow \\ \text{node2} \Longrightarrow \\ \text{node3} \Longrightarrow\end{array}\begin{bmatrix} \boldsymbol{Y}_{aA}^{ji} & \boldsymbol{Y}_{aB}^{ji} & \boldsymbol{Y}_{aC}^{ji} \\ \boldsymbol{Y}_{bA}^{ji} & \boldsymbol{Y}_{bB}^{ji} & \boldsymbol{Y}_{bC}^{ji} \\ \boldsymbol{Y}_{cA}^{ji} & \boldsymbol{Y}_{cB}^{ji} & \boldsymbol{Y}_{cC}^{ji} \end{bmatrix} \begin{bmatrix} \boldsymbol{V}_i^A \\ \boldsymbol{V}_i^B \\ \boldsymbol{V}_i^C \end{bmatrix} = \begin{bmatrix} \boldsymbol{F}_a^j \\ 0 \\ -\boldsymbol{F}_c^j \end{bmatrix}.$$

The column vector of unknowns has sinusoidal effective values at the velocity-holor nodes; these are designated as $\boldsymbol{V}_i^A$ and $\boldsymbol{V}_i^C$. The column vector of knowns consists of sinusoidal effective values of the force-holor inputs. The force-holor effective value-entering node A due to the force-holor inputs is merely $\boldsymbol{F}_a^j$. There is no force-holor input entering node B exactly and thus there is a zero in the known column vector at node B. For node C one may observe that the force-holor effective value, F_a, acts away from the node. As a result, the force-holor effective value is $-F_c$.

One may now determine the admittance-holor matrix. Matrix component $\boldsymbol{Y}_{aA}^{ji}$ represents the admittance holor from terminal A ($\boldsymbol{V}_i^A$) with all the other terminals grounded ($\boldsymbol{V}_i^B = \boldsymbol{V}_i^C = 0$). This is the parallel combination of b_1, ωm_1 and b_3, so that:

$$\boldsymbol{Y}_{aA}^{ji} = \begin{Vmatrix} b_1 + d_3 & -m_1\omega \\ \omega m_1 & b_1 + b_3 \end{Vmatrix}.$$

In a similar way, $\boldsymbol{Y}_{bB}^{ji}$ represents the equivalent admittance holor from terminal B ($\boldsymbol{V}_i^B$) with all the other terminals grounded ($\boldsymbol{V}_i^A = \boldsymbol{V}^C = 0$):

$$\boldsymbol{Y}_{bB}^{ji} = \begin{Vmatrix} b_3 & k/\omega \\ -k/\omega & b_3 \end{Vmatrix},$$

and $\boldsymbol{Y}_{cC}^{ji}$ applies to terminal C ($\boldsymbol{V}_i^C$) with the other terminals grounded

$$\left(\boldsymbol{V}_i^A = \boldsymbol{V}_i^B = 0\right):$$

$$\boldsymbol{Y}_{cC}^{ji} = \begin{Vmatrix} b_2 & -\omega m_2 + k/\omega \\ \omega m_2 - k/\omega & b_2 \end{Vmatrix}.$$

The other admittance holors are merely the negative of all the common terms. For instance, to find $\boldsymbol{Y}_{aB}^{ji}$ and $\boldsymbol{Y}_{bA}^{ji}$, there is only one common term, b_3, in

$$\boldsymbol{Y}_{aB}^{ji} = \boldsymbol{Y}_{bA}^{ji} = \begin{Vmatrix} -b_3 & 0 \\ 0 & -b_3 \end{Vmatrix}.$$

To find $\boldsymbol{Y}_{bC}^{ji}$ and $\boldsymbol{Y}_{cB}^{ji}$, the only common term in $\boldsymbol{Y}_{bB}^{ji}$ and $\boldsymbol{Y}_{cC}^{ji}$ is k/ω, and thus

$$\boldsymbol{Y}_{bC}^{ji} = \boldsymbol{Y}_{cB}^{ji} = \begin{Vmatrix} 0 & k/\omega \\ -k/\omega & 0 \end{Vmatrix}.$$

Since there are no common terms between $\boldsymbol{Y}^{aA}$ and $\boldsymbol{Y}^{cC}$, the admittance holors

$$\boldsymbol{Y}_{aC}^{ji} = \boldsymbol{Y}_{cA}^{ji} = \begin{Vmatrix} 0 & 0 \\ 0 & 0 \end{Vmatrix}.$$

The complete hypermatrix is therefore:

$$\begin{bmatrix} \begin{Vmatrix} b_1+b_3 & -\omega m_1 \\ \omega m_1 & b_1+b_3 \end{Vmatrix} & \begin{Vmatrix} -b_3 & 0 \\ 0 & -b_3\end{Vmatrix} & \begin{Vmatrix}0&0\\0&0\end{Vmatrix} \\ \begin{Vmatrix}-b_3&0\\0&-b_3\end{Vmatrix} & \begin{Vmatrix} b_3 & k/\omega \\ -k/\omega & b_3\end{Vmatrix} & \begin{Vmatrix}0 & k/\omega\\ -k/\omega & 0\end{Vmatrix} \\ \begin{Vmatrix}0&0\\0&0\end{Vmatrix} & \begin{Vmatrix}0&k/\omega\\-k/\omega&0\end{Vmatrix} & \begin{Vmatrix} b_2 & -\omega m_2 + k/\omega \\ \omega m_2 - k/\omega & b_2\end{Vmatrix}\end{bmatrix} \begin{bmatrix}\begin{Vmatrix}V_c^A\\V_s^A\end{Vmatrix}\\ \begin{Vmatrix}V_c^B\\V_s^B\end{Vmatrix}\\ \begin{Vmatrix}V_c^C\\V_s^C\end{Vmatrix}\end{bmatrix} = \begin{bmatrix}\begin{Vmatrix}F_a^c\\F_a^s\end{Vmatrix}\\ \begin{Vmatrix}0\\0\end{Vmatrix}\\ \begin{Vmatrix}-F_c^c\\-F_c^s\end{Vmatrix}\end{bmatrix}.$$

To express the effective values at any terminal in terms of the basic mechanical commutation matrixer components and the known input effective values, one must use determinants, for example:

$$V_i^A = \begin{Vmatrix}V_c^A\\V_s^A\end{Vmatrix} = \frac{\left|\begin{matrix} \begin{Vmatrix}F_a^c\\F_a^s\end{Vmatrix} & \begin{Vmatrix} -b_3 & 0 \\ 0 & -b_3\end{Vmatrix} & \begin{Vmatrix}0&0\\0&0\end{Vmatrix} \\ \begin{Vmatrix}0&0\\0&0\end{Vmatrix} & \begin{Vmatrix} b_3 & k/\omega \\ -k/\omega & b_3\end{Vmatrix} & \begin{Vmatrix}0 & k/\omega\\ -k/\omega & 0\end{Vmatrix} \\ \begin{Vmatrix}-F_c^c\\-F_c^s\end{Vmatrix} & \begin{Vmatrix}0&k/\omega\\-k/\omega&0\end{Vmatrix} & \begin{Vmatrix} b_2 & -\omega m_2 + k/\omega \\ \omega m_2 - k/\omega & b_2\end{Vmatrix}\end{matrix}\right|}{\left|\begin{matrix} \begin{Vmatrix} b_1+b_3 & -\omega m_1 \\ \omega m_1 & b_1+b_3 \end{Vmatrix} & \begin{Vmatrix} -b_3 & 0 \\ 0 & -b_3\end{Vmatrix} & \begin{Vmatrix}0&0\\0&0\end{Vmatrix} \\ \begin{Vmatrix}-b_3&0\\0&-b_3\end{Vmatrix} & \begin{Vmatrix} b_3 & k/\omega \\ -k/\omega & b_3\end{Vmatrix} & \begin{Vmatrix}0 & k/\omega\\ -k/\omega & 0\end{Vmatrix} \\ \begin{Vmatrix}0&0\\0&0\end{Vmatrix} & \begin{Vmatrix}0&k/\omega\\-k/\omega&0\end{Vmatrix} & \begin{Vmatrix} b_2 & -\omega m_2 + k/\omega \\ \omega m_2 - k/\omega & b_2\end{Vmatrix}\end{matrix}\right|},$$

and alternatively, if one evaluates the determinants

$$V_i^A = \frac{\left[\begin{Vmatrix} b_3 & k/\omega \\ -k/\omega & b_3\end{Vmatrix}\boldsymbol{D} - \begin{Vmatrix}0&k/\omega\\-k/\omega&0\end{Vmatrix}\begin{Vmatrix}0&k/\omega\\-k/\omega&0\end{Vmatrix}\right]\begin{Vmatrix}F_c^a\\F_s^a\end{Vmatrix} + \left[\begin{Vmatrix}-b_3&0\\0&-b_3\end{Vmatrix}\begin{Vmatrix}0&k/\omega\\-k/\omega&0\end{Vmatrix}\right]\begin{Vmatrix}-F_c^c\\-F_s^c\end{Vmatrix}}{\left[\begin{Vmatrix} b_3 & k/\omega \\ -k/\omega & b_3\end{Vmatrix}\boldsymbol{D} - \begin{Vmatrix}0&k/\omega\\-k/\omega&0\end{Vmatrix}\begin{Vmatrix}0&k/\omega\\-k/\omega&0\end{Vmatrix}\right]\begin{Vmatrix} b_1+b_3 & -\omega m_1 \\ \omega m_1 & b_1+b_3 \end{Vmatrix} - \left[\begin{Vmatrix}(-b_3)^2 & 0\\ 0 & (-b_3)^2\end{Vmatrix}\boldsymbol{D}\right]},$$

where

$$\boldsymbol{D} = \begin{Vmatrix} b_2 & -\omega m_2 + \dfrac{k}{\omega} \\ \omega m_2 - \dfrac{k}{\omega} & b_2 \end{Vmatrix}.$$

One may find expressions for $\boldsymbol{V}_i^B$ and $\boldsymbol{V}_i^C$ in a similar way.

12.9 Damper–mass mechanical commutation matrixer

Figure 12.12(a) shows a **damper–mass** (*b–m*) mechanical homogeneous continuous dynamical system with two dampers and a mass. Either of the two impedance-holor mechanical commutation matrixers shown in figures 12.12(b) and (c) may represent this mechanical homogeneous continuous dynamical system, showing that the force-holor that acts to accelerate the mass is the same in both mechanical commutation matrixers.

In the mechanical commutation matrixer shown in figure 12.12(b), the mesh holor equation hypermatrix has the form:

$$\begin{bmatrix} Z_{ij}^{Aa} & Z_{ij}^{An} \\ Z_{ij}^{Na} & Z_{ij}^{Nn} \end{bmatrix} \begin{bmatrix} F_A^j \\ F_a^j \end{bmatrix} = \begin{bmatrix} V_i^A \\ 0 \end{bmatrix},$$

where the energy source has no contribution to mesh (F_a^j) and contributes a velocity-holor rise (V_i^A) to mesh (F_A^j).

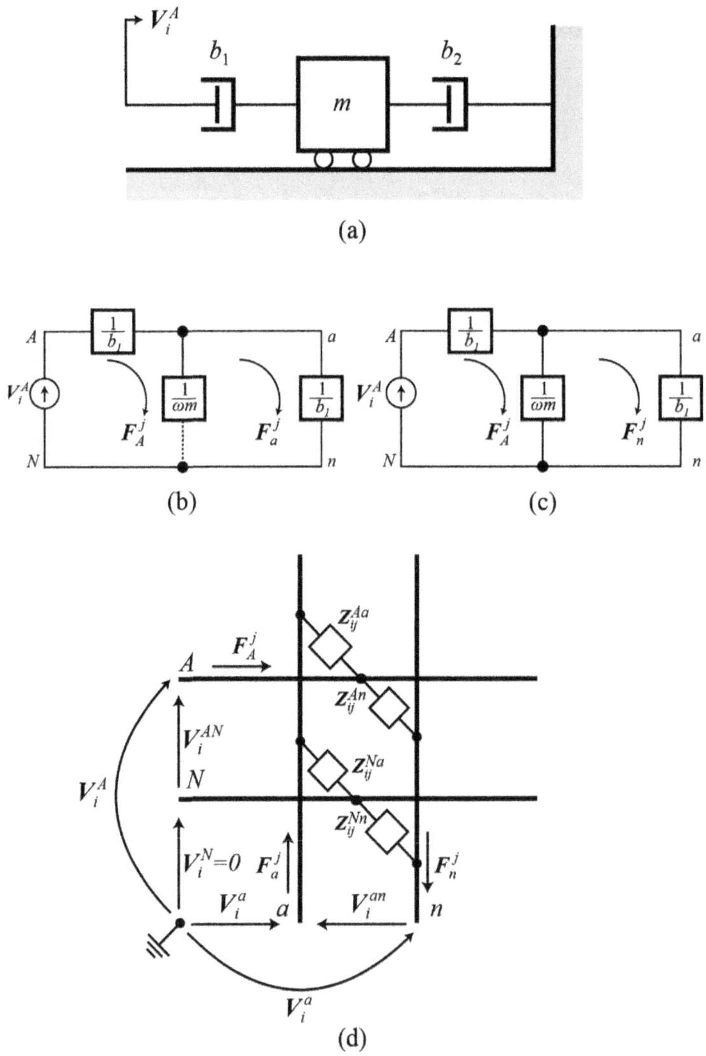

Figure 12.12. Mechanical commutation matrixer.

The sum of the impedance holor around the meshes yields:

$$\boldsymbol{Z}_{ij}^{Aa} = \begin{Vmatrix} 1/b_1 & 1/\omega m \\ -1/\omega m & 1/b_1 \end{Vmatrix}, \quad \boldsymbol{Z}_{ij}^{Nn} = \begin{Vmatrix} 1/b_2 & 1/\omega m \\ -1/\omega m & 1/b_2 \end{Vmatrix},$$

and the common term is $1/\omega m$, so that

$$\boldsymbol{Z}_{ij}^{An} = \boldsymbol{Z}_{ij}^{Na} = \begin{Vmatrix} 0 & -\dfrac{1}{\omega m} \\ \dfrac{1}{\omega m} & 0 \end{Vmatrix}.$$

Therefore the matrix becomes:

$$\begin{bmatrix} \begin{Vmatrix} 1/b_1 & 1/\omega m \\ -1/\omega m & 1/b_1 \end{Vmatrix} & \begin{Vmatrix} 0 & 1/\omega m \\ -1/\omega m & 0 \end{Vmatrix} \\ \begin{Vmatrix} 0 & 1/\omega m \\ -1/\omega m & 0 \end{Vmatrix} & \begin{Vmatrix} 1/b_2 & 1/\omega m \\ -1/\omega m & 1/b_2 \end{Vmatrix} \end{bmatrix} \begin{bmatrix} \begin{Vmatrix} F_A^c \\ F_A^s \end{Vmatrix} \\ \begin{Vmatrix} F_a^c \\ F_a^s \end{Vmatrix} \end{bmatrix} = \begin{bmatrix} \begin{Vmatrix} V_c^A \\ V_s^A \end{Vmatrix} \\ \begin{Vmatrix} 0 \\ 0 \end{Vmatrix} \end{bmatrix}.$$

In this mechanical commutation matrixer (figure 12.12(b)) the force holor applied to the mass is $\boldsymbol{F}_M^j = \boldsymbol{F}_A^j - \boldsymbol{F}_a^j$.

One must therefore solve the matrix for $\boldsymbol{F}_A^j$ and $\boldsymbol{F}_a^j$ and then subtract the results.

$$\boldsymbol{F}_A^j = \begin{Vmatrix} F_A^c \\ F_A^s \end{Vmatrix} = \frac{\begin{Vmatrix} \begin{Vmatrix} V_c^A \\ V_s^A \end{Vmatrix} & \begin{Vmatrix} 0 & \frac{1}{\omega m} \\ -\frac{1}{\omega m} & 0 \end{Vmatrix} \\ \begin{Vmatrix} 0 \\ 0 \end{Vmatrix} & \begin{Vmatrix} \frac{1}{b_2} & \frac{1}{\omega m} \\ -\frac{1}{\omega m} & \frac{1}{b_2} \end{Vmatrix} \end{Vmatrix}}{\begin{Vmatrix} \begin{Vmatrix} \frac{1}{b_1} & \frac{1}{\omega m} \\ -\frac{1}{\omega m} & \frac{1}{b_1} \end{Vmatrix} & \begin{Vmatrix} 0 & \frac{1}{\omega m} \\ -\frac{1}{\omega m} & 0 \end{Vmatrix} \\ \begin{Vmatrix} 0 & \frac{1}{\omega m} \\ -\frac{1}{\omega m} & 0 \end{Vmatrix} & \begin{Vmatrix} \frac{1}{b_2} & \frac{1}{\omega m} \\ -\frac{1}{\omega m} & \frac{1}{b_2} \end{Vmatrix} \end{Vmatrix}}$$

$$= \frac{\begin{Vmatrix} \dfrac{1}{b_2} & \dfrac{1}{\omega m} \\ -\dfrac{1}{\omega m} & \dfrac{1}{b_2} \end{Vmatrix} \begin{Vmatrix} V_c^A \\ V_s^A \end{Vmatrix}}{\begin{Vmatrix} \dfrac{1}{b_1} & \dfrac{1}{\omega m} \\ -\dfrac{1}{\omega m} & \dfrac{1}{b_1} \end{Vmatrix} \begin{Vmatrix} \dfrac{1}{b_2} & \dfrac{1}{\omega m} \\ -\dfrac{1}{\omega m} & \dfrac{1}{b_2} \end{Vmatrix} - \begin{Vmatrix} 0 & \dfrac{1}{\omega m} \\ -\dfrac{1}{\omega m} & 0 \end{Vmatrix} \begin{Vmatrix} 0 & \dfrac{1}{\omega m} \\ -\dfrac{1}{\omega m} & 0 \end{Vmatrix}},$$

$$
\boldsymbol{F}_a^j = \begin{Vmatrix} F_a^c \\ F_a^s \end{Vmatrix} = \frac{\begin{Vmatrix} \begin{Vmatrix} \frac{1}{b_1} & \frac{1}{\omega m} \\ -\frac{1}{\omega m} & \frac{1}{b_1} \end{Vmatrix} & \begin{Vmatrix} V_c^A \\ V_s^A \end{Vmatrix} \\ \begin{Vmatrix} 0 & \frac{1}{\omega m} \\ -\frac{1}{\omega m} & 0 \end{Vmatrix} & \begin{Vmatrix} 0 \\ 0 \end{Vmatrix} \end{Vmatrix}}{\begin{Vmatrix} \begin{Vmatrix} \frac{1}{b_1} & \frac{1}{\omega m} \\ -\frac{1}{\omega m} & \frac{1}{b_1} \end{Vmatrix} & \begin{Vmatrix} 0 & \frac{1}{\omega m} \\ -\frac{1}{\omega m} & 0 \end{Vmatrix} \\ \begin{Vmatrix} 0 & \frac{1}{\omega m} \\ -\frac{1}{\omega m} & 0 \end{Vmatrix} & \begin{Vmatrix} \frac{1}{b_2} & \frac{1}{\omega m} \\ -\frac{1}{\omega m} & \frac{1}{b_2} \end{Vmatrix} \end{Vmatrix}}
$$

$$
= \frac{\begin{Vmatrix} 0 & \frac{1}{\omega m} \\ -\frac{1}{\omega m} & 0 \end{Vmatrix} \begin{Vmatrix} V_c^A \\ V_s^A \end{Vmatrix}}{\begin{Vmatrix} \frac{1}{b_1} & \frac{1}{\omega m} \\ -\frac{1}{\omega m} & \frac{1}{b_1} \end{Vmatrix} \begin{Vmatrix} \frac{1}{b_2} & \frac{1}{\omega m} \\ -\frac{1}{\omega m} & \frac{1}{b_2} \end{Vmatrix} - \begin{Vmatrix} 0 & \frac{1}{\omega m} \\ -\frac{1}{\omega m} & 0 \end{Vmatrix} \begin{Vmatrix} 0 & \frac{1}{\omega m} \\ -\frac{1}{\omega m} & 0 \end{Vmatrix}},
$$

and thus

$$
\boldsymbol{F}_M^j = \boldsymbol{F}_A^j - \boldsymbol{F}_a^j = \frac{\begin{Vmatrix} b_2 & 0 \\ 0 & b_2 \end{Vmatrix} \begin{Vmatrix} V_c^A \\ V_s^A \end{Vmatrix}}{\begin{Vmatrix} \frac{1}{b_1} & \frac{1}{\omega m} \\ -\frac{1}{\omega m} & \frac{1}{b_1} \end{Vmatrix} \begin{Vmatrix} \frac{1}{b_2} & \frac{1}{\omega m} \\ -\frac{1}{\omega m} & \frac{1}{b_2} \end{Vmatrix} - \begin{Vmatrix} 0 & \frac{1}{\omega m} \\ -\frac{1}{\omega m} & 0 \end{Vmatrix} \begin{Vmatrix} 0 & \frac{1}{\omega m} \\ -\frac{1}{\omega m} & 0 \end{Vmatrix}} +
$$

$$
- \frac{\begin{Vmatrix} b_2 & 0 \\ 0 & b_2 \end{Vmatrix} \begin{Vmatrix} V_c^A \\ V_s^A \end{Vmatrix}}{\begin{Vmatrix} \frac{1}{b_1} & 0 \\ 0 & \frac{1}{b_1} \end{Vmatrix} \begin{Vmatrix} \frac{1}{b_2} & 0 \\ 0 & \frac{1}{b_2} \end{Vmatrix} + \begin{Vmatrix} 0 & \frac{1}{\omega m} \\ -\frac{1}{\omega m} & 0 \end{Vmatrix} \left[\begin{Vmatrix} \frac{1}{b_1} & 0 \\ 0 & \frac{1}{b_1} \end{Vmatrix} + \begin{Vmatrix} \frac{1}{b_2} & 0 \\ 0 & \frac{1}{b_2} \end{Vmatrix} \right]}.
$$

Let us solve the same problem using the mechanical commutation matrixer shown in figure 12.12(c).

The mesh holor equations by inspection are:

$$
\begin{bmatrix} \begin{Vmatrix} \frac{1}{b_1} & 0 \\ 0 & \frac{1}{b_1} \end{Vmatrix} + \begin{Vmatrix} \frac{1}{b_2} & 0 \\ 0 & \frac{1}{b_2} \end{Vmatrix} & \begin{Vmatrix} -\frac{1}{b_2} & 0 \\ 0 & -\frac{1}{b_2} \end{Vmatrix} \\ \begin{Vmatrix} -\frac{1}{b_2} & 0 \\ 0 & -\frac{1}{b_2} \end{Vmatrix} & \begin{Vmatrix} \frac{1}{b_2} & \frac{1}{\omega m} \\ -\frac{1}{\omega m} & \frac{1}{b_2} \end{Vmatrix} \end{bmatrix} \begin{bmatrix} \begin{Vmatrix} F_A^c \\ F_A^s \end{Vmatrix} \\ \begin{Vmatrix} F_n^c \\ F_n^s \end{Vmatrix} \end{bmatrix} = \begin{bmatrix} \begin{Vmatrix} V_c^A \\ V_s^A \end{Vmatrix} \\ \begin{Vmatrix} 0 \\ 0 \end{Vmatrix} \end{bmatrix}.
$$

Note that this matrix is different from the matrix determined for the mechanical commutation matrixer shown in figure 12.12(b). In this mechanical commutation matrixer, the force holor applied to the mass is $\boldsymbol{F}_n^j = \begin{Vmatrix} F_n^c \\ F_n^s \end{Vmatrix}$ alone. Thus if one solves for $\boldsymbol{F}_n^j$ one obtains:

$$\boldsymbol{F}_M^j = \boldsymbol{F}_n^{\;j} = \begin{Vmatrix} F_n^c \\ F_n^s \end{Vmatrix} = \frac{\begin{vmatrix} \begin{Vmatrix} \frac{1}{b_1} & 0 \\ 0 & \frac{1}{b_1} \end{Vmatrix} + \begin{Vmatrix} \frac{1}{b_2} & 0 \\ 0 & \frac{1}{b_2} \end{Vmatrix} & \begin{Vmatrix} V_c^A \\ V_s^A \end{Vmatrix} \\ \begin{Vmatrix} -\frac{1}{b_2} & 0 \\ 0 & -\frac{1}{b_2} \end{Vmatrix} & \begin{Vmatrix} 0 \\ 0 \end{Vmatrix} \end{vmatrix}}{\begin{vmatrix} \begin{Vmatrix} \frac{1}{b_1} & 0 \\ 0 & \frac{1}{b_1} \end{Vmatrix} + \begin{Vmatrix} \frac{1}{b_2} & 0 \\ 0 & \frac{1}{b_2} \end{Vmatrix} & \begin{Vmatrix} -\frac{1}{b_2} & 0 \\ 0 & -\frac{1}{b_2} \end{Vmatrix} \\ \begin{Vmatrix} -\frac{1}{b_2} & 0 \\ 0 & -\frac{1}{b_2} \end{Vmatrix} & \begin{Vmatrix} \frac{1}{b_2} & \frac{1}{\omega m} \\ -\frac{1}{\omega m} & \frac{1}{b_2} \end{Vmatrix} \end{vmatrix}}$$

$$= \frac{\begin{Vmatrix} \frac{1}{b_2} & 0 \\ 0 & \frac{1}{b_2} \end{Vmatrix} \begin{Vmatrix} V_c^A \\ V_s^A \end{Vmatrix}}{\left[\begin{Vmatrix} \frac{1}{b_1} & 0 \\ 0 & \frac{1}{b_1} \end{Vmatrix} + \begin{Vmatrix} \frac{1}{b_2} & 0 \\ 0 & \frac{1}{b_2} \end{Vmatrix} \right] \begin{Vmatrix} \frac{1}{b_2} & \frac{1}{\omega m} \\ -\frac{1}{\omega m} & \frac{1}{b_2} \end{Vmatrix} - \begin{Vmatrix} \frac{1}{b_2} & 0 \\ 0 & \frac{1}{b_2} \end{Vmatrix} \begin{Vmatrix} \frac{1}{b_2} & 0 \\ 0 & \frac{1}{b_2} \end{Vmatrix}},$$

or

$$F^M = \frac{\begin{Vmatrix} \frac{1}{b_2} & 0 \\ 0 & \frac{1}{b_2} \end{Vmatrix} \begin{Vmatrix} V_c^A \\ V_s^A \end{Vmatrix}}{\begin{Vmatrix} \frac{1}{b_1} & 0 \\ 0 & \frac{1}{b_1} \end{Vmatrix} \begin{Vmatrix} \frac{1}{b_2} & 0 \\ 0 & \frac{1}{b_2} \end{Vmatrix} + \left[\begin{Vmatrix} \frac{1}{b_1} & 0 \\ 0 & \frac{1}{b_1} \end{Vmatrix} + \begin{Vmatrix} \frac{1}{b_2} & 0 \\ 0 & \frac{1}{b_2} \end{Vmatrix} \right] \begin{Vmatrix} 0 & \frac{1}{\omega m} \\ -\frac{1}{\omega m} & 0 \end{Vmatrix}}.$$

Moreover, the results are the same in both mechanical commutation matrixers.

References

Beer F P and Johnston E R Jr 1984 *Vector Mechanics for Engineers—Statics and Dynamics* (New York: McGraw-Hill)

Neimark J I and Fufaiev N A 1972 *Dynamics of non-holonomic systems.* (Providence, RI: American Mathematical Society) (Translated from Russian ed, 1967)

Chapter 13

Fluidic homogeneous continuous dynamical systems as fluidic commutation matrixers

Windmills, which are used in the great plains of Holland and North Germany to supply the want of falling water, afford another instance of the action of velocity. The sails are driven by air in motion—by wind.

Hermann von Helmholtz

13.1 Fluidic-transmission line

Let us consider one of the basic fluidic commutation matrixer components—the fluidic-transmission line. Figure 13.1(a) shows a **mechano-fluidic** (MF) pump.

How are we to describe this physical commutation matrixer component mathematically? To answer this question we must first decide what we wish to know about the passage of the fluid to a transmission line. One thing we want to know is the rate at which liquid is passing through the fluidic-transmission line. We call this the flow, and for a fluidic-commutation matrixer, it is designated as q. Another thing we might need to know is the size of the MF pump required to produce the flow. Instead of physical size, the pressure often indicates the capability of an MF pump, p_A, it can develop.

If we were able to relate the pressure, $p_{A,}$, to the flow, q, we would have a type of mathematical model of the fluidic transmission line. But can we do this? We cannot, because the flow through the fluidic-transmission line also depends on the pressure, p_a, at the end of the fluidic-transmission line.

If p_a were equal to p_A, there would be no flow through the fluidic-transmission line. Thus the mathematical model must relate the flow, q, to the pressure difference, $p_A - p_a$.

doi:10.1088/978-0-7503-1350-6ch13

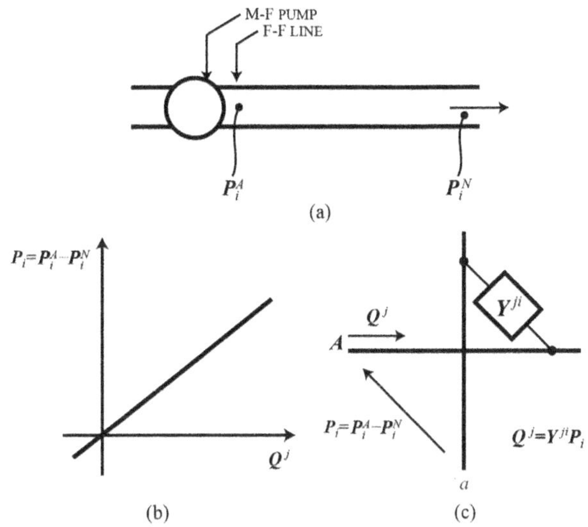

Figure 13.1. Fluidic-transmission line fluidic-commutation matrixer component.

These quantities, flow and pressure difference, are the signal holors for fluidic commutation matrixers. Flow is the energy-transfer holor (V^j) and pressure difference is the energy-potential-difference holor (F_i).

To determine the mathematical model of the fluidic-transmission line, one could perform some experiments on the fluidic-transmission line in which the pressure difference was changed and the resulting flow measured.

Figure 13.1(b) shows some typical results for data points. As the pressure difference increases, the flow increases. The mathematical model, of course, must hold for an infinite number of pressure differences—flow combinations—not just the three data points.

13.2 Node-to-datum holor equations for the fluidic commutation matrixer

The node-to-datum holor equations are usually referred to as node holor equations. The method is a very convenient way of formulating holor equations when the desired output is of the energy-potential-difference holor type. When we apply the node-to-datum method, the coefficients turn out simplest if admittance holors are used. Accordingly, consider the fluidic homogeneous continuous dynamical system shown in figure 13.2.

Suppose it is the analogous fluidic commutation matrixer for a particular fluidic homogeneous continuous dynamical system that has flow-source holors, Q^A and Q^a. One seeks to find the equilibrium holor equations that describe the pressure holors, P_A and P_a.

One begins by applying equation (10.1) (KCL) to nodes A and a.

$$\text{node } A \rightarrow Q_1^j + Q_2^j = Q_A^j, \tag{13.1a}$$

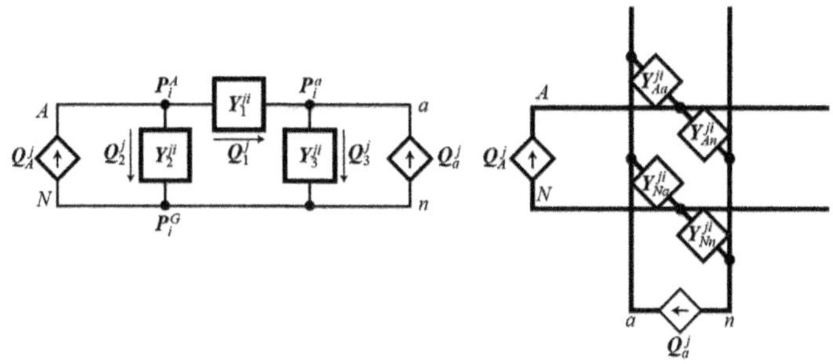

Figure 13.2. Analogous fluidic homogeneous continuous dynamical system involving admittance holors.

$$\text{node } a \rightarrow \boldsymbol{Q}_3^j - \boldsymbol{Q}_1^j = \boldsymbol{Q}_a^j, \tag{13.1b}$$

where the flow holors away from the mode are positive on the left-hand side of the holor equation and negative on the right-hand side.

Once all the node holor equations are written (in this case the two equations (13.1a) and (13.1b), one formulates the fluidic-component relations. The three fluidic-component relations are:

$$\left(\boldsymbol{P}_i^A - \boldsymbol{P}_i^a\right)\boldsymbol{Y}_1^{ji} = \boldsymbol{Q}_1^j \tag{13.2a}$$

$$\left(\boldsymbol{P}_i^A - \boldsymbol{P}_i^G\right)\boldsymbol{Y}_2^{ji} = \boldsymbol{Q}_2^j \tag{13.2b}$$

$$\left(\boldsymbol{P}_i^a - \boldsymbol{P}_i^G\right)\boldsymbol{Y}_3^{ji} = \boldsymbol{Q}_3^j. \tag{13.2c}$$

Now, if equations (13.2) are substituted into equations (13.1) and one notes that the reference pressure holor, $\boldsymbol{P}_i^{\mathrm{G}}$, is zero, the result is as follows.

For node A:

$$\left(\boldsymbol{Y}_1^{ji} + \boldsymbol{Y}_2^{ji}\right)\boldsymbol{P}_i^A - \boldsymbol{Y}_1^{ji}\boldsymbol{P}_i^a = \boldsymbol{Q}_A^j. \tag{13.3a}$$

For node a:

$$\boldsymbol{Y}_1^{ji}\boldsymbol{P}_{\mathrm{i}}^{\mathrm{A}} + \left(\boldsymbol{Y}_1^{ji} + \boldsymbol{Y}_3^{ji}\right)\boldsymbol{P}_i^a = \boldsymbol{Q}_{\mathrm{a}}^j. \tag{13.3b}$$

Let us consider the individual terms in these two holor equations and try to interpret them.

Node A: $(\boldsymbol{Y}_1^{ji} + \boldsymbol{Y}_2^{ji})\boldsymbol{P}_i^A$—total flow holor away from node A due to $\boldsymbol{P}_i^A$ only, with all other pressure holors set to zero;

$\boldsymbol{Y}_1^{ji}\boldsymbol{P}_i^a$—total flow holor away from node a due to $\boldsymbol{P}_i^a$ only, with all other pressure holors set to zero;

$\boldsymbol{Q}_A^j$—total flow holor into node A from all flow-holor sources connected to node A.

Node a: $(\boldsymbol{Y}_1^{ji} + \boldsymbol{Y}_3^{ji})\boldsymbol{P}_i^a$—total flow holor away from node a due to $\boldsymbol{P}_i^a$, with all other pressure holors set to zero;
$\boldsymbol{Y}_1^{ji}\boldsymbol{P}_i^A$—total flow holor away from node a due to $\boldsymbol{P}_i^A$, with all other pressure holors set to zero.

At this stage, it is valuable to note the coefficients in the holor equations, and the matrix form of the holor equations is useful for this purpose.

$$\begin{matrix}\text{node } A \Longrightarrow \\ \text{node } a \Longrightarrow\end{matrix}\begin{bmatrix} \boldsymbol{Y}_1^{ji} + \boldsymbol{Y}_2^{ji} & -\boldsymbol{Y}_1^{ji} \\ -\boldsymbol{Y}_1^{ji} & \boldsymbol{Y}_1^{ji} + \boldsymbol{Y}_3^{ji} \end{bmatrix}\begin{bmatrix} \boldsymbol{P}_i^A \\ \boldsymbol{P}_i^a \end{bmatrix} = \begin{bmatrix} \boldsymbol{Q}_A^j \\ \boldsymbol{Q}_a^j \end{bmatrix}. \tag{13.4}$$

In this array of holor admittances, the terms on the main diagonal are positive and all the others are negative. This is true in general for node-to-datum equilibrium holor equations.

The general form is:

$$\begin{bmatrix} \boldsymbol{Y}_{Aa}^{ji} & \boldsymbol{Y}_{An}^{ji} \\ \boldsymbol{Y}_{Na}^{ji} & \boldsymbol{Y}_{Nn}^{ji} \end{bmatrix}\begin{bmatrix} \boldsymbol{P}_i^A \\ \boldsymbol{P}_i^a \end{bmatrix} = \begin{bmatrix} \boldsymbol{Q}_A^j \\ \boldsymbol{Q}_a^j \end{bmatrix}. \tag{13.5}$$

Note that $\boldsymbol{Y}_{Aa}^{ji}$ is the sum of the admittance holor of the positive branches connected to node A and $\boldsymbol{Y}_{Nn}^{ji}$ is the sum of admittance holors at node a. $\boldsymbol{Y}_{An}^{ji}$ is the negative of the admittance holor of the branch connecting nodes A and a. If every mechanical commutation matrixer component connecting two nodes is 'bilateral' (i.e. the pressure/flow relationship is the same for flow in both senses of direction), $\boldsymbol{Y}_{An}^{ji} = \boldsymbol{Y}_{Na}^{ji}$ or $\boldsymbol{Y}^{jk} = \boldsymbol{Y}^{kj}$, and the matrixer is symmetrical. With some practice, it is a simple matter to write mnemonically the equilibrium holor equations of many physical heterogeneous continuous dynamical hypersystems directly in this form (i.e. 'by inspection' of the physical commutation matrixer), thus reducing the likelihood of error. One will clarify the procedure in the examples given in section 13.3.

13.3 Loop and mesh holor equations for the fluidic commutation matrixer

The loop and mesh holor equations method is a convenient means of formulating holor equations when the unknowns are of the energy-transfer-holor type. This method favours the use of impedance-holor parameters.

Let us consider, therefore, the fluidic commutation matrixer shown in figure 13.3. The fluidic commutation matrixer represents a particular fluidic homogeneous continuous dynamical system that has input pressure-source holors of sinusoidal type with effective values of pressure-source holors $\boldsymbol{P}_i^A$ and $\boldsymbol{P}_i^a$.

One wants to find holor equations that describe the sinusoidal flow holors through all the impedance-holor components. In this case, there are three distinct impedance holors and three distinct flow holors. However, only two of these are

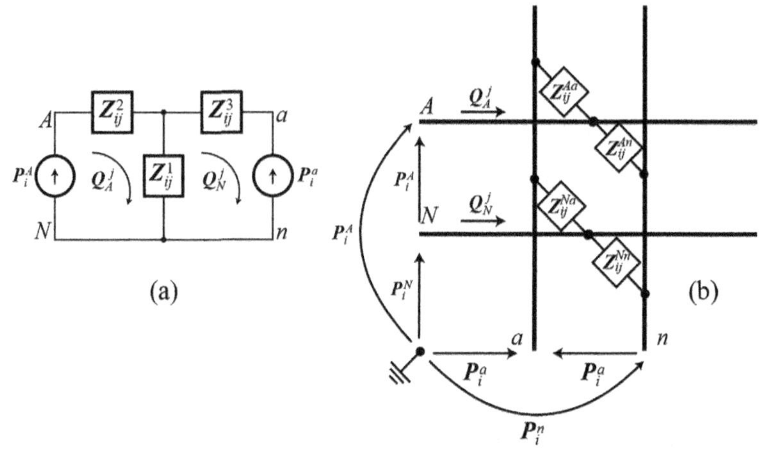

Figure 13.3. Analogous physical commutation matrixer (fluidic commutation matrixer) involving impedance holors.

independent. For instance, if $\boldsymbol{Q}_A^j$ is the flow holor through $\boldsymbol{Z}_{ij}^2$, and $\boldsymbol{Q}_a^j$ is the flow holor through $\boldsymbol{Z}_{ij}^3$, then the flow holor through $\boldsymbol{Z}_{ij}^1$ is $\boldsymbol{Q}_A^j - \boldsymbol{Q}_a^j$.

The flow holors $\boldsymbol{Q}_A^j$ and $\boldsymbol{Q}_a^j$ are the mesh flow holors and they are the most convenient choice for the unknowns, since one may apply equation (10.1b) around the closed mesh.

If one applies equation (10.1b) around meshes A and a:

$$\text{mesh } A \Longrightarrow \left(\boldsymbol{P}_i^A - \boldsymbol{P}_i^y\right) + \left(\boldsymbol{P}_i^y - \boldsymbol{P}_i^g\right) = \left(\boldsymbol{P}_i^A - \boldsymbol{P}_i^g\right), \tag{13.6a}$$

$$\text{mesh } a \Longrightarrow \left(\boldsymbol{P}_i^g - \boldsymbol{P}_i^y\right) + \left(\boldsymbol{P}_i^y - \boldsymbol{P}_i^a\right) = \left(\boldsymbol{P}_i^g - \boldsymbol{P}_i^a\right). \tag{13.6b}$$

The meshes are the open spaces in the impedance-holor fluidic commutation matrixer.

One always takes the flow holor in each mesh in a clockwise sense of direction. One may now substitute the fluidic commutation matrixer component relations into the mesh equations (13.6a) and (13.6b) to obtain:

$$\text{mesh } A \Longrightarrow \boldsymbol{Z}_{ij}^2\boldsymbol{Q}_A^j + \boldsymbol{Z}_{ij}^1\left(\boldsymbol{Q}_A^j - \boldsymbol{Q}_a^j\right) = \boldsymbol{P}_i^A, \tag{13.7a}$$

$$\text{mesh } a \Longrightarrow \boldsymbol{Z}_{ij}^1\left(\boldsymbol{Q}_a^j - \boldsymbol{Q}_A^j\right) + \boldsymbol{Z}_{ij}^3\boldsymbol{Q}_a^j = \boldsymbol{P}_i^a. \tag{13.7b}$$

Alternatively, if one groups the flow holors together in equation (13.7) the result is:

$$\text{mesh } A \Longrightarrow \left(\boldsymbol{Z}_{ij}^1 + \boldsymbol{Z}_{ij}^2\right)\boldsymbol{Q}_A^j - \boldsymbol{Z}_{ij}^1\boldsymbol{Q}_a^j = \boldsymbol{P}_i^A, \tag{13.8a}$$

$$\text{mesh } a \Longrightarrow \boldsymbol{Z}_{ij}^1\boldsymbol{Q}_A^j + \left(\boldsymbol{Z}_{ij}^1 + \boldsymbol{Z}_{ij}^3\right)\boldsymbol{Q}_a^j = \boldsymbol{P}_j^a. \tag{13.8b}$$

One may interpret the terms in equation (13.8) as follows.

In mesh A: $(\boldsymbol{Z}_{ij}^1 + \boldsymbol{Z}_{ij}^2)\boldsymbol{Q}_A^j$—the sum of the pressure-holor drops across each impedance holor in a sense of the flow-holor direction and due to $\boldsymbol{Q}_A^j$ only; all the other flow holors are set to zero;

$\boldsymbol{Z}_{ij}^1\boldsymbol{Q}_a^j$—the sum of the pressure-holor drops across impedance holors, in a sense of direction $\boldsymbol{Q}_A^j$, but due to flow holor $\boldsymbol{Q}_a^j$; all the other flow holors are set to zero;

In mesh a: $-\boldsymbol{Z}_{ij}^1\boldsymbol{Q}_A^j$—the sum of the pressure-holor drops in mesh a in a sense of flow direction due to $\boldsymbol{Q}_A^j$; all the other holor flows are set to zero;

$(\boldsymbol{Z}_{ij}^1 + \boldsymbol{Z}_{ij}^3)\boldsymbol{Q}_2^j$—the sum of the pressure-holor drops due to holor flow $\boldsymbol{Q}_a^j$; all the other flow holors are set to zero.

In matrix form, equation (13.8) may be written as:

$$\begin{matrix}\text{mesh } A \Longrightarrow \\ \text{mesh } a \Longrightarrow\end{matrix}\begin{bmatrix} \boldsymbol{Z}_{ij}^1 + \boldsymbol{Z}_{ij}^2 & -\boldsymbol{Z}_{ij}^1 \\ -\boldsymbol{Z}_{ij}^1 & \boldsymbol{Z}_{ij}^1 + \boldsymbol{Z}_{ij}^3 \end{bmatrix}\begin{bmatrix} \boldsymbol{Q}_A^j \\ \boldsymbol{Q}_a^j \end{bmatrix} = \begin{bmatrix} \boldsymbol{P}_i^{\mathrm{A}} \\ \boldsymbol{P}_i^{\mathrm{a}} \end{bmatrix}. \tag{13.9}$$

In the impedance-holor array, the terms on the main diagonal are positive and all others are negative. This is true for mesh holor equations when the assumed senses of flow direction are all clockwise or all counter clockwise.

The general form for the mesh holor equations is:

$$\begin{bmatrix} \boldsymbol{Z}_{ij}^{Aa} & \boldsymbol{Z}_{ij}^{An} \\ \boldsymbol{Z}_{ij}^{Na} & \boldsymbol{Z}_{ij}^{Nn} \end{bmatrix}\begin{bmatrix} \boldsymbol{Q}_A^j \\ \boldsymbol{Q}_a^j \end{bmatrix} = \begin{bmatrix} \boldsymbol{P}_i^A \\ \boldsymbol{P}_i^a \end{bmatrix}. \tag{13.10}$$

The fluidic-commutation-matrixer component $\boldsymbol{Z}_{ij}^{Aa}$ is the sum of all the impedance holors in mesh A and $\boldsymbol{Z}_{ij}^{Nn}$ is the sum of all the impedance holors in mesh a. The impedance holors $\boldsymbol{Z}_{ij}^{An}$ and $\boldsymbol{Z}_{ij}^{Na}$ are equal and are the negative of all the common terms in $\boldsymbol{Z}_{ij}^{Aa}$ and $\boldsymbol{Z}_{ij}^{Nn}$, provided that all the circulating flow holors are taken in the same sense of direction (e.g. all clockwise).

Part III

Physical Matrixers as Physical Commutators

IOP Publishing

Mechatronics

Dynamical systems approach and theory of holors

B T Fijalkowski

Chapter 14

Mechanical commutation matrixer commutators for conventional DC and AC magneto-mechano-dynamical electrical machines

Everything should be made as simple as possible, but not simpler.'

Albert Einstein

14.1 Introduction

The 'Gramme machine' or 'Gramme dynamo' is an AC–DC commutator synchronous generator designed by Belgian inventor Zénobe Gramme in 1871. It was the first **magneto-mechano-dynamical** (MMD) electrical machine to be used for industrially generated DC electrical energy. His innovation was to use many armature windings, wound on a doughnut shaped armature and commutated (switched) with a many-segmented mechanical ring commutator, to rectify (smooth) the output AC waveforms, generating nearly constant DC electrical energy. Figure 14.1 shows a hand-cranked physical model from the 1870s, designed for use in laboratories. It used the laminated permanent magnets invented by Jamin (Fontaine 1878).

The phenomenon of unipolar electrical-current conductivity had already been applied in 1883, when the British physicist, W Ritchie, constructed an MMD electrical machine, i.e. an AC–DC/DC–AC commutator dynamotor (figure 14.2), with a rotary **mechanical commutation matrixer** (MCM) commutator, i.e. an MCM AC–DC/DC–AC commutator, which can be taken for the prototype of a contemporary mechanical ring- or disc-type polysectorial MCM AC–DC/DC–AC commutator.

MCM AC–AC and AC–DC/DC–AC commutators are characteristic components that are common to a large group of DC electrical machines. MCM AC–AC and AC–DC/DC–AC commutators (figure 14.3) are conventionally made by assembling hard-drawn copper sectors (bars or segments) interleaved with 0.7–2.0 mm sheet mica or micanite, with these separators being 'undercut' by about 1 mm.

Figure 14.1. The Gramme dynamo (Fontaine 1878).

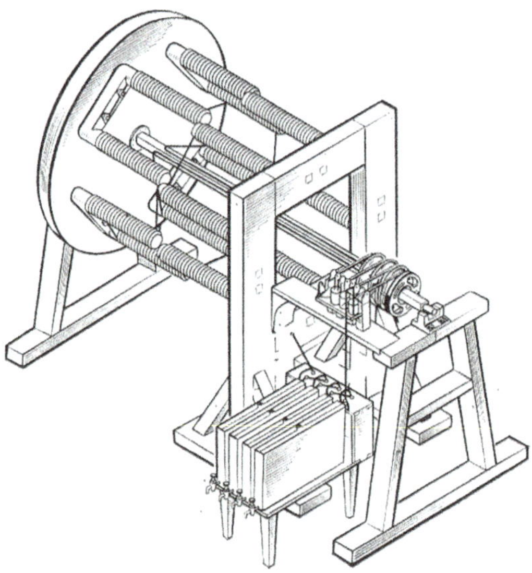

Figure 14.2. Ritchie's MCM AC–DC/DC–AC commutator dynamotor (Veselovsky and Kopylov 1984).

The stationary or movable brushes of a suitable carbon/graphite content are mounted in boxes with spring loading to hold them against the MCM commutator surface, with a medium to strong pressure depending on the application. The circumferential brush width is typically 2–3 sector widths (10–20 mm) and about 30 mm axially. One brush-arm per pole is employed, except for in certain four-pole wave-wound MCM AC–DC/DC–AC commutator dynamotors (generator/motors), which have two brush-arms in adjacent positions to facilitate maintenance.

The MCM AC–DC/DC–AC commutator is a necessary part of a DC commutator electrical machine, while the use of an MCM AC–AC commutator made possible AC commutator motors that have certain valuable features. Finally, MCM AC–DC/DC–AC commutators are used in single-armature converters, the function of which is to convert AC–DC or DC–AC. Thus, the first MCM with 'mechanical-contact' electrical valves became an MCM connection of sliding contacts, setting up, for example, collector phosphorus bronze springs or carbon/graphite brushes, sliding, respectively, on the surface of the collector split-rings or hard-drawn copper sectors (bars or segments). Rotary sliding-contacts, **mechano-electrical** (ME) AC–DC commutators, termed 'MIMO MCM DC–AC commutators' by the author (figures 14.4–14.8)—which are widely used in MCM AC–DC commutator generators, according to the dynamical systems approach on physical heterogeneous continuous simple dynamical hypersystems, that is, MMD electrical machines—are a series of copper sectors (bars or segments), insulated from one another and forming matrixer-input primary conductive collectors of an MCM AC–DC commutator.

The copper sectors, i.e. matrixer-input primary conductive collectors, are connected to the commutating sections, i.e. phase-winding terminals, of an armature winding mounted on the rotor of an electrical machine. Stationary carbon brushes are placed on the MCM DC–AC commutator cylindrical surface with copper sectors.

Carbon brushes forming the matrixer-output secondary conductive collectors of an MCM commutator dynamical system are connected by means of sliding-contacts, i.e. electrical valves (copper-sector/carbon-brush contacts), to the copper sectors, i.e. the matrixer-input primary conductive collectors of an MCM DC–AC commutator.

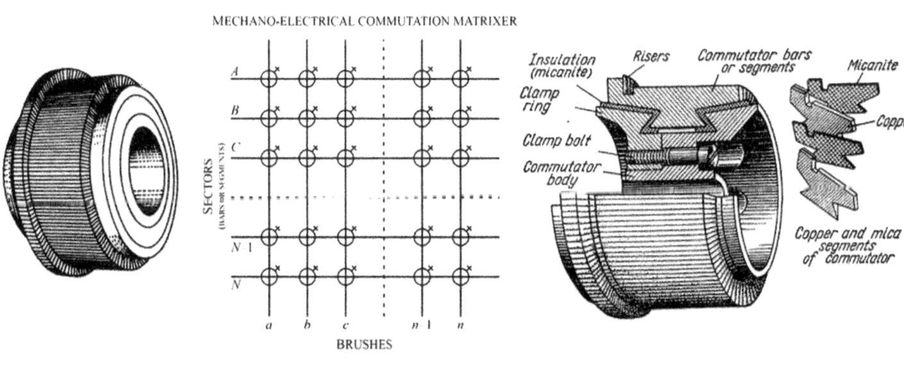

Figure 14.3. Section through an MCM AC–AC and/or AC–DC/DC–AC commutator (Kuznetsov 1960, Kasatkin and Perekalin 1970, Fijalkowski 1973).

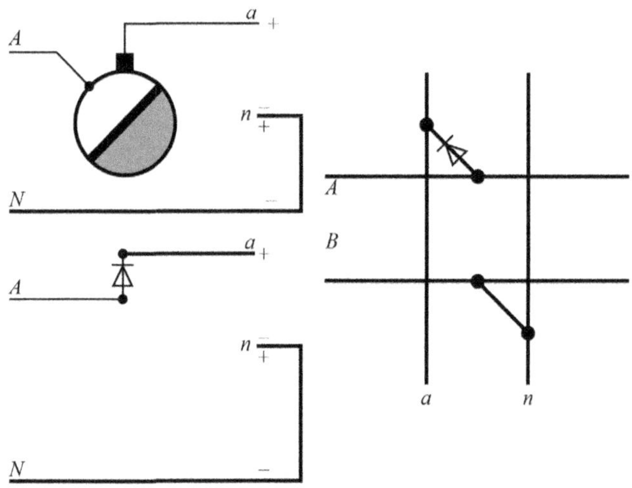

(a)

Figure 14.4. The single-phase, half-wave MCM AC–DC commutator, with a single homogeneous cathode commutation group and a neutral, acting as a one-way MCM AC–DC rectifier.

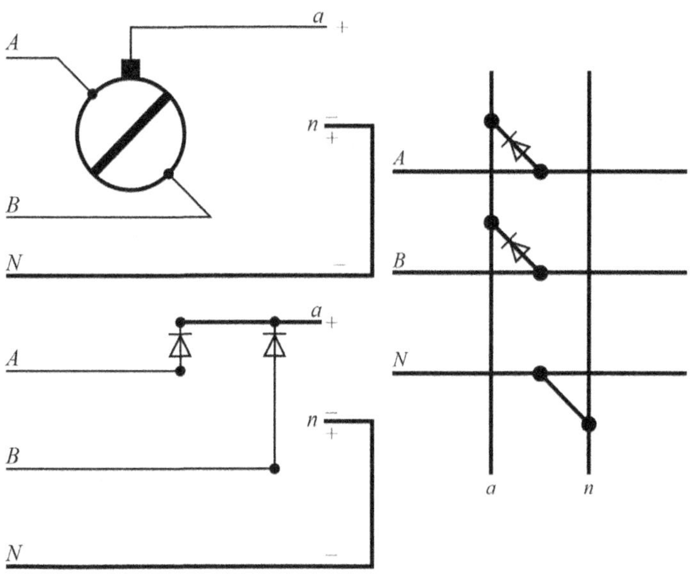

(b)

Figure 14.5. The two-phase, half-wave MCM AC–DC commutator, with a single homogeneous cathode commutation group and a neutral, acting as a one-way MCM AC–DC rectifier.

By means of these sliding contacts acting as electrical valves, it is possible to connect the phase windings of the rotary armature winding in the electrical machine to an external electrical commutation matrixer dynamical system.

The operating principle of the MIMO ECM AC–DC/DC–AC commutator will be considered in more detail later on. The MIMO MCM AC–DC/DC–AC

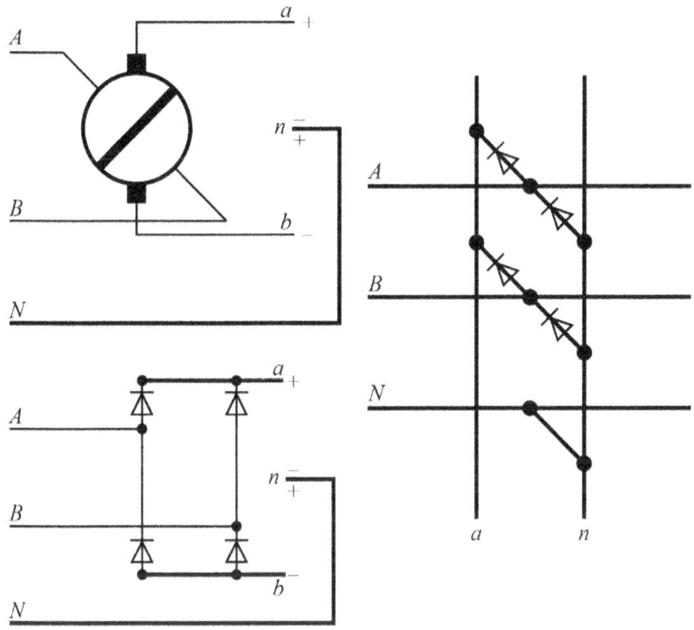

(c)

Figure 14.6. The two-phase full-wave MCM AC–DC commutator, with double heterogeneous cathode and anode commutation groups and a neutral, acting as a double-way MCM AC–DC rectifier.

commutator was in the past, and to date still is, a necessary component of AC–DC/DC–AC commutator dynamotors.

In addition, the use of an MIMO MCM AC–AC and/or AC–DC/DC–AC commutator made possible yesterday and today's polyphase MCM AC–AC and/ or AC–DC/DC–AC commutator electrical machines.

Finally, MIMO MCM AC–DC commutators have been and are used in single-armature MCM AC–DC commutator rotary converters, the function of which is to convert three-phase alternating voltages to direct voltage, or vice versa.

On the other hand, MCM AC–AC and AC–DC/DC–AC commutator dynamotors are quite expensive to manufacture compared with AC and DC commutatorless machines, while they demand special care and are not so rugged.

MCM AC–AC and AC–DC/DC–AC commutator dynamotors are not as maintenance-free as AC and DC commutatorless dynamotors, due to brush and MCM commutator wear.

The roundness of the MIMO MCM AC–AC and AC–DC/DC–AC commutator is the most important design parameter to reduce the brush noise of the MCM AC–DC/DC–AC commutator dynamotor. The roundness of the MIMO MCM AC–AC or AC–DC/DC–AC commutator needs to be maintained in order to maintain the brush noise under a certain level throughout the life span of the MCM AC–AC and AC–DC/DC–AC commutator dynamotor.

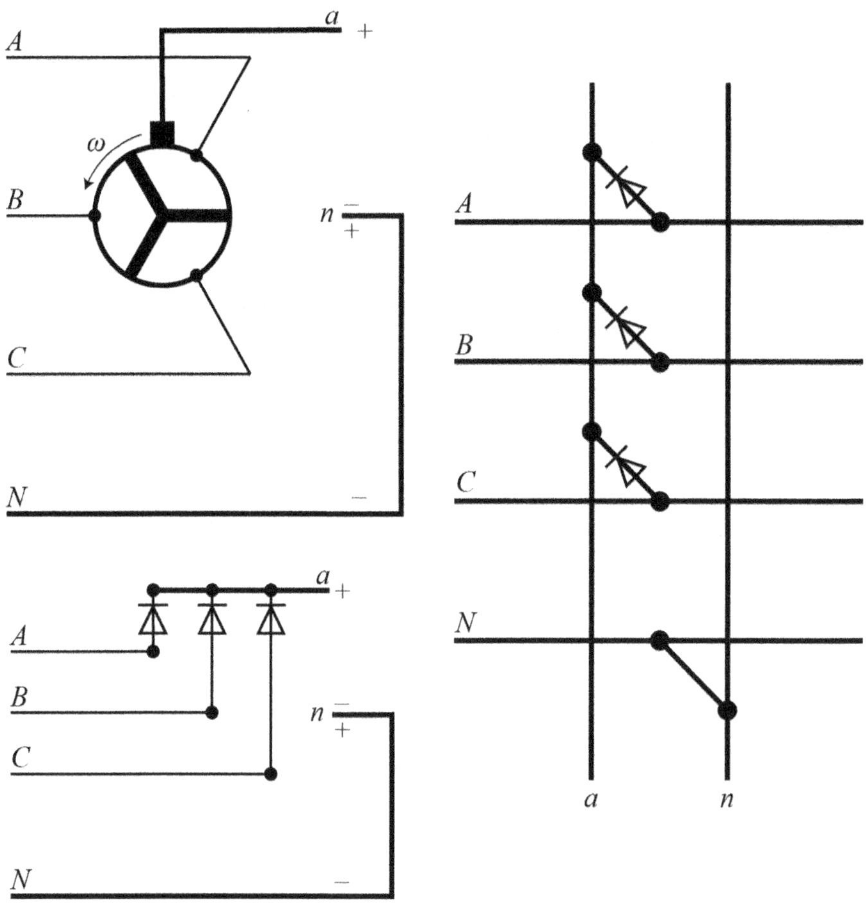

(d)

Figure 14.7. The three-phase, half-wave MCM AC–DC commutator, with a single homogeneous cathode commutation group and a neutral, acting as a one-way MCM AC–DC rectifier.

AC and DC commutatorless electrical machines do not permit smooth and economical control of voltage and current (for MMD generation) or torque and speed control (for MMD motoring) of the kind that is necessary for some forms of industrial and traction propulsion complex dynamical hypersystems.

The use of the integral MCM commutator in MCM AC–AC and AC–DC/DC–AC commutator dynamotors makes it possible to produce variable-voltage and variable-current or variable-torque and variable-speed dynamotors with flat or drooping voltage–current and torque–speed characteristics and with the power factor in some of them brought to unity.

All of these useful characteristics of MCM AC–AC and AC–DC/DC–AC commutator dynamotors are achieved at the expense of complexity and the extra

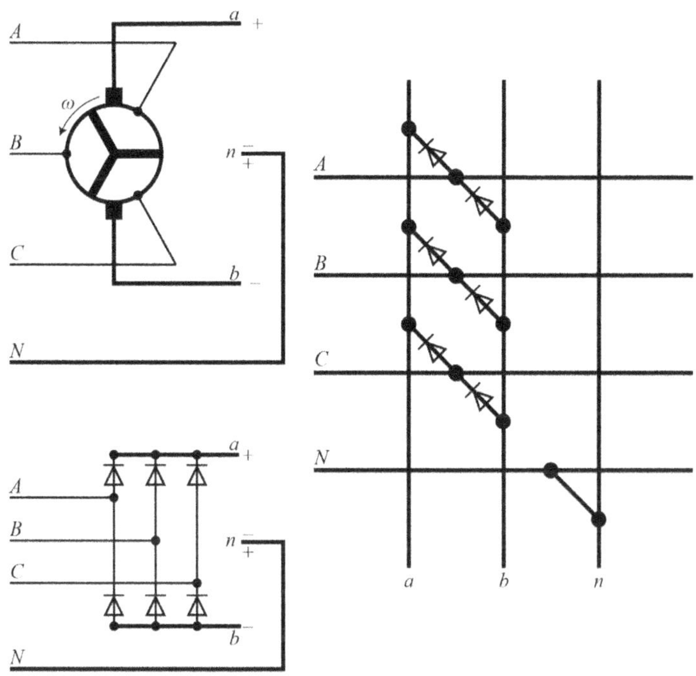

(e)

Figure 14.8. The three-phase full-wave MCM AC–DC commutator, with double heterogeneous cathode and anode commutation groups and a neutral, acting as a double-way MCM AC–DC rectifier.

cost of the MMD electrical machine, with reduced reliability and worse operating conditions.

The main difficulty in building MCM AC–AC and AC–DC/DC AC commutator dynamotors is to obtain satisfactory ME commutation of alternating and direct currents, especially under starting conditions.

In order to make an MCM AC–AC commutator dynamotor suitable for AC supply it is necessary to adapt its magnetic circuit to an alternating magnetic field, reduce the reactance of the windings to a minimum and, finally, improve the conditions of the contacted commutation in the MCM AC–AC commutator dynamotors. The first two requirements may be considered as being suitably fulfilled at the expense of considerable complexity and the increased cost of the MMD electrical machine. As for securing good commutation conditions, nothing that has been done in this direction has satisfied scientists, who do not think that the problem has been properly resolved and are looking for a better solution.

14.2 MCM AC–DC/DC–AC commutator dynamotors

Industry has used MCM AC–DC/DC–AC commutator dynamotors (generators/ motors) for years because of their outstanding voltage and/or current as well as torque and/or speed control capabilities, down to essentially zero (0) revolutions per second (rps), and because of their good starting and low-speed torque. MCM AC–DC/DC–AC commutator dynamotors are usually much more expensive than standard AC commutatorless dynamotors, and the high-current part (the 'armature') of the MCM AC–DC/DC–AC commutator dynamotor is on the rotor. In getting from the DC terminals into the armature, these high currents must pass through brushes gearing on an MCM AC–DC/DC–AC commutator, a rotating AC–DC rectifier/DC–AC inverter.

Wear on the MCM AC–DC/DC–AC commutator and the brushes adds substantially to maintenance costs and reduces reliability. MCM AC–DC/DC–AC commutator motors have the further disadvantage that large DC power supplies are generally not available unless especially provided for.

In the last few years, cheaper and higher power electronic-control equipment has been developed, which converts AC–DC right at the dynamotor.

Thus, small and moderate-sized MCM AC–DC/DC–AC commutator motors can be supplied in effect right from the AC mains, eliminating the need for special power supplies. Such an electronic controller can also readily adjust input DC voltage to the MCM AC–DC/DC–AC commutator motor. Typical applications are propulsion motors in electric traction railroads or automotive vehicles, aircraft and ships, and also electrical drills and hand tools.

The use of MCM AC–DC/DC–AC commutators gives the electrical machines several valuable properties, of which the first is the possibility of AC–DC rectifying the polyphase AC induced voltages in the rotating armature winding into a single-phase DC induced voltage at the carbon brushes, and/or DC–AC inverting the DC induced voltage at the carbon brushes into polyphase AC induced voltages in the rotating armature winding.

The generation of an AC induced voltage in the armature winding of an MCM AC–DC commutator generator is exactly the same as in an AC commutatorless generator. As the armature winding rotates, an alternating current is produced in it. Some rotating AC–DC rectifier must be inserted between the armature winding and the terminals so that the terminal AC induced voltage can always act upon the external load in the same sense of direction.

This rotating AC–DC rectifier must reverse the connection of the winding conductors to the external circuit at the instant when the AC induced voltage of the conductors is zero and changing its sense of direction. This rotating AC–DC rectifier is termed by the author the 'MCM AC–DC commutator', and the rotating DC–AC inverter is termed the 'MCM DC–AC commutator'. The Achilles' heel for an MCM AC–DC/DC–AC commutator dynamotor is its MCM commutator. Consistently transferring DC from a rotating shaft to the stationary point, or, vice versa, from a stationary point to the rotating shaft, requires materials that can carry DC yet withstand friction and arcing. MCM commutators require periodic maintenance, decrease dynamotor life and limit the maximum speed at which the dynamotor can turn.

14.3 Schrage MCM AC–AC commutator motor

The AC commutatorless, hyposynchronous wound-rotor motor, frequency converter and adjustable-ratio transformer were first combined into a single structure by K H Schrage of Sweden to provide a complete, self-contained, adjustable-speed EM motor (Laughton and Say 1985). The electrical machine consists of a special phase-wound motor, whose ‘rotor’ acts as the ‘primary’ (it is connected to the three-phase electrical energy source through a set of slip rings) and the ‘stator’, which performs as the ‘secondary’. This is just the reverse of the functions of the stator and rotor windings in the standard AC commutatorless, hyposynchronous wound-rotor motor. In addition, there is a third winding that functions as an electromagnetic link between the primary and the secondary and as a source of slip-frequency voltage for the secondary. This so-called MCM commutator ‘tertiary’ winding is placed directly over the primary and is connected to an MCM AC–AC commutator mounted on the shaft end opposite the slip rings. Instead of being connected to a resistance controller, the secondary winding is wired to three pairs of carbon brushes that ride on the MCM AC–AC commutator. Speed control is effected by moving a lever or turning a handwheel that shifts the three pairs of carbon brushes. Thus, the recovery of slip energy can be achieved in a single electrical machine by incorporating an MCM AC–AC commutator in the rotor. ‘Schrage’ (rotor-fed) and ‘doubly fed’ (stator-fed) MCM AC–AC commutator motors have some commercial importance. Both have a ‘shunt’ speed–torque characteristic and can operate both above and below synchronous speed. A three-phase MCM AC–AC commutator motor with a ‘series’ characteristic is also available.

Summing up, the Schrage MCM AC–AC commutator motor has its primary winding on the rotor, connected to the supply through slip rings and carbon brushes. The rotor also carries a low-voltage MCM commutator tertiary winding with conductor located in the same slots as, and above, those of the primary. The secondary winding is on the stator. The primary and secondary windings are similar to those of the conventional AC commutatorless induction motor. The brush gear comprises two movable rockers, each fitted with three carbon-brush spindles per pair of poles. The two rockers are geared together, each being fitted with a toothed segment. The two segments mesh with pinions fitted to a short shaft to which either a handwheel or a small pilot actuator is connected. The gearing is so arranged that the movement of the handwheel or pilot actuator causes the carbon-brush rockers to move in opposite senses of direction.

The carbon brushes attached to each rocker move over separate portions of the MCM commutator surface to enable these brushes to be placed ‘in line’ or to be moved in either sense of direction, so that more or fewer commutator sectors are included between a brush on one rocker and the corresponding carbon brush on another rocker. The bus-rings of each rocker are connected to the secondary (stator) winding as in figure 14.9, which represents a two-pole electrical machine.

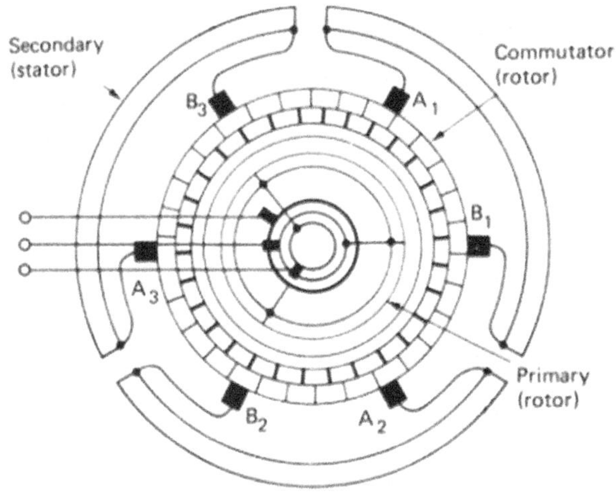

Figure 14.9. Schrage MCM AC–AC commutator motor (Laughton and Say 1985).

As the primary winding is on the rotor, AC induced voltages at the brushes of slip frequency are induced in the secondary (stator) winding. The AC induced voltage at the carbon brushes is also of this frequency.

The AC induced voltage induced in each coil of the MCM tertiary commutator winding is constant at all speeds, and therefore the AC induced voltages injected via the carbon brushes into the secondary winding are proportional to the number of MCM commutator sectors included between the corresponding carbon brushes on the two rockers, i.e. between the brushes connected to a particular phase of the secondary. Thus when these carbon brushes are in line, the secondary winding has no AC induced voltage injected into it and the MCM AC–AC commutator motor will run with its natural slip. When the carbon brushes are moved so that the injected AC induced voltage opposes current, the angular velocity is reduced (slip positive); for the opposite movement, the angular velocity is increased (slip negative).

In the Schrage MCM AC–AC commutator motor with a synchronous angular velocity ω_s and a carbon brush-separation electrical angle θ, the no-load angular velocity is

$$\omega = \omega_s \left(1 - k \sin \frac{1}{2}\theta\right),$$

where k is a constant, depending on the number of turns in the secondary and commutator windings.

Because of its simplicity of operation and the ease with which a large stepless angular-velocity range can be obtained (above and below synchronism), Schrage MCM AC–AC commutator motors have been and are still used for many moderately small industrial applications.

14.4 Exemplary applications of mechanical commutation matrixer (MCM) commutators

14.4.1 AC commutatorless motor—an MCM AC–DC commutator generator for converting AC–DC

In order to obtain DC electrical energy, one can first convert AC electrical energy into mechanical energy by means of an AC commutatorless EM motor, and then convert the mechanical energy back into electrical energy (but this time in the form of DC) by means of an MCM AC–DC commutator generator. The necessary **motor-generator** (MG) system (Ward Leonard system) must, therefore, consist of two electrical machines, connected together mechanically (figure 14.10).

A valuable advantage of this system is the complete electrical independence of the DC mains from the AC mains; in many practical applications, it is very important to be able to control, over a wide range, the DC voltage delivered by the MG system. This system is, for example, an essential part of the **generator-motor** (GM) system, called the 'rotary dynamo electrical Ward Leonard system', which was, in the past, used in cases where it was necessary to regulate the speed of an actuating mechanism over a very wide range. The disadvantage of the rotary Ward Leonard system is, above all, its low efficiency, since the losses in the unit are made up of the losses in both electrical machines.

MG systems, in which a change in a relatively small control input-signal power produces a significant change in a large controlled output-signal power of the system, are called 'dynamoelectrical amplifiers'. Of course, the dynamoelectrical amplifier must receive electrical energy from some energy source whose power is essentially controlled by the input-signal power. In this sense, a separately excited MCM AC–DC commutator generator can serve as a dynamoelectrical amplifier; a small change in electrical energy applied to the exciting winding produces a greater change in the electrical energy delivered to the external circuit. However, the amplification in such an MG system is relatively small, of the order of 30 to 40, and

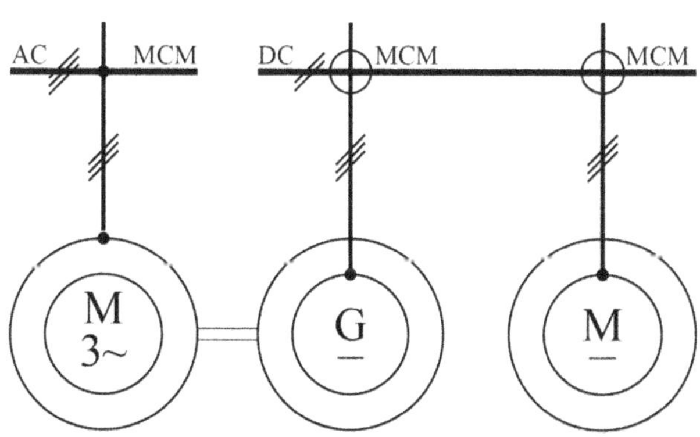

Figure 14.10. The AC commutatorless motor—an MCM AC–DC commutator generator for converting AC–DC (Ward Leonard system).

the variation process proceeds slowly because of large inductance in the exciting winding.

14.4.2 MCM AC–DC commutator amplidyne

The development of industrial electrification required the production of more up-to-date dynamoelectrical amplifiers on the basis of separately excited MCM AC–DC commutator generators. The previously widely used dynamoelectrical amplifier also known as the 'MCM AC–DC commutator amplidyne' is basically a separately excited MCM AC–DC commutator generator, but the basic magnetic flux which determines the load voltage is produced by the armature magnetic field. This method of exciting the main magnetic field was, in the past, used in several special-purpose MCM AC–DC commutator generators (for railway trains, welding, lighting, etc). These electrical machines are called 'cross-field excitation' electrical machines. Figure 14.11 shows the basic schematic diagram of an MCM AC–DC commutator amplidyne.

In contrast to normal MCM AC–DC commutator generators, its magnetic-field system is usually a non-salient magnetic pole with slots of differing sizes for several exciting windings, control windings and a compensating winding.

The MCM AC–DC commutator amplidyne operates as a generator with a constant rotational speed, and its prime mover is usually an AC commutatorless motor. One of the exciting windings is connected across some small DC energy

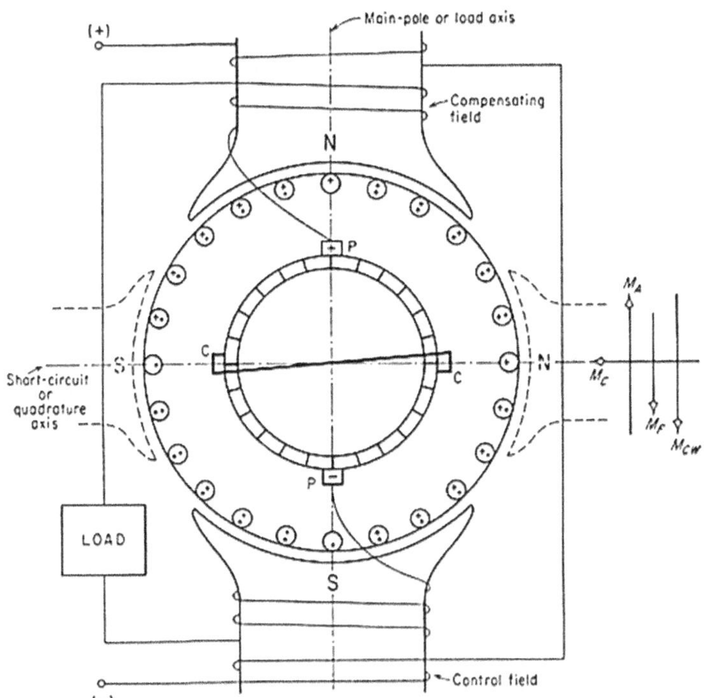

Figure 14.11. MCM AC–DC commutator amplidyne with basic control and compensating fields (Siskind 1963).

source and the input-signal voltage at its terminals produces a relatively small exciting current. The latter excites a relatively small longitudinal magnetic flux, which links the armature winding. The first pair of carbon brushes placed on the ring commutator across the cross axis of the amplidyne are short-circuited, and so the small DC induced voltage produces a considerable current in the armature winding. A second pair of carbon brushes is situated on the ring commutator on the vertical axis of the electrical machine, but the DC induced voltage does not cause any potential difference between them.

The armature current passes through the armature winding and produces a relatively large cross magnetic flux; the magnetising force of this current is large, and because of the non-salient magnetic-pole construction of the MCM AC–DC commutator amplidyne frame, the reluctance is a salient magnetic-pole DC electrical machine. The cross-magnetic flux induces a second relatively large DC induced voltage in the armature winding. This DC induced voltage produces a DC output voltage between the brushes placed on the longitudinal axis, resulting in a certain output current being produced in the controlled load circuit of the MCM AC–DC commutator amplidyne.

The controlled electrical energy at the amplidyne output will be many times greater than the input electrical energy. Of course, a certain resultant armature current is produced in the armature winding, but it is convenient on the basis of the superposition principle to consider the armature current and output current separately. It is necessary that the magnetising force of the output current should not change the longitudinal magnetic flux. In order to ensure this, in the frame slots of the amplidyne there is (in addition to the interpole commutating winding) a compensation winding situated on the frame of the amplidyne and connected in series with the current circuit, the function of which is to compensate for the magnetising force of the output current.

At present, the field of application of the MCM AC–DC commutator amplidyne is very narrow. To date, they have been employed in old installations for controlling reversing rolling mills, cold rolling mills, blast-furnaces and mine winders, and for metal-cutting and paper machines, etc.

14.4.3 MCM AC–DC/DC–AC ring-commutator single-armature frequency converter

In the past, a natural development of the idea of converting AC electrical energy directly into DC electrical energy was to do away with the mechanical link and perform the conversion in one electrical machine. This was first achieved in a rotary dynamoelectrical MCM AC–DC/DC–AC ring-commutator single-armature frequency converter, which is also known as an MCM AC–DC/DC–AC commutator dynamotor. This is an MCM AC–DC/DC–AC commutator generator with slip rings on the extended shaft at the end opposite to the MCM commutator. These slip rings are connected to equidistant points of the armature winding in the sense of rotation direction. Carbon brushes are placed on the rings and are connected across an AC voltage (figure 14.12).

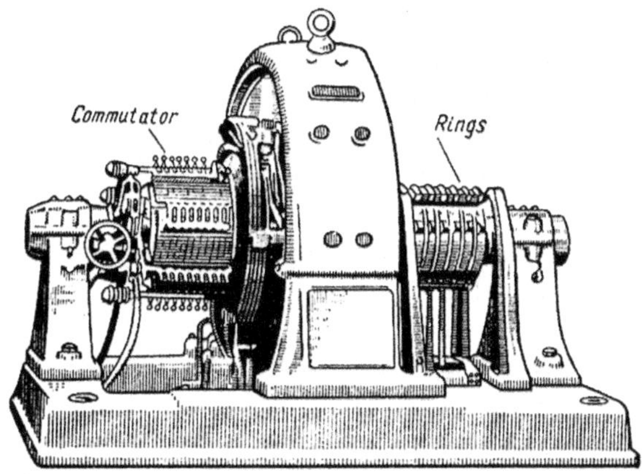

Figure 14.12. MCM AC–DC ring-commutator single-armature frequency converter (Kasatkin and Perekalin 1970).

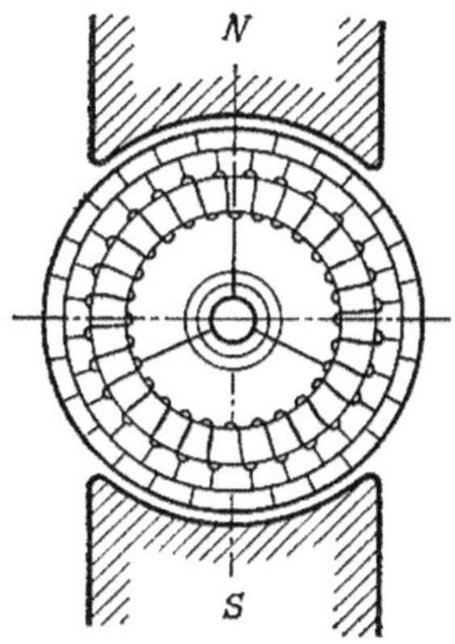

Figure 14.13. Circuit of a three-phase MCM AC–DC ring-commutator single-armature frequency converter (Kasatkin and Perekalin 1970).

The number of slip rings depends on the number of phases of AC supply that must be converted. For a single-phase supply, two slip rings are necessary, for a three-phase supply, three are required, and for five-phase supply, five are needed.

The diagram in figure 14.13 shows a three-phase MCM AC–DC/DC–AC ring-commutator single-armature frequency converter; for simplicity, it is two magnetic poles and has a ring armature winding.

The points at which the slip rings are connected to the armature winding in a three-phase MCM AC–DC ring-commutator single-armature converter must be spaced $2\pi/3$ electrical radians apart.

If the armature winding consists of $2a$ parallel branches, each pair of parallel branches, in succession round the armature winding, must be divided into three parts by the points of construction of the slip rings. It follows that a point of the armature winding is connected to each slip ring. Thus, relative to the AC side, the armature

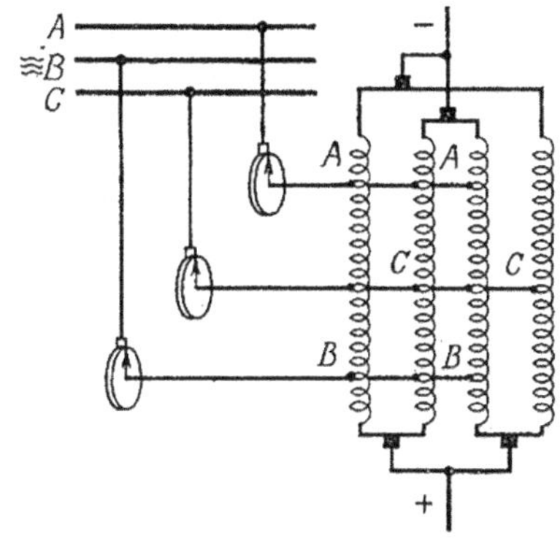

Figure 14.14. Circuit connections of the windings of a slip-ring MCM AC–DC ring-commutator single-armature frequency converter (Kasatkin and Perekalin 1970).

winding is connected in delta (Δ), each side of which consists of a parallel branch (figure 14.14).

In converting AC to DC, the MCM AC–DC/DC–AC ring-commutator single-armature frequency converter receives AC on the slip-ring side and delivers DC from the MCM AC–DC commutator side; it operates as an AC commutatorless synchronous motor with respect to the AC mains, and as an MCM AC–DC commutator generator with respect to the DC mains. If the DC side of the MCM AC–DC/DC–AC ring-commutator single-armature converter is disconnected and the excitation left connected, one obtains an AC synchronous electrical machine with stationary magnets and rotating armature.

It can operate as an AC commutatorless synchronous motor at the expense of electrical energy taken from the AC mains, and as an AC synchronous generator if it is driven by some prime mover.

MCM AC–DC/DC–AC ring-commutator single-armature converters are also used for converting DC into AC, acting as a rotating DC–AC inverter.

In this case, it operates as an MCM DC–AC commutator motor with respect to the DC mains and an AC commutatorless synchronous generator with respect to the AC mains. If the AC side is disconnected, the electrical machine will be able to operate as an MCM DC–AC commutator motor at the expense of the electrical energy taken from the DC mains, and as an MCM AC–DC commutator generator if it is driven by some prime mover.

The armature current in the armature winding of the MCM AC–DC/DC–AC ring-commutator single-armature frequency converter may be regarded as the result of a superimposition of the alternating and direct currents, consequently, the actual current in the winding will be the result of the alternating and direct currents. The armature current reacts with the magnetic field of the poles and produces a corresponding torque.

A great advantage of the MCM AC–DC/DC–AC ring-commutator single-armature frequency converter is the fact that the armature winding is heated less by the resultant current than would be the case if there were only alternating or direct current flowing through it. The more phases there are in the MCM AC–DC/DC–AC ring-commutator single-armature frequency converter, the smaller the resultant current. When it is operating as a polyphase MCM AC–DC/DC–AC ring-commutator single-armature frequency converter, the electrical machine has considerably smaller losses than it does when operating as an MCM AC–DC commutator generator. This fact makes it possible to obtain more electrical energy from an MCM AC–DC/DC–AC ring-commutator single-armature frequency converter on the DC side than from an MCM AC–DC commutator generator of the same size.

In the past, MCM AC–DC/DC–AC commutator single-armature frequency converters were widely used as high-, medium- and low-power rotating DC–AC inverters for converting DC to AC. This was required in DC supply systems for powering AC devices and radio apparatus (e.g. in aircraft and automotive vehicles), as well in high-power, high-frequency MCM DC–AC commutator generators for long-wave radio stations (e.g. in the old radio station in Nauen, Germany).

A few years ago, the high-power MCM AC–DC ring-commutator single-armature frequency converters were also used in DC supply systems for urban electrical traction substations for tramways and trolleybuses (e.g. in the old electrical traction substation in Szczecin, West Pomerania, Poland).

14.4.4 MCM DC–AC/AC–DC ring-commutator single-armature frequency converter and adjustable-ratio EE transformer connected to an AC commutatorless wound-rotor motor for rotational-speed-control purposes

Something that is important for speed-control by frequency variation is the need for an injected voltage whose frequency must always be the same as the slip frequency. Since the rotor frequency increases progressively as the speed drops below or rises above synchronism, the equipment supplying the injected voltage must automatically adjust itself to generate slip-frequency AC induced voltage as the main AC commutatorless wound-rotor motor changes speed.

In one fundamental design, the electrical machine that supplies the adjustable voltage for the phase-wound rotor is somewhat similar to a rotary converter, except that the stator is a laminated core with no winding, acting merely as part of a good magnetic circuit. The rotor contains the usual DC armature winding that is connected to an MCM commutator at one end and slip rings at the other (Siskind 1963).

As illustrated in figure 14.15, the brushes riding on the slip rings are connected to the three-phase power lines through a set of adjustable-ratio EE transformers, while another set of brushes, spaced $2\pi/3$ electrical radians apart, rests on the MCM commutator and is wired to the wound-rotor slip rings.

The special electrical machine is coupled to the main AC commutatorless wound-rotor motor and therefore always operates at the same speed as the latter.

To understand why the slip-ring frequency of the MCM DC–AC/AC–DC single-armature converter always matches the slip frequency of the AC commutatorless wound-rotor motor, it is necessary to recognise the following conditions of operation.

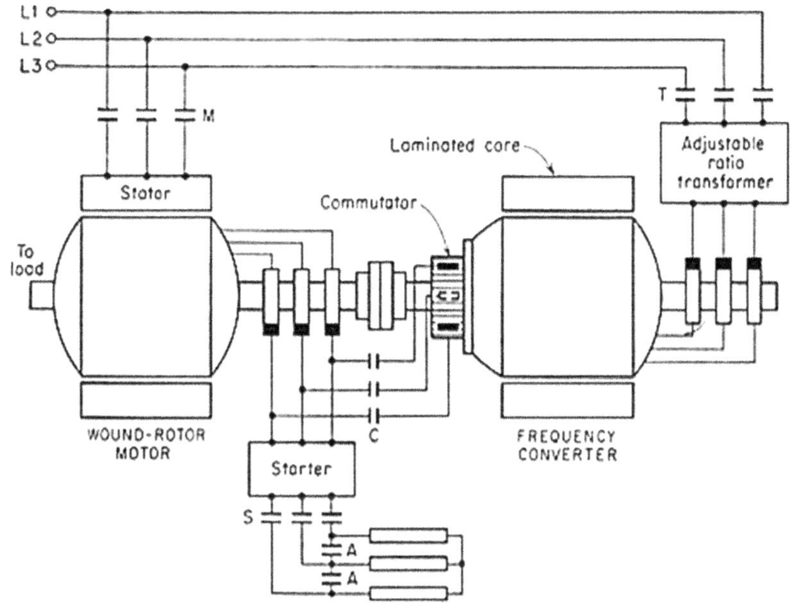

Figure 14.15. MCM DC–AC ring-commutator single-armature frequency converter and adjustable-ratio EE transformer connected to an AC commutatorless wound-rotor motor for rotational-speed-control purposes (Siskind 1963).

- Line frequency currents enter the AC end of the MCM DC–AC/AC–DC single armature frequency converter to set up a revolving magnetic field that rotates at a synchronous rotational speed with respect to the armature-winding conductors.
- If the rotor is at rest, the magnetic field thus produced revolves in space at a synchronous rotational speed.
- If the rotor is turned at a synchronous rotational speed in a sense of direction that is opposite to the motion of the magnetic field with respect to the moving conductors, the magnetic field is stationary in space. Under this condition, the magnetic field is, in effect, rotated backward as rapidly as it is moved forward.
- At any rotational speed intermediate between rest and synchronism, the magnetic field revolves at a rotational speed that is the difference between synchronism and the rotational speed of the rotating electrical machine.

It follows, then, that the frequency at the MCM commutator brushes is always identical to that of the interconnected wound-rotor slip rings. In particular, the MCM commutator-brush frequency and the slip frequency are both f (cps or Hz) when the rotor is at rest, and the two frequencies are zero (0, cps or Hz) when the rotor turns at synchronous rotational speed in a sense of direction opposite to the rotation of the magnetic field with respect to its own conductors.

As in all rotary converters, the voltage at the MCM commutator end always bears a very definite ratio to the impressed AC induced voltage at the slip-ring end.

This means that, with an adjustable-voltage EE transformer to supply electrical energy to the slip-ring end, the correct voltage, for the desired rotational speed, will be available at the MCM commutator.

The Scherbius method permits the operator to control the rotational speed of a constant torque application over an extremely wide range above and below synchronism and to do so steplessly.

Figure 14.13 illustrates the arrangement of the electrical machines and auxiliary equipment and shows how the control circuit interconnects the various units. The propulsion mechatronical control system, the so-called 'rotary dynamoelectrical Scherbius system' is put into operation by first bringing the main AC commutator-less wound-rotor motor up to rotational speed. As previously explained, this is done by closing the main contactor's line contacts *M*, and then closing the starting contacts *S* and the accelerating contacts *A*. The slip-ring brushes are next transferred to the MCM commutator brushes by closing the *C* contacts and opening the *S* and *A* contacts. Finally, the *T* contacts are closed to connect the adjustable-ratio EE transformer to the slip rings of the MCM DC–AC/AC–DC single armature frequency converter, and manual settings are made at the EE transformer to provide the required injected voltage for the desired rotational-speed adjustment.

Although this scheme was rarely used as described in the past, it is basic to all other propulsion mechatronical control systems in which an AC commutatorless wound-rotor motor is the main drive and voltage injection in the secondary is behind some of the advanced propulsion mechatronical control systems presently in use, as well as the special Schrage MCM AC–AC commutator motor discussed in this chapter.

14.4.5 MCM DC–AC/AC–DC ring-commutator single-armature frequency converter and MCM DC–AC commutator motor coupled to the shaft of the AC commutatorless wound-rotor motor for speed-control purposes

For variable-speed propulsion mechatronic control systems requiring very large AC commutatorless wound-rotor motors, from about several megawatts (MW) of power, the wound-rotor is preferred. When used in such installations, it is, however, associated with auxiliary equipment, which unlike resistance controllers, returns much of the slip-frequency electrical energy of the rotor to the system. One method of doing this was described and illustrated in figure 14.15.

Another variable-speed propulsion mechatronic control system that was frequently employed in the past in, for instance, steel-mill drives, dredges and other heavy-machinery loads whose rotational speeds must be adjustable over extremely wide ranges, is the so-called 'rotary dynamo-electrical Kraemer system', two variations of which are used to convert the slip frequency electrical energy from the main AC commutatorless wound-rotor motor into useful electrical energy.

In one, the constant-power propulsion mechatronical control system, DC electrical energy from the MCM DC–AC/AC–DC ring-commutator single-armature frequency converter is supplied to an MCM DC–AC commutator motor coupled to the shaft of the main AC commutatorless wound-rotor motor. In the other, the constant-torque propulsion mechatronical control system, the DC output of the MCM DC–AC/AC–DC ring-commutator single-armature frequency converter ultimately drives an AC commutatorless synchronous or hyposynchronous (induction) generator that feeds back into the AC supply lines.

Figure 14.16 shows the wiring connections for the constant-power propulsion mechatronical control system, called the rotary dynamoelectrical Kraemer system. With the MCM DC–AC/AC–DC ring-commutator single-armature frequency-converter's contactor *C* open, the unloaded main AC commutatorless wound-rotor motor is brought up to rotational speed in the usual way with the rotor-resistance starter. Contacts *S* and *A* are then opened, normal excitation is applied to the frequency converter, and contact *C* is quickly closed to impress slip-frequency voltage across its slip rings to permit the electrical machine to rotate slowly.

The magnetic field of the MCM DC–AC commutator motor is excited next, whereupon that unit, which is coupled to the main AC commutatorless wound-rotor motor and running at nearly synchronous rotational speed, generates a DC voltage that is impressed across the MCM commutator end of the MCM DC–AC/AC–DC ring-commutator single-armature frequency converter. Because a definite voltage ratio always exists between the AC and DC ends of the MCM DC–AC/AC–DC ring-commutator single-armature frequency converter, a certain

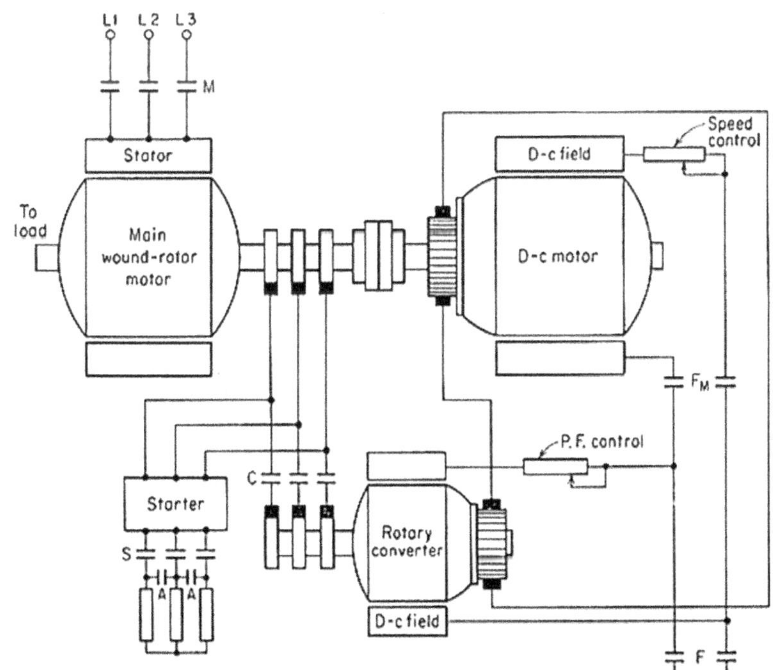

Figure 14.16. Rotary dynamoelectrical Kraemer system for rotational-speed control for constant-power operation (Siskind 1963).

slip-frequency potential appears at the slip ring and is injected into the rotor of the main AC commutatorless wound-rotor motor. The latter therefore slows down, automatically adjusting its own input current for the particular load that was applied after starting.

With the rotational speed of the main AC commutatorless wound-rotor motor reduced and its slip frequency and voltage increased, the MCM ring-commutator single-armature frequency converter speeds up to deliver electrical energy to the armature of the MCM DC–AC commutator motor, which in turn converts it to mechanical energy that aids the main AC commutatorless wound-rotor motor in driving the load. The set now stabilises at the adjusted rotational speed for the particular load, delivering a mechanical-energy output equivalent to the mechanical-energy input, minus the losses.

To decrease the rotational speed of the main drive, the excitation of the MCM DC–AC commutator motor must be increased. This raises the DC induced voltage of that electrical machine, causes the voltage at the slip rings to increase, and results in a higher injected slip-frequency potential, which brings about the desired drop in the rotational speed. The reverse is true when the excitation is reduced.

In fact, if the excitation to the MCM DC–AC commutator motor is reduced to zero (0) when the main AC commutatorless wound-rotor motor operates with small slip, and is then reversed with slightly increased excitation, the drive will run at synchronous rotational speed. Under this condition, the MCM DC–AC/AC–DC ring-commutator single-armature frequency converter stops, the slip frequency is zero (0), and the main AC commutatorless wound-rotor motor becomes an AC commutatorless synchronous motor with DC excitation. Further increasing the reverse excitation causes the main AC commutatorless wound-rotor motor to operate above synchronous rotational speed, and the sense of the rotation direction of the MCM DC–AC/AC–DC ring-commutator single–armature frequency converter reverses. For rotational speeds above synchronism, the DC electrical machine becomes an MCM AC–DC commutator generator, imposes an additional load on the main AC commutatorless wound-rotor motor, and feeds its electrical energy through the MCM DC–AC/AC–DC ring-commutator single–armature frequency converter into the rotor of the main AC commutatorless wound-rotor motor.

A useful feature of the propulsion mechatronical control system is the fact that it is possible to control the power-factor by adjusting the excitation of the rotary dynamoelectrical MCM DC–AC/AC–DC ring-commutator single–armature frequency converter; correction up to about 0.95 is generally practical through over-excitation.

Compared with other propulsion mechatronic control systems for heavy power drives—notably the rotary dynamoelectrical Scherbius system, which uses such special types of equipment as ohmic drop exciters, AC commutatorless hyposynchronous (induction) electrical machines, EE transformers and MCM DC–AC commutator motors, the rotary dynamoelectrical Kramer system is economical. This is because the MCM DC–AC commutator motors and rotary dynamoelectrical DC–AC/AC–DC ring-commutator single–armature frequency converters were in the past, for the most part, similar to those of the standard design.

References

Fijalkowski B 1973 Kierunki rozwoju napedu elektrycznego lokomotyw (Development trends of the electric drive of locomotives) *Materiały konferencyjne—I Krajowa Konferencja—Pojazdy Szynowe* (Krakow: Instytut Pojazdow Szynowych Politechniki Krakowskiej) pp 109–28 (In Polish)

Fontaine H 1878 *Electric Lighting: A Practical Treatise* (Translated from French by Higgs P.) (London: E and FN Spon) p 86

Kasatkin A and Perekalin M 1970 *Basic electrical engineering* (Moscow: Mir Publishers) p 480

Kuznetsov M 1960 *Fundamentals of Electrical Engineering* (Moscow: Peace Publishers)

Laughton M A and Say M G 1985 *Electrical Engineer's Reference Book* (London: Butterworth)

Siskind Ch S 1963 *Electrical control systems in industry* (New York: McGraw-Hill)

Vesolovsky O N and Kopylov I P 1984 U istokov elektromekhaniki [K 150-letiyu elektrodvigatelya Yakobi i principa obratimosti] *Elektrotekhnika* (In Russian)

IOP Publishing

Mechatronics
Dynamical systems approach and theory of holors
B T Fijalkowski

Chapter 15

Electrical commutation matrixer commutators for modern DC and AC magneto-mechano-dynamical electrical machines

A scientific truth does not triumph by convincing its opponents and making them see the light, but rather because its opponents eventually die and a new generation grows up that is familiar with it.

Max Planck

15.1 Introduction

The world must solve important challenges to convert electrical energy and transform information for mechatronic control systems in an efficient way. Examples of this are transportation, renewable energies and industrial processing applications. These problems can be solved using energy converters based on modern semi- and/or superconductive **electrical commutation matrixer** (ECM) AC–AC, AC–DC–AC, AC–DC/DC–AC, DC–DC and DC–AC–DC commutator applications in **magneto-mechano-dynamical** (MMD) electrical machines.

The ideal ECM commutator in many of these applications may have the following characteristics:

- sinusoidal input and output currents;
- operation with unity power factor;
- regeneration capability;
- compact design with a good power-to-mass ratio.

Semi- and/or superconductive ECM AC–AC, AC–DC–AC, AC–DC/DC–AC, DC–DC and DC–AC–DC commutators (matrix converter) can fulfil all these characteristics and this is the reason for the tremendous interest in this topology.

In the last decade, many advances in the development of this topology have been presented, including industrial applications up to megawatt level. The use of the semi- and/or super-conductive ECM AC–AC, AC–DC–AC, AC–DC/DC–AC, DC–DC and DC–AC–DC commutator (matrix converter) in real applications and the challenges that these applications present are very topical and important.

In view of the fact that sources of conventional energy are gradually being depleted, the general public is increasingly interested in discussions about novel types of future-oriented energy-saving industrial and traction propulsion complex dynamical hypersystems. These energy-saving complex dynamical hypersystems have already become an alternative to be taken seriously. The development of today's industrial and traction propulsion complex dynamical hypersystems is closely tied up with developments in electronic technology.

The MMD electrical machines (dynamotors) currently in use have remained unchanged since their conception before the turn of the 20th century; new electrical dynamotor concepts have only gained marginal acceptance.

Today, there are good possibilities available to select optimal industrial and traction propulsion complex dynamical hypersystems with due consideration being paid to performance, lifetime costs, environment, influence on the storage battery, etc. This **very advanced propulsion** (VAP) technology is based on two important disciplines—**electrical-integrated-matrixer** (EIM) commutator and **electrical-integrated-circuit** (EIC) processor techniques. Some of the results achieved higher performance and other operational benefits. The **microcontroller unit** (MCU) gives the necessary conditions for further improvements, such as diagnostics.

Continued **research and development** (R&D) work on ECM commutators will lead to greater competitiveness for AC–AC and AC–DC/DC–AC commutator dynamotors (generators/motors) for industrial and traction propulsion complex dynamical hypersystems.

Recent years have not only witnessed rapid progress in 'microelectronics' (i.e. the area of electronics dealing with the design, realisation and application of various low-power EICs made in a solid (continuous medium) or on its surface), but also in 'macroelectronics' (i.e. the area of electronics dealing with the design, realisation and application of various high-power ECMs made in a solid (continuous medium) or on its surface).

Above all, it refers to the developments that have taken place in the field of MCUs (i.e. low-power EICs with a **central processing unit** (CPU), memory and **input/output** (I/O) capability, made in the shape of individual chips containing **very large scale of integration** (VLSC) EICs) and ECM commutators (i.e. electronic commutators containing **standard scale integration** (SSI) and **medium-scale integration** (MSI) ECMs). The latter, implemented by the author in 1973, have been widely popularised, particularly in the field of industrial and traction propulsion complex dynamical hypersystems (Fijalkowski 1973, 1985, 1987, 1988, 2010, 2011).

Over 40 years ago, the author started R&D works on mono- and poly-crystalline as well as amorphous semiconductor electronic commutators with uni- and/or bipolar electrical valves, which he called for the first time 'ECM commutators', and these aroused the interest of the greatest world consortia. These consortia are aware of the fact that the author is one of the few scientific men who initialised the avalanche of concepts, the consequences of which are difficult to foresee. It is obvious that these concepts will not only change contemporary integrated power electronics (now termed by the author 'macroelectronics') to a certain degree, but also modern industrial and traction propulsion complex dynamical hypersystems and, consequently, newly designed energy-saving electrical machines. The scale of these changes remains unknown for the time being. Behind these concepts hides the harbinger of the new universe—the 'universe of macroelectronics'.

Is the new universe—the 'universe of power electronics'—going to be better than the existing one? This is unknown. Certainly, it offers new possibilities that are difficult not to take advantage of.

Power electronics and industrial and traction propulsion complex dynamical hypersystems are again queuing for the results achieved in solid-state physics laboratories.

The scientific revolution in the field of power electronics has taken place very rapidly, because stereotypical ideas in this area have been overcome.

All progress is stimulated by thinking that is free from dogmas that are opposed to challenges to existing concepts. It is also known that the best practical solutions are born out of ingenious theory. 'There is nothing more practical than a good theory' said Ludwig Boltzmann (1844–1906).

The concept of a generalised intelligent **multi-input/multi-output** (MIMO) ECM AC–AC and/or AC–DC/DC–AC commutator with discrete and/or continuous uni- and/or bipolar electrical valves was born due to the mathematical hypermatrix notation of the dynamical systems approach for the formation methods of mathematical models of ECM AC–AC and/or AC–DC/DC–AC commutator dynamotors (Fijalkowski 1973, 1985, 1987).

MIMO ECM AC–AC and/or AC–DC/DC–AC commutators are the key to overcoming the contactless (brushless), i.e. sparkless commutation (changing-over the way of electrical current flow), and so the static conversion of one kind of electrical energy into another one, with the application of the phenomena inducing uncontrolled or controlled electrical current conductivity (carrying of electrical charges by positive or negative carriers into a definite medium under the action of electrical field).

MIMO ECM AC–AC and/or AC–DC/DC–AC commutators have introduced the era of newly designed energy-saving AC–AC and AC–DC/DC–AC commutator dynamotors, enabling continuous (stepless) and contactless control of their current and voltage (during generating), or their torque and speed (during motoring).

These ECM AC–AC and/or AC–DC/DC–AC commutators with uni- and/or bipolar electrical valves constitute a successive breakthrough in the development of ECM AC–AC and AC–DC\DC–AC commutator dynamotors, and multiply their possibilities enormously. There are fantastic perspectives that will probably become reality as soon as in the current century. However, the scientific sense of concepts conceived for the first time by this author forty years ago does not consist of this. It is hope that attracted the attention of the greatest world consortia. The most essential value of this new concept was the achievement of a cardinal breakthrough in how one thinks about the static conversion of one kind of electrical energy into another one, with the aid of continuous uni- and/or bipolar electrical valves—a change of paradigm regarding the kind of thinking that is standard in the field of power electronics. In order to understand it, it is necessary to look into the past.

The physical science of electricity and electrical machinery is one of the youngest branches of physical knowledge. Electrical machine building is actually above 100 years old. However, in this relatively short period a tremendous amount of R&D work has been carried out, which has revolutionised the technical and economic features of modern industry, traction and our social lives.

The construction of different types of **magneto-mechano-dynamical** (MMD), **magneto-fluido-dynamical** (MFD), **magneto-plasmo-dynamical** (MPD), **electro-mechano-dynamical** (EMD), **electro-fluido-dynamical** (MFD) and **electro-plasmo-dynamical** (EPD) electrical machines is a subject for a special course of study and is beyond the scope of this textbook. Here the author considers only this fundamental principle underlying the MMD conversion of mechanical energy into electrical energy and vice versa.

In the dynamical systems approach to MMD electrical machines, one classifies them as physical heterogeneous continuous simple dynamical hypersystems, i.e. **mechano-electrical** and/or **electro-mechanical** (ME/EM) heterogeneous continuous simple dynamical hypersystems (Fijalkowski 1985).

An MMD conversion involves the interchange of energy between a mechanical continuous dynamical system and an electrical continuous dynamical system, or vice versa, through the medium of coupling magnetic-field energy, and utilises the effect of electromagnetic induction and the mechanical action of the coupling magnetic-field energy for its accomplishment.

The process is essentially reversible, except that a small amount of the thermal energy of a thermal continuous dynamical system is lost as heat.

Thus, MMD electrical machines are the principal rotary and linear converters of mechanical energy into electrical energy, and/or electrical energy into mechanical energy, and they are also the principal rotary converters of one form of electrical energy into another form, differing in voltage, current and sometimes frequency.

When the MMD conversion takes place from mechanical into electrical energy, the MMD electrical machine is called an 'ME generator' (or an 'ME dynamo').

When electrical energy is converted into mechanical energy, the MMD electrical machine is called an 'EM motor'.

When bilateral energy MMD conversion is necessary, the MMD electrical machine is called an 'ME/EM dynamotor'.

The constant improvements in the design of MMD electrical machines have made possible many new practical applications, and have provided the impulse for further progress, while fostering the most diverse uses of electrical energy.

This explains why MMD electrical machines were given great attention by scientists and engineers, and why MMD electrical machines quickly attained technical perfection of design.

15.2 Status and trends

The properties of unipolar electrical-current conductivity, i.e. the nonlinearity of the current–voltage characteristics of polycrystalline semiconductors, were discovered by Adams and Day in the 1880s. In 1883, Fritsch described a construction of selenium electrical valves, and these were not applied on a large scale until the end of the 1920s. In 1926, Grondahl discovered cuprous-oxide electrical valves, which became popular in the 1930s. At the same time, titanium dioxide electrical valves with an admissible working temperature of up to 473 K were also applied. In 1882, Jamin and Manevrieu published a work on the properties of unipolar electrical-arc conductivity in a mercury vapour, i.e. mercury-arc or ionic electrical valves, which were broadly applied at the beginning of the 1930s. Fleming invented the first electron diode in 1904, which he named the 'kenotron'. In parallel to polyvalvular (multi-anode) mercury-arc electrical valves, called 'mutators', the manufacturing technique for 'gasotrons' and, a little later, 'thyratrons' expanded. In the 1950s 'ignitrons' and 'exitrons' were also applied on a large scale. The development of the manufacturing technology for monocrystalline semiconductors, and then germanium and silicon unipolar electrical valves, took place in the 1940s (uncontrolled discrete electrical valves—germanium and silicon diodes) and 1950s (controlled discrete electrical valves—germanium and silicon triodes: transistors, **insulated-gate bipolar transistors** (IGBT), **metal-oxide-semiconductor field-effect transistors** (MOSFET), thyristors, trigistors, **magnetic-field controlled thyristors** (MCT), etc—as well as germanium and silicon tetrodes: spacistors, tetristors, etc). In addition to silicon unipolar electrical valves and silicon bipolar electrical valves with single or two control gates—including, for example, bipolar field transistors and photo-transistors, as well as symistors (symmetrical thyristors), photo-symistors and bipolar trigistors (**gate-turn-off** (GTO) thyristors)—**reverse-blocking insulated-gate bipolar transistors** (RB IGBT) have also been applied. Progress in the recognition of the electrical properties of glassy amorphous materials is leading to a series of promising new applications for amorphous semiconductors in macro- and microelectronics, and thus in the field of discrete or continuous uni- and bipolar (uni- and bilateral) electrical valves.

In 1958 S R Ovshinsky discovered the phenomenon of switching over in amorphous semiconductors, i.e. a rapid changeover of electrical-current conductivity when the applied electrical field attains its critical value. In the case of chalcogenide glasses, nearly the same phenomenon appeared in electrical field with an intensity of 10^5 V cm^{-1}, with the changeover time yielding a few nanoseconds (Holmberg and Shaw 1974), while the electrical-current conductivity can be changed approximately 10^6 times. The results obtained were not published before 1970 (Fijalkowski 1988c), while references for them can be found in the monograph (Fijalkowski 1987). However, some technological difficulties, as well as military applications, contributed to the limitation of their initial development.

More than 50 years have passed since Ovshinsky demonstrated the 'ovonic' semiconductor device (the name is a contraction of 'OVshinsky electrONIC'), which was the first electrical valve made from amorphous semiconductors.

For 19 years, although the ovonic existed and was in operation, nobody precisely knew the principle of its operation. There have been several well-known cases in the history of electronics where semiconductor devices have been in common use before their operating principles were exactly known and their theoretical basis created. This was the case for the protoplast of the galena detector, as well as the best polycrystalline semiconductor electrical valve in the past—the selenium rectifier.

It was not until the 1970s that three physicists—P W Anderson and J van Vleck of the USA and N F Mott of the UK—worked out new fundamentals for the theory and technology of amorphous semiconductors, which showed that unipolar electrical-current conductivity could not only exist in mono- and polycrystalline semiconductors, but also in amorphous semiconductors. The theory gained common approval, and its creators were honoured with the Nobel Prize for physics in 1977. If everybody had assented to a theory that had been formulated earlier (Seitz 1940; Kittel 1956; Mott and Davis 1971), which anticipated that unipolar electrical-current conductivity could only exist in unipolar mono- and polycrystalline semiconductors, nobody would have an inkling about some of the technologies that humanity could aspire to. Fortunately, a group of scientists doubted the correctness of this theory (Mott and Davis 1971). Due to their obstinacy, it is known today that the earlier theory was only valid for mono- and polycrystalline semiconductors.

All the aforementioned discrete uni- and bipolar electrical valves were or are used in MIMO ECM AC–AC, AC–DC–AC and AC–DC/DC–AC commutators with non-integrated ECMs (Fijalkowski 1973, 1982, 1985, 1987). In the field of macroelectronics, they are named by the author 'polyvalvular MIMO ECM AC–AC and/or AC–DC/DC–AC commutators'. It is obvious that in the field of power electronics these polyvalvular MIMO ECM commutators are still called 'static converters'. The same thing also took place in the field of MCM AC–AC and AC–DC/DC–AC commutator dynamotors, which in the field of macroelectronics are named by the author 'polysectorial MIMO MCM AC–AC and/or AC–DC/DC–AC commutators' (Fijalkowski 1973, 1985, 1987).

Less expensive, lighter and smaller than electrical machines' polysectorial MIMO MCM AC–AC and/or AC–DC/DC–AC commutators, polyvalvular MIMO ECM AC–AC and/or AC–DC/DC–AC commutators lie at the very heart of the conversion and control of electrical energy, which in turn lie at the heart of making that energy useful. From domestic appliances to space-faring and automotive vehicles, the applications of macroelectronics are virtually limitless. Until now, however, the same could not be said for access to up-to-date reference books devoted to power electronics (see the references and bibliography).

MIMO ECM AC–AC, AC–DC–AC, AC–DC/DC–AC, DC–DC and DC–AC–AC commutators are ECMs of controlled semiconductor electrical valves that directly connect each matrixer-input phase to each matrixer-output phase, especially without any intermediate DC link. The MIMO ECM AC–AC, AC–DC–AC, AC–DC/DC–AC, DC–DC or DC–AC–AC commutator is usually fed at the matrixer-input side by a voltage source and it is connected to inductive and/or capacitive loads at the matrixer-output side. The main advantage of the ECM commutator is the absence of bulky reactive components, which are subject to ageing and reduce the MIMO ECM AC–AC, AC–DC–AC, AC–DC/DC–AC, DC–DC and DC–AC–AC commutator reliability. One significant advantage of this arrangement is that the ECM commutator does not carry current to the rotor, which eliminates the brushes and their wear-related drawbacks.

Furthermore, MIMO ECM AC–AC, AC–DC–AC, AC–DC/DC–AC, DC–DC and DC–AC–AC commutators provide uni- and/or bilateral (bipolar) power flow, nearly sinusoidal matrixer-input and matrixer-output waveforms, a controllable matrixer-input power factor, a regenerative capability and the origin function to generate magnitude-frequency controllable matrixer-output voltages. Therefore, ECM AC–AC, AC–DC/DC–AC and DC–DC commutators have received considerable attention as an advantageous alternative to an ECM AC–DC–AC commutator acting in **voltage-source inverter** (VSI) topology, or an ECM DC–AC–DC commutator acting in **voltage-source chopper** (VSC) ECM topology.

The author showed in 1973 in Fijalkowski (1973) that each static converter or MIMO MCM AC–AC, AC–DC–AC, AC–DC/DC–AC, DC–DC and DC–AC–DC commutator of electrical machines cannot only be considered theoretically, but must also be made practically as polyvalvular MIMO ECM AC–AC and AC–DC/DC–AC and DC–DC commutators with built-in discrete and/or continuous unipolar electrical valves called 'thyristors' (see figure 15.1), but he did not expect then that the results of this would have any practical consequences in the future.

In addition, the author showed once more in 1982 (Fijalkowski 1982) that each MIMO MCM AC–AC and/or AC–DC/DC–AC commutator of electrical machines cannot only be considered theoretically, but must also be made practically as MIMO ECM AC–AC, AC–DC–AC, AC–DC/DC–AC, DC–DC and DC–AC–DC

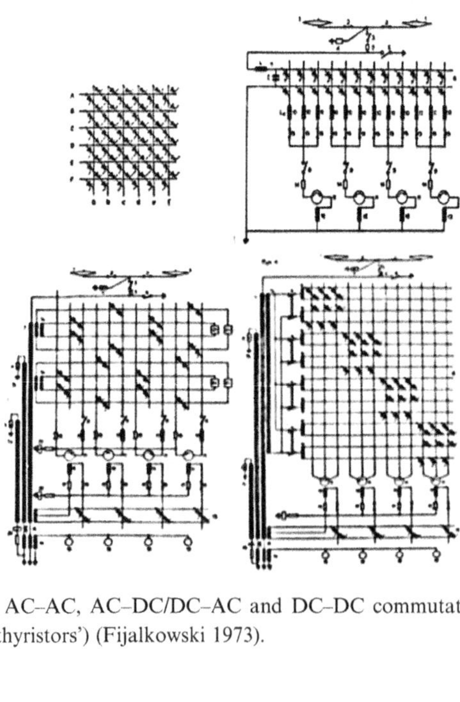

Figure 15.1. MIMO ECM AC–AC, AC–DC/DC–AC and DC–DC commutators with built-in continuous unipolar electrical valves ('thyristors') (Fijalkowski 1973).

commutators with built-in discrete and/or continuous bipolar, triggered and quenched electrical valves called 'trigistors' (see figure 15.2).

The development of **three-phase by three-phase** (3 × 3) ECM AC–AC commutators (called 'direct frequency converters with nine electrical valves') was started by Zagorski (1969), and the development of each and every one of the $m \times n$ ECM AC–AC, AC–DC–AC, AC–DC/DC–AC, DC–DC and DC–AC–DC commutators (with $m \times n$ bipolar, triggered and quenched electrical valves) was started afterwards by Fijalkowski (1973, 1982, 1985, 1987), while the development of 3 × 3 ECM AC–AC commutators (called 'matrix converters with nine bipolar electrical valves') continued soon after, as performed by Gyugyi and Pelly (1976), Cheron *et al* (1985), and Alesina and Venturini (1989), who proposed the basic principles of their operation.

The 3 × 3 ECM AC–AC commutators are superior to the up-to-now applied ECM AC–AC commutators (called 'thyristor cycloconverters with eighteen or thirty-six unipolar electrical valves') in being able to produce unity **power factor** (PF)

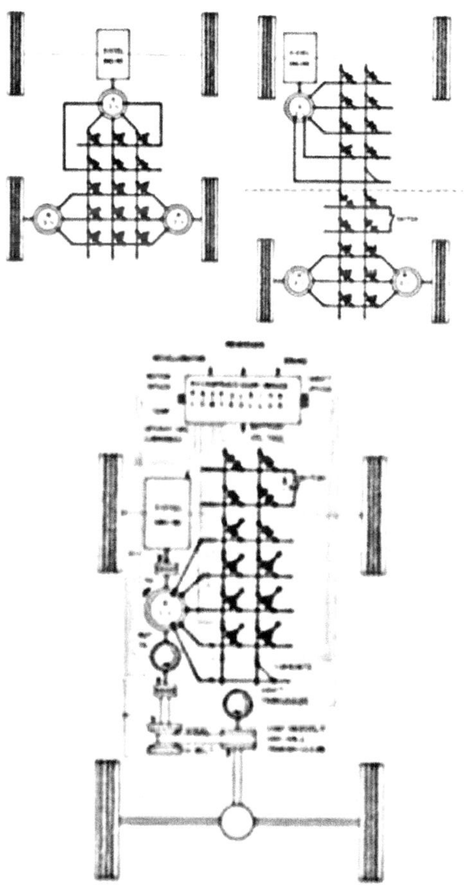

Figure 15.2. MIMO ECM AC–AC, AC–DC/DC–AC and DC–DC commutators with built-in discrete and/or continuous bipolar, triggered and quenched electrical valves ('trigistors') (Fijalkowski 1982).

line current. However, compared to PWM voltage-fed AC–DC–AC ECM commutators, the parts count is significantly higher. Consequently, the R&D works in this field show two trends (see the references and bibliography).

On the one hand, there was the need for reliable discrete and continuous uni- and/or bipolar electrical valves (electronic switches). On the other hand, the initial modulation strategy was unsuitably in favour of ultra-modern ECM solutions, allowing higher energy (voltage) transfer ratios and better power (current) quality.

The performance of MIMO ECM AC–AC, AC–DC–AC, AC–DC/DC–AC, DC–DC and DC–AC–DC commutators is strictly dependent upon the **pulse-width modulation** (PWM) strategy employed to control uni- and/or bipolar, triggered and quenched electrical valves. Since the introduction of MIMO ECM commutators, numerous modulation methods have been developed.

In Alesina and Venturini (1989) and Tenti *et al* (1992), off-line global duty-ratio functions are mathematically derived, which can be modulated by a carrier-frequency signal. This modulation method necessitates a formidable number of complicated computations at the implementation stage.

The **space vector PWM** (SVPWM) for an ECM commutator explores a more systematic approach to understand the operation of the ECM (Huber and Borojevic 1995; Casadei *et al* 2002). However, the SVPWM is far from spontaneous and requires look-up tables with the *a priori* prepared and stored commutation patterns. The carrier-frequency based PWM is the latest modulation strategy for ECM AC–AC commutators (Yoon and Sul 2006; Li *et al* 2008, 2009a, 2009b).

The carrier-frequency based PWM employs the carrier-frequency and reference signals and can be implemented without complicated computations and look-up tables.

The modulation method described in Yoon and Sul (2006), however, necessitates proper offset voltages and discontinuous carrier-frequency signals, which implies relatively indirect understanding, and cannot be applied to ECM topologies with a neutral connection between the matrixer-input and matrixer-output neutrals.

In Li *et al* (2008), a carrier-frequency-based PWM strategy, termed the **direct duty-ratio PWM** (DDPWM), is presented for the three-phase to three-phase (3×3) ECM AC–AC commutator, which does not require the reference offset and employs a continuous triangular carrier-frequency waveform. The desired matrixer-output phase voltages can be synthesised by utilising the matrixer-input phase voltages based on the per-output-phase average concept over one commutation period, i.e. per carrier cycle. A maximum gain of 0.866 can be obtained simply by injecting third harmonic components into matrixer-output voltage references in the three-phase electrical energy system.

In Li *et al* (2009a), the DDPWM method is extended in order to use it for the various ECM commutator topologies, e.g. three-phase to one-phase 3×1 or 3×2 ECM AC–AC and/or AC–DC/DC–AC commutators, three-phase to two-phase 3×2 or 3×3 ECM AC–AC commutators, three-phase to three-phase 3×3 or 3×4 ECM AC–AC commutators, three-phase to five-phase 3×5 or 3×6 ECM AC–AC commutators, and five-phase to five-phase 5×5 or 5×6 ECM AC–AC commutators.

In Li *et al* (2009b), a systematic method to generate the reference signals in various ECM commutators is presented. Before the DDPWM is used to synthesise the matrixer-output voltages, the per-phase matrixer-output voltage references should be given at each commutation time. The global minimum of the matrixer-input voltage range is used to determine the matrixer-output voltage magnitude. The feasibility and validity of the proposed DDPWM method applied in various EMC commutators was verified with simulation and experiment results. When, over 40 years ago, the author had the idea of MCM and ECM commutators realised on discrete and continuous uni- and/or bipolar electrical valves, only a few people believed in the feasibility of the R&D programme started by him. It is clear that amorphous semiconductor electrical valves will be the basis of the next great advance in macro-, meso-, micro- and nanoelectronics.

The author intends to show why the field of power electronics is presently in a state of crisis, and therefore historically ripe for a new basic approach. A characteristic component of a large group of MMD electrical machines is an integral MCM or ECM AC–AC and/or AC–DC/DC–AC commutator that is in the dynamical systems approach for MMD electrical machines, i.e. a physical heterogeneous continuous dynamical hypersystem is classified as a **generalised commutation matrixer** (GCM) (Fijalkowski 1973, 1985).

The concept of designing fully static power electronic electrical-valve ECM AC–DC/DC–AC commutators—named 'ECM commutators' by the author—for the conversion of polyphase alternating voltages that are constant in both amplitude and frequency, or direct voltages that are constant in magnitude, to polyphase alternating voltages, modulated in both amplitude and frequency in order to be able to control the torque and speed of ECM AC–AC and AC–DC/DC–AC commutator dynamotors, is by no means new (see figures 15.3–15.10). If one studies the older literature on static converters, one can observe that such static power electronic electrical-valve ECM AC–DC/DC–AC commutators were described as far back as the early 1890s.

The chronological background of MIMO ECM AC–DC commutators dates back to the end of the 19th century, when in 1895 (date of application) the notable Polish electrician, Charles (Karol Franciszek) Pollak (honorary doctor of Warsaw University of Technology), was issued with UK (24,398 AD 1895) and German

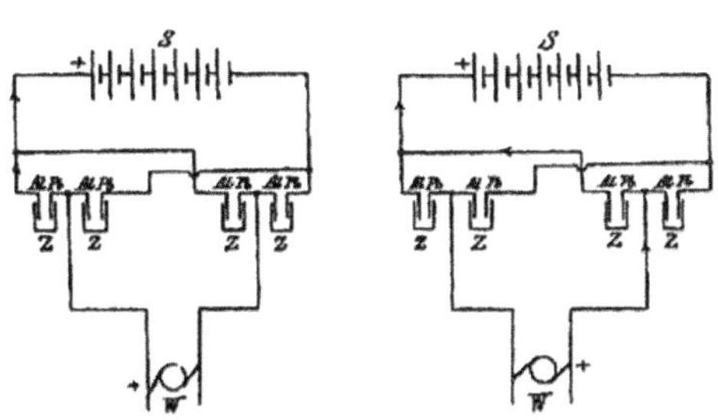

Figure 15.3. The Pollak bridge rectifier with four aluminium electrical valves (Pollak 1895, 1896).

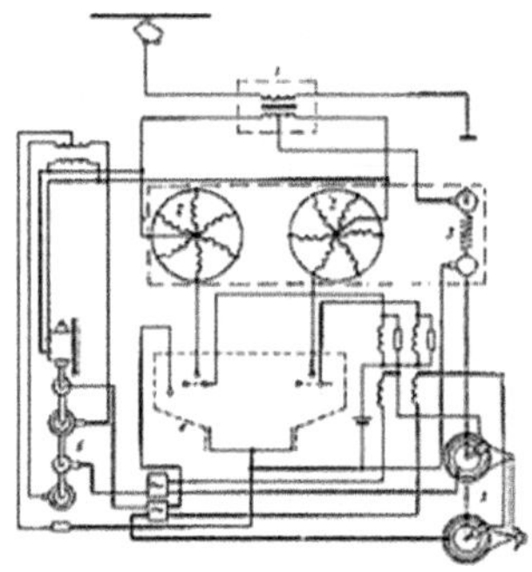

Figure 15.4. A single-phase electrical commutation matrixer (ECM) AC–AC commutator motor: 1—an EE transformer; 2—a stator winding of the traction EM motor; 3—a rotor winding; 4—a mercury-arc controlled electrical valve; 5—a rotary-commutator; and 6—a control system (Kern 1938).

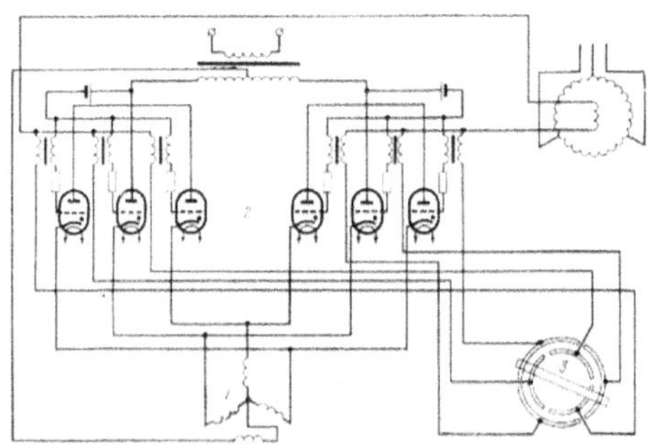

Figure 15.5. A single-phase electrical commutation matrixer (ECM) AC–AC commutator motor: 1—a stator winding; 2—thyratrons; and 3—rotary-commutator (Willis 1933).

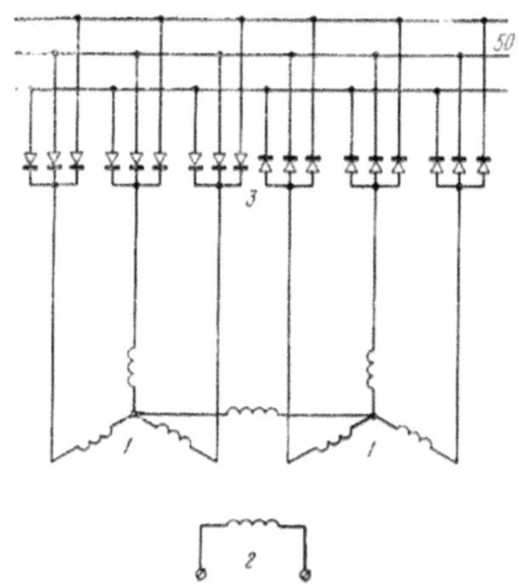

Figure 15.6. A three-phase electrical commutation matrixer (ECM).

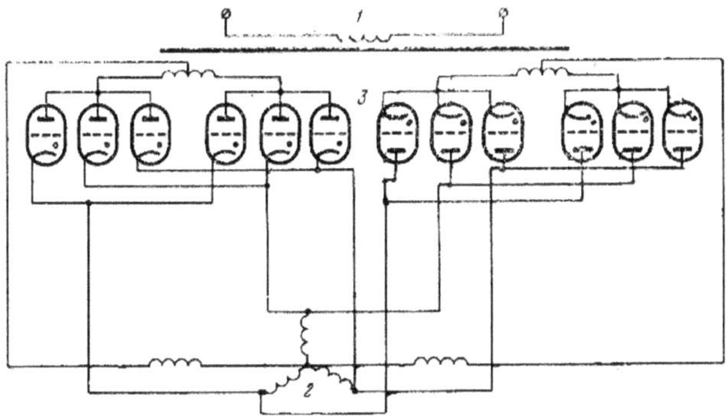

Figure 15.7. A single-phase electrical commutation matrixer (ECM) AC–AC commutator motor: 1—an EE transformer; 2—a stator winding; and 3—thyratrons (Zavalishin and Vegner 1936).

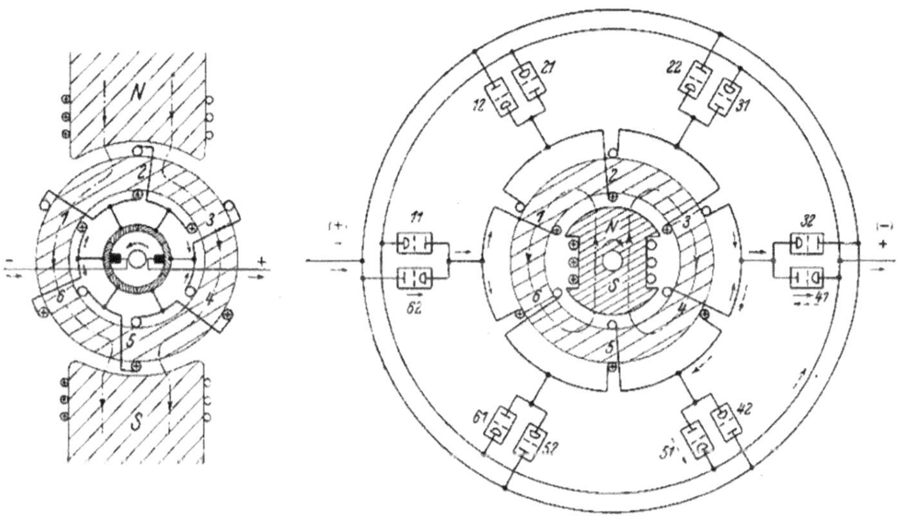

Figure 15.8. A 6 × 2/2 × 6 mechanical commutation matrixer (ECM) AC–DC/DC–AC commutator generator/motor (left) and a 6 × 2/2 × 6 electrical commutation matrixer (ECM) AC–DC/DC–AC commutator generator/motor (right) (Schilling 1940).

(DRP 96564) patents for an aluminium electrical valve bridge rectifier (Pollak 1895, 1896).

Descriptions of the patents included a scheme of the Pollack bridge rectifier (figure 15.3) as a single-phase full-wave MIMO ECM AC–DC commutator in a single-phase double-way connection, now known as the Graetz bridge rectifier. These descriptions were published in the *Elektronische Zeitung* (1897) with notation from the editor that at that time Leo Graetz was working on rectifiers with a similar principle of operation. However, the solution of Leo Graetz was published in 1897, a year and a half after Karol Pollak's first patent had been issued (Strzelecki and Benysek 2008).

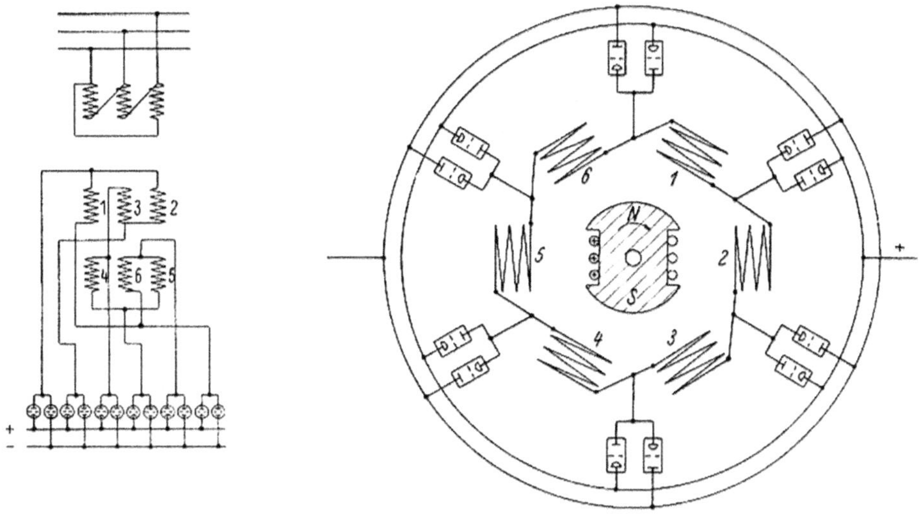

Figure 15.9. A 6 × 2/2 × 6 electrical commutation matrixer (ECM) AC–DC/DC–AC commutator generator/motor (right) compared with a 6 × 2/2 × 6 electrical commutation matrixer (ECM) AC–DC/DC–AC rectifier/inverter (left) (Schilling 1940).

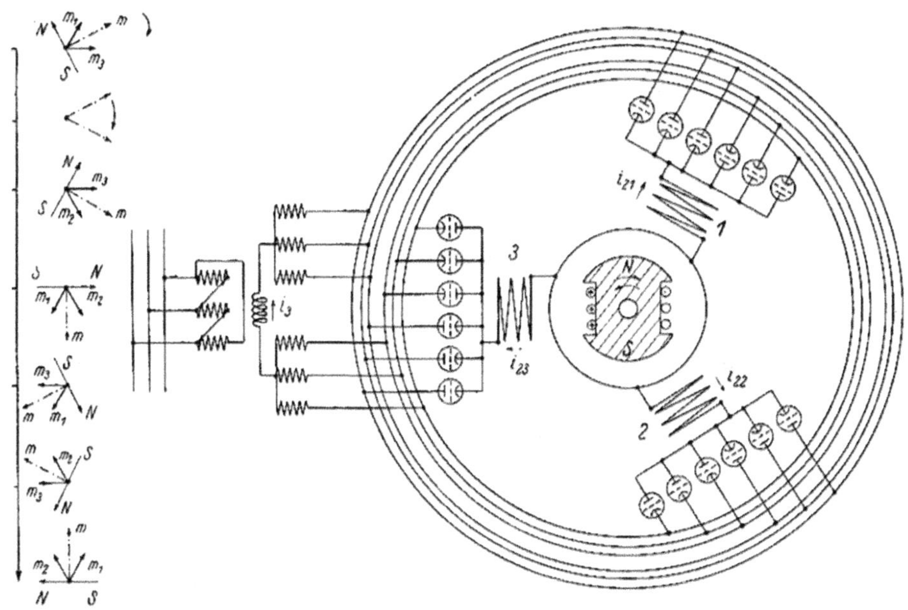

Figure 15.10. A 2(2 × 3) electrical commutation matrixer (ECM) AC–AC commutator generator/motor (Schilling 1940).

In 1901, the Belorussian academician Vladimir Fedorovich Mitkevich (1872–1951) published on two-phase and three-phase half-wave ECM AC–DC commutators in a two- and/or three-phase single-way connection (now known as Mitkevich rectifiers) (Neiman 1972). Soon after, in 1924, the Russian academician Andrei

Nikolaevich Larionov (1889–1963) published on a three-phase full-wave ECM AC–DC commutator in a three-phase double-way connection (now known as the Larionov bridge rectifier) (Anon 1950). Other subsequent inventions that are of importance to ECM AC–AC and AC-DC/DC–AC commutators (see figures 15.4–15.10) were developed in the 1930s (Kloninger 1932; Willis 1933; Alexanderson and Mittag 1934; Kern 1938; Schilling 1940).

Improved ECM AC–AC and AC–DC/DC–AC commutators were frequently proposed during the 1940s and 1950s (Schilling 1940; Rémy 1954). However, such ECM AC–AC and AC–DC/DC–AC commutators did not gain wide acceptance and most of the concepts probably did not leave the laboratory. The fact that this was so must naturally be attributed to the characteristics of the electrical valves available to the scientists and engineers in those days—electronic tubes and mercury-arc electrical valves of different kinds. For technical and economic reasons, it was not until the 1960s that scientists could solve this problem by introducing the **semiconductor-controlled rectifier** (SCR) (later called the 'thyristor') into the power electronics world. In France, the first ECM commutator for an MMD electrical machine (probably the first in the world) was conceived by Kloninger (1932), while in the UK the first complete ECM commutator based on thyratron electrical valves was developed by Alexanderson and Mittag (1934).

AC–AC commutator motor: 1—a stator winding; 2—a rotor winding; and 3—thyratrons (Alexanderson 1931).

The author started R&D work on ECM AC–DC/DC–AC commutator dynamotors (figure 15.11) for high-power EM propulsion systems in the early 1960s (Fijalkowski 1962, 1964).

The 1960s witnessed rapid progress in ECM high-power electronics (called 'macroelectronics' by the author), above all when it came to the developments that took place in the field of bipolar, triggered and quenched electrical valves such as trigistor, **gate-turn-off** (GTO) thyristors and/or bipolar transistor ECMs.

New applications for macroelectronics have thus evolved and established applications have advanced further. This textbook briefly presents some of the progress and new developments that occurred within the macroelectronics field over the previous century.

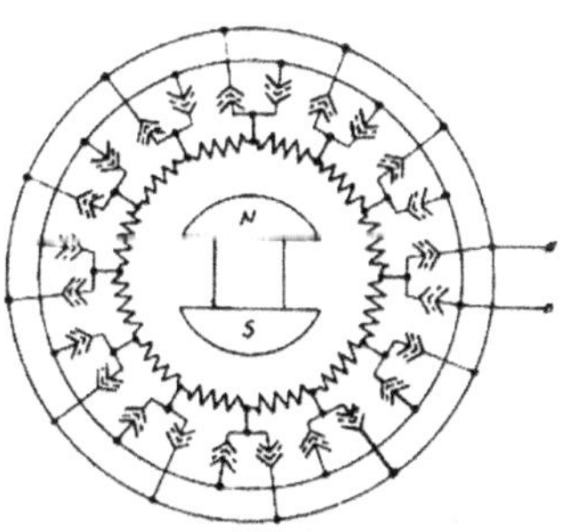

Figure 15.11. A 2 × 12/12 × 2 electrical commutation matrixer (ECM) AC–DC/DC–AC commutator dynamotor (Fijalkowski 1964).

By definition, macroelectronics is the specific application of electronics using ECM commutators (called **multi-input/multi-output** (MIMO) ECM commutators by the author) to the electrical power field, mainly through the new concept of ECM AC–AC and AC–DC/DC–AC commutator dynamotors for industrial and traction EM propulsion systems. It followed the developments that have taken place within macro- and microelectronics as a whole.

The introduction of the ECM AC–AC, AC–DC–AC, AC–DC/DC–AC, DC–DC and/or DC–AC–DC commutator in combination with macro- and microelectronic based technology has been the most significant factor in the tremendous improvement of the power handling capacity, voltage and current data, etc, of the new-concept power-handling components, i.e. ECM commutators.

At the same time, it has been possible to fully exploit the rapid integration of components by miniaturising the static high-power MIMO ECM AC–AC, AC–DC–AC, AC–DC/DC–AC, DC–DC and/or DC–AC–DC commutators in MMD electrical machines as well as low-power (small signal) electronic computer processors, i.e. microprocessors in the higher- and lower-level controllers included in macro- and microelectronics equipment (or ECM AC–AC, AC–DC–AC, AC–DC/DC–AC, DC–DC and/or DC–AC–DC commutators, as they are usually called), and in this way the ECM AC–AC, AC–DC–AC, AC–DC/DC–AC, DC–DC and/or DC–AC–DC commutator has evolved from a rather unqualified integral static high-power ECM AC–AC, AC–DC–AC, AC–DC/DC–AC, DC–DC and/or DC–AC–DC commutator into EM propulsion process equipment that incorporates qualified **electrical integrated circuits** (EIC) for control, monitoring, operation and protection.

Up to now, the use of analogue EIC techniques has predominated, but the trend towards digital EIC techniques is both justified for cost reasons and essential in this case—for example, with very advanced ECM AC–AC, AC–DC–AC, AC–DC/DC–AC, DC–DC and/or DC–AC–DC commutators of the **sinusoidal pulse-width-modulated** (SPWM) type.

In the 1970s, as ECM AC–AC, AC–DC–AC, AC–DC/DC–AC, DC–DC and/or DC–AC–DC commutators were developed and put into practical use, demand grew for fast-switching thyristor ECMs with high performance, such as higher blocking voltage, larger current capability and shorter turn-off times (Fijalkowski 1973).

The current trends in aerospace avionics and automotive vetronics are towards intelligent electronically commutated (reciprocal and rotational) EM actuators, namely electrical machines and other digital devices with much higher performance and greater compactness, and not only at the interconnection stage.

The increasing complexity of intelligent electronically commutated actuators and digital devices is leading to large chip sizes, and the greater degree of integration means that the power dissipated by the working chip is also rising. Increasingly, the tendency is to package several chips together in the form of a multichip module, using techniques such as wire bonding, **tape automated bonding** (TAB) and flip chip bonding to reduce the inter-connection length and hence the overall size.

There ought to be continuing demand for high-performance intelligent MIMO ECM AC–AC, AC–DC–AC, AC–DC/DC–AC, DC–DC and/or DC–AC–DC commutators of far greater complexity than before.

Artificial **neural networks and fuzzy logic** (NF) are increasingly being incorporated into intelligent MIMO ECMs to provide robust and effective control in a wide range of aerospace and automotive applications. The method is based on many-valued-logic, which enables general principles and expert knowledge to be used to provide control rules and procedures. The intrinsic nonlinearity and variability of operating conditions make NF control an ideal method for this area.

The author is actively pursuing the integration of matrixery and circuitry on the same transparent substrate as the intelligent MIMO ECM AC–AC, AC–DC–AC, AC–DC/DC–AC, DC–DC and/or DC–AC–DC commutators and NF controllers, respectively.

The most obvious advantage of integrated matrixery and circuitry on the same transparent substrate is the reduction of the number of mono- and polycrystalline 'work-horse' components, as well as amorphous Si or GaAs super- and/or semiconductors, fast-switching discrete and/or continuous uni- and/or bipolar electrical valves, e.g. diodes, triodes, tetrodes, etc, namely **charge-injection transistors** (CHINT), **complementary metal-oxide-semiconductor** (CMOS) transistors, **hetero-junction bipolar transistors** (HBT), **high-electron injection transistors** (HEMT), **metal-semiconductor field-effect transistors** (MESFET), **metal-insulator-semiconductor FETs** (MISFET), **metal-oxide-semiconductor FETs** (MOSFET), **organic FETs** (OFET), **junction-gate field-effect transistor** (JFET or JUGFET), bipolar **quantum interference transistors** (QUIT), **static induction transistors** (SIT), bipolar **super-conductor-base hot electron transistors** (SUPER-HET), **insulated-gate bipolar transistors** (IGBT), **gate-turn-off** (GTO) thyristors, **light-triggered-and-quenched** (LTQ) thyristors, **MOS controlled thyristors** (MCT), **reverse-blocking insulated-gate bipolar transistors** (RB IGBT), ovonics, etc, and also external connections. Discrete electrical valves suffer from strongly temperature-dependent performance.

The author therefore carried out a detailed programme of research into intelligent MIMO ECM commutators using continuous electrical valves in which the electrical-current conduction is much less temperature-dependent than in discrete electrical valves. The external connections are also one of the dominant causes of MIMO ECM commutator failure. They will be lighter and more compact because of their proportion of size and mass. It is difficult to estimate the cost and reliability of these ECM commutators, as they are not yet in manufacture.

Although the R&D and initial manufacture costs are high, the leading electronic manufacturers will be investing significantly in this key technology and in intelligent MIMO ECM AC–AC, AC–DC–AC, AC–DC/DC–AC, DC–DC and/or DC–AC–DC commutators that are capable of truly inaudible (>20 kHz) operation. NF controllers will soon extend their application from aerospace avionics and automotive vetronic intelligent electronically commutated EM actuators and other digital device markets into other areas and they have every chance of completely revolutionising the ECM universe.

Recently, attention has been focused on the bipolar **reverse-conducting** (RC) GTO thyristor, MCT, MOSFET and IGBT ECM, called the **RC trigistor** ECM, as fast-switching semiconductor controlled electrical-valve ECMs, which may feasibly overcome these difficulties. The electrical-valve ECMs proved to be successful for practical ECM AC–AC, AC–DC–AC, AC–DC/DC–AC, DC–DC and/or DC–AC–DC commutator applications, and important application data were obtained.

These bipolar trigistor and bipolar transistor ECMs have been used in ECM AC–AC, AC–DC–AC, AC–DC/DC–AC, DC–DC and/or DC–AC–DC commutators for industrial and traction propulsion systems and have the advantage of making them more compact, less expensive and of lighter mass than the unipolar **reverse-blocking** (RB) thyristor with freewheeling diode ECMs.

The Nobel Prize in Physics 2010 was awarded jointly to Andre Geim and Konstantin Novoselov 'for groundbreaking experiments regarding the two-dimensional material graphene'[1] (the most promising and versatile material at present). Graphene has attracted attention as a promising two-dimensional honeycomb lattice composed of sp^2-bonded carbon-atoms material for very advanced ECM AC–AC, AC–DC–AC, AC–DC/DC–AC, DC–DC and/or DC–AC–DC commutators, including high-speed **single-electron transistors** (SET), which play a central role in LSI technology.

The SET is based on the idea of controlling the number of electrons in a quantum dot using its electrical potential. The energy consumption of SET is thought to be very small, because the unit of information is a single electron.

Graphene is a one-atom-thick layer of carbon atoms arranged in a hexagonal lattice. It is the building block of graphite (which is used, among others things, for pencil tips), but graphene is a remarkable substance on its own, with a multitude of astonishing properties that have repeatedly earned it the title 'wonder material'.

Graphene is the thinnest material known to man, at one atom thick, and is also incredibly strong—about 200 times stronger than steel. On top of that, graphene is an excellent conductor of heat and electricity and has interesting light absorption abilities. It is truly a material that could change the world, with unlimited potential for integration in almost every industry.

Graphene has a potential advantage in that quantum dots can be formed easily just by cutting out small regions via etching, because graphene begins with a thickness of only a single atomic layer. However, controllable manufacture with a precision of several nanometers still remains a challenge. For example, the SET commutation (switching) performance at room temperature is not yet well reproducible due to the lack of precise manufacture technology.

Therefore, SET is expected to be an electrical valve (electronic switch) that will be used widely in the near future, and much research effort has been put into it.

Today, the author is designing ECM commutators, which, compared with the first thyristor ECM AC–AC, AC–DC–AC, AC–DC/DC–AC, DC–DC and/or DC–AC–DC commutators, will be designed for a 40 × higher ECM AC–AC, AC–DC–AC, AC–DC/DC–AC, DC–DC and/or DC–AC–DC commutator power rating. The integral ECM commutators now being conceived and developed for current projects of ECM AC–AC and AC–DC/DC–AC commutator dynamotors belong to the third generation. The first two generations were air-cooled and air-circulated

[1] www.nobelprize.org/nobel_prizes/physics/laureates/2010/

thyristor ECM AC–AC, AC–DC–AC, AC–DC/DC–AC, DC–DC and/or DC–AC–DC commutators, while oil has been selected as the coolant for the third generation.

The change to oil cooling was a natural consequence of the fact that ECM AC–AC, AC–DC–AC, AC–DC/DC–AC, DC–DC and/or DC–AC–DC commutators show considerably superior performance. Further, oil-cooling systems have substantially lower inherent losses, which offers significant gains with the rapid escalation of the evaluation costs.

The performance and efficiency of MIMO ECM commutators and MMD electrical machines (ME/EM dynamotors and EE transformers) is limited by thermal constraints in the cooling ECM channels and MMD electrical-machine windings. These constraints can be reduced somewhat by circulating liquid through the liquid-cooling ECM channels, or air through the air gaps in the MMD electrical machines. However, such approaches necessitate the addition of pumping mechanisms and are limited by the thermal conductivity of air and liquid.

Recently, a novel approach has been taken to cool MIMO ECM commutators and MMD electrical machines, using **nano-magneto-rheological** (NMR) or **giant-electro-rheological** (GER) fluids as the cooling media.

The aim is to take advantage of the thermal and magnetic or electrical characteristics of NMR or GER fluids, respectively, such that not only will the NMR or GER fluid circulate through the MIMO ECM commutators and MMD electrical machines, but the existing thermal and magnetic or electrical fields in the MIMO ECM commutators and MMD electrical machines will provide the energy required to pump the NMR or GER fluid. The cooling mechanism will therefore not require additional moving mechanical parts, MF pumps, MP compressors, or sensors. The behaviour of the NMR or GER fluid will provide the method to pump and cool the MIMO ECM commutators and MMD electrical machines (thus increasing the MIMO ECM commutators and MMD electrical machines' potential energy density and efficiency).

It is generally the need for a high short-circuit capability and low losses rather than the performance during normal operation that determines, e.g. the choice of trigistor cathode area for the ECM AC–AC, AC–DC–AC, AC–DC/DC–AC, DC–DC and/or DC–AC–DC commutator's electrical valves.

A high short-circuit current capability allows one to optimise the armature-winding reactance concerning reactive power consumption, MMD electrical machine costs, and losses.

Today, there are no plans for MIMO ECM AC–AC, AC–DC–AC, AC–DC/DC–AC, DC–DC and/or DC–AC–DC commutators where the data suggest that parallel connection of electrical valves will be necessary.

The bipolar trigistors with a cathode area exceeding 100 cm^2 now available give a single chain of trigistors connected in series. This results in high reliability and minimal complexity, costs and losses. The triggering pulses for MIMO ECM AC–AC, AC–DC–AC, AC–DC/DC–AC, DC–DC and/or DC–AC–DC commutator electrical valves, where a large number of electrical valves are connected in series to obtain a sufficient voltage rating, must be transmitted by non-galvanic means. For instance, for this purpose, one uses light guides, which transmit light pulses for the electrical-valves' **electronic control unit** (ECU) at earth potential to the individual electrical valves. Because the electrical valves require an electrical triggering pulse, the conversion from light to electrical current pulses is performed in an ECU by each electrical valve (electronic switch).

15.3 Physical and mathematical models of a generalised ECM AC–AC and/or AC–DC/DC–AC commutator

15.3.1 Introduction

In the existing theory of ECM AC–AC and AC–DC/DC–AC commutator dynamotors, both the rotary and linear roles, which in these MMD electrical machines are played by the sliding-contact **mechano-electrical commutation matrixer** (MCM) AC–AC and/or AC–DC/DC–AC commutator, are completely devalued. Up to now, the theory for a generalised **electrical commutation matrixer** (ECM) commutator of MMD electrical machines had also not been worked out and its mathematical model in the author's dynamical systems approach for MMD electrical machines and matrix notation had not been formulated. A complex variety of phenomena taking place with the changeover of the sense of current direction polarity in individual commutating sections, i.e. the phase coils of a polyphase armature winding short circuited through the carbon brush, i.e. the matrixer-input primary collectors or a rotary sliding-contact MCM AC–AC and/or AC–DC/DC–AC commutator, have merely been considered (Kostenko and Piotrovsky 1968).

As mentioned above, up to now, the theory of the MCMs of MCM AC–AC and AC–DC/DC–AC commutator dynamotors, both rotary and linear, has not been worked out. The mathematical model formulation to fill this theoretical gap was first carried out by the author (Fijalkowski 1973, 1982, 1985, 1987, 1988) and will be presented below.

The generalised ECMs of MMD electrical machines (figure 15.12, both rotary and linear), sliding-contact MCM AC–AC and/or AC–DC/DC–AC commutators, and solid-state semi- or superconductor electrical-valve ECM AC–AC and/or AC–DC/DC–AC commutators always act as static frequency and phase-number changers, converting one form of electrical energy of (e.g. with the voltage holor $\boldsymbol{U}_i^P$, the frequency f^P and the phase number m^P) into another form of electrical energy (e.g. with the voltage holor $\boldsymbol{U}_i^s$, the frequency f^s and the phase number m^s) without losses. At the same time, they take into account the properties of the uni- or bipolar current-conduction of electrical valves.

For instance, sliding contacts (copper-bar–carbon-brush contacts) form the components of a sliding-contact MCM AC–AC and/or AC–DC/DC–AC commutator, or solid-state semi- and/or superconductor electrical valves, both monocrystal and amorphous, form the components of an MIMO ECM.

In the simplest way, the generalised ECM AC–AC and/or AC–DC/DC–AC commutator of MMD electrical machines, both rotary sliding-contact MCM AC–AC and/or AC–DC/DC–AC commutators and static ECM AC–AC and/or AC–DC/DC–AC commutators, can be presented as generalised ECM AC–AC and/or AC–DC/DC–AC commutators (figure 15.12) consisting of crossed-over but non-galvanically connected (by means of electrical valves) current-conducting matrixer-input and matrixer-output conductive collectors.

In this generalised ECM, the horizontal collectors (matrixer rows) *A, B, C, …, N* are called the 'matrixer-input' or 'primary collectors', and the vertical collectors (matrixer columns) *a, b, c, …, n* are called the 'matrixer-output' or 'secondary collectors'. Discrimination of the matrixer-input or primary collectors is relative, since at an evident moment of time electrical energy can be both absorbed and recovered through the same collectors. This generalised ECM AC–AC and/or AC–DC/DC–AC commutator may only be connected with two other electro-energetic systems, namely the armature winding of an MMD electrical machine and the AC or DC supply line, by means of self-collectors.

Considering an ECM AC–AC and/or AC–DC/DC–AC commutator of MMD electrical machines as the ME/EM dynamical hypersystem, it is necessary to take into account its cooperation with the armature winding of an MMD electrical machine by means of either voltage sources or current-holor sources led to the matrixer-input and matrixer-output conductive collectors.

The generalised ECM AC–AC and/or AC–DC/DC–AC commutator of MMD electrical machines is the MCM or ECM with the matrixer-input or primary collectors *A, B, C, …, N* connected to the machine's polyphase armature winding terminals, which are connected as poly-star or polygon (e.g. wye or delta), and the matrixer-output or secondary conductive collectors *a, b, c, …, n* connected to the AC or DC supply line. The individual collectors of this ECM are connected to each other by means of uni- and/or bipolar electrical valves.

The impedance of such a real electrical valve is nonlinear, but the impedance of such an ideal electrical valve in the turn-on state is equal to zero, while in the turn-off state it is equal to infinity.

The most obvious difference between a generalised ECM and other types of MCMs and ECMs is the electrical-valve action. In contrast to digital ECMs, the electrical valves do not indicate a logic level. Control is effected by determining the times at which the electrical valves should turn-on. Whether there is just one electrical valve or a large group of them, there is a complexity limit: if a converter has P primary collectors (inputs) and s secondary conductive collectors (outputs), even the densest possible collection of electrical valves would have a single electrical valve between each input conductive collector (column line) and each output conductive collector (row line). The $P \times s$ electrical valves in the ECM AC–AC and/or AC–DC/DC–AC commutator can be arranged according to their connections.

The pattern suggests an ECM, as shown in figure 15.12. MIMO ECM AC–AC, AC–DC/DC–AC and/or DC–DC commutators fall into two classes.

- **Direct ECM AC–AC, AC–DC/DC–AC and/or DC–DC commutators.** In these ECMs, energy-storage components are only connected to the ECM at the primary (input) and secondary (output) of its collectors (ports). The electrical energy-storage components effectively become part of the energy source or load. For instance, an ECM AC–DC commutator (rectifier) with an external low-pass filter is an example of a direct ECM AC–DC commutator. In the literature, only direct 3×3 ECM AC–AC commutators are called 'matrix converters'.

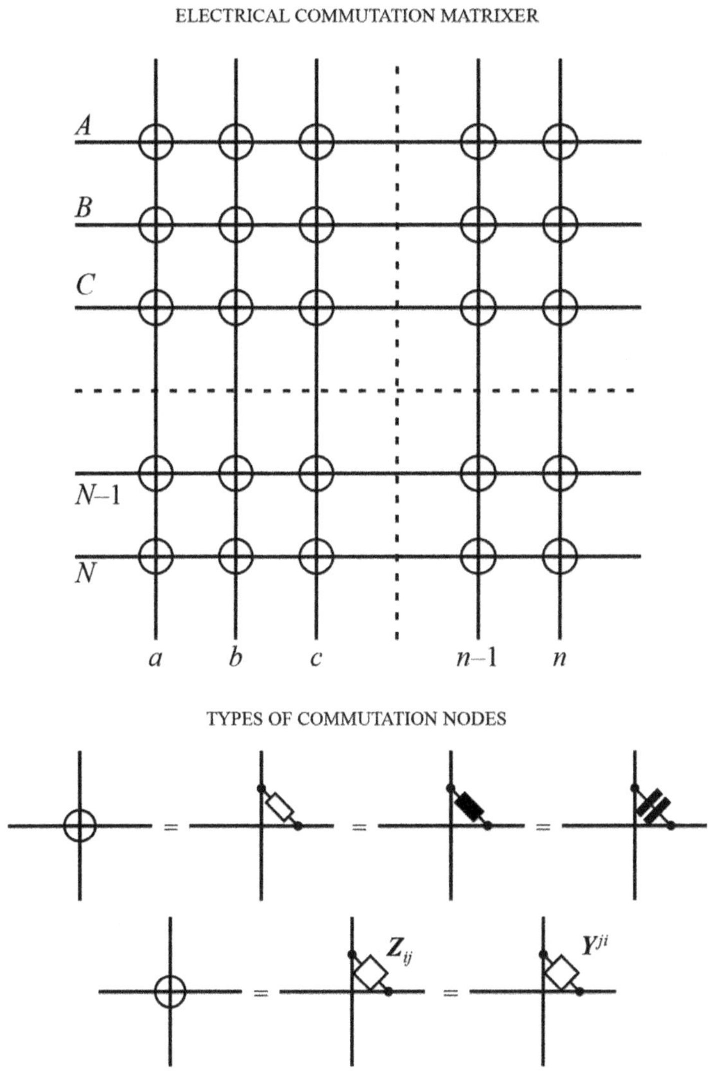

Figure 15.12. Generalised ECM AC–AC, AC–DC/DC–AC and/or DC–DC commutator (Fijalkowski 1987).

- **Indirect ECM AC–DC–AC and/or DC–AC–DC commutators,** also termed **embedded electrical commutators.** These ECM AC–DC–AC and/or DC–AC–DC commutators, for instance, the polarity-reverser example, have energy-storage components connected within the ECM structure. There are usually a very small number of energy-storage components. Indirect ECM AC–DC–AC and/or DC–AC–DC commutators are most commonly analysed as a cascade connection of direct ECM AC–AC and/or AC–DC/DC–AC commutators with the energy-storage in between.

From such a dynamical systems approach for MMD electrical machines with a generalised MIMO ECM AC–AC, AC–DC–AC, AC–DC/DC–AC, DC–DC and/or DC–AC–DC commutator (Fijalkowski 1973, 1982, 1985, 1987, 1988), in referring to AC commutatorless MMD electrical machines with which this generalised MIMO ECM AC–AC, AC–DC–AC, AC–DC/DC–AC, DC–DC and/or DC–AC–DC commutator cooperates and establishes an ME/EM simple dynamical hypersystem, two different mathematical models may be formulated, and therefore, two different types of matrix notation can be obtained for the generalised MIMO ECM commutator's holor equations.

Taking into account ECM excitations (forcings) in the form of the voltage-source holor matrix, and taking as the ECM responses the external mesh-current holor matrix, the mathematical model of a generalised MIMO ECM AC–AC, AC–DC–AC, AC–DC/DC–AC, DC–DC and/or DC–AC–DC commutator of MMD electrical machines has been formulated in the dynamical systems approach. The matrix notation is as follows (figure 15.13):

$$\begin{aligned} \boldsymbol{E}_i &= \left\| \boldsymbol{E}_i^{PT} \quad \boldsymbol{E}_i^{sT} \right\|^T \\ &= \left\| \boldsymbol{E}_i^{ANT} \quad \boldsymbol{E}_i^{BAT} \quad \boldsymbol{E}_i^{CBT} \quad \cdots \quad \boldsymbol{E}_i^{N(N-1)T} \quad \boldsymbol{E}_i^{anT} \quad \boldsymbol{E}_i^{baT} \quad \boldsymbol{E}_i^{cbT} \quad \cdots \quad \boldsymbol{E}_i^{n(n-1)T} \right\|^T. \end{aligned} \tag{15.1}$$

are the responses of the multi-port MIMO generalised MIMO ECM, the column hypermatrices or the hypervectors of the outer-mesh current holors,

$$\begin{aligned} \boldsymbol{I}^j &= \left\| \boldsymbol{I}_P^{jT} \boldsymbol{I}_s^{jT} \right\|^T \\ &= \left\| \boldsymbol{I}_{AN}^{j\;T} \quad \boldsymbol{I}_{BA}^{j\;T} \quad \boldsymbol{I}_{CB}^{j\;T} \quad \cdots \quad \boldsymbol{I}_{N(N-1)}^{j\;T} \quad \boldsymbol{I}_{an}^{j\;T} \quad \boldsymbol{I}_{ba}^{j\;T} \quad \boldsymbol{I}_{cb}^{j\;T} \quad \cdots \quad \boldsymbol{I}_{n(n-1)}^{j\;T} \right\|^T. \end{aligned} \tag{15.2}$$

A set of the differential equations of dynamics, i.e. the mathematical model of the multi-port MIMO generalised ECM as the electrical heterogeneous continuous dynamical hypersystems may be formulated in the dynamical systems approach and holor notation as follows:

$$\boldsymbol{E}_i = \boldsymbol{Z}_{ij}\boldsymbol{I}^j, \tag{15.3}$$

where $i \times j$ hypermatrix $\boldsymbol{Z}_{ij}$ has a dimension of the impedance holor.

To denote the state variables associated with the primary (input) conductive collectors by an index P, and those associated with the secondary (output) conductive collectors by an index s, a set of the differential equations of dynamics, i.e. the mathematical model of the multi-port generalised MIMO ECM may be expressed as follows:

$$\left\| \begin{matrix} \boldsymbol{E}_i^P \\ \boldsymbol{E}_i^s \end{matrix} \right\| = \left\| \begin{matrix} \boldsymbol{Z}_{ij}^{PP} & \boldsymbol{Z}_{ij}^{Ps} \\ \boldsymbol{Z}_{ij}^{sP} & \boldsymbol{Z}_{ij}^{ss} \end{matrix} \right\| \left\| \begin{matrix} \boldsymbol{I}_P^j \\ \boldsymbol{I}_s^j \end{matrix} \right\|, \tag{15.4a}$$

or in detail as

$$\left\| \begin{array}{c} \boldsymbol{E}_i^{AB} \\ \boldsymbol{E}_i^{BC} \\ \boldsymbol{E}_i^{C(N-1)} \\ \vdots \\ \boldsymbol{E}_i^{NA} \\ \boldsymbol{E}_i^{ab} \\ \boldsymbol{E}_i^{bc} \\ \boldsymbol{E}_i^{c(n-1)} \\ \vdots \\ \boldsymbol{E}_i^{na} \end{array} \right\| = \left\| \begin{array}{cccccccccc} \boldsymbol{Z}_{ij}^{AA} & \boldsymbol{Z}_{ij}^{AB} & \boldsymbol{Z}_{ij}^{AC} & & \boldsymbol{Z}_{ij}^{AN} & \boldsymbol{Z}_{ij}^{Aa} & \boldsymbol{Z}_{ij}^{Ab} & \boldsymbol{Z}_{ij}^{Ac} & & \boldsymbol{Z}_{ij}^{An} \\ \boldsymbol{Z}_{ij}^{BA} & \boldsymbol{Z}_{ij}^{BB} & \boldsymbol{Z}_{ij}^{BC} & \cdots & \boldsymbol{Z}_{ij}^{BN} & \boldsymbol{Z}_{ij}^{Ba} & \boldsymbol{Z}_{ij}^{Bb} & \boldsymbol{Z}_{ij}^{Bc} & \cdots & \boldsymbol{Z}_{ij}^{Bn} \\ \boldsymbol{Z}_{ij}^{CA} & \boldsymbol{Z}_{ij}^{CA} & \boldsymbol{Z}_{ij}^{CC} & & \boldsymbol{Z}_{ij}^{CN} & \boldsymbol{Z}_{ij}^{Ca} & \boldsymbol{Z}_{ij}^{Cb} & \boldsymbol{Z}_{ij}^{Cc} & & \boldsymbol{Z}_{ij}^{Cn} \\ & \vdots & & \ddots & \vdots & & \vdots & & \ddots & \vdots \\ \boldsymbol{Z}_{ij}^{NA} & \boldsymbol{Z}_{ij}^{NB} & \boldsymbol{Z}_{ij}^{NC} & \cdots & \boldsymbol{Z}_{ij}^{NN} & \boldsymbol{Z}_{ij}^{Na} & \boldsymbol{Z}_{ij}^{Nb} & \boldsymbol{Z}_{ij}^{Nc} & \cdots & \boldsymbol{Z}_{ij}^{Nn} \\ \boldsymbol{Z}_{ij}^{aA} & \boldsymbol{Z}_{ij}^{aB} & \boldsymbol{Z}_{ij}^{aC} & & \boldsymbol{Z}_{ij}^{aN} & \boldsymbol{Z}_{ij}^{aa} & \boldsymbol{Z}_{ij}^{ab} & \boldsymbol{Z}_{ij}^{ac} & & \boldsymbol{Z}_{ij}^{an} \\ \boldsymbol{Z}_{ij}^{bA} & \boldsymbol{Z}_{ij}^{bB} & \boldsymbol{Z}_{ij}^{bC} & \cdots & \boldsymbol{Z}_{ij}^{bN} & \boldsymbol{Z}_{ij}^{ba} & \boldsymbol{Z}_{ij}^{bb} & \boldsymbol{Z}_{ij}^{bc} & \cdots & \boldsymbol{Z}_{ij}^{bn} \\ \boldsymbol{Z}_{ij}^{cA} & \boldsymbol{Z}_{ij}^{cB} & \boldsymbol{Z}_{ij}^{cC} & & \boldsymbol{Z}_{ij}^{cN} & \boldsymbol{Z}_{ij}^{ca} & \boldsymbol{Z}_{ij}^{cb} & \boldsymbol{Z}_{ij}^{cc} & & \boldsymbol{Z}_{ij}^{cn} \\ & \vdots & & \ddots & \vdots & & \vdots & & \ddots & \vdots \\ \boldsymbol{Z}_{ij}^{nA} & \boldsymbol{Z}_{ij}^{nB} & \boldsymbol{Z}_{ij}^{nC} & \cdots & \boldsymbol{Z}_{ij}^{cN} & \boldsymbol{Z}_{ij}^{na} & \boldsymbol{Z}_{ij}^{nb} & \boldsymbol{Z}_{ij}^{nc} & \cdots & \boldsymbol{Z}_{ij}^{nn} \end{array} \right\| \left\| \begin{array}{c} \boldsymbol{I}_{AB}^{j} \\ \boldsymbol{I}_{BA}^{j} \\ \boldsymbol{I}_{CB}^{j} \\ \vdots \\ \boldsymbol{I}_{N(N-1)}^{j} \\ \boldsymbol{I}_{an}^{j} \\ \boldsymbol{I}_{ba}^{j} \\ \boldsymbol{I}_{cb}^{j} \\ \vdots \\ \boldsymbol{I}_{n(n-1)}^{j} \end{array} \right\| . \tag{15.4b}$$

The coefficients

$$\boldsymbol{Z}_{kl} = \frac{\partial \boldsymbol{E}_{k(k-1)}}{\partial \boldsymbol{I}^{l(l-1)}} \bigg| \boldsymbol{I}^{j(j-1)} \ (j \neq l) \tag{15.5}$$

have a dimension of the impedance holor.

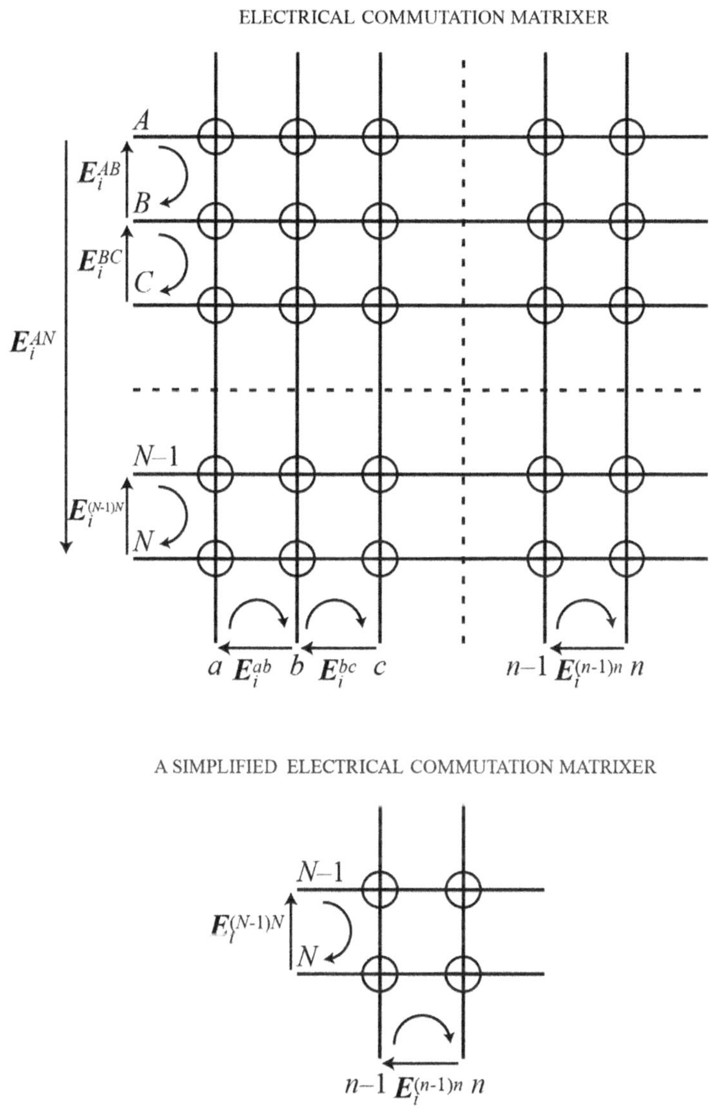

Figure 15.13. A multi-port generalised MIMO ECM with voltage-source holors (physical model I) (Fijalkowski 1987).

To adapt a mathematical model of the multi-port generalised MIMO ECM to the theory of the electrical heterogeneous continuous dynamical hypersystems' demands, one refers to a simplified and very general mathematical model of the multi-port generalised MIMO ECM, derived from the physical model shown in figure 15.12.

In order to fulfil the demands of physical energy conversion of one kind of electrical energy with the voltage-source holor $\boldsymbol{E}_i^P$, the current holor $\boldsymbol{I}_P^j$, the frequency f_P and the number of phases m_P of a primary electrical heterogeneous continuous dynamical hypersystem of the multi-port generalised MIMO ECM into another kind of electrical energy with the voltage-source holor E_i^s, the current holor I_s^j, the frequency f_s and the number of phases m_s of a secondary electrical continuous dynamical system of the multi-port generalised physical commutation matrixer—as well as the free passage of electrical energy between these dynamical hypersystems—they ought to be connected by means of the multi-port generalised MIMO ECM's uni- and/or bipolar passive and active electrical components and devices.

A physical model of the multi-port generalised MIMO ECM (figure 15.14) fulfils the aforementioned demands. Uni- and bipolar passive and active electrical components and devices connect both electrical heterogeneous continuous dynamical hypersystems of the multi-port generalised MIMO ECM and they make free passage of electrical energy between them possible.

In the general case, this MIMO ECM can connect many electrical heterogeneous continuous dynamical hypersystems. In this textbook we consider a case where there are only two electrical heterogeneous continuous dynamical hypersystems—the primary and secondary. This is sufficient to understand the multi-port MIMO ECM concept. Both electrical heterogeneous continuous dynamical hypersystems of the multi -port generalised MIMO ECM may be considered to be autonomous, i.e. when disconnected from the multi-port generalised MIMO ECM, they have determined potentials on their terminals. They have voltage-source holors, current-source holors and electrical components (e.g. inductances, capacitances and resistances), as well as discrete or continuous uni- and/or bipolar physical valves. Assuming that the excitations of a multi-port generalised MIMO ECM are given in the form of the column hypermatrices or hypervectors of current-source holors flowing in the primary (input) and secondary (output) conductive collectors (figure 15.14):

$$I^j = \left\| I_P^{jT} I_s^{jT} \right\|^T = \left\| I_A^{jT} \quad I_B^{jT} \quad I_C^{jT} \quad \cdots \quad I_N^{jT} \quad I_a^{jT} \quad I_b^{jT} \quad I_c^{jT} \quad \cdots \quad I_n^{jT} \right\|^T. \tag{15.6}$$

are the responses of the multi-port MIMO ECM, the column hypermatrices or the hypervectors of the node voltage holors

$$V_i = \left\| V_i^{PT} V_i^{sT} \right\|^T = \left\| V_i^{AT} \quad V_i^{BT} \quad V_i^T \quad \cdots \quad V_i^{NT} V_i^{aT} \quad V_i^{bT} \quad V_i^{cT} \quad \cdots \quad V_i^{nT} \right\|^T. \tag{15.7}$$

A set of the differential equations of dynamics, i.e. the mathematical model of the multi-port generalised MIMO ECM as the electrical heterogeneous continuous dynamical hypersystems, may be formulated in the dynamical systems approach and matrix notation as follows:

$$I^j = Y^{ji} V_i, \tag{15.8}$$

where $n \times m$ hypermatrix Y^{ji} has a dimension of the admittance holor.

Node voltage holors, consisting of electrical components of the hypermatrix V_i, are measured with regard to the reference node, which is located on the outside of the multi-port generalised MIMO ECM. For example, in a star connection of the multi-phase generalised electrical heterogeneous continuous dynamical hypersystem, the reference nodes may be a null point of the multi-phase star connection.

To denote the state variables associated with the primary (input) collectors by an index P, and those associated with the secondary (output) collectors by an index s, the set of differential equations of dynamics, i.e. the mathematical model of the multi-port generalised MIMO ECM, may be expressed as follows:

$$\left\| \begin{matrix} \boldsymbol{I}_P^j \\ \boldsymbol{I}_s^j \end{matrix} \right\| = \left\| \begin{matrix} \boldsymbol{Y}_{PP}^{ji} & \boldsymbol{Y}_{Ps}^{ji} \\ \boldsymbol{Y}_{sP}^{ji} & \boldsymbol{Y}_{ss}^{ji} \end{matrix} \right\| \left\| \begin{matrix} \boldsymbol{V}_i^P \\ \boldsymbol{V}_i^s \end{matrix} \right\|, \tag{15.9a}$$

or in detail as

$$\left\| \begin{matrix} \boldsymbol{I}_A^j \\ \boldsymbol{I}_B^j \\ \boldsymbol{I}_C^j \\ \vdots \\ \boldsymbol{I}_N^j \\ \boldsymbol{I}_a^j \\ \boldsymbol{I}_b^j \\ \boldsymbol{I}_c^j \\ \vdots \\ \boldsymbol{I}_n^j \end{matrix} \right\| = \left\| \begin{matrix} \boldsymbol{Y}_{AA}^{ji} & \boldsymbol{Y}_{AB}^{ji} & \boldsymbol{Y}_{AC}^{ji} & & \boldsymbol{Y}_{AN}^{ji} & \boldsymbol{Y}_{Aa}^{ji} & \boldsymbol{Y}_{Ab}^{ji} & \boldsymbol{Y}_{Ac}^{ji} & & \boldsymbol{Y}_{An}^{ji} \\ \boldsymbol{Y}_{BA}^{ji} & \boldsymbol{Y}_{BB}^{ji} & \boldsymbol{Y}_{BC}^{ji} & \cdots & \boldsymbol{Y}_{BN}^{ji} & \boldsymbol{Y}_{Ba}^{ji} & \boldsymbol{Y}_{Bb}^{ji} & \boldsymbol{Y}_{Bc}^{ji} & \cdots & \boldsymbol{Y}_{Bn}^{ji} \\ \boldsymbol{Y}_{CA}^{ji} & \boldsymbol{Y}_{CA}^{ji} & \boldsymbol{Y}_{CC}^{ji} & & \boldsymbol{Y}_{CN}^{ji} & \boldsymbol{Y}_{Ca}^{ji} & \boldsymbol{Y}_{Cb}^{ji} & \boldsymbol{Y}_{Cc}^{ji} & & \boldsymbol{Y}_{Cn}^{ji} \\ & \vdots & & \ddots & \vdots & & \vdots & & \ddots & \vdots \\ \boldsymbol{Y}_{NA}^{ji} & \boldsymbol{Y}_{NB}^{ji} & \boldsymbol{Y}_{NC}^{ji} & \cdots & \boldsymbol{Y}_{NN}^{ji} & \boldsymbol{Y}_{Na}^{ji} & \boldsymbol{Y}_{Nb}^{ji} & \boldsymbol{Y}_{Nc}^{ji} & \cdots & \boldsymbol{Y}_{Nn}^{ji} \\ \boldsymbol{Y}_{aA}^{ji} & \boldsymbol{Y}_{aB}^{ji} & \boldsymbol{Y}_{aC}^{ji} & & \boldsymbol{Y}_{aN}^{ji} & \boldsymbol{Y}_{aa}^{ji} & \boldsymbol{Y}_{ab}^{ji} & \boldsymbol{Y}_{ac}^{ji} & & \boldsymbol{Y}_{an}^{ji} \\ \boldsymbol{Y}_{bA}^{ji} & \boldsymbol{Y}_{bB}^{ji} & \boldsymbol{Y}_{bC}^{ji} & \cdots & \boldsymbol{Y}_{bN}^{ji} & \boldsymbol{Y}_{ba}^{ji} & \boldsymbol{Y}_{bb}^{ji} & \boldsymbol{Y}_{bc}^{ji} & \cdots & \boldsymbol{Y}_{bn}^{ji} \\ \boldsymbol{Y}_{cA}^{ji} & \boldsymbol{Y}_{cB}^{ji} & \boldsymbol{Y}_{cC}^{ji} & & \boldsymbol{Y}_{cN}^{ji} & \boldsymbol{Y}_{ca}^{ji} & \boldsymbol{Y}_{cb}^{ji} & \boldsymbol{Y}_{cc}^{ji} & & \boldsymbol{Y}_{cn}^{ji} \\ & \vdots & & \ddots & \vdots & & \vdots & & \ddots & \vdots \\ \boldsymbol{Y}_{nA}^{ji} & \boldsymbol{Y}_{aB}^{ji} & \boldsymbol{Y}_{nC}^{ji} & \cdots & \boldsymbol{Y}_{cN}^{ji} & \boldsymbol{Y}_{na}^{ji} & \boldsymbol{Y}_{nb}^{ji} & \boldsymbol{Y}_{nc}^{ji} & \cdots & \boldsymbol{Y}_{nn}^{ji} \end{matrix} \right\| \left\| \begin{matrix} \boldsymbol{V}_i^A \\ \boldsymbol{V}_i^B \\ \boldsymbol{V}_i^C \\ \vdots \\ \boldsymbol{V}_i^N \\ \boldsymbol{V}_i^a \\ \boldsymbol{V}_i^b \\ \boldsymbol{V}_i^c \\ \vdots \\ \boldsymbol{V}_i^n \end{matrix} \right\|. \tag{15.9b}$$

The coefficients

$$\boldsymbol{Y}^{kl} = \frac{\partial \boldsymbol{j}^k}{\partial \boldsymbol{U}_l} \bigg| \boldsymbol{U}_i = 0 \; (i \neq l) \tag{15.9c}$$

have a dimension of the admittance holor.

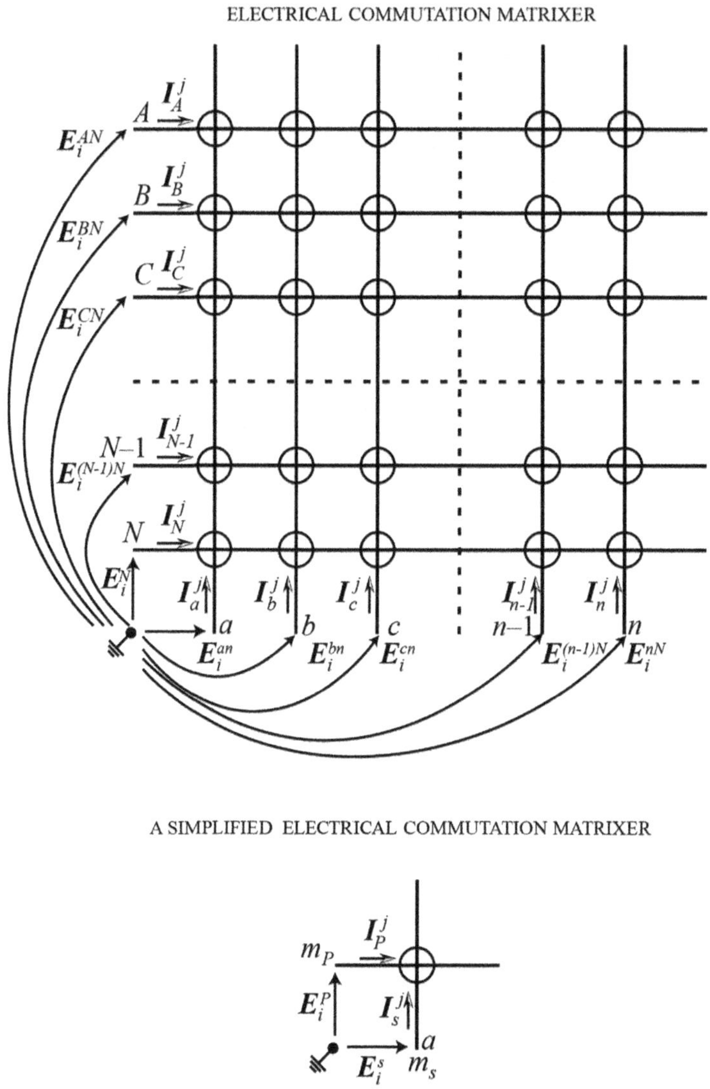

Figure 15.14. A multi-port generalised MIMO ECM with current-source holors (physical model I) (Fijalkowski 1987).

Summing up, a hybrid mathematical model of the multi-port generalised MIMO ECM may be written as follows:

$$
\begin{bmatrix}
\mathbf{E}_i^{AN}\\
\mathbf{E}_i^{BA}\\
\mathbf{E}_i^{CB}\\
\vdots\\
\mathbf{E}_i^{N(N-1)}\\
\mathbf{E}_i^{an}\\
\mathbf{E}_i^{ba}\\
\mathbf{E}_i^{cb}\\
\vdots\\
\mathbf{E}_i^{n(n-1)}\\
\mathbf{J}_A^j\\
\mathbf{J}_B^j\\
\mathbf{J}_C^j\\
\vdots\\
\mathbf{J}_N^j\\
\mathbf{J}_a^j\\
\mathbf{J}_b^j\\
\mathbf{J}_c^j\\
\vdots\\
\mathbf{J}_n^j
\end{bmatrix}
=
\left[\begin{array}{cccccccccc|cccccccccc}
\mathbf{Z}_{ij}^{AA} & \mathbf{Z}_{ij}^{AB} & \mathbf{Z}_{ij}^{AC} & & \mathbf{Z}_{ij}^{AN} & \mathbf{Z}_{ij}^{Aa} & \mathbf{Z}_{ij}^{Ab} & \mathbf{Z}_{ij}^{Ac} & & \mathbf{Z}_{ij}^{An} & 0 & 0 & 0 & & 0 & 0 & 0 & 0 & & 0\\
\mathbf{Z}_{ij}^{BA} & \mathbf{Z}_{ij}^{BB} & \mathbf{Z}_{ij}^{BC} & \cdots & \mathbf{Z}_{ij}^{BN} & \mathbf{Z}_{ij}^{Ba} & \mathbf{Z}_{ij}^{Bb} & \mathbf{Z}_{ij}^{Bc} & \cdots & \mathbf{Z}_{ij}^{Bn} & 0 & 0 & 0 & \cdots & 0 & 0 & 0 & 0 & \cdots & 0\\
\mathbf{Z}_{ij}^{CA} & \mathbf{Z}_{ij}^{CA} & \mathbf{Z}_{ij}^{CC} & & \mathbf{Z}_{ij}^{CN} & \mathbf{Z}_{ij}^{Ca} & \mathbf{Z}_{ij}^{Cb} & \mathbf{Z}_{ij}^{Cc} & & \mathbf{Z}_{ij}^{Cn} & 0 & 0 & 0 & & 0 & 0 & 0 & 0 & & 0\\
& \vdots & & \ddots & \vdots & & \vdots & & \ddots & & & \vdots & & \ddots & \vdots & & \vdots & & \ddots & \vdots\\
\mathbf{Z}_{ij}^{NA} & \mathbf{Z}_{ij}^{NB} & \mathbf{Z}_{ij}^{NC} & \cdots & \mathbf{Z}_{ij}^{NN} & \mathbf{Z}_{ij}^{Na} & \mathbf{Z}_{ij}^{Nb} & \mathbf{Z}_{ij}^{Nc} & \cdots & \mathbf{Z}_{ij}^{Nn} & 0 & 0 & 0 & \cdots & 0 & 0 & 0 & 0 & \cdots & 0\\
\mathbf{Z}_{ij}^{aA} & \mathbf{Z}_{ij}^{aB} & \mathbf{Z}_{ij}^{aC} & & \mathbf{Z}_{ij}^{aN} & \mathbf{Z}_{ij}^{aa} & \mathbf{Z}_{ij}^{ab} & \mathbf{Z}_{ij}^{ac} & & \mathbf{Z}_{ij}^{cn} & 0 & 0 & 0 & & 0 & 0 & 0 & 0 & & 0\\
\mathbf{Z}_{ij}^{bA} & \mathbf{Z}_{ij}^{bB} & \mathbf{Z}_{ij}^{bC} & \cdots & \mathbf{Z}_{ij}^{bN} & \mathbf{Z}_{ij}^{ba} & \mathbf{Z}_{ij}^{bb} & \mathbf{Z}_{ij}^{bc} & \cdots & \mathbf{Z}_{ij}^{bn} & 0 & 0 & 0 & \cdots & 0 & 0 & 0 & 0 & \cdots & 0\\
\mathbf{Z}_{ij}^{cA} & \mathbf{Z}_{ij}^{cB} & \mathbf{Z}_{ij}^{cC} & & \mathbf{Z}_{ij}^{cN} & \mathbf{Z}_{ij}^{ca} & \mathbf{Z}_{ij}^{cb} & \mathbf{Z}_{ij}^{cc} & & \mathbf{Z}_{ij}^{cn} & 0 & 0 & 0 & & 0 & 0 & 0 & 0 & & 0\\
& \vdots & & \ddots & \vdots & & \vdots & & \ddots & \vdots & & \vdots & & \ddots & \vdots & & \vdots & & \ddots & \vdots\\
\mathbf{Z}_{ij}^{nA} & \mathbf{Z}_{ij}^{nB} & \mathbf{Z}_{ij}^{nC} & \cdots & \mathbf{Z}_{ij}^{cN} & \mathbf{Z}_{ij}^{na} & \mathbf{Z}_{ij}^{nb} & \mathbf{Z}_{ij}^{nc} & \cdots & \mathbf{Z}_{ij}^{nn} & 0 & 0 & 0 & \cdots & 0 & 0 & 0 & 0 & \cdots & 0\\
\hline
0 & 0 & 0 & & 0 & 0 & 0 & 0 & & 0 & \mathbf{Y}_{AA}^{ji} & \mathbf{Y}_{AB}^{ji} & \mathbf{Y}_{AC}^{ji} & & \mathbf{Y}_{AN}^{ji} & \mathbf{Y}_{Aa}^{ji} & \mathbf{Y}_{Ab}^{ji} & \mathbf{Y}_{Ac}^{ji} & & \mathbf{Y}_{An}^{ji}\\
0 & 0 & 0 & \cdots & 0 & 0 & 0 & 0 & \cdots & 0 & \mathbf{Y}_{BA}^{ji} & \mathbf{Y}_{BB}^{ji} & \mathbf{Y}_{BC}^{ji} & \cdots & \mathbf{Y}_{BN}^{ji} & \mathbf{Y}_{Ba}^{ji} & \mathbf{Y}_{Bb}^{ji} & \mathbf{Y}_{Bc}^{ji} & \cdots & \mathbf{Y}_{Bn}^{ji}\\
0 & 0 & 0 & & 0 & 0 & 0 & 0 & & 0 & \mathbf{Y}_{CA}^{ji} & \mathbf{Y}_{CA}^{ji} & \mathbf{Y}_{CC}^{ji} & & \mathbf{Y}_{CN}^{ji} & \mathbf{Y}_{Ca}^{ji} & \mathbf{Y}_{Cb}^{ji} & \mathbf{Y}_{Cc}^{ji} & & \mathbf{Y}_{Cn}^{ji}\\
& \vdots & & \ddots & \vdots & & \vdots & & \ddots & \vdots & & \vdots & & \ddots & \vdots & & \vdots & & \ddots & \vdots\\
0 & 0 & 0 & \cdots & 0 & 0 & 0 & 0 & \cdots & 0 & \mathbf{Y}_{NA}^{ji} & \mathbf{Y}_{NB}^{ji} & \mathbf{Y}_{NC}^{ji} & \cdots & \mathbf{Y}_{NN}^{ji} & \mathbf{Y}_{Na}^{ji} & \mathbf{Y}_{Nb}^{ji} & \mathbf{Y}_{Nc}^{ji} & \cdots & \mathbf{Y}_{Nn}^{ji}\\
0 & 0 & 0 & & 0 & 0 & 0 & 0 & & 0 & \mathbf{Y}_{aA}^{ji} & \mathbf{Y}_{aB}^{ji} & \mathbf{Y}_{aC}^{ji} & & \mathbf{Y}_{aN}^{ji} & \mathbf{Y}_{aa}^{ji} & \mathbf{Y}_{ab}^{ji} & \mathbf{Y}_{ac}^{ji} & & \mathbf{Y}_{an}^{ji}\\
0 & 0 & 0 & \cdots & 0 & 0 & 0 & 0 & \cdots & 0 & \mathbf{Y}_{bA}^{ji} & \mathbf{Y}_{bB}^{ji} & \mathbf{Y}_{bC}^{ji} & \cdots & \mathbf{Y}_{bN}^{ji} & \mathbf{Y}_{ba}^{ji} & \mathbf{Y}_{bb}^{ji} & \mathbf{Y}_{bc}^{ji} & \cdots & \mathbf{Y}_{bn}^{ji}\\
0 & 0 & 0 & & 0 & 0 & 0 & 0 & & 0 & \mathbf{Y}_{cA}^{ji} & \mathbf{Y}_{cB}^{ji} & \mathbf{Y}_{cC}^{ji} & & \mathbf{Y}_{cN}^{ji} & \mathbf{Y}_{ca}^{ji} & \mathbf{Y}_{cb}^{ji} & \mathbf{Y}_{cc}^{ji} & & \mathbf{Y}_{cn}^{ji}\\
& \vdots & & \ddots & \vdots & & \vdots & & \ddots & \vdots & & \vdots & & \ddots & \vdots & & \vdots & & \ddots & \vdots\\
0 & 0 & 0 & \cdots & 0 & 0 & 0 & 0 & \cdots & 0 & \mathbf{Y}_{nA}^{ji} & \mathbf{Y}_{aB}^{ji} & \mathbf{Y}_{nC}^{ji} & \cdots & \mathbf{Y}_{cN}^{ji} & \mathbf{Y}_{na}^{ji} & \mathbf{Y}_{nb}^{ji} & \mathbf{Y}_{nc}^{ji} & \cdots & \mathbf{Y}_{nn}^{ji}
\end{array}\right]
\begin{bmatrix}
\mathbf{I}_{AB}^j\\
\mathbf{I}_{BA}^j\\
\mathbf{I}_{CB}^j\\
\vdots\\
\mathbf{I}_{N(N-1)}^j\\
\mathbf{I}_{an}^j\\
\mathbf{I}_{ba}^j\\
\mathbf{I}_{cb}^j\\
\vdots\\
\mathbf{I}_{n(n-1)}^j\\
\mathbf{V}_i^A\\
\mathbf{V}_i^B\\
\mathbf{V}_i^C\\
\vdots\\
\mathbf{V}_i^N\\
\mathbf{V}_i^a\\
\mathbf{V}_i^b\\
\mathbf{V}_i^c\\
\vdots\\
\mathbf{V}_i^n
\end{bmatrix}. \quad (15.10)
$$

$$\boldsymbol{Z}_{kl} = \left.\frac{\partial \boldsymbol{E}_{k(k-1)}}{\partial \boldsymbol{I}^{l(l-1)}}\right|_{\boldsymbol{I}^{j(j-1)}} (j \neq l) \text{ and } \boldsymbol{Y}^{kl} = \left.\frac{\partial \boldsymbol{J}^{k}}{\partial \boldsymbol{V}_{l}}\right|_{\boldsymbol{V}_i=0} (i \neq l) \tag{15.11}$$

have dimensions of the impedance or admittance holors, respectively.

Similar to the mathematical models of physical heterogeneous continuous dynamical hypersystems of generalized MIMO ECMs shown in figures 15.13 and 15.14, consisting of a set of differential equations of dynamics (15.4) and (15.9), is the possible formulation of the mathematical model of physically continuous dynamical systems of generalized MIMO ECMs (figure 15.14) acting as voltage-source ECM commutators (matrix converters) with an arbitrary number N of primary collectors (inputs), equal to the number m of physical commutation matrixer phases, and an arbitrary number n of secondary collectors (outputs).

Assuming that the voltage-source holors applied across the primary collectors (inputs) $A, B, C, \dots, N$ of a generalised MIMO ECM commutator (matrix converter) are given in the form of the voltage-source holor column matrix or vector (input holors), then

$$\boldsymbol{E}_i^P = \left\| \boldsymbol{E}_i^{AT} \quad \boldsymbol{E}_i^{BT} \quad \boldsymbol{E}_i^{CT} \quad \cdots \quad \boldsymbol{E}_i^{N-1T} \quad \boldsymbol{E}_i^{NT} \right\|^T \tag{15.12}$$

is the primary electrical state variables (inputs), and a voltage-source holor, applied across the secondary collectors (outputs) $a, b, c, \dots, n$ of a generalised MIMO ECM commutator (matrix converter), in the form of the voltage-source holor matrix or vector (output holors)

$$\boldsymbol{E}_s^j = \left\| \boldsymbol{E}_a^{jT} \quad \boldsymbol{E}_b^{jT} \quad \boldsymbol{E}_c^{jT} \quad \cdots \quad \boldsymbol{E}_{n-1}^{j\ T} \quad \boldsymbol{E}_n^{jT} \right\|^T \tag{15.13}$$

is the secondary electrical state variables (output holors).

A mathematical model of the electrical heterogeneous continuous dynamical hypersystem of a generalised MIMO ECM commutator (matrix converter) acting as a voltage-source ECM commutator, making the outputs dependent on the inputs, in the dynamical systems approach and matrix notation may be written in the form:

$$\boldsymbol{E}_s^j = \boldsymbol{C}_{sP}^{ji} \boldsymbol{E}_i^P, \tag{15.14}$$

or in detail as

$$\left\| \begin{matrix} E_a^j \\ E_b^n \\ E_c^i \\ \vdots \\ E_{n-1}^j \\ E_n^j \end{matrix} \right\| = \left\| \begin{matrix} C_{aA}^{ji} & C_{aB}^{ji} & C_{aC}^{ji} & & C_{a(N-1)}^{ji} & C_{aN}^{ji} \\ C_{bA}^{ji} & C_{bB}^{ji} & C_{bC}^{ji} & \cdots & C_{b(N-1)}^{ji} & C_{bN}^{ji} \\ C_{cA}^{ji} & C_{cB}^{ji} & C_{cC}^{ji} & & C_{c(N-1)}^{ji} & C_{cN}^{ji} \\ & \vdots & & \ddots & \vdots & \\ C_{(n-1)A}^{ji} & C_{(n-1)B}^{ji} & C_{(n-1)C}^{ji} & \cdots & C_{(n-1)(N-1)}^{ji} & C_{(n-1)N}^{ji} \\ C_{nA}^{ji} & C_{nB}^{ji} & C_{nC}^{ji} & & C_{n(N-1)}^{ji} & C_{nN}^{ji} \end{matrix} \right\| \cdot \left\| \begin{matrix} E_i^A \\ E_i^B \\ E_i^C \\ \vdots \\ E_i^{N-1} \\ E_i^N \end{matrix} \right\| \qquad (15.15)$$

However, a mathematical model of the electrical continuous dynamical system of a generalised MIMO ECM commutator (matrix converter) acting as a voltage-source ECM commutator, making the inputs dependent on the outputs, may be written in the form:

$$\left\| E_i^P \right\| = \left\| C_{ij}^{Ps} \right\| \left\| E_s^j \right\|, \qquad (15.16)$$

or in detail as

$$\left\| \begin{matrix} E_i^A \\ E_i^B \\ F_i^C \\ \vdots \\ E_i^{N-1} \\ E_i^N \end{matrix} \right\| = \left\| \begin{matrix} C_{ij}^{Aa} & C_{ij}^{Ab} & C_{ij}^{Ac} & & C_{ij}^{A(n-1)} & C_{ij}^{An} \\ C_{ij}^{Ba} & C_{ij}^{Bb} & C_{ij}^{Bc} & \cdots & C_{IJ}^{B(n-1)} & C_{IJ}^{Bn} \\ C_{IJ}^{Ca} & C_{IJ}^{Cb} & C_{IJ}^{Cc} & & C_{IJ}^{C(n-1)} & C_{IJ}^{Cn} \\ & \vdots & & \ddots & \vdots & \\ C_{IJ}^{(N-1)a} & C_{IJ}^{(N-1)b} & C_{IJ}^{(N-1)c} & \cdots & C_{IJ}^{(N-1)(n-1)} & C_{IJ}^{(N-1)n} \\ C_{IJ}^{Na} & C_{IJ}^{Nb} & C_{IJ}^{Nc} & & C_{IJ}^{N(n-1)} & V_{IJ}^{Nn} \end{matrix} \right\| \left\| \begin{matrix} E_a^j \\ E_b^j \\ E_c^j \\ \vdots \\ E_{n-1}^j \\ E_n^j \end{matrix} \right\|, \qquad (15.17)$$

where $\boldsymbol{C}_{kL}^{JI}$ and $\boldsymbol{C}_{ij}^{KI}$ are the voltage-source coefficients of the commutation nodes kL and Kl, respectively, which may be expressed as follows:

$$\boldsymbol{C}_{kL}^{JI} = \left.\frac{\partial \boldsymbol{E}_k^j}{\partial \boldsymbol{E}_i^L}\right|_{\boldsymbol{E}_i^l = 0} = \begin{cases} +1 \stackrel{\Delta}{\rightarrow} \boldsymbol{E}_k^j = +\boldsymbol{E}_i^L \\ 0 \stackrel{\Delta}{\rightarrow} \boldsymbol{E}_k^j = 0 \\ -1 \stackrel{\Delta}{\rightarrow} \boldsymbol{E}_k^j = -\boldsymbol{E}_i^L \end{cases} .$$

and

$$\boldsymbol{C}_{ij}^{Kl} = \left.\frac{\partial \boldsymbol{E}_i^K}{\partial \boldsymbol{E}_l^j}\right|_{\boldsymbol{E}_i^j = 0} = \begin{cases} -1 \stackrel{\Delta}{\rightarrow} \boldsymbol{E}_i^K = -\boldsymbol{E}_l^j \\ 0 \stackrel{\Delta}{\rightarrow} \boldsymbol{E}_i^K = 0 \\ +1 \stackrel{\Delta}{\rightarrow} \boldsymbol{E}_i^K = +\boldsymbol{E}_l^j. \end{cases}$$

In the case of the generalized MIMO ECM with bipolar (symmetrical) electrical valves, the voltage-source coefficients of the commutation nodes $\boldsymbol{C}_{kL}^{ji} = \pm 1$ i $\boldsymbol{C}_{ij}^{Kl} = \mp 1$ represent an ideal electrical-valve whose impedance equals zero, and which is incorporated between the primary collector L and the secondary collector k, or between the secondary collector l and the primary collector K, which corresponds to a turn-on state of the electrical valves with a sense of the current flow direction (polarisation) that is dependent on a sign of these coefficients. However, for voltage-source coefficients of the commutation nodes $\boldsymbol{C}_{kL}^{ji} = 0$ and $\boldsymbol{C}_{ij}^{Kl} = 0$, an ideal electrical-valve impedance equals infinity, which corresponds to a turn-off state of the electrical valves.

Analogically, a mathematical model of the physical heterogeneous continuous dynamical system of a generalised MIMO ECM AC–AC and/or AC–DC/DC–AC commutator (matrix converter) shown in figure 15.14, as a current-source ECM commutator with an arbitrary number N of primary collectors (inputs), equals a number m of physical phases and an arbitrary number n of secondary collectors (outputs).

Assuming that an energy-transfer holor's sources, flowing in the primary collectors (inputs) $A, B, C, \ldots, N$ of an MIMO ECM commutator (matrix converter), are given in the form of the current-source holor column matrix or vector (input holors), then

$$\boldsymbol{I}_i^P = \left\| \boldsymbol{I}_i^{AT} \quad \boldsymbol{I}_i^{BT} \quad \boldsymbol{I}_i^{CT} \quad \cdots \quad \boldsymbol{I}_i^{N-1T} \quad \boldsymbol{I}_i^{NT} \right\|^T \tag{15.18}$$

are the primary electrical state variables (inputs), and the secondary current-source holors of a generalised MIMO ECM commutator, flowing in the secondary collectors (outputs) *a, b, c, …, n* of an MIMO ECM, in the form of the secondary current-source holors (output holors)

$$\boldsymbol{I}_s^j = \left\| \boldsymbol{I}_a^{jT} \; \boldsymbol{I}_b^{jT} \; \boldsymbol{I}_c^{jT} \; \cdots \; \boldsymbol{I}_{n-1}^{j\;T} \; \boldsymbol{I}_n^{jT} \right\|^T \tag{15.19}$$

are the secondary physical state variables (inputs).

A mathematical model of the physical heterogeneous continuous dynamical hypersystem of a generalized MIMO ECM commutator (matrix converter) acting as a current-source ECM commutator, making the outputs dependent on the inputs, in the dynamical systems approach and a matrix notation may be written in the form:

$$\boldsymbol{I}_s^j = \boldsymbol{C}_{sP}^{ji} \boldsymbol{I}_i^P, \tag{15.20}$$

or in detail as

$$\left\| \begin{matrix} \boldsymbol{I}_a^j \\ \boldsymbol{I}_b^j \\ \boldsymbol{I}_c^j \\ \vdots \\ \boldsymbol{I}_{n-1}^j \\ \boldsymbol{I}_n^j \end{matrix} \right\| = \left\| \begin{matrix} \boldsymbol{C}_{aA}^{ji} & \boldsymbol{C}_{aB}^{ji} & \boldsymbol{C}_{aC}^{ji} & & \boldsymbol{C}_{a(N-1)}^{ji} & \boldsymbol{C}_{aN}^{ji} \\ \boldsymbol{C}_{bA}^{ji} & \boldsymbol{C}_{bB}^{ji} & \boldsymbol{C}_{bC}^{ji} & \cdots & \boldsymbol{C}_{b(N-1)}^{ji} & \boldsymbol{C}_{bN}^{ji} \\ \boldsymbol{C}_{cA}^{ji} & \boldsymbol{C}_{cB}^{ji} & \boldsymbol{C}_{cC}^{ji} & & \boldsymbol{C}_{c(N-1)}^{ji} & \boldsymbol{C}_{cN}^{ji} \\ & \vdots & & \ddots & \vdots & \\ \boldsymbol{C}_{(n-1)A}^{ji} & \boldsymbol{C}_{(n-1)B}^{ji} & \boldsymbol{C}_{(n-1)C}^{ji} & & \boldsymbol{C}_{(n-1)(N-1)}^{ji} & \boldsymbol{C}_{(n-1)N}^{ji} \\ \boldsymbol{C}_{nA}^{ji} & \boldsymbol{C}_{nB}^{ji} & \boldsymbol{C}_{nC}^{ji} & \cdots & \boldsymbol{C}_{n(N-1)}^{ji} & \boldsymbol{C}_{nN}^{ji} \end{matrix} \right\| \left\| \begin{matrix} \boldsymbol{I}_i^A \\ \boldsymbol{I}_i^B \\ \boldsymbol{I}_i^C \\ \vdots \\ \boldsymbol{I}_i^{N-1} \\ \boldsymbol{I}_i^N \end{matrix} \right\|. \tag{15.21}$$

However, the mathematical model of an electrical continuous dynamical system of the generalised MIMO ECM commutator (matrix converter) shown in figure 15.15, acting as a current-source ECM commutator, making the inputs dependent on the outputs, in the dynamical systems approach and matrix notation may be expressed in the form:

$$\boldsymbol{I}_i^P = \boldsymbol{C}_{ij}^{Ps} \boldsymbol{I}_s^j, \tag{15.22}$$

or in detail as

$$\begin{Vmatrix} I_i^A \\ I_i^B \\ I_i^C \\ \vdots \\ I_i^{N-1} \\ I_i^N \end{Vmatrix} = \begin{Vmatrix} C_{ij}^{Aa} & C_{ij}^{Ab} & C_{ij}^{Ac} & & C_{ij}^{A(n-1)} & C_{ij}^{An} \\ C_{ij}^{Ba} & C^{Bb} & C_{ij}^{Bc} & \cdots & C_{ij}^{B(n-1)} & C_{ij}^{Bn} \\ C_{ij}^{Ca} & C_{ij}^{Cb} & C_{ij}^{Cc} & & C_{ij}^{C(n-1)} & C_{ij}^{Cn} \\ & \vdots & & \ddots & \vdots & \\ C_{ij}^{(N-1)a} & C_{ij}^{(N-1)b} & C_{ij}^{(N-1)c} & \cdots & C_{ij}^{(N-1)(n-1)} & C_{ij}^{(N-1)n} \\ C_{ij}^{Na} & C_{ij}^{Nb} & C_{ij}^{Nc} & & C_{ij}^{N(n-1)} & C_{ij}^{Nn} \end{Vmatrix} \begin{Vmatrix} I_a^j \\ I_b^j \\ I_c^j \\ \vdots \\ I_{n-1}^j \\ I_n^j \end{Vmatrix}, \tag{15.23}$$

where C_{kL}^{ji} i C_{ij}^{Kl} are the current-source coefficients of the commutation nodes kL and Kl, respectively.

To recapitulate, it may be affirmed that an action of a generalised MIMO ECM commutator as a current-source ECM commutator (figure 15.15) may be presented by means of its mathematical model, consisting of equations of the unholonomic constraints of current-source holors in matrix notation ((15.24) and (15.25)). For thus-formulated equations of the unholonomic constraints of current-source holors in matrix notation, the isochronic variations of the generalised coordinates are not already independent of each other, and that is why it is effective to apply the dynamical systems approach in the methodology for formulating the mathematical models of MIMO ECM commutators (matrix converters) in the matrix notation of the Euler–Lagrange second-order differential equation of dynamics. In the case of an ideal bipolar electrical valve, the voltage-holor coefficient of a commutating junction, moreover, concerns the electrical-valve resistance connected between the primary collector P and the secondary collector s. In this case, the electrical-valve resistance is equal to zero, which corresponds to the turn-on state of an electrical valve with the sense of the current direction dependent upon the sign of this voltage-source coefficient. On the other hand, the voltage-holor coefficient of a commutating junction regards the electrical-valve resistance as equal to infinity, which corresponds to the turn-off state (blocking) of an electrical valve.

As future spacecraft grow in size, multiple electrical energy sources may be required to supply the electrical distribution network. To supply the various electrical loads, these energy sources must be combined to derive an ECM DC–AC–DC commutator with several different load voltages. Such an ECM is important from a reliability standpoint, since the electrical loads can still be supplied even if some of the electrical energy sources fail.

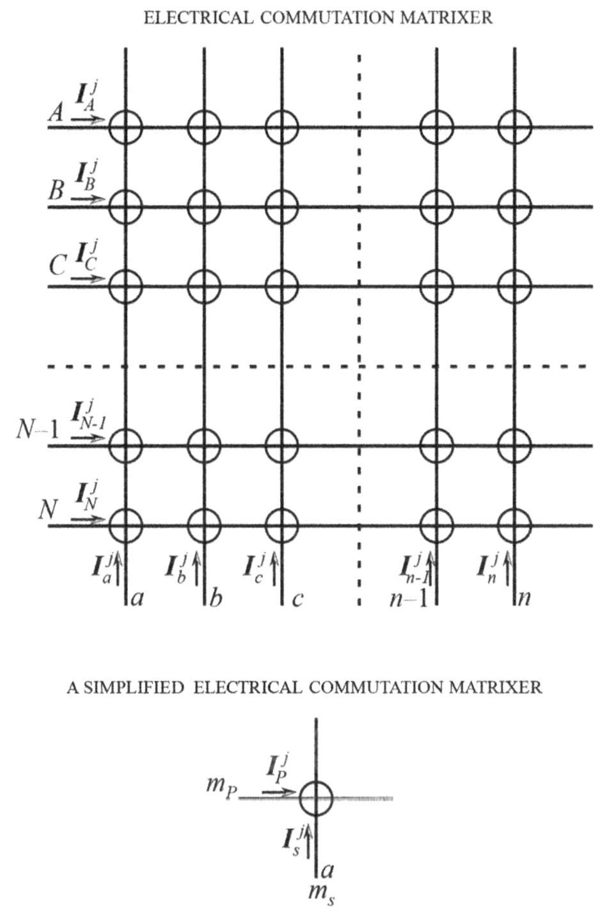

Figure 15.15. Physical model of the generalised MIMO ECM AC–AC and/or AC–DC/DC–AC commutator of an AC and/or DC MMD electrical machine (Fijalkowski 1987).

One method of implementing such an ECM DC–AC–DC commutator, where several DC–AC inverters are supplied from different electrical energy sources to drive a common AC bus, may also be indicated by figure 15.15. Because of their high output impedance, these DC–AC inverters share the electrical load in a predictable manner, and they provide an ECM DC–AC–DC commutator that can be designed to operate in a short circuit and thus provide a fault-tolerant system. These DC–AC inverters are operated synchronously through a common **microcontroller unit** (MCU) and regulated by monitoring one of the AC–DC rectifier outputs. The output DC voltage is regulated in the usual manner by variation of the operating frequency. The use of a common AC bus appears to be very advantageous for those applications where multiple ECM outputs must be driven from multiple DC inputs. This type of ECM DC–AC–DC commutator provides the high efficiency of a resonant circuit, and at higher power levels it can still be operated in the 25 kHz range to minimise the mass of the magnet. For example, in a **personal computer**

(PC) energy supply, there are commonly five (5) separate DC loads, and the ECM is 2×10.

A small number of applied ECM commutators have more than 36 electrical valves, and most designs use fewer than 12. An ECM is a means to organise ECM AC–AC, AC–DC–AC, AC–DC/DC–AC, DC–DC and/or DC–AC–AC commutators for a given application.

It also helps to focus efforts into three major spheres. Each of these spheres must be addressed effectively in order to create a useful macro- and microelectronic continuous dynamical system.

- **The hardware sphere**—create an ECM; this involves the selection of appropriate super- and/or semiconductor electrical valves and the auxiliary elements that drive and protect them.
- **The software sphere**—activate the ECM to achieve the desired AC–AC, AC–DC–AC, AC–DC/DC–AC, DC–DC and/or DC–AC–DC conversion; all operational decisions are implemented by adjusting the electrical-valve timing.
- **The interface sphere**—insert energy-storage components to provide the low-pass filters or intermediate energy-storage necessary to meet the application requirements; unlike with most passive filter applications, lossless active filters with simple structures are required.

15.3.2 Hybrid electrical commutation matrixer commutators

The hybrid ECM AC–AC, AC–DC–AC, AC–DC/DC–AC, DC–DC and/or DC–AC–DC commutator, realised in continuous uni- and/or bipolar electrical valves, establishes a mediate component amidst the polyvalvular ECM AC–AC, AC–DC–AC, AC–DC/DC–AC, DC–DC and/or DC–AC–DC commutators, also realised on continuous uni- and/or bipolar electrical valves. The reactive components of ECM, and so primary (matrixer input) and secondary (matrixer output) collectors, i.e. the connections between the continuous uni- and/or bipolar electrical valves, are made on ceramic or glassy bases using vacuum evaporation, printing or injection casting techniques.

The active elements of ECMs, and so continuous uni- and/or bipolar (bilateral) electrical valves—diodes, triodes (transistors, thyristors, trigistors, etc) or tetrodes (spacistors, tetristors, etc)—without packages are soldered or pressed into the ECM manufactured in this manner.

The hybrid ECM AC–AC, AC–DC–AC, AC–DC/DC–AC, DC–DC and/or DC–AC–DC commutator has a number of essential advantages, especially when it is made using the thick-film technique:

- relatively short design and production preparation time;
- moderate cost of production, even in small series;
- possibility of integrating continuous uni- and/or bipolar electrical valves and protection circuits;

- high overcurrent capability;
- high over-voltage capability;
- high output power.

By way of example, rotary ECM AC–DC/DC–AC commutators, which are designed to be built into the rotor of the ECM AC–DC/DC–AC commutator dynamotor and work in the most severe conditions, are now most frequently made using the hybrid technique.

15.3.3 Monolithic electrical commutation matrixer commutators

The monolithic ECM AC–AC, AC–DC–AC, AC–DC/DC–AC, DC–DC and/or DC–AC–DC commutator, realised on continuous uni- and/or bipolar electrical valves, is an intelligent MIMO ECM commutator in which all the active elements (continuous uni- and/or bipolar electrical valves) are made in a single wafer of the crystalline or amorphous material in the shape of single chips. Thus, the technique has assured high packing density for the active components.

According to the number of individual single active components integrated on the single wafer of crystalline or amorphous material, ECMs with an SSI (comprising several to 36 continuous uni- and/or bipolar electrical valves) and ECMs with an MSI (from 24 to hundreds of continuous uni- and/or bipolar electrical valves) can be distinguished.

The monolithic intelligent MIMO ECM AC–AC, AC–DC–AC, AC–DC/DC–AC, DC–DC and/or DC–AC–DC commutator, realised on mono-crystalline semi-conductor continuous uni- and/or bipolar electrical valves, used in the static converter technique and ECM AC–AC, AC–DC–AC, AC–DC/DC–AC, DC–DC and DC–AC–DC commutator dynamotors, show far more limitations, namely:

- difficulties with the integration of continuous uni- and/or bipolar electrical valves;
- limited resistance against over-voltages;
- low cost only for series production on a large scale (100 000 ECMs per annum).

Of late, in hybrid and monolithic ECM AC–AC, AC–DC–AC, AC–DC/DC–AC, DC–DC and DC–AC–DC commutators, two basic types of ECMs have been used: unipolar and/or bipolar.

The author announced the expansion of the pioneering family of compact intelligent MIMO ECM AC–AC, AC–DC–AC, AC–DC/DC–AC, DC–DC and DC–AC–DC commutators with integral press-pack ECMs.

These novel macroelectronic devices have recently shown a continuous electrical valve voltage rating of 8 kV and a current rating of 2.5 kA. The compact matrixer-built package design gives the new macroelectronic device unrivalled megawatt power density, with the advantages of high reliability and efficiency.

The unique matrixer-built hermetic ceramic package and internal construction provide double-sided cooling for the electrical valves, resulting in superior thermal power cycling and power ratings when compared to alternative module technologies.

The compact matrixer-built, fully-hermetic package design is mechanically compatible with cooling systems, including total immersion systems such as oil, currently employed for 3 or 4 kA GTO thyristors, this feature allowing for either retrofitting or design continuation of existing systems.

The high current rating allows the conversion of old bipolar static converter drives to state-of-the-art ECM technology, without the need to replace the basic mechanical structure. Adopting a refurbishment approach to component obsolescence is both economical and environmentally friendly, as it reduces the consumption of materials and energy compared to the replacement of equipment.

The high power density of compact intelligent MIMO ECM AC–AC, AC–DC–AC, AC–DC/DC–AC, DC–DC and DC–AC–DC commutators with integral presspack ECMs, and their unique matrixer-built mechanical construction, make them ideal for new designs—not only in MCM AC–AC and AC–DC/DC–AC commutator dynamotors, but also in green power applications, such as the matrix converters normally used for wind and/or wave power generation in the high megawatt range.

The novel ECM AC–AC, AC–DC–AC, AC–DC/DC–AC, DC–DC and DC–AC–DC commutator dynamotors with intelligent self-protecting compact MIMO ECMs offer size and mass reduction, together with improved reliability, among others benefits.

ECM AC–AC, AC–DC–AC, AC–DC/DC–AC, DC–DC and DC–AC–DC commutator dynamotors are highly integrated and provide key functions, including overcurrent, over-temperature and short-circuit protection, and supply **ultra-violet** (UV) lock functions.

The novel ECM AC–AC, AC–DC–AC, AC–DC/DC–AC, DC–DC and DC–AC–DC commutator dynamotors are high-performance MMD electrical machines with built-in current-sense electrical valves in the ECM stage, which is housed in a thin and compact matrixer-built package, thanks to the insulated base-plate with high thermal conductivity. They are the first ECMs capable of truly inaudible (>20 kHz) operation.

The high reliability and exceptional thermal ruggedness are also well suited to the demands of high-power ECM AC–AC, AC–DC–AC, AC–DC/DC–AC, DC–DC and DC–AC–DC commutator dynamotors for industrial and traction propulsion mechatronical control systems, e.g. for both high rail and main line applications such as high-speed **electrical multiple units** (EMU). For instance, full-diffusion technology with loose mono- and/or poly-silicon wafers may be used to manufacture **MOS-controlled thyristor** (MCT) type intelligent MIMO ECM commutators. Up to now, neutron-doped silicon of the so-called 'float-zone' type has mainly been used. Neutron doping means that the silicon rod is irradiated with neutrons in a nuclear reactor prior to being sliced into wafers. A number of silicon atoms are then converted into phosphorus atoms and the material becomes weakly *n*-doped.

The major advantage, however, is that after heat treatment the silicon rod will have a very homogeneous resistivity. This is valid for achieving high-voltage capability for the ECMs. To obtain an acceptable yield in the ECM manufacture, i.e. the number of MCT-type ECMs fulfilling the specifications in relation to the number of MCT-type ECMs manufactured, the environment must be extremely clean.

Diffusion technology means that the silicon wafer is coated with gallium, aluminium or phosphorus. These substances are then allowed to diffuse into the silicon wafer in diffusion furnaces under high temperature. In certain ECM manufacturing stages, only parts of the silicon wafer surface are coated for subsequent diffusion. For this purpose, photolithography is utilised. After completion of the diffusion and the coating of the aluminium contacts, the silicon wafers are sliced into ECM chips with the aid of high-power lasers. Depending on the silicon wafer area of the chips, the intelligent MIMO ECMs, e.g. for AC–AC commutator dynamotors (generators/motors), incorporate 3 × 3 or 3 × 4 ECMs, and AC–DC/DC–AC commutator dynamotors incorporate 3 × 2/2 × 3 ECMs. By controlling the lifetime of the charge carriers, it is possible to determine the position of the MCT-type ECM on the recovery charge–minimum on-state voltage drop curve. For this purpose, gold irradiation is applied, which provides considerably closer tolerances. With electron irradiation, the ECM components are bombarded with high-energy electrons (>10 MeV) from electron guns.

On the basis of the large volume of ECM commutators (static converter) that were manufactured in the 20th century (primarily diode- or thyristor-type MIMO ECMs for AC–AC, AC–DC–AC, AC–DC/DC–AC, DC–DC and/or DC–AC–DC commutators), it was possible to introduce more reliable dimensioning criteria. In combination with closer MIMO ECM manufacturing tolerances, above all, this substantially raised the maximum reverse voltage/off-state voltage. This lowered the on-state voltage drop, and as a consequence reduced the on-state charge and thus the high power losses in protective circuits around the intelligent MIMO ECM AC–AC, AC–DC–AC, AC–DC/DC–AC, DC–DC and/or DC–AC–DC commutators.

The author also recently worked out a concept for an ECM commutator with a bipolar ECM, realised on super- and semiconductor hybrid continuous bipolar electrical valves, based on the use of heterojunctions, i.e. a contact of amorphous and mono- and/or polycrystalline materials. The properties of these heterojunctions seem to be for various reasons competitive in relation to structures made from mono- and/or poly-crystalline semiconductors.

A concept for the bipolar ECM, realised on continuous bipolar power transistors of the *n–p–n* type—in which the emitters will be made from chalcogenide glass and the bases will be made from monocrystalline silicon—or continuous bipolar heterojunctions and dielectrically insulated gates in field-effect transistors, is being worked out.

In addition, there is a concept for a bipolar ECM, realised on continuous bipolar power **superconductor-base hot electron transistors** (SUPER-HET) on the base of GaAs/Nab/Ins, continuous bipolar power **quantum interference transistors** (QUIT), or continuous bipolar power and very high-speed **static induction transistors** (SIT) and/or graphene **single electron transistors** (SET) with low power-delay product and very high transconductance. Bipolar ECMs of that type can be used, owing to specific conditions of the electrical-current super- and/or semi-conductivity, as basic trivalent-logic functions of the ECM AC–AC, AC–DC–AC, AC–DC/DC–AC, DC–DC and/or DC–AC–DC commutator, which seem to be the simplest and most effective static converters for conversion of one kind of electrical energy into another. This trivalent-logic is based on the Lukasiewicz–Tarski algebra.

Semi- and/or superconductive ECM AC–AC, AC–DC–AC, AC–DC/DC–AC, DC–DC and/or DC–AC–DC commutators are now particularly indispensable in modern EM actuator engineering, and for energy conversion purposes, mainly in **fly-by-wire** (FBW) aircraft and **ride-by-wire** (RBW) or **x-by-wire** (XBW) automotive vehicles. Compared with previous polysectorial (polysegmental) MCM commutators and polyvalvular ECM AC–AC, AC–DC–AC, AC–DC/DC–AC, DC–DC and/or DC–AC–DC commutators (static converters), they are more practice-oriented, especially from the point of view of operating convenience and the space requirements of ultra-modern super- and/or semiconductor ECM techniques.

Numerous semi- and/or superconductive ECM AC–AC, AC–DC–AC, AC–DC/DC–AC, DC–DC and/or DC–AC–DC commutators have been developed for modern MMD electrical machines and power applications. When selecting these commutators, various criteria have to be borne in mind, i.e. the cost of the semi- and/or superconductive ECM AC–AC, AC–DC–AC, AC–DC/DC–AC, DC–DC, or DC–AC–DC commutator, the driver requirements, the switching and network, mounting and wiring needs, together with the ECM commutator's electrical characteristics and performance.

The cost of semi- and/or superconductive ECM AC–AC, AC–DC–AC, AC–DC/DC–AC, DC–DC and/or DC–AC–DC commutators, when compared to that of the non-MCM wiring requirements, has been reducing and will continue to reduce significantly.

The development aims for semi- and/or superconductive ECM AC–AC, AC–DC–AC, AC–DC/DC–AC, DC–DC and/or DC–AC–DC commutators are: to manufacture ECM commutators that have uncomplicated drivers, and to reduce circuitry and produce a simplified method for the ECM wiring and mounting in a realistic process.

Newly developed insulated ECMs meet the requirements for modern MMD electrical machines and power applications.

Insulated ECMs reduce mounting costs:

- ECM wiring through printed circuits;
- ECM wiring using copper bus-bars;
- a common heat-sink for all ECM components.

Insulated ECMs limit RFI phenomena:

- extremely low parasitic capacitance;
- the heat-sink is grounded to the ECM's shield and chassis.

Insulated ECMs increase safety:

- internal insulation of 2.5 kV (50 HZ, 1 min);
- safer construction, no mica insulation, as is needed for MCMs;
- very low thermal resistance between the junction and case of electrical valves.

The semi- and/or superconductive ECM AC–AC, AC–DC–AC, AC–DC/DC–AC, DC–DC and/or DC–AC–DC commutators for modern MMD electrical machines and power applications are specifically designed to reduce the number of components, especially electrical valves.

In the case of very high-power semi- and/or superconductive ECM AC–AC, AC–DC–AC, AC–DC/DC–AC, DC–DC and/or DC–AC–DC commutators for modern MMD electrical machines and power applications, the use of continuous electrical valves permits a reduction in parasitical inductance and an improvement in cooling characteristics.

The semi- and/or superconductive ECM AC–AC, AC–DC–AC, AC–DC/DC–AC, DC–DC and/or DC–AC–DC commutators for modern MMD electrical machines and power applications operate at high-current values and, due to their high-current gain, can be controlled by low-power gate or base drivers. Their switching safe operating areas, namely **reverse-biased safe-operating areas** (RBSOA), allow repeatable operation without use of snubbers.

For high-performance ECM AC–AC, AC–DC–AC, AC–DC/DC–AC, DC–DC and/or DC–AC–DC commutator dynamotors, ECM-mode power supplies, storage-battery chargers and welding equipment, the main requirements are:

- noiseless operation;
- large band-pass;
- a reduction in the size and mass of the passive components.

The result is the choice of switching frequencies $f \geqslant 15\,\text{kHz}$, or even more for amorphous electrical valves, e.g. ovonics.

The ECM AC–AC, AC–DC–AC, AC–DC/DC–AC, DC–DC and/or DC–AC–DC commutator has a number of advantages over conventional power converter technologies when it is used on an AC bus system. As higher temperature power semi- or superconductive electrical valves, such as silicon carbide ones, become available, the technology will become a good candidate for power conversion in applications where volume and mass are at a premium. For example, the monolithic 3×3 ECM commutator with a bipolar ECM, realised on only nine (9) continuous **reverse-blocking insulated-gate bipolar transistors** (RB IGBT), or **metal-oxide semiconductor field-effect transistors** (MOSFET), or bipolar symmetrical trigistors, that is, **gate turn-off** (GTO) symmetrical thyristors (symistors) in a 3×3 ECM connection, is shown in figure 15.16.

This monolithic 3×3 ECM AC–AC commutator makes the bilateral flow of secondary electrical energy possible during generating (absorbing) as well as during motoring (driving) of the AC–AC commutator dynamotor with variable angular velocity. Six operating modes are possible in this MMD electrical machine.

Potentially attractive features of the monolithic 3×3 ECM, such as the absence of DC link energy storage with reactive components, high power density, high quality of input and output waveforms, are leading to continued research efforts aimed at translating their promise into reality (Fijalkowski 1973, 1982, 1985; Wheeler *et al* 2004).

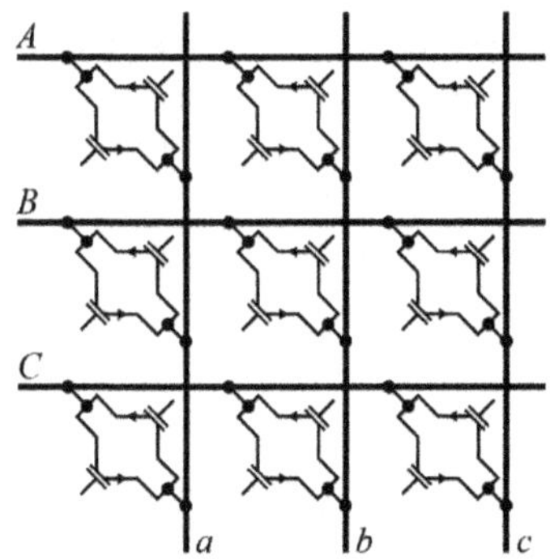

Figure 15.16. The monolithic 3 × 3 ECM AC–AC commutator with nine (9) RB IGBTs.

Although monolithic 3 × 3 ECMs were first introduced as a class of frequency and phase-number changer realised using controlled electrical valves (electronic switches) by Fijalkowski (1973, 1982, 1988) and then later by Gyugyi and Pelly (1976), significant progress began with the introduction of high-frequency synthesis, as proposed by Venturini (1980).

Although the initial approach was limited to a voltage transfer ratio of one half, improved modulation strategies with a voltage transfer ratio of $\sqrt{3/2}$ were published on by Alesina and Venturini (1989). Further along, indirect modulation methods based on a fictitious DC link concept were proposed (Rodriguez 1984; Alesina and Venturini 1989; Oyama *et al* 1989; Neft and Schauder 1988a; 1988b; Ziogas *et al* 1986). With the exception of the programmed commutation (switching) pattern in (Ziogas *et al* 1986), none of these modulation strategies for monolithic 3 × 3 ECMs based on high-frequency synthesis have been able to break the upper bound given by $\sqrt{3/2}$ p.u.

Since AC–AC commutator motor output power is proportional to the square of the applied voltage, this limitation leads to reduced output power, limited to 3/4 of the AC–AC commutator motor rating, when fed from a monolithic 3 × 3 ECM AC–AC commutator operating from the line nominally rated at AC–AC commutator motor voltage. Otherwise, a custom-designed AC–AC commutator motor designed for the appropriate output voltage level would be required. This remains one of the many roadblocks limiting the viability of the monolithic 3 × 3 ECM AC–AC commutator.

A low commutation (switching) frequency six-step modulation strategy for frequency conversion using 3 × 3 ECM AC–AC commutators capable of providing more than unity voltage transfer factor was presented by Wang and Venkataramanan (2006).

The modulation strategy may be more easily developed on the basis of the operation of the dual monolithic 3 × 2 plus 2 × 3 ECM topology with a fictitious DC link by using commutation-matrixer holor multiplication of these two cascaded connected ECMs (see sub-section 10.6.1), as previously introduced by the author of this book (Fijalkowski 1967, 1982, 1985, 1987).

However, the approach may also be applicable for the monolithic 3 × 3 ECM (nine-electrical valve) topology, as may be demonstrated through appropriate topological mapping.

Figure 15.17(a) illustrates the 3 × 3 ECM using three monolithic **single-pole-triple-valvular** (SPTV) commutation groups *Va*, V_b, *Vc* with nine (9) RB IGBTs. Figure 15.17(b) illustrates the reduced electrical valve (switch) two monolithic SPTV commutation groups *Vp* and *Vn* with six (6) RB IGBTs and three monolithic **single-pole-double-valvular** (SPDV) commutation groups *Va*, V_b, *Vc* with six (6) RB IGBTs. The two realisations are labeled as conventional ECM and indirect ECM, respectively. If the commutation function of the RB IGBTs, as illustrated in figure 15.17, is defined to be $\boldsymbol{C}_{sP}^{ji}$, the input-output transposition properties of the ECM can be expressed as

$$\boldsymbol{E}_s^j = \boldsymbol{C}_{sP}^{ji}\boldsymbol{E}_i^P, \qquad (15.24)$$

or in detail as

$$\begin{Vmatrix} \boldsymbol{E}_a^j \\ \boldsymbol{E}_b^j \\ \boldsymbol{E}_c^j \end{Vmatrix} = \begin{Vmatrix} \boldsymbol{C}_{aA}^{ji} & \boldsymbol{C}_{aB}^{ji} & \boldsymbol{C}_{aC}^{ji} \\ \boldsymbol{C}_{bA}^{ji} & \boldsymbol{C}_{bB}^{ji} & \boldsymbol{C}_{bC}^{ji} \\ \boldsymbol{C}_{cA}^{ji} & \boldsymbol{C}_{cB}^{ji} & \boldsymbol{C}_{cC}^{ji} \end{Vmatrix} \begin{Vmatrix} \boldsymbol{E}_i^A \\ \boldsymbol{E}_i^B \\ \boldsymbol{E}_i^C \end{Vmatrix}. \qquad (15.25)$$

However, a mathematical model of an electrical continuous dynamical system of the 3 × 3 ECM AC–AC commutator shown in figure 15.17, acting as a current-source ECM AC–AC commutator, making the input holors dependent on the output holors, in the dynamical systems approach and matrix notation may be expressed in the form:

$$\boldsymbol{I}_P^j = \boldsymbol{C}_{Ps}^{ji}\boldsymbol{I}_i^s, \qquad (15.26)$$

or in detail as

$$\begin{Vmatrix} \boldsymbol{I}_A^j \\ \boldsymbol{I}_B^j \\ \boldsymbol{I}_C^j \end{Vmatrix} = \begin{Vmatrix} \boldsymbol{C}_{Aa}^{ji} & \boldsymbol{C}_{Ab}^{ji} & \boldsymbol{C}_{Ac}^{ji} \\ \boldsymbol{C}_{Ba}^{ji} & \boldsymbol{C}_{Bb}^{ji} & \boldsymbol{C}_{Bc}^{ji} \\ \boldsymbol{C}_{Ca}^{ji} & \boldsymbol{C}_{Cb}^{ji} & \boldsymbol{C}_{Cc}^{ji} \end{Vmatrix} \begin{Vmatrix} \boldsymbol{I}_i^a \\ \boldsymbol{I}_i^b \\ \boldsymbol{I}_i^c \end{Vmatrix}, \qquad (15.27)$$

where

$$\left\|\boldsymbol{E}_s^j\right\| = \left\|\boldsymbol{E}_a^j \quad \boldsymbol{E}_b^j \quad \boldsymbol{E}_c^j\right\|^T;$$

$$\left\|\boldsymbol{E}_i^P\right\| = \left\|\boldsymbol{E}_i^A \quad \boldsymbol{E}_i^B \quad \boldsymbol{E}_i^C\right\|^T; \ \left\|\boldsymbol{I}_P^j\right\| = \left\|\boldsymbol{I}_A^J \quad \boldsymbol{I}_B^J \quad \boldsymbol{I}_C^J\right\|^T; \ \left\|\boldsymbol{I}_i^s\right\| = \left\|\boldsymbol{I}_I^a \quad \boldsymbol{I}_I^b \quad \boldsymbol{I}_i^c\right\|^T$$

$\boldsymbol{C}_{sP}^{ji}$ is the voltage-source holor commutation matrixer with the elements being the commutation functions of various throws shown in figure 15.17(a):

$$\boldsymbol{C}_{sP}^{ji} = \begin{Vmatrix} \boldsymbol{C}_{aA}^{ji} & \boldsymbol{C}_{aB}^{ji} & \boldsymbol{C}_{aC}^{ji} \\ \boldsymbol{C}_{bA}^{ji} & \boldsymbol{C}_{bB}^{ji} & \boldsymbol{C}_{bC}^{ji} \\ \boldsymbol{C}_{cA}^{ji} & \boldsymbol{C}_{cB}^{ji} & \boldsymbol{C}_{cC}^{ji} \end{Vmatrix}.$$

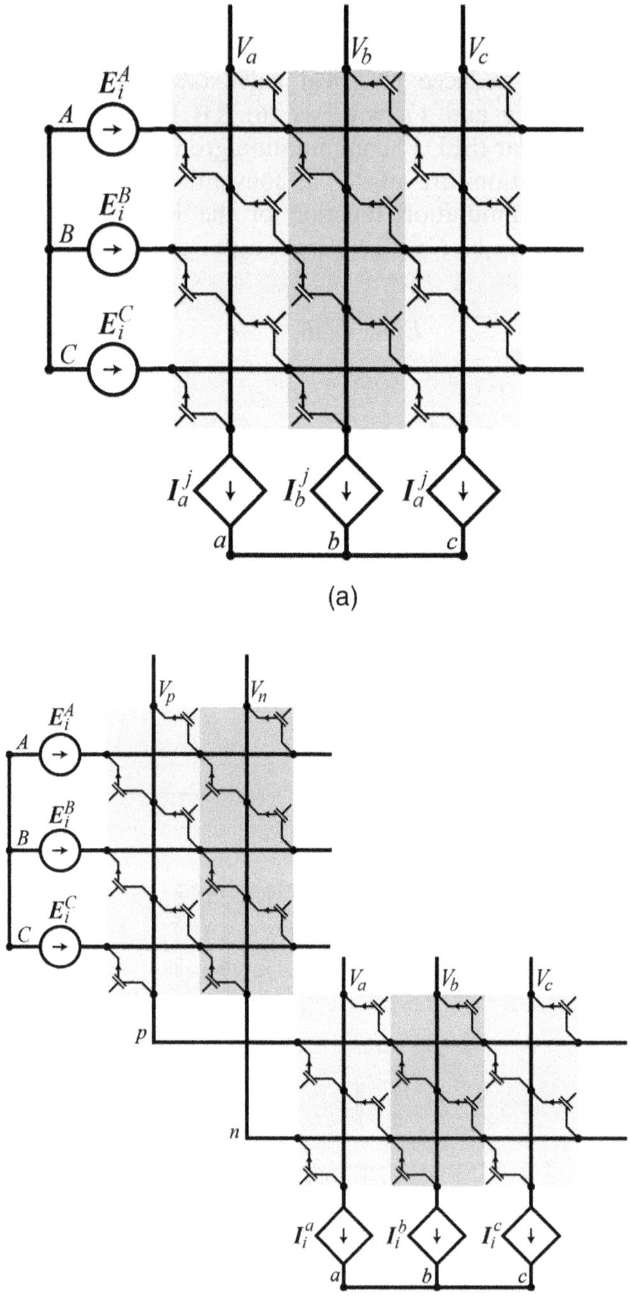

Figure 15.17. Representation of the monolithic 3 × 3 ECM AC–AC commutator (a) and the monolithic 3 × 2 plus 2 × 3 ECM AC–DC–AC commutator (b).

C^{ji}_{Ps} is the current-source holor commutation matrixer with the elements being the commutation functions of various throws shown in figure 15.17(a):

$$C^{ji}_{Ps} = \begin{Vmatrix} C^{ji}_{Aa} & C^{ji}_{Ab} & C^{ji}_{Ac} \\ C^{ji}_{Ba} & C^{ji}_{Bb} & C^{ji}_{Bc} \\ C^{ji}_{Ca} & C^{ji}_{Cb} & C^{ji}_{Cc} \end{Vmatrix} = \begin{Vmatrix} C^{ji}_{aA} & C^{ji}_{aB} & C^{ji}_{aC} \\ C^{ji}_{bA} & C^{ji}_{bB} & C^{ji}_{bC} \\ C^{ji}_{cA} & C^{ji}_{cB} & C^{ji}_{cC} \end{Vmatrix}^T .$$

C^{JI}_{kL} are the voltage-source coefficients of the commutation nodes kL, which may be expressed as follows:

$$C^{JI}_{kL} = \left.\frac{\partial E^j_k}{\partial E^L_i}\right|_{E^I_i=0} = \begin{cases} +1 \overset{\Delta}{\rightarrow} E^j_k = +E^L_i \\ 0 \overset{\Delta}{\rightarrow} E^j_k = 0 \\ -1 \overset{\Delta}{\rightarrow} E^j_k = -E^L_i \end{cases} .$$

C^{Kl}_{ij} are the current-source coefficients of the commutation nodes Kl, which may be expressed as follows:

$$C^{Kl}_{ij} = \left.\frac{\partial E^K_i}{\partial E^j_l}\right|_{E^j_i=0} = \begin{cases} -1 \overset{\Delta}{\rightarrow} E^K_i = -E^j_l \\ 0 \overset{\Delta}{\rightarrow} E^K_i = 0 \\ +1 \overset{\Delta}{\rightarrow} E^K_i = +E^j_l \end{cases} .$$

The input–output transposition properties of the 3 × 2 plus 2 × 3 EMC AC–DC–AC commutator can be expressed as

$$E^J_s = C^{ji}_{sP(E)} C^{sP(I)}_{ij} E^{P}_{i}, \qquad (15.28)$$

or in detail as

$$\begin{Vmatrix} E^j_a \\ E^j_b \\ E^j_c \end{Vmatrix} = \begin{Vmatrix} C^{ji}_{pA} & C^{ji}_{nA} \\ C^{ji}_{pB} & C^{ji}_{nB} \\ C^{ji}_{pC} & C^{ji}_{nC} \end{Vmatrix} \begin{Vmatrix} C^{ap}_{ij} & C^{bp}_{ij} & C^{cp}_{ij} \\ C^{an}_{ij} & C^{bn}_{ij} & C^{cn}_{ij} \end{Vmatrix} \begin{Vmatrix} E^A_i \\ E^B_i \\ E^C_i \end{Vmatrix}, \qquad (15.29)$$

where $C^{sP(I)}_{ij}$ is the commutation matrixer of the current-source 2 × 3 EMC DC–AC commutator connected with the stiff voltage holors. $C^{ji}_{sP(E)}$ is the commutation matrixer of the voltage-source 3 × 2 EMC AC–DC commutator connected with the stiff current holors. They are given by

$$C^{ji}_{sP(E)} = \begin{Vmatrix} C^{ji}_{pA} & C^{ji}_{nA} \\ C^{ji}_{pB} & C^{ji}_{nB} \\ C^{ji}_{pC} & C^{ji}_{nC} \end{Vmatrix}$$

and

$$C_{ij}^{sP(I)} = \begin{Vmatrix} C_{ij}^{ap} & C_{ij}^{bp} & C_{ij}^{cp} \\ C_{ij}^{an} & C_{ij}^{bn} & C_{ij}^{cn} \end{Vmatrix}.$$

The input–output transposition properties of the 3 × 2 EMC AC–DC commutator can be expressed as

$$E_{s(E)}^{j} = C_{sP(E)}^{ji} E_{i}^{P(E)}, \tag{15.30}$$

or in detail as

$$\begin{Vmatrix} E_{p(E)}^{j} \\ E_{n(E)}^{j} \end{Vmatrix} = \begin{Vmatrix} C_{pA}^{ji} & C_{nA}^{ji} \\ C_{pB}^{ji} & C_{nB}^{ji} \\ C_{pC}^{ji} & C_{nC}^{ji} \end{Vmatrix} \begin{Vmatrix} E_{i}^{A(E)} \\ E_{i}^{B(E)} \\ E_{i}^{C(E)} \end{Vmatrix}, \tag{15.31}$$

and the input–output transposition properties of the 2 × 3 EMC DC–AC commutator can be expressed as

$$E_{i}^{s(I)} = C_{ij}^{sP(I)} E_{P(I)}^{j}, \tag{15.32}$$

or in detail as

$$\begin{Vmatrix} E_{i}^{a(I)} \\ E_{i}^{b(I)} \\ E_{i}^{c(I)} \end{Vmatrix} = \begin{Vmatrix} C_{ij}^{ap} & C_{ij}^{bp} & C_{ij}^{cp} \\ C_{ij}^{an} & C_{ij}^{bn} & C_{ij}^{cn} \end{Vmatrix} \begin{Vmatrix} E_{p(I)}^{j} \\ E_{n(I)}^{j} \end{Vmatrix}. \tag{15.33}$$

However, a mathematical model of an electrical continuous dynamical system of the 3 × 2 plus 2 × 3 ECM AC–DC–AC commutator shown in figure 15.17, acting as a current-source ECM AC–DC–AC commutator, making the input holors dependent on the output holors, in the dynamical systems approach and matrix notation may be expressed in the form:

$$I_{s}^{j} = C_{Ps(I)}^{ji} C_{ij}^{Ps(E)} I_{i}^{P}, \tag{15.34}$$

or in detail as

$$\begin{Vmatrix} I_{A}^{j} \\ I_{B}^{j} \\ I_{C}^{j} \end{Vmatrix} = \begin{Vmatrix} C_{pa}^{ji} & C_{na}^{ji} \\ C_{pb}^{ji} & C_{nb}^{ji} \\ C_{pc}^{ji} & C_{nc}^{ji} \end{Vmatrix} \begin{Vmatrix} C_{ij}^{Ap} & C_{ij}^{Bp} & C_{ij}^{Cp} \\ C_{ij}^{An} & C_{ij}^{Bn} & C_{ij}^{Cn} \end{Vmatrix} \begin{Vmatrix} I_{i}^{a} \\ I_{i}^{b} \\ I_{i}^{c} \end{Vmatrix}. \tag{15.35}$$

The input–output transposition properties of the 3 × 2 EMC AC–DC commutator can be expressed as

$$I_{s(I)}^{j} = C_{Ps(I)}^{ji} I_{i}^{P(I)},$$

or in detail as

$$\begin{Vmatrix} I^{j}_{p(I)} \\ I^{j}_{n(I)} \end{Vmatrix} = \begin{Vmatrix} C^{ji}_{pa} & C^{ji}_{na} \\ C^{ji}_{pb} & C^{ji}_{nb} \\ C^{ji}_{pc} & C^{ji}_{nc} \end{Vmatrix} \begin{Vmatrix} I^{a(I)}_{i} \\ I^{b(I)}_{i} \\ I^{c(I)}_{i} \end{Vmatrix},$$

and the input–output transposition properties of the 2 × 3 EMC DC–AC commutator can be expressed as

$$I^{P(E)}_{i} = C^{Ps(E)}_{ij} I^{j}_{s(E)},$$

or in detail as

$$\begin{Vmatrix} I^{A(E)}_{i} \\ I^{B(E)}_{i} \\ I^{C(E)}_{i} \end{Vmatrix} = \begin{Vmatrix} C^{Ap}_{ij} & C^{Bp}_{ij} & C^{Cp}_{ij} \\ C^{An}_{ij} & C^{Bn}_{ij} & C^{Cn}_{ij} \end{Vmatrix} \begin{Vmatrix} I^{j}_{p(E)} \\ I^{j}_{n(E)} \end{Vmatrix},$$

where

$$\begin{aligned}
\left\| E^{j}_{s(E)} \right\| &= \left\| E^{j}_{p(E)} \quad E^{j}_{n(E)} \right\|^{T} \\
\| E^{P(E)}_{i} \| &= \left\| E^{A(E)}_{i} \quad E^{B(E)}_{i} \quad E^{C(E)}_{i} \right\|^{T} \\
\left\| E^{j}_{s(I)} \right\| &= \left\| E^{j}_{a(I)} \quad E^{j}_{b(I)} \quad E^{j}_{c(I)} \right\|^{T} \\
\| E^{P(I)}_{i} \| &= \left\| E^{p(I)}_{i} \quad E^{n(I)}_{i} \right\|^{T} \\
\left\| E^{j}_{s(E)} \right\| &= \left\| E^{j}_{p(I)} \quad E^{j}_{n(I)} \right\|^{T} \\
\left\| I^{j}_{s(I)} \right\| &= \left\| I^{j}_{A(I)} \quad I^{j}_{B(I)} \quad I^{j}_{C(I)} \right\|^{T} \\
\| I^{P(I)}_{i} \| &= \left\| I^{p(I)}_{i} \quad I^{n(I)}_{i} \right\|^{T} \\
\| I^{P(E)}_{i} \| &= \left\| I^{A(E)}_{i} \quad I^{B(E)}_{i} \quad I^{C(E)}_{i} \right\|^{T} \\
\left\| I^{j}_{s(E)} \right\| &= \left\| I^{j}_{p(E)} \quad I^{j}_{n(E)} \right\|^{T}
\end{aligned}$$

$C^{ji}_{Ps(E)}$ is the commutation-matrixer holor of the voltage-source 3 × 2 EMC AC–DC commutator connected with the stiff voltage holors and $C^{ji}_{Ps(I)}$ is the commutation-matrixer holor of the current-source 2 × 3 EMC DC–AC commutator connected with the stiff current holors. They are given by

$$C^{ji}_{Ps(I)} = \begin{Vmatrix} C^{ji}_{pa} & C^{ji}_{na} \\ C^{ji}_{pb} & C^{ji}_{nb} \\ C^{ji}_{pc} & C^{ji}_{nc} \end{Vmatrix}$$

and

$$C^{Ps(E)}_{ij} = \begin{Vmatrix} C^{Ap}_{ij} & C^{Bp}_{ij} & C^{Cp}_{ij} \\ C^{An}_{ij} & C^{Bn}_{ij} & C^{Cn}_{ij} \end{Vmatrix}.$$

By matrix multiplication of the commutation-matrixer holor of the voltage-source 3 × 2 EMC AC–DC commutator and the commutation-matrixer holor of the current-source 2 × 3 EMC DC–AC commutator, we may receive mathematical confirmation that a product of the commutation-matrixer holors of the 3 × 2 plus 2 × 3 EMC commutator is equal to the commutation-matrixer holor of the 3 × 3 EMC AC–AC commutator.

$$\begin{aligned} C^{ji}_{sP(E)} C^{sP(I)}_{ij} &= \begin{Vmatrix} C^{ji}_{pA} & C^{ji}_{nA} \\ C^{ji}_{pB} & C^{ji}_{nB} \\ C^{ji}_{pC} & C^{ji}_{nC} \end{Vmatrix} \begin{Vmatrix} C^{ap}_{ij} & C^{bp}_{ij} & C^{cp}_{ij} \\ C^{an}_{ij} & C^{bn}_{ij} & C^{cn}_{ij} \end{Vmatrix} \\ &= \begin{Vmatrix} C^{ji}_{pA}C^{ap}_{ij} + C^{ji}_{nA}C^{an}_{ij} & C^{ji}_{pA}C^{bp}_{ij} + C^{ji}_{nA}C^{bn}_{ij} & C^{ji}_{pA}C^{cp}_{ij} + C^{ji}_{nA}C^{cn}_{ij} \\ C^{ji}_{pB}C^{ap}_{ij} + C^{ji}_{nB}C^{an}_{ij} & C^{ji}_{pB}C^{bp}_{ij} + C^{ji}_{nB}C^{bn}_{ij} & C^{ji}_{pB}C^{cp}_{ij} + C^{ji}_{nB}C^{cn}_{ij} \\ C^{ji}_{pC}C^{ap}_{ij} + C^{ji}_{nC}C^{an}_{ij} & C^{ji}_{pC}C^{bp}_{ij} + C^{ji}_{nC}C^{bn}_{ij} & C^{ji}_{pC}C^{cp}_{ij} + C^{ji}_{nC}C^{cn}_{ij} \end{Vmatrix} \\ &= \begin{Vmatrix} C^{ji}_{aA} & C^{ji}_{aB} & C^{ji}_{aC} \\ C^{ji}_{bA} & C^{ji}_{bB} & C^{ji}_{bC} \\ C^{ji}_{cA} & C^{ji}_{cB} & C^{ji}_{cC} \end{Vmatrix}. \end{aligned}$$

Similarly, by matrix multiplication of the commutation-matrix holor of the current-source 2 × 3 EMC AC–DC commutator and the commutation-matrix holor of the voltage-source 3 × 2 EMC DC–AC commutator, we may receive mathematical confirmation that a product of the commutation-matrix holors of the 3 × 2 plus 2 × 3 EMC commutator is also equal to the commutation-matrix holor of the 3 × 3 EMC AC–AC commutator.

$$C_{Ps(I)}^{ji} C_{ij}^{Ps(E)} = \left\| \begin{matrix} C_{pa}^{ji} & C_{na}^{ji} \\ C_{pb}^{ji} & C_{nb}^{ji} \\ C_{pc}^{ji} & C_{nc}^{ji} \end{matrix} \right\| \left\| \begin{matrix} C_{ij}^{Ap} & C_{ij}^{Bp} & C_{ij}^{Cp} \\ C_{ij}^{An} & C_{ij}^{Bn} & C_{ij}^{Cn} \end{matrix} \right\|$$

$$= \left\| \begin{matrix} C_{pa}^{ji}C_{ij}^{Ap} + C_{na}^{ji}C_{ij}^{An} & C_{pa}^{ji}C_{ij}^{Bp} + C_{na}^{ji}C_{ij}^{Bn} & C_{pa}^{ji}C_{ij}^{Cp} + C_{na}^{ji}C_{ij}^{Cn} \\ C_{pb}^{ji}C_{ij}^{Ap} + C_{nb}^{ji}C_{ij}^{An} & C_{pb}^{ji}C_{ij}^{Bp} + C_{nb}^{ji}C_{ij}^{Bn} & C_{pb}^{ji}C_{ij}^{Cp} + C_{nb}^{ji}C_{ij}^{Cn} \\ C_{pc}^{ji}C_{ij}^{Ap} + C_{nc}^{ji}C_{ij}^{An} & C_{pc}^{ji}C_{ij}^{Bp} + C_{nc}^{ji}C_{ij}^{Bn} & C_{pc}^{ji}C_{ij}^{Cp} + C_{nc}^{ji}C_{ij}^{Cn} \end{matrix} \right\|$$

$$= \left\| \begin{matrix} C_{Aa}^{ji} & C_{Ab}^{ji} & C_{Ac}^{ji} \\ C_{Ba}^{ji} & C_{Bb}^{ji} & C_{Bc}^{ji} \\ C_{Ca}^{ji} & C_{Cb}^{ji} & C_{Cc}^{ji} \end{matrix} \right\|,$$

where

$$\left\| \begin{matrix} C_{aA}^{ji} & C_{aB}^{ji} & C_{aC}^{ji} \\ C_{bA}^{ji} & C_{bB}^{ji} & C_{bC}^{ji} \\ C_{cA}^{ji} & C_{cB}^{ji} & C_{cC}^{ji} \end{matrix} \right\| = \left\| \begin{matrix} C_{Aa}^{ji} & C_{Ab}^{ji} & C_{Ac}^{ji} \\ C_{Ba}^{ji} & C_{Bb}^{ji} & C_{Bc}^{ji} \\ C_{Ca}^{ji} & C_{Cb}^{ji} & C_{Cc}^{ji} \end{matrix} \right\|^{T}$$

Commutation functions and voltage/current holors

Let the three phase matrixer-input-voltage holors and three-phase matrixer-output-current holors be assumed to be balanced stiff AC holor quantities and expressed as

$$\begin{aligned} E_A &= E\angle\alpha & I^a &= I\angle\beta \\ E_B &= E\angle\alpha + 2\pi/3 & I^b &= I\angle\beta + 2\pi/3 \\ E_C &= E\angle\alpha - 2\pi/3 & I^c &= I\angle\beta - 2\pi/3, \end{aligned} \tag{15.36}$$

where E is the voltage-source holor amplitude; I is the current-source holor amplitude; α_P is the phase angle of the voltage-source holor, given by $\alpha_P = \omega_P t + \alpha_{P0}$; and β_s is the phase angle of the current-source holor, given by $\beta_s = \omega_s t + \beta_{s0}$.

Let the modulation functions for the current-source 3 × 2 ECM and voltage-source 2 × 3 ECM be defined as

$$\begin{aligned} M_A &= 1\angle\alpha & M^a &= 1\angle\beta \\ M_B &= 1\angle\alpha + 2\pi/3 & M^b &= 1\angle\beta + 2\pi/3 \\ M_C &= 1\angle\alpha - 2\pi/3 & M^c &= 1\angle\beta - 2\pi/3. \end{aligned} \tag{15.37}$$

The displacement angle φ_P on the current-source 3 × 2 ECM primary side and the displacement angle φ_s on the voltage-source 2 × 3 ECM secondary side are defined as

$$\varphi_P = \alpha_{P0} - \beta_{P0}, \ \varphi_s = \alpha_{s0} - \beta_{s0}. \tag{15.38}$$

The commutation (switching) functions for the various throws of the current-source 3 × 2 ECM and voltage-source 2 × 3 ECM are determined by equations (15.37) and (15.38), respectively.

$$
\begin{aligned}
C_{pA}^{ji} &= \begin{cases} 1 & \text{if}(\boldsymbol{M}_A > \boldsymbol{M}_B) \wedge (\boldsymbol{M}_A > \boldsymbol{M}_C) \\ 0 & \text{otherwise} \end{cases} \\
C_{pB}^{ji} &= \begin{cases} 1 & \text{if}(\boldsymbol{M}_B > \boldsymbol{M}_C) \wedge (\boldsymbol{M}_B > \boldsymbol{M}_A) \\ 0 & \text{otherwise} \end{cases} \\
C_{pC}^{ji} &= \begin{cases} 1 & \text{if}(\boldsymbol{M}_C > \boldsymbol{M}_A) \wedge (\boldsymbol{M}_C > \boldsymbol{M}_B) \\ 0 & \text{otherwise} \end{cases} \\
C_{nA}^{ji} &= \begin{cases} 1 & \text{if}(\boldsymbol{M}_A < \boldsymbol{M}_B) \wedge (\boldsymbol{M}_A < \boldsymbol{M}_C) \\ 0 & \text{otherwise} \end{cases} \\
C_{nB}^{ji} &= \begin{cases} 1 & \text{if}(\boldsymbol{M}_B < \boldsymbol{M}_C) \wedge (\boldsymbol{M}_b < \boldsymbol{M}_a) \\ 0 & \text{otherwise} \end{cases} \\
C_{nC}^{ji} &= \begin{cases} 1 & \text{if}(\boldsymbol{M}_C < \boldsymbol{M}_A) \wedge (\boldsymbol{M}_C < \boldsymbol{M}_B) \\ 0 & \text{otherwise} \end{cases}
\end{aligned}
\tag{15.39}
$$

$$
\begin{aligned}
\boldsymbol{C}_{ij}^{ap} &= \begin{cases} 1 \ \text{if} \ \boldsymbol{M}_A > 0 \\ \quad ; \boldsymbol{C}_{ij}^{an} = 1 - \boldsymbol{C}_{ij}^{ap} \\ 0 \ \text{otherwise} \end{cases} \\
\boldsymbol{C}_{ij}^{bp} &= \begin{cases} 1 \ \text{if} \ \boldsymbol{M}_B > 0 \\ \quad ; \boldsymbol{C}_{ij}^{bn} = 1 - \boldsymbol{C}_{ij}^{bp} \\ 0 \ \text{otherwise} \end{cases} \\
\boldsymbol{C}_{ij}^{cp} &= \begin{cases} 1 \ \text{if} \ \boldsymbol{M}_C > 0 \\ \quad ; \boldsymbol{C}_{ij}^{cn} = 1 - \boldsymbol{C}_{ij}^{cp}. \\ 0 \ \text{otherwise} \end{cases}
\end{aligned}
\tag{15.40}
$$

The commutation functions for the current-source 3 × 2 ECM and the voltage-source 2 × 3 ECM are depicted in figure 15.18.

With this choice of commutation functions, the resulting output-voltage holors and input-current holors achieve unity displacement factor operation ($\cos \varphi = 1$) at the input and output simultaneously.

Summary

Conventional modulation strategies for ECM commutators based on trigonometric $\boldsymbol{\alpha} - \boldsymbol{\beta}$ or $\boldsymbol{d} - \boldsymbol{q}$ transformations and **space-vector** (SV) techniques have their voltage transfer capability limited at about 0.866 p.u. This leads to a derating of AC–AC commutator motor capability by 25% in the realisation of variable frequency electromechanical drives using **off-the-shelf** (OTS) line-fed commutatorless AC motors, or requires custom-designed AC–AC commutator motors for operation at

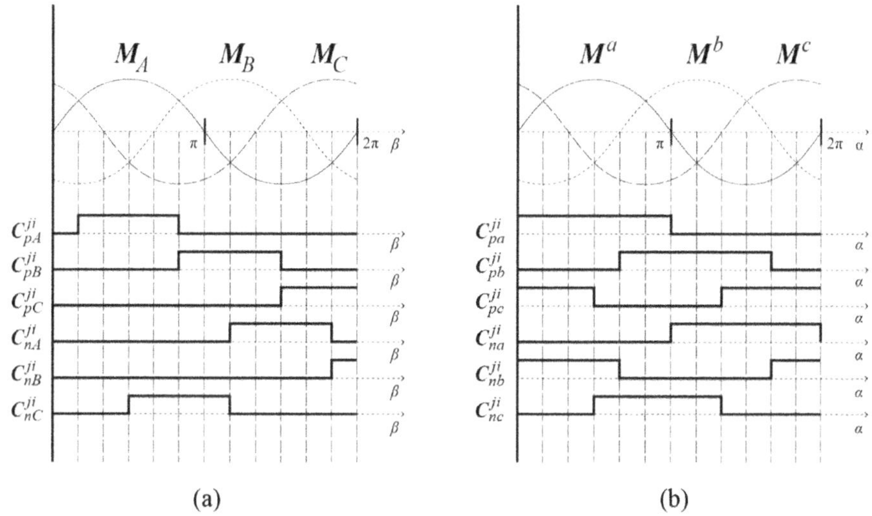

Figure 15.18. Waveforms of various commutation functions for the current-source 3 × 2 ECM AC–DC commutator (a) and the voltage-source 2 × 3 ECM DC–AC commutator (b) for unity power factor operation at both sets of matrixer terminals or both matrixer ports.

reduced voltage. Thus, a six-step variable-frequency modulation strategy for ECM commutators that are capable of providing more than a 1 p.u. fundamental component of output voltage is presented.

In this sub-section a new modulation/control strategy for the 2 × 3 plus 3 × 2 ECM commutator featuring up to unity voltage transfer ratio is presented. If such a low-frequency modulation approach is combined with proper filter design and/or multi-pulse techniques, practical designs may be realized for applications using electrical valves as RB IGBTs.

The combination of the presented modulation with conventional **pulse-width modulation** (PWM) provides independent specification of power factor at both the matrixer-input and matrixer-output ports of the MCM commutator. Furthermore, the presented approach makes it possible to develop the limits of conventional scalar and vector PWM approaches in the transition region between linear operation and six-step operation.

A six-step operation that is suitable for the conventional nine-electrical-valve topology and the reduced electrical valve (switch) dual monolithic ECM topologies is presented, and it is capable of unity displacement-factor operation ($\cos\varphi = 1$) at the matrixer input and matrixer output simultaneously. It is demonstrated that by varying the commutation (switching) angles, control of the fundamental component of output-voltage holor and input-current holor is possible, with concomitant generation of reactive power at the matrixer-input and matrixer-output terminals of the monolithic ECM AC–DC–AC commutator, with bipolar (bilateral) power flow.

15.4 MCM and ECM AC–AC and AC–DC/DC–AC commutator dynamotors—a basic application

Macro- and microelectronics technology led to the change-over from today's power electronics static converter torque and speed control systems for MCM DC–AC commutator motors and AC commutatorless motors to tomorrow's ECM AC–AC and DC–AC commutator motors.

In the past, power electronics technology led to the change-over from the rotary dynamoelectric Ward Leonard torque and speed control systems for MCM DC–AC commutator motors to static-converter Ward Leonard torque and speed control systems. This commenced in the 1930s with the introduction of mercury-arc converters and in the 1950s with the introduction of transductor-controlled **semiconductor rectifiers** (SR), i.e. power diodes.

Today, **semiconductor controlled rectifiers** (SCR), i.e. thyristor converters for MCM DC–AC commutator motor advanced propulsion (traction dynamical hypersystems) have matured, and this technology is well established in many applications. Because of their simple and robust design, AC commutatorless motors, and then all the AC hyposynchronous (induction) squirrel-cage-rotor motors invented by the Polish electrician, M O Dolivo-Dobrovolsky (1862–1919), in 1888, are preferable to the MCM DC–AC commutator motor for many torque and speed control applications, particularly in difficult and hazardous environments (Hoseman *et al* 1999). However, torque and speed control for AC commutatorless motors have only been a dream for many years. It is true that this technique has existed for a long time and indeed manufacturers have also been manufacturing static converters (static frequency changers) for AC MCM commutatorless motors for today's industrial and traction propulsion complex dynamical hypersystems.

In recent years, the cost trend for ECM commutators has now made it increasingly interesting from the commercial viewpoint to use ECM commutators more widely, in ECM AC–AC and AC–DC/DC–AC commutator dynamotors for variable-speed industrial and traction propulsion complex dynamical hypersystems. This trend is still in its infancy and this technique should experience its definite breakthrough during the coming years.

The design of ECM commutators is closely tied to the complexity of the MMD electrical machine. Beyond doubt, electrical energy conversion is technically the most advanced today. It involves the static conversion of electrical energy.

It has been, and still is, a challenge for the application of newly designed energy-saving ECM AC–AC and AC–DC/DC–AC commutator dynamotors (generators/motors), particularly if industrial and traction propulsion complex dynamical hypersystems are considered. It is consequently not an exaggeration to state that ECM AC–AC and AC–DC/DC–AC commutator dynamotors are technically in advance of those for the other MMD electrical machines.

The author has been very active during recent years in the field of ECM AC–AC and AC–DC/DC–AC commutator dynamotors. Viewed from a historical perspective, the author has kept up with the technical developments in this area and on many occasions has quickly and successfully exploited new technical advances. Figures 15.19 and 15.20 show some important landmarks schematically.

As far back as 1881, the electrical machinery industry supplied the first MCM AC–DC/DC–AC commutator dynamotors for industrial and traction propulsion complex dynamical hypersystems.

The technology was based on ME energy conversion. Development then continued with the introduction of mercury-arc and semiconductor converters. The first fully static ECM commutators and ECM AC–AC and AC–DC/DC–AC commutator dynamotors were launched in the early 1930s (figures 15.16 and 15.17).

Recently, there has been considerable interest in the use of ECM AC–AC, AC–DC–AC, AC–DC/DC–AC, DC–DC and DC–AC–DC commutator technology for industrial and traction propulsion mechatronical control systems, as well as for aerospace and automotive applications.

The aerospace and automotive industries have also shown considerable interest in ECM AC–AC, AC–DC–AC, AC–DC/DC–AC, DC–DC and DC–AC–DC commutators due to the inherent potential advantages they have in aircraft and automotive applications. This textbook will give an in-depth introduction to the full range of ECM AC–AC, AC–DC–AC, AC–DC/DC–AC, DC–DC and DC–AC–DC commutator technologies and investigate the suitability and design of ECM AC–AC, AC–DC–AC, AC–DC/DC–AC, DC–DC and DC–AC–DC commutators.

The potential advantages of ECM AC–AC, AC–DC–AC, AC–DC/DC–AC, DC–DC and DC–AC–DC commutator technology will be examined, and the factors that have so far limited commercial exploitation of the ECM will be discussed (see the references and bibliography).

The book will use appropriate practical results from various prototypical MIMO ECM AC–AC, AC–DC–AC, AC–DC/DC–AC, DC–DC and DC–AC–DC commutators to illustrate the topics under discussion. These examples will focus on

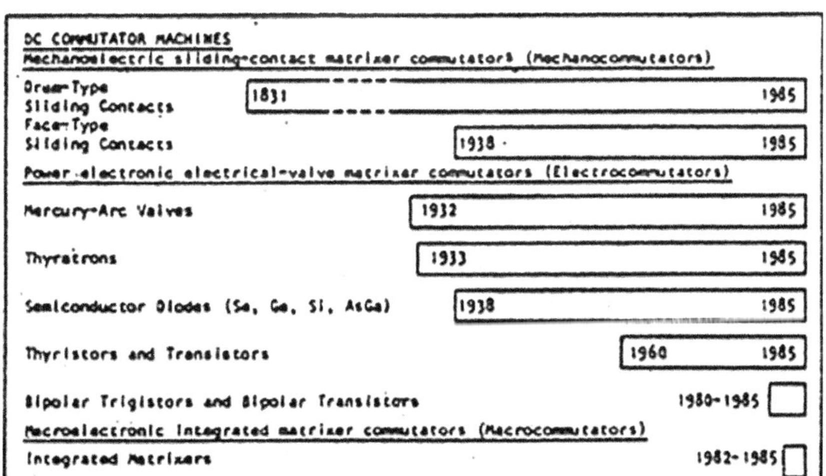

Figure 15.19. Some development milestones in the field of MCM AC–DC/DC–AC commutator dynamotors (Fijalkowski 1988).

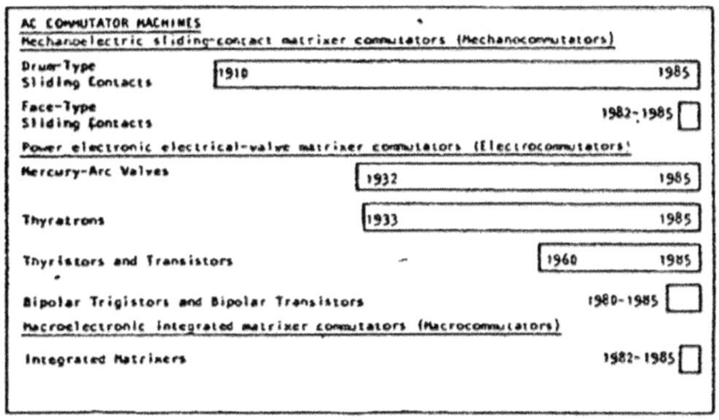

Figure 15.20. Some development milestones in the field of MCM AC–AC commutator motors (Fijalkowski 1988).

application areas where the advantages of the ECM AC–AC, AC–DC–AC, AC–DC/DC–AC, DC–DC and DC–AC–DC commutator carry a large premium, for example industrial and traction propulsion mechatronical control systems, especially for aerospace and automotive applications.

The textbook is aimed at anybody interested in the area of MIMO ECM AC–AC, AC–DC–AC, AC–DC/DC–AC, DC–DC and DC–AC–DC commutators. It will start with a basic introduction to the topology and lead into in-depth discussion on the critical research developments and topical issues.

Today, macro- and microelectronics technology has become widely accepted for new-concept integral ECM AC–AC and AC–DC/DC–AC commutator dynamotors. Such technology has resulted in considerably improved industrial and traction propulsion complex dynamical hypersystems, providing fast, smooth and precise torque and speed control. For instance, for railway and automotive vehicles this means better slip controls, which have made it possible to take out high tractive efforts close to the limit of adhesion.

15.5 New-concept ECM AC–AC and AC–DC/DC–AC commutator dynamotors

The development of modern ECM AC–AC and AC–DC/DC–AC commutator dynamotors is closely connected with the further development of macro- and microelectronics technology. Moreover, it is necessary to pay attention to the interesting fact that contemporary MCM AC–AC and AC–DC/DC–AC commutator dynamotors have not changed their basic construction and appearance since the late 1980s. R&D work on ECM AC–AC and AC–DC-/DC–AC commutator dynamotors initially concentrated on designing their details (component parts) and a deeper understanding of the physical phenomena appearing during their working.

Modern ECM AC–AC and AC–DC/DC–AC commutator dynamotors with ECM commutators have up to now been used very rarely, mainly in aviation and automotive industrial and traction propulsion complex dynamical hypersystems (Fijalkowski 1985b).

Another factor that contributed to such a state of affairs in power electronics was the fact that in the 1940s the rapid development of mercury-arc converters took place, followed by the development of thyristor converters in the 1950s.

Thus, scientists were only interested on a large scale in the development of electronic commutators, i.e. polyvalvular ECM AC–AC and/or AC–DC/DC–AC commutators with unipolar ECMs, realised on continuous power diodes and thyristors—up to now called 'static converters' in the field of power electronics.

The new-concept integral ECM AC–AC and/or AC–DC/DC–AC commutator dynamotors conceived by the author can be used, e.g. in advanced cradled dynamometer simple continuous dynamical systems for testing combustion engines, and electrical, fluidic and pneumatic machines, as well as other rotary machines, etc, while industrial and traction propulsion complex dynamical hypersystems operate in **sinusoidal-pulse width modulation** (SPWM) mode with integral ECM commutators, just like conventional MCM AC–AC and AC–DC/DC–AC commutator dynamotors.

A dynamometer that is designed to be driven is called an 'absorption dynamometer'. A dynamometer that can either drive or absorb is called a 'universal dynamometer'. A dynamometer can also be used to determine the torque and power required to operate a driven machine, for example, a **mechano-fluidic** (MF) pump or a **mechano-pneumatic** (MP) compressor. In that case, a motoring or driving dynamometer is used.

A motoring dynamometer acts as an MCM AC–AC or a DC–AC commutator motor that drives the equipment being tested. It must be able to drive the combustion engine, EM, FM or PM motor, or other rotating prime mover, at any angular speed and develop any level of torque that the test requires. Only torque and angular speed can be measured directly.

In most dynamometers, power is not measured directly, it must be computed from the torque and angular-speed holors, or force and linear-velocity holors, according to the following holor product:

$$\boldsymbol{P}^{\kappa} = \boldsymbol{T}_i \otimes \boldsymbol{\Omega}^j = \gamma^{\kappa i}_{..j} \boldsymbol{T}_i \boldsymbol{\Omega}^j,$$

or

$$\boldsymbol{P}^{\kappa} = \boldsymbol{F}_i \otimes \boldsymbol{V}^j = \gamma^{\kappa i}_{..j} \boldsymbol{F}_i \boldsymbol{V}^j,$$

where $\boldsymbol{P}^{\kappa}$ is the power holor (W), $\boldsymbol{T}_i$ is the torque holor (Nm), $\boldsymbol{\Omega}^j$ is the angular-velocity holor (rad s^{-1}), $\boldsymbol{F}_i$ is the force holor (N) and $\boldsymbol{V}^j$ is the linear-velocity holor (m s^{-1}).

A growing feature of industrial and traction dynamical systems is the sense of the power-flow direction in the ECM commutator of the MMD electrical machine being arbitrary, that is, it can be used for both EM motoring and ME generation (regenerative braking). The change to the sense of the power-flow direction can take place without any reconnections. Other characteristics that can be achieved with industrial and traction propulsion dynamical systems are a power factor that can be controlled in an optimum way and low machine armature-winding phase current harmonics. These characteristics are perhaps more important than the industrial and traction propulsion complex dynamical hypersystem as such.

Combined with reasonable forecasts for the ever more widespread application of new-concept integral ECM AC–AC and AC–DC/DC–AC commutator dynamotors in industrial and traction propulsion complex dynamical hypersystems, constant attention has been devoted to studying and updating these initial new-concept integral ECM AC–AC and AC–DC/DC–AC commutator dynamotors and integral lower- and higher-level microprocessor controllers, as broken down into the following integral MMD electrical machine types (figures 15.21 and 15.22).

- Three- or two-phase ECM AC–AC commutator hypersynchronous orthogonal-two-phase wound-rotor dynamotor (triply fed brushed MMD electrical machine)—figures 15.21(a).
- Three- or two-phase ECM AC–AC commutator hypersynchronous orthogonal-two-phase wound-rotor dynamotor (triply fed brushless MMD electrical machine)—figures 15.21(b).
- Three- or two-phase ECM AC–AC commutator hypersynchronous three-phase wound-rotor dynamotor (doubly fed brushed MMD electrical machine)—figures 15.21(c).
- Three- or two-phase ECM AC–AC commutator hypersynchronous three-phase wound-rotor dynamotor (doubly fed brushless MMD electrical machine)—figures 15.21(d).
- Three- or two-phase ECM AC–AC commutator synchronous orthogonal-two-phase wound-rotor electromagnetically excited dynamotor (triply fed brushed MMD electrical machine)—figure 15.21(e).

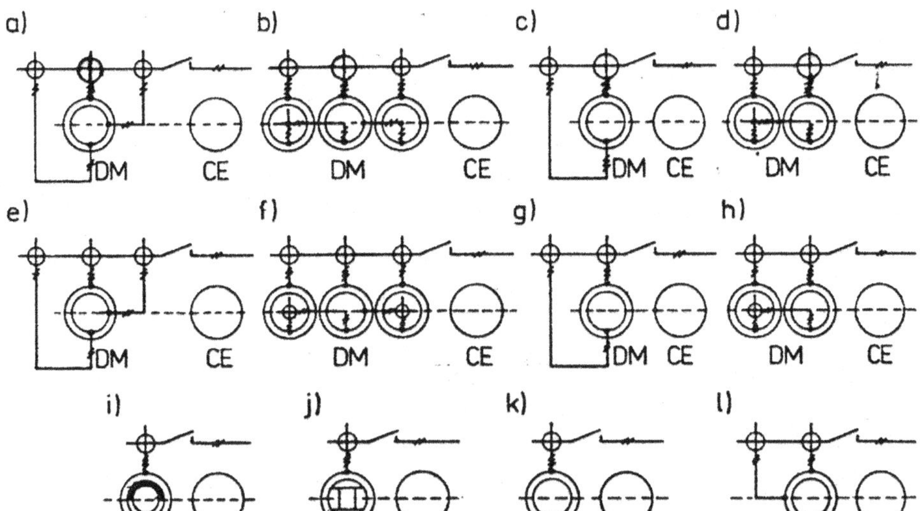

Figure 15.21. Three- or two-phase ECM AC–AC commutator MMD electrical machines (Fijalkowski 1987).

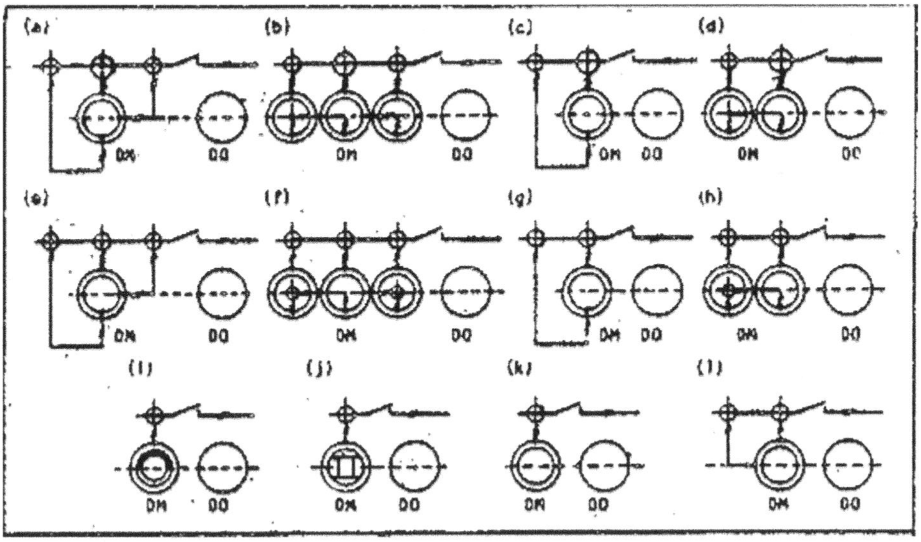

Figure 15.22. Single-phase ECM AC–AC and/or AC–DC/DC–AC commutator MMD electrical machines (Fijalkowski 1987).

- Three- or two-phase ECM AC–AC commutator synchronous orthogonal-two-phase wound-rotor electromagnetically excited dynamotor (triply fed brushless MMD electrical machine)—figure 15.21(f).
- Three- or two-phase ECM AC–AC commutator synchronous single-phase wound-rotor electromagnetically excited dynamotor (doubly fed brushed MMD electrical machine)—figure 15.21(g).
- Three- or two-phase ECM AC–AC commutator synchronous single-phase wound-rotor electromagnetically excited dynamotor (doubly fed brushless MMD electrical machine)—figure 15.21(h).
- Three- or two-phase ECM AC–AC commutator synchronous magneto-electrically excited dynamotor (doubly fed brushless MMD electrical machine)—figure 15.21(i).
- Three- or two-phase ECM AC–AC commutator synchronous reluctance dynamotor (singly fed brushless MMD electrical machine)—figure 15.21(j).
- Three- or two-phase ECM AC–AC commutator hyposynchronous (induction) squirrel-cage-rotor dynamotor (singly fed brushless MMD electrical machine)—figure 15.21(k).
- Three- or two-phase ECM AC–AC commutator hyposynchronous (induction) orthogonal-two-phase stator and hollow-rotor dynamotor (doubly fed brushless MMD electrical machine)—figure 15.21(l).
- Single-phase ECM AC–AC and/or AC–DC/DC–AC commutator hyper-synchronous orthogonal-two-phase wound-rotor dynamotor (triply fed brushed MMD electrical machine)—figure 15.22(a).

- Single-phase ECM AC–AC and/or AC–DC/DC–AC commutator hypersynchronous orthogonal-two-phase-wound-rotor dynamotor (triply fed brushless MMD electrical machine)—figure 15.22(b).
- Single-phase ECM AC–AC and/or AC–DC/DC–AC commutator hypersynchronous three-phase-wound-rotor dynamotor (doubly fed brushed MMD electrical machine)—figure 15.22(c).
- Single-phase ECM AC–AC and/or AC–DC/DC–AC commutator hypersynchronous three-phase-wound-rotor dynamotor (doubly fed brushless MMD electrical machine)—figure 15.22(d).
- Single-phase ECM AC–AC and/or AC–DC/DC–AC commutator hypersynchronous orthogonal-two-phase-wound-rotor electromagnetically excited dynamotor (triply fed brushed MMD electrical machine)—figure 15.22(e).
- Single-phase ECM AC–AC and/or AC–DC/DC–AC commutator synchronous orthogonal-two-phase wound-rotor electromagnetically excited dynamotor (triply fed brushless MMD electrical machine)—figure 15.22(f).
- Single-phase ECM AC–AC and/or AC–DC/DC–AC commutator synchronous single-phase wound-rotor electromagnetically excited dynamotor (doubly fed brushed MMD electrical machine)—figure 15.22(g).
- Single-phase ECM AC–AC and/or AC–DC/DC–AC commutator synchronous single-phase wound-rotor electromagnetically excited dynamotor (doubly fed brushless MMD electrical machine)—figure 15.20(h).
- Single-phase ECM AC–AC and/or AC–DC/DC–AC commutator synchronous single-phase wound-rotor magnetoelectrically excited dynamotor (doubly fed brushless MMD electrical machine)—figure 15.22(i).
- Single-phase ECM AC–AC and/or AC–DC/DC–AC commutator synchronous reluctance dynamotor (singly fed brushless MMD electrical machine)—figure 15.22(j).
- Single-phase ECM AC–AC and/or AC–DC/DC–AC commutator hyposynchronous (induction) squirrel-cage-rotor dynamotor (singly fed brushless MMD electrical machine)—figure 15.22(k).
- Single-phase ECM AC–AC and/or AC–DC/DC–AC commutator hyposynchronous (induction) orthogonal-two-phase wound-stator and hollow-rotor dynamotor (doubly fed brushless MMD electrical machine)—figure 15.22(l).

These new-concept integral MMD electrical machines with ECM commutators will have a fully laminated stator and all the electrical insulation will be of class H.

The material utilisation, which is the ratio between the mass and the output power, will be better for the new-concept ECM AC–AC and AC–DC/DC–AC commutator dynamotors compared with the preceding MMD electrical machines. This improvement will be achieved without having to increase either the electrical stresses or the relative level of the material used. This means that, for example, the margin for a harmful temperature rise in connection with overloads is at least as great as for the older designs.

ECM AC–AC and AC–DC/DC–AC commutator dynamotors with their accessories and above all intelligent MMD electrical machines represent one of the main themes of this textbook. Concepts such as artificial intelligence and microcomputer capacity are factors that are increasingly putting their imprint on our world. The use of industrial and traction propulsion complex dynamical hypersystems with integral ECM AC–AC and AC–DC/DC–AC commutator dynamotors is growing rapidly. This is largely due to the increasingly severe demands being made on these dynamical hypersystems.

Today's and tomorrow's traction propulsion complex dynamical hypersystems must be capable of being adjusted to cope with the varying operation conditions and handle a wide range of current and voltage/torque and speed controls.

In addition, they must be flexible to meet operation modifications, where new operating conditions might be needed in existing and traction propulsion complex dynamical hypersystems. To meet these demands, integral ECM AC–AC and AC–DC/DC–AC commutator dynamotors are becoming more advanced and their range of accessories is growing. These dynamotors will be available in the near future as complete 'function packages' for different applications. Their accessories will include flexible microprocessor-based ECM commutator controllers and a new artificial intelligence dynamical system.

Advantages and disadvantages of integral ECM AC–AC and AC–DC/DC–AC commutator dynamotors

Here follows a short list of the advantages and disadvantages of integral ECM AC–AC and AC–DC/DC–AC commutator dynamotors.

Some of the advantages are as follows.

- Variable/adjustable voltage and/or current, as well as torque and/or speed controls, which may increase the efficiency of the load. Increased efficiency leads to reduced electrical energy costs.
- Easy installation, leading to reduced installation costs.
- Easy commissioning, leading to reduced commissioning costs.
- No space is required for an ECM AC–AC or AC–DC/DC–AC commutator cabinet.
- Reduced EMC problems, both radiated and line-carried, due to the containment of the ECM and the cable from the ECM AC–AC and/or AC–DC/DC–AC commutator to the dynamotor.
- Reduced stock inventory, since one integral ECM AC–AC and AC–DC/DC–AC commutator dynamotor can replace, for example, several AC induction motors with different pole numbers for a certain torque.

Some of the disadvantages are as follows.

- Novel integral ECM AC–AC and AC–DC/DC–AC commutator dynamotor concept scares potential buyers.
- Slightly more expensive to buy than an AC or a DC dynamotor with a separate static converter.
- The AC–AC and/or AC–DC/DC–AC commutator's ECM is exposed to the same environment as the dynamotor, regarding vibrations and heat.

Modern industrial and traction propulsion complex dynamical hypersystems with an integral ECM AC–AC or AC–DC–AC commutator dynamotor as their chief and only component are generally built up as hierarchical physical heterogeneous continuous complex dynamical hypersystems (figure 15.23).

The dynamotor's microprocessor-based ECM commutator controllers (lower-level controllers) communicate with higher-level propulsion controllers, which handle the coordination within the industrial and traction propulsion complex dynamical hypersystem and, in their turn, the propulsion controllers communicate with the ECM commutator controllers. Further down in the hierarchy is a relatively small volume of data for the different ECM AC–AC and AC–DC/DC–AC commutator dynamotors' ECM commutator controllers. On the other hand, greater demands are frequently made on the signal-processing rate than in the higher-level propulsion controllers. The need for the signal-processing rate in the control functions decreases at the higher level.

In the case of higher-level control and monitoring with operator presentation, the microcomputer must be able to handle large volumes of data, while the demand made on the processing rate decreases.

There are three main function categories in each type of process computer system:

- **process control** (PC) functions;
- **man–machine communication** (MMC) functions;
- **computer system** (CS) functions.

Figure 15.21 shows the main functions and signal routes of the industrial or traction propulsion complex dynamical hypersystem. The control equipment is

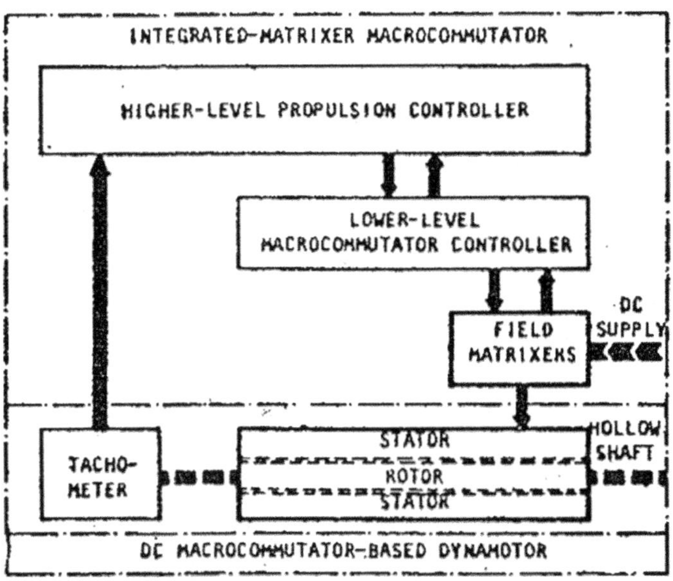

Figure 15.23. Main functions of an ECM AC–AC and or AC–DC–AC commutator dynamotor (Fijalkowski 1985).

divided into a 'higher-level propulsion controller' and a 'lower-level static commutator'. Both sets of control equipment are located in the ECM commutator compartment.

The ECM commutator controller incorporates circuits for controlling and monitoring the armature and field ECMs. In principle, the input signal to the ECM commutator controller is a torque reference (set point), which is compared with a torque signal obtained by means of suitable control of the armature and field ECMs.

The propulsion controller incorporates circuits for controlling and monitoring the object driven by the integral ECM AC–AC or AC–DC/DC–AC commutator dynamotor. The control mode is generally speed control, but also position or some other variables; for example, tension, pressure, flow and loop may also be controlled. The reference (set point) for the controlled variable may be formed in the propulsion controller or may be obtained from the higher-level industrial or traction process commander (control system). Today's microprocessor-based ECM commutator controllers consist of microcomputers of varying degree of complexity working in real time. They are frequently linked in hierarchical structures to satisfy different functional requirements in the best possible way. Modern industrial and traction propulsion complex dynamical hypersystems impose demands on advanced data communication methods.

The operator controls and supervises his/her control process with the aid of an MMC system, which essentially consists of multi-colour **video display units** (VDU) and keyboards. These VDUs reproduce in a quick, correct and easily understandable way what is happening in the integral ECM AC–AC and AC–DC/DC–AC commutator dynamotors, i.e. in the industrial and traction propulsion complex dynamical hypersystems. This highlights the current requirements and demands for new-concept integral MMD electrical machines with ECM AC–AC and/or AC–DC/DC–AC commutators.

Placement of the ECM AC–AC and/or AC–DC/DC–AC commutators

Some of the possibilities may place the ECM AC–AC and/or AC–DC/DC–AC commutator on top of the integral AC–AC and/or AC–DC/DC–AC commutator dynamotor, while some of them also allow it to be placed on the left or right side of the dynamotor.

Placement at the top conserves the footprint of the integral ECM AC–AC and/or AC–DC/DC–AC commutator dynamotor, but increases the height. The ECM AC–AC and/or AC–DC/DC–AC commutator can also be placed at the non-drive end. This placement extends the axial length of the ECM AC–AC and/or AC–DC/DC–AC commutator dynamotor, but maintains the existing height and reduces the risk of harmful vibrations. In this case, the AC–AC and/or AC–DC/DC–AC commutator's ECM is better protected from heat, especially the rising heat of the integral ECM AC–AC and/or AC–DC/DC–AC commutator dynamotor after it is turned off.

Modern integral ECM AC–AC and/or AC–DC/DC–AC commutator dynamotors should have ECMs, together with single-chip microcontrollers (designed for control and monitoring), built into the newly designed terminal boxes of integral MMD electrical machines.

15.6 Exemplary applications of electrical commutation matrixer commutators

15.6.1 A 2 × 2 ECM AC–DC/DC–AC commutator

The roles of ECM AC–DC/DC–AC commutators in realistic applications are major. A 2 × 2 ECM AC–DC/DC–AC commutator, e.g. covers all possible cases with a single-port input energy source and a two-terminal load.

The ECM AC–DC/DC–AC commutator is commonly drawn with a single homogeneous anode or cathode commutation group, or two heterogeneous anode and cathode commutation groups, i.e. the well-known H-bridge converter (rectifier/inverter) shown in figure 15.24.

15.6.2 A 3 × 3, 3 × 5 or 5 × 5 ECM AC–AC commutator

This book, among other subjects, especially concerns 3 × 3, 3 × 5 and/or 5 × 5 ECM AC–AC commutator dynamotors as alternative **voltage-source inverter** (VSI) AC motor drives (see figure 15.25).

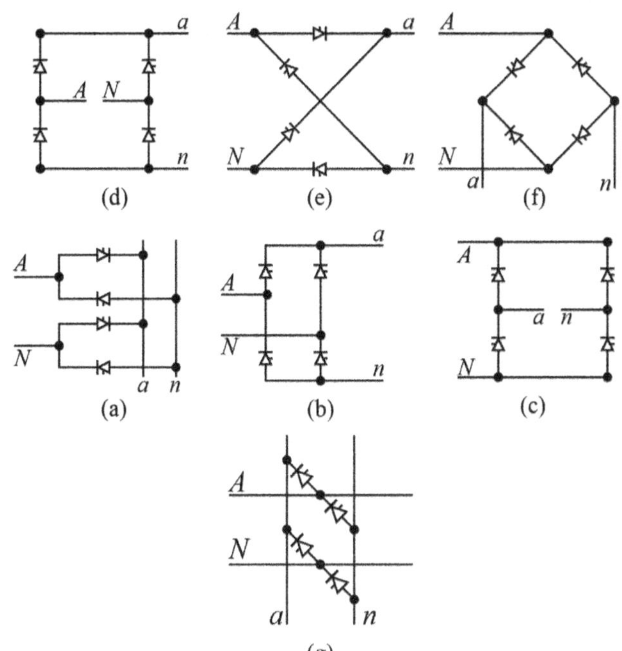

Figure 15.24. A 2 × 2 ECM AC–DC/DC–AC commutator with a single homogeneous anode or cathode commutation group, or two heterogeneous anode and cathode commutation groups (H-bridge configuration).

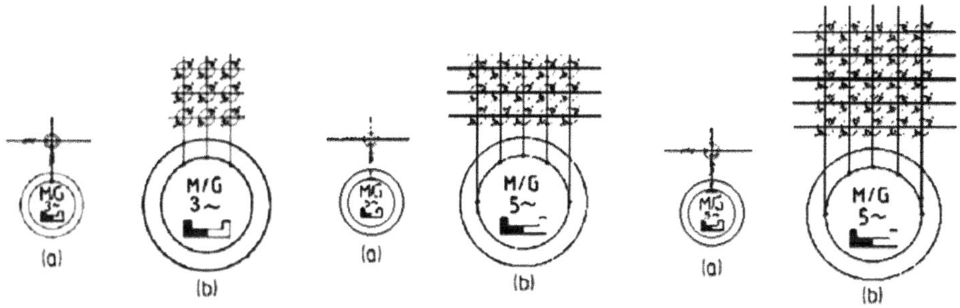

Figure 15.25. The 3 × 3, 3 × 5 and 5 × 5 ECM AC–AC commutator dynamotors for steered, motorised and/or generatorised wheels (SM&GW) (Fijalkowski 1987, 1982, 1985, 1988, 1997).

The 3 × 3, 3 × 5 and 5 × 5 ECM commutators are direct ECM AC–AC commutators with no DC link. The lack of reactive components in the DC link is one of the salient advantages of the 3 × 3, 3 × 5 and 5 × 5 ECM commutators.

Furthermore, the 3 × 3, 3 × 5 and/or 5 × 5 ECM features full four-quadrant operation and sinusoidal ECM input current. The ECM output voltage is limited to 87% of the ECM input voltage.

The 3 × 3, 3 × 5 or 5 × 5 ECM AC–AC commutator needs nine (9), 15 or 25 bipolar electrical valves, respectively, to connect two three- or five-phase voltage systems in all possible combinations.

Most R&D work about 3 × 3, 3 × 5 and 5 × 5 ECMs has so far concerned the modulation and control of the ECM AC–AC commutator dynamotor. The implemented 3 × 3 ECM AC–AC commutator shows a reduction of the reactive elements by more than a factor of 12 compared to a standard VSI inverter AC motor drive at the same power rating.

The power flow through the 3 × 3, 3 × 5 or 5 × 5 ECM AC–AC commutator is bipolar and it draws sinusoidal input current in the generative as well as the regenerative mode. The input power-power factor is above 0.98. It may be concluded that VSI converter AC motor drives will become standard in the near future. For example, in civil and/or military **all-electric vehicles** (AEV) and **hybrid-electric vehicles** (HEV), mass and size reduction is critical. High-temperature operation is desirable to ease the problem of the thermal management of 3 × 3, 3 × 5 and/or 5 × 5 ECM AC–AC commutator dynamotors with single cooling heat-sinks (Fijalkowski 1988b, 1988e, 1988g, 1997).

It is believed that the 3 × 3, 3 × 5 or 5 × 5 ECM AC–AC commutator dynamotor may have significant advantages over the conventional ECM AC–DC–AC commutator dynamotor in all of these areas, since it is possible to eliminate the DC link capacitor, especially if an AC primary power system is used.

An interesting example of the design and construction of a snubberless 150 kVA experimental 3 × 3 ECM AC–AC commutator induction dynamotor, built using 600 A, 1400 V continuous IGBTs (Powerex CM600HA-28H) is shown in figure 15.26 (Wheeler *et al* 2004).

The 150 kVA experimental 3 × 3 ECM AC–AC commutator is controlled with a microprocessor (Infineon SAB80C 167). Despite the relatively low-performance

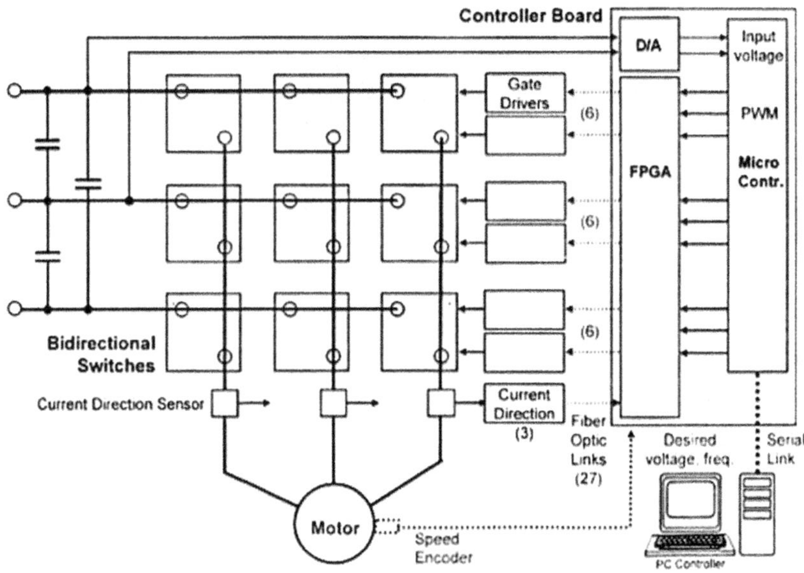

Figure 15.26. A snubberless 150 kV experimental 3 × 3 ECM AC–AC commutator induction motor (Wheeler *et al* 2004).

control platform, it has been possible to achieve full closed-loop vector control of the ECM AC–AC commutator induction motor load. From the output-voltage and output-current waveforms of the 3 × 3 ECM AC–AC commutator induction motor shown in figure 15.27, it can be noted that the output power of the 3 × 3 ECM AC–AC commutator is 150 kVA (Wheeler *et al* 2004).

It is very important to minimise the parasitic inductance within the ECM AC–AC commutator. In this ECM AC–AC commutator, the parasitic inductance has been minimised using laminate input power planes with distributed input filter capacitance.

Another interesting example of the design and construction of a static Scherbius system with a 3 × 3 ECM AC–AC commutator wound-rotor induction dynamotor, which is used to control the rotor-side currents, is shown in figure 15.28 (Altun 2005).

In such a static Scherbius system, the slip power is taken from the rotor circuit and given to the grid through the 3 × 3 ECM AC–AC commutator. The static Scherbius system combines a 3 × 3 ECM AC–AC commutator and a commutator-less AC wound-rotor induction dynamotor in a slip energy recovery EM propulsion mechatronic control system, as shown in figure 15.28. This static Scherbius system is inherently capable of bilateral power flow and also offers virtually sinusoidal input currents, without the harmonics usually associated with present commercial ECM AC–AC and AC–DC–AC commutators, the so-called 'cycloconverters' and/or 'inverters'.

In the static Scherbius system, a 3 × 3 ECM AC–AC commutator has been used to extract the slip energy from the rotor into the mains in the hyposynchronous

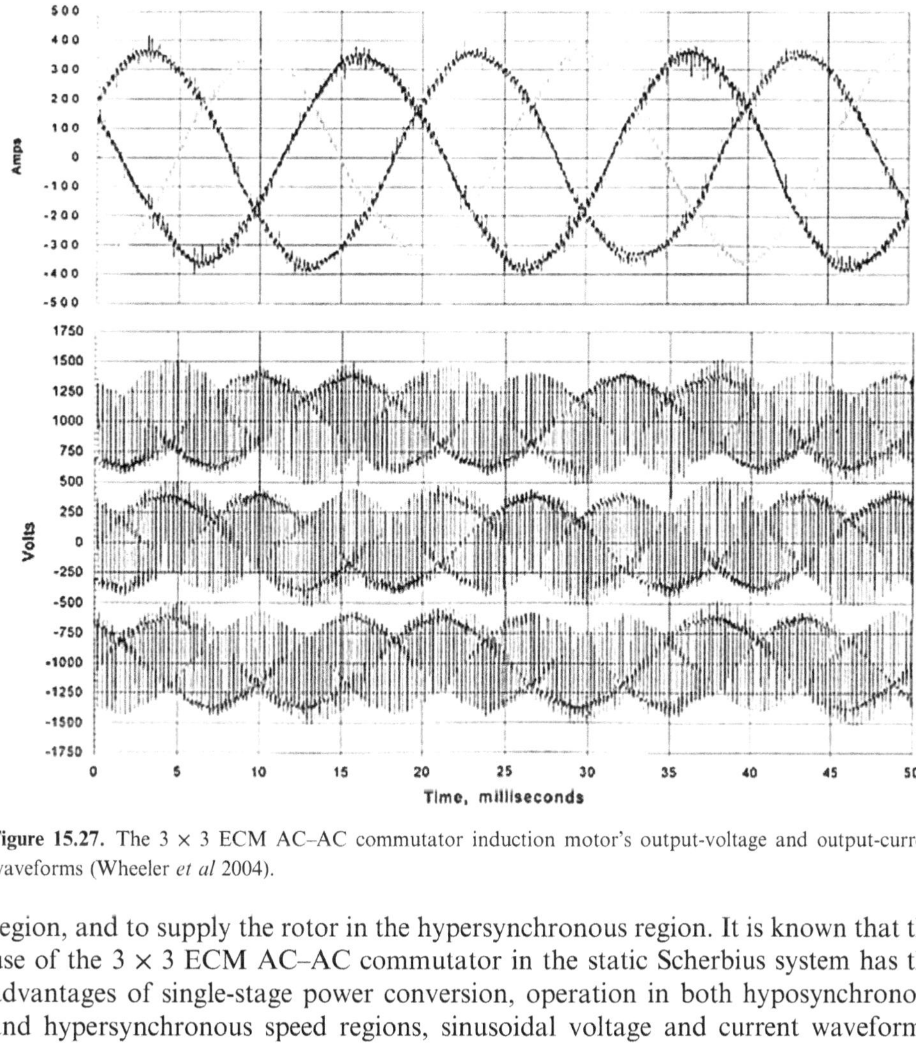

Figure 15.27. The 3 × 3 ECM AC–AC commutator induction motor's output-voltage and output-current waveforms (Wheeler *et al* 2004).

region, and to supply the rotor in the hypersynchronous region. It is known that the use of the 3 × 3 ECM AC–AC commutator in the static Scherbius system has the advantages of single-stage power conversion, operation in both hyposynchronous and hypersynchronous speed regions, sinusoidal voltage and current waveforms, and simpler microelectronic gate-control drivers than other ECM AC–AC and/or AC–DC–AC commutators (cycloconverters and/or inverters).

At present, the static Scherbius system has an important place in wind-power generation applications, where a variable-speed, constant-frequency static Scherbius system produces electrical energy for a wide range of wind speeds.

15.6.3 A 3 × 2 ECM AC–DC commutator

A slightly more complicated example of a 3 × 2 ECM AC–DC commutator generator (alternator), with a two heterogeneous anode and cathode commutation groups, i.e. the well-known three-phase bridge rectifier, is shown in figure 15.29 (Fijalkowski 1987).

There are three possible primary collectors (inputs) and two secondary collectors (outputs) of the DC circuit, which gives a 3 × 2 ECM.

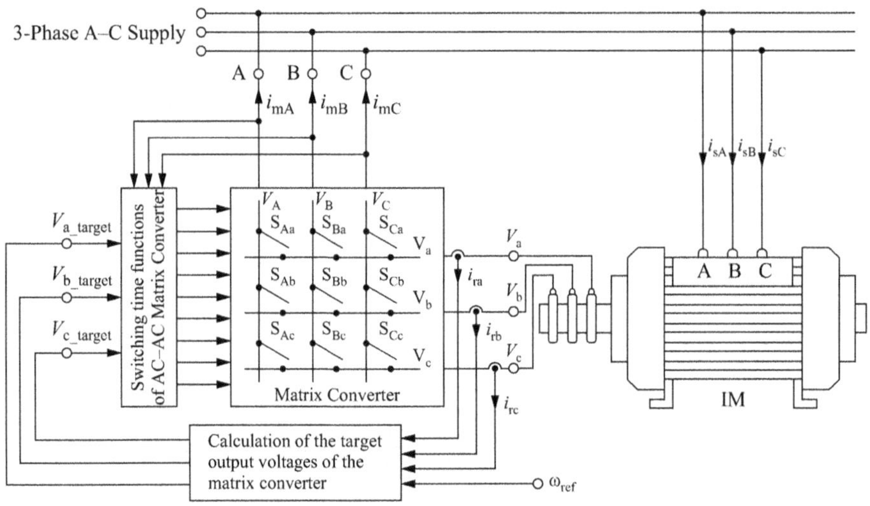

Figure 15.28. Slip energy recovery static Scherbius system with a 3 × 3 ECM AC–AC commutator that is a matrix converter (adapted from Altun 2005).

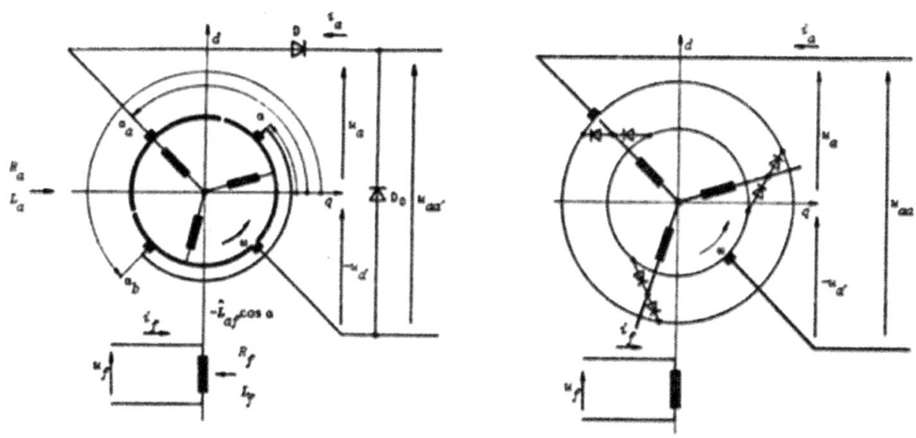

Figure 15.29. A 3 × 2 electrical commutation matrixer (ECM) AC–DC commutator generator (right) compared with a 3 × 2 mechanical commutation matrixer (MCM) AC–DC commutator generator (left) (Fijalkowski 1987, Tutaj 1996).

Because ECMs may be used in ECM AC–AC and AC–DC/DC–AC commutator dynamotors, two important recent achievements may now be applied with great success: contactless (sparkless) commutation and contactless (brushless) electromagnetic excitation.

15.6.4 A 3 × 3 ECM AC–AC commutator

Three-phase ECM AC–AC commutator hypersynchronous dynamotors (figure 15.30) with three-phase wound rotors consist of the ECM commutator with a bipolar ECM, realised on only nine (9) continuous **reverse-blocking insulated-gate bipolar transistors**

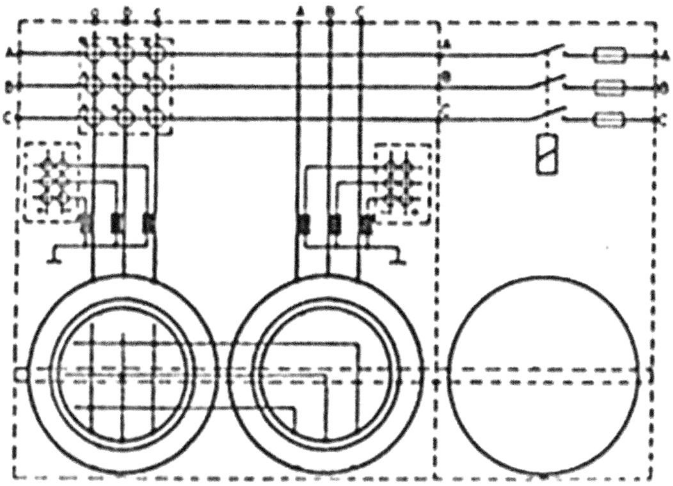

Figure 15.30. Three-phase ECM AC–AC commutator hypersynchronous dynamotor with a 3 × 3 ECM commutator and three-phase AC wound rotors (Fijalkowski 1988).

(RB IGBT), **metal-oxide semiconductor field-effect transistors** (MOSFET) or bipolar symmetrical trigistors, that is, **gate turn-off** (GTO) symmetrical thyristors (symistors) in a 3 × 3 ECM connection (Fijalkowski 1988).

This 3 × 3 ECM AC–AC commutator (figure 15.31) also makes the bilateral flow of secondary electrical energy possible during generation (absorbing) and motoring (driving) with the hypersynchronous angular velocity.

Six operating modes are possible in this MMD electrical machine.

15.6.5 A single-phase ECM AC–AC and/or AC–DC/DC–AC commutator

The single-phase ECM AC–AC and/or AC–DC/DC–AC commutator synchronous (electromagnetically excited) dynamotor (figure 15.32) with two three-phase armature windings, one of which is wye-connected, and another of which is delta-connected, consists of a non-rotary main ECM commutator with two bipolar ECMs, realised on 12 continuous bipolar trigistors in a 2 s(2 × 3)/2 s(3 × 2) ECM connection in series, and a rotary auxiliary ECM commutator with a unipolar ECM, realised on six (6) continuous diodes in a 2 × 3 ECM connection. It has a contactless (brushless) electromagnetic excitation.

The single-phase ECM AC–AC and/or AC–DC/DC–AC commutator synchronous (magnetoelectrically excited) dynamotor (figure 15.33) consists of an ECM commutator with a single or two bipolar ECMs, realised on six (6) or 12 continuous bipolar trigistors in a 2 × 3 ECM connection or a 2 s(2 × 3) ECM connection in series, by means of permanent-magnet poles (for example, alloys of neodymium, iron and boron).

The single-phase ECM AC–AC and/or AC–DC/DC–AC commutator hyposynchronous (induction) dynamotor (figure 15.34) with a squirrel-cage rotor, consists of

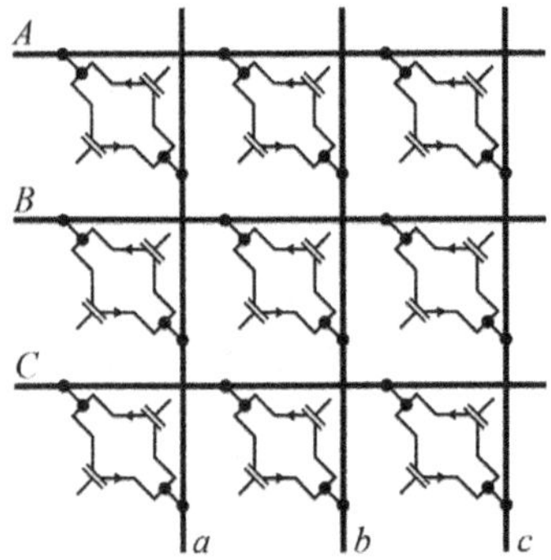

Figure 15.31. Three-phase 3 × 3 ECM AC–AC commutator with RB IGBT.

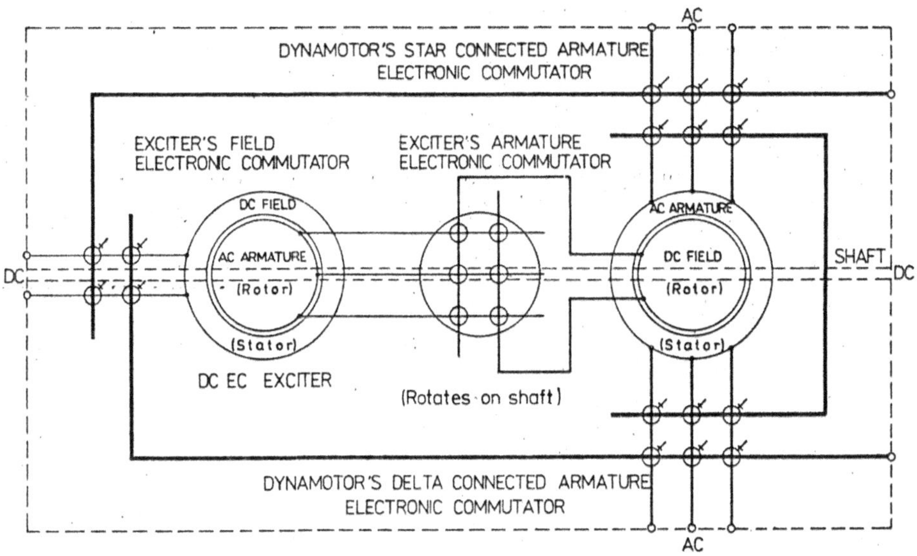

Figure 15.32. A single-phase ECM AC–AC and/or DC–AC/AC–DC commutator synchronous (electromagnetically excited) dynamotor with a 2 s(2×3)/2 s(3×2) ECM commutator (Fijalkowski 1988, Tutaj 2012).

an ECM commutator with a single or two bipolar ECMs, realised on six (6) or 12 continuous bipolar trigistors in a 2 × 3 ECM connection or a 2 s(2 × 3) ECM connection in series.

15.6.6 A 2 × 3/3 × 2 ECM DC–AC/AC–DC commutator and 2 × 2 ECM DC–DC commutator

An advantage of the hybrid-electric **drive-by-wire** (DBW **two-wheel-drive** (2WD) and/or **four-wheel-drive** (4WD) propulsion mechatronical control system is that a novel form of **external combustion engine** (ECE) or **internal combustion engine** (ICE) can be used, designed specifically for high-efficiency constant engine-shaft angular velocity and power operation and minimum emission (Fijalkowski 1997).

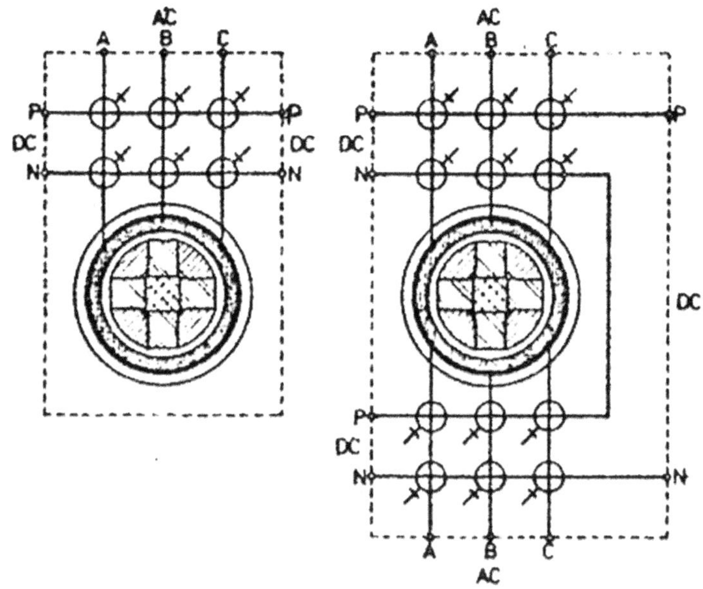

Figure 15.33. A single-phase ECM AC–AC and/or AC–DC/DC–AC commutator synchronous (magneto-electrically excited) dynamotor with a 2 × 3/3 × 2 or 2 s(2 × 3)/2 s(3 × 2) ECM commutator (Fijalkowski 1988).

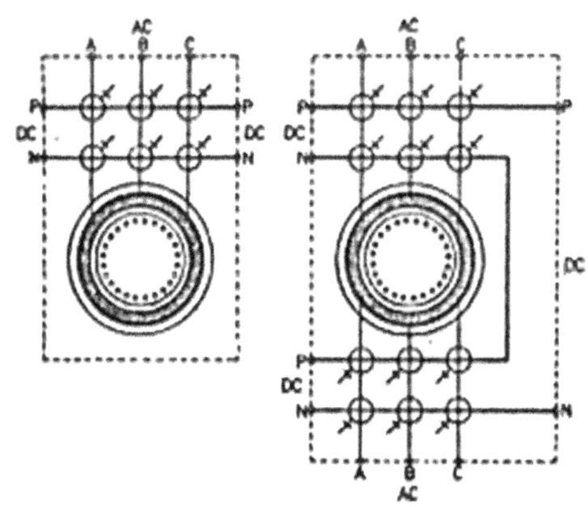

Figure 15.34. A single-phase ECM AC–AC and/or AC–DC/DC–AC commutator hyposynchronous (induction) dynamotor with a 2×3/3×2 or a 2(2×3)/2(3×2) ECM commutator and squirrel-cage rotor (Fijalkowski 1988).

The DBW 4WD propulsion mechatronical control system using an EM/ME **steered, motorised and/or generatorised wheel** (SM&GW) with ECM DC–AC/AC–DC commutator wheel-hub dynamotors will eliminate the need for axles and drive shafts, which will allow a separate suspension for each independently sprung SM&GW (figure 15.35). The rotating housing is made in the form of a wheel hub, designed to fit standard rim sizes on the market. The ECM DC–AC/AC–DC

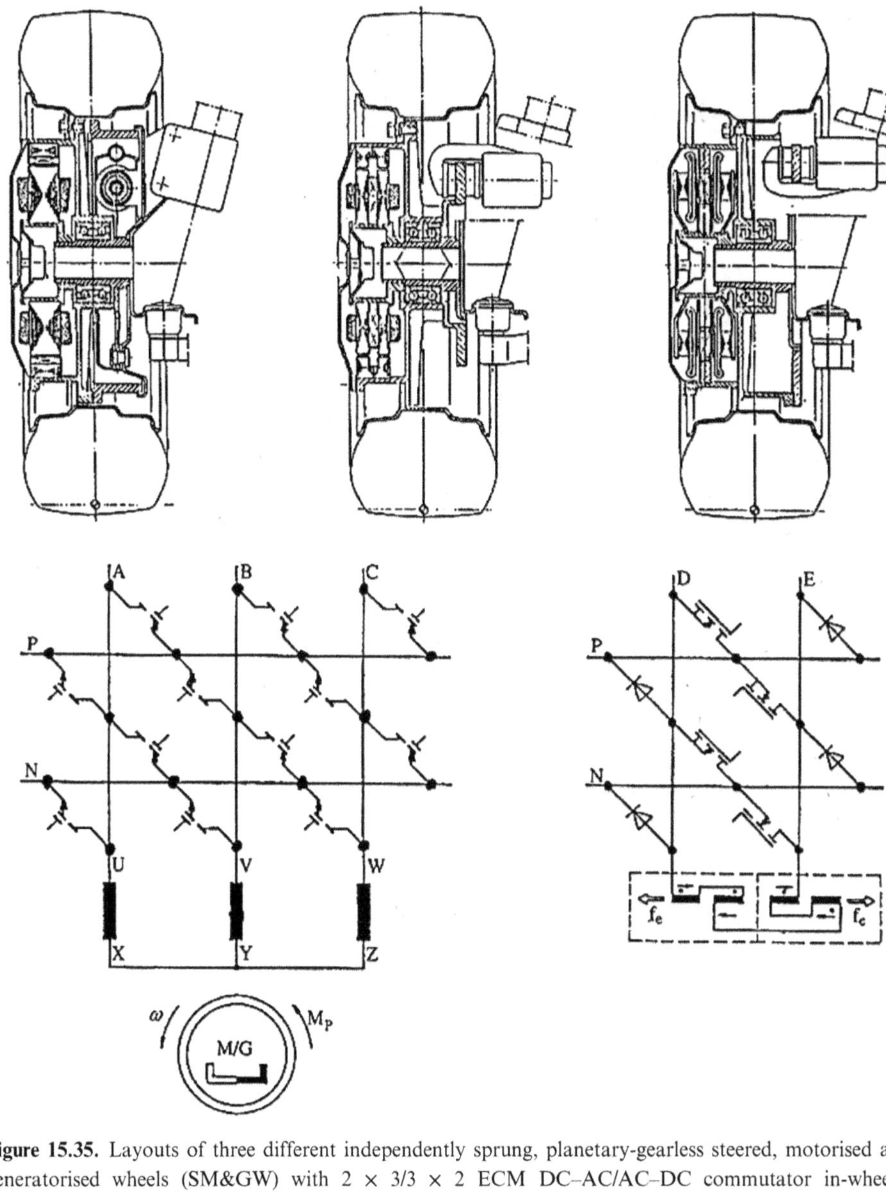

Figure 15.35. Layouts of three different independently sprung, planetary-gearless steered, motorised and/or generatorised wheels (SM&GW) with 2 × 3/3 × 2 ECM DC–AC/AC–DC commutator in-wheel-hub dynamotors and tubular, linear 2 × 2 ECM DC–DC commutator wheel-drum or wheel-disc brake-actuator motors (Fijalkowski 1997).

commutator wheel-hub motor's transmission offers complete freedom in the design of the **hybrid-electric vehicle** (HEV).

A DBW 2WD and/or 4WD propulsion mechatronical control system may provide the HEV with good climbing ability (up to $\pi/4$ rad on firm surfaces). The two SM&GWs on each side of the HEV's hull-wheeled unit may have an independent **absorb-by-wire** (ABW) **four-wheel-absorbed** (4WA) suspension.

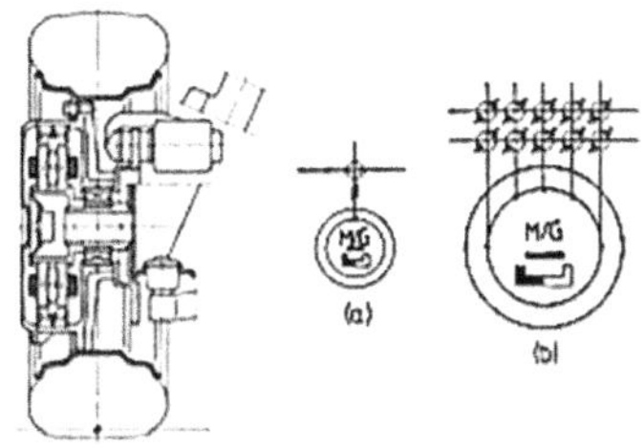

Figure 15.36. Layout of the steered, motorised and/or generatorised wheel (SM&GW) (A) and the physical model of the ECM DC–AC/AC–DC commutator reluctance and IPM magnetoelectrically excited in-wheel-hub motor/generator (B) (Fijalkowski 1984, 1990).

15.6.7 A 2 × 5/5 × 2 ECM DC–AC/AC–DC commutator

The two front and the two rear SM&GWs (figure 15.36) are driven and/or absorbed individually by four 2 × 5/5 × 2 ECM DC–AC/AC–DC commutator reluctance and IPM magnetoelectrically excited in-wheel-hub dynamotors, respectively, and the angular velocity of each SM&GW may be arbitrarily controlled by a driver-vehicle and a terrain-vehicle real-time expert system, incorporating a mathematical model following **fuzzy-logic** (FL) programmable and **neural-network** (NN) learning motion mechatronical control (Fijalkowski 1990, 2010).

15.6.8 A 2 × 3 ECM DC–AC/AC–DC commutator and 2 × 2 DC–DC commutator

The author considered a novel mobility and steerability concept for HEVs with EM differentials. The use of these novel triad hybrid mobility and triad hybrid steerability enhancing concept DBW 2WD and/or 4WD propulsion mechatronical control systems for HEV (figure 15.37) opens up wide possibilities for improving fossil and non-fossil fuel economy, cutting initial and whole-life DBW 4WD propulsion costs, protecting the environment and improving the distribution of terrain thrust (gross tractive effort), as well as keeping the net motion resistance of the DBW 4WD propulsion and the SM&GW sinkage (rut depth) low—not only by increasing the velocity of travel, but also by decreasing the rolling resistance of all the SM&GWs.

HEVs may be able to retain a sufficient level of mobility even if they lose one or more SM&GWs. Using EM differential torque and/or angular-velocity controls of the outer SM&GWs, as well as current and/or voltage controls of the inner SM&GWs, can increase the lateral motion control effect, especially when recovering from braking with the inner SM&GWs acting as the ECM AC–DC commutator in-wheel-hub generators, because the front gravitational forces on the HEVs become greater than the respective rear ones. At the same creep, this leads to greater horizontal (longitudinal and lateral) forces.

The DBW 4WD propulsion mechatronical control systems used on HEVs satisfy nearly all the same requirements that the running gear systems used on conventional vehicles have (Fijalkowski 1986, 1990, 1994, 2010):

- to apply an EM 'single-shaft' DBW 4WD propulsion mechatronical control system to a complete number of SM&GWs;

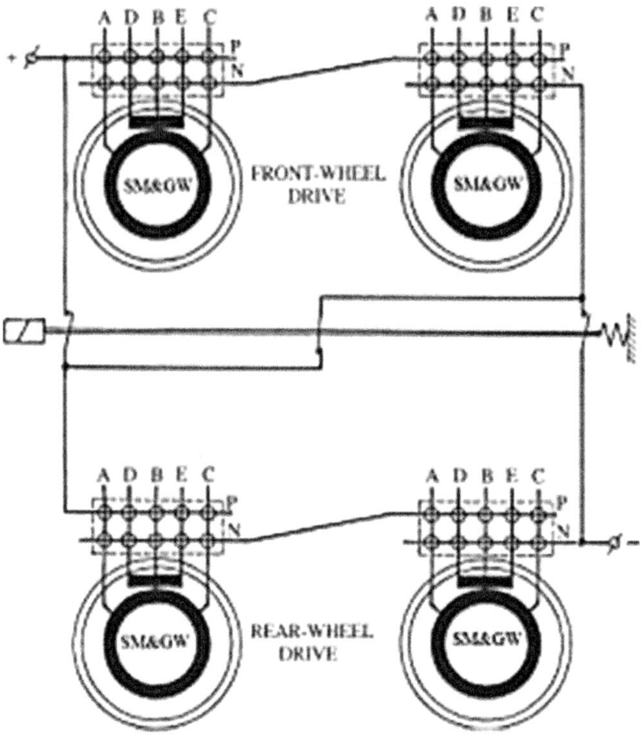

Figure 15.37. Elementary wiring connection for the axleless, full-time DBW 4WD transmission arrangement with the torque proportioning front-wheel drive (FWD) and rear-wheel drive (RWD), as well as centre-wheel drive (CWD) electro-mechanical (EM) differentials (Fijalkowski 1997, 2010).

- to allow a positive propelling (driving) torque to the outer side of the curve and a negative dispelling (braking) torque to the inner side, achieving their maximum value for pivot-skid steering;
- to occupy the minimum volume within the space envelope of the HEV;
- to distribute the mass of the HEV over a relatively spacious ground surface or soil area.

The requirement of the first feature may contribute to a very good soil performance for HEVs. The third feature may tend to conflict with requirement of the first feature.

15.6.9 A 2 s(2 × 5)/2 s(5 × 2) ECM DC–AC/AC–DC commutator

The layout of a particularly interesting planetary-gearless SM&GW with a wire-mesh tyre for the **Martian Roving Vehicle** (MRV) with a 2 s(2 × 5)/2 s(5 × 2) ECM DC–AC/AC–DC commutator reluctance and a magneto-electrically excited (unwound MSI single ring-shape outer-rotor and wound and IPM twin ring-shaped inner stator) in-wheel-hub motor/generator is shown in figure 15.38 (Fijalkowski 2001a, 2001b).

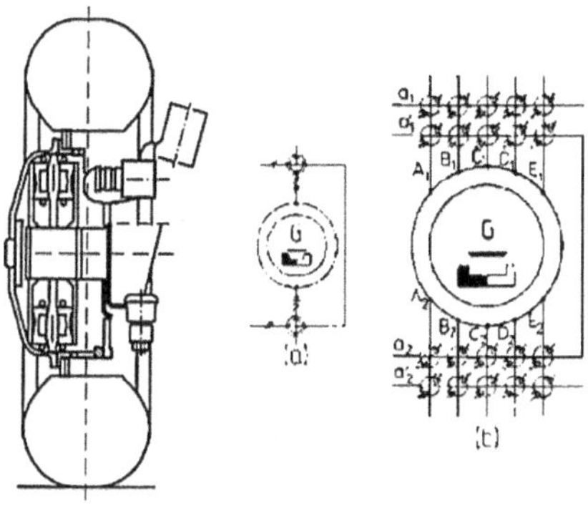

Figure 15.38. Planetary-gearless steered, motorised and/or generatorised wheel (SM&GW) with a 2 s(2 × 5)/2 s (5 × 2) ECM DC–AC/AC–DC commutator reluctance and a magnetoelectrically excited in-wheel-hub motor/generator (Fijalkowski 2001b).

It is evident that there are no windings on the **mild soft iron** (MSI) single ring-shaped outer-rotor, while there are no slip rings and there is no MCM DC–AC/AC–DC commutator with sliding-copper-segments and carbon brushes. What is not evident is the nature of the twin ring-shaped inner-stator five-phase windings. If the twin ring-shaped inner-stator is a six-pole, five-phase structure with a large number of 'teeth', then the single ring-shaped outer-rotor will also have teeth and it will be magnetised so that the south pole occupies one-half of the periphery, while the north pole occupies the other half. Both members afford a large number of opportunities for positional lock-up of the single ring-shaped outer rotor. Although the twin ring-shaped inner-stator only has six poles, and the single ring-shaped outer-rotor only has three poles, the presence of teeth on both members affords a large number of opportunities for positional lock-up of the single ring-shaped outer-rotor.

The 2 s(2 × 5)/2 s(5 × 2) ECM DC–AC/AC–DC commutator reluctance and the magnetoelectrically excited in-wheel-hub motor/generator with different pole numbers on the unwound MSI single ring-shaped outer-rotor, and the wound and IPM ring-shaped inner-stator, can provide traction if the stator five-phase winding phase-coils are sequentially energised. Here the unwound MSI single ring-shaped outer-rotor is made to 'chase' the sequentially switched wound and IPM twin ring shaped inner-stator's poles. As soon as magnetic alignment occurs, the attracting pole is de-energised. A variation of this basic scheme employs Hall-effect elements to directly sense the position of either the single ring-shaped outer-rotor itself, or a suitably magnetised ring on the shaft.

15.7 Electrical commutation matrixer commutators—a look into the future

The ECM commutator (conceived by the author in 1988) represents the 'superconductive ECM commutator' in which all the active (continuous bipolar electrical valves) and reactive elements (primary and secondary collectors, i.e. connections) of the ECM are manufactured in a super- and/or semiconducting single wafer of amorphous material in the shape of single chips (Fijalkowski 1988).

In 1987, the Nobel Prize in the field of physics was awarded to the Swiss physicist, K A Müller, and the German physicist, J G Bednorz, who both worked for IBM in Zürich. In early 1987, they discovered the properties of electrical-current conductivity (a rapid loss of resistance) in barium, lanthanum, copper and oxygen compounds at a temperature of 35 K. A little later, the American physicist, P Chu, who followed the path indicated by Müller and Bednorz, went beyond the 'magic' limit of the nitrogen boiling point temperature of 77 K.

Today, the electrical-current superconductivity phenomenon has already been registered even at the ambient temperature of the natural (human) environment. The methods for approaching electrical-current superconductivity have changed completely. Amorphous materials may be used similarly to mono- and polycrystalline conductors (e.g. metals), non-conductors and superconductors (Kazmerski 1980).

The electrical-current conductivity of amorphous germanium and silicon can be controlled easily by means of adequate doping, and the remaining properties of these materials are in principle analogous to those existing in their mono- and polycrystalline equivalents. Thus, it is possible to perform junctions of the *p–n* type from amorphous germanium and silicon, which constitute the basic elements of the majority of semiconductor electrical valves.

Cheap and easy to attain, amorphous material can be used for the construction of both active (super- and/or semiconductor amorphous continuous bipolar electrical valves) and reactive elements (superconductor amorphous primary and secondary collectors, i.e. connections).

If the transistor revolution took place due to the discovery of the properties of mono- and polycrystalline semiconductors, the next revolution in electronics will make use of semiconductor amorphous materials.

From the point of view of industrial and traction technology, amorphous materials are characterised above all by their enormous heterogeneity of physical properties and by the fact that they require no complex technological processes.

Amorphous materials are usually manufactured in the shape of glasses by means of rapid quenching of the material from the liquid phase in the shape of thin films, which are vacuum evaporated on adequate quenched substrates. These technologies are distinctly competitive compared to the manufacturing technologies for mono- and polycrystalline materials.

ECM AC–AC and AC–DC/DC–AC commutators with bipolar ECMs, of which primary and secondary collectors are realised on ceramic superconductors, are contributing to one more revolution in the history of both macroelectronics and ECM AC–AC and AC–DC/DC–AC commutator dynamotors. Although the ECM AC–AC and AC–DC/DC–AC commutator with a bipolar ECM, realised on continuous bipolar ovonics, is made of glassy material that does not have a crystalline structure, it is a super- and/or semiconductor static converter, which provides a solution to the problem of its cooling.

Continuous bipolar ovonics are characterised by the short time-lag of the transition between the on state (conduction-state) of the electrical valve and its off state (non-conductive-state) of 0.1 μs. The ratio of the off-state to the on-state resistance yields 10^7 and more.

The triggering signal applied to the triggering gate may have an arbitrary polarisation. Continuous bipolar ovonics can operate at a frequency below 500 kHz and at an ambient temperature range up to 573 K (Ovshinsky 1958; Young 1961; Cohen *et al* 1969; Ovshinsky 1970; Holmberg and Shaw 1974; Ovshinsky and Adler 1978; Kazmerski 1980).

Continuous bipolar ovonics, under the influence of some control voltage, the value of which depends upon the chemical composition and dimensions of the glassy wafer, undergo a metamorphosis from a resistor with a resistance of 10^7 Ω into a conductor with a resistance of 10^{-2} Ω.

After the metamorphosis into a conductor, a relatively high electrical current starts to flow in the ovonic. As long as the electrical current is flowing, the glassy wafer remains a conductor. It maintains this ability without any change, even when the electrical current changes its instantaneous value over a small range. However, a single power pulse opposite to the primary is sufficient for the amorphous conductor (without crystalline structure) to undergo a metamorphosis back into a resistor with a high resistance.

Since the electrical current practically does not flow in this resistor, the heat is not released in it. However, if the ovonic metamorphosises itself into a conductor and conducts the electrical current, its resistance is so insignificant that the quantity of heat that is released can be neglected.

In this way, the dream of electronics scientists and engineers to make a great number of continuous electrical valves on the single carrying wafer and to realise in practice uni- and/or bipolar ECMs for the ECM commutators becomes realisable.

15.8 Conclusion

Macroelectronics is the study of ECM commutators for the conversion and control of electrical energy. The technology is a critical part of energy infrastructure and supports almost all important electrical applications. For macroelectronics design, one considers only those matrixers and devices that, in principle, introduce no loss and can achieve near-perfect reliability. The two key characteristics of high efficiency and high reliability are implemented with ECMs, supplemented with energy storage. ECMs in turn can be organised as electrical-valve matrixers. This facilitates their analysis and design.

In a macroelectronic continuous dynamical system, the three primary challenges are the hardware problem of implementing an ECM, the software problem of deciding how to operate that ECM, and the interface problem of removing unwanted distortion and providing the user with the desired clean electrical-energy source. The hardware is implemented using special types of power super- and/or semiconductors. These include several types of SiC and AsGa diodes, *n–p–n* and *p–n–p* transistors, especially MOSFETs, IGBTs, RB IGBT, OT IGBTs and HV IGBTs, and several types of thyristor, especially SCRs, PPTs, PPCSTs and GTOs. The software problem can be represented in terms of commutation (switching) functions. The frequency, duty ratio and phase of the commutation functions are available for operational purposes. The interface problem is addressed by means of lossless filters. Most often, these are lossless LC passive filters to smooth out ripples or reduce harmonics. More recently, active filters have been applied to make dynamical corrections in power conversion waveforms.

For the time being, it is difficult to pronounce optimistic opinions on the prospects for practical applications of ECM commutators. The author believes that scientists should not promise users too much. As an example to put users into an optimistic mood, one ought to mention the invention of the transistor by J Bardeen, W G Brattain and W B Shockley in 1948. At that time, people assumed that the application of transistors resolved itself into the replacement of electron lamps, especially triodes. Nobody could foresee that soon after this invention the unprecedented miniaturisation of electronics and the creation of EICs with larger scales of integration would follow. In the author's opinion, such a thing can occur with ECM commutators, but for the time being it is impossible to foresee some of the future practical applications.

The results of the invention of ECM commutators and the R&D work carried out in this field were that solid-state physics began life anew. Not long ago, people believed that in the field of power electronics everything had already been invented, whereas this is only half-true. Recently, there has been enormous demand for R&D work in the field of solid-state physics.

Scientific research concerning ECM and commutators will be transformed into a great race by scientific teams from great world consortia and its results, which will revolutionise the universe of macroelectronics, may be expected at any moment. Very similar races by scientific teams have already taken place today in the field of solid-state physics, where a series of discoveries concerning high-temperature electric-current conductivity in materials of a new kind took place.

These particular experiments terminated in the invention of ECM commutators with bipolar ECMs, realised on continuous bipolar ovonics; hybrid transistors with heterojunctions; SUPER-HETs with a superconductive base; QUITs; and very high-speed SITs.

These ECM commutators will be used in newly designed energy-saving MMD electrical machines with the application of high-temperature superconductors, for example, made from Y–Cu–Ba–O ceramics. Superconductivity is an exciting and rapidly developing technology, especially since the recent discovery of high T_C materials.

The basic technology and performance of IECs based on Nb Josephson junctions were implemented, and then followed by an assessment of the prospects for IECs incorporating high-T_C materials. These include uni- and bipolar semiconductor triodes (three-terminal electrical valves) and hybrid semiconductor/superconductor electrical valves, with the challenge being to compete successfully with current semiconductor technology, which is itself developing rapidly. ECM commutators are now indispensable in modern MMD, MEM, MFD and MPD electrical machine engineering, and for energy converting processes in industrial and traction propulsion complex dynamical hypersystems. Compared with the previous polyvalvular technology, ECM commutators (static converters) are more practice-oriented, especially from the point of view of operating convenience and the space requirements of ultra-modern semiconductor/superconductor ECM techniques.

A major step forward has been taken within electrical machinery technology with the development of macroelectronic commutators, called 'ECM commutators' by the author. The ECM commutator is a new product family based on the ECM and the EIC computer controller. The latest advances made within the field of macroelectronic components, such as powerful ECM commutators, and in the field of 64-bit, 128-bit and 256-bit microprocessors, have been exploited during the R&D work.

Components in the form of hardware and software have been developed with the aid of these new macro-, meso-, micro- and nanoelectronic components, and adapted to new modern control system structures.

Today, the ECM on its own may be used for all the new-concept integral ECM AC–AC. AC–DC–AC, AC–DC/DC–AC, DC–DC and DC–AC–DC commutator dynamotor's industrial and traction propulsion complex dynamical hypersystems, previously covered by several earlier modular dynamical hypersystems. The method used to build industrial and traction propulsion complex dynamical hypersystems has varied over the years, depending on the types of components that were available.

A changeover from extremely distributed industrial and traction propulsion complex dynamical hypersystems, where each object had its own ECU, to modern integrated industrial and traction propulsion complex dynamical hypersystems took place. The torque and speed control functions were grouped together in ECM commutator dynamotors to provide central access to all information.

The world must solve major challenges to transform and control electrical energy in an efficient way. Examples of this can be seen in aerospace, automotive and railway transportation applications, and in renewable energy and industrial processing applications. These problems can be solved using ECM AC–AC, AC–DC–AC, AC–DC/DC–AC, DC–DC and DC–AC–DC commutators based on modern power super- and/or semiconductor amorphous electrical valves.

The ideal ECM AC–AC, AC–DC–AC, AC–DC/DC–AC, DC–DC and DC–AC–DC commutator in many of these applications might have the following characteristics:

- compact design with a good power-to-mass ratio;
- regeneration capability;
- operation with unity power factor;
- sinusoidal input and output currents.

ECM AC–AC, AC–DC–AC, AC–DC/DC–AC, DC–DC and DC–AC–DC commutators can fulfil all of these characteristics, and this is the reason for the tremendous interest in this matrix topology.

Over the last decade, many advances in the development of this matrix topology were presented, including industrial applications up to the megawatt (MW) level.

The use of ECM AC–AC, AC–DC–AC, AC–DC/DC–AC, DC–DC and DC–AC–DC commutators in real applications and the challenges that these applications present is very topical and important.

This book will present the most recent advances with topics to the macro-, meso-, micro- and nanoelectronics community. These include, but not limited to, the following.

- ECM AC–AC, AC–DC–AC, AC–DC/DC–AC, DC–DC and DC–AC–DC commutators for aerospace, automotive and railway transportation applications, and renewable energy and industrial applications.
- ECM AC–AC, AC–DC–AC, AC–DC/DC–AC, DC–DC and DC–AC–DC commutator derived matrix topologies.
- New super- and/or semiconductor amorphous electrical valves for use in ECM AC–AC, AC–DC–AC, AC–DC/DC–AC, DC–DC and DC–AC–DC commutators.
- Power quality issues and ECM AC–AC, AC–DC–AC, AC–DC/DC–AC and DC–DC and DC–AC–DC commutator reliability and stability.
- Implementation of **artificial intelligence** (AI) current commutation strategies in application examples.
- New control/modulation methods for ECM AC–AC, AC–DC–AC, AC–DC/DC–AC, DC–DC, DC–AC–DC, DC–DC or DC–AC–DC commutator applications, including **field-oriented control** (FOC), **space vector modulation** (SVM), **direct torque control** (DTC), **passively based control** (PBC) and **predictive control** (PC).
- Comprehensive comparative evaluation of the ECM AC–AC and AC–DC/DC–AC commutator concepts versus the ECM AC–DC–AC or DC–AC–DC commutator concepts, with DC or AC link energy storage components, respectively.

However, the practical applications of the 3 × 3 ECM AC–AC, commutator, as of now, are very limited.

The main reasons are:

- non-availability of bilateral fully controlled monolithic electrical valves capable of high-frequency operation;
- complex control law implementation;
- an intrinsic limitation of the output-input voltage ratio;
- commutation (Case 2000) and protection of the switches.

To date, the electrical valves are assembled from existing continuous devices, resulting in increased costs and complexity, and only experimental ECMs with a capacity well below 1 MVA have been built. However, with advances in device technology, it is hoped that these problems will eventually be solved, and 3 × 3 ECM AC–AC super or static commutators will not only replace naturally commutated cyclo-converters in all the applications, but will also take over from the PWM rectifiers/inverters.

In Mahlein *et al* (1999), it was shown that with **space-vector** (SV) PWM control using over-modulation, the voltage transfer ratio may be increased to 1.05 at the expense of more harmonics and large filter capacitors.

Improvements in devices and advances in control concepts have led to steady improvements in ECMs and macroelectronic continuous dynamical systems. This is driving a tremendous expansion of their application. For instance, PCs would be unwieldy and inefficient without macroelectronic DC energy supplies. Portable communication devices and computers would be impractical. High-performance HVDC transmission links, ME/EM generator/motor (voltage/torque) controls, lighting systems and a wide range of industrial controls depend on macroelectronics.

In the near future, one can expect strong growth in automotive applications, in DC energy supplies for communication systems, in portable applications and in high-end ECMs for advanced microprocessors.

During the next generation, we will reach a time when almost all electrical energy is processed through macroelectronics somewhere in the path from generation to end use.

References

Alesani A and Venturini M G B 1989 Analysis and design of optimum-amplitude nine-switch-direct AC–AC converters *IEEE Trans. Power Electron.* **4** 101–12

Alexanderson E F W and Mittag A H 1934 The 'Thyratron' Motor *Electr. Eng.* **53** 1817

Altun H 2005 Application of a matrix converter in a slip energy recovery drive system *Eng. Modelling* **18** 3–4, 69–80

Anon 1950 Professor AN Larionov k 60-letiiu so dnia rozhdeniia i 30-letiiu nauchno-pedagogicheskoi deiatelnosti *Elektrichestvo* **9** (In Russian)

Casadei D, Serra G, Tani A and Zarri L 2002 Matrix converter modulation strategies: a new general approach on space-vector representation of the switch state *IEEE Trans. on Ind. Electron.* **49**

Case M J 2000 The commutation process in the matrix converter *Conf. Proc. EPE-PEMC 2000 (Kosice, Slovakia)* **2** 109–12

Cheron Y, Foch H and Perin A 1985 An improved direct frequency changer using power transistors *Proc. First European Conf. on Power Electronics and Applications* **1** 1.123–8

Cohen M H, Fritzsche H and Ovshinsky S R 1969 *Phys. Rev. Lett.* **22** 1065

Fijalkowski B 1962 *Sterowanie maszyny wyciagowej pradu stalego z przeksztaltnikami rtęciowymi (Control of the DC mine winder using mercury-arc converters)* Zeszyty naukowe Politechniki Szczecinskiej, Nr 36, Prace Monograficzne, Nr 9 (Politechnika Szczecinska: Szczecin) p 59 (In Polish)

Fijalkowski B 1964 Sterowany przeksztaltnikowy naped elektryczny i perspektywy jego rozwoju (Converter drive and perspectives of its development) *Zeszyty Naukowe Politechniki Szczecinskiej* Elektryka VI, Nr 53, 1964, pp 6–34 (In Polish)

Fijalkowski B 1967 Zastosowanie rachunku holorowego w teorii pradow przemiennych (Application of a holor calculus in the theory of alternating currents) *Przeglad elektrotechniczny (Electrotechnics Review)* Rok XLIII, Zeszyt 2, Warszawa (In Polish)

Fijalkowski B 1973 Kierunki rozwoju napedu elektrycznego lokomotyw (Development trends of the electric drive of locomotives) *Materiały konferencyjne—I Krajowa Konferencja—Pojazdy*

Szynowe (Krakow: Instytut Pojazdow Szynowych Politechniki Krakowskiej) pp 109–128 (In Polish)

Fijalkowski B 1984 Electronic commutator AC/DC motor-driven tracked all-terrain vehicles with extremely high mobility *Proc. 8th Int. ISTVS Conf.: The performance of off-road vehicles and machines (Churchill College, Cambridge University, 5–11 August 1984) J. Terramech* 1045–63

Fijalkowski B 1985a New concept MACRO- and MICRO-electronics cradled dynamometer systems for testing of combustion engines *Proc. ISATA 85: Int. Symp. on Auto-motive Technology and Automation* (Graz, Austria) **2** 587–616

Fijalkowski B 1986 Future hybrid electromechanical very advanced propulsion systems for civilian wheeled and tracked all-terrain vehicles with extremely high mobility *Proc. 1986 EVS-8: The Eight International Electric Vehicle Symp. (Washington DC)* pp 428–43

Fijalkowski B 1987 *Modele matematyczne wybranych lotniczych i motoryzacyjnych mechano-elektro-termicznych dyskretnych nadsystemów dynamicznych* (*Mathematical Models of Selected Aviation and Automotive Continuous Dynamical Hyperystems*) (Krakow: Monografia 53 Politechnika Krakowska imienia Tadeusza Kościuszki) p 276 (In Polish)

Fijalkowski B 1988a Choice of hybrid propulsion systems—wheeled city and urban vehicle—tracked all-terrain vehicle. Part II *Electr. Vehicle Dev.* **7** 31–4

Fijalkowski B 1988b Odnochipnyi makrokommutator—Tehnologiia v stadii razvitiia—Uzhe nie teoriia, no eshche ne promyshlennost' *Zbornik Prednasok z VIII. celostatnej konferencje so zahranicnou ucastou Elektricke Pohony A Vykonova Elektronika* (Kosice, Kosicka Bela, Czecho-Slovakia: I. DIEL) pp 182–94 (In Russian)

Fijalkowski B 1988c Single-chip static-commutator—an indispensable module of new concept AC and/or DC dynamotors for automotive very advanced propulsion systems *Proc. EVS9: The 9th Int. Electric Vehicle Symposium* (*Toronto, 13–16 November 1988*) EVS88-031 1-6

Fijalkowski B 1988d Amorphous and polycrystalline semiconductor single-chip super-commutator—an indispensable module of new concept AC and/or DC electrical machines *Proc. Int. Conf. and Intensive Tutorial Course on Semiconductor Materials (New Delhi)* 8–16

Fijalkowski B 1990 Very advanced propulsion spheres for high-speed tracked vehicles *Proc. of the 10th Int. Conf. of the International Society for Terrain-Vehicle Systems (ISTVS) (Kobe)* **3** 783–97

Fijalkowski B 1994 Front- or rear-wheel-drive and inter-wheel-drive electromechanical differentials for full time 4WD × 4WB × 4WA × 4WS supercars with steerable & motorized wheels *Proc. ISATA 94: Int. Dedicated Conf. on ELECTRIC, HYBRID & ALTERNATIVE FUEL VEHICLES and SUPERCARS (ADVANCED ULTRALIGHT HYBRIDS) in conjunction with the 27th Int. Symp. on Automotive Technology and Automation, Aachen (Germany, 31 October–4 November 1994)* Paper No. 94UL024 681–8

Fijalkowski B T 1982 Trigistor frequency changer-controlled toothed gearless propulsion systems for electric and hybrid vehicles *Proc. Europe's Int. Conf. on Electric Road Vehicle Systems DRIVE ELECTRIC (Amsterdam)* 554–67

Fijalkowski B T 1985b On the new concept hybrid and bimodal vehicles for the 1980s and 1990s *Proc. DRIVE ELECTRIC Italy '85 (Sorrento)* pp 4.04.2–8

Fijalkowski B T 1985c On the new concept MMD electrical machines with integral macro-electronic commutators—Development for the future *Proc. First European Conf. on POWER ELECTRONICS AND APPLICATIONS (Brussels, 16–18 October 1985)* **2** pp 3.377–3.384

Fijalkowski B T 1997 Intelligent Automotive systems: development in full-time chassis motion spheres for intelligent vehicles *Advanced Vehicles and Infra-structure Systems—Computer Applications, Control and Automation* ed C O Nwagboso (New York: Wiley) pp 125–142

Fijalkowski B T 2001a Articulated triad Martian roving vehicle *The Fourth Int. Mars Society Convention (Stanford, CA)* (Lakewood, CO: The Mars Society Inc.)

Fijalkowski B T 2001b A concept of the All-Electric Pressurized Articulated Triad Martian Roving Vehicle *On To Mars 2* ed R M Zubrin and F Crossman (Lakewood, CO: The Mars Society)

Fijalkowski B T 2010 *Automotive Mechatronics: Operational and Practical Issues.— Vol 1 (International Series on Intelligent Control, and Automation Science and Engineering vol. 47)* (Berlin: Springer)

Fijalkowski B T 2011 *Automotive Mechatronics: Operational and Practical Issues.— Vol. 2 (International Series on Intelligent Control, and Automotive Science and Engineering vol 52)* (Berlin: Springer)

Gyugyi L and Pelly B R 1976 *Static Power Frequency Changers: Theory, Performance and Application* (New York: Wiley)

Holmberg S N and Shaw M P 1974 *Proc. of the 5th Int. Conf. Amorphous Liquid Semiconductors* (London: Taylor and Francis)

Hoseman G, Krolikowski L, Raatz E, Shapilov E D, Turek-Kwiatkowska L and Vieselovsky O N 1999 Zhizn i delyatelnost Mikhaila Dolivo-Dobrovolskogo—vydaiushchegosia inzhenera elektrika evropeiskogo masshtaba *Proc. of the 4th Int. Conf. on Unconventional Electromechanical and Electrical Systems (St Petersburg)* **1** 59–70 (In Russian)

Huber L and Borojevic D 1995 Space vector modulated three-phase to three-phase matrix converter with input power factor correction *IEEE Trans. Ind. Appl.* **31** 1234–46

Kazmerski L L (ed) 1980 *Polycrystalline and amorphous thin films and devices* (New York: Academic)

Kern E 1938 Der Dreiphasenstromrichtermotor und seine Steuerung bei Betrieb als Umkehrmotor *ETZ* 59 467 (In German)

Kittel Ch 1956 *Introduction to Solid State Physics* (New York: Wiley)

Kloninger M J 1932 Applications des redresseurs á grilles polarisées. Séance S.F.E. 6 Fevrier *Bulletin S.F.E. Fev.* (In French)

Kostenko M and Piotrovsky L 1968 *Electrical Machines* 2 vols (Moscow: Mir Publishers) vol 1 chs 5 and 6 pp 201–67; vol 2 ch 27 649–54

Li Y and Choi N-S 2009a Carrier based pulse width modulation for matrix converter *Conf. Record of IEEE Applied Power Electronics Conf.* 1709–15

Li Y, Choi N-S and Han B-M 2009b DDPWM based control of matrix converters. *Power Electron.* **9** 535–43

Li Y, Choi N S, Han B-M, Kim K-M, Lee B and Park J-H 2008 Direct duty ratio pulse width modulation method for matrix converter *Int. J. Control Autom. Syst.* **6** 660–9

Mahlein J, Simon O and Braun M 1999 A matrix converter with space vector control enabling overmodulation *Conf. Proc. EPE 99 (Lausanne)* pp 1–11

Mott M F and Davis E A 1971 *Electronic Processes in Mono-crystalline Materials* (London and New York: Oxford University Press)

Neft C L and Schauder C D 1988a Theory and design of a 30-hp matrix converter *Conf. Record of the 1988 Industry Applications Society Annual Meeting (Pittsburgh, PA)* (IEEE Cat. No. 88CH 2565-0)

Neft C L and Schauder C D 1988b Theory and design of a 30-hp matrix converter *Proc. of the IEEE Industrial Application Society Annual Meeting (Pittsburgh, PA)* pp 248–53

Neiman L R 1972 Akademik Vladimir Fedorovich Mitkevich ego trudy i progressivnye idei. K 100-letiiu so dnia rozhdeniya *Elektrichestvo* **8** (In Russian)

Ovshinsky S R 1958 *Phys. Rev. Lett.* **21** 1450

Ovshinsky S R 1970 *J. Non-Cryst. Solids* **2** 99

Ovshinsky S R and Adler D 1978 *Contemp. Phys.* **10** 100

Oyama J, Higuchi T, Yamada E, Koga T and Lipo T 1989 New control strategy for matrix converter *Record of the 20th Annual IEEE Power Electronics Specialists Conf.—PESC '89 (Milwaukee, WI)* Cat. No.89 CH2721-9 pp 360–7

Pollak Ch 1895 English Patent (24,398 A.D)

Pollak Ch 1896 *Elektrische Flüssigkeitskondensator mit Aluminium-elektroden*. Kaiserliches Patentamt, Patentschrift No 92564, Klasse 21: Elektriche Apparate, Patentirt im Deutschen Reiche vom 14. Januar 1896 ab. Ausgaben den 19. Mai 1897.

Rémy E 1954 Le collecteur magnétique *Bulletin E.S.E.,Nõel 1954, et - Dynamo sans balais. Revue court-circuit de l'amicale E.S.E* (In France)

Rodriguez J 1984 A new control technique for AC–AC converter *Proc. of the 3rd IFAC Symp. on Control in Power Electronics and Electrical Drives (Lausanne)* p 203

Schilling W 1940 *Die Wechselrichter und Umrichter—Ihre Berechnung und Arbeitweise*Verlag von R Oldenbourg) (In German)

Seitz P 1940 *Modern Theory of Solids* (New York: McGraw-Hill)

Strzelecki R M and Benysek G (ed) 2008 *Power Electronics in Smart Electrical Energy Networks* (Berlin: Springer)

Tenti P, Malesani L and Rossetto L 1992 Optimum control of N-input K-output matrix converters *IEEE Trans. Power Electron.* **7** 707–13

Tutaj J 1996 Estymacja jakosci energii samochodowych komutatorowych pradnic mechano-elektrycznych pradu stalego z komutatorami elektronicznymi (Energy quality estimation of DC mechano-electrical generators with electronic commutators) *PhD Thesis* (Promotor: Bogdan Fijalkowski, Automotive Vehicles and Combustion Engines Institute, Cracow University of Technology, Krakow) (In Polish)

Tutaj J 2012 *Ujecie systemowe dynamiki wielofunkcyjnego pradnico-rozrusznika silnika spalino-wego pojazdu samochodowego (Dynamical Systems Approach of a Multifunctional Generator/ Starter for the Combustion Engine of an Automotive Vehicle)* Scientific Editor: Bogdan Fijalkowski, Seria Mechanika, Monografia, 409, Wydawnictwo Naukowe Politechniki Krakowskiej im. Tadeusza Kosciuszki, Krakow (In Polish)

Wang B and Venkataramanan G 2006 Six Step Modulation of Matrix Converter with Increased Voltage Transfer Ratio *Proc. of the 37th IEEE Power Electronics Specialists Conf. (Jeju, Korea)* **49** 930–6

Wheeler P W, Clare J C, Katsis D, Empringham L, Bland M and Podlesak T 2004 Design and construction of a 150 kVA matrix converter induction motor drive *Proc. of the 2nd IEE Int. Conf. on Power Electronics Machines and Drives (Edinburgh)* **498**

Willis C H 1933 A study of the thyratron commutation motor *Gen. Elect. Rev.* 76

Yoon Y-D and Sul S K 2006 Carrier-based modulation technique for matrix converter *IEEE Trans. Power Electron.* **21** 169–703

Young L 1961 *Anodic Oxide Films* (New York: Academic)

Zagorsky V T 1969 Tekhniko-ekonomicheskiye pokazateli neposredstvennykh preobrazo-vateley chastoty s prinuditelnoy kommutaciey *Elektrichestvo* pp 35–41 (In Russian)

Zavalishin D A and Vegner O G 1936 Ionnye skhemy ventilnykh dvigateley *Elektrichestvo* **3** 1936 (In Russian)

Ziogas P D, Khan S I and Rashid M H 1986 Analysis and design of forced commutated cycloconverter structures with improved transfer characteristics *IEEE Trans. Ind. Electron.* **33** 271

IOP Publishing

Mechatronics
Dynamical systems approach and theory of holors
B T Fijalkowski

Chapter 16

Electrical commutation matrixer keyboards for computers

People who are really serious about software should make their own hardware.
Alan Kay

Information: the negative reciprocal value of probability.
Claude Shannon

16.1 Introduction

In this chapter, **electrical commutation matrixer** (ECM) keyboards for computers are introduced. Computers are electronic machines that take input from a human user, process the given input and generate output in the form of useful information. ECM keyboards are the most commonly used input devices. One can use ECM keyboards to type data and text and execute commands (see figure 16.1).

Figure 16.2 illustrates a typical laptop computer ECM keyboard layout, where all the common keys found on the large desktop computer ECM keyboard have been miniaturised to laptop size. The design is intended to let human users type without the risk of friction from their fingers rubbing together.

The layout of the keys was borrowed from the typewriter with a number of new keys added. Various ECM keyboards have different extra keys. An ECM keyboard for a Windows computer has different extra keys than one for an Apple computer.

Laptop computers have fewer keys, but have an additional 'Fn' key that can be combined with other keys to replace the missing ones. ECM keyboards can be customised for other languages, too.

ECM keyboards (as shown in figures 16.1 and 16.2) are the most common input device. Several kinds of ECM keyboards are available, but they resemble each other, with minor variations. The ECM keyboard in most common use is the QWERTY

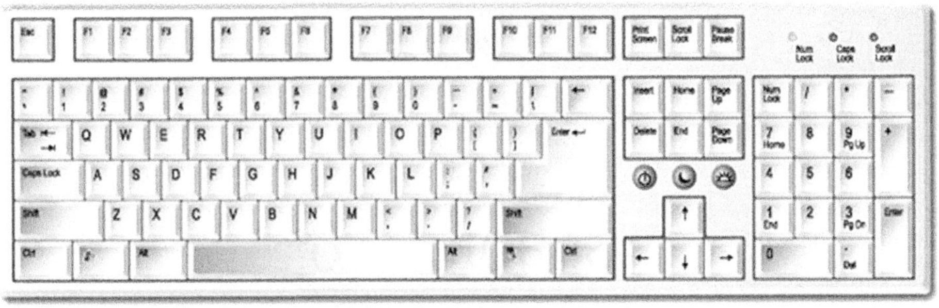

Figure 16.1. Standard MS Windows keyboard layout—Windows Natural Keyboard.

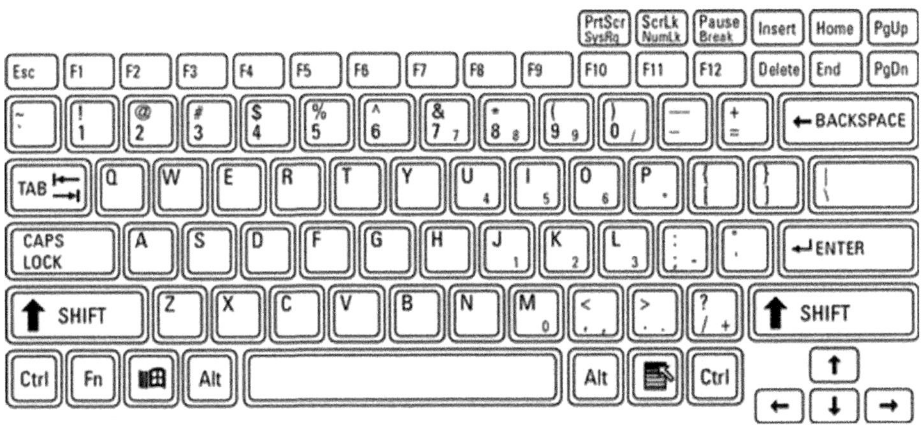

Figure 16.2. Typical laptop keyboard layout.

board. Generally, a standard keyboard has 104 keys. In these ECM keyboards, the cursor control keys are duplicated to allow easier use of the numeric pad.

A virtual ECM keyboard is actually a key-in device, roughly the size of a fountain pen, which uses highly advanced laser technology, to project a full-sized ECM keyboard onto a flat surface. Since their invention, computers have undergone rapid miniaturisation. Disks and components grew smaller in size, and the only component that remained the same for decades was the ECM keyboard. Since the miniaturisation of a conventional ECM keyboard is very difficult, one opts for a virtual ECM keyboard. Here, a camera tracks the finger movements of the typist to obtain the correct keystroke. A virtual ECM keyboard is one that a human user operates by typing on or within a wireless or optical detectable surface or area rather than by depressing physical keys.

ECM keyboards have changed very little in layout since their introduction. In fact, the most common change has simply been the natural evolution of adding more keys that provide additional functionality. There are various types of ECM keyboards, such as the 101-key enhanced keyboard, the 104-key Windows keyboard, the 82-key Apple standard keyboard and the 108-key Apple extended keyboard. Portable computers, such as laptops, quite often have custom ECM keyboards that have slightly different key arrangements from a standard ECM keyboard. In addition, many system manufacturers add specialist buttons to the standard layout. A standard ECM keyboard consists of the following commutation groups of electrical valves acting as keys.

- **Alphanumerical keys.** The alphanumeric or 'typewriter' keys include the number keys and alphabet keys. These keys are arranged in the same style as in normal typewriters, popularly known as the 'QWERTY' layout.
- **Functional keys.** Arranged in a row on the top of the ECM keyboard, they help perform specific tasks, such as searching for a file or refreshing a web page.
- **Central keys.** Used for controlling the movement of the cursor and screen display, they include arrow keys and modifier keys, such as 'ALT', 'CTRL' 'SHIFT'.
- **Numerical keypad.** Located on the right side of the keyboard, this looks like a calculator's keypad.
- **Special purpose keys.** 'Delete' or 'Del'; 'End'; Escape'; 'Home'; 'Insert' or 'Ins'; 'Page Down' or 'PgDn'; 'Page Up or 'PgUp'; 'Pause'; 'Print Screen'; 'Spacebar'; and 'Tab'.

Many changes to the conventional ECM keyboard design are attempts to make them safer or easier to use. For example, some human users have associated increased ECM keyboard use with repetitive stress injuries like carpal tunnel syndrome, although scientific studies have produced conflicting results. They are conventional ECM keyboards.

- **Das keyboard** is a completely black keyboard with weighted keys that require more pressure from a human user's strongest fingers and less pressure from the weaker ones.
- The **Virtual Laser Keyboard** projects a representation of an ECM keyboard onto a flat surface. When used successfully, a human user's fingers pass through the beam of **infrared** (IR) light above the projected surface and a sensor interprets this as a keystroke.
- The **True-Touch Roll-Up keyboard** is flexible and can be rolled up to fit in a backpack or bag.
- The **Ion Illuminated keyboard** uses light-emitting diodes or electroluminescent film to send light through the keys or the spaces between keys.
- The **Optimus keyboard** has **organic light-emitting diodes** (OLED) in the keys. Users can change which letter, command or action each key represents, and the OLED can change to display the new information.

With the exception of the Virtual Laser Keyboard, which has its own sensing system, each of these ECM keyboards uses the same technology as conventional ECM keyboards do to communicate with the computer. We shall look at that technology next.

16.2 How electrical commutation matrixer keyboards for computers work

A computer ECM keyboard is a hardware device that functions in accordance with the instructions made by the human user. It comprises ECMs, electrical valves (switches) and processors that help in transferring keystroke messages to the computer.

In the contemporary technology-driven world, everyone is aware of computing and the use of computer ECM keyboards. Everyone knows that the ECM keyboard is an input device that functions in accordance with the instructions of the human user.

Computer ECM keyboards are used to perform various tasks, such as typing on a word processor or text editor, accessing menus and playing games. Although pressing a key corresponds to typing a letter, sign or symbol, it may also represent computer commands. For example, pressing the 'F5' key is a command for refreshing. For some characters, it is necessary to press and hold more than one key at the same time.

Speaking about the history of computer ECM keyboards, they were modeled in the 1940s based on the technology of the typewriter keyboard. Similar to a typewriter keyboard, a computer keyboard consists of a set of keys engraved with symbols and letters. However, the productivity of working on a computer ECM keyboard is much greater than that for working a typewriter keyboard.

In general, there are 80–110 keys on a computer keyboard. The keys may vary depending upon the brand and the type of **operating system** (OS). Nevertheless, the shape, size and spacing of the keys are almost the same for all ECM keyboards. In addition, the layout/arrangement of the keys that represent letters, signs and symbols is the same, and this is referred to as QWERTY.

The working of a computer ECM keyboard can be compared to that of a miniature computer. Inside the keyboard there is a metallic plate, a circuit board (key ECM) and a processor, and these are responsible for transferring information from the ECM keyboard to the computer.

When one looks at all the extras and options that are available for modern computer ECM keyboards, it can be hard to believe that their original design came from mechanical typewriters that did not even use electricity. Now, one can buy ergonomic ECM keyboards that bear little resemblance to the flat, rectangular models with ordinary square keys.

Some flashier ECM keyboards light up, roll up or fold up, and others offer options for programming one's own commands and shortcuts. But no matter how many bells and whistles they offer, most ECM keyboards operate using similar technology. They use electrical valves acting as keys to translate a human user's keystrokes into signals that a computer can understand. In this chapter the author will explore ECM keyboard technology, together with different key layouts, options and designs.

16.3 Electrical commutation matrixer keyboard fundamentals

An ECM keyboard's primary function is to act as an input device. Using an ECM keyboard, a human user can type a document, use keystroke shortcuts, access menus, play games and perform a variety of other tasks. ECM keyboards can have different electrical valves acting as keys, depending on the manufacturer, the OS they are designed for and whether they are attached to a desktop computer or are part of a laptop.

But, for the most part, these keys, also called 'keycaps', are the same size and shape from keyboard to keyboard. They are also placed at a similar distance from one another in a similar pattern, no matter what language or alphabet the keys represent.

Most ECM keyboards have between 80 and 110 keys, including: typing alphanumerical keys; a numerical keypad; functional keys; and control keys. The typing alphanumerical keys include the letters of the alphabet, generally laid out in the same pattern that is used for typewriters. According to legend, this layout, known as 'QWERTY' for its first six letters, helped keep mechanical typewriters' metal arms from colliding and jamming as users typed. Some people question this story, but whether it is true or not, the QWERTY pattern had been standard for a long time when computer keyboards were first developed. ECM keyboards can also use a variety of other typing alphanumerical key arrangements. The most widely known is 'Dvorak', named after its creator, August Dvorak.

The Dvorak layout places all of the vowels on the left side of the ECM keyboard, and the most common consonants on the right. The most commonly used letters are all found along the home row. The home row is the main row where one places one's fingers when one begins typing. Users who prefer the Dvorak layout say it increases their typing speed and reduces fatigue. Other layouts include: 'ABCDE', 'XPeRT', 'QWERTZ' and 'AZERTY'. Each is named after the first keys in the pattern.

The QWERTZ and AZERTY arrangements are commonly used in Europe. The numerical keypad is a more recent addition to the computer ECM keyboard. As the use of computers in business environments increased, so did the necessity for speedy data entry. Since a large part of the data was made up of numbers, a set of 17 keys, arranged in the same configuration found on adding machines and calculators, was added to the ECM keyboard.

In 1986, IBM further extended the fundamental ECM keyboard with the addition of functional and control keys. Applications and OSs can assign specific commands to the functional keys. Control keys provide cursor and screen control. Four arrow keys arranged in an inverted 'T' formation between the typing alphanumerical keys and numerical keypad move the cursor on the screen in small increments.

The Windows ECM keyboard added some extra control keys: two (2) Windows or Start keys and an Application key. Apple keyboards, on the other hand, have Command (also known as 'Apple') keys.

An ECM keyboard developed for Linux users features Linux-specific hot keys, including one marked with 'Tux' the penguin—the Linux logo/mascot. An ECM keyboard is a lot like a miniature computer. It has its own processor and circuitry that carries information to and from that processor. A large part of this circuitry makes up the key ECM keyboard.

The key ECM keyboard is a grid of circuits underneath the keys. In all ECM keyboards (except for capacitive ones, which we shall discuss in the next section), each circuit is broken at a point below each key. When one presses a key, it presses an electrical valve, completing the circuit and allowing a tiny amount of current to flow through.

The mechanical action of the electrical valve causes some vibration, called 'bounce', which the processor filters out. If one presses and holds a key, the processor recognises this as the equivalent of pressing a key repeatedly. When the processor finds a circuit that is closed, it compares the location of that circuit on the key ECM keyboard to the character map in its **read-only memory** (ROM).

A character map is basically a comparison chart or look-up table. It tells the processor the position of each key in the ECM keyboard and what each keystroke or combination of keystrokes represents. For example, the character map lets the processor know that pressing a key by itself corresponds to a small letter 'a', but the 'Shift' and 'a' keys pressed together correspond to a capital 'A'.

A computer can also use separate character maps, overriding the one found in the ECM keyboard. This can be useful if a user is typing in a language that uses letters that do not have English equivalents on an ECM keyboard with English letters. Users can also set their computers to interpret their keystrokes as though they were typing on a Dvorak keyboard, even though their actual keys are arranged in a QWERTY layout. In addition, OSs and applications have ECM keyboard accessibility settings that let human users change their keyboard's behaviour to adapt to disabilities.

16.4 Electrical commutation matrixer keyboard's electrical valves

ECM keyboards use a variety of electrical valve (switch) technologies. Capacitance electrical valves acting as keys are considered to be non-mechanical because they do not physically complete a circuit like most other ECM keyboard technologies. Instead, current constantly flows through all the parts of the key ECM keyboard. Each key is spring-loaded and has a tiny plate attached to the bottom of it. When one presses a key, it moves this plate closer to the plate below it. As the two plates move closer together, the amount of current flowing through the ECM keyboard changes. The processor detects the change and interprets it as a key press for that location.

Capacitive electrical-valve ECM keyboards are expensive, but they have a longer life than any other keyboard. In addition, they do not have problems with bounce, since the two surfaces never come into actual contact. All of the other types of electrical valves used in ECM keyboards are mechanical in nature. Each provides a different level of audible and tactile response—the sounds and sensations that typing creates.

Depending upon the working principle, there are twelve main types of keys, namely, metal contact, capacitance, metal dome, rubber dome, membrane, buckling spring, roll-up, Hall effect, reed, foam component, laser and photo-optics. Let us discuss the functioning of these key electrical and optical valves (switches) in brief.

- **Metal contact.** A metal (hard) contact key is attached to a metallic plate that helps in connecting the circuit board. When the metal key is pressed, it pushes a metallic plate, which in turn touches the metallic portion of the circuit plate. This overall process of completing a circuit results in a circuit flow, allowing the transfer of the message to the **central processing unit** (CPU), which is further transmitted to the software. Metal ECM keyboards are growing less common. Metal electrical valves (switches) simply have a spring-loaded key with a strip of metal on the bottom of the plunger. When the key is pressed, the metal strip connects the two parts of the circuit. This ECM keyboard technology has good tactile response, making satisfyingly audible 'clicks', and is inexpensive to manufacture. The simplest switch mechanism employs mechanical contacts that are brought together as the key plunger is depressed. An example of a mechanical-contact key electrical valve (switch) is shown in figure 16.3.

Mechanical-contact ECM keyboards use separate complete electrical valves (switches) underneath every key. Each electrical valve (switch) is composed of a base, a spring and a stem. Depending on the shape of the stem, such ECM keyboards have varying actuation and travel distances. Depending on the resistance of the spring, the key requires different amounts of pressure to actuate. Mechanical-contact ECM keyboards allow for the removal and replacement of keycaps. Mechanical-contact ECM keyboards have proven to be popular amongst the PC gaming community and are gaining in popularity among typists. Although low in cost, one problem with mechanical contacts is that they suffer from contact bounce, which can generate false signals. This problem usually necessitates the use of appropriate denounce circuitry. Another disadvantage of mechanical-contact

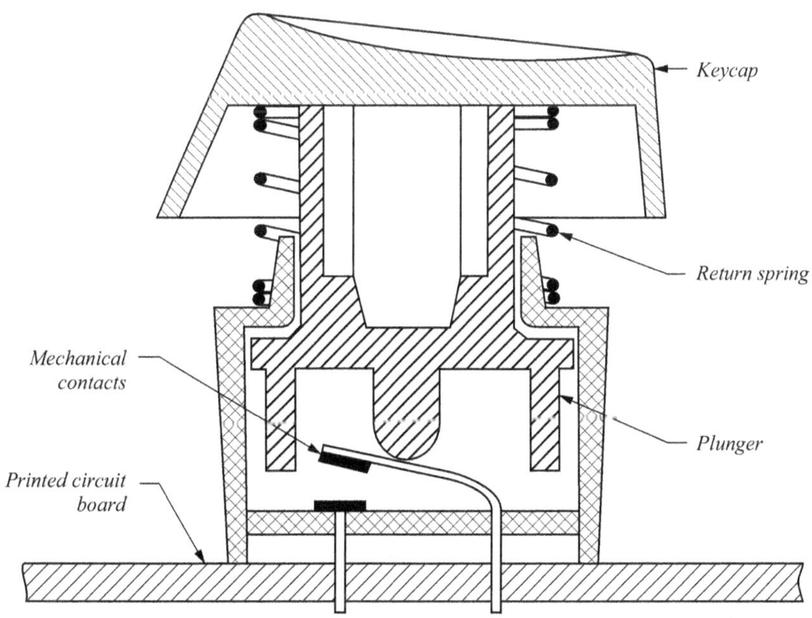

Figure 16.3. An example of a mechanical-contact key electrical valve (switch) (adapted from McLoughlin 2011).

electrical valves (switches) is that contacts can be contaminated by particles or oxidation. Such contamination results in increased contact resistance, causing unreliable operation or even total switch failure. The other problem is that the contacts tend to wear out or corrode faster than on ECM keyboards that use other technologies. There is no barrier that prevents dust or liquids from coming into direct contact with the circuitry of the key ECM. Key life is in the range of 5–10 million actuations, but the survival rate often drops rapidly above 5 million operations (McLoughlin 2011).

- **Capacitance electrical valves (switches).** One of the most popular ECM keyboard technologies uses capacitance changes to signal key activation. In this type of keyboard, pressing the key changes the capacitance of a pattern of capacitor pads. Unlike in 'dome switch' ECM keyboards, the pattern consists of two D-shaped capacitor pads for each electrical valve (switch), printed on a **printed circuit board** (PCB) and covered by a thin, insulating film of solder mask, which plays the role of a dielectric. The mechanism of the capacitance electrical valves (switches) is very simple, compared to that of mechanical ones. Its movable part ends with a flat foam element (with dimensions near to those of an aspirin tablet) finished with aluminum foil below. The opposite side of the electrical valve (switch) is a PCB with capacitor pads. When a key is pressed, the foil clings tightly to the surface of the PCB, forming a daisy chain of two capacitors between contact pads and itself separated with thin solder mask, thus 'shorting' the contact pads with an easily detectable drop of capacitive reactance between them. Usually this permits a pulse or pulse train to be sensed. The keys do not need to be pressed fully to be fired on, which enables some typists to work faster. In its most common form, the capacitance commutation (switching) mechanism employs a foam pad that is bonded to the bottom of the key plunger. Figure 16.4 illustrates an example of a capacitance key electrical valve (switch). A metal plunger is fixed on the underside of a capacitive key, which helps in activating the circuit flow. When a capacitive key is pressed, the metal plunger applies gentle pressure to the circuit board.

  The pressure is identified by the computer and the circuit flow is initiated, resulting in the transfer of information from the circuit to the currently installed software. A sheet of conductive foil is attached to the bottom of the foam pad. As a key is pressed, the foil, which acts as a capacitive plate, is brought into contact with a printed circuit board mounted under the array of key plungers. The increase in capacitance as the foil contacts the circuit board passes a low-level signal onto detector circuits, and this signals key activation. Because capacitance switching does not depend on the closing of a circuit through low resistance contacts, capacitance keyboards exhibit high reliability. Additional circuits and AC signal are needed to sense electrical valve (switch) closure. Typical key life characteristics range from 10–50 million keystrokes (McLoughlin 2011).
- **Metal dome electrical valve (switch).** One popular electrical valve (switch) technology is the dome electrical valve (switch). Dome electrical valve

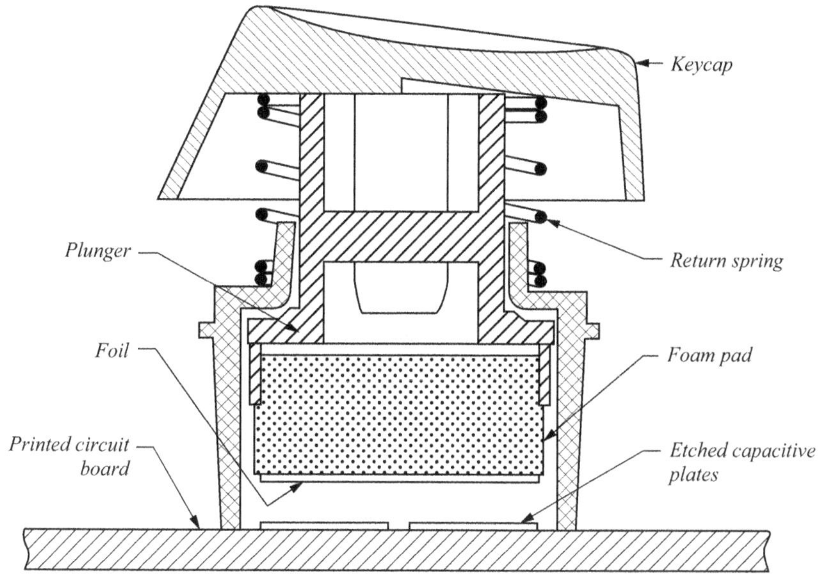

Figure 16.4. An example of a capacitance key electrical valve (switch) (adapted from McLoughlin 2011).

(switch) ECM keyboards are a hybrid of flat-panel membrane and mechanical ECM keyboards. They bring two circuit board traces together under a rubber or silicone keypad using either metal 'dome' electrical valves (switches) or polyurethane formed domes. The metal-dome electrical valves (switches) are formed pieces of stainless steel that, when compressed, give the human user a crisp, positive tactile feedback. These metal types of dome electrical valves (switches) are very common, are usually reliable for over 5 million cycles, and can be plated in either nickel, silver or gold. Dome electrical valve (switch) ECM keyboards are also called direct electrical valve (switch) ECM keyboards. As shown in the cross-section of figure 16.5, in a metal dome electrical valve (switch), the metal dome provides a spring action, and when pressed, it snaps to the closed position (McLoughlin 2011).

This provides some travel and good tactile response. Metal dome electrical valves (switches) are available as a single push button and in an ECM keyboard. A combination of membrane and metal dome technologies is widely used. This combines the low cost and environmental immunity of the membrane technology with the tactile response of the metal dome electrical valves (switches)

- **Rubber dome electrical valves (switches).** These are very common. They use small, flexible rubber domes, each with a hard carbon center. When the user presses a key, a plunger on the bottom of the key pushes down against the dome, and the carbon center presses against a hard, flat surface beneath the key ECM. As long as the key is held, the carbon center completes the circuit. When the key is released, the rubber dome springs back to its original shape, forcing the key back up to its at-rest position. The rubber dome electrical

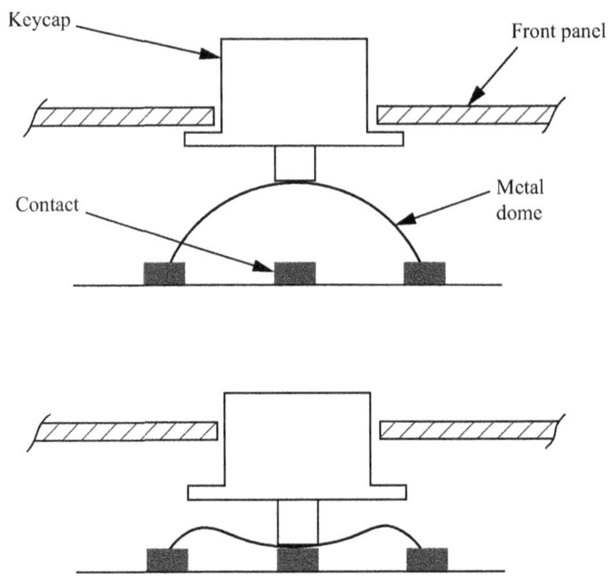

Figure 16.5. An example of a metal dome key electrical valve (switch) (adapted from McLoughlin 2011).

valves (switches), most commonly referred to as polydomes, are formed polyurethane domes where the inside bubble is coated in graphite. While polydomes are typically cheaper than metal domes, they lack the crisp snap of the metal domes and usually have a lower life specification. Polydomes are considered very quiet, but purists tend to find them 'mushy' because the collapsing dome does not provide as much positive response as metal domes. For either metal or polydomes, when a key is pressed, it collapses the dome, which connects the two circuit traces and completes the connection to enter the character. The pattern on the PC board is often gold-plated. Both are common switch technologies and are used in mass-market ECM keyboards today. This type of electrical valve (switch) technology is most commonly used in handheld controllers, mobile phones, automotive vehicles, consumer electronics and medical devices. Rubber dome electrical-valve ECM keyboards are inexpensive, have pretty good tactile response and are fairly resistant to spills and corrosion because of the rubber layer covering the key ECM.

- **Membrane.** Rather than having an electrical valve (switch) for each key, membrane ECM keyboards use a continuous membrane that stretches from one end to another. A pattern printed in the membrane completes the circuit when one presses a key. Some membrane ECM keyboards use a flat surface printed with representations of each key rather than keycaps. There are two types of membrane-based ECM keyboards, flat-panel membrane ECM keyboards and full-travel membrane ECM keyboards. Flat-panel membrane ECM keyboards are most often found on appliances like microwave ovens or photocopiers. A common design consists of three layers. The top layer (and the one the user touches) has the labels printed on its front and conductive stripes printed on the back. Under this it has a spacer layer, which holds the front and back layer apart so that they do not normally make electrical contact. The back layer has conductive stripes printed perpendicularly to those of the front layer. When placed together, the stripes form a grid. When a user pushes down at a particular position, his/her finger pushes the front layer down through the spacer layer to close a circuit at one of the intersections of the grid. This indicates to the computer or ECM keyboard control processor that a particular button has been pressed. Generally, flat-panel membrane ECM keyboards do not have much of a 'feel', so many machines that use them issue a beep or flash a light when the key is pressed. They are often used in harsh environments where water- or leak-proofing is desirable. Although they were used in the early days of the personal computer (on the Sinclair ZX80 and ZX81 and the Atari 400), they have been supplanted by the more tactile dome and mechanical switch keyboards. Full-travel membrane-based ECM keyboards are the most common computer ECM keyboards today. They have one-piece plastic keytop/switch plungers which press down on a membrane to actuate a contact in an ECM. Although essentially a metal electrical valve (switch) technology, full travel membrane ECM keyboards present a low-cost and reliable keyboard alternative. An ECM of full travel (approximately 4 mm) key button assembly is mounted above a three-layer sandwich of circuit materials that comprise the commutation (switching) mechanisms for all the keys on the ECM keyboard (see figure 16.6).

The top and bottom layers of the sandwich are thin sheets of flexible film (typically, clear polymer) with a conductive silver composition deposited onto facing sides of the flexible film. The silver composition is 'printed' in a pattern that forms the electrical contacts for each key station and all the electrical interconnections. Between the conductive layers is a third layer that provides a gap between the contacts through holes in the layer. The entire three-layer switch assembly is bonded to a sheet metal base plate. When a key is pressed, a switch closure is accomplished by deflecting the top film layer through the hole in the separation layer so that the conductive material on the top film layer comes in contact with the conductive material on the bottom layer. Typical key life now exceeds 10 million keystrokes (McLoughlin 2011). Unfortunately, this type of ECM keyboard also suffers from switch bounce. Membrane ECM keyboards do not have good tactile response, and without additional mechanical components they do not make the clicking sound that some human users like to hear when they are typing. However, they are generally inexpensive to make.

- **Buckling spring.** Many typists prefer buckling spring ECM keyboards. The buckling spring mechanism[1] atop the electrical valve (switch) is responsible

[1] Expired US patent 4,118,611.

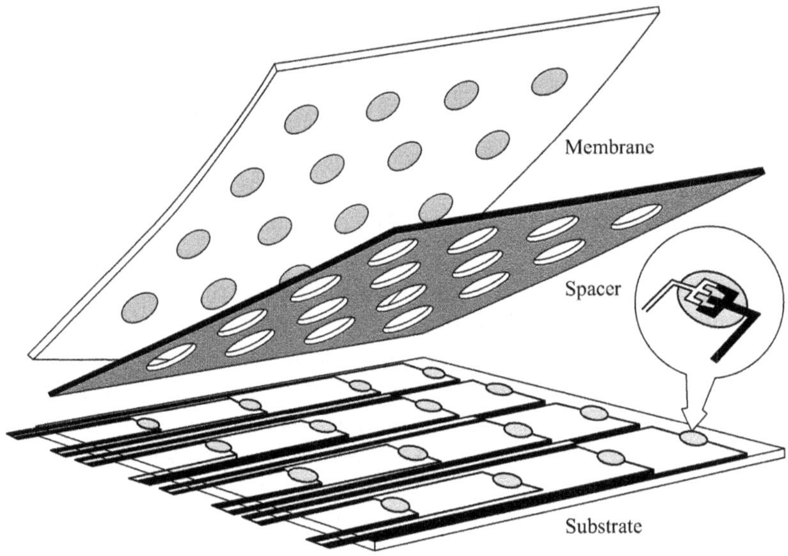

Figure 16.6. Exploded view of a full travel membrane ECM keyboard showing the unitised frame, space layer, adhesive layers and base plate (adapted from McLoughlin 2011).

for the tactile and aural response of the keyboard. This mechanism controls a small hammer that strikes a capacitive or membrane switch.

- **Roll-up.** Some ECM keyboards are designed using flexible materials that can roll up in a moderately tight bundle. Normally, the external materials are either silicone or polyurethane. Although manufacturers may claim that roll-up ECM keyboards are foldable, they cannot be folded without damaging the membrane that holds the circuitry. Typically, they are completely sealed in rubber, making them watertight like membrane ECM keyboards. Like membrane ECM keyboards, they are reported to be very hard to get used to, as there is little tactile feedback, and silicone will tend to attract dirt, dust and hair.
- **Hall effect.** Hall-effect ECM keyboards use **interior permanent magnets** (IPM) and Hall-effect sensors instead of an actual electrical valve (switch). When a key is depressed, it moves an IPM, which is detected by the solid-state sensor. In Hall-effect ECM keyboards, a magnetic transducer (typically an IC) is mounted near an IPM, which is affixed to the key plunger (see figure 16.7).

  Depression of a key moves the IPM closer to the magnetic transducer, causing the transducer to output a voltage change, signaling key activation. Because the commutation (switching) mechanism employs solid-state electronics, Hall-effect ECM keyboards exhibit extremely high reliability. Key life of more than 100 million actuations per key is typical (McLoughlin 2011). These ECM keyboards are extremely reliable, and are able to accept millions of keystrokes before failing. They are used for ultra-high reliability

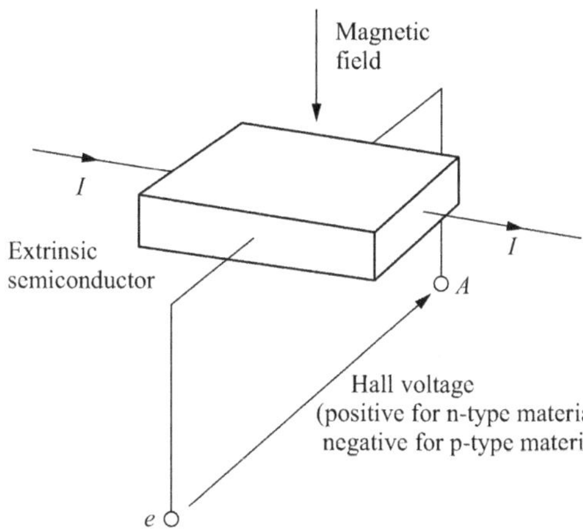

Figure 16.7. Hall effect (adapted from McLoughlin 2011).

applications, in locations like nuclear power plants or aircraft cockpits. They are also sometimes used in industrial environments. These ECM keyboards can easily be made totally waterproof. They also resist large amounts of dust and contaminants. Because an IPM and sensor are required for each key, as well as custom control electronics, they are very expensive. In addition, because individual key electrical valves (switches) can be sealed, Hall effect ECM keyboards are often used in military, marine, industrial, and aerospace and automotive applications, where reliability is a critical factor. The disadvantages of Hall-effect electrical valves (switches) are their relatively high cost and the large size of the individual key electrical valve (switch) modules.

- **Reed electrical valves (switches).** These are essentially mechanical contacts; the problems caused by contamination are eliminated by enclosing the electrical valve (switch) contacts in a sealed capsule, which is usually made of glass. Depression of a key moves an IPM close to the encapsulated reed electrical valves (switches). The attraction of the IPM on the reed electrical valves (switches) causes them to be brought together until the contacts touch. Because the sealed capsule prevents contamination of the electrical valve (switch) contacts, the long-term reliability is generally good. Key life of 50 million operations is possible (McLoughlin 2011). The biggest disadvantage of reed electrical valves (switches) is their high cost relative to other technologies.
- **Foam element.** These ECM keyboards are growing less common. The foam element electrical valve (switch) is basically the same design as that used for metal ECM keyboards, but with a small piece of spongy foam between the bottom of the plunger and the metal strip, providing a better tactile response. This ECM keyboard technology has good tactile response, make satisfyingly audible 'clicks', and is inexpensive to manufacture. The problem is that the contacts tend to wear out or corrode faster than on ECM keyboards that use

other technologies. In addition, there is no barrier that prevents dust or liquids from coming into direct contact with the circuitry of the key ECM.

- **Laser.** A laser projection device approximately the size of a computer mouse projects the outline of the ECM keyboard keys onto a flat surface, such as a table or desk. This type of laser ECM keyboard is portable enough to be used easily with PDAs and cellphones, and many models have retractable cords and wireless capability. However, sudden or accidental disruption of the laser will register unwanted keystrokes. Also, if the laser malfunctions, the whole unit becomes useless, unlike conventional ECM keyboards, which can be used even if a variety of parts (such as the keycaps) are removed. This type of ECM keyboard can be frustrating to use, since it is susceptible to errors, even in the course of normal typing, and its complete lack of tactile feedback makes it even less user-friendly than the lowest quality membrane ECM keyboards.
- **Photo-optics.** This is for photo-optic ECM keyboards, light-responsive ECM keyboards, photo-electric ECM keyboards and optical key actuation detection technology. Optical ECM keyboard technology was introduced in 1962 by Harley E Kelchner for use in a typewriter machine with the purpose of reducing the noise generated by actuating the typewriter keys. Optical ECM keyboard technologies utilise light-emitting devices and photo sensors to optically detect actuated keys. Most commonly, the emitters and sensors are located at the perimeter, mounted on a small PCB. The light is directed from one side of the keyboard interior to the other, and it can only be blocked by the actuated keys. Most optical ECM keyboards require at least two beams (most commonly a vertical beam and a horizontal beam) to determine the actuated key. Some optical keyboards use a special key structure that blocks the light in a certain pattern, allowing only one beam per row of keys (most commonly a horizontal beam). The mechanism of the optical ECM keyboard is very simple—a light beam is sent from the emitter to the receiving sensor, and the actuated key blocks, reflects, refracts or otherwise interacts with the beam, resulting in an identified key. Some earlier optical ECM keyboards were limited in their structure and required special casing to block external light, while no multi-key functionality was supported and the design was limited to a thick rectangular case. The advantages of optical keyboard technology are that it offers a truly waterproof keyboard, is resilient to dust and liquids, and uses about 20% PCB volume, compared with membrane or dome electrical valve (switch) ECM keyboards, significantly reducing electronic waste.

In all the key types, the circuit signals the processor to read and/or identify the character that has been pressed. For example, in a metal (hard) contact key, the processor reads that pressing the ‘Shift’ and ‘a’ keys at the same time corresponds to ‘A’. Hence, the letter, sign or symbol is displayed on the screen accordingly. Releasing the pressed key breaks the circuit flow, after which the key retains its original position. The communication between the computer ECM keyboard and the main computer is bilateral, meaning that messages or information can be sent to each other.

Different manufacturers have used these standard ECM keyboard technologies, and a few others, to create a wide range of unconventional ECM keyboards. We shall look at some of these unconventional ECM keyboards in the next section.

16.5 Unconventional electrical commutation matrixer keyboards

Many of the modifications to conventional ECM keyboard designs are an attempt to make them safer or easier to use. For example, some human users have associated increased keyboard use with **repetitive stress injuries** (RSI) like carpal tunnel syndrome, although scientific studies have produced conflicting results.

Ergonomic ECM keyboard designs are intended to keep a human user's hands in a more natural position while they are typing in an attempt to prevent injuries. While these ECM keyboards can certainly keep human users from holding their hands in a 'praying mantis' position, studies disagree on whether they actually prevent injury.

The simplest ergonomic ECM keyboards look like conventional ones that have been divided down the middle, keeping a human user's hands further apart and aligning the wrists with the forearms. More complex designs place the two halves of the keyboard at varying angles to one another and to the surface on which the ECM keyboard rests. Some go even further, placing the two halves of the ECM keyboard on the armrests of chairs, or making them completely perpendicular to the desk surface. Others, like the 'Datahand', do not look much like ECM keyboards at all. Some modifications, while not necessarily ergonomic, are designed to make ECM keyboards more portable, more versatile, or just 'cooler'. Some unconventional ECM keyboards are designed to reduce fatigue for RSI, or for one-handed use. e.g. the Maltron left-handed keyboard shown in figure 16.8.

Figure 16.8. The Maltron left-handed keyboard.

16.6 Virtual electrical commutation matrixer keyboards

Since their invention, computers have undergone rapid miniaturisation, from being a 'space saver' to 'as tiny as your palm'. Disks and components grew smaller in size, but one component remained the same for decades—the ECM keyboard. The miniaturisation of ECM keyboards proved to be a nightmare for users.

Human users of PDAs and smartphones are annoyed by the tiny size of the keys. The innovative new virtual ECM keyboard uses advanced technologies to project a full-sized computing ECM keyboard onto any surface. This device has become the solution for mobile computer users who prefer to touch-type rather than cramping over tiny keys. Typing information into mobile devices usually feels about as natural as a linebacker riding a Big Wheel. The virtual ECM keyboard is a way to eliminate finger cramping. All that is needed to use the ECM keyboard is a flat surface. Using laser technology, a bright red image of a keyboard is projected from a device such as a handheld. Detection technology based on optical recognition allows users to tap the images of the keys, so the virtual ECM keyboard behaves like a real one. It is designed to support any typing speed.

Theoretically, with either approach, the ECM keyboard can be in space and the human user can type by moving fingers through the air. The regular QWERTY keyboard layout is provided. Several products have been developed that by 'virtual ECM keyboard' mean one that has been put on a display screen as an image map. In some cases, the ECM keyboard can be customised. Depending on the product, the human user (who may be someone who is unable to use a regular ECM keyboard) can use a touch screen or a mouse to select the keys. A virtual ECM keyboard is one that a user operates by typing (moving fingers) on or within a wireless or optical-detectable surface or area rather than by depressing physical keys. In one technology, the ECM keyboard is projected optically onto a flat surface and, as the human user touches the image of a key, the optical device detects the stroke and sends it to the computer. In another technology, the ECM keyboard is projected onto an area and selected keys are transmitted as wireless signals using short-range Bluetooth technology. With either approach, a virtual keyboard makes it possible for the human user of a very small smartphone or a wearable computer to have full ECM keyboard capability. The advantages of the virtual ECM keyboard are as follows:

- keyboard layouts can be changed by software, allowing for 'foreign' or alternative keyboard layouts;
- ability to minimise the risk of repetitive strain injuries;
- flat or large typing surfaces are unnecessary;
- speed of text entry;
- accuracy;
- portability.

16.7 Canesta keyboard

The Canesta Keyboard IR Light Source (IR-CK100) plays a critical role in any Canesta Keyboard Perception Chipset implementation by enabling Canesta's patented electronic perception technology. The module collects the IR laser diode and all the associated optics into a single concise package, making it easy to integrate into a final product.

The IR Light Source operates by emitting a beam of infrared light. This light beam is designed to overlap the area on which the Canesta Keyboard Pattern Projector (PP-CK100) displays the ECM keyboard layout, so the human user's fingers are illuminated by the infrared light beam. The Canesta Keyboard Sensor Module (SM-CK100) detects the finger movement and the typing activity is resolved into the appropriate keystrokes or mouse actions.

16.8 Samsung's Scurry keyboard

Samsung's Scurry is also a wearable ECM keyboard. It works by attaching motion sensors to each finger. It does not detect muscle movement, but rather uses gyroscopic technology to detect angular movements of fingers through space. This approach works better. Demonstrations on the show floor were far more impressive than its competitor's performance in terms of accuracy. However, the device is too bulky. Nonfunctional prototypes of the final product are much smaller. The ECM keyboard demonstrated is wired. The manufacturers say that by the time their products become commercially available they will support the wireless Bluetooth protocol. In addition, the product requires specialised software for handheld devices.

16.9 Exemplary electrical commutation matrixer keyboard

Computer ECM keyboards include control circuitry to convert key presses into key codes (usually scancodes) that the computer's electronics can understand. The key electrical valves (switches) are connected via the printed circuit board in an $X \times Y$ ECM, where a voltage is provided sequentially to the Y lines and, when a key is depressed, detected sequentially by scanning the X lines. The first computer ECM keyboards were for mainframe computer data terminals and used discrete electronic parts. The first ECM keyboard microprocessor was introduced in 1972 by General Instruments, but keyboards have been using the single-chip 8048 microcontroller variant since it became available in 1978.

The ECM keyboard commutation (switching) matrixer is wired to its inputs, it converts the keystrokes to key codes, and, for a detached keyboard, sends the codes down a serial cable (the keyboard cord) to the main processor on the computer motherboard. This serial ECM keyboard cable communication is only bilateral to the extent that the computer's electronics controls the illumination of the caps lock, number lock and scroll lock lights.

When pressing an ECM keyboard key, the key contacts may 'bounce' against each other for several milliseconds before they settle into metal (hard) contact. When released, they bounce some more until they revert to the uncontacted state. If the computer were watching for each pulse, it would see many keystrokes for what the human user thought was just one. To resolve this problem, the processor in an ECM keyboard (or computer) 'debounces' the keystrokes by aggregating them across time to produce one 'confirmed' keystroke.

The keyboard one is interfacing with here is an ECM keyboard. This ECM keyboard is designed with a particular rows and columns. These rows and columns are connected to the microcontroller through its ports of the microcontroller. One normally uses and $X \times Y$ ECM keyboard. So only two ports of the microcontroller can be easily connected to the rows and columns of the ECM keyboard. Whenever a key is pressed, a row and a column gets shorted through that pressed key and all the other keys are left open. When a key is pressed only a bit in the port goes high. What indicates to the microcontroller that the key has been pressed? It is identified through this high on the bit key in the corresponding column. Once one is sure that one of the keys in the keyboard has been pressed, one's next aim is to identify that key. To do this, one first checks for a particular row and then checks the corresponding column in the ECM keyboard.

To check the row of the pressed key in the ECM keyboard, one of the rows is made high by making one bit in the output port of the microcontroller high. This is done until the row is revealed. Once one obtains the row, the next task is to ascertain the column of the pressed key. The column is detected through the contents in the input ports with the help of a counter. The content of the input port is rotated with carry until the carry bit is set.

The scan ECM lies beneath the keys, as shown in figure 16.9. Whenever a key is depressed and released, the control code at the time of the event is sensed, and this is then translated by the microcontroller into K-scan code. When a key is depressed, a make code is produced, and when a key is released, a break code is produced.

Figure 16.10 shows the key ECM for the 107 standard keyboard with power management keys (power, wake and sleep).

Figure 16.11 shows the key ECM for a multimedia keyboard with function keys.

Figure 16.12 shows a block diagram for an exemplary ECM keyboard. The solution only includes the key button inputs and LED indicator outputs.

Figure 16.13 shows the schematic of the exemplary ECM keyboard, where:

- J1 is used for USB connection;
- J2 is used for PS/2 mouse connection (for future development);
- J3 and J7 are used for in-circuit programming;
- an eight row $\times$ 18 column key ECM is implemented;
- eight rows are implemented in PTA[7:0];

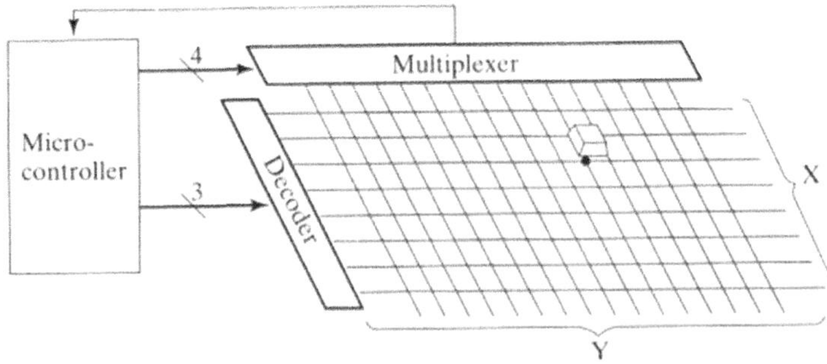

Figure 16.9. Scan electrical commutation matrixer.

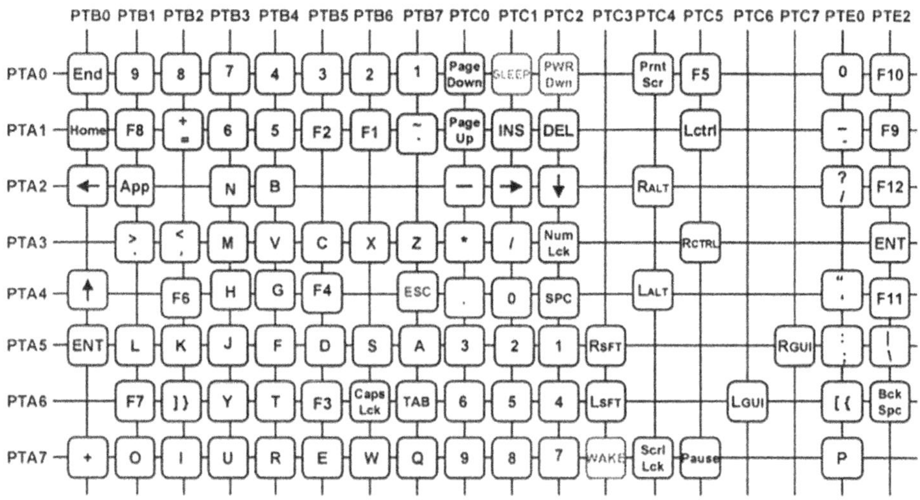

Figure 16.10. 107-key ECM (Lau 2006).

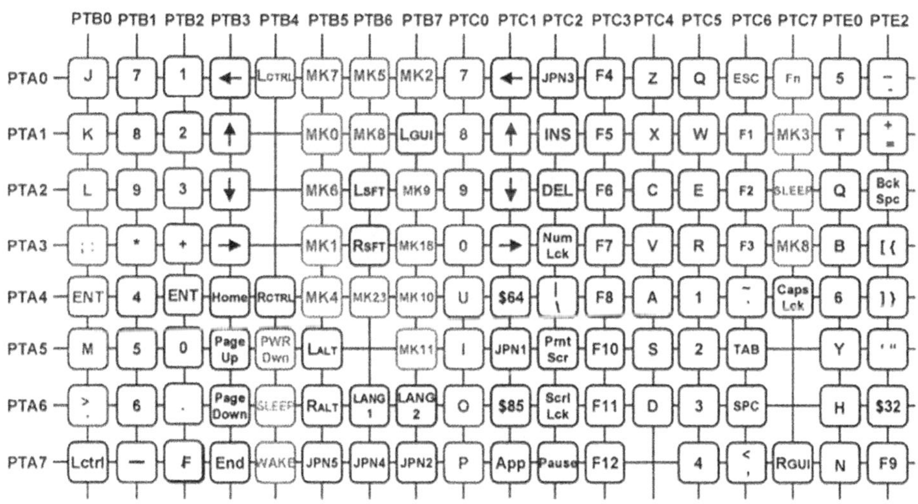

Figure 16.11. Multimedia key ECM (Lau 2006).

- 18 columns are implemented in PTB[7:0], PTC[7:0], PTE0 and PTE2;
- keyboard LEDs:
  - scroll lock at PTD2;
  - caps lock at PTD3;
  - number lock at PTD4.

The 'intellect' of the exemplary ECM keyboard is the commutation matrixery within the unit that handles the processing of keystrokes and exchanges information

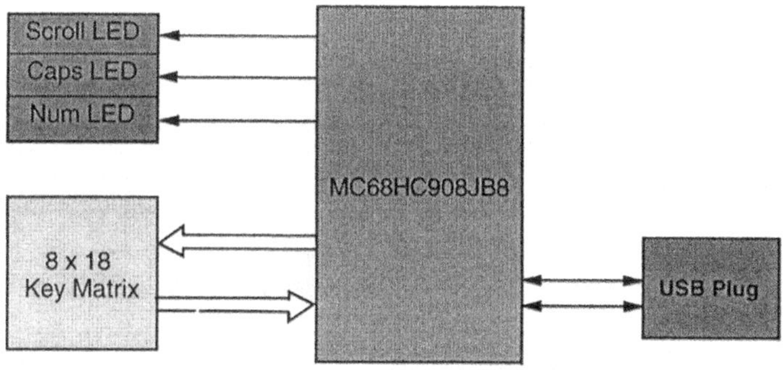

Figure 16.12. Block diagram of the exemplary ECM keyboard (Lau 2006).

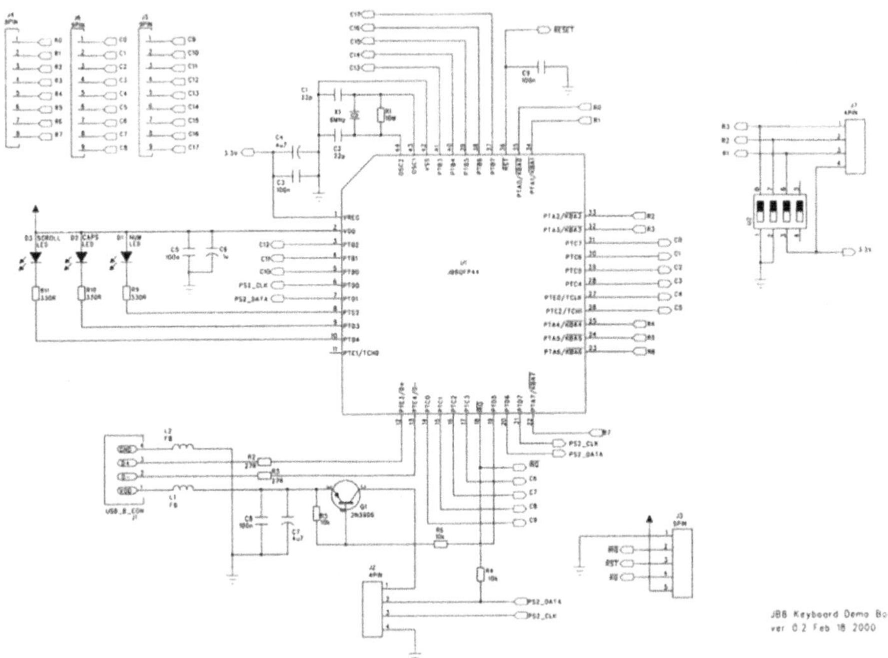

Figure 16.13. Exemplary ECM keyboard schematics (Lau 2006).

with the host system. While ECM keyboards are fairly simple devices, one might be surprised at just how complex the internal commutation matrixery in an ECM keyboard is (figure 16.14).

In fact, the ECM keyboard has within it small versions of several components that one finds within the computer as a whole. It has its own microprocessor, which 'runs the show', so to speak. This is, of course, a tiny **central processor unit** (CPU), not a fully fledged CPU, like the main computer hardware uses. There is also some **read-only memory** (ROM) that runs this small processor, similar to the system BIOS code on the motherboard.

Figure 16.14. Internal commutation matrixer board from a typical ECM keyboard PCB. The chip is the ECM keyboard controller. Note the LEDs sticking up from the circuit board in this design; they 'poke through' holes in the keyboard case (Lau 2006).

Programmable ECM keyboards also contain some EEPROM memory to hold programming information; this functions as if it were a tiny hard disk, in a way. This internal commutation matrixery has two fundamental tasks. First, it is responsible for sensing the raw signals created by the key electrical valves (switches) as they make electrical contact, and then translating them into signals that can be sent to the computer. Due to the electrical characteristics of keystrokes, this is not as simple as it sounds. Second, it handles the actual interfacing between the ECM keyboard and the rest of the host system. The interface between the computer and the ECM keyboard is a fully bilateral serial communications line, with its own special protocols and commands.

16.10 Conclusions

As one types, the processor in the ECM keyboard analyses the key ECM and determines what characters to send to the computer. It maintains these characters in its memory buffer and then sends the data. Many ECM keyboards connect to the computer through a cable with a PS/2 or **universal serial bus** (USB) connector. Laptops use internal connectors. Regardless of which type of connector is used, the cable must carry power to the ECM keyboard, and it must carry signals from the ECM keyboard back to the computer. Wireless ECM keyboards, on the other hand, connect to the computer through **infrared** (IR), **radio frequency** (RF) or Bluetooth connections. IR and RF connections are similar to what one finds in a remote control. Regardless of which sort of signal they use, wireless ECM keyboards require a receiver, either built-in or plugged into the USB port, to communicate with the computer. Since they do not have a physical connection to the computer, wireless ECM keyboards have an AC power connection or use, e.g. graphene Li-ion or Mg-ion storage batteries for power. In the field of storage batteries, conventional storage-battery electrode materials (and prospective ones) are significantly improved when they are enhanced with graphene. Graphene can make storage batteries that are light, durable and suitable for high-capacity energy storage, while it also shortens charging times. Whether it is through a cable or wireless, the signal from the ECM keyboard is monitored by the computer's ECM keyboard controller. This is an **integrated circuit** (IC) that processes all of the data that come from the ECM keyboard and forwards them to the OS. When the OS is notified that there are data from the ECM keyboard, it checks to see if the ECM keyboard data are a system level command. A good example of this is 'Ctrl-Alt-Del' on a Windows computer, which reboots the system. Then, the OS passes the ECM keyboard data onto the current application.

The application determines whether the ECM keyboard data are a command, like 'Alt-f', which opens the File menu in a Windows application. If the data are not a command, the application accepts them as content, which can be anything from typing a document, to entering a URL, to performing a calculation. If the current application does not accept the ECM keyboard data, it simply ignores the information. This whole process, from pressing the key to entering content into an application, happens almost instantaneously. Virtual ECM keyboards are projected images of the real thing that let typists compose their sentences on any flat surface. They are inching closer to store shelves. The ECM keyboard will be convenient for users who travel so they can easily access information through web pages. This will be efficient for individuals who have physical impairments. This technology will provide a keyboard that does not require force to activate the keys. It will serve people with physical impairments better.

Of the various devices available for transmitting information from humans to computers, the electrical valve (switch) is the simplest form of input device. However, the standard ECM keyboard is certainly the most generally applied. It is clearly the device of choice for applications involving significant amounts of textual input. It is also the computer input technology most affected by historical precedent. The data entry ECM keyboard achieved some degree of design standardisation long before its linkage to the computer. While this linkage has had a substantial effect on the design of current ECM keyboards, the functionality, layout, mechanics and dimensions of these devices are still in many ways a reflection of 19th-century technology.

References

Lau D 2006 USB and PS/2 Multimedia Keyboard Interface. Designer Reference Manual. M68HC08 Microcontrollers, DRM014, Rev. 1.0, 4/2006, Freescale Semiconductor Inc., Hong Kong 34 p

McLoughlin I 2011 *Computer Architecture: an embedded approach* (New York: McGraw-Hill)

Chapter 17

Programmable-logic and/or generic-logic electrical commutation matrixers for digital devices

You have to learn the rules of the game. And then you have to play better than anyone else.

Albert Einstein

17.1 Introduction

In this chapter, **programmable-logic electrical commutation matrixer** (PECM) and/or **generic-logic electrical commutation matrixers** (GECM) are introduced. The emphasis is on using PECMs to implement combinational logic. The sequential logic applications of PECMs are discussed. PECMs and GECMs can be used to replace SSI and MSI digital devices in most types of applications, resulting in fewer parts and lower costs (Broesch 1991; Floyd 1994; Carter 1997; Sharma 1998).

Programming software, such as the **advanced Boolean expression language** (ABEL) and the **compiler for universal programmable logic** (CUPL), are used to implement logic design in PECMs.

Once one is familiar with one PECM programming language, the others are easy to learn because of their similarities. For instance, ABEL can be used to progress any type of PECM and is therefore a matrix-independent language. ABEL is run on a computer connected to a language-independent ECM programmer into which the PECM is inserted (Pellerin and Holley 1994).

17.2 Programmable-logic electrical-commutation-matrixers and classifications

All PECM consists of programmable matrixers, which are essentially grids of conductors that form rows and columns with a fusible link at each cross point. PECMs can be either fixed or programmable.

The earliest type of **programmable logic matrix** (PLM), which dates back to the 1960s, has an **electrical commutation matrixer** (ECM) with an electrical valve as a diode at each cross point of the ECM.

17.2.1 The OR electrical commutation matrixer

The original diode ECM evolved into the integrated OR ECM, which consists of an ECM of OR gates (that produce a HIGH output when one or more inputs are HIGH) connected to a PECM with fusible links at each cross point of a row and column, as shown in figure 17.1(a).

The OR ECM is programmed by means of electrical valves, i.e. blowing fuses to eliminate selected variables from the output functions, as illustrated in figure 17.1(b) for a specific case. For each input to an OR gate, only one fuse is left intact in order to connect the desired variable to the gate input. Once a fuse is blown, it cannot be reconnected.

17.2.2 The AND electrical commutation matrixer

This type of ECM consists of AND gates (that produce HIGH output only when all of the inputs are HIGH) connected to a PECM with fusible links at each cross point, as shown in figure 17.2(a). Like the OR ECM, the AND ECM is programmed by means of electrical valves, i.e. blowing fuses to eliminate variables from the output function, as illustrated in figure 17.2(b).

For each input to an AND gate, only one fuse is left intact in order to connect the desired variable to the gate input. Also like the OR ECM, the AND ECM with fusible links is one-time programmable.

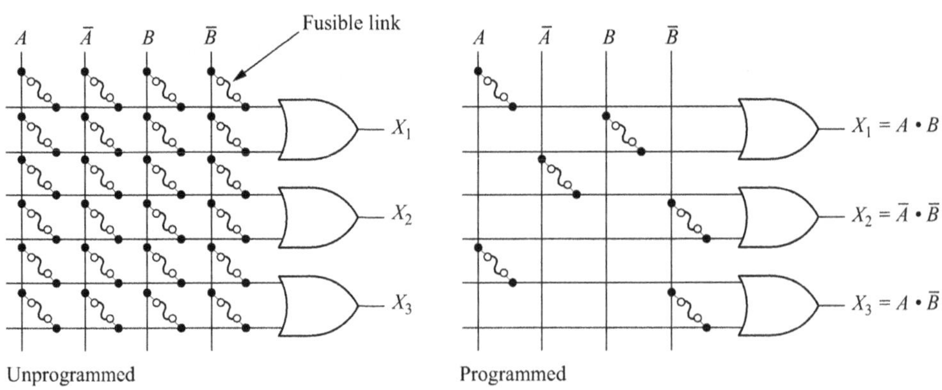

Figure 17.1. An example of a basic OR electrical commutation matrixer (adapted from Floyd 1994).

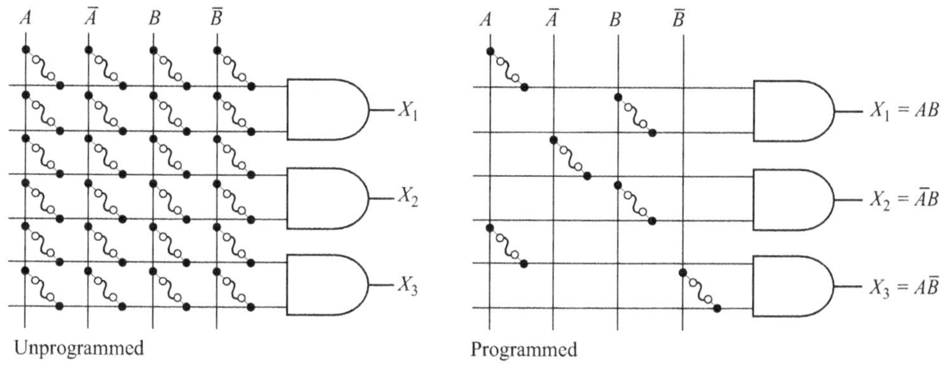

Figure 17.2. An example of a basic AND electrical commutation matrixer (adapted from Floyd 1994).

17.2.3 Classifications of programmable-logic electrical commutation matrixers

PECMs are classified according to their architecture, which is basically the internal functional arrangement of the elements that give a PECM its particular operating characteristics.

- **Programmable Read-Only Memory (PROM).** The PROM consists of a set of fixed (non-programmable) AND gates connected as a decoder and a programmable OR ECM. It is used primarily as an addressable memory and not as a logic matrix because of the limitations imposed by the fixed AND gates.
- **Programmable Logic Matrix (PLM).** The PLM is a PECM that consists of a programmable AND ECM and a programmable OR ECM. The PLM was developed to overcome some of the limitations of PROM. The PLM is also called **field-programmable logic array** (FPLA) because the user in the field, not the manufacturer, programs it.
- **Programmable Matrix Logic (PML).** The PML is an ECM that was developed to overcome certain disadvantages of the PLM, such as longer delays due to the additional fusible links that result from using two programmable matrixers and more circuit complexity. The basic PML consists of a programmable AND ECM and a fixed OR ECM with output logic. The PML is the most common on-time programmable logic matrix and is implemented with bipolar **transistor-transistor logic** (TTL) or **emitter-coupled logic** (ECL) technology.
- **Generic Matrix Logic (GML).** A more recent development in GE CMs is the GML. The GML, like the PML, has a programmable AND ECM and a fixed OR ECM with programmable output logic. The two main differences between GML and PML matrixers are (a) the GML is reprogrammable and (b) the GML has programmable output configurations.

17.3 Programmable ROMs (PROMs and EPROMs)

A ROM contains permanently or semi-permanently stored data, which can be read from the memory but either cannot be changed at all or cannot be changed without special equipment. A ROM stores data that are used repeatedly in system applications, such as tables, conversions, or programmed instructions for system initialisation and operations. ROMs retain stored data when the power is off and are therefore non-volatile memory.

PROM, or programmable ROM, is the type in which the user with the aid of specialised instruments electronically stores the data. PROMs are basically the same as mask ROMs, once they have been programmed. The difference is that PROMs come from the manufacturer unprogrammed and are custom programmed in the field to meet the user's needs.

PROMs use some type of fusing process to store bits, in which a memory link is burned open or left intact to represent a binary zero (0) or a binary one (1). The fusing process is irreversible; once a PROM has been programmed, it cannot be changed. Figure 17.3 illustrates a **metal-oxide semiconductor** (MOS) PROM matrix with fusible links.

The fusible links are manufactured into the PROM between the source of each cell's transistor and its column line.

In the programming process, a sufficient current is injected through the fusible link to burn it open to create a stored binary zero (0). The link is left intact for a stored binary one (1).

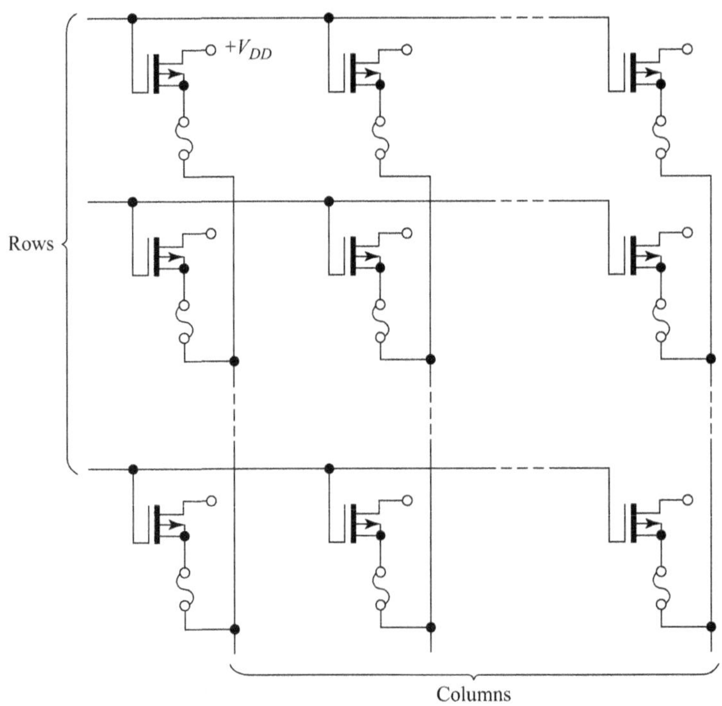

Figure 17.3. MOS PROM matrixer with fusible links (adapted from Floyd 1994). (All drains are commonly connected to V_{DD}.)

A PROM is normally programmed by inserting it into a special instrument called a 'PROM programmer'. Basically, the programming is accomplished by the simplified setup. An address is selected by the electrical valves on the address lines, and then a pulse is applied to those output lines corresponding to bit locations where zeros (0) are to be stored. These pulses blow the fusible links, thus creating the desired bit pattern. The next address is then selected and the process is repeated. This sequence is actually done automatically by a software-driven PROM programmer.

Unlike an ordinary PROM, an **erasable PROM** (EPROM) can be reprogrammed if an existing program in the memory matrixer is erased first. An EPROM uses an NMOSFET matrixer with an isolated gate structure.

The isolated transistor gate has no electrical connections and can store an electrical charge for indefinite periods. The data bits in this type of matrix are represented by the presence or absence of a stored gate charge.

Two basic types of EPROMs are the **ultraviolet** (UV) light EPROM, i.e. the UV EPROM, and the electrically erasable PROM, i.e. the E^2PROM.

One can recognise the UV EPROM matrix by the transparent quartz lid on the package. The isolated gate in the **field-effect transistor** (FET) or an ultra-violet EPROM is 'floating' within an oxide insulating material.

The programming process causes electrons to be removed from the floating gate. Erasure is done by exposure of the memory matrix chip to high-intensity UV radiation through the quartz window on top of the package. The positive charge stored on the gate is neutralised after several minutes to an hour of exposure time.

Electrically erasable complementary metal-oxide semiconductor (E^2C MOS) can be both erased and programmed with electrical pulses. Since it can be both electronically written into and electronically erased, the E^2PROM can be rapidly programmed and erased in matrix for reprogramming.

Two types of E^2PROMs are the floating-gate MOS and the **metal nitride-oxide silicon** (MNOS). The application of a voltage on the control gate in the floating gate structure permits the storage and removal of charge from the floating gate.

17.4 Programmable matrix logic (PML)

The PML and the GML are the most common EPMs used for logic implementation. The PML consists of a programmable ECM of AND gates that connects to a fixed EPM of OR gates. This structure allows any **sum-of-products** (SOP) logic expression with a defined number of variables to be implemented. Note that any logic function can be expressed in SOP forms.

The basic structure of a PML is illustrated in figure 17.4 for two input variables and one output, although most PMLs have many inputs and many outputs.

As we know, a programmable matrix is essentially a grid of conductors forming rows and column with a fusible link at each cross point. Each fused cross point of a row and column is called a 'cell' and is the programmable component of a PML.

Each row is connected to the input of an AND gate and each column is connected to an input variable or its complement. By using the presence or absence of fused

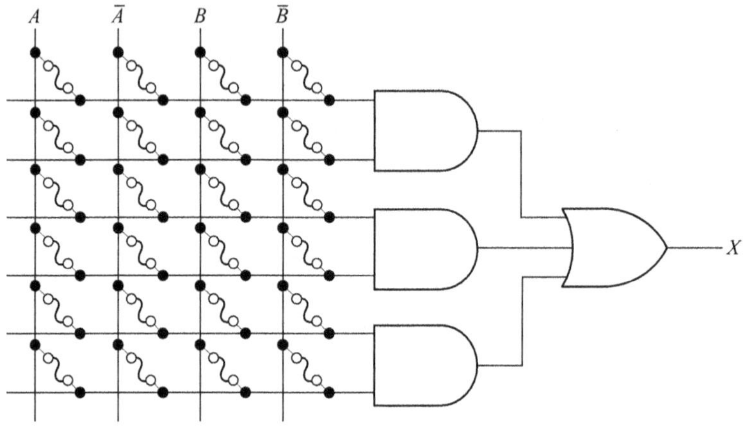

Figure 17.4. Basic structure of programmable matrix logic (PML) (adapted from Floyd 1994).

connections created by programming, any combination of input variables or complements can be applied to an AND gate to form any desired product term.

17.5 Generic matrix logic (GML)

The GML in its basic form is an EPM with a reprogrammable AND ECM, a fixed OR ECM and programmable output logic. The GML can be reprogrammed repeatedly because it uses E^2CMOS technology instead of bipolar technology and fusible links.

Just as in a PML, this structure allows any SOP logic expression with a defined number of variables to be implemented.

The basic structure of a GML is illustrated in figure 17.5 for two input variables and one output, although most GMLs have many inputs and many outputs.

The reprogrammable EGM is essentially a grid of conductors forming rows and columns with an E^2CMOS cell at each cross point, rather than a fuse as in PML. These cells are shown as blocks in figure 17.5. Each row is connected to the input of an AND gate, and each column is connected to an input variable or its complement.

By programming each E^2CMOS cell to be either 'on' or 'off', any combination of input variables or complements can be applied to an AND gate to form any desired product term.

A cell that is on effectively connects its corresponding row and column, and a cell that is off disconnects the row and column.

The cells can be electrically erased and reprogrammed. A typical E^2C-MOS cell can retain its programmed state for 25 years or more.

E^2CMOS technology is based on a combination of **complementary metal-oxide semiconductor** (CMOS) and **non-complementary metal-oxide semiconductor** (NMOS) technologies and is used in GML ECMs such as those above discussed and other PLMs.

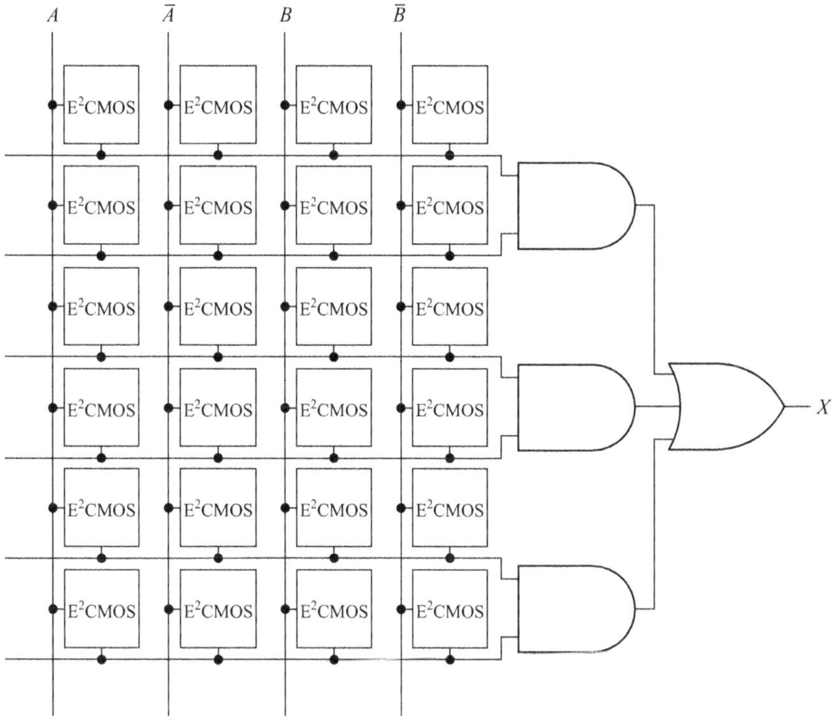

Figure 17.5. Basic E²CMOS matrix structure of generic matrixer logic (GML) (adapted from Floyd 2000).

An E²CMOS cell is built around a MOS transistor with a floating gate that is externally charged or discharged by a small programming current. A physical model of this type of cell is shown in figure 17.6.

When the flowing gate is charged to a positive potential by removing electrons, the sense transistor is turned on, storing a binary zero (0). When the floating gate is charge to a negative potential by placing electrons on it, the sense transistor is turned off, storing a binary one (1).

The control gate controls the potential of the floating gate. The pass transistor isolates the sense transistor from the EGM during read and write operations that use the word and bit lines. The cell is programmed by applying a programming pulse to either the control gate or the bit line of a cell that has been selected by a voltage on the word line. During the programming cycle, the cell is erased by applying a voltage to the control gate to make the floating gate negative. This leaves the sense transistor in the off state, storing a binary one (1). A write pulse is applied to the bit line of a cell in which a binary zero (0) is to be stored. This will charge the floating gate to a point where the sense transistor is on, storing a binary (0). The bit stored in the cell is read by sensing the presence or absence of a small cell current in the bit line. When a binary one (1) is stored, there is no cell current because the sense transistor is off. When is a binary zero (0) is stored, there is a small current because the sense transistor is on. Once a bit is stored in a cell, it will remain indefinitely unless the cell is erased or a new bit is written into the cell.

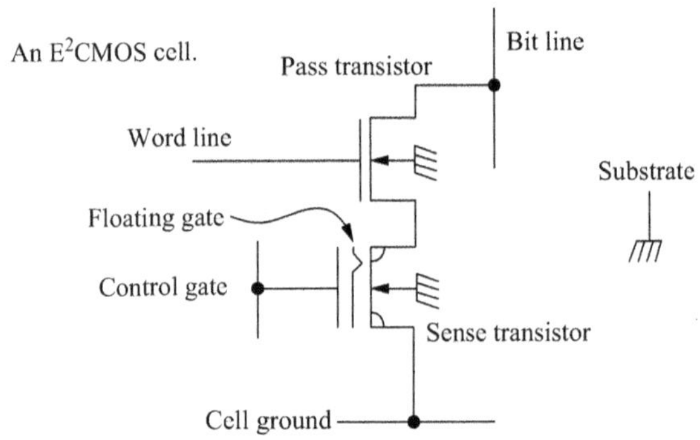

Figure 17.6. An E^2CMOS cell (adapted from Floyd 2000).

17.6 Electrical commutation matrixer crossbar

A type of mid-19th-century telegraph-exchange mechanical ECM crossbar consisted of a grid of vertical and horizontal brass bars with a hole at each intersection. The operator inserted a brass pin to connect one telegraph line to another. An ECM crossbar consisted of metal bars associated with each matrixer input and matrixer output, together with some means of controlling the movable contacts (electrical valves) at each cross point.

In the later part of the 20th century, these literal ECM crossbars declined and the term came to be used symbolically for rectangular array ECMs in general. The development of reed-relay switch (electrical valve) cross points was crucial to the advancement of ECM crossbar implemented digital switching.

Modern 'electrical-commutation-matrixer crossbars' are usually implemented with semiconductor technology. An important emerging class of optical crossbars is being implemented with technology.

An ECM crossbar (also termed an 'electrical commutation matrixer cross point') is an ECM connecting multiple matrixer inputs to multiple matrixer outputs in a matrix manner. Originally, the term was used literally, for an ECM controlled by a grid of crossing metal bars, but later it was broadened to describe an ECM's electrical valves in general. It is one of the principal commutation architectures, together with a rotating switch, memory switch and a crossover switch. An ECM crossbar is an assembly of individual electrical valves between multiple matrixer inputs and multiple matrixer outputs. The electrical valves are arranged in an ECM. If the ECM crossbar has m matrixer inputs and n matrixer outputs, then a crossbar has an ECM with $m \times n$ cross points or places where the 'bars' cross (see figure 17.7).

At each crosspoint is an electrical valve; when turned on, it connects one of the m matrixer inputs to one of the n matrixer outputs. A given crossbar is a single-layer non-blocking switch. Collections of crossbars can be used to implement multiple layer and/or blocking electrical valves. An ECM crossbar is also termed a **coordinate electrical commutation matrixer** (CECM).

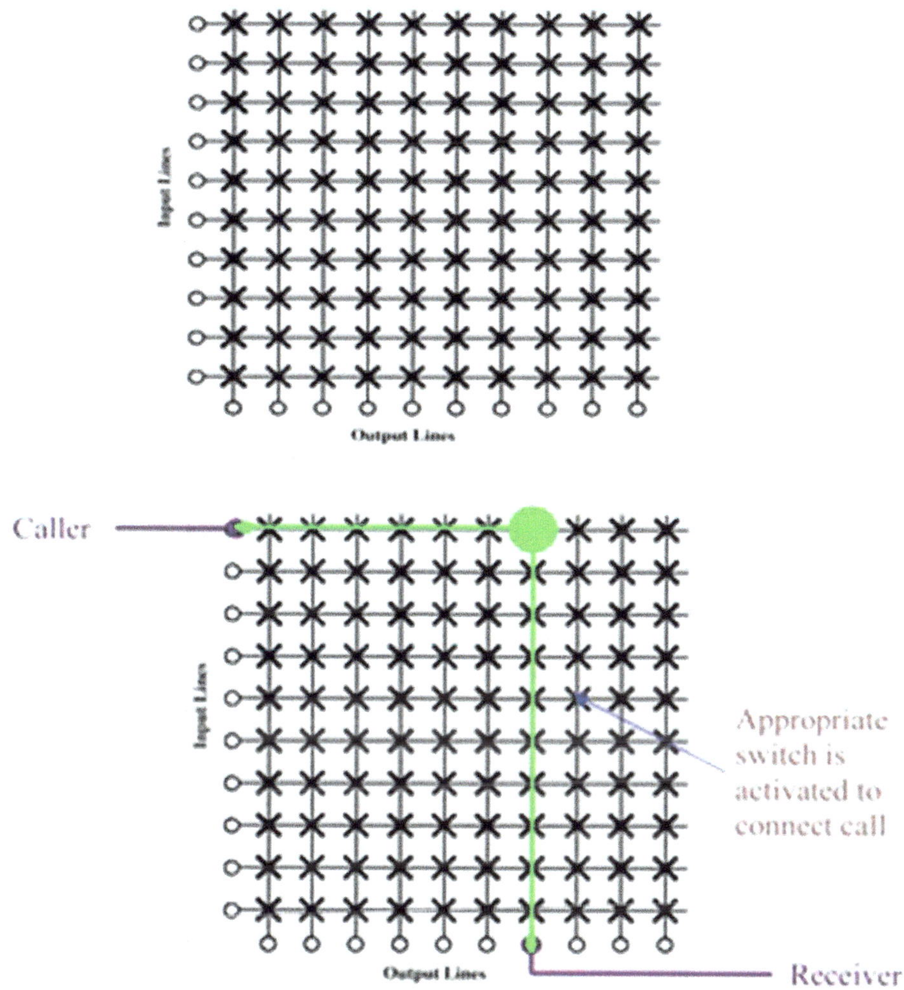

Figure 17.7. Electrical commutation matrixer crossbar (Ogorzalek 2007).

ECM crossbars are characterised by the fact that (Ogorzalek 2007):

- the number of cross points grows as the square of number of stations;
- loss of cross points prevents connection;
- there is inefficient use of cross points:
    - all stations are connected, but only a few cross points are in use;
- they are non-blocking.

ECM crossbars are most commonly used in information processing applications such as telephony and circuit commutation, but they are also used in applications such as mechanical sorting machines with inputs.

The ECM crossbar is also used in some semiconductor memory digital devices implemented with nanotechnology. Here the bars are extremely thin metal wires and the electrical valves are fusible links. The fuses are blown or opened using high voltage and they are read using low voltage. Such digital devices are called 'programmable read-only memory' (Chen *et al* 2003). Mouttet (2008) presented a nanoscale ECM crossbar implementation of an adding circuit being used as an alternative to logic gates for computation.

17.7 Electrical commutation matrixer tactile sensor

A modern 'electrical commutation matrixer tactile sensor' of sensing devices consists of tactile sensors, which are used for several purposes, such as shape recognition, contact point determination and pressure/force measurement.

A number of ECM tactile sensors have been proposed in the literature, with several different solutions concerning their realisable features—optical, piezoresistive, piezoelectric and so on; see, for example, (Pugh 1986; Howe and Cutkosky 1989; Nicholls and Lee 1989) for an overview of the technologies and applications.

ECM tactile sensors have in particular been implemented in automation and robotics. Currently, as with force sensors, ECM tactile sensors are commercially available devices. Doubtless, they are the most universally implemented sensorial ECMs for industrial grippers, even though they are often used as advanced on/off devices to certify whether a grasp or contact condition occurs (Siciliano and Khabit 2008). Usually, they consist of an ECM (array) of sensing devices (figure 17.8(a)). Each sensing device is usually referred to as a taxel (from tactile device), and the whole set of information is termed a tactile image (figure 17.8(b)). The key objective of this set of sensors is to measure the pressure map over the sensing area.

In general, the types of information that may be obtained from an ECM tactile sensor are (Siciliano and Khabit 2008):

- **contact**: this is the most simple information given by the sensor, and concerns the presence or absence of a contact;
- **force**: each sensing device provides information related to the amount of locally applied force, which can be used in several ways for successive elaborations;
- **simple geometrical information**, i.e. the position of the contact area and the geometrical shape of the contact itself (planar, circular, and so on);
- **the main geometrical features of the object**: by proper elaboration of the data from the taxels, it is possible to deduce the type of object in contact with the sensor, for example, a sphere or a cylinder (data relative to the three-dimensional shape);
- **mechanical properties**, such as the friction coefficient or roughness, while the thermal properties of the object may also be measured by a tactile sensor;
- **slip condition**, i.e. the relative movement between the object and the sensor.

Several technologies have been adopted for the design of tactile sensors, ranging from the piezoresistive through the magnetic to the optical. Among the most common, one can mention (Siciliano and Khabit 2008):

- resistive and conductive effects;
- electromagnetic effects;
- capacitive effects;

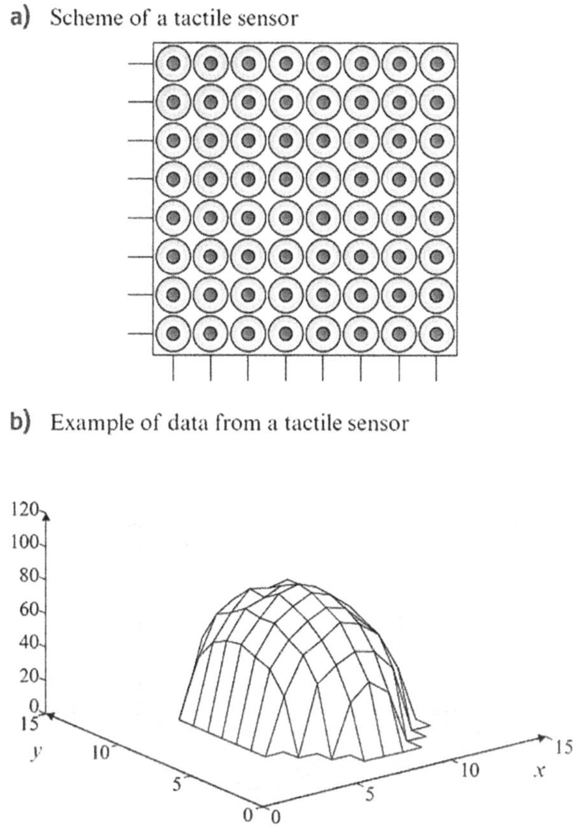

Figure 17.8. An electrical commutation matrixer (ECM) tactile sensor: (a) scheme and (b) example data (Siciliano and Khabit 2008).

- piezoelectric effects;
- optical effects;
- mechanical methods.

Each of these technologies has positive and negative aspects. Common drawbacks, however, include the size of these sensors, which are usually quite large in comparison with the available space, and the necessity of a large number of electrical connections.

17.8 Electrical commutation matrixer seven-segment display

A common type of display device that uses **light-emitting diodes** (LED) is the electrical commutation matrixer seven-segment display. Combinations of the segments form ten decimal digits, as illustrated in figure 17.9 (Floyd 2000).

Each segment in the electrical commutation matrixer display is an LED. By forward-biasing selected combinations of segments, any decimal digit and a decimal point can be formed. Two types of LED electrical commutation matrixers are the common anode and the common cathode, as shown in figure 17.10.

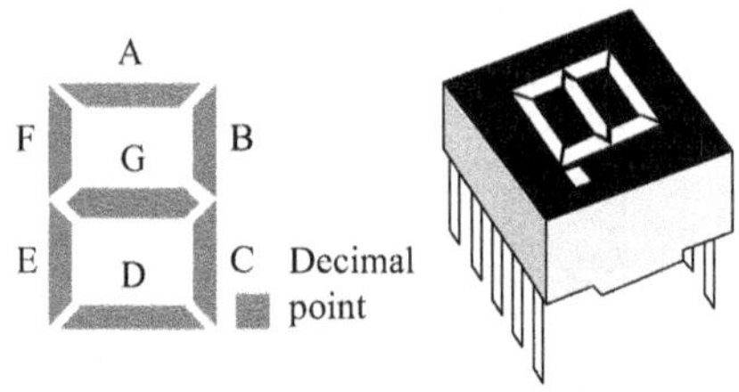

Figure 17.9. LED seven-segment arrangement and typical device. (adapted from Floyd 2000)

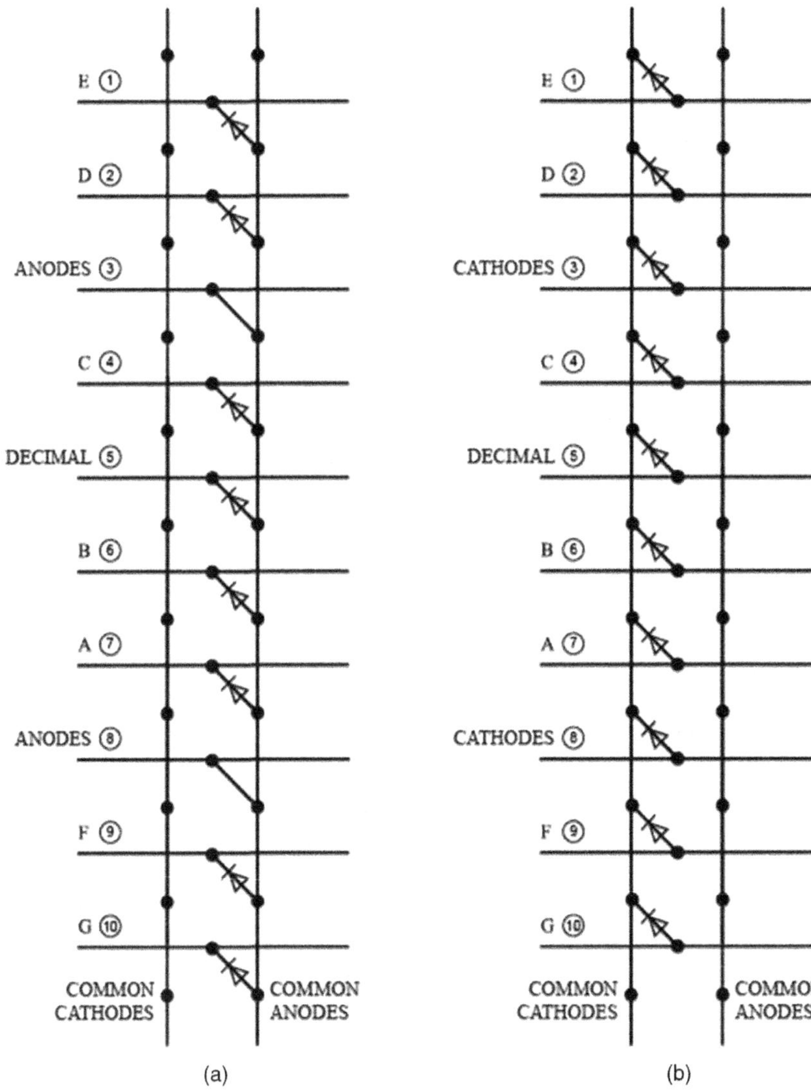

Figure 17.10. Electrical commutation matrixer seven-segment display. (a) Common-anode commutation group. (b) Common-cathode commutation group.

Standard LEDs are also used for indication lamps and readout displays on a wide variety of instruments, ranging from consumer appliances through to scientific apparatus.

17.9 Fluidic commutation matrixer combinational-chemistry microreaction

In the near future, scientists and engineers face the task of developing designs for both **micro-electro-mechanical systems** (MEMS) and **nano-electro-mechanical systems** (NAEMS). These can be standalone systems, like chemical microplants. Combinational chemistry is an important new methodology developed by researchers in the pharmaceutical industry to reduce the time and costs associated with producing effective and competitive new drugs.

Figure 17.11 shows a state-of-the art chemical microplant as a fluidic commutation matrixer combinational-chemistry microreaction (Koivo 2012).

The production of drugs, chemicals and other substances takes place at much smaller plants, chemical microplants, which produce less waste, use less energy and cause less environmental damage. Such inventions will grow in number, but to make strong advances, this microfluidic-matrix domain is in need of vigorous efforts in physical and mathematical modelling and simulation (Koivo 2012).

The control of **electro-osmotic flows** (EOF) in microfluidic electrical valves has been achieved by using a new generation of **flow field effect transistors** (FFET) with integrated **polarisable interfaces** (PI). This new type of microfluidic electrical valve was compared to more conventional metal-isolator-electrolyte FFETs.

Using a fluidic Wheatstone bridge configuration, precise control of the liquid flow rate was obtained with very small voltages (<500 mV). By controlling the EOF independently of the external electrical field, electrophoretic filter effects could be observed. The ability of scientists and engineers to integrate such fluidic transistors in serial or parallel architectures opens up new routes for high-selectivity biomolecule sorting using programmable chip matrixers.

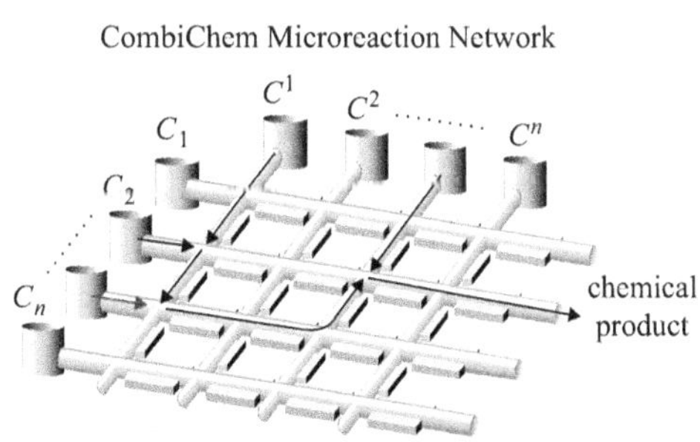

Figure 17.11. Artist's impression of a microfluidic electrical commutation matrixer based on flow FETs (adapted from Koivo 2012).

17.10 Exemplary applications of programmable-logic electrical commutation matrixer for digital devices

17.10.1 PCM implementing a 2048 × 8 EPROM

Its operation is representative of that of other typical EPROMs. As the logic symbol in figure 17.12 shows, this EPM has 2049 addresses ($20^{11} = 2048$), each with eight (8) bits. Note that the eight outputs are tri-state ($\overline{V}$).

To read from the memory, the output enable input ($\overline{OE}$) must be LOW and the power-down/program ($\overline{CE}/PGM$) input must be LOW. To erase the stored data, the EPM is exposed to high-intensity UV light through the transparent lid. As in most EPROMs after erasure, all the bits are binary ones (1). To program the EPM, a high DC voltage is applied to V_{PP}, and $\overline{OE}$ is HIGH.

The eight (8) data bits to be programmed into a given address are applied to the outputs O_0 through O_7, and the address is selected on inputs A_0 through A_{10}. Next, a 10 ms to 55 ms HIGH-level pulse is applied to the $\overline{CE}/PGM$ input. The addresses can be programmed in any order.

Another example of EPM implementing an EPROM is shown in figure 17.13, where it is programmed by blowing a fuse link to create a binary zero (0).

17.10.2 PCM implementing a sum-of-products (SOP) expression

In its simplest form, each cell in a basic AND ECM consists of a fusible link connecting a row and a column, as represented in figure 17.14.

When a connection between a row and column is required, the fuse is left intact. When no connection between a row and column is required, the fuse is blown open

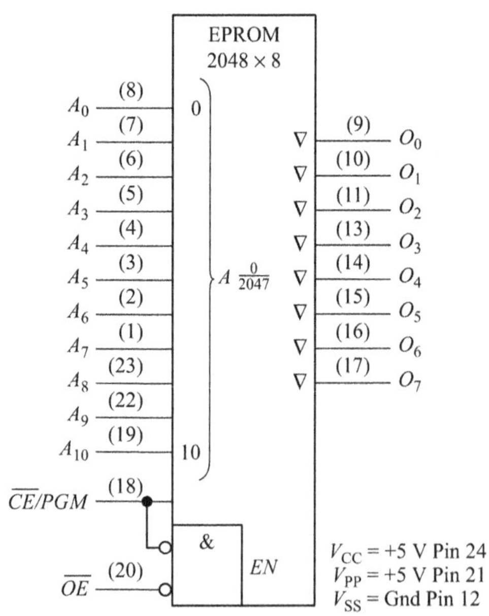

Figure 17.12. The logic symbol for a 2094 × 8 EPROM (adapted from Floyd 2000).

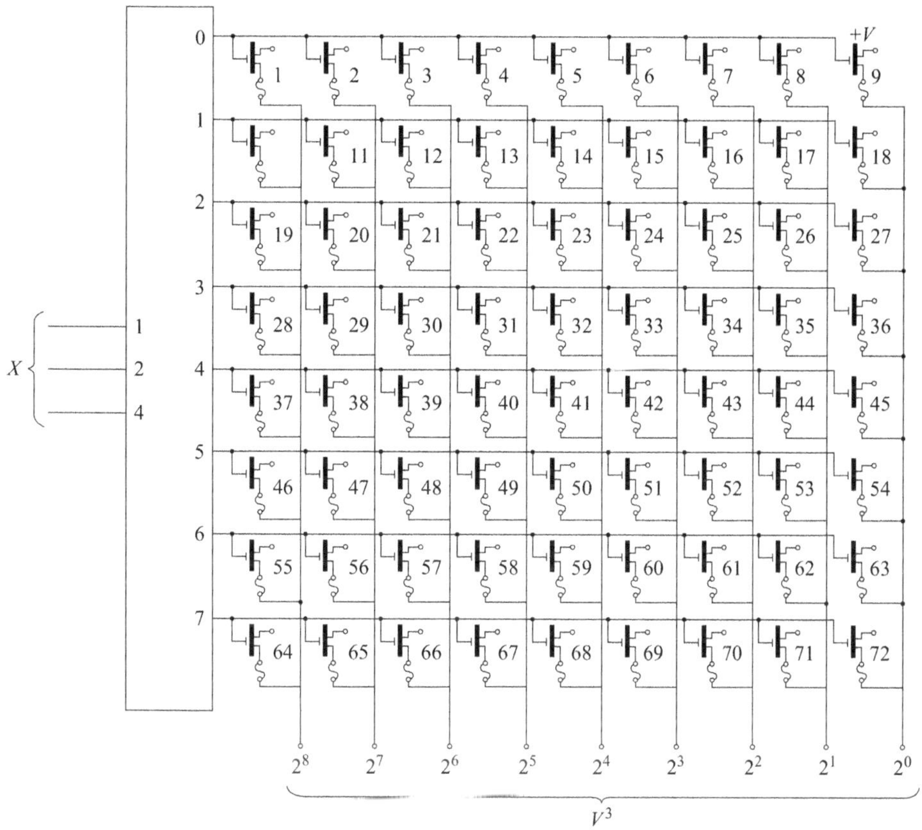

Figure 17.13. The EPROM which is programmed by blowing a fuse link (adapted from Floyd 2000).

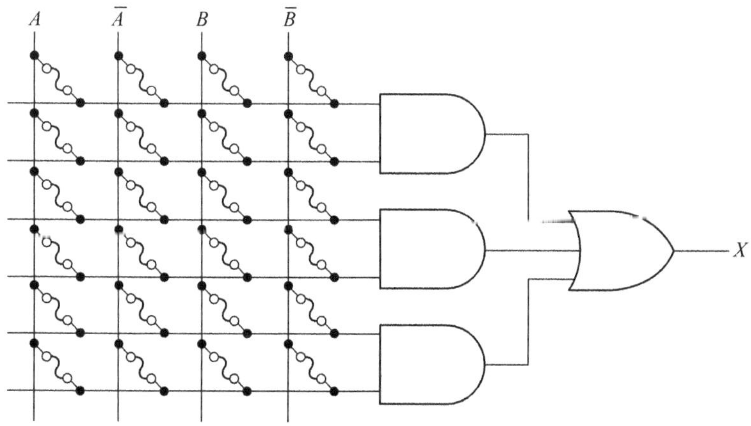

Figure 17.14. Basic structure of a PML (adapted from Floyd 2000).

during the programming process. As one can see, the fusible links are left intact to connect the desired variables or their complements to the appropriate AND gate inputs.

As an example, a simple PML matrixer is programmed as shown in figure 17.15 so that the product term AB is produced by the top AND gate, $A\overline{B}$ is produced by the middle AND gate and $\overline{AB}$ is produced by the bottom AND gate.

The fusible links are opened when a variable or its component is not used in a given product term. The final output from the OR gate is the SOP expression,

$$X = AB + A\overline{B} + \overline{AB}.$$

The programmed polarity output in a PML matrixer operates by selectively opening or leasing intact a cell on one input of an XOR gate, causing the gate to invert its other input or not. See figure 17.16.

The selected matrixer inputs determine the combinational or registered mode, matrixer input or matrixer output and inverted or non-inverted matrixer output. See figure 17.17.

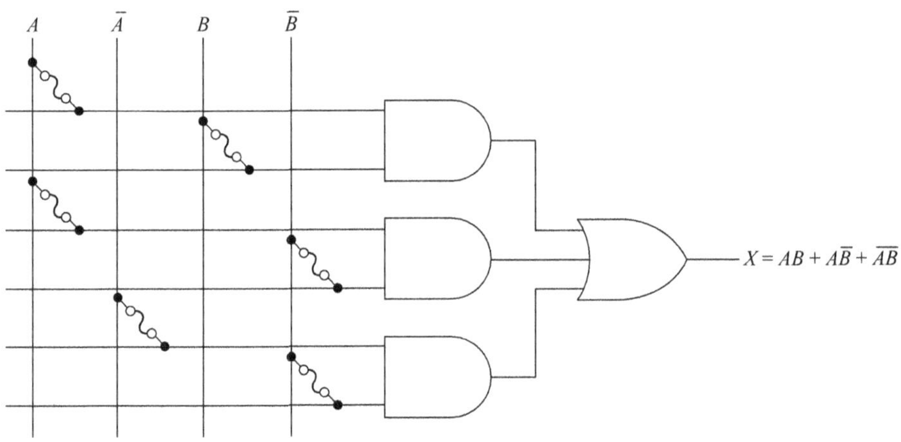

Figure 17.15. PML implementation of a SOP expression $X = AB + A\overline{B} + \overline{AB}$ (adapted from Floyd 2000).

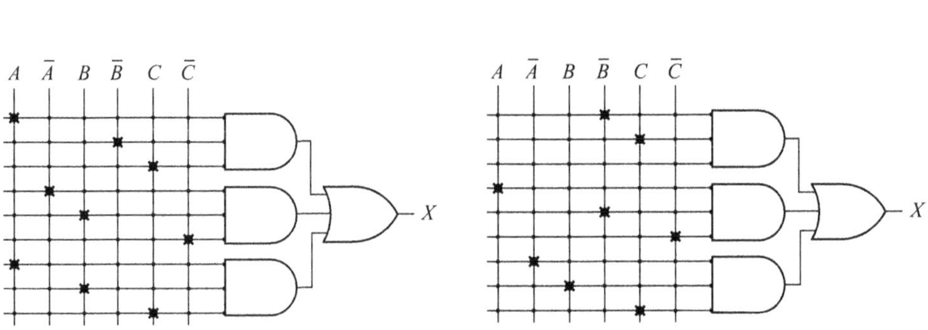

Figure 17.16. PML implementation of SOP expressions (adapted from Floyd 2000).

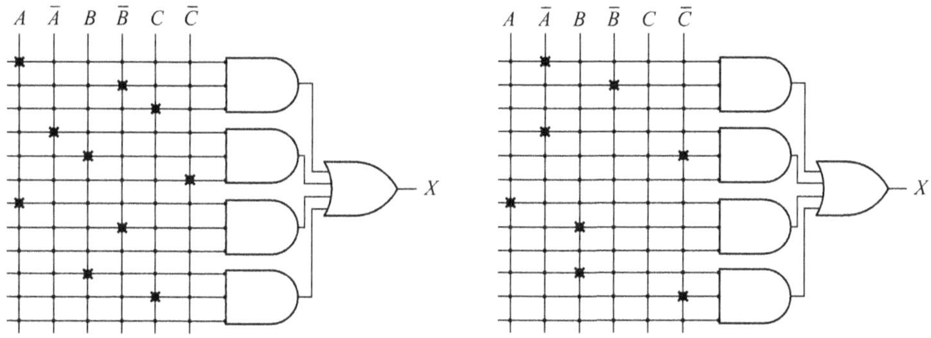

(a) $X = A\bar{B}C + \bar{A}B\bar{C} + A\bar{B} + BC$ (b) $X = \bar{A}\bar{B} + \bar{A}\bar{C} + AB + B\bar{C}$

Figure 17.17. PML implementation of SOP expressions (adapted from Floyd 2000).

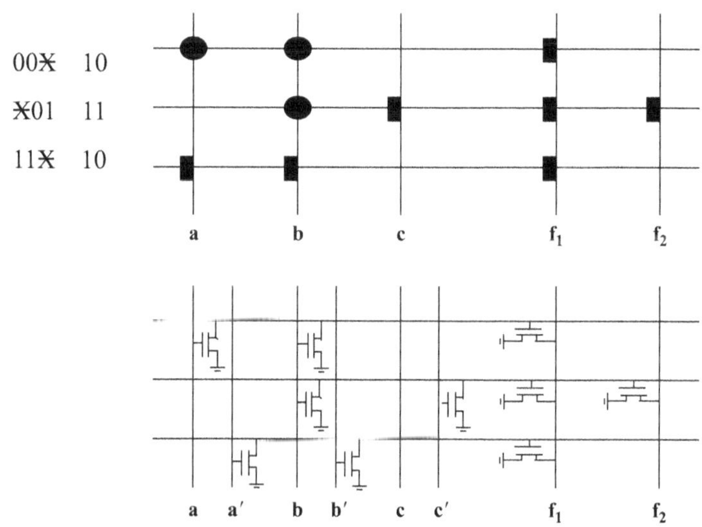

Figure 17.18. PLM implementation of the expressions $\mathbf{f_1 = a'b' + b'c + ab}$ and $\mathbf{f_2 = b'c}$ (adapted from De Micheli 2007).

Figure 17.18 shows a PLM implementation of the expressions $\mathbf{f_1 = a'b' + b'c + ab}$ and $\mathbf{f_2 = b'c}$ (De Micheli 2007).

17.10.3 GML implementing a sum-of-products (SOP) expression

As an example, a simple GML EDM is programmed as shown in figure 17.19 so that the product term $\bar{A}B$ is produced by the top AND gate, AB is produced by the middle AND gate and $\overline{AB}$ is produced by the bottom AND gate.

As shown, the E^2CMOS cells are on to connect the desired variables or their complements to the appropriate AND gate inputs. The E^2CMOS cells are off where a variable or its complement is not used in a given product term. The final output from the OR gate is an SOP expression.

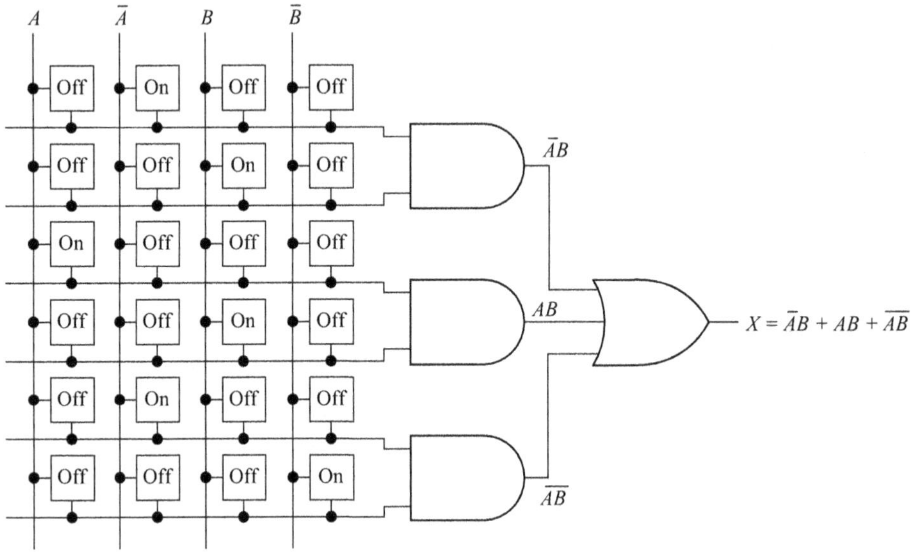

Figure 17.19. GML implementation of a SOP expression $X = \bar{A}B + AB + \overline{AB}$ (adapted from Floyd 2000).

17.10.4 Electrical commutation matrixer seven-segment display to put on show '5'

Figure 17.20 shows how to connect the electrical commutation matrixer seven-segment display shown in figure 17.18(a) to put on show '5'. The maximum value of the continuous forward current for each LED is 30 mA and a +5 V source is used.

The maximum forward current for each LED is 30 mA and a +5 V_{DC} voltage source is to be used.

17.11 Conclusion

PMLs are programmed by leasing specific fusible links intact and blowing open all the others. GMLs are programmed in a similar way, except that the E^2CMOS cells are turned on or off. The logic functions to be implemented determine which cells are affected. In order to program a PML or GML, the following items are required: a computer, programming software and a PML or GML programmer.

As mentioned earlier, there are several software packages for implementing logic design in PMLs and DMLs. ABEL and CUPL are two commonly used **hardware description languages** (HDL). Because both languages are similar and produce the same results in terms of programming a PCM or GCM, use is often a matter of preference and availability.

When considering tactile sensors, it is useful to begin by considering the fundamental physical quantities that can only be sensed through contact with the environment. The most important quantities measured with touch sensors are shape and force. Each of these may be measured as either an average quantity for some part of the robot or as a spatially resolved, distributed quantity across a contact area.

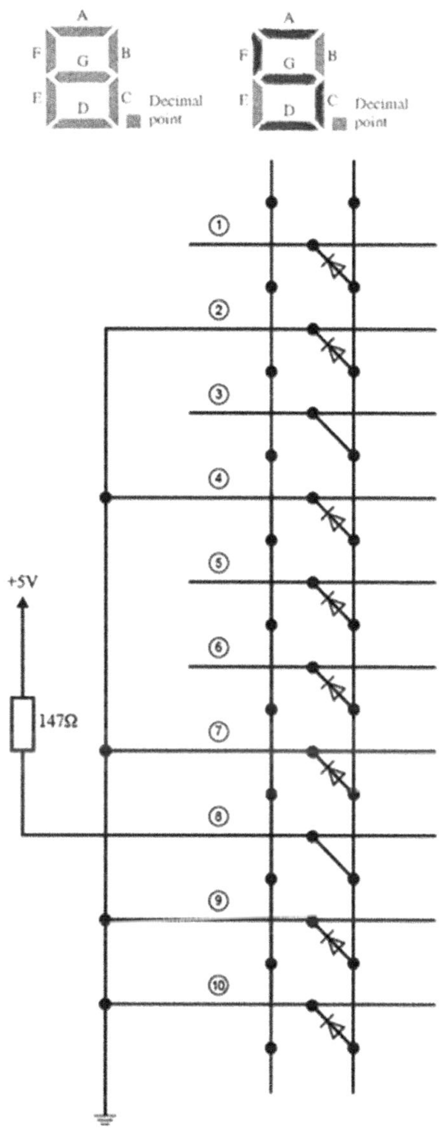

Figure 17.20. The ECM seven-segment display to put on show '5'.

PECMs and DECMs are fundamental to modern flat-panel displays. Thin-film-transistor LCDs have a transistor at each cross point, and so including an ECM crossbar as part of their structure might be considered.

For video switching in home and professional theatre applications, an ECM crossbar is used to make the output of multiple video appliances available to every monitor or every room throughout a building simultaneously.

In a typical installation, all the video sources are located on an equipment rack, and they are connected as matrixer inputs. Where central control of the commutation matrixer is practical, a typical rack-mounted ECM crossbar offers front-panel buttons to allow the manual connection of matrixer inputs to matrixer outputs. An example of such a use might be in a so-called 'sports bar', where numerous programs are displayed simultaneously.

In order to accomplish this, a sports bar would ordinarily need to purchase a separate cable or satellite subscription for each display for which independent control is desired. The ECM crossbar enables the signals to be re-routed on a whim, thus allowing the establishment to purchase only those subscriptions needed to cover the total number of unique programs viewed anywhere in the building.

Such ECM crossbars are used in high-end home theatre applications. The video sources that are typically shared include set-top cable/satellite receivers and DVD changers, while the same concept also applies to audio devices. The matrixer outputs are wired to televisions in individual rooms.

The ECM crossbar may be controlled (via an Ethernet or RS-232 serial connection) by a whole-house automation controller (such as those made by AMX, Crestron, or Control4), which provides the user interface that enables the user in each room to select which appliance to watch. The actual user interface varies according to system brand, and might include a combination of on-screen menus, touch-screens and handheld remote controls.

The ECM crossbar is necessary to enable users to select the program they wish to watch from the same room they will watch it from; otherwise it would be necessary for them to walk to the equipment rack.

The special ECM crossbars used in distributing satellite TV signals are called 'multimatrixers'. ECM crossbar technology opens up the opportunity to use computational resources to achieve automatic commutation or switching.

References

Broesch J D 1991 *Practical Programmable Circuits: a Guide to PLDs, State Machines, and Microcontrollers* (San Diego, CA: Academic Press)

Carter J W 1997 *Digital Programming with Programmable Logic Devices* (Upper Saddle River, NJ: Prentice Hall) p 418

Chen Y, Jung G-Y, Ohlberg D A A, Li X, Stewart D R, Jeppesen J O, Nielsen K A and Stoddart J F 2003 Nanoscale molecular-switch cross bar circuits *Proc. of the 2003 Nano-technology* **14** 462–8

De Micheli 2007 Two-level Logic Synthesis and Optimization. Presentation Centre SI: <Two-Level Exact Minimization.ppt>, Integrated Systems Centre, EPF Lausanne, Switzerland

Floyd Th L 1994 *Digital Fundamentals* (New York: Macmillan)

Floyd Th L 2000 *Digital Fundamentals* (Upper Saddle River, NJ: Prentice Hall)

Howe R D and Cutkosky M R 1989 Sensing skin acceleration for texture and slip perception *Proc. IEEE Int. Conf. Robotics and Automation (Scottsdale, AZ)*

Koivo H 2012 Challenging paradigms in modelling and simulation. *Whitepaper 104*, Theme (T.4), Helsinki University of Technology, Control Engineering Laboratory, FIN-02015 HUT, FINLAND http://www.imserv.com/pdf/whitepapers/white-paper_104.pdf

Mouttet B 2008 Logicless computational architectures with nano-scale crossbar arrays. *NSTI Nanotech 2008 Conf.*

Nicholls H R and Lee M H 1989 A survey of robot tactile sensing technology *Int. J. Robot. Res.* **8** 3–30

Ogorzalek M J 2007 Telecommunication Technologies, Week 11, Circuit Switching—Packet Switching *Presentation EIE 325 Telecommunication Technologies* (Dept. of EIE, Hong Kong Polytechnical University)

Pinto S F, Silva J F, Pellerin D and Holley M 1994 *Digital Design Using ABEL* (Upper Saddle River, NJ: Prentice Hall)

Pugh A 1986 *Robot Sensors: Tactile and Non-Vision* vol 2 (Berlin: Springer)

Sharma A K 1998 *Programmable Logic Handbook: PLDs, CPLDs, and FPGAs* (New York: McGraw-Hill)

Siciliano B and Khabit O (ed) 2008 *Springer Handbook of Robotics* (Berlin: Springer)

Chapter 18

Nano-magneto-rheological fluido-mechanical commutation matrixers for internal combustion engines

18.1 Introduction

The automotive world is moving towards **hybrid-electric vehicles** (HEV) in every respect other than actual conventional automotive vehicles. Enthusiasm for **all-electric vehicles** (AEV), a feature of the early 1990s, has declined a little. It is now certain that HEVs are within reach of the advanced **hybrid power source** (HPS), which is an advanced **internal combustion engine** (ICE) without **crankshaft and connecting rod (conrod) mechanisms** (C^2M), conceived and developed by the author (Fijalkowski 1986).

As a result, the generation changes in C^2M, which acts as **linear reciprocating motion to rotary motion** (LM–RM/RM–LM) and vice versa (**nano-magneto-rheological** (NMR) **fluido-mechanical commutation matrixer** (FCM) LM–RM/RM–LM commutator), are taking place throughout the world.

The author will focus on the main innovations and trends for tri-mode **hybrid-electric** (HE) **all-wheel-driven** (AWD) **drive-by-wire** (DBW) propulsion and **all-wheel-braked** (AWB) **brake-by-wire** (BBW) dispulsion mechatronic control systems, as well as a two- or four- or even five-stroke thermodynamic cycle, twin-opposed-piston, **crankless** reciprocating ICE, termed the 'Fijalkowski engine', with a **nano-magneto-rheological fluid** (NMRF) FCM LM–RM/RM–LM and/or RM–RM commutator.

Despite the recent emphasis on fuel cells, ICEs will continue to be the **chemo-thermo-fluido-mechanical** (Ch-Th-FM) energy converters of choice in automotive applications for the next few decades. However, the decades ahead are likely to be characterised by a fragmentation of the engine market, with the emergence of many competing technologies and a blurring of the distinction between engine

doi:10.1088/978-0-7503-1350-6CH018

families. The driving forces behind these technological developments are, and will continue to be, dominated by energy efficiency, emission regulations and drivability, coupled with great emphasis on reliability, manufacturability, cost, diagnostics, serviceability and recyclability. Emerging technologies in sensors, actuators and advanced mechatronical control systems are opening up many parallel paths, and the future of the ICE is likely to be shaped by the economic and regulatory climate on a global scale.

This section provides an overview of emerging trends, opportunities and challenges in ICE technology. Most conventional ICEs are reciprocating ones. Pistons that move back and forth inside enclosed cylinders power reciprocating IC engines. The LM of the pistons is converted to RM by a conventional crankshaft or geared crankshaft, bent axis, swash plate, wobble plate, transition arm or cam mechanism to convert LM to RM, or vice versa, to perform useful work such as turning the wheels of an automotive vehicle or the rotors of a helicopter.

The crankshaft, sometimes colloquially abbreviated to 'crank', is the part of an ICE that converts the LM of the pistons into the RM of the crankshaft. It typically connects to a flywheel to reduce the pulsation characteristic of the two- or four-stroke thermodynamic cycle, and sometimes to a torsion or vibration damper at the opposite end to reduce the torsion vibrations that are often caused along the length of the crankshaft by the cylinders furthest from the output end acting on the torsion elasticity of the metal.

The crankshaft is subjected to various forces, but it needs to be checked in two positions. Firstly, failure may occur at the position of maximum bending. In such a condition, the failure is due to bending and the pressure in the cylinder is maximal. Secondly, the crank may fail due to twisting, so the crankpin needs to be checked for shear at the position of maximal twisting. The pressure at this position is not maximal pressure, but a fraction of the maximal pressure. Thus, the most important mechanical motions of the ICE are the LM of the pistons and connecting rods (conrods) and the RM of the crankshaft. The latter converts the LM of the pistons into the RM of the crankshaft. Then, the RM can be output to the driving wheels of the automotive vehicle through the clutch, transmission and shafts. During the combustion and expansion processes, the high cylinder pressure pushes the piston motion to **bottom dead centre** (BDC). During compression and intake/exhaust processes, the inertial of the flywheel and the cylinder pressure of other cylinders drive the piston motion to **top dead centre** (TDC).

The novel ICE without a C^2M that the author would like to offer is based on another law of physics. The invention called the 'nano-magneto-rheological mechatronical commutator internal combustion engine', more concisely termed the 'Fijalkowski engine' elsewhere (Pinto-Silva 1987; De Falco 1989), is based on a completely novel propulsion engineering solution that has no analogies in the world. Furthermore, with the elimination of the conventional C^2M, its friction and mass are substantially reduced, which results in even greater mechanical energy output.

The Fijalkowski engine, as conceived in the 1980s and first presented and published on in October 1986 at the 8th International Electric Vehicle Symposium in Washington, DC (Fijalkowski 1986), has only one or more moving parts, the piston-rod assembly. It may be constructed of one or more pairs of directly opposed pistons with opposed cylinders and heads, and one or more IPM arrays fixed to the piston rod between them, or driven by an NMR FCM LM–RM/RM–LM and/or RM–RM commutator, which replaces the conventional C^2M. The piston-rod assembly may shuttle back and forth in a straight line from compression–ignition to compression–ignition in its opposing cylinders.

Mechanical energy from the combustion and expansion of a fuel mixture is concentrated in a straight line, and because the pistons do not experience angular loading in the cylinders (as seen in conventional crankshaft reciprocating ICE configurations), friction is substantially reduced, which translates into greater mechanical energy output.

The Fijalkowski engine has major advantages; it can burn a large variety of liquid-fossil as well as gaseous-nitrogen fuels, and has good potential for depollution.

The new technology, which was discovered by the author (Fijalkowski 1986), is a highly efficient method for converting LM to RM in advanced reciprocating ICEs, **mechano-fluidic** (MF) pumps or compressors, and **mechano-vacuum** (MV) pumps, as well as **fluido-mechanical** (FM) and **pneumo-mechanical** (PM) motors. This new method may also provide an effective way to vary the capacity of MF pumps and MP compressors, as well as the compression ratio of diesel and Otto engines.

The specific result is an advanced opposed piston, opposed cylinder compression ICE that is much more fuel-efficient than conventional ICEs and may therefore save users millions of dollars in fuel costs each year. Taking into account the instantaneous ICE friction, torque is decisive in recognising which ICE components are contributing the most towards the overall frictional losses of the ICE. The attenuation of frictional losses for such components may also rationalise the fuel consumption and increase the power output of the ICE. Therefore, one of our aims ought to be the development of a tool for reckoning the instantaneous friction torque of the C^2M.

The invention that is being offered is from the area of propulsion engineering, specifically the area of the opposed-piston, opposed-cylinder ICE, which is widespread. It is a popularly known fact that ICEs have a C^2M, except for rotary ICEs.

The C^2M has a crankshaft, which is a key part of that mechanism. Manufacturing and servicing that part is costly. According to various sources, about 30% of the cost of the entire ICE comes from the crankshaft alone. The presence of a C^2M causes the reciprocating ICE to vibrate, and this leads to many technical problems. Therefore, attempts to create an ICE without a C^2M have been made for quite a while, however, to date the majority of ICEs still use them. So-called rotary ICEs are an exception (e.g. Wankel engines), and they are used extremely rarely because of their difficult servicing procedures and complicated structure. The necessity of separating the lubrication system will lead to designs with outside scavenging, such as mechanical blowers, and maybe a design solution similar to the Fijalkowski engine can be found.

Figure 18.1 shows the design animation of the Fijalkowski engine's NMR FCM LM–RM/RM–LM commutator. The four pairs of opposing pistons of a 'boxer' ICE arrangement are solidly connected to each other as well as to an NMR

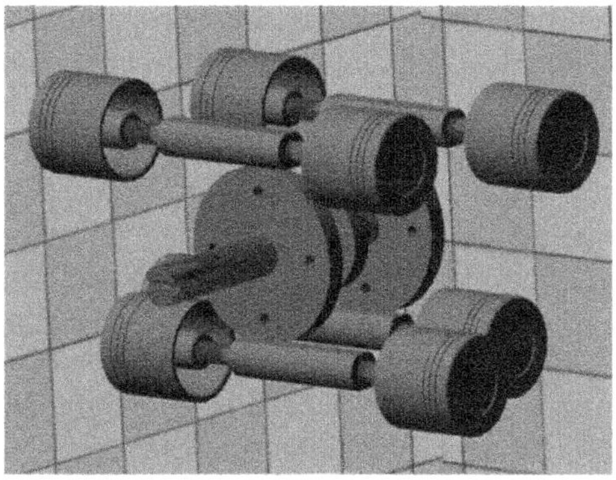

Figure 18.1. Design of the Fijalkowski engine's NMR FCM LM–RM/RM–LM commutator.

mechatronical commutator by means of round piston rods. There are separating walls between the NMR FCM LM–RM/RM–LM commutator case and the underside of the pistons. Because the piston rods are guided and are supported by bearings in the separation wall, the pistons hardly touch the cylinder walls, there are no side forces, and there is no piston tipping at the dead centres. As is the case with the conventional two-stroke thermodynamic cycle reciprocating ICE, the bottom side of the pistons may function as charge pumps, one for each cylinder. Because the ceramic pistons do not need to be lubricated, the combustion air will remain free of lubricating oily fluid.

As can be seen in this Fijalkowski engine, the two-stroke thermodynamic cycle may allow us to place great expectations in novel non-metallic composites such as fibreglasses (e.g. Kevlar and Technora) or ceramic materials. Ceramics are assumed to construct a highly loaded piston that will be thermally stable. It seems possible that cylinders constructed from ceramics will develop less thermal energy (heat) distortion and less wear. The components of the Fijalkowski engine that may be constructed from ceramic materials are: pistons; piston rings (if any); piston rods' mover (forcer) blocks inside the NMR FCM LM–RM/RM–LM and RM–RM commutator; cylinder sleeves; and exhaust channels. There are many advantages in the design of the Fijalkowski engine.

Because there is no gudgeon pin or short skirt, the piston will only be about half the mass of its conventional counterpart. This in turn means that the primary vibration forces may be reduced by about 50%. Since the pistons may be connected rigidly, because of the absence of a gudgeon pin, the angular motion of the sliding followers (piston rods), and therefore the secondary vibration forces, may be eliminated. This all means that such an unconventional reciprocating ICE with fewer cylinders than a conventional one does not have unacceptable levels of vibration. Because the combustion-gas forces act perpendicularly to the opposed pistons, their resultant (as the total outcome of these forces) in the Fijalkowski engine may always be reacted to by pure compression in the sliding followers (piston rods). This and the reduction of lateral acceleration forces owing to the light mass of the opposed pistons may virtually eliminate piston slaps. The value of the inertia force will be smaller than the value of the combustion-gas force between the rings and the bores at all times, while their maximum values do not occur simultaneously—the lateral components of the inertia force are zero at the end of each stroke.

Consequently, ring flutter should not occur and the rate of wear for the rings and bores should be low. By virtue of the shortening of the piston skirt, the oily-fluid drag (if any) ought to be about 5% of that of a conventional piston. Therefore, the cold cranking (starting) of crankless ICEs should be easier, and so a smaller **starting, lighting and ignition** (SLI) **chemo-electrical/electro-chemical** (ChE/ECh) storage battery and an ECM DC–AC commutator starter motor would be needed.

In addition, the length of the cylinder bore to be machined might be only about two-thirds of that required conventionally. Many of these advantages would be of special benefit for HEVs, and the author has also designed two- or four- or even five-stroke thermodynamic cycle versions with triangular, rectangular and oval twin-opposed pistons, instead of circular twin-opposed pistons, which may also be utilised.

The Fijalkowski engine concept presented below may operate on the **high-pressure direct injection** (HDI) and constant-volume four-stroke thermodynamic cycle principle. The twin-opposed pistons move almost symmetrically in their horizontal ICE cylinders, and each of their LMs may be converted into the RM of the ICE's output shaft by means of a pair of sliding followers (piston conrods) with their bearings, and the main shaft's NMR rotary ratchets, and the NMR rotary clutch, with the NMR rotary ratchets' mechanical harmonic oscillators shifted at π rad. This engine has a simple and low-cost supercharging system, which may have a compact volume and the freedom to position its components. The HEV may have lower values of under-bonnet (under-hood) temperature.

The Fijalkowski engine may have small dimensions, but it will produce much more power than a conventional reciprocating IC engine. With small cylinder sizes (three times smaller than a conventional reciprocating IC engine's), this particular crankless reciprocating IC engine may develop two–three times more power than a conventional one. This crankless reciprocating IC engine may be described as 'square'—a term indicating that the bore and stroke are equal. This engine may have a high velocity, and it may have no C^2M, but it may have only piston rods and pistons, and it may be successfully utilised in technologies and products that necessitate high-velocity small-size crankless reciprocating ICEs.

18.2 The Fijalkowski engine concept

An advanced **hybrid power source** (HPS), termed the Fijalkowski engine, has been preferred to other solutions with conventional crankshaft reciprocating ICEs. This HPS, that is, the Fijalkowski engine (see figure 18.2), with its silent 'four-in-one', electronically commutable, magneto-electrically excited, high-temperature superconductor, brushless ECM AC–DC/DC–AC commutator, **interior permanent magnet** (IPM) flywheel/damper generator/motor, may provide **chemo-thermo-fluido-mechano-electrical** (Ch-Th-F-ME) or Ch-Th-FM energy conversion for HE AWD DBW propulsion.

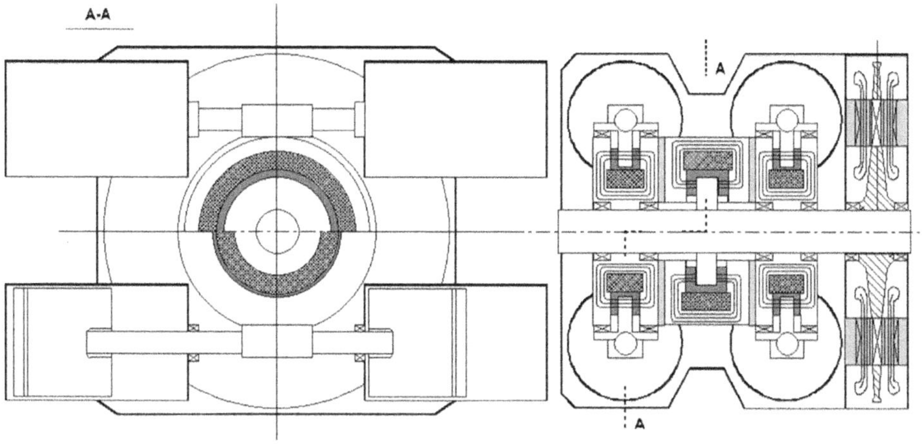

Figure 18.2. Layout of the Fijalkowski engine with its NMR FCM LM–RM/RM–LM and RM–RM commutator and two-, four-, or even five-stroke thermodynamic cycle rigidly connected twin-opposed piston configuration (A) Transverse section and (B) longitudinal section (Fijalkowski 1986, 1988d, 1989, 1999b, 2006).

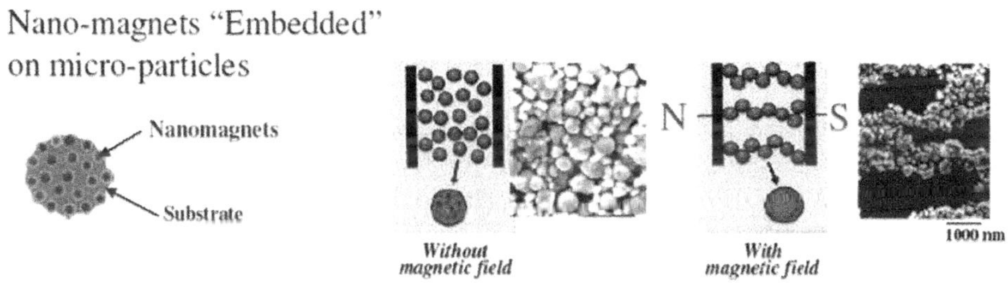

Figure 18.3. Nano-magneto-rheological fluid (NMRF) action. Credit: Centro Richerche, FIAT (Folonari 2006).

At the very heart of this novel HPS is a **nano-magneto-rheological fluid** (NMRF). This is essentially a suspension of micro-sized, magnetically polarisable nano-magnets embedded on micro-particles in oily fluid or some other liquid. Under normal operating conditions, NMRF is a free-flowing liquid with a consistency similar to that of ICE oily fluid (Folonari 2006).

Exposure to a magnetic field, however, may transform the NMRF into a nearly solid state in milliseconds. Just as quickly, the NMRF may return to its liquid state with the removal of the magnetic field. The degree of change in an NMRF is proportional to the magnitude of the applied magnetic field. When it is subjected to a magnetic field, an NMRF actually develops yield strength and behaves as a Bingham solid. The flow ratio of NMRFs may vary with the application of a magnetic field, as shown in figure 18.3 (Folonari 2006).

The advantages of working with nano-magnets as compared to micro-particles are as follows:

- reduced wear;
- reduced aggregation;

- reduced power consumption;
- faster response time.

The applications may be as follows:
- high-speed NMR mechanical components;
- active NMR shock absorbers (dampers);
- NMR rotary ratchets;
- NMR brakes;
- NMR clutches.

The change may appear as a very large change in effective viscosity. NMRF is similar to a **giant-electro-rheological fluid** (GERF), but it is 20–50 times stronger (Wen *et al* 2005, Novak 2005). It may also operate directly from a low-voltage (14 V_{DC} or 28 V_{DC}, or even 42 V_{DC}) automotive onboard **electrical energy distribution** (E^2D) system, while it is far less sensitive to contaminants and extremes in temperature. Applied to a novel NMR mechatronical commutator prime mover, that is, the Fijalkowski engine, NMR technology may provide flexible mechatronical control capabilities in designs that may be far less complicated and more reliable than conventional crankshaft reciprocating ICEs.

With an efficient low-voltage E^2D system and a response time of a few milliseconds, two fully controllable NMR rotary ratchets, which are real bipolar mechanical valves in a novel NMR mechatronical commutator prime mover, offer one of the most effective means available for interfacing mechanical components with an HE AWD DBW propulsion mechatronical control system (Fijalkowski 1986, 1997, 2011).

The simple design of the NMR rotary ratchets, which are real NMR fully controllable bipolar mechanical valves with few moving parts, translates into high reliability and quiet operation.

NMR rotary ratchets are real NMR rotary-ratchet composite mechanical harmonic oscillators provided with NMR fully controllable bipolar mechanical valves, operating with a mechatronical catch or a pawl that prevents the outer composite mechanical harmonic oscillator from slipping back and allows them to move in only a single sense of rotary direction (rotation).

They are smooth acting, with proportional brakes that are more compact and require substantially less power to operate than magnetic-hysteresis or eddy-current ones. Their simplicity and ease of mechatronical control makes them a cost-effective choice for a novel NMR prime mover. They may be capable of high torque and low velocity, and can be used in lock/unlock applications.

The author has focused his attention on the complexity as well as the physical mass and size of the conventional IC engine's gudgeon-pin arrangement, the close tolerances entailed in their manufacture and the time spent on assembly. The author conceived and developed an advanced HPS, termed the Fijalkowski engine, that is nearly conventional except that its pistons and their sliding followers (piston rods) are rigidly connected (see the transverse section in figure 18.2(A)).

The Fijalkowski engine in question may have eight ICE cylinders that are arranged co-axially with the ICE output shaft and rotate the latter by means of two NMR rotary ratchets and a single NMR rotary clutch on it, and sliding followers (piston rods). The two pistons of each pair of ICE cylinders are of the opposed type and are therefore dynamically balanced. The Fijalkowski engine's pistons longitudinal motion, which is LM, may be converted to RM by unconventional cylindrical NMR rotary ratchets, with sliding followers (piston rods), mounted in the housing, engaging sinusoidal harmonic oscillators around the shanks of the piston assembly units, while splines couple the pistons' rotation to an output shaft passing through the axle of housing, as shown in figures 18.2 and 18.4. If further development proves to be satisfactory, it could have a considerable impact on the Fijalkowski engine design. For vehicular applications, high thermal efficiency is important for good fuel economy, but it is also important that the power plant is simple and compact.

Preliminary design optimization studies resulted in an advanced NMR mechatronical commutator, four-stroke thermodynamic cycle, reciprocating ICE concept with a relatively high thermal efficiency. This IC engine has not generally been favoured for HEV applications in the past. However, use of a direct-coupled automotive silent four-in-one composite flywheel damper onboard **mechano-electrical** (ME) generator/starter **electro-mechanical** (EM) motor, with macro- and microelectronic control of the electrical energy transmission, may make the constant IC engine's output shaft angular velocity running feasible, maintaining near-optimal ICE conditions over a wide power range.

The Fijalkowski engine, that is, the crankless reciprocating IC engine with reciprocating twin-opposed pistons, may have an output shaft connected to an NMR rotary clutch with MRF rotary ratchets, and several radially located opposed pistons engaging NMR rotary ratchets' mechanical harmonic oscillators, mechatronically controlled so that their power strokes balance the opposing pistons (see figure 18.4).

The crankless reciprocating ICE may include housing, and an output main shaft radially connected to the NMR rotary ratchets. Eight ICE cylinders in the housing may be spaced about the NMR rotary clutch's axis of rotation, so that each cylinder may be coupled to a sliding follower (piston rod) that engages the NMR rotary

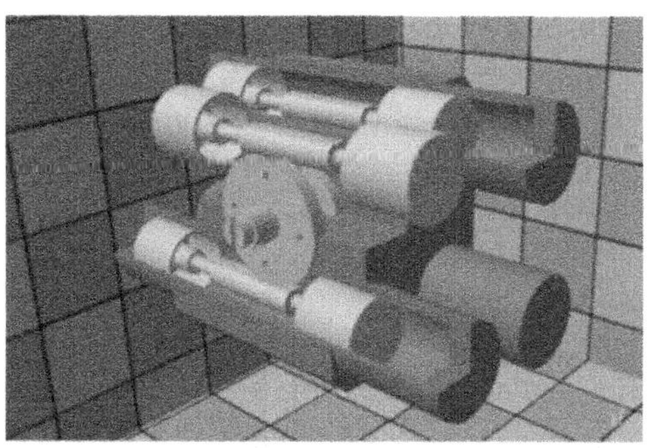

Figure 18.4. Design of the Fijalkowski engine.

ratchets so that the piston's movement between the BDC may drive it round, thus imparting torque through the NMR rotary clutch to the output shaft.

The Fijalkowski engine's thermodynamic cycle may be mechatronically controlled so that the power stroke of any one piston precedes that of the one that is the most radially distant from that one and the approaching TDC. This may keep the crankless ICE well balanced.

The Fijalkowski engine may have low friction vibration, overall mass and combustion pressure, resulting in high efficiency, low NO_x emissions, low component wear, and cool and quiet running.

The Fijalkowski engine will be an unconventional ICE, with eight ICE cylinders of approximately the same length as conventional IC engine cylinders, giving the same power output as a conventional boxer IC engine, and eight or 12 opposed pistons with sliding followers (piston rods) and a rotating output shaft, four-in-one, automotive silent, electronically commutable, magnetoelectrically excited, high-temperature superconductor, brushless ECM AC–DC/DC–AC commutator, composite-flywheel/damper onboard generator/starter motor that can convert the mechanical energy supplied to the opposed pistons by the combustion of the usual hydrogen or hydrocarbons directly into electrical energy.

The Ch-Th-M part of the Fijalkowski engine may have the following advantages:

- use of the constant-volume four-stroke thermodynamic cycle principle;
- use of automotive mechatronics to ensure the starting, synchronization and delivery of usable mechanical energy, and the ability to ignite either of the opposed pistons on many short circular ICE cylinders;
- use of ICE output shafts with four-in-one generator/motor rotor inertia to return the opposed pistons;
- use of sliding-follower/output-shaft NMR FCM LM–RM/RM–LM and RM–RM commutators to replace the C^2M, which may permit the elimination of the inlet (intake) and outlet (exhaust) valves, pushrods, lifters, camshafts, timing gear and timing belt or chain required in conventional ICEs; thus it might require fewer parts and it might be simpler to manufacture, while lowering the consumption of metal reserves, and it may not require a separate ICE flywheel.

Then there is the ME/EM part of the Fijalkowski engine, that is, the four-in-one, automotive silent, electronically commutable, magnetoelectrically excited, high-temperature, superconductor, brushless-type ECM AC–DC/DC–AC commutator, composite-flywheel/damper on-board generator/starter motor (see figure 17.20). It may have two 2×5-phase, pentagon-connected, wire-wound (slotted-core), disc-shaped outer stators and an unwound IPM disc-shaped inner rotor (also see figure 18.5).

Armature alternating voltages, which are real ECM AC–DC commutator-input alternating voltages, can be rectified to armature direct voltages, that is, ECM AC–DC commutator-output direct voltages, or alternatively, to armature direct voltages, that is, ECM DC–AC commutator-input direct voltages. The latter can be inverted to armature alternating voltages, which are real ECM DC–AC commutator

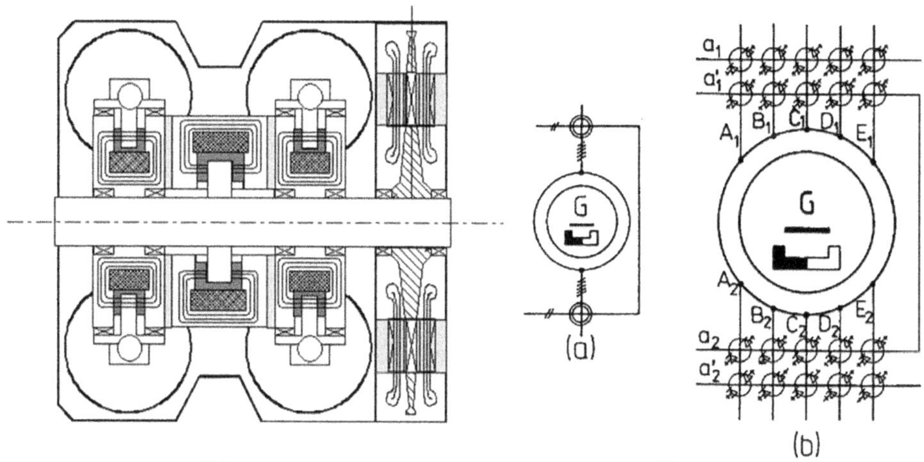

(A) (B)

Figure 18.5. Layout of the Fijalkowski engine. (A) The silent four-in-one, electronically commutable, magnetoelectrically excited, high-temperature superconductor, brushless ECM AC–DC/DC–AC commutator. (B) IPM flywheel/damper generator/motor.

output alternating voltages, modulated in both amplitude and frequency in **sinusoidal pulse-width modulation** (SPWM) mode to produce armature sinusoidal phase-current waveforms in the electrical energy, or vice versa, in advantageous conditions with regard to mass and volume as well as efficiency.

The Fijalkowski engine may be designed to perform three key functions: reduce fuel consumption by 20% compared to previous reciprocating IC engines, increase drivability and minimize exhaust emissions.

The first objective may be achieved by implementing the latest fuel HDI technology—**common rail** (CR)—which involves mechatronically controlled high-pressure injection. The Fijalkowski engine architecture may also be designed to reduce fuel consumption.

The second objective is to persuade consumers to forget about the Fijalkowski engine's poor image. The HDI may provide 50% additional torque at low values of the engine's output shaft angular velocity and add more power while reducing noise and vibration. Superior environmental performance may also be achieved.

The Fijalkowski engine may be one of the cleanest reciprocating ICEs on the automotive market, while offering considerable potential for further development in the future. The longitudinal section of the NMR reciprocating prime mover, that is, a crankless, reciprocating, constant-volume, two- or four-stroke thermodynamic cycle, eight-cylinder, opposed IC engine, called the Fijalkowski engine, is shown in figure 18.6.

Its transmission, by which power may be transmitted from the ICE to the live axle(s) in an HEV, may represent a transmission drum, which consists of an outer NMR rotary ratchet NMR FCM LM–RM/RM–LM commutator (see figure 18.7), acting as a mechatronical LM–RM rectifier (during ICE operating action), or an NMR FCM RM–LM inverter (during ICE cranking action), a floating composite mechanical

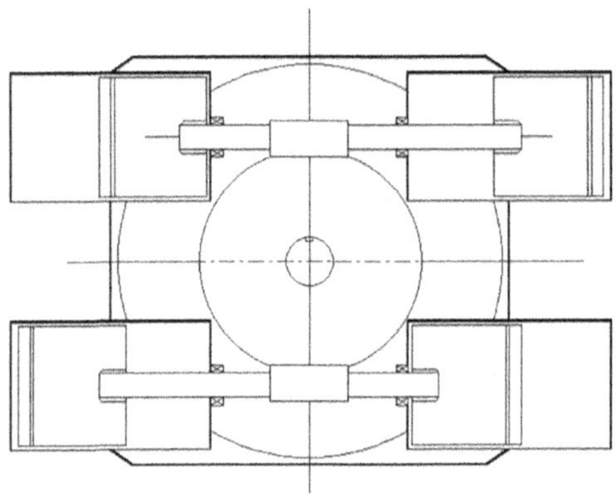

Figure 18.6. Longitudinal section of the Fijalkowski engine.

harmonic oscillator, and an inner NMR rotary ratchet's composite mechatronical common drum.

The reciprocating motion of two parallel acting IC engine pistons' straight piston rods of the crankless, constant-volume, four-stroke, thermodynamic cycle, eight-cylinder opposed reciprocating ICE may be transferred tangentially to a pair of NMR rotary ratchets, which are a real NMR rotary ratchet's fully controlled bipolar mechanical valves.

The NMR rotary ratcheting action may result from locking of the NMR rotary ratchet's composite mechanical harmonic oscillator when a magnetic-field exciter's current pulses solidify the NMRF. This may cause the inner NMR rotary ratchet composite mechanical common drum to rotate with uniform RM (see figures 17.18–18.8).

The uniform RM may in turn be transferred to a central NMR rotary clutch's composite mechanical common hub by a continuously variable-velocity NMR rotary clutch. The latter controls the mechanical energy throughout (see figure 17.18).

The layout of the Fijalkowski engine's NMR FCM LM–RM/RM–LM and RM–RM NMR commutators is shown in figure 18.7.

Mechanical energy is mechatronically controlled by variation of the magnetic-field-exciter current of the outer NMR rotary-clutch's composite mechanical common hub's NMR layer, or by current-pulse controlled 'quick passage' of the outer-clutch composite mechanical common hub from pushing to dragging current pulses.

A very advanced novel HPS that is the Fijalkowski engine (see figures 18.1–18.8) supplies reciprocating mechanical energy to a pair of its straight piston rods, which in turn supply reciprocating mechanical energy tangentially to a pair of NMR rotary-ratchet composite mechanical harmonic oscillators, that is the NMR rotary-ratchets' fully controlled bipolar mechanical valves.

Synchronized current pulses of the NMR rotary-ratchet magnetic-field-exciter latch the presently forward RM outer rotary-ratchet's composite mechanical harmonic oscillator to an inner NMR rotary-ratchet's composite mechanical common drum, while the alternate outer NMR rotary-ratchet composite mechanical harmonic oscillator slips backward to reset.

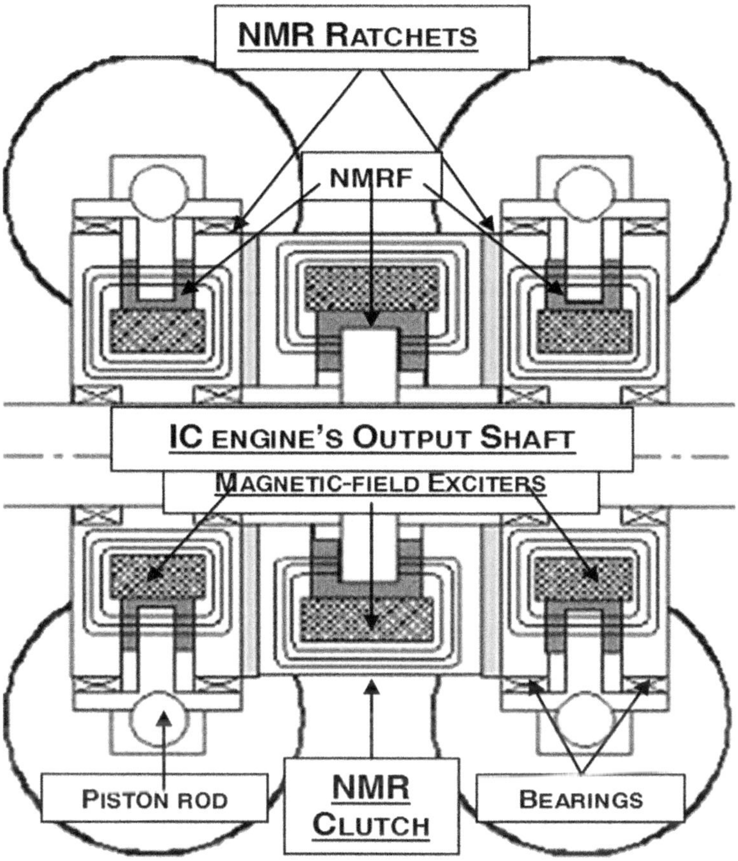

Figure 18.7. A transmission drum that consists of an outer NMR rotary ratchet NMR FCM LM–RM/RM–LM commutator acting as a floating composite mechanical harmonic oscillator, and an inner NMR rotary ratchet composite mechatronic common drum.

Upon resetting, the situation is reversed, and the previously latched NMR rotary ratchet deactivates and slips backward to reset while the previously slipping NMR rotary ratchet is reactivated to latch onto a forward stroke of the straight piston rod.

In this manner, the sum of the two alternating NMR rotary ratchets action provides continuous forward RM to the central NMR rotary-clutch's composite mechanical common hub (see figures 18.7 and 18.8).

18.2.1 Magnetic-field-exciter's electric-current sequencing for rotary motion

The reversal of RM can be realised easily by swapping the NMR rotary ratchets' magnetic-field-exciter's electric-current-pulse sequencing, as shown in figure 18.9. Thus, the sense of RM direction (rotation) is changed by skipping one pulse on the left-hand NMR rotary-ratchet's fully controllable bipolar mechanical valve to allow the left-hand NMR rotary-ratchet composite mechanical flywheel cylinder to slip, while one NMR rotary-ratchet's magnetic-field-exciter's electric-current pulse is

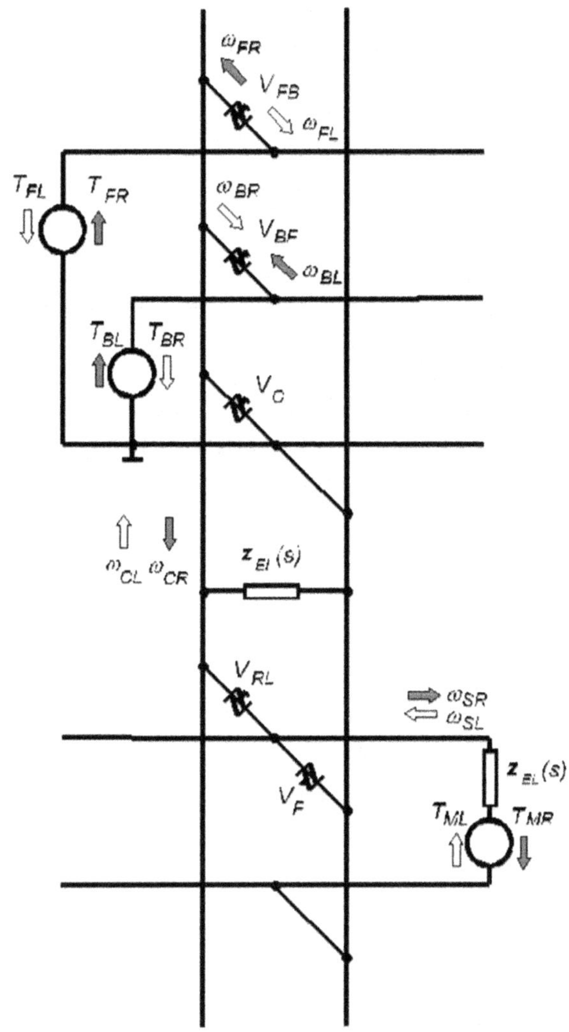

Figure 18.8. Layout of the Fijalkowski engine's NMR FCM LM–RM/RM–LM and RM–RM commutators. T_{FR}, T_{FL}— NMR rotary-ratchet composite harmonic oscillators' forward torques for the right- and left-hand IC engine output-shaft rotations, respectively; T_{BR}, T_{BL}—NMR rotary ratchet composite harmonic oscillators' backward torques for right- and left-hand IC engine main-shaft rotations, respectively; T_{MR}, T_{ML}—ECM AC–DC/DC–AC commutator flywheel/damper generator/motor electromagnetic torques for right- and left-hand IC engine main-shaft rotations, respectively; ω_{FR}, ω_{FL}—NMR rotary-ratchet composite harmonic oscillators' forward angular velocities for right- and left-hand IC engine main-shaft rotations, respectively; ω_{BR}, ω_{BL}—NMR rotary-ratchet composite harmonic oscillators' backward angular velocities for right- and left-hand IC engine output shaft rotations, respectively; ω_{CR}, ω_{CL}—NMR rotary-ratchets and clutch composite mechanical common drum angular velocities for right- and left-hand IC engine output shaft rotations, respectively; and ω_{SR}, ω_{SL}—NMR rotary-ratchets and clutch composite common drum angular velocities for right- and left-hand IC engine output shaft rotations, respectively.

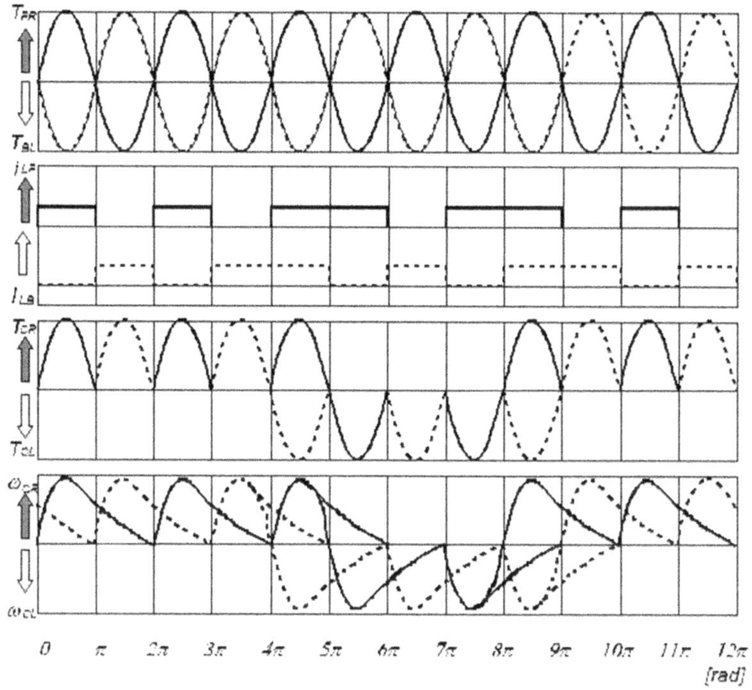

Figure 18.9. Layout of the Fijalkowski engine's NMR FCM LM–RM/RM–LM and RM–RM commutation waveforms for its NMR FCM LM–RM/RM–LM and RM–RM commutator: T_{FR}, T_{FL}—NMR rotary-ratchet's composite mechanical harmonic oscillators' forward torques for right- and left-hand IC engine output shaft rotations, respectively; T_{BR}, T_{BL}—NMR rotary ratchet's composite mechanical harmonic oscillators' backward torques for right- and left-hand IC engine output-shaft rotations, respectively; T_{MR}, T_{ML}—ECM AC–DC/DC–AC commutator IPM flywheel/damper generator/motor electromagnetic torques for right- and left-hand IC engine output-shaft rotations, respectively; ω_{FR}, ω_{FL}—NMR rotary-ratchet composite mechanical harmonic oscillators' forward angular velocities for right- and left-hand IC engine output-shaft rotations, respectively; ω_{BR} ω_{BL}—NMR rotary-ratchet composite mechanical harmonic oscillators' backward angular velocities for right- and left-hand IC engine output-shaft rotations, respectively; ω_{CR}, ω_{CL}—NMR rotary-ratchets and clutch composite mechanical common drum angular velocities for right- and left-hand IC engine output-shaft rotations, respectively; and ω_{SR}, ω_{SL}—NMR rotary-ratchets and clutch composite mechanical common drum angular velocities for right- and left-hand IC engine output-shaft rotations, respectively.

added to the right-hand NMR rotary-ratchet's fully controllable mechanical valve to latch the right-hand rotary-ratchet's composite mechanical common drum to the backward RM. The NMR rotary-ratchet's magnetic-field-exciter electric-current-pulse sequence is in the opposite phase and the transmission in the drag mode supplies reverse torque to the central NMR rotary-clutch's composite hub.

18.2.2 Engine output-shaft angular velocity control

The inner NMR rotary-ratchet's composite mechanical common drum contains the central NMR rotary-ratchet's composite mechanical common hub and is separated from its NMR layer by means of the diamagnetic-material layer. Transmission of torque between the inner NMR rotary-ratchet's composite mechanical common drum and a separate clutch magnetic-field-exciter's electric-current-pulse driver achieves the central NMR rotary-ratchet's composite mechanical common hub. During acceleration, one of two methods for controlling the torque transmission may be taken into account.

Either the NMR rotary-clutch's magnetic-field-exciter's applied voltage may be progressively raised to solidify the NMR, or the NMR rotary-clutch's magnetic-field-exciter's electric-current-pulse control techniques may be used to vary the electric-current-pulse width and/or the frequency of the force impulses between the inner NMR rotary-ratchet's composite mechanical common drum and the central NMR rotary-ratchet's composite mechanical common hub by means of the RM–RM NMR mechatronical commutator acting as a bipolar four-quadrant RM–RM NMR mechatronical chopper, with the **pulse-width-modulation** (PWM), and/or the **frequency modulation** (FM).

18.2.3 Engine output-shaft deceleration and reverse

The action for slowing the Fijalkowski engine comprises sequential operations.

Firstly, by switching off the applied control voltage, the NMR layer between the inner NMR rotary-ratchet's composite mechanical common drum and the output central NMR rotary-clutch's composite common hub (see figures 17.4 and 17.5) is deactivated.

Secondly, the electric-current-pulse sequence will trip shortly to give reverse RM to the inner NMR rotary-ratchet's composite mechanical common drum.

Thirdly, the NMR rotary-ratchet's composite mechanical common drum will be progressively reconnected to the output central NMR rotary-clutch's composite mechanical common hub by one of the methods mentioned above to transmit reverse torque to decelerate the Fijalkowski engine. This introduces another unique feature of the advanced NMR mechatronical commutator, twin-opposed-piston-type reciprocating ICE. If the deceleration torque continues after the reciprocating IC engine has come to a halt, the Fijalkowski engine may begin to move backwards. This can facilitate parking and HEV manoeuvring. It should be pointed out here that the conventional AWB BBW dispulsion mechatronical control system might have to be retained, at least for emergency stopping.

18.3 Fijalkowski engine cooling

The performance and efficiency of conventional ICEs and ECEs, as well as MMD electrical machines (ME/EM dynamotors and EE transformers), is limited by thermal constraints in the cooling channels and windings. These constraints can be reduced somewhat by circulating liquid through liquid cooling channels or air through the air gaps in the ICEs and ECEs and the MMD electrical machines. However, such approaches necessitate the addition of pumping mechanisms and are limited by the thermal conductivity of air or liquid.

A novel approach has been taken to cooling the Fijalkowski engine, using ferrofluids as the cooling media. The aim is to take advantage of the thermal and magnetic characteristics of ferrofluids, such that not only will the ferrofluid circulate through the Fijalkowski engine, but the existing thermal and magnetic fields in the Fijalkowski engine will provide the energy required for pumping the ferrofluid. The cooling mechanism will therefore not require any additional moving mechanical parts, pumps, compressors or sensors.

The behaviour of the ferrofluid will provide the pumping and cooling method for the Fijalkowski engine (thus increasing the Fijalkowski engine's potential energy density and efficiency). The necessary challenge has been the synthesis of ferrofluid magnetic nanoparticles that possess the required ferrofluidic magnetic and thermal properties. There ought to be particular focus on the chemical and biological synthesis of ferrofluid magnetic nanoparticles that exhibit the necessary properties.

18.4 The Fijalkowski engine advantages versus conventional internal combustion engines

As compared to conventional reciprocating ICEs, the Fijalkowski engine has the following advantages.

- It does not have a C^2M (there is no need to utilise the ICE's oily fluid lubricant), crankcase, cooling jacket or coolant; the demands made on the metal or ceramics that the block of cylinders are made of are thereby lowered.
- It does not have a radiator, an MP fan, an MF pump, a cylinder head, an overhead camshaft or related auxiliary systems.
- The torque of the efficient force in it may be increased many times over without increasing the piston's power stroke; for example, if a conventional reciprocating ICE has 1.5 revs during idle strokes in the working range, there are no idle strokes in it at all; the torque may be increased sharply in it because of the jet of the exhausts.
- It will experience almost none of the vibrations that present conventional reciprocating ICEs with a C^2M have—which eventually destroy the drive-shaft and weaken the joints in HEVs.
- Without the C^2M, it may be almost perfectly balanced, reducing the bearing loads to a minimum.
- Without the side loading common to conventional reciprocating ICEs, mechanical efficiency may be improved by 5–8%, as demonstrated by estimates, according to its inventor.
- It may show a threefold increase in power-to-mass over diesel engines, and it may be run in reverse as an MP compressor or even an MF pump.

What is wrong with the highly developed conventional C^2M? Nothing at all is wrong with the crankshaft. The problem with the mechanism lies in the statement that the piston connecting rods flip from side to side with each cycle, in effect shortening their length in the process.

If one imagines that the crank is extremely smooth, the 'big-end' bearing may be described as a sinusoidal motion, in view of the fact that the 'flipping' has altered it. To illustrate this effect, the Fijalkowski engine's sinusoidal harmonic piston motion

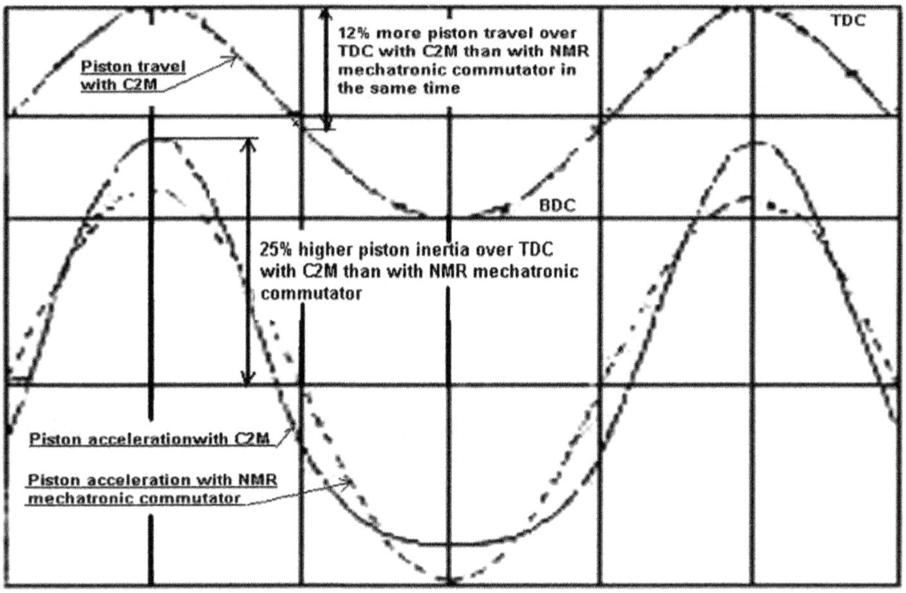

Figure 18.10. The Fijalkowski engine's sinusoidal harmonic piston motion compared with the conventional reciprocating ICE's crankshaft-driven piston motion.

is compared with the conventional reciprocating ICE's crankshaft-driven piston motion in figure 18.10.

Are the conventional C^2M mechanisms really so defective? Not if drivers do not necessitate higher values of the ICE's output crankshaft angular velocity, and they may take steps to prevent continued dangerous resonance.

Increased values of the IC engine's crankshaft angular velocity mean increased ICE power output—without an increase in size and mass. For example, the average crank reciprocating ICE is approaching its potential at 12 000 rpm, and the average crankless reciprocating ICE with the NMR FCM LM–RM/RM–LM commutator, that is, the Fijalkowski engine, may reach 20 000 rpm.

The NMR ratchet mechanism is efficient, it is smooth and it is easy to engineer. It may reduce the number of significant moving parts, it does not necessitate high-pressure oily fluid lubrication, it is less expensive, it requires less skill and it takes precise sinusoidal harmonic motion as the C^2M (see figure 18.10), so it does not generate spurious harmonics, and the piston movement is more efficient, a plus that may rise with the crankless reciprocating IC engine's output-shaft angular velocity to become very significant at 20 000 rpm[1].

Controllable NMRF stems are interesting because of their ability to provide simple, quiet, rapid-response (within few milliseconds) interfaces between mechatronical controls and mechanical systems. The ability of a controllable NMRF to be utilised directly as an NMR ratchet's fast-acting, fully controlled bipolar mechanical

[1] NB Although 20 000 rpm seems very fast by modern reciprocating IC engine standards, the Fijalkowski engine can be expected to reach 40 000 rpm with further development.

valves (with no moving parts) in an NMR FCM LM–RM/RM–LM commutator is one of the principal motivating factors for the development of such NMRFs.

Fijalkowski engines in the 20 000–40 000-rpm range may be used for stationary applications such as **distributed generation** (DG), **combined heat and power** (CHP) and shaft power, as well as other stationary power plant applications.

Let us look at the negatives. It looks funny, seems to be the most common reaction, but the author may skip that one. It does necessitate oily-fluid slides for lateral stabilisation. This is because a piston must have a little 'play', that is to say, it must not scrape the piston walls, and because the rings that may be employed to effect this gap without loss of compression have to allow for thermal energy (heat) expansion, a tiny amount of slap may occur at mid-stroke, just when the piston rods alter their sense of motion direction. This may be dampened out adequately by the slides (which may be fed constantly with oily fluid), except when the dreaded resonance occurs. However, because the assembly may have a very simple and fundamental resonant frequency, and because the whole design of the crankless reciprocating ICE, MF pump, MP compressor, or FM or PM motor may recognise the necessity of avoiding this problem, the NMR ratchet mechanism may be equipped with resonance dampers—the universally accepted best method of dealing with unavoidable resonance problems.

A quick look at the sketches (see figures 18.1 and 18.4) may reveal all. What the author might have thought of as a crankless reciprocating ICE's combustion chamber with an air inlet and exhaust gas outlets, a fuel injector and a couple of spark plugs (with an excess of air), may also be a tuneable damper. There may be a theoretical resonance, hazard, but by tweaking the liquid fuel input and the exhaust port trimming and duration, resonance amplification may be avoided.

The conventional C^2M is a relatively simple mechanical device to build and, with its soft bearings and high-pressure NMR fluid cooling system, it is made to work well enough for most applications. On the other hand, the advanced NMR FCM LM–RM/RM–LM commutator has to be precision-engineered, with significant attention given to stress relief and cooling.

18.5 Conclusions

A generational change in C^2M is taking place throughout the world. The author has performed R&D work on an advanced NMR FCM LM–RM/RM–LM and RM–RM commutator, reciprocating IC engine, that is, a two- or four- or even five-stroke thermodynamic cycle, twin-opposed piston, crankless, reciprocating IC engine (the 'Fijalkowski engine'), for over 20 years (Fijalkowski 1986). In most conventional reciprocating IC engines, the reciprocating motion of the pistons drives a rotating crankshaft. The notable exception to this general configuration, the Fijalkowski engine, may utilise an NMR FCM LM–RM/RM–LM commutator instead of the **mechano-mechanical commutation matrixer** (MCM) LM–RM/RM–LM commutators of the C^2M. Compared with earlier MCM LM–RM/RM–LM commutators, NMR FCM LM–RM/RM–LM commutators are more practice-oriented, especially from the viewpoint of operating convenience and the space requirements of ultra-modern and innovative NMR mechatronical techniques. NMR FCM LM–RM/RM–LM and/or LM-LM or RM–RM commutators adapt the characteristics of the mechanical energy to the operational requirements.

The objective of this chapter is to obtain feedback through exciting interest from HPS designers who are working in the automotive, aerospace, marine and other industries.

The Fijalkowski engine may be much more practical than conventional reciprocating IC engines and are likely to be manufactured by the aerospace, automotive, marine and other industries in the not-too-distant future. Such is the hope of the author, who is working towards this goal by utilising several innovative mechatronical techniques that incorporate artificial intelligence.

References

De Falco F 1989 Materiale rotabile e trazione *ATTI Relazioni Generali, 1-mo Convegno Internazionale 'LA FERROVIE NEI TRASPORTI DEGLI ANII 2000* (Bologna *12–14 April*) pp 39–50

Fijalkowski B 1986 Future hybrid electromechanical very advanced propulsion systems for civilian wheeled and tracked all-terrain vehicles with extremely high mobility *Proc. 1986 EVS-8: The Eight International Electric Vehicle Symposium* (Washington DC *20–23 October*) pp 428–43

Fijalkowski B 1988 Unconventional internal combustion engines for automotive vehicles *Motor Vehicles and Motors '88, 3–5 October 1988 (also in Motorna Vozila - Motori Saopstenja* (XIV-83) - Kragujevac, November 1988) pp 265–74

Fijalkowski B 1989 The novel means of very advanced propulsion for 2000s thermo-electric high-speed trains. *Proc. Int. Congress The role of railways in the year 2000* (Bologna *12–14 April 1989*) pp 425–31

Fijalkowski B T 1997 Intelligent automotive systems: development in full-time chassis motion spheres for intelligent vehicles *Advanced Vehicles and Infra-structure Systems—Computer Applications, Control and Automation* ed C O Nwagboso (New York: Wiley) pp 125–42

Fijalkowski B T 1999a A novel concept of the air independent propulsion and/or dispulsion for attack submarines *Proc. of the 4th UEES'99: Int. Conf. on Unconventional Electromechanical and Electrical Systems* (St Petersburg *21–24 June 1999*) vol 2 pp 357–62

Fijalkowski B T 1999b A novel automotive MRF or ERF reciprocating prime mover *Proc. Barcelona 1999 European Automotive Congress—Vehicle Systems Technology for the Next Century* (Barcelona *30 June–2 July 1999*) Paper No. STA99P410

Fijalkowski B T 2006a An advanced reciprocating internal combustion engine without crankshaft and connecting rod mechanism *Proc. of the ASME Internal Combustion Engine Division 2006 Fall Technical Conference – ICEF2006* (Sacramento, CA *5–8 November*) ICE2006-1519

Fijalkowski B T 2011 *Automotive Mechatronics: Operational and Practical Issues* Volume II (International Series on Intelligent Control, and Automotive Science and Engineering vol 52) (Berlin: Springer)

Folonari C V 2006 Nanotechnology for automotive suspensions *Presentation to Centro Richerche FIAT (Turin)* 1–19

Novak R 2005 'Stiff' fluid could soon put the brakes on cars *New Scientist* 11 October 2005 p 23

Pinto-Silva J 1987 Non-conventional land transportation systems *J. Terramech.* **24** 153–57

Wen W, Huang W, Yang S, Lu K and Sheng P 2003 The giant electro-rheological effect in suspension of nanoparticles *Nat. Mater.* **2** 727–30

Mechatronics
Dynamical systems approach and theory of holors
B T Fijalkowski

Chapter 19

Conclusion and future trends

19.1 Concluding remarks

The title of this book requires some explanation.

- 'Mechatronics' is the integration of physics—mechanics, fluidics, electrics and electronics, etc—and computer technologies into the research and development of physical heterogeneous continuous complex and/or simple dynamical hypersystems (Fijalkowski 2010, 2011). Thus, mechatronics consists of the synergistic combination of different physical disciplines like mechanics, fluidics, electrics and electronics, as well as informatics, etc. Synergy is a very creative and therefore dynamic process, far more so than the usual co-operation of the different physical disciplines, and even more so than a close integration of machine hardware and software. Mechatronics offers new solutions and unprecedented flexibility in transportation systems, industrial production processes, and aerospace, automotive and traction components, etc. Its economic success is based on functional integration, i.e. the multiple use of mechano-mechanical (MM), fluido-mechanical (FM) and/or electromechanical (EM) actuators (machines); the decentralisation of intelligence into physical heterogeneous continuous dynamical hypersystems, e.g. aerospace, automotive and/or traction single-shaft propulsion mechatronical control systems; and the inherent options for sensorless self-monitoring and system protection.
- The 'dynamical systems approach' is a holistic description of observable physical reality. *The whole of the dynamical systems approach is not composed of the sum of its elementary units—the multiplicity of elementary units is a result of the diversification of the whole.*
- The idea of a *holor* (from the Greek $\delta\lambda o\sigma$, as in the English *holistic*, pertaining to an entity or a whole) seems to have been originated by Wallis (1673) to

doi:10.1088/978-0-7503-1350-6ch19

describe a mathematical entity that is made up of one or more *merates* (from the Greek $\mu\acute{\varepsilon}\rho o\sigma$, part of an entity) or *coordinates* (independent quantities) of the holor. It includes complex numbers, scalars, vectors, matrices, tensors, quaternions and other hypernumbers. The works of Grassmann (1844, 1894) strongly influenced the subsequent development of holors, including the works of Gibbs (1881) and Heaviside (1893, 1925). Holors, thus defined, have been known for centuries, but each has been developed more or less independently, accompanied by separate nomenclature and theory (Moon and Spencer 1986). The 'theory of holors' demonstrates how these complicated subjects can be made simple by using a single notation that applies to all holors, both tensors and non-tensors. Moon and Spencer (1965, 1986) consider all possible types of holors and develop holor algebra and holor calculus in the most general sense. The theory of holors should establish a method by which students and teachers can learn vector and tensor analysis via a uniform treatment (Moon and Spencer 1963, 1965, 1986; Fijalkowski 1967).

The physical heterogeneous continuous dynamical hypersystem as a whole may be concerned or it may be divided into physical homogeneous continuous dynamical systems on the grounds of which form of energy is concerned (e.g. kinetic energy: radiant, thermal, motion, sound and electrical homogeneous continuous dynamical systems; potential energy: chemical, nuclear, stored mechanical, fluidic, and electrical and gravitational homogeneous continuous systems). Next, they may be divided as physical homogeneous continuous dynamical hyposystems and, as follows, they may be divisible as functional and structural physical homogeneous continuous dynamical components.

The author's intention in writing this textbook is to submit 'generalised physical commutation matrixer holor analyses' in the easiest practicable dynamical systems approach (systems thinking). The author supposes that this can be best realised through the application of physical commutation matrixers and the functional and structural physical homogeneous continuous dynamical components.

Nearly all mechanical and electrical engineering students have been educated concisely regarding physical commutation matrixers during their university physics courses. In view of this, the physical model represents an exceptional structure with which to form a junction between static and dynamical analysis.

It is probably particularly important that the generalised physical commutation matrixer concept lends itself systematically to analogy. Owing to the application of the physical commutation matrixer method, one may confirm that it is practicable to consider the area under discussion in mechatronics (kinetic energy—radiant, thermal, motion, sound and electrical—as well as potential energy—chemical, nuclear, stored mechanical, fluidic, electrical and gravitational). For instance, the analogy between charging an electrical capacitor and saturating a fluidic container of water is made evident by equating the physical and mathematical models of these physical processes. Naturally, one might have identified this analogy instinctively. In spite of this, one becomes aware that these physical processes are also comparable to the variation in velocity of a mass when a force is relevant. This is not practically so evident. As a result, the physical commutation matrixer concept confirmation is exceptionally convenient in the absence of physical situations in various physical heterogeneous continuous dynamical hyposystems for a generalised physical heterogeneous continuous dynamical hyposystem from which a universal type of solution may be completed.

A new definition of the 'physical commutation matrixer', based on matrix-interconnection physical heterogeneous continuous dynamical hypersystem theory concepts, is proposed. The definition may be used to evaluate physical commutation matrixers, which perform multivalent logical functions, and continuously operating physical commutation matrixers. The serial and parallel connections of two physical commutation matrixers and simplified formulas are defined, which are valid for high component physical commutation matrixers.

A comparison is made between results arrived at using the new and conventional definitions. It is found that the results differ little in the case of equal probability output letters. For instance, a physical commutation matrixer is a general term referring to a physical heterogeneous continuous dynamical hypersystem or part of a physical heterogeneous continuous dynamical hypersystem of matrixer conductive parts and their matrix interconnections through which an energy-transfer holor is intended to flow.

A physical commutation matrixer; i.e. a configuration of physically (i.e. electrically, magnetically, optically, or radiationally) connected dynamical components or analogue and/or digital devices was made up of active and passive physical, functional and structural dynamical components or an assemblage of physical, functional and structural dynamical components and their matrix interconnected row and column conductive collectors.

Physical commutation matrixer theory includes the study of all aspects of matrixers, including analysis, design and application. In it, the fundamental quantities are the 'energy-potential-difference holors' (i.e. generalised force holors) between various points, the 'energy-transfer holors' (i.e. generalised velocity holors) flowing in a number of row and column conductive collectors, and the parameters, which describe the passive physical, functional and structural dynamical components. Other important physical commutation matrixer quantities, such as power, energy and time constants, may be computed from the fundamental physical variables.

This textbook provides a comprehensive and careful development of the physical commutation matrixer base and the use of holor theory formulations of analytical methods. No attempt is made to provide the required mathematical tools: matrix theory, the functions of a complex variable, Laplace transforms, numerical methods and programming. However, some references in these areas have been included in the bibliography.

Holor analysis has been shown beyond doubt to be of prime importance in physics and engineering. Every current scientist and engineer must be systematically well acquainted with the symbolism and methods of manipulation. Because both the nomenclature and routine manipulation are quite uncomplicated and may be learned in a short time, the question emerges as to whether an independent one-term subject on holors is adequate for a programme of scientific study. If the subject was taught on the experiential level, it might be integrated into one of the necessary physics subjects, such as electrics and electronics, magnetics, mechanics, fluidics, photonics, etc. The treatment presented in this interdisciplinary textbook differs from the old school one by including invariance in the theory and providing general definitions that are sensible in all coordinate systems. In this approach, holor calculus and matrix analysis is endowed with a concrete logical foundation. Irrespective of the forward-thinking aspects of the textbook, one does not sense that the treatment is too difficult for ordinary readers.

19.2 Future work

Further investigation of physical heterogeneous dynamical hypersystems, as presented in this textbook, may build on the promising cyber-physical heterogeneous dynamical hypersystems. They are integrations of computation, networking and physical processes. Computers and networks monitor and control the physical processes, with feedback loops where physical processes affect computations, and vice versa. The societal and economic potential of such dynamical hypersystems is vastly greater than has been realised, and major investments are being made worldwide to develop this technology. The technology builds on the discipline of embedded dynamical hypersystems, computers and software embedded in devices whose principle mission is not computation, such as aircraft, automotive vehicles, medical devices and scientific instruments. Cyber-physical heterogeneous dynamical hypersystems integrate the dynamics of the physical processes with those of the software and networking, providing abstractions and modeling, design and analysis techniques for the integrated whole (Podelski and Bogomolov 2012; Lee and Seshia 2013; Sanfelice 2014).

The combination of networking and information technology with engineered physical heterogeneous dynamical hypersystems and associated services has enabled an innovative generation of 'smart dynamical hypersystems'. These cyber-physical heterogeneous dynamical hypersystems combine distributed sensing, monitoring, actuation and control networks. They have inter-operable dynamical systems, hyposystems and component integration, very advanced analytics, and user interfaces featuring customised degrees of independence, which can facilitate adaptive, predictive and mutual optimisation of cyber-physical heterogeneous dynamical hypersystem performance over the entire life cycle of a machine or device (e.g. design, construction, operation/use, maintenance, and service). The existing and emerging applications are creating exciting opportunities, especially for the aerospace and automotive industries. Currently, **unmanned aircraft systems** (UAS) are being exploited for search and rescue, surveillance, mapping, crop spraying and environmental monitoring. Moreover, **unmanned ground vehicles** (UGV) have very diverse applications, such as security and mine clearance (Cranfield University 2016).

Cyber-physical heterogeneous continuous dynamical hypersystems integrate analogue and/or digital devices, interfaces, networks, computer systems and the like with the natural and man-made physical world. The naturally interrelated and heterogeneous combination of behaviors in these dynamical systems makes their analysis and design a challenging task. The safety and reliability specifications imposed in cyber-physical applications, which are normally translated into stringent robustness standards, aggravate the matter. Unfortunately, the state-of-the-art tools for system analysis and design cannot cope with the intrinsic complexity of cyber-physical heterogeneous dynamical hypersystems. Tools that are suitable for the analysis and design of cyber-physical heterogeneous dynamical hypersystems must allow for a combination of the physical or continuous dynamics and the cyber or computational components, as well as being able to handle a variety of modes of perturbation, such as exogenous disturbances, time delays and dynamical hypersystem failures (Sanfelice 2014). For instance, cyber-physical heterogeneous dynamical hypersystems demonstrate the applicability of cyber-mode control algorithms to physical heterogeneous dynamical hypersystems without significant design effort being required, and underline their advantage of high robustness. The remaining chattering effects, which decrease cyber-mode position control accuracy, may be reduced by means of faster hardware or sophisticated observers and disturbance estimators. As a result, a challenging research tendency is, for instance, the investigation of an adaptive disturbance estimator for the flexible shaft system that does not require information about the structure of a disturbance or the frequency of periodical perturbations. The performance of the aforementioned cyber-mode position control of an AC–AC or AC–DC–AC commutator asynchronous (induction) motor may also be improved by implementing better observers for magnetic flux in addition to angular velocity. Furthermore, an estimator that provides information about the unknown load may lead to more accurate control performance.

Cyber-physical heterogeneous dynamical hypersystems can be described as smart dynamical hypersystems that encompass computational (i.e. hardware and software) and physical homogeneous dynamical systems, hyposystems and components, which are seamlessly integrated and interact closely to sense the changing state of the real world (Lu *et al* 2015; Anon 2013). Unlike more conventional embedded dynamical hypersystems, a cyber-physical heterogeneous dynamical hypersystem is usually designed to be a network (whole) of interacting dynamical systems, hyposystems and components, with physical inputs and outputs, rather than a standalone device (Anon 2008).

Cyber-physical heterogeneous dynamical hypersystems are becoming popular in many application areas: power networks, the aerospace and automotive industries, manufacturing, healthcare, critical infrastructure, etc. However, with its popularity and extensive application in critical and important applications, cyber-physical heterogeneous dynamical hypersystems are becoming increasingly vulnerable and more of a target for cyber-physical attacks (Wang *et al* 2010). Hackers can launch malicious attacks on power networks and transportation systems (Mills 2009) and be able to hack into wireless medical devices implanted in the human body (Leavitt 2014). More and more security vulnerabilities are being found in all kinds of cyber-physical heterogeneous dynamical hypersystems (Fletcher and Liu 2011). The security of cyber-physical heterogeneous dynamical hypersystems has become a concern of utmost importance in the research and systems design of cyber-physical heterogeneous dynamical hypersystems.

Thus, a cyber-physical heterogeneous dynamical hypersystem integrates computing, communication and storage capabilities with the monitoring and/or control of entities in the physical world dependably, safely, securely, efficiently and in real time (Mahapatra 2016).

- Cyber-physical heterogeneous dynamical hypersystems will be an exciting development area for the coming decades;
- they involve multi-disciplinary research-and-development (R&D) work:
    - they involve high-confidence software;
- cyber-physical heterogeneous dynamical hypersystems have the potential to change the way people interact with their surroundings;
- future applications for cyber-physical heterogeneous dynamical hypersystems are only limited by the human imagination;
- affordability and ease of application will drive adoption.

Cyber-physical heterogeneous dynamical systems are expected to play a major role in the R&D of future physical heterogeneous dynamical systems, with new capabilities that far exceed today's levels of autonomy, functionality, usability, reliability and cyber security. R&D advances can be accelerated by close collaboration between academic disciplines in computation, communication, control, and other engineering and computer science disciplines, coupled with grand challenge applications.

The author's selected recommendations for R&D in cyber-physical heterogeneous dynamical systems are as follows.

- Standardised abstractions and architectures that permit modular design of cyber-physical heterogeneous dynamical systems are urgently needed.
- Cyber-physical heterogeneous dynamical system applications involve components that interact through a complex, coupled physical environment. Reliability and security pose particular challenges in this context—new frameworks, algorithms and tools are required.
- Future cyber-physical heterogeneous dynamical systems will require hardware and software components that are highly dependable, reconfigurable and, in many applications, certifiable—and trustworthiness must also extend to the system level.

In future work, the author will focus on cyber-physical heterogeneous dynamical hypersystems in many application areas: power networks, the aerospace and automotive industries, manufacturing, healthcare and critical infrastructure—as well as electro-mechanical (EM) drives, etc—and measure their performance to compare against theoretical holor analysis.

References

Anon 2008 University of California at Berkeley, Electrical Engineering and Computer Sciences, Technical Report No. UCB/EECS-2008, http://eecs.berkeley,edu/Pubs/TechRpts/2008/EECS-2008-8.htm

Anon 2013 Foundations for innovation in cyber physical systems workshop summary report, http://nist.gov/el/upload/CPS-WorkshopReport-1-30-13-Final.pdf

Anon 2016 Cranfield University Centre for Cyber-physical Systems. 2016, http://CentreforCyber-physicalSystems.htm

Fijalkowski B 1967 Zastosowanie rachunku holorowego w teorii pradow przemiennych (Application of a holor calculus in the theory of alternating currents). *Przeglad elektrotechniczny (Electrotechnics Review)*, Rok XLIII, Zeszyt 2, Warszawa 1967 (In Polish)

Fijalkowski B T 2010 *Automotive Mechatronics: Operational and Practical Issues*—Volume 1 (International Series on Intelligent Control, and Automation Science and Engineering vol 47) (Berlin: Springer)

Fijalkowski B T 2011 *Automotive Mechatronics: Operational and Practical Issues*—Volume 2 (International Series on Intelligent Control, and Automotive Science and Engineering vol 52) (Berlin: Springer)

Fletcher K K and Liu X F 2011 Security requirements analysis, specification, prioritization, and policy development in cyber-physical systems *5th Int. Conf. on Secure Soft-ware Integration & Reliability Improvements Companion* (SSIFI-C) pp 106–13

Gibbs J W 1881 *Elements of Vector Analysis* (privately printed, New Haven, pp. 17–50, 1881; pp. 50–90, 1884) 1906 *The scientific papers of J Willard Gibbs* (London: Longmans, Green and Co) vol. II p 17

Grassmann H 1844 Die linealoe Ausdehnungslehre, ein neuer Zweig der Mathematik (Otto Wigand, Leipzig, 1844), Die Ausdehnungslehre (Enslin, Berlin, 1862)

Grassmann H 1894 Grassmann's gesammelte math U. phys. Werke, 3 vols (Leipzig: B G Teubner)

Heaviside O 1893 *Electromagnetic Theory* (London: Electrician Printing and Pub. Co., London), 3 vols; (New York: Dover). Heaviside's treatment is given in vol I, Ch III, pp 132–305: 'The elements of vectorial algebra and analysis'

Heavside O 1925 Electrical Papers (Boston, MA: The Copley publishers)

Leavitt N 2014 Researchers Fight to keep implanted medical devices safe from hackers *Computer* **43** 11–4

Lee E A and Seshia S A 2013 *Introduction to Embedded Systems—A Cyber-Physical Systems Approach* http://leeshesia.org/releases/leesgesia_DigitalV1_08.pdf

Lu T , Zhao J, Zhao L, Li Y and Zhang X 2015 Towards a framework for assuring cyber physical system security *Int. J. Security Appl.* **9** 25–40

Mahapatra R N 2016 Cyber-physical systems: issues and challenges (presenation adapted from NSF workshops) *Third Edition of Int. Conf. on Wireless Networks & Embedded Systems—WECON 2011* sls 1–50

Mills E 2009 Hackers broke into FAA air traffic contro system *The Wall Street Journal* A6

Moon P H and Spencer D E 1963 A new mathematical representation of alternating currents *Tensor* p 110

Moon P H and Spencer D E 1965 *Vectors* (Princeton, NJ: van Nostrand)

Moon P H and Spencer D E 1986 *Theory of Holors. A Generalization of Tensors* (Cambridge: Cambridge University Press) (Digitally printed, 2005)

Podelski A and Bogomolov S 2012 *Lectures notes.* http://informatik,uni-freiburg.de/teaching/532012/cps-hm

Sanfelice R G 2014 *Introduction to Cyber-physical Systems* (CMPE 142) Fall 2014 (Santa Cruz, CA: University of California)

Wallis J 1685 Letter to Collins, 6 May 1673 Treatise of algebra (London, 1685) p 264; Cajori F 1912 Historical notes on the graphic representation of imaginaries before the time of Wessel Am. Math. Mon. 19 167

Wang E K, Ye Y, Xu X, Yiu S M, Hui I C K and Chou K P 2010 Security Issues and challenges for cyber physical systems *2010 IEEE/ACM Conf. on Green Computers and Communications & 2010 IEEE/ACM Conf. on Cyber Physical and Social Computing* pp 733–8

Glossary

Actuator—The component of an open-loop or closed-loop mechatronical control system that connects the electronic control unit (ECU) with the process; the actuator consists of a mechano-electrical (MM) commutation matrixer (MCM) or an electrical commutation matrixer (ECM) AC–AC, AC–DC–AC or DC–AC commutator motor and a final-control unit; positioning electrical signals are converted to mechanical output.

All-electrical vehicle (AEV)—Any ground automotive vehicle whose original source of energy is electrical power, such as an electrical automobile or electrical locomotive.

Angular velocity holor is defined as the rate of change of the angular displacement holor and is a vector quantity (more precisely, a pseudovector) that specifies the angular speed (rotational speed) of an object and the axis about which the object is rotating. The SI unit of angular velocity is radians per second (rad-s^{-1}). Angular velocity holor is usually represented by the symbol omega ($\boldsymbol{\Omega}^{j}$). The sense of direction of the angular velocity holor is perpendicular to the plane of rotation, in a sense of direction that is usually specified by the 'right-hand rule'.

Artificial intelligence (AI) is the synthetic intelligence exhibited by machines or software, and the domain of computer science that develops machines and software with human-like intelligence.

Application specific integrated circuit (ASIC)—An integrated circuit (IC) designed for a custom requirement, frequently a gate array or single-chip microprocessor or programmable logic device.

Application specific integrated matrixer (ASIM)—An integrated matrixer (IM) designed for a custom requirement, frequently a gate array or single-chip static-commutator or electrical commutation matrixer (ECM) AC–AC, AC–DC–AC, AC–DC/DC–AC or DC–DC commutator.

doi:10.1088/978-0-7503-1350-6ch20

Cartesian dualism—Developed by René Descartes, this states that there are two basic forms of substance: mental and material. According to Descartes's philosophy, the mental does not have extension in space, and the material cannot think. Also called substance dualism, it is credited with giving rise to much thought concerning the well-known mind-body dilemma.

Cartesian reductionism—The idea of reductionism has existed since the ancient Greeks, while René Descartes, the 17th-century French philosopher, was the first to formally state the concept. He stated that the world is like a machine composed of many smaller parts, and that it can be understood by taking apart and studying the parts before learning how they all fit into the whole.

Chemo-electrical/electro-chemical storage battery (ChE/ECh)Self-contained ChE/ECh cell/cells or an accumulator system that converts chemical energy into electrical energy in a reversible process.

Complexity is the property of a real-world dynamical hypersystem that is perceptible in the inability of any one formalism to denote all its properties. It necessitates that one finds different means of interacting with physical heterogeneous continuous dynamical hypersystems. Different in the sense that when one formulates adequate physical and mathematical models, the formal physical heterogeneous continuous dynamical hypersystems required to describe each particular aspect are not derivable from each other.

Computer-aided systems engineering (CASE)—A tool designed to help a systems analyst complete development tasks.

Connecting rod—Any straight link that transmits motion or power from one linkage to another within a mechanism, especially linear to rotary (LM–RM), as in a reciprocating internal combustion engine (ICE) or mechano-pneumatic (MP) compressor.

Constraint—A restriction on the natural degrees of freedom (DoF) of a physical heterogeneous continuous dynamical hypersystem, the number of constraints is the difference between the number of natural DoF and the number of actual DoF.

Cost function—Can refer to: in mathematical optimisation, the loss function, a function to be minimised; in artificial neural networks, the function to return a number representing how well the neural network performed to map training examples to correct output.

Crankshaft—A shaft about which a crank rotates.

Current holor—A current holor, also known as an electrical energy-transfer holor, is a flow of electrical charge. In electrical commutation matrixers (ECM) this charge is often carried by moving electrons in wires, i.e. ECM rows and columns. It can also be carried by ions in an electrolyte, or by both ions and electrons, as in a plasma. The SI unit for measuring a current is the ampere (A), which is the flow of electrical charges through a surface at the rate of one coulomb per second ($C\ s^{-1}$).

Cybernetics is the study of human/machine interaction guided by the principle that numerous different types of systems can be studied according to the principles of feedback, control and communication. The field has a quantitative component, inherited from feedback control and information theory, but it is primarily a qualitative, analytical tool—one might even say a philosophy of technology. Cybernetics is characterized by a tendency to universalise the notion of feedback, seeing it as the underlying principle of the technological world. The study of systems that are open to energy but closed to information and control systems that are tight is called cybernetics. Cybernetics takes as its domain the design or discovery and application of the principles of regulation and communication. Cybernetics does not treat things, but rather ways of behaving.

Cyber-physical heterogeneous continuous complex and/or simple dynamical hypersystems have collections of various cyber-physical dynamical systems, hyposystems and components so connected or related as to perform a particular function not performable by the cyber-physical dynamical systems, hyposystems and components alone, and they must encompass computational (i.e. hardware and software) and physical homogeneous dynamical systems, hyposystems and components, seamlessly integrated and interacting closely to sense the changing state of the real world. These cyber-physical heterogeneous dynamical hypersystems combine distributed sensing, monitoring, actuation and control networks.

d'Alambert principle—The principle that the resultant of the external forces and the kinetic reaction acting on a body equals zero.

Darcy's Law—The law that the rate at which a fluid flows through a permeable substance per unit area is equal to the permeability, which is only a property of the substance through which the fluid is flowing, times the pressure drop per unit length of flow, divided by the viscosity of the fluid.

Deterministic mathematical model of the physical heterogeneous continuous dynamical hypersystem is a model whose description is possible in the form of the function relationships between the inputs and outputs of the physical heterogeneous continuous dynamical hypersystem.

Developmental systems approach is a perspective that has been adopted by increasing numbers of scientists and engineers since it emerged in the twentieth century. Developmental systems approach into the methodology of physical models of the physical heterogeneous continuous dynamical hypersystems, in addition to the dynamical systems approach has met with indifference not only from some theorists on the research side of mathematical models, but also some practitioners making experimental studies. It still omnipotently predominates in the excellence as well the physical model, as the modelled existing or projected physical heterogeneous continuous dynamical hypersystem, the dynamical systems approach in the methodology of a formulation of mathematical models, creating in researchers who are occupied in analytical studies the illusion of also steering the physical models. The developmental systems approach in the methodology of the excellence of physical models did not bring about the panacea on the routine for creating the physical model. This requires a considerably *stronger* impulse for the creation of the macro-conditions for computerised analytical studies of mathematical models. If the theorists—systems scientists and engineers—and researchers of mathematical models are interested in their computerised analytical studies as well as their verification (checking) while taking into account the feedback of an algorithm for the formulation of a mathematical model, then sooner or later, they must look towards having a repertoire of methods to improve their physical models. This is an opportunity to recognise the developmental systems approach in the methodology for improving physical models—the idea of improving the physical model, and the comprehensive trouble-shooting processes for dynamical and static problems, which are of interest to researchers. What, therefore, is this idea, and what might be the terms of its usefulness? The basic concept in a glossary of the developmental systems approach in the methodology to create excellent physical models is 'change'. Mathematical models continuously improve due to the effect of the alteration of physical models. Lack of alteration means they are inadequate and unfit. Changes to physical models are therefore in the world of mathematical models, and are important functions of the control of their adequacy.

Dynamical systems approach—Offers a holistic from whole to local approach for describing physical reality; produces a comprehensive image of physical reality; produces accurate predictions with straightforward mathematics and clear logic.

Dynamical universe (DU) theory is a holistic description of observable physical reality. It is a unifying theory converting space–time in variable coordinates into dynamic space in absolute coordinates. The dynamic universe (DU) theory relies on an overall zero-energy balance in space and the conservation of the total energy in interactions in space. *The whole in the DU is not composed of the sum of the elementary units—the multiplicity of elementary units is a result of the diversification of the whole.* Many physical processes in the Universe are highly nonlinear and coupled in a complex way. A deep understanding of the DU often requires detailed numerical simulations, coupled with sophisticated data analysis and visualisation.

Electrical commutation matrixer (ECM) AC–DC commutator—This commutator is an ECM AC–DC rectifier for a rotary or transversal AC–DC commutator generator, the commutator electronically commutes (switches) the armature windings so that the resultant induced source AC armature phase-voltages always act with the same sense of voltage polarisation; this requires a reversal of the armature winding connection every π rad; the induced source AC armature phase-voltages are mechanically rectified to the induced source DC armature voltage via bipolar electrical valves, e.g. diodes, MOS-FETs, IGBTs or MCTs.

Electrical commutation matrixer (ECM) DC–AC commutator—This commutator is an ECM DC–AC inverter for a rotary or transversal DC–AC commutator motor or actuator, the commutator electronically commutes (switches) the armature windings so that the resultant torque or force always acts in the same sense of rotary or transverse direction, respectively; this requires a reversal of the armature winding connection every π rad; the DC armature supply is via bipolar electrical valves, e.g. MOSFETs, IGBTs or MCTs.

Electrical capacitance is the ability of a body to store an electrical charge. Any object that can be electrically charged exhibits capacitance. A common form of energy storage device is the parallel-plate capacitor. In a parallel plate capacitor, capacitance is directly proportional to the surface area of the conductor plates and inversely proportional to the separation distance between the plates.

Electrical conductance—The inverse quantity of electrical resistance is electrical conductance, the ease with which an electric current passes. The SI unit of electrical conductance is measured in siemens (S) or mhos (℧).

Electrical inductance—In electromagnetism and electronics, inductance is the property of a conductor by which a change in current flowing through it 'induces' (generates) an electromotive force (induced voltage) in both the conductor itself (self-inductance) and in any nearby conductors (mutual inductance).

Electrical resistance is the repulsion of a current within an electrical homogeneous continuous electrical system. It explains the relationship between voltage (the amount of electrical pressure) and the current (flow of electricity), e.g. the electrical resistance of an electrical conductor is the opposition to the passage of an electric current through that conductor. Electrical resistance shares some conceptual parallels with the mechanical notion of friction. The SI unit of electrical resistance is the ohm (Ω).

Electrics is the technology that generally deals with the application of electricity, electronics and electromagnetism. This branch of physics first became distinct in the latter half of the 19th century, after the introduction of the electrical telegraph, the telephone and electrical power distribution and use. Subsequently, broadcasting and recording media made electronics part of everyday life.

Electro-mechanical (EM)—Pertaining to an electro-mechanical (EM) motor or actuator or device, EM heterogeneous continuous dynamical hypersystem, or EM process that is electromagnetically or electrostatically actuated or controlled.

Engine—A machine in which power is applied to do work by the conversion of various forms of energy into mechanical torque, force and motion.

Enthalpy—The sum of the internal energy of a thermal homogeneous continuous dynamical system plus the product of the system's volume multiplied by the pressure exerted on the system by its surroundings. Also known as heat content, sensible heat or total heat.

Entropy—Function of the state of a thermal homogeneous continuous dynamical system whose change in any differential reversible process is equal to the heat absorbed by the system from its surroundings divided by the absolute temperature of the system. Also known as thermal charge.

Eulerian coordinates—Any system of coordinates in which the properties of a fluid are assigned to points in space at each given time, without an attempt to identify individual fluid parcels from one time to the next; a sequence of synoptic charts is a Eulerian representation of the data.

Eulerian correlation—The correlation between the properties of a flow at various points in space at a single instant of time. Also known as synoptical correlation.

Eulerian equation—A mathematical representation of the motions of a fluid in which the behaviour and the properties of the fluid are described at fixed points in a coordinate system.

Euler method—A method of studying fluid motion and the mechanics of deformable bodies in which one considers volume elements at fixed locations in space, across which material flows, the Euler method is in contrast to the Lagrangian method.

Extended finite element method (XFEM) is a numerical technique based on the generalized finite element method (GFEM) and the partition of unity method (PUM). It extends the classical finite element method (FEM) approach by enriching the solution space for solutions to differential equations with discontinuous functions.

External combustion engine (ECE)—An engine that operates by the energy of external combustion of a fuel.

External energy-potential-difference holor—An energy-potential-difference holor exerted on a physical heterogeneous continuous dynamical hypersystem or on some physical homogeneous continuous dynamical systems, hyposystems and components by an agency outside the physical heterogeneous continuous dynamical hypersystem.

External work—The work done by a physical heterogeneous continuous dynamical hypersystem in expanding against energy-potential-difference holors exerted from outside.

Final-control unit—The second or last stage of an actuator to control mechanical output.

Finite elements method (FEM) is a numerical technique for finding approximate solutions to boundary value problems for partial differential equations. It uses the subdivision of a whole problem domain into simpler parts, called finite elements, and variational methods from the calculus of variations to solve the problem by minimising an associated error function.

Flow—The forward continuous movement of a fluid, such as gases, vapours, or liquids, through closed or open channels or conduits.

Flow-rate—Also known as the rate of flow holor. (i) Time necessitated for a given quantity of flowable material to flow a measured distance. (ii) Weight or volume of flowable material flowing per unit time.

Flow resistance—Any factor within a conduit or channel that impedes the flow of fluid, such as surface roughness or sudden bends, contractions, or expansions. Also known as viscosity.

Friction flow—Fluid flow in which a significant amount of mechanical energy is dissipated into heat by the action of viscosity.

Friction torque—The torque that is produced by frictional forces and opposes rotational motion, such as that associated with journal or sleeve bearings in machines.

Fluid is a substance (a continuum) that continuously deforms the application of a shear, no matter how small the shear stress may be, and includes both liquids and gases. A fluid cannot sustain a shear when at rest. Liquids are nearly incompressible, but gases are highly compressible. Liquids are distinguished from gases by orders of magnitude differences in their density, absolute viscosity and bulk modulus.

Fluid density—The mass of a fluid per unit volume.

Fluid dynamics—The science of fluids in motion.

Fluid mechanics—The science concerned with fluids, either at rest or in motion, and dealing with pressures, velocities and accelerations in the fluid, including fluid deformation and compression or expansion.

Fluid statics—The determination of the pressure intensities and forces exerted by fluids at rest.

Fluidic capacitance—The volume of fluid in a channel can change just because of a change in pressure: this is either due to fluid compressibility or channel elasticity. Fluidic capacitors are one of two types of fluidic energy storing dynamical components in fluidic homogeneous continuous dynamical systems. Capacitance in a fluidic homogeneous continuous dynamical system comes from continuous fluido-fluidic (FF) accumulators, but also from the fluid itself if it is compliant. It is essential to consider fluid compliance in pneumatic homogeneous continuous dynamical systems, but it generally does not play a significant role in the physical models of hydraulic homogeneous continuous dynamical systems unless there is significant trapped air causing spongy behaviour. The capacitance of the fluid is captured by its bulk modulus property. The idea of the capacitance of fluid trapped in a cylinder can be expanded to estimate the capacitance of a plug of fluid in a hose or pipe, which in turn can be used in a dynamic model. One application of a physical model involving fluid compression is to understand the hydro-mechanical (HM) hammer, which is impact loading caused by sudden changes in flow, such as when a fluidic valve is commuted (switched) from on to off. One has already seen that a fluid itself, whether a liquid or gas, exhibits fluid compliance due to its compressibility. Certain fluidic machines or devices may also introduce flow compliance into a fluidic homogeneous continuous dynamical system, even if the fluids are absolutely incompressible.

Fluidic equation of motion—One of the set of fluido-dynamic equations representing the application of Newton's second law (NSL) of motion to a fluidic homogeneous continuous dynamical system, the total acceleration of an individual fluid particle is equated to the sum of the forces acting on the particle within the fluid.

Fluidic inertance or inductance—The second type of energy-storing dynamical component is fluidic inertance or inductance. In mechanical homogeneous continuous dynamical systems, mass and rotary inertia often dominate dynamical-system behavior and must be physically and mathematically modelled. In fluidic homogeneous continuous dynamical systems, the inertia of the fluid is generally insignificant and it is usually ignored in dynamical system physical models. This is because in hydraulic homogeneous continuous dynamical systems, the pressures are so high that inertial forces can be neglected, and in pneumatic homogeneous continuous dynamical systems, the mass of air is so low that inertial forces can also be neglected. When analysing the high-frequency behaviour of a fluidic homogeneous continuous dynamical system, for example, with the sudden on–off commutation (switching) of fluidic valves that causes transients in fluid flow, fluidic inertance should be included in the physical model.

Fluidic resistance—The force exerted by a gas or liquid opposing the motion of a body through it. Like mechanical friction and electrical resistance, this fluidic dynamical component performs an energy-dissipation function. The dissipation of fluidic energy into thermal energy (heat) occurs in all fluidic machines (FM motors, MF pumps and compressors) or devices to some extent. Thus, fluidic resistors are any dynamical component that resists flow. Another way of looking at fluidic resistors is: any dynamical component that causes a pressure drop when fluid flows through the component. Fluidic resistors include fluidic valves, filters, hoses, pipes and fittings.

Fluidic friction—Conversion of mechanical energy in fluid flow into thermal energy.

Fluidics is the technology of using the flow characteristics of liquid or gas to operate a fluidic homogeneous continuous dynamical system. One of the newest of the control technologies, fluidics has in recent years come to compete with mechanical and electrical homogeneous continuous dynamical systems.

Fluido-mechanical (FM)—Pertaining to a fluido-mechanical (FM) motor or actuator or device, FM heterogeneous continuous dynamical hypersystem, or FM process that is fluido-dynamically or fluido-statically actuated or controlled.

Force holor—That influence on a mechanical homogeneous continuous dynamical system that causes it to accelerate, quantitatively it is a mechanical energy-potential-difference holor, equal to the mechanical homogeneous continuous dynamical system's rate of change of momentum holor.

Fuzzy-logic (FL)—Software design based upon a reasoning node rather than a fixed mathematical algorithm, a fuzzy logic design allows the mechatronic engineer to participate in the software design because the fuzzy language is linguistic and built upon easy-to-comprehend fundamentals.

Generalised admittance holor is a measure of how easily a physical heterogeneous continuous dynamical hypersystem will allow an energy-transfer holor to flow. It is defined as the holor inverse of generalized impedance holor, that is the holor reciprocal of the generalised impedance holor. Thus, the generalised admittance holor is the energy-transfer holor-to-energy-potential-difference holor ratio, and it conventionally carries SI units of siemens (S), formerly called mhos (℧).

Generalised coordinates—A set of variables used to specify the positional orientation of a physical heterogeneous continuous dynamical hypersystem, in principle defined in terms of the Cartesian coordinates of the physical heterogeneous continuous dynamical hyposystem's particles and of the time to some convenient manner, the number of such coordinates equals the number of degrees of freedom (DoF) of the physical heterogeneous continuous dynamical hyposystem. Also known as Lagrangian coordinates.

Generalised equation of motion—Equation that specifies the coordinates of particles as functions of time. Also known as a differential equation of dynamics, or one of several such equations, from which the coordinates of particles as functions of time can be obtained if the initial positions and velocities of the particles are known.

Generalised energy-potential-difference holor—The generalised energy-potential-difference holor corresponding to a generalised coordinate is the ratio of the virtual work done in an infinitesimal virtual displacement, which alters those coordinates and no others.

Generalised energy-transfer holor—The derivative with respect to time of one of the generalised coordinates of a particle. Also known as the Lagrangian generalised energy-transfer holor.

Generalised finite element method (GFEM) is a direct extension of the standard finite element method (SFEM, or FEM), which makes it possible to attain the accurate solution of physical and engineering problems in complex domains that might be practically impossible to solve using the FEM.

Generalised impedance holor is the measure of the opposition that a physical heterogeneous continuous dynamical hypersystem presents to an energy-transfer holor when an energy-potential-difference holor is applied. In quantitative terms, it is the holor ratio of the energy-potential-difference holor to the energy-transfer holor in a physical heterogeneous continuous dynamical hypersystem. The impedance holor extends the concept of the resistance holor to physical heterogeneous continuous dynamical hypersystems, and possesses both magnitude and phase, unlike the resistance holor, which only has magnitude. When a physical heterogeneous discrete dynamical hypersystem is driven with a step function, there is no distinction between the impedance holor and the resistance holor; the latter can be thought of as an impedance holor with a zero phase angle. The magnitude of the impedance holor is the holor ratio of the energy-potential-difference holor amplitude to the energy-transfer holor amplitude. The phase of the impedance is the phase shift by which the energy-transfer holor lags the energy-potential difference holor. In general, the generalised impedance will be a holor, with the same SI units as the physical resistance holor, for which the SI unit is the ohm (Ω).

Generalised momentum—If q_j ($j = 1, 2, 3,\ldots$) are the generalised coordinates of a classical physical heterogeneous continuous dynamical hypersystem, and $\mathcal{L}$ is its Lagrangian, the momentum conjugate to q_j is $p_j = \partial\mathcal{L}/\partial q_j$. Also known as canonical momentum or conjugate momentum.

Graphene is a remarkable substance with a multitude of astonishing properties that have repeatedly earned it the title of 'wonder material'. Graphene is the thinnest material known to man at one atom thick, and it is also incredibly strong—about 200 times stronger than steel. On top of that, graphene is an excellent conductor of heat and electricity and has interesting light-absorption abilities. It is truly a material that could change the world, with unlimited potential for integration in almost any industry.

Hamilton–Ostrogradsky principle—A variational principle that states that the path of a conservative physical heterogeneous continuous dynamical hypersystem in configuration space between two configurations is such that the integral of the Lagrangian function over time is a minimum or maximum relative to near paths between the same end points and taking the same time.

Hamilton's equations of motion—A set of first-order, highly symmetrical equations describing the motion of a classical dynamical system, namely $\dot{q}_j = \partial H/\partial p_j$. $\dot{p}_j = -\partial H/\partial p_j$, where q_j ($j = 1, 2,\ldots$) are the generalised coordinates of the system, p_j is the momentum conjugate to q_j and H is the Hamiltonian. Also known as canonical equations of motion.

Hamiltonian function—A function of the generalised coordinates and momenta of a physical heterogeneous continuous dynamical hypersystem, equal in value to the sum over the coordinates of the product of the generalised momentum corresponding to the coordinate, and the coordinate's time derivative, minus the Lagrangian of the physical heterogeneous continuous dynamical hypersystem; it is numerically equal to the total energy of the Lagrangian and does not depend on time explicitly, the equations of motion of the physical heterogeneous continuous dynamical hypersystem are determined by the functional dependence of the Hamiltonian on the generalised coordinates and momenta.

Heat—Thermal energy is transmitted due to a temperature difference between the source from which the energy is coming and a sink towards which the energy is going, other types of energy in transit are called work.

Heat conduction—The flow of thermal energy through a substance from a higher to a lower temperature region. Also known as heat diffusion.

Heat convection—The transfer of thermal energy by actual physical movement from one location to another of a substance to which thermal energy is stored. Also known as thermal convection.

Heat flow—Heat thought of as thermal energy flowing from one substance to another, quantitatively, the amount of heat transferred in a unit time. Also known as heat transmission.

Heat radiation—Also known as thermal radiation, this is electromagnetic radiation generated by the thermal motion of charged particles in matter. All matter with a temperature greater than absolute zero emits thermal radiation. When the temperature of the body is greater than absolute zero, interatomic collisions cause the kinetic energy of the atoms or molecules to change. Thermal radiation is one of the fundamental mechanisms of heat transfer. Heat transfer through radiation takes place in the form of electromagnetic waves that are mainly in the infrared region. Radiation emitted by a body is a consequence of the thermal agitation of its composing molecules.

Heat transfer describes the exchange of thermal energy between physical heterogeneous continuous dynamical hypersystems, depending on the temperature and pressure, by dissipating heat. Thermal homogeneous discrete dynamical systems which are not isolated may decrease in entropy. Most objects emit infrared thermal radiation near room temperature. The fundamental modes of heat transfer are conduction or diffusion, convection, advection and radiation.

Holism—Asserts that everything exists in relationship, in a context of connection and meaning—and that any change or event causes realignment, however slight, throughout the entire pattern. The whole is greater than the sum of its parts and this means that the whole comprises a pattern of relationships that are not contained by the parts, but ultimately define them. Holism stands in stark opposition to the method of reductionism, which holds that analysis, dissection and strict definition are the tools for understanding reality. Holism asserts that phenomena can never be fully understood in isolation; it asserts that reductionism can only give us a partial view of anything it dissects. Holism is difficult to pin down precisely, because by its very nature it embraces paradox, mystery and contradiction. The holistic world view has its roots in the new physics, systems thinking and ecology, holism and perennial philosophy.

Holonomic constraints—An integrable set of differential equations of dynamics which describe the restrictions on the motion of a physical heterogeneous continuous dynamical hypersystem, a function relating several variables in the form $f(x_1, \ldots x_n) = 0$, in the optimisation of physical problems.

Holonomic physical heterogeneous continuous dynamical hypersystem—A physical heterogeneous continuous dynamical hypersystem in which the constraints are such that the original coordinates can be expressed in terms of independent coordinates and possibly also time.

Holor—A holor (hoʊlər; Greek ὅλος *whole*) is a mathematical entity that is made up of one or more independent quantities ('merates', as they are called in the theory of holors). Complex numbers, scalars, vectors, matrices, tensors, quaternions and other hypercomplex numbers are kinds of holor.

Hybrid-electrical vehicle (HEV) an automotive vehicle that uses two or more distinct power sources. However, one of them ought to be a secondary power source with the possibility of energy storage. The term most commonly refers to hybrid electrical vehicles (HEVs), which combine an internal combustion engine (ICE) or external combustion engine (ECE) and one or more electro-mechanical/mechano-electrical (EM/ME) dynamotors. However, other mechanisms to capture and utilise energy are included. A hybridisation of a conventional automotive vehicle and an electrical vehicle combines an ICE or ECE propulsion system with an electrical propulsion system. The presence of the electrical power train is intended to achieve either better specific fuel consumption (SFC) than a conventional automotive vehicle or better performance.

Internal combustion engine (ICE)—A prime mover in which the fuel is burned within the engine and the products of combustion serve as thermodynamic fluid, as with gasoline and diesel engines. An engine that operates by the energy of internal combustion of a fuel.

Kinetic energy—The energy which a physical heterogeneous continuous dynamical hypersystem possesses because of the motion in classical physics equal to one-half of the physical heterogeneous continuous dynamical hypersystem's mass and/or capacitance times the square of its energy-transfer variable.

Kirchhoff's first law (KFL)—This law is also known as Kirchhoff's point rule, or Kirchhoff's junction rule (or nodal rule). The principle of conservation of electrical charge implies that at any commutation node (junction) in an electrical commutation matrixer (ECM), the holor sum of the current holors flowing into that commutation node is equal to the holor sum of the current holors flowing out of that commutation node, or the holor sum of the current holors in an ECM of conductors (rows and columns) meeting at a point is zero.

Kirchhoff's second law (KSL)—This law is also known as Kirchhoff's loop rule, Kirchhoff's mesh rule, or Kirchhoff's second rule. The principle of conservation of energy implies that the holor sum of the electrical energy-potential-difference holors (voltage holors) around any closed electrical commutation matrixer (ECM) is zero, or, more simply, the holor sum of the electromotive force (emf) holors in any closed loop is equivalent to the holor sum of the potential-drop holors in that loop, or the holor sum of the products of the resistances of the ECM conductors (rows and columns) and the current holors in them in a closed loop is equal to the total emf holor available in that loop.

Lagrange–Hamilton theory—The formalised study of continuous systems in terms of field variables where a Lagrangian density function and a Hamiltonian density function are introduced to produce equations of motion.

Lagrange's equations—Equation of dynamics of a physical heterogeneous continuous dynamical hypersystem for which a classical (non-quantum mechanical) physical description is possible, and which relates the kinetic energy of the physical heterogeneous continuous dynamical hypersystem to the generalised coordinates, the generalised energy-potential-difference variables and the time. Also known as the Lagrangian equations of dynamics.

Lagrangian—The difference between the kinetic energy and the potential energy of a physical heterogeneous continuous dynamical hypersystem, expressed as a function of generalised coordinates and energy-transfer variables from which Lagrange's equations can be derived. Also known as the kinetic potential Lagrange function.

Lagrangian density—For a dynamical system of fields or continuous media, a function of the fields, of their time and space derivatives, and the coordination and time, whose integral over space is the Lagrangian.

Lagrangian function—The function which measures the difference between the kinetic and the potential energy of a physical heterogeneous continuous dynamical hypersystem.

Lagrangian method—A method of studying of fluid motion and the mechanics of deformable bodies in which one considers volume elements which are carried along with the fluid or body, and across whose boundaries material does not flow, in contrast to the Euler method.

Machine—A combination of rigid or resistant bodies that have definite motions and are capable of performing useful work.

Magnetic reluctance is a concept used in the analysis of magnetic homogenous continuous dynamical systems. It is analogous to electrical resistance in an electrical homogeneous continuous dynamical system, but rather than dissipating electrical energy it stores magnetic energy. In likeness to the way an electrical field causes an electrical current to follow the path of least resistance, a magnetic field causes magnetic flux to follow the path of least magnetic reluctance. It is a scalar, extensive quantity, akin to electrical resistance. The units for magnetic reluctance are inverse henries (H^{-1}). Also known as magnetic resistance.

Magnetic resistance—See magnetic reluctance.

Magnetomotive force—In physics, the magnetomotive force (mmf), is a quantity appearing in the equation for the magnetic flux in a magnetic homogeneous continuous dynamical system, sometimes known as Hopkinson's law: $M = R\,\Phi$, where Φ is the magnetic flux and R is the magnetic reluctance of the homogeneous continuous dynamical system. It can be seen that the mmf plays a role in this equation that is analogous to that of the voltage V in Ohm's law: $V = R\ I$. The mmf is analogous to electromotive force (emf), i.e. the difference in electrical potential, or voltage, between the terminals of a source of electricity, e.g. a ChE/ECh storage battery from which no current is being drawn, since it is the cause of magnetic flux in a magnetic homogeneous continuous dynamical system, i.e. (i) $M = N\ I$, where N is the number of turns in the coil and I is the electrical current through the magnetic homogeneous continuous dynamical system, and (ii) $M = R\ \Phi$, where Φ is the magnetic flux and R is the magnetic reluctance, and (iii) $M = H\ l$, where H is the magnetising force (the strength of the magnetising field) and l is the mean length of a solenoid or the circumference of a toroid.

Mass-flow holor—Also known as the mass-transfer holor and bulk-flow holor, it is the measure of the movement of material matter. In physics, mass flow occurs in open fluidic homogeneous continuous dynamical systems and is often measured as occurring when moving across a certain boundary characterised by its cross-sectional area and a flow-rate holor. In fluidics it may also be a flow of fluids in a tube or vessel of a certain diameter. A bulk transfer of particles of matter in a characterised type of flow is also known as bulk flow.

Mass-velocity holor—The mass flow-rate holor of a fluid divided by the cross-sectional area of the enclosing chamber or conduit, The SI unit is kilogram per second and square metres (kg s^{-1} m^2).

Mathematical model of the physical heterogeneous continuous dynamical hypersystem is adequate to a physical process, functional and structural homogeneous continuous dynamical systems, hyposystems and components and superimposed constraints.

Mechanical commutation matrixer (MCM) AC–DC commutator—This commutator is a mechano-electrical commutation matrixer (MCM) AC–DC rectifier for a rotary AC–DC commutator generator, the commutator mechano-electrically (ME) commutes (switches) the armature windings so that the resultant induced source AC armature phase-voltages always act with the same sense of voltage polarisation; this requires a reversal of the armature winding connection every π rad; the induced source AC armature phase-voltages are mechano-electronically rectified to the induced source DC armature voltage via bipolar mechano-electrical valves, i.e. commutator segments that contact the carbon brushes.

Mechanical commutation matrixer (MCM) DC–AC commutator—This commutator is a mechano-electrical commutation matrixer (MCM) DC–AC inverter for a rotary DC–AC commutator motor or actuator, the commutator mechano-electrically commutes (switches) the armature windings so that the resultant torque always acts in the same sense of rotary direction; this requires a reversal of the armature winding connection every π rad; the DC armature supply is via bipolar mechano-electrical valves, i.e. carbon brushes that contact the commutator segments.

Mechanics (Greek μηχανική) is concerned with the behavior of physical bodies when subjected to forces or displacements, and the subsequent effects of the bodies on their environment. It is a branch of classical physics that deals with particles that are either at rest or are moving at velocities significantly less than the speed of light. It can also be defined as a branch of science that deals with the motion of and forces on objects.

Mechanism—The part of a machine which contains two or more pieces so arranged that the motion of one compels the motion of the others.

Mechano-electrical (ME)—Pertaining to a mechano-electrical (ME) generator or device, ME heterogeneous continuous dynamical hypersystem, or ME process that is mechanically driven or controlled.

Mechano-fluidic (MF)—Pertaining to a mechano-fluidic (MF) pump or compressor or device, MF heterogeneous continuous dynamical hypersystem, or MF process that is fluidically driven or controlled.

Microcomputer unit (MCU)—A semiconductor device that has a central processing unit (CPU), memory and input/output (I/O) capability on the same chip.

Motion—A continuous change of position of a body.

Mechano-fluidic (MF) pump—A machine that draws a fluid into itself through an entrance port and forces the fluid out through an exhaust port.

Mechano-pneumatic (MP) compressor—A mechano-pneumatic (MP) machine that increases the pressure of a gas or vapour by increasing the gas density and delivering the gas or vapour against the connected pneumatic homogeneous continuous dynamical system resistance.

Newtonian mechanics—A mechanical homogeneous continuous dynamical system based upon Newton's laws of motion in which mass and energy are considered as separate, conservative, mechanical properties, in contrast to their treatment in relativistic mechanics.

Newton's first law—The law that a particle not subjected to external forces remains at rest or moves with constant velocity in a straight line. Also known as the first law of motion, or Galileo's law of inertia.

Newtonian paradigm—Newton gave us three 'laws of motion', which were intended to describe the motion of the planets. It turned out that these laws could be applied in a seemingly perfectly general way. This broader application has been the foundation of the modern scientific method and will be referred to here as the Newtonian paradigm. At the centre of this paradigm is 'dynamics'. Dynamics is the way the laws of motion get applied.

Newton's second law—The law that the acceleration of a particle is directly proportional to the resultant external force acting on the particle and is inversely proportional to the mass of the particle. Also known as the second law of motion.

Newton's third law—The law that, if two particles interact, the force exerted by the first particle on the second particle (called the action force) is equal in magnitude and opposite in sense of direction to the force exerted by the second particle on the first particle (called the reaction force). Also known as the law of action and reaction, or the third law of motion.

Partition of unity method (PUM)—Can be regarded as a generalisation of the classic finite element methods (FEM). Instead of a mesh, the construction of a global approximation space and the evaluation of the discretisation use a much more versatile set of patches, the so-called cover. The method enables the inclusion of arbitrary local function spaces and is particularly well suited for adaptive refinements with respect to spatial resolution and reproducing the quality of the function space.

Physical heterogeneous continuous complex and/or simple dynamical hypersystems—Have collections of various physical dynamical systems, hyposystems and components so connected or related as to perform a particular function not performable by the dynamical systems, hyposystems and components alone.

Physical model of the physical heterogeneous continuous dynamical hypersystem is a model in which passing physical processes are equivalent to each other when being described by mathematical relationships.

Pneumatics—Fluid statics and behaviour in closed pneumatic homogeneous continuous dynamical systems when the fluid is a gas.

Potential energy—The capacity to do the work that a physical heterogeneous continuous dynamical hypersystem has by virtue of its position or configuration.

Pressure-drop holor—The difference in pressure holor between two points in a fluidic homogeneous continuous dynamical system, usually caused by frictional resistance to a fluid flowing through a conduit, filter media, or other flow-conducting, fluidic homogeneous continuous dynamical system.

Pressure holor—A type of stress which is exerted uniformly in all senses of directions; its measure is the force holor exerted on per unit area.

Prime mover—The component of a power plant that transforms energy from the thermal or the fluidic form to the mechanical form.

Principle of virtual work—The principle that the total work done by all energy-potential-difference variables acting on a physical heterogeneous continuous dynamical hypersystem in static equilibrium is zero for any infinitesimal generalised coordinate from equilibrium that is consistent with the constraints of the physical heterogeneous continuous dynamical hypersystem. Also known as the virtual work principle.

Protocol—The rules governing the exchange of information (data) between networked elements.

Pulse-width modulation (PWM)—The precise and timely creation of negative and positive waveform edges to achieve a waveform with a specific frequency and duty cycle.

Ratchet—A wheel operating with a catch or a pawl so as to rotate in only a single sense of direction.

Ratchet coupling—A coupling between two shafts that uses a ratchet to allow the driven shaft to be turned in one sense of direction only and also to permit the driven shaft to overrun the driving shaft.

Semicustom MCU—A microcontroller unit (MCU) that incorporates normal MCU elements plus application-specific peripheral devices, such as higher-power port outputs, special times units, etc; mixed semiconductor technologies, such as high-density CMOS (HCMOS) and bipolar analogues, are available in a semicustom MCU, generally the HCMOS is limited to 10 V_{DC}, whereas the bipolar-analogue is suitable to 60 V_{DC}.

Specific fuel consumption (SFC)—The mass flow rate of fuel required to produce a unit of power or thrust, e.g. kilograms per kilowatt-hour (kg kWh^{-1}).

Statistical mathematical model of the physical heterogeneous continuous dynamical hypersystem is a model that is formulated by taking into consideration the statistical parameters and the probability expansion function.

Statistical mechanics—Statistical mechanics is a branch of physics that uses probability theory to study the average behaviour of a mechanical homogeneous continuous dynamical system where the state of the system is uncertain. Statistical mechanics is a collection of mathematical tools that are used to fill this disconnection between the laws of mechanics and the practical experience of incomplete knowledge.

Systems thinking approach involves shifting one's attention from the dynamical parts to the whole, from objects to relationships, from structures to processes, from hierarchies to networks. It also includes shifts of emphasis from the rational to the intuitive, from analysis to synthesis, from linear to nonlinear thinking. By contrast to reductionism, systems thinking is a style of thought and reasoning that seeks to understand a whole dynamical hypersystem by examining the physical heterogeneous continuous dynamical hypersystem as a whole instead of disassembling and studying the dynamical parts, i.e. the physical homogeneous continuous dynamical systems, hyposystems and components. While some scientists prefer to use one of the two styles of thinking to the exclusion of the other, it is more common to use whichever style fits a given situation. Quite simply, some situations call for systems thinking, while others require a closer look at the dynamical parts of the physical heterogeneous continuous dynamical hypersystem and are better suited to reductionism. Summing up, systems thinking is an approach to integration that is based on the belief that the dynamical parts, i.e. the physical homogeneous continuous dynamical systems, hyposystems and components of a physical heterogeneous continuous dynamical hypersystem, will act differently when isolated from the hypersystem's environment or other parts of the hypersystem. Standing in contrast to positivist and reductionist thinking, systems thinking sets out to view physical heterogeneous continuous dynamical hypersystems in a holistic manner. Consistent with systems philosophy, systems thinking concerns an understanding of a physical heterogeneous continuous dynamical hypersystem by examining the linkages and interactions between the physical homogeneous continuous dynamical systems, hyposystems and components that comprise the whole of the physical heterogeneous continuous dynamical hypersystem. When one encounters situations that are complex and messy, then systems thinking can help one to understand the situation systemically. This helps one to see the full-size image from which one may identify multiple leverage points that can be addressed to support constructive change. It also helps one to see the connectivity between physical homogeneous continuous dynamical systems, hyposystems and components in the situation, so as to support joined-up actions.

Theoretical reductionism states that all theories in a field are part of a larger theory with a broader scope. In theory, this supports the idea of the existence of a 'grand unified theory' of physics that combines quantum physics with other observed phenomena.

Thermal capacitance—The ratio of the entropy added to a body to the resulting rise in temperature.

Thermal conductance—The amount of heat transmitted by a material divided by the difference in temperature of the surfaces of the material.

Thermal conductivity—The heat flow through a surface per unit area per unit time, divided by the negative of the rate of change of temperature with distance to a direction perpendicular to the surface. Also known as heat conductivity.

Thermal inductance—The product of temperature difference and time divided by entropy flow.

Thermal energy potential difference is the difference between the thermodynamic temperatures of two points.

Thermal resistance—A measure of a body's ability to prevent heat from flowing through it, equal to the difference between the temperatures of opposite faces of the body divided by the rate of heat flow. Also known as heat resistance.

Thermal resistivity—The reciprocal of thermal conductivity.

Thermics—Also known as thermal physics, this is the combined study of thermodynamics, statistical mechanics and kinetic theory. This umbrella-subject is typically designed for physics students and functions to provide a general introduction to each of the three core heat-related subjects. Others, however, define thermics loosely as a summation of thermodynamics and statistical mechanics.

Thermodynamics is a branch of physics concerned with heat and temperature and their relation to energy and work. It defines the macroscopic variables, such as internal energy, entropy, and pressure, which partly describe a body of matter or radiation. It states that the behaviour of these variables is subject to general constraints that are common to all materials, not the peculiar properties of particular materials. These general constraints are expressed in the four laws of thermodynamics.

Torque holor—(i) For a single force holor, the cross product of a holor from some reference point to the point of application of the force holor with the force holor itself. Also known as the moment of force holor or rotation moment holor. (ii) For several force holors, the holor sum of the torque holors (first definition) associated with each of the force holors.

Transfer-function holor—The mathematical relationship between the output holor of a mechatronical control system and its output holor for a linear system; it is the output holor divided by the input holor under conditions of zero initial energy stored. Also known as the transmittance holor.

Transfer-matrix holor—The generalisation of the concept of a transfer-function holor to a multivariable mechatronical control system; it is the matrix whose product with the vector representing the input holors yields the vector representing the output holors.

Unholonomic physical heterogeneous continuous dynamical hypersystem—A physical heterogeneous continuous dynamical hypersystem that is subject to restraints of such a nature that the physical heterogeneous continuous dynamical hypersystem cannot be described by independent coordinates; examples are generalised commutation matrixer (GCM) commutators, e.g. in mechanical homogeneous continuous dynamical systems are a rolling hoop, or an ice skate which must point along its path.

Vapour is a gas at a temperature below the critical temperature, so that it can be liquefied by compression, without lowering the temperature.

Velocity holor—The time rate of change of position of a body; it is a vector quantity with a sense of direction and magnitude. Also known as the linear velocity holor.

Voltage holor—Also known as the electrical energy-potential-difference holor, electrical tension holor or electrical pressure holor (denoted V_i) and measured in SI units of electrical potential: volts (V) or joules per coulomb (J C^{-1}). It is the electrical energy-potential-difference holor between two electrical commutation matrixers' (ECM) separated terminals, i.e. between ECM rows and columns, or the difference in electrical potential energy of a unit charge transported between two ECM separated terminals.

Volumetric flow-rate holor—In fluidics-in particular, fluid dynamics and hydrometry-the volumetric flow-rate holor (also known as the volume flow-rate holor, the rate of fluid-flow holor or the volume velocity holor) is the volume of fluid that passes through a given surface per unit time. The SI unit is cubic meters per second (m^3 s^{-1}). It is usually represented by the symbol Q^j. The volumetric flow-rate holor should not be confused with the volumetric flux holor, as defined by Darcy's law and represented by the symbol q^j, with SI units of m^3/m^2 s, that is, (m s^{-1}). The integration of a flux over an area gives the volumetric flow rate.

Lightning Source UK Ltd.
Milton Keynes UK
UKOW07n1538260117
292944UK00004B/32/P